Model Neural Networks and Behavior

Model Neural Networks and Behavior

EDITED BY
ALLEN I. SELVERSTON
University of California, San Diego
La Jolla, California

Springer Science+Business Media, LLC

Library of Congress Cataloging in Publication Data

Main entry under title:

Model neural networks and behavior.

Includes bibliographical references and index.
1. Neuropsychology. 2. Neural circuitry. 3. Animal behavior. I. Selverston, Allen I.
QP360.M57 1985 599′.051 85-6351

ISBN 978-1-4757-5860-3 ISBN 978-1-4757-5858-0 (eBook)
DOI 10.1007/978-1-4757-5858-0

Originally published by Plenum Press, New York in 1985.
Softcover reprint of the hardcover 1st edition 1985

Dedicated to Graham Hoyle, whose uncompromising principles and love of comparative neurophysiology were a constant source of inspiration to his students, colleagues, and all those interested in model neural networks and behavior.

Contributors

THOMAS W. ABRAMS, Center for Neurobiology and Behavior, College of Physicians and Surgeons, Columbia University, and Howard Hughes Medical Institute, New York, New York 10032

EDMUND A. ARBAS, The Biological Laboratories, Harvard University, Cambridge, Massachusetts 02138

HUGO ARÉCHIGA, Department of Physiology and Biophysics, Center of Investigation and of Advanced Studies of the IPN, Mexico, D.F.

MICHAEL J. BASTIANI, Department of Biological Sciences, Stanford University, Stanford, California 94305

BARBARA BELTZ, Harvard Medical School, Department of Neurobiology, Boston, Massachusetts 02115

PAUL R. BENJAMIN, M.R.C. Neurophysiology Research Group, School of Biology, University of Sussex, Falmer, Brighton, Sussex, United Kingdom

MALCOLM BURROWS, Department of Zoology, University of Cambridge, Cambridge, United Kingdom

RONALD L. CALABRESE, The Biological Laboratories, Harvard University, Cambridge, Massachusetts 02138

JOHN A. CONNOR, Department of Molecular Biophysics, AT&T Bell Laboratories, Murray Hill, New Jersey 07974 and Department of Physiology and Biophysics, University of Illinois, Urbana, Illinois 61801

WALTER COSTELLO, Yale University, Department of Biology, New Haven, Connecticut 06511

W. JACKSON DAVIS, The Thimann Laboratories and The Long Marine Laboratories, University of California at Santa Cruz, Santa Cruz, California 95064

MICHAEL S. DEKIN, Department of Physiology and Biophysics, The University of Iowa, Iowa City, Iowa 52242

CHRISTOPHER J. H. ELLIOTT, M.R.C. Neurophysiology Research Group, School of Biology, University of Sussex, Falmer, Brighton, Sussex, United Kindgom

GRAHAM P. FERGUSON, M.R.C. Neurophysiology Research Group, School of Biology, University of Sussex, Falmer, Brighton, Sussex, United Kingdom

UBALDO GARCÍA, Department of Physiology and Biophysics, Center of Investigation and of Advanced Studies of the IPN, Mexico, D.F.

ALAN GELPERIN, AT&T Bell Laboratories, Murray Hill, New Jersey 07974, and Department of Biology, Princeton University, Princeton, New Jersey 08544

PETER A. GETTING, Department of Physiology and Biophysics, The University of Iowa, Iowa City, Iowa 52242

SILVIO GLUSMAN, Harvard Medical School, Department of Neurobiology, Boston, Massachusetts 02115

MAURICE GOLA, Institute of Neurophysiology and Psychophysiology, C.N.R.S., Marseilles, France

COREY S. GOODMAN, Department of Biological Sciences, Stanford University, Stanford, California 94305

MICHAEL GOY, Harvard Medical School, Department of Neurobiology, Boston, Massachusetts 02115

RONALD HARRIS-WARRICK, Harvard Medical School, Department of Neurobiology, Boston, Massachusetts 02115

JOHN G. HILDEBRAND, Department of Biological Sciences, Columbia University, New York, New York 10027

PHILIP HOCKBERGER, Department of Molecular Biophysics, AT&T Bell Laboratories, Murray Hill, New Jersey 07974 and Department of Physiology and Biophysics, University of Illinois, Urbana, Illinois 61801

SCOTT L. HOOPER, Biology Department, Brandeis University, Waltham, Massachusetts 02254

J. J. HOPFIELD, AT & T Bell Laboratories, Murray Hill, New Jersey 07974 and Divisions of Chemistry and Biology, California Institute of Technology, Pasadena, California 91125

LILY YEH JAN, Department of Physiology, University of California, San Francisco, California 94143

YUH NUNG JAN, Department of Physiology, University of California, San Francisco, California 94143

MICHAEL JOHNSTON, Harvard Medical School, Department of Neurobiology, Boston, Massachusetts 02115

S. B. KATER, Department of Biology, University of Iowa, Iowa City, Iowa 52242

EDWARD A. KRAVITZ, Harvard Medical School, Department of Neurobiology, Boston, Massachusetts 02115

WILLIAM B. KRISTAN, JR., Department of Biology, University of California, San Diego, California 92093

SASCHA DU LAC, Neuroscience Program, Stanford University, Stanford, California 94305

MARGARET LIVINGSTONE, Harvard Medical School, Department of Neurobiology, Boston, Massachusetts 02115

EVE MARDER, Biology Department, Brandeis University, Waltham, Massachusetts 02254

EARL MAYERI, Department of Physiology, University of California, San Francisco, California 94143

RICHARD E. MCCAMAN, Section of Neuropharmacology, Division of Neurosciences, Beckman Research Institute of the City of Hope, Duarte, California 91010

JOHN P. MILLER, Department of Zoology, University of California, Berkeley, California 94720

MAURICE MOULINS, Laboratory of Comparative Neurobiology, CNRS and University of Bordeaux I, Arcachon, France

FRÉDÉRIC NAGY, Laboratory of Comparative Neurobiology, CNRS and University of Bordeaux I, Arcachon, France

JOYCE K. ONO, Section of Neuropharmacology, Division of Neurosciences, Beckman Research Institute of the City of Hope, Duarte, California 91010

MICHAEL O'SHEA, University of Geneva, Laboratory of Neurobiology, CH-1211 Geneva 4, Switzerland

KEIR G. PEARSON, Department of Physiology, University of Alberta, Edmonton, Alberta, Canada

R. MELDRUM ROBERTSON, Department of Biology, McGill University, Montreal, Quebec, Canada

LEONARDO RODRÍGUEZ-SOSA, Department of Physiology and Biophysics, Center of Investigation and of Advanced Studies of the IPN, Mexico, D.F.

BARRY S. ROTHMAN, Department of Physiology, University of California, San Francisco, California 94143

LAWRENCE SALKOFF, Yale University, Department of Biology, New Haven, Connecticut 06511

MARK SCHAEFER, Department of Biological Sciences, Stanford University, Stanford, California 94305

RICHARD H. SCHELLER, Department of Biological Sciences, Stanford University, Stanford, California 94305

THOMAS SCHWARZ, Harvard Medical School, Department of Neurobiology, Boston, Massachusetts 02115

ALLEN I. SELVERSTON, Department of Biology, University of California, San Diego, La Jolla, California 92093

KATHLEEN KING SIWICKI, Harvard Medical School, Department of Neurobiology, Boston, Massachusetts 02115

D. W. TANK, AT&T Bell Laboratories, Murray Hill, New Jersey 07974

JOHN B. THOMAS, Yale University, Department of Biology, New Haven, Connecticut 06511

JAMES W. TRUMAN, Department of Zoology, University of Washington, Seattle, Washington 98195

JANIS C. WEEKS, Department of Entomology, University of California, Berkeley, Berkeley, California 94720

DAVID A. WEISBLAT, Department of Zoology, University of California, Berkeley, California 94720

ROBERT J. WYMAN, Yale University, Department of Biology, New Haven, Connecticut 06511

BIRGIT ZIPSER, Cold Spring Harbor Laboratory, Cold Spring Harbor, New York 11724

ROBERT S. ZUCKER, Department of Physiology-Anatomy, University of California, Berkeley, California 94720

Preface

The most conspicuous function of the nervous system is to control animal behavior. From the complex operations of learning and mentation to the molecular configuration of ionic channels, the nervous system serves as the interface between an animal and its environment. To study and understand the fundamental mechanisms underlying the control of behavior, it is often both necessary and desirable to employ biological systems with characteristics especially suitable for answering specific questions. In neurobiology, many invertebrates have become established as model systems for investigations at both the systems and the cellular level.

Large, readily identifiable neurons have made invertebrates especially useful for cellular studies. The fact that these neurons occur in much smaller numbers than those in higher animals also makes them important for circuit analysis. Although important differences exist, some of the questions that would be technically impossible to answer with vertebrates can become experimentally tractable with invertebrates.

The principal purpose of this volume is to present an overview of the remarkable progress being made using these model systems. Yet invertebrates can in fact be approached from two overlapping points of view. A zoologist may look at the functioning of an insect nervous system because he is interested in insects. To him the system being studied is not a model but the real thing. However, from a neurobiological point of view the system that best allows one to solve a particular problem is a model system, and this is the sense in which we will use this sometimes overworked term.

When the use of invertebrate material became widespread in the early 1950s, several important issues were raised. The first was whether or not the use of invertebrates would really solve all of the important problems in a matter of a few years. The second was whether or not the data gleaned from these preparations had any relevance to "real" nervous systems. The answer to the first question became clear in a short time: The "simple" invertebrate nervous systems were extremely complex and there would not be any shortcuts. The question of relevance, which still seems to trouble many psychologists, has for the most part been answered affirmatively. As in most biological systems, more general principles may be elucidated at the cellular than at the systems level. Nevertheless, important insights are being achieved at both levels, and chapters containing information derived from studies at both levels are included in this volume.

The contributed chapters have been grouped into thematic areas, beginning with seven chapters dealing with neural circuitry. These are followed by sections on development, learning and plasticity, neurotransmitters and neuromodulators, and cellular and membrane biophysics, and ending with the relatively new fields of neurogenetics and molecular neurobiology.

Neural Circuitry

Behavior is the result of the interactions of neurons within networks. Some interactions produce motor programs, and a central question is how exactly they do it. A good place to start then might be an analysis of the neural circuits that generate motor programs. Considerable progress has been made in the study of rhythmic behaviors, where preparations utilizing a few large identifiable neurons are capable of generating the basic oscillatory drive. It is becoming clear as a result of these studies that considerably more than a knowledge of the participating neurons and their synaptic relationships is necessary to explain the underlying mechanisms. This is made clear in the first chapter by Getting and Dekin. They describe the current status of work on the swimming pattern generator for the marine mollusk *Tritonia.* This normally slow-moving animal undergoes a series of dorsal and ventral flexions that propel it through the water when properly stimulated. The premotor pool of interneurons that underlies the behavior has been almost completely worked out. However, the network may not be as "hard-wired" as was previously thought; instead it may be composed of several different subcircuits whose individual expression is dependent upon both the type of initial stimulus and the internal state of the animal. Getting and Dekin call this arrangement a "polymorphic" network, and we will return to this new concept of network plasticity in some of the other chapters dealing with motor systems.

Much of the original impetus for the study of oscillatory neural networks derived from the work of Wilson and his colleagues on the flight system of the locust. This work led to the idea that a completely deafferented ganglion could nevertheless produce a reasonable motor pattern and laid to rest the hypothesis that rhythmic behaviors were generated as a result of sensory feedback. The chapter by Robertson and Pearson is illuminating because it not only makes a start at unraveling the flight-pattern-generating circuitry but also discusses the role of sensory feedback and its importance in controlling the motor behavior.

The lobster pyloric central pattern generator (CPG) is well described and can serve to study the cellular basis of modulatory action on pattern-generating networks. Miller and Selverston describe the mechanisms underlying the generation of the pyloric rhythm, which were determined using a new dye-sensitized photoinactivation technique. As with most of the CPG circuits, the pyloric rhythm uses both cellular and network properties synergistically to form a three-phase cycle.

Many of the cellular properties can be shown to be the result of extrinsic tonic input, which causes the release of local hormones. These substances, along with several newly discovered putative modulators, are discussed more fully in a later chapter by Marder and Hooper. The chapter by Moulins and Nagy discusses the concept of hierarchically arranged oscillators and particularly the role of inputs to the lobster stomatogastric pattern generators. As with the case of *Tritonia,* a specific input, called the anterior pyloric modulator, can in effect rewire a known circuit by altering the strengths of synaptic and electrical connections. The result is to produce a marked change in the output of the isolated, deafferented system. The following chapter, by Calabrese and Arbas, presents a further variation on this theme using the leech heartbeat system. Here there is both a central and a peripheral (myogenic) oscillator, which operate in parallel. Each can be modulated independently by sensory feedback or by specific identified serotonergic neurons. Extensive knowledge of the pattern-generating circuit can now suggest specific mechanisms for these modulatory drives. Another preparation that has been exploited for circuit analysis is the snail feeding system. Benjamin and his colleagues have been pioneers in the description of this complex network, which, like the others, employs mainly inhibitory chemical synapses to cause the pattern-generating neurons to fire in the correct sequence. The oscillation itself could result either from the synaptic connections or from the large number of electrically coupled cells feeding excitation back on themselves.

One of the most important ideas to emerge from the study of these small circuits is that much of the transmission between neurons is graded, i.e., occurring without the presence of all-or-none action potentials. An example of the integrative effects of such nonspiking transmission in the insect central nervous system is given in the chapter by Burrows. Here, the complex pathways that integrate sensory information and convert it to motor output are seen to utilize a large amount of graded transmission. In the system Burrows describes, it may be possible to ascribe to the graded release the function of shaping the motor outflow, while the "spikers" provide the first stage of integration for the sensory inflow.

Development

New anatomical techniques, many of which emerged to take specific advantage of large invertebrate cells, have permitted workers to make enormous progress in the study of neuronal development. No other system has to deal with as much complexity as does the nervous system during development. Individual neurons must find their correct targets, often over long distances and often in competition with many other neurons. The questions of when this process begins during embryogenesis and how it is controlled by the genome and by other cells are some of the most fundamental ones in neurobiology. Hildebrand looks at the problem of intercellular control of development using the sensory pathways from the antenna to the brain of the moth *Manduca.* He and his colleagues have shown that the wiring and synaptic connectivity within the antennal lobes develop and become consolidated before sensory input is present. However, if the primordial antennal disk is removed, a stunted and displaced antennal lobe results. It appears therefore that in this model system the inputs are necessary to organize and sustain the normal dendritic morphology and function, but not for the survival of the neurons.

Some of the most remarkable progress in understanding how axonal growth occurs during development has come from the study of grasshopper embryos. Here large identifiable cells have permitted Goodman and his colleagues to study the mechanisms underlying this growth. Their results suggest that the axon pathways are labeled and direct the filopodia of growth cones by the presence of surface molecules that cause selective adhesion. The chapter by Bastiani, du Lac, and Goodman also presents data suggesting that many of the same neurons in *Drosophila* are homologous, so that future studies may be able to include a genetic analysis of the system as well.

The question of the embryonic origins of specific identified cells is examined in the chapter by Weisblat and Kristan. Here serotonin-containing cells can be identified early and their initial stages of differentiation studied.

Another type of axon growth occurs following axotomy in adult preparations, and Kater's laboratory has been extremely successful looking at such growth in the snail *Helisoma.* What has become clear from this work is that there is considerably more plasticity present in the formation of neural connections than was previously thought—novel connections can be routinely formed by interfering with the normal regrowth process. Furthermore, the mechanism underlying this plasticity is now open to investigation because the system can be operated in culture as well as in the animal.

Learning and Plasticity

As stated by Abrams in his opening chapter for the section on learning and plasticity, "mechanisms underlying associative learning might be more analyzable in invertebrate preparations." The presence of defined circuits and identifiable neurons offers the promise of isolating the changes in the nervous system that are responsible for a learned behavioral modification. Abrams brings us up to date on the preparation that has received the most intense scrutiny—*Aplysia.* Moving away from habituation, one of the simplest forms of learning, to associative conditioning, the work presented here presents a new theory of how such a complicated phenomenon may occur; it is both elegant and testable.

A much more sophisticated preparation for learning studies may turn out to be the common garden slug *Limax maximus.* The chapter by Gelperin and his coworkers describes how the many types of paradigms that have been successfully used show how similar is learning in this simple system to vertebrate learning. This system is remarkable in that an isolated lip–brain preparation can also show associative learning, and an animal that has learned to make an associative choice *in vivo* can show retention of the response *in vitro.* The authors also present a new computer simulation that models learning in this animal and which promises to be of enormous help in planning future experiments.

This section closes with a discussion of motivation, learning, and choice in another mollusk, *Pleurobranchia,* by Davis. As with *Limax,* both intact and semiisolated preparations can be used to study the changes that occur during learning. However, Davis has also been able to examine the role of motivation in this process and presents some interesting ideas of how the two closely related phenomena may interact.

Neurotransmitters and Neuromodulators

The chapters dealing with neurotransmitters and neuromodulators show how well and how quickly the advances that have recently taken place in this area using vertebrate material can be incorporated into the simpler invertebrate systems. This is not to say that fundamental research in this area has been neglected. Studies on transmitter-receptor mechanisms, transmitter identification, and the role of peptides have been underway for some time. However, the incorporation of the new immunological techniques has had a major impact on investigations of invertebrate systems, just as it has on work with vertebrate systems.

Mayeri and Rothman describe the complexities of the bag cell hormones responsible for egg laying in *Aplysia.* This is in fact a multitransmitter neural system, and the effects these hormones have on the total behavioral repertoire appear to be much more complex than was previously thought.

Ono and McCaman also use *Aplysia* as a model system for examining the role of chemically mediated synaptic responses in small neuronal networks. Taking advantage of the large cell sizes, they have been able to perform chemical analyses on single identifiable cell bodies as well as to examine the effects of specific agonists and antagonists that can differentiate between different receptor–iontophore complexes for the same transmitter.

We have already introduced the stomatogastric pyloric system in covering previous chapters, in which the importance of conventional synaptic input for controlling motor output was discussed. In the chapter by Marder and Hooper, we see how putative neuromodulatory substances have long-term effects on the output patterns and therefore the behavior. The advantage of the pyloric system for this analysis is that the synaptic connectivity and the burst-generating mechanisms are so well worked out that it is possible to look at the cellular substrates for the action of the neuromodulators.

Kravitz and his colleagues wonder about the roles of some of these substances in their chapter entitled "The Well-Modulated Lobster." They not only describe the biosynthesis and distribution of three important compounds—serotonin, octopamine, and proctolin—but also describe the action of these compounds on muscle and motoneurons. They also document the fact that these compounds appear to produce their behavioral effects via activation of the central motor programs for two well characterized behaviors, abdominal flexion and extension.

Another good model system for the study of neuromodulatory action is the crustacean eyestalk, in which a small peptide called neurodepressing hormone is being examined by Aréchiga and his co-workers. This peptide has depressing effects on most types of neurons, possibly by acting on a sodium pump mechanism.

The substance is also interesting because it has been shown to be released in a circadian fashion, coinciding with the general activity phases of the animal.

In insects, circulating peptide and steroid hormones play a key role throughout the life of the animal and, once more, they serve as excellent models with which to study hormone action on the central nervous system at the cellular level. Truman and Weeks describe their work on the ecdysis system of insects, in which they have found that ecdysal hormone can modify neural circuits so that they can be used for different tasks.

Proctolin, already mentioned in several preceding chapters, may have a direct effect on insect muscles, being released as a co-transmitter by the motoneurons. O'Shea presents evidence that, in addition to its many known actions, proctolin may also be acting as an excitatory transmitter.

Cellular and Membrane Biophysics

One of the most useful ways in which large molluscan neurons have served as models is for the analysis of ionic currents underlying various forms of neuronal excitability. Gola, for example, examines the effect of different calcium and potassium currents on the electrical behavior of various snail neurons and how the multitude of fast and slow currents can affect the integrating ability of a single cell within a circuit.

Calcium and cyclic AMP can both act as internal messengers, since both normally exist at low levels within the cell. The coupling between calcium and cyclic AMP forms the basis of the chapter by Connor and Hockberger. They describe how conditions that simulate action potential trains cause calcium to enter gastropod neurons, leading to high concentrations near the cell membrane and thus influencing calcium-dependent processes.

Finally, Zucker presents a residual calcium hypothesis for facilitation, which can be tested using giant molluscan cell bodies and synapses. His evidence, which includes increased spontaneous release (which would result if there was a build-up of residual calcium), provides the best evidence to date that this mechanism is the cause of synaptic facilitation.

Neurogenetics and Molecular Neurobiology

This volume concludes with a section covering four new approaches to solving some of the more intractable problems of neuroscience—the links between the genes and the nervous system. We begin with Zipser's work on identifying the antigenic markers on identifiable leech neurons. These markers give an identity not only to individual cells but also to specific groups of cells as well. It is enticing to think that these molecules may also serve as cell recognition sites during development, and this possibility is put forward.

Another approach, taken by Scheller, is to use molecular techniques to determine how neuropeptides control behavior. As presented in the chapter by Mayeri, egg laying in *Aplysia* is controlled by a group of neuropeptides released by the bag cells. Scheller has been able to clone the DNA making these peptides and by means of restriction mapping and nucleotide sequencing has identified a family of 5–9 genes encoding several of the peptides. Why so many genes and peptides are necessary to control a relatively simple stereotyped behavior is now open to analysis.

Drosophila, the animal that has been a standard for genetic analysis, is now also being turned to by neurobiologists, who hope to use the vast store of catalogued mutants to solve some fundamental problems. Wyman's group, for example, has been using the escape jump behavior of *Drosophila* because the neurons involved are the largest in the animal and because it is possible to isolate jumpless mutants. These animals can then be subjected to anatomical and physiological studies to find the source of the disrupted behavior.

Finally, the Jans hope to find the gene for the voltage-sensitive potassium channel by using transposable elements to mark and clone the *Shaker* locus, which appears to contain the structural gene for the channel. *Shaker* mutations have been induced by hybrid dysgenesis caused by the insertion of transposable p factors into the *Shaker* locus. A cloned p element can then be used to isolate the DNA contiguous with the p element. Since there are no high-affinity toxins or antibodies against the potassium channels, *Drosophila* genetics and the *Shaker* locus may be a viable method of finding the gene for this important channel protein.

An overview such as the one presented in this volume must necessarily omit a large amount of good work. The comparative approach to the study of nervous systems is a popular one and the chapters which follow are meant to convey a sampling of some of the most fruitful areas under investigation. Nevertheless, they do reflect my own biases and for that I must assume full responsibility.

Allen I. Selverston

La Jolla, California

Contents

PART II. DEVELOPMENT

PART III. LEARNING AND PLASTICITY

PART IV. NEUROTRANSMITTERS AND NEUROMODULATORS

Model Neural Networks and Behavior

Part I
Neural Circuitry

1

Tritonia Swimming

A Model System for Integration within Rhythmic Motor Systems

Peter A. Getting and Michael S. Dekin

1. Introduction

Over the past several decades, the functional organization of motor systems, and in particular those controlling rhythmic movements, has been viewed in the context of two conceptual hypotheses. The observation that many rhythmic motor patterns and behaviors could persist in the absence of phasic sensory feedback led to the idea of a central pattern generator (CPG). The CPG is envisioned as a group of central neurons that generates a sequence of temporally and spatially coordinated activity. It is now clear that most, if not all rhymic behaviors have as their basis a central pattern generator (Delcomyn, 1980). The second major hypothesis has been the concept of the "command" neuron or system. This idea is founded in the work of Wiersma and Ikeda (1964), who observed that stimulation of certain neurons in the crayfish could elicit rhythmic movements of the swimmerets. Despite attempts to define a command neuron by explicit criteria (Kupfermann and Weiss, 1978), the term is most commonly used to describe neurons that, when active, will "turn on" some recognizable, coordinated behavior. The overall organization of rhythmic motor systems can be represented as a series of "black boxes" representing the command and CPG function (Fig. 1) (Grillner, 1977). In this scheme, initiating stimuli would activate an appropriate command neuron or set of command neurons that in turn would activate the central pattern generator for a particular rhythmic behavior. The command neuron could be viewed as largely "permissive" in that its function would be to select and "turn on" the appropriate

Peter A. Getting and Michael S. Dekin • Department of Physiology and Biophysics, The University of Iowa, Iowa City, Iowa 52242.

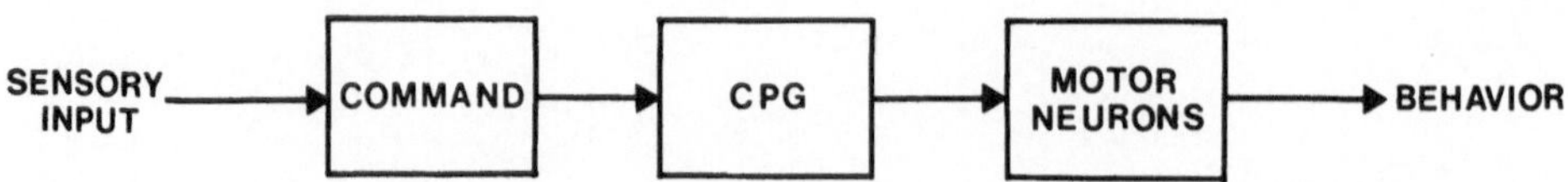

FIGURE 1. Functional organization of a rhythmic motor system represented as a block diagram.

pattern generator. According to this view, the command would not be involved directly in the generation of the pattern, but would activiate a pre-existing CPG. In the extreme, one could imagine that each different movement involving a common pool of motor neurons would be generated by a different CPG each activated by a different set of commands.

Considerable effort has been directed toward seeing how the command and CPG functions are implemented at the cellular and synaptic levels. Criteria have been established for identifying both command (Kupfermann and Weiss, 1978) and CPG neurons (Getting *et al.*, 1980). Application of these criteria to a variety of invertebrate preparations suggests that the functional organization of rhythmic motor systems may not be as simple as implied by Fig. 1. First, single neurons may subserve more than one function. For example, motor neurons may be part of the CPG (Selverston *et al.*, 1977; Heitler and Mulloney, 1978). In addition, command neurons may receive feedback from the CPG that they activate (Gillette *et al.*, 1980). These findings suggest that the progressive flow of information from left to right as depicted in Fig. 1 may not be an appropriate description of these systems. In this chapter, we present evidence that the escape swimming system of the marine mollusk *Tritonia diomedea* is not organized in such a linear fashion. Our evidence suggests that this CPG is not "prewired" to produce only one activity pattern, but is multifunctional.

2. Escape Swimming in *Tritonia*

The sea slug *T. diomedea* escapes from predators by making a series of alternating ventral and dorsal flexions called swimming (Williows, 1967; Hume *et al.*, 1982). The neural control of swimming in *Tritonia* has been reviewed recently (Getting, 1983c), so this material will be summarized here only briefly. Figure 2 shows the activity pattern recorded from two classes of pedal ganglion neurons, the dorsal flexion neurons (DFN) and the ventral flexion neurons (VFN). The DFN and VFN provide the major motor drive underlying the dorsal and ventral flexions, respectively, of a swim. The upper trace in Fig. 2 is the output from a position monitor placed on the tail of the animal. Downward deflections represent ventral flexions. Each flexion movement of the swim is preceded at a constant latency by a burst in the corresponding flexion neuron type (Hume *et al.*, 1982). The basic swim pattern recorded from the flexion neurons is an alternation in the bursts between the DFN and the VFN with a swim always ending on a DFN burst. Three additional types of efferent neuron have been identified, but their role in swim-

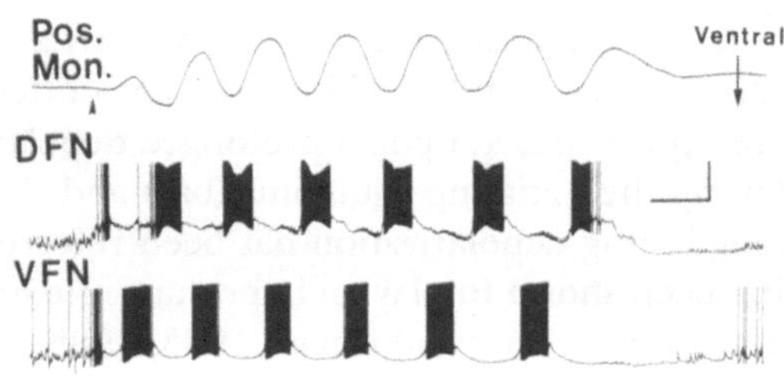

FIGURE 2. Burst pattern recorded intracellularly from a dorsal flexion neuron (DFN) and a ventral flexion neuron (VFN) during an escape swim in a whole animal preparation. The swim was initiated by placing a few crystals of sea salt on the epithelium at the arrow below the top trace. The alternating flexion movements of the swim are indicated by a position monitor (Pos. Mon.) placed on the tail of the animal (see Hume *et al.*, 1982, for details). Vertical scale, 40 mV; time scale, 5 sec.

ming is not clear (Hume *et al.*, 1982). The central orgin of the swim motor pattern was demonstrated by Dorsett *et al.* (1969, 1973). Despite total isolation of the cerebral-pleural-pedal ganglion complex, the DFN and VFN can still produce, in response to a brief electrical stimulation of a peripheral nerve, the alternating burst pattern typical of a swim.

The CPG for swimming has been localized to at least four groups of premotor interneurons (Getting, 1977, 1981, 1983a,b; Getting *et al.*, 1980; Taghert and Willows, 1978). Figure 3 shows simultaneous intracellular recordings from a representative member of each of the four interneuron populations during a swim sequence initiated in an isolated brain preparation. The swim pattern is characterized by alternating bursts between the dorsal swim interneurons (DSI) and the high-frequency bursts in two classes of ventral swim interneurons (VSI-A and VSI-B). The

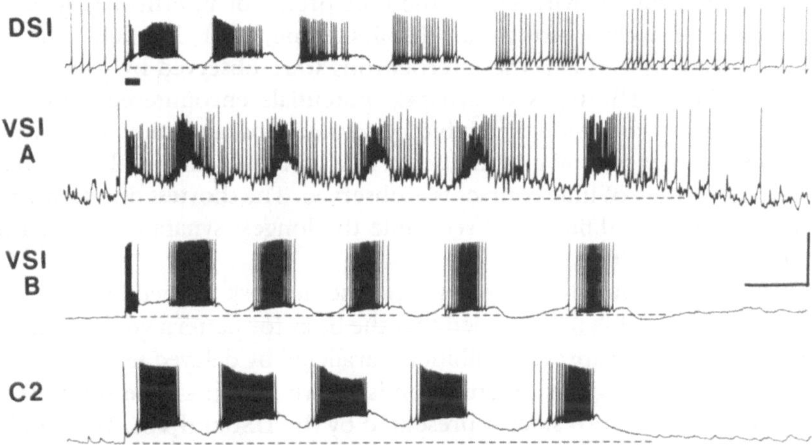

FIGURE 3. Burst pattern recorded intracellularly from a member of the four premotor interneuron groups. The swim sequence was initiated in an isolated brain preparation by electrical stimulation of a peripheral nerve during the bar below the top trace. The dotted lines indicate resting potential. DSI, dorsal swim interneurons; VSI-A, ventral swim interneurons (class A); VSI-B, ventral swim interneurons (class B); C2, cerebral cell 2. Vertical scale, 50 mV (DSI, VSI-B, and C2), 25 mV (VSI-A); time scale, 5 sec.

bursts of cerebral cell 2 (C2) always start after the DSI burst and overlap the transition from the DSI to VSI bursts. The dotted lines of Fig. 3 indicate that the bursts are superimposed upon a prolonged depolarization that is largest immediately following the initiating stiumulus (bar) and that decays as the swim progresses. This long-lasting depolarization has been referred to as the RAMP depolarization and has been shown to play an important role in maintaining swim activity (Lennard *et al.*, 1980; Getting and Dekin, 1983, 1985). The direct participation of these interneurons in pattern generation has been demonstrated by resetting experiments in which advancing or delaying the normal burst onset time causes a permanent time shift in the remaining swim cycles (Getting *et al.*, 1980; Getting, 1983b). Each of the four premotor interneuron types makes monosynaptic connections to the DFN and VFN (Hume and Getting, 1982a; Getting, 1983b). Integration of these synaptic drives with the intrinsic membrane properties of the flexion neurons serves to shape the final motor pattern (Hume and Getting, 1983b). No direct feedback has been found from the flexion neurons to the premotor pattern-generating interneurons (Hume and Getting, 1982a; but see Taghert and Willows, 1978, for exception). The overall organization of the *Tritonia* swim system, therefore, appears to follow the scheme depicted in Fig. 1 with a premotor pattern-generating network that enforces its activity onto a set of somatic motor neurons.

3. How Does the Swim CPG Work?

In an effort to uncover mechanisms for the generation of the swim pattern, the monosynaptic connectivity among the four premotor interneuron groups has been studied extensively (Getting *et al.*, 1980; Getting, 1981, 1983a,b). The monosynaptic connections between the interneurons were observed by pairwise intracellular recordings. The types of synaptic potentials encountered were diverse including excitatory connections, inhibitory connections, and multiaction synapses that combine both excitation and inhibition in a variety of temporal sequences. In addition, a wide range of time courses was observed. The shortest duration synaptic potential ranged from 0.65 to 0.8 sec while the longest synaptic potential had a duration in excess of 20 sec.

Using a digital computer to reconstruct the network formed by the C2, DSI, and VSI-A, Getting (1983a) concluded that the basis for pattern generation among these interneurons is reciprocal inhibition paralleled by delayed excitation. A simplified diagram depicting this organization is shown in Fig. 4. The swim oscillator is thought to consist of two sides represented by the DSI and VSI. Within each of these groups, synergists are coupled by reciprocal excitation. The monosynaptic interaction between DSI and VSI is reciprocal inhibition. The monosynaptic inhibitory pathway from DSI to VSI is paralleled by a polysynaptic excitatory pathway using the C2 interneuron. The events leading to the sequence of bursts during a swim cycle can be summarized as follows. A swim cycle is thought to begin when the DSI are depolarized and begin to fire (Lennard *et al.*, 1980). When the DSI fire, they inhibit the VSI and begin to excite the C2. Once the C2 reach threshold,

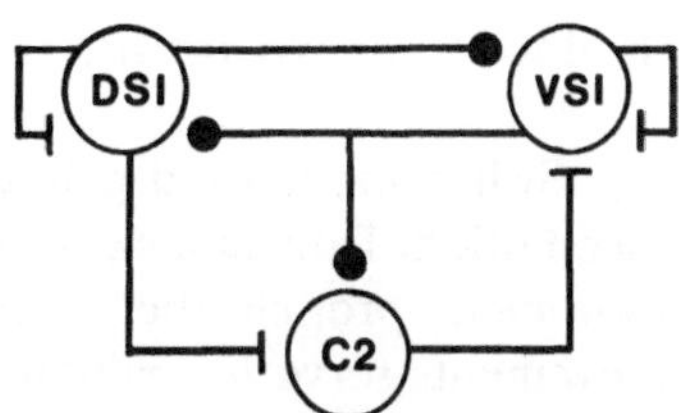

FIGURE 4. Simplified diagram showing reciprocal inhibition between the DSI and VSI paralleled by excitation via C2. The excitatory effect of the C2 synapse onto VSI is delayed (see text for details). T-bars, excitatory connections; solid dots, inhibitory connections.

the DSI and C2 fire coactively for a short period of time. During this time, VSI is receiving inhibition from the DSI, but excitation from C2. Summation of the excitatory input from C2 eventually causes VSI to start firing. Once VSI starts firing, it inhibits both the DSI and C2, terminating their bursts. At this point, VSI is no longer receiving excitation from C2, so the VSI spike frequency declines as the residual excitation dissipates. The decrease in VSI firing rate releases the DSI from inhibition to complete a cycle. Cycling will continue as long as the firing rate of DSI upon release from inibition is sufficient to excite C2 to threshold. This excitation is apparently provided by RAMP depolarization (Lennard *et al.*, 1980; Getting and Dekin, 1983, 1984).

For the network of Fig. 4 to oscillate, there must be a delay between the monosynaptic inhibition of VSI and the polysynaptic excitation. Two different mechanisms for delayed excitation have been found for the two classes of VSI (VSI-A and VSI-B). For VSI-A, the synaptic connection from C2 consists of a multiaction, inhibitory–excitatory synapse (Getting, 1980). When C2 fires in a burst, the action of this synapse is intially to inhibit VSI. After a delay of 1–1.5 sec, the action of this synapse switches to become excitatory despite ongoing activity in C2 (Getting, 1983a). The VSI-B receive only excitation from C2. The delay for this cell is mediated by an intrinsic membrane property of the VSI-B. Getting (1983b) has shown that the VSI-B possess a fast, transient, potassium current termed A-current by Connor and Stevens (1971). When the VSI-B are depolarized, this potassium current is activated and counteracts the depolarizing action of excitatory synaptic potentials. But A-current inactivates with time, so after a period of 1–2 sec, A-current no longer has a significant effect and the cell is released to respond to the full magnitude of the excitatory synaptic drive.

Several conclusions can be drawn from these studies. First, the CPG for swimming is located in a set of premotor interneurons with little or no involvement of the motor neurons directly in pattern generation. Second, pattern generation within the premotor network emerges as a property of the network as a whole. No single cell nor synapse is capable, in isolation, of producing the oscillatory burst pattern. The ability of this network to generate patterned activity depends on the interaction of both synaptic connectivity and the intrinsic cellular properties of each neuron. Third, the major synaptic mechanism for pattern generation appears to be reciprocal inhibition paralleled by delayed excitation. It is interesting to note that the delayed excitation is asymmetrical in that it is mediated only from DSI to VSI and not vice versa. Likewise, changes in cycle period are asymmetrically distributed with 80% of the increase in period occurring during the DSI bursts, whereas VSI bursts remain relatively constant (Lennard *et al.*, 1980).

4. How Is Swimming Initiated and Maintained?

Swimming is used primarily as an escape mechanism rather than a mode of locomotion. During a swim, the animal may tumble and twist so that each flexion movement propels the animal in a different direction. The alternating flexion movements serve to maintain buoyancy allowing the animal to be swept away by water currents. This escape response can be initiated by epithelial contact with the tube feet of predatory star fish or by a variety of noxious chemical stimuli including application of sea salt crystals. In isolated brain preparations, the neural correlate of swimming can be initiated by a brief (1–2 sec) electrical stimulation of peripheral nerves (Doesett *et al.*, 1969). Thus, a transient sensory stimulus may release a complex behavior lasting a minute or more. A second major thrust of our research has been focused on neural events and mechanisms underlying the initiation and maintenance of swimming.

The initiation of swimming takes placed in several stages (Willows and Hoyle, 1969). Immediately upon epithelial stimulation, a local and general withdrawal reflex occurs. During this time, all external body extremities are withdrawn and the animal contracts so as to remove the body wall from the stimulus. The next stage, called preparation, consists of a longitudinal elongation of the body and a flattening of the posterior mantle and anterior oral veil to form paddlelike areas. The first movement of a swim is a rapid ventral flexion that propels the animal off the substrate. Swimming proceeds as a series of alternating dorsal and ventral flexions, always ending on a dorsal flexion. Swimming is always preceded by a reflexive withdrawal phase, but reflexive withdrawals are not always superceded by swimming. Tactile stimulation of the epithelium can evoke a reflexive withdrawal without swimming.

The neural correlate of reflexive withdrawal and swimming can be observed in the premotor interneurons (Getting, 1977). In a quiescent, nonswimming animal, the C2 and VSI-B are normally silent while the DSI and VSI-A fire at a rate of 1–3 spikes per sec. Upon stimulation, the DSI and VSI are excited to fire coactively during the reflexive withdrawal phase. The C2 neuron typically fires a single spike at the onset of the stimulation and then remains silent until late in the reflexive withdrawal phase when it starts its first burst. The onset of activity in C2 marks the beginning of the swim phase. An important event in the initiation and maintenance of swimming appears to be the establishment of the RAMP depolarization (Lennard *et al.*, 1980). Since the bursts of a swim ride on the RAMP depolarization, we have tried depolarizing the interneurons to see which ones could initiate or maintain swimming. Tonic depolarization of either VSI (A or B) does not result in generation of the swim pattern. This result is consistent with the model presented in Fig. 4 since there is no pathway for the excitation of VSI to spread to other interneurons. Tonic depolarization of C2 or DSI can elicit a swim (Fig. 4) (Getting, 1977; Taghert and Willows, 1978; Lennard, *et al.*, 1980). The ability of tonic depolarization of DSI to elicit swim activity is not surprising. Excitation of DSI can pass to C2 and thus to the VSI (Fig. 4). The observation that C2 can also elicit swim activity is more surprising and was not predicted on the basis of a computer reconstruction of the DSI–C2–VSI–A network (Getting, 1983a).

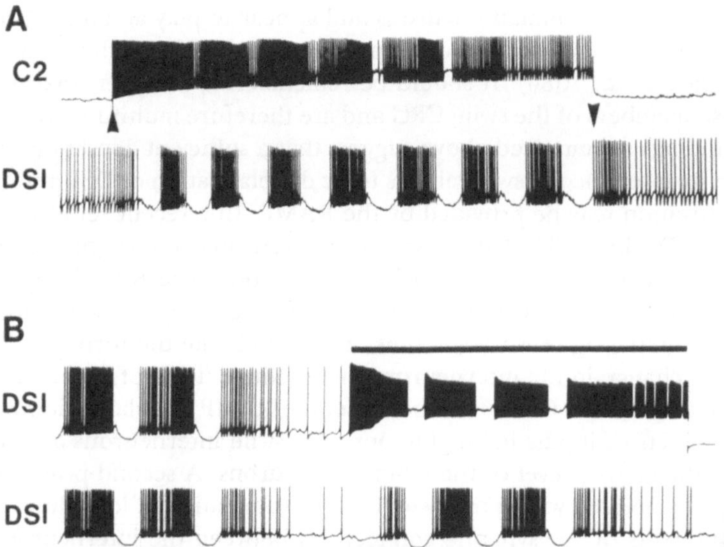

FIGURE 5. Tonic depolarization of C2 or DSI can elicit pattern generation. (A) Intracellular recordings from a C2 and a DSI. Between the arrows, a constant depolarizing current was injected into C2. The recurrent bursts in the DSI indicate that the swim pattern was generated for the duration of the depolarization. (Adapted from Getting, 1977.) (B) Intracellular recordings from two DSI. The two bursts at the left are the last bursts of a swim sequence initiated by electrical stimulation of a peripheral nerve. After the termination of the normal swim sequence, the DSI in the upper trace was depolarized with constant current for the duration marked by the bar. Three additional swim cycles were elicited by depolarization of only one of the six known DSI. (Adapted from Lennard *et al.*, 1980.) Vertical scale, 50 mV; time scale, 5 sec.

One may also ask whether or not activity in a particular neuron is necessary for the initiation and maintenance of swim activity. Cerebral cell 2 is necessary for both the initiation and maintenance of swimming (Getting, 1977; Lennard *et al.*, 1980). If C2 spike activity is blocked by hyperpolarization prior to an initiating stimulus, then swimming cannot be evoked. If C2 is held hyperpolarized until after the initiating stimulus, swim activity will start only when C2 is released. In addition, if C2 is hyperpolarized during a swim, then pattern generation is curtailed. Similar experiments on the DSI have not been performed because this group of interneurons contains six cells. It has proved extremely difficult to penetrate all six neurons simultaneously and hyperpolarize all of them during a swim. As an alternative, we have functionally removed first one, then two, then three or four DSI by hyperpolarization and looked at the effect on the number of swim cycles produced by a constant initiating stimulus. As more DSI were functionally removed from the network (i.e., fewer active DSI remained), the number of swim cycles declined monotonically. Extrapolation of the data suggested that if five or more of the DSI could be removed, no swim activity would be produced. Kupfermann and Weiss (1978) have suggested criteria for identifying cells as command neurons. A command neuron must be active during a behavior, it must be necessary for the behavior, and it must be sufficient to initiate the behavior. By these criteria, both C2 and

DSI would qualify as command neurons and appear to play an integral role in the initiation and maintenance of swimming (Getting, 1977; Taghert and Willows, 1978; Lennard *et al.*, 1980). It should be remembered, however, that the C2 and DSI are also members of the swim CPG and are therefore multifunctional neurons.

The findings summarized above suggest that a sufficient condition for the initiation and maintenance of swimming is tonic depolarization of DSI or C2 and that this depolarization may be provided by the RAMP. In a recent set of experiments (Getting and Dekin, 1983, 1985), we have investigated the origin of the RAMP depolarization in the various premotor interneurons. The RAMP depolarization could arise from two sources. First, the RAMP might represent the long-lasting effects of the initiating sensory stimulus. This could take the form of a persistent conductance change in the interneurons or repetitive firing of excitatory neurons presynaptic to the swim CPG. In either case, the RAMP depolarization would represent the effects of inputs impinging onto the swim interneurons and would not depend on the activity level of the swim interneurons. A second possibility is that the RAMP is generated within the swim CPG by summation of long duration, excitatory components of the synaptic connections between the interneurons. In this case, generation of the RAMP would depend on the level of activity in the swim CPG. The relative contribution of these two mechanisms was measured by comparing the RAMP produced by a constant initiating stimulus with and without cyclic activity within the swim CPG. Swim activity was blocked by hyperpolarization of both C2. The magnitude and time course of the RAMP was quantified using the voltage clamp technique to measure the current underlying the RAMP depolarization. The interneurons could be divided into two categories. For the C2 and the flexion neurons, the RAMP is nearly abolished when cyclic activity within the swim CPG is blocked. For these cells, the RAMP depolarization appears to be generated from activity within the swim CPG. The primary source for the RAMP depolarization in C2 resides in the summation of excitatory inputs from the DSI.

The RAMP in DSI and VSI-B arises from two sources (Fig. 6). At early times in a swim (first 10–15 sec), the RAMP in DSI is independent of activity within the CPG and represents inputs impinging onto the network. At longer times, the RAMP is substantially reduced by blocking swim activity within the CPG. These findings suggest that the initiating input is addressed primarily to the DSI and VSI, but not the C2. The DSI in particular play a pivotal role in the initiation and maintenance of swimming. It is these cells that are depolarized by the initiating stimulus and that when depolarized can initiate and maintain pattern generation.

The neuronal pathway and mechanisms for the generation of the early, external component of the RAMP in DSI are not known. The latter component of the RAMP in DSI is dependent on activity within the CPG and may therefore represent the summation of synaptic inputs from other interneurons within the network. Each DSI receives synaptic excitation from only two known sources within the CPG. Cerebral cell 2 produces a dual action excitatory-inhibitory postsynaptic potential in DSI (Getting, 1977, 1981). The duration of the excitatory component, however, is only 1–2 sec and cannot account for the long-lasting RAMP depolarization. The second source of excitation to the DSI are monosynaptic excitatory connections between the DSI (Getting, 1981). Because these inter-DSI connections

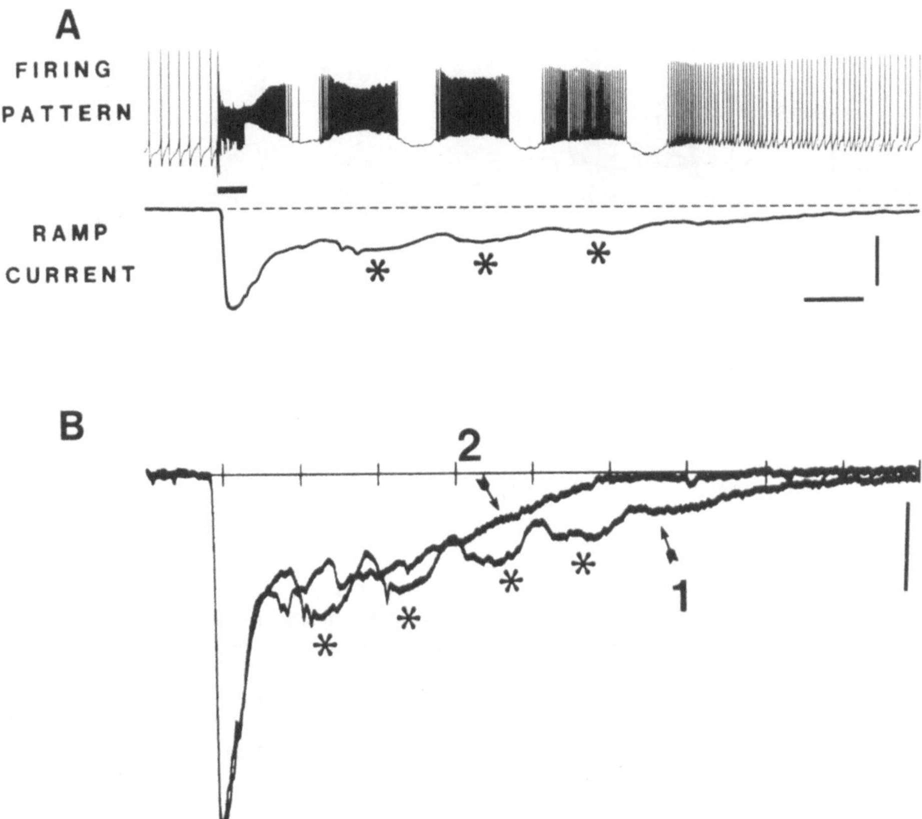

Figure 6. RAMP depolarization in DSI. (A) shows the firing pattern recorded intracellularly from one DSI (upper trace) and the membrane current recorded from a neighboring DSI under voltage clamp (lower trace) (hold potential, −80 mV). In response to stimulation of a peripheral nerve during the bar, the unclamped DSI depolarized and produced the typical swim burst pattern. The membrane current of the clamped DSI rose rapidly to an inward peak during the stimulus and then decayed during the remainder of the swim. Each burst in the unclamped DSI is correlated with a slow inward current transient in the clamped DSI (marked by an asterisk). This pattern of inward current during a swim sequence is referred to as the RAMP current. Vertical scale, 30 mV (upper trace), 6 nA (lower trace); time scale, 3 sec. (B) Superposition of two RAMP currents recorded from a DSI during a control swim sequence (trace 1) and when the C2 were hyperpolarized to block swim activity (trace 2). The RAMP currents have been normalized to the same peak amplitude. The inward current transients associated with each swim cycle are marked by the asterisks in the control swim. For the first 10–15 sec following the nerve stimulus, the two RAMP currents are similar in time course. At later times the RAMP current recorded with C2 inhibited (2) decayed more rapidly than the control RAMP (1). Vertical scale, 1.5 nA; time marks, 5 sec; holding potential, −80 mV. (Adapted from Getting and Dekin, 1985.)

may contribute to the RAMP, we have investigated the nature of the interaction among the DSI in more detail. Figure 7A illustrates the functional interaction between DSI in a quiescent, nonswimming preparation. When one DSI is driven, it causes inhibition of the other ipsilateral and contralateral DSI. In fact, when any DSI is driven in a quiescent preparation, it causes inhibition of all other DSI. This

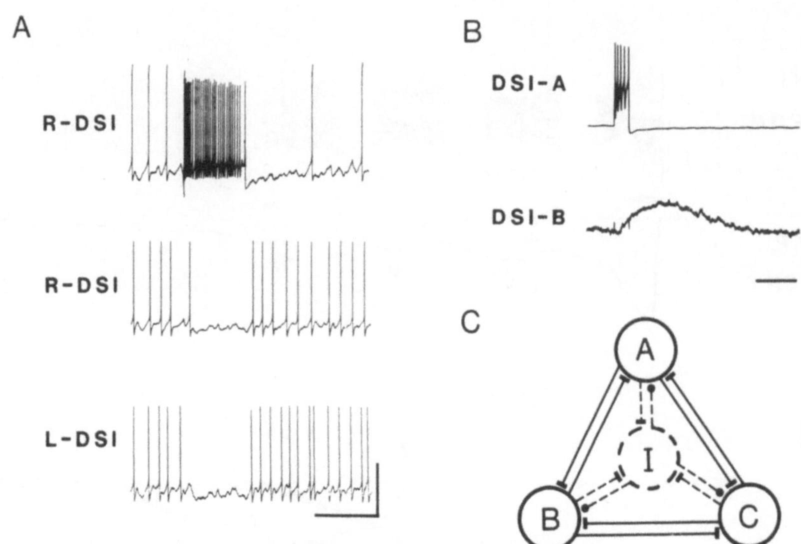

FIGURE 7. Interactions between DSI. (A) Driving a right DSI (R-DSI, upper trace) caused inhibition of all other DSI both ipsilaterally (R-DSI, middle trace) and contralaterally (L-DSI, lower trace). Vertical scale, 40 mV; time scale, 5 sec. (B) Monosynaptic excitatory connections are revealed between ipsilateral DSI when the brain is bathed in high-calcium, high-magnesium sea water which blocks polysynaptic pathways. Vertical scale, 30 mV (upper trace), 3 mV (lower trace); time scale, 3 sec. (C) Circuit diagram showing the proposed connections of the unidentified I-neuron. This cell would mediate polysynaptic inhibition in parallel to the monosynaptic excitatory pathway between DSI. (From Getting and Dekin, 1985.)

inhibition could be mediated by either direct monosynaptic or polysynaptic pathways. To test these possibilities, the brain was bathed in high-calcium, high-magnesium sea water (calcium and magnesium concentration raised to 2.5 times normal). This solution blocks polysynaptic pathways, but leaves monosynaptic potentials largely unaffected (Getting, 1981). Under these conditions, monosynaptic, excitatory connections are observed between the three ipsilateral DSI (Getting, 1981) (Fig. 7B). To reconcile these divergent observations, we have proposed the network shown in Fig. 7C (Getting and Dekin, 1985). Each DSI is shown to excite its ipsilateral mates via monosynaptic connections. In addition, each DSI excites an inhibitory interneuron (I-neuron) that in turn inhibits the DSI. If the inhibition mediated via the polysynaptic I-neuron pathway were stronger than the monosynaptic excitation, then driving one DSI would result in inhibition of all others. If, on the other hand, the I-neuron pathway were blocked (e.g., as in high-calcium, high-magnesium sea water), reciprocal excitatory interactions would be observed.

During a swim, the six DSI fire coactively at high frequency. Inhibitory synaptic interactions between the DSI would appear inconsistent with this firing pattern. We tested the nature of the DSI interactions during a swim by driving one DSI while observing the firing pattern of a second DSI. An increase in the firing rate of one DSI caused an increase in the firing rate of other ipsilateral DSI (Getting and Dekin, 1985). This observation suggests that during a swim, the DSI–DSI interactions are dominated by the monosynaptic excitation and that the I-neuron is probably either silent or greatly reduced in excitability. Thus, the synaptic interactions between DSI can be "switched" from inhibition in a quiescent, nonswimming preparation to excitation during swimming. A major difference between a quiescent and a swimming preparation is the presence of recurrent bursts in the C2 interneurons. Figure 8 shows that C2 is capable of modulating the sign of the synpatic interactions between DSI (Getting and Dekin, 1985). The upper trace shows inhibition of one DSI when another was driven for the duration of the bar above the trace. When a DSI and C2 were driven simultaneously, excitation rather than inhibition was observed (middle trace). It is possible that the excitation could be mediated by a direct pathway from C2 to the DSI (Getting, 1977, 1981). The lower trace shows that when C2 is driven alone, the excitation of DSI is less than that when DSI and C2 are driven togehter. These results are consistent with the hypothesis that C2 inhibits the I-neuron. Thus, when C2 is silent, the synaptic interaction between DSI would be dominated by polysynaptic inhibition, but when C2 is active, the synaptic interaction would change to excitation.

Figure 9 shows an overall network diagram incorporating our current understanding of the synaptic organization of premotor interneurons. The known monosynaptic connections are shown as solid lines. Functional pathways which may be polysynaptic or monosynaptic are shown as dotted lines. Although the I-neuron has not been identified, its connections can be inferred from the effects of driving

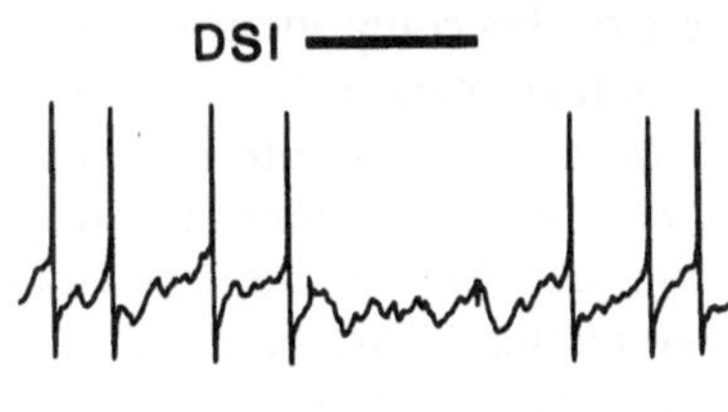

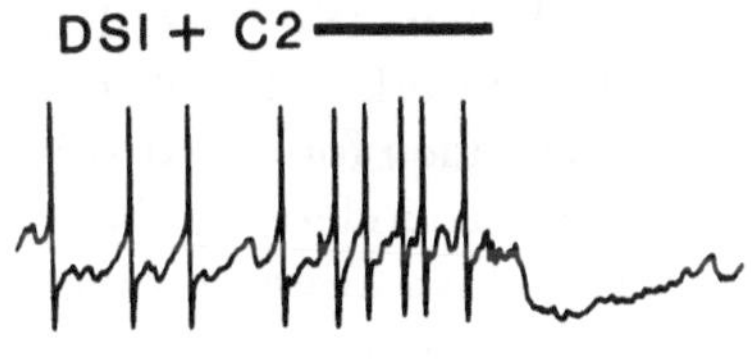

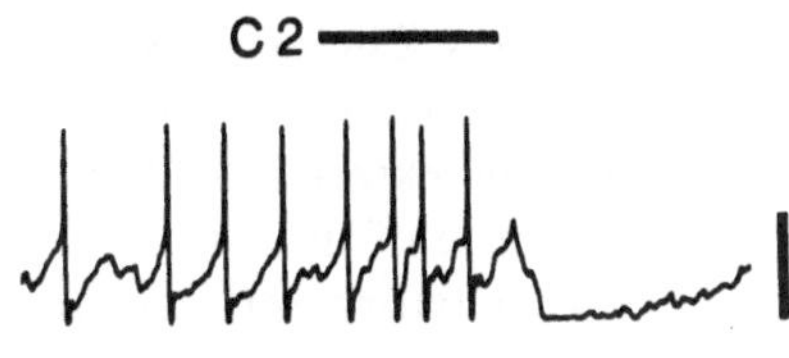

FIGURE 8. Modulation of DSI–DSI interactions by activity in C2. Each trace shows an intracellular recording from the same DSI. In the top trace a second DSI was driven for the duration marked by the bar and caused inhibition of the recorded DSI. When a DSI and C2 were driven together (middle trace), the recorded DSI was excited during the period of drive. The effect of C2 activity alone is illustrated in the bottom trace. Vertical scale, 30 mV; time scale, 3 sec. (From Getting and Dekin, 1985.)

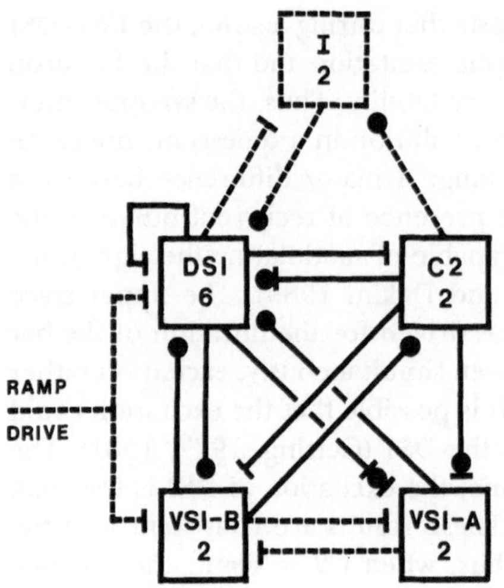

FIGURE 9. Network diagram summarizing the known monosynaptic (solid lines) and postulated pathways (dotted) among the premotor interneurons. T-bars, excitatory synapse; filled dot, inhibitory synapse; multiaction synapses are shown as mixed symbols. The temporal sequence of action is given by the order of symbols going from the presynaptic to the postsynaptic cell. For example, the synapse from C2 to DSI is excitatory, then inhibitory. The number of known cells for each group is given in the corresponding box.

DSI and C2 (Fig. 7 and 8). The I-neuron is excited by DSI and in turn inhibits the DSI. The I-neuron is also inhibited by C2. The lower part of the circuit (C2, DSI, VSI-A and B) can form a pattern generating network for swimming as discussed above (see Fig. 4).

The RAMP depolarization underlying a swim has two components. An external component that lasts 10–15 sec is shown impinging upon the DSI and VSI. The source of this component is currently unknown. At longer times the RAMP depolarization appears to be generated from within the network. A good candidate for the source of this internal component is the mutually excitatory connection between the DSI.These are shown in Fig. 9 as a recurrent excitatory pathway among the DSI. The monosynaptic excitatory postsynaptic potentials (EPSP) generated by this recurrent pathway have durations between 7–9 sec (Getting, 1981). The interburst interval between DSI bursts during a swim ranges from 1–3 sec (Lennard *et al.,* 1980). Thus, the excitation produced by one DSI burst would outlast the period of inhibition between bursts and could help provide the depolarization for the next DSI burst. If the DSI–DSI interactions were dominated by excitation, the network of Fig. 9 could be self-regenerative: each swim cycle providing the depolarization for a next swim cycle.

The events underlying the initiation of a swim can be summarized as follows. Sensory input is addressed primarily to the DSI and VSI in the form of an external RAMP depolarization. If the DSI fire sufficiently, they will excite C2 to threshold, which in turn will inhibit the I-neuron. When the I-neuron is inhibited, the DSI–DSI interactions switch from inhibition to excitation and the network adopts a pattern-generating mode to produce a swim. The discovery of the I-neuron and its modulation by C2 helps to explain why tonic depolarization provided to either DSI or C2 causes swim activity. Depolarization of DSI by current injection simply substitutes for the external RAMP depolarization. Tonic depolarization of C2 would cause inhibition of the I-neuron which in turn would change the DSI–DSI interactions to excitatory, once again allowing the network to adopt its pattern-generating mode.

If the DSI–DSI interactions are excitatory during a swim, one might ask why does not a swim, once initiated, go on forever? Several mechanisms can be identified that contribute to the termination of a swim. If the loop gain from one DSI through another and back to the first were less than one, then the recurrent excitatory connections between the DSI would not be strong enough to maintain the RAMP depolarization. Thus, the level of excitation within the network would slowly decrease over time. The loop gain among the DSI may be decreased further by the presence of spike frequency adaptation in the DSI (Getting, 1983a). In addition, C2 makes a multiaction excitatory–inhibitory connection onto DSI (Getting, 1980). The inhibitory component has a duration of over 20 sec. Because this inhibitory component is so long, it will sum over many cycles causing a progressive decrease in the excitability of DSI. When the DSI frequency at the beginning of a new cycle is insufficient to excite C2 to threshold, the I-neuron would come back on causing inhibition between the DSI and thus termination of the swim. In summary, the network depicted in Fig. 9 appears to adopt at least two configurations. If C2 is off, the DSI–DSI interactions are dominated by inhibition and the network is stable in a nonpattern-generating mode. Any increase in the frequency of one DSI would cause inhibition of the other DSI, thus tending to return the network to the baseline activity. If C2 is firing, the network adopts a second quasi-stable configuration with the I-neuron inhibited and the DSI–DSI interactions dominated by excitation. In this mode, the network will produce the swim pattern for as long as C2 remains active. When C2 fails to fire, the network reverts to the original stable nonpattern-generating state.

5. The "Polymorphic" Network Concept

From what we currently know about the cellular and synaptic organization of the *Tritonia* swim system, it cannot be described by the conceptual organization presented in Fig. 1. For example, it is difficult to ascribe the command function to a single neuron or group. Both the C2 and DSI qualify as command neurons by the criteria of Kupfermann and Weiss (1978), but they are also part of the swim CPG. One might also be tempted to ascribe a command function to the I-neurons as well in that inhibition of the I-neuron would release swim activity by disinhibition of the DSI. The command function appears to be an emergent property of the network as a whole. It is not a funciton of a single cell, but a process that emerges as a consequence of multiple synaptic interactions within the network shown in Fig. 9. This same network also forms the pattern generator for swimming.

A second issue is that the *Tritonia* swim system does not conform to the linear flow of information from command to CPG depicted in Fig. 1. In the swim system, the command and CPG functions are merged inseparably into a single network of neurons. Davis 1976 proposed a number of alternative organizations for motor systems involving various kinds of feedback between the command and CPG elements. Unfortunately, none of these are particularly applicable to the *Tritonia* swim system. The major problem with these schemes is that the commands are viewed as "permissive," serving to turn on a pre-existing CPG network. Perhaps the command function should be viewed as also being "instructive." An instructive command would serve not only to select and activate the motor system, but also to organize the network into an appropriate configuration to generate a particular motor pattern. The *Tritonia* swim network provides an illustrative example. Sensory input is addressed to a network of neurons (Fig. 9) that can, under appropriate conditions, be organized into a pattern generator for swimming. To do so, however, requires changing the sign of the DSI–DSI interactions from inhibition to excitation. This is accomplished by activation of C2 which in turn inhibits the I-neurons, thus reorganizing the network into a configuration to generate the swim pattern.

The inadequacy of previous schemes to incorporate our findings on the organization of the *Tritonia* swim system has led us to propose an alternative concept for the organization of the motor system. Our ideas are based on two major observations. The first is that a group of premotor interneurons can be multifunctional and participate in more than one behavior. The second observation is that inputs can reorganize the motor network into different configurations, each subserving a different function or behavior. In other words, commands may be instructive as well as permissive.

We have incorporated these ideas into a concept that we call the "polymorphic" network (Fig. 10). To discuss this concept, we will first need to define a few terms. We define a network as an ensemble of neurons interconnected by monosynaptic connections. A network is therefore defined on anatomical rather than functional criteria. Under this definition, terms such as "swim network" or "feeding network" are misnomers, for a network should not be defined in terms of its function. A polymorphic network is a network that can be organized into mulitple states or configurations called circuits. Each circuit is defined by a set of functional interactions between the neurons, and each circuit may subserve a particular behavioral function. These ideas are illustrated in Fig. 10, which shows a polymorphic network with four circuit states. Each circuit may involve the entire set of neurons within the network or some subset. Also illustrated is the idea that each circuit can be transformed into the others so that the network may adopt any one of its four different states. Although this is the most general case, in reality some transitions may not be possible.

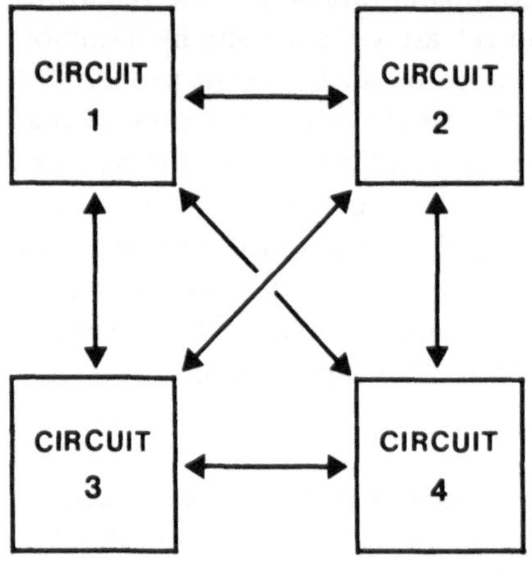

FIGURE 10. Diagrammatic representation of a polymorphic network with four different circuit configurations. Each circuit is postulated to produce a different motor pattern. In the general case shown, all transitions between the four circuits are possible.

Perhaps one of the best ways to illustrate the usefulness of the polymorphic network idea is to apply it to the *Tritonia* swim system. Figure 9 illustrates our current understanding of the premotor network controlling the DFN and VFN. This network contains 14 known premotor interneurons divided into five groups: DSI, VSI-A, VSI-B, C2, and I-neuron. These interneurons are interconnected in a complex pattern of monosynaptic pathways, many of which are reciprocal and involve both excitatory, inhibitory, and multiaction synapses. In addition, each of the interneurons has selective access to the motor pool, the DFN and VFN. For example, the VSI-A inhibit the DFN, but make no connections to the VFN (Hume and Getting, 1982a). The VSI-B excite the VFN and make multiaction synapses onto the DFN (Getting, 1983b). Thus, a variety of motor patterns could be generated by a single set of premotor interneurons if the interneurons were activated in different spatial and temporal coordination patterns. With reference to the polymorphic network idea, each patttern of activity within the interneurons would represent the expression of a different circuit.

For the *Tritonia* swim system, we currently know of two circuit configurations which the network of Fig. 9 can adopt. The first circuit is used in reflexive withdrawals and appears to be the preferred state for the network. In this state, the C2 neuron is silent and the DSI–DSI interactions are characterized by inhibition mediated via the I-neuron. Excitatory input from epithelial mechanoreceptors and chemoreceptors is addressed to both the DSI and VSI causing coactivation of these two cell types. During reflexive withdrawals, then, the DSI and VSI act synergistically to coactivate the flexion neurons. If the sensory-evoked activity in DSI is strong and prolonged, the DSI will depolarize C2 to threshold. When C2 start firing, the network adopts a second circuit configuration that produces the swim pattern. In this case, the I-neuron is inhibited and the DSI–DSI interactions are dominated by excitation. In addition, the DSI and VSI now act as antagonists to activate the flexion neurons in an alternating burst pattern. As discussed in Section 4, a swim will continue as long as C2 is active and maintains the network in this pattern-generating configuration. We currently know of only two circuit configurations for this network of premotor interneurons. This is largely because their activity has been observed only during two behaviors, reflexive withdrawals and swimming. It is entirely possible that this same set of interneurons activates the flexion neurons during other behaviors, such as feeding.

By what mechanisms could a network be switched into its various circuit configurations? The pattern of activity displayed by a neuronal circuit depends on a complex interaction between the nature of the inputs to the system, the nature of the synaptic interactions among the neurons, and the intrinsic membrane properties of the neurons themselves (Byrne, 1980a,b,c; Miller and Selverston, 1982a,b; Getting, 1981, 1983a,b). An alteration in any one of these general properties could result in a change in the spatial and temporal coordination pattern of activity within the network. Two of these mechanisms appear to be operating in the *Tritonia* network. First is the way in which sensory input is addressed to the network. If the sensory stimulus is weak and of very short duration, only reflexive withdrawals will be evoked. This is because C2 will not be depolarized to threshold. If the stimulus is intense and of long duration (1–2 sec), C2 will be depolarized to threshold and a swim evoked. One could easily imagine other modes of addressing sensory input to this network that could produce different coordination patterns among the interneurons. For example, if C2 were inhibited, the other interneurons could be activated in almost any pattern without evoking a swim. The second mechanism operating is a change in the functional synaptic interactions within the network. A dramatic example is the reversal in the sign of the DSI–DSI interactions during swimming. One could imagine more subtle changes involving modulation (both increases and decreases) in the strength of synaptic connections. Such a modulation appears to be mediated by dopamine in the lobster stomatogastric system (Marder and Eisen, 1984). A third possible mechanism is a modulation of the intrinsic membrane properties. The somas and dendrites of central neurons are known to contain numerous voltage- and time-dependent ionic conductances (Adams *et al.*, 1980). Many of these currents can be modulated by transmitterlike agents (see Nicoll, 1982). For example, M-current and spike frequency adaptation in bullfrog sympathetic neurons can be modulated by acetylcholine (Brown and Adams, 1979; Adams *et al.*, 1982). Inputs to the lobster stomatogastric system can induce "plateau potentials" (Russell and Hartline, 1978). Alterations in instrinsic membrane properties need not involve transmitterlike substances, but may come about simply by changes in membrane potential. We have shown that neurons in nucleus tractus solitarius of the guinea pig can show substantial changes in their repetitive firing properties by modulating the depth of hyperpolarization preceding a depolarization (Dekin and Getting, 1983). Thus, a number of mechanisms are currently known that could serve to reorganize a network into its various circuit configurations. It is likely that still additional mechanisms will be forthcoming.

Whether or not the polymorhpic network idea is a viable concept for explaining the organization of motor systems in general remains to be seen. We have evidence in *Tritonia* that a single network of premotor interneurons participates in the generation of two different motor patterns and that the functional interactions within the network can be reorganized into several circuit configurations. The application of this idea to other systems will depend on identifying the relevant interneuronal pools and on characterizing mechanisms for configuring these interneurons into circuits.

ACKNOWLEDGMENTS. We thank Drs. Michael O'Donovan and Eve Marder for many helpful and illuminating discussions. The research presented in this chapter was funded in part by NIH research grant NS17328 to Dr. Peter A. Getting. We also thank Mrs. Michelle Lopez for typing the manuscript.

References

Adams, D. J., Smith, S. M., and Thompson, S. H., 1980, Ionic currents in molluscan soma, *Annu. Rev. Neurosci.* **3**:141–167.

Adams, P. R., Brown, D. A., and Constanti, A., 1982, M-currents and other potassium currents in bullfrog sympathetic neurons, *J. Phyiol. (London)* **330**:537–572.

Brown, D. A., and Adam, P. R., 1979, Muscarinic suppression of a novel voltage-sensitive K-current in a vertebrate neuron, *Nature (London)* **283**:673–676.

Byrne, J. H., 1980a, Analysis of ionic conductance mechanisms in motor cells mediating inking behavior in *Aplysia californica*, *J. Neurophysiol.* **43:**630–650.

Byrne, J. H., 1980b, Quantitative aspect of ionic conductance mechanisms contributing to firing pattern of motor cells mediating inking behavior in *Aplysia californica*, *J. Neurophysiol.* **43:**651–668.

Byrne, J. H., 1980c, Neural circuit for inking behavior in *Aplysia californica*, *J. Neurophysiol.* **43:**896–911.

Connor, J. A., and Stevens, C. F., 1971, Voltage clamp studies of a transient outward current in gastropod neural somata, *J. Physiol. (London)* **213:**21–30.

Davis, W. J., 1976, Organizational concepts in the central motor networks of invertebrates, in: *Neural Control of Locomotion* (R. M. Herman, S. Grillner, P. S. G. Stein, and D. G. Stuart, eds.), Plenum Publishing Corporation, New York, pp. 265–292.

Dekin, M. S., and Getting, P.A., 1983, Delayed excitation in neurons of the nucleus tractus solitarius studied *in vitro*, *Neurosci. Abstr.* **9:**677.

Delcomyn, F., 1980, Neural basis of rhythmic behavior in animals, *Science* **210:**492–498.

Dorsett, D. A., Willows, A. O. D., and Hoyle, G., 1969, Centrally generated nerve impulse sequences determining swimming behavior in *Tritonia*, *Nature (London)* **244:**711–712.

Dorsett, D. A., Willows, A. O. D., and Hoyle, G., 1973, The neuronal basis of behavior in *Tritonia*, IV. The central origin of a fixed action pattern demonstrated in the isolated brain, *J. Neurobiol.* **4:**287–300.

Getting, P. A., 1977, Neural organization of escape swimming in *Tritonia*, *J. Comp. Physiol.* **121:**325–342.

Getting, P. A., 1981, Mechanisms of pattern generation underlying swimming in *Tritonia*, I. Neuronal network formed by monosynaptic connections, *J. Neuroophysiol.* **46:**65–79.

Getting, P. A., 1983a, Mechanisms of pattern generation underlying swimming in *Tritonia*, II. Network reconstruction, *J. Neurophysiol.* **49:**1017–1035.

Getting, P. A., 1983b, Mechanisms of pattern generation underlying swimming in *Tritonia*, III. Intrinsic and synaptic mechanisms for delayed excitation, *J. Neurophysiol.* **49:**1036–1050.

Getting, P. A., 1983c, Neural control of swimming in *Tritonia*, in: *Neural Origin of Rhythmic Movements* (A. Roberts and B. L. Roberts, eds.), Cambridge University Press, Cambridge, England, pp. 89–128.

Getting, P. A., and Dekin, M. S., 1983, Maintenance of *Tritonia* swimming by reciprocal excitation, *Neurosci. Abstr.* **9:**541.

Getting, P. A., and Dekin, M. S., 1985, Mechanisms of pattern generation underlying swimming in *Tritonia*, IV. Gating of a central pattern generator, *J. Neurophysiol.* (in press).

Getting, P. A., Lennard, P. R., and Hume, R. I., 1980, Central pattern generator mediating swimming in *Tritonia*, I. Identification and synaptic interations, *J. Neurophysiol.* **44:**151–164.

Gillette, R., Kovac, M. P., and Davis, W. J., 1978, Command neurons in *Pleurobranchaea* receive synaptic feedback from motor network they excite, *Science* **199:**798–801.

Grillner, S., 1977, On the neural control of movement—A comparison of different rhythmic behaviors, in: *Function and Formation of Neural Systems* (G. S. Stent, ed.), Dahlem Konferenzen, Berlin, pp. 197–224.

Heitler, W. J., and Mulloney, B., 1978, Crayfish motor neurons are an integral part of the swimmeret central oscillator, *Soc. Neurosci. Abstr.* **4:**381.

Hume, R. I., and Getting, P. A., 1982a, Motor organization of *Tritonia* swimming, II. Synaptic drive to flexion neurons from premotor interneurons, *J. Neurophysiol.* **47:**75–90.

Hume, R. I., and Getting, P. A., 1982b, Motor organization of *Tritonia* swimming, III. Contribution of intrinsic membrane properties to flexion neuron burst formation, *J. Neurophysiol.* **47:**91–102.

Hume, R. I., Getting, P. A., and Del Beccaro, M. A., 1982, Motor organization of *Tritonia* swimming, I. Quantitative analysis of swim behavior and flexion neuron firing patterns, *J. Neurophysiol.* **47:**60–74.

Kupfermann, I., and Weiss, K. R., 1978, The command neuron concept, *Behav. Brain Sci.* **1:**3–39.

Lennard, P. R., Getting, P. A., and Hume, R. I., 1980, Central pattern generator mediating swimming in *Tritonia*, II. Initiation, maintenance, and termination, *J. Neurophysiol.* **44:**165–173.

Marder, E., and Eisen, J. S., 1984, A mechanism for the production of phase shift in a pattern generator, *J. Neurophysiol.* **51:**1375–1393.

Miller, J. P., and Selverston, A. I., 1982a, Mechanisms underlying pattern generation in lobster stomatogastric ganglion as determined by selective inactivation of identified neurons, II. Oscillatory properties of pyloric neurons, *J. Neurophysiol.* **48:**1378–1391.

Miller, J. P., and Selverston, A. I., 1982b, Mechanisms underlying pattern generation in lobster stomato gastric ganglion as determined by selective inactivation of identified neurons, IV. Network properties of pyloric system, *J. Neurophysiol.* **48:**1416–1432.

Nicoll, R. A., 1982, Neurotransmitters can say more than just "yes" or "no," *Trends Neurosci.* **5:**369–374.

Russell, D. F., and Hartline, D. K., 1978, Bursting neural network: A reexamination, *Science* **200:**453–456.

Selverston, A. I., Russell, D. F., Miller, J. P., and King, D. G., 1977, The stomatogastric nervous system: Structure and function of a small neural network, *Prog. Neurobiol.* **7:**215–290.

Taghert, P. H., and Willows, A. O. D., 1978, Control of a fixed action pattern by single, central neurons in the marine mollusk, *Tritonia diomedea, J. Comp. Physiol.* **123:**253–259.

Weeks, J. C., and Kristan, W. B., Jr., 1978, Initiation, maintenance, and modulation of swimming in the medicinal leech by the activity of a single neurone, *J. Exp. Biol.* **77:**71–88.

Wiersma, C. A. G., and Ikeda, K., 1964, Interneurons commanding swimmeret movements in the crayfish, *Procambarus clarkii* (Girard), *Comp. Biochem. Physiol.* **12:**509–525.

Willows, A. O. D., 1967, Behavioral acts elicited by stimulation of single, identifiable brain cells, *Science* **157:**570–574.

Willows, A. O. D., and Hoyle, G., 1969, Neuronal network triggering a fixed action pattern, *Science* **166:**1549–1551.

2

Neural Networks Controlling Locomotion in Locusts

R. MELDRUM ROBERTSON AND KEIR G. PEARSON

1. Introduction

Recently there has been considerable increase in interest in the organization and functioning of nerve cells in the central nervous system of insects. It is now clear that the neural control of many simple behaviors in these animals can be analyzed using modern intracellular recording and staining techniques, and that insects offer attractive preparations for determining the events associated with neuronal development (Chapter 9, this volume). In some large orthopterans (locusts, crickets) and cockroaches, substantial progress has now been made toward understanding the nervous control of flying, jumping, respiration, walking, and predator avoidance (Robertson and Pearson, 1982, 1983; Pearson *et al.*, 1980; Burrows, 1982; Pearson, 1976; Westin and Ritzmann, 1982) and toward elucidating integrative events in auditory and tactile sensory systems (Wohlers and Huber, 1982; Romer *et al.*, 1981; Siegler and Burrows, 1983). Moreover, these animals have provided useful preparations for the analysis of graded transmitter release (Burrows, 1981), the modulatory influences of biogenic amines (Evans and O'Shea, 1978), and the physiological action of neuropeptides (Adams and O'Shea, 1983).

In this chapter we review only a small part of the field of insect neurobiology, namely our own work on the nervous control of jumping and flight in the locust. We believe that the cellular analysis of these two behaviors has revealed some general features of insect motor systems. These are discussed in the final section of the chapter.

R. MELDRUM ROBERTSON • Department of Biology, McGill University, Montreal, Quebec, Canada.
KEIR G. PEARSON • Department of Physiology, University of Alberta, Edmonton, Alberta, Canada.

2. The Locust Jump

The motor program for the jump consists of three phases: (1) a rapid flexion of both hindleg tibia leading to the two tibia being locked in full flexion, (2) a short period of cocontraction of the hindleg extensor and flexor tibia muscles (lasting about 500 msec) during which time energy for the jump is stored in elastic elements of the leg, and (3) a sudden inhibition of flexor activity that allows the extensor muscles to shorten and the transfer of stored energy to rapid extension movements of both hindlegs (Heitler and Burrows, 1977a). The neuronal circuitry controlling these three phases of the jump is centered around two pairs of large thoracic interneurons, the C-neurons and the M-neurons (Figs. 1 and 2) (see also Pearson, 1983). The C-neurons function to lock both tibia into full flexion and the M-neurons trigger the jump by terminating the cocontraction phase.

Each C-neuron makes monosynaptic excitatory connections to the fast extensor tibiae motoneuron and the flexor tibiae motoneurons of one hindleg (Pearson and Robertson, 1981). These connections are very powerful so that a single spike in a C-neuron is capable of synchronously activating the fast extensor and flexor motoneurons of one hindleg. When the tibia is close to full flexion, this synchronous pattern of activation results in the tibia flexing rapidly and locking into full flexion. Flexion movement occurs despite extensor activity because the flexor muscles enjoy a considerable mechanical advantage close to full flexion (Heitler, 1974).

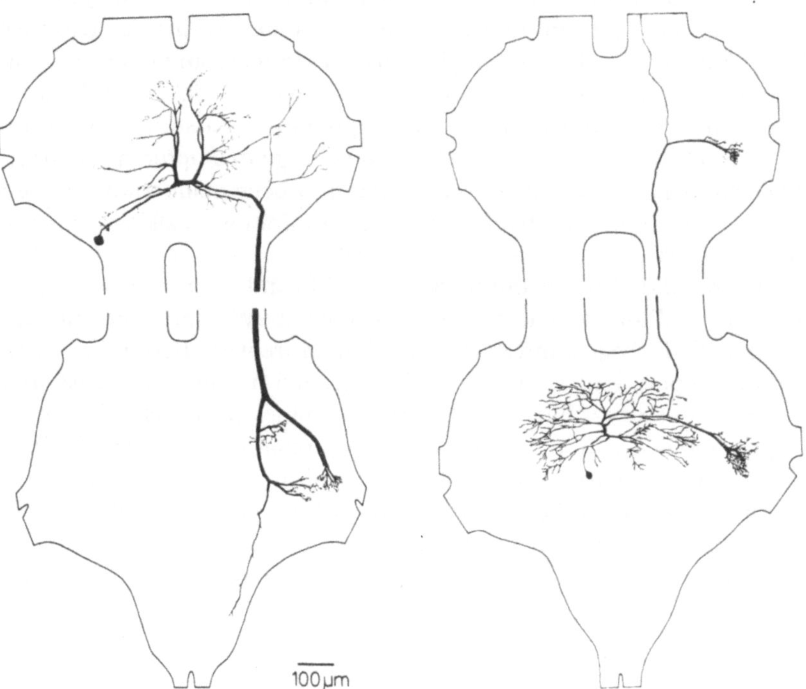

FIGURE 1. Drawing of the structure of a C-neuron (left) and an M-neuron (right) in the mesothoracic (top) and metathoracic (bottom) ganglia of the locust. (From Pearson, 1983.)

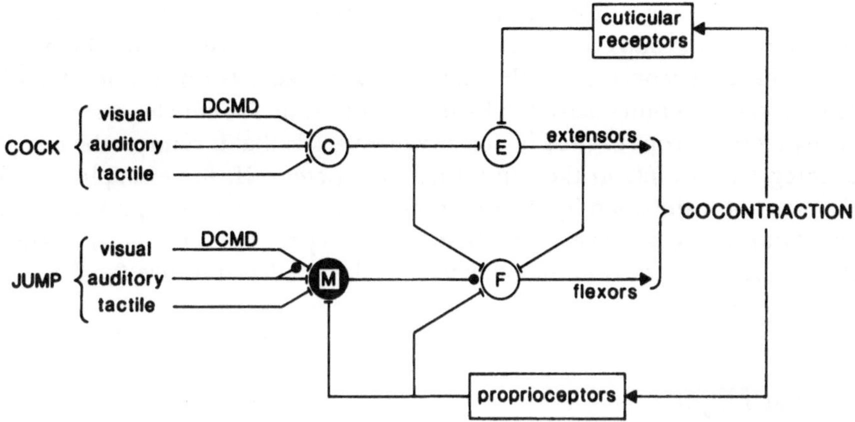

FIGURE 2. Diagram summarizing the main features of the neural circuits for jumping in the locust. C, C-neuron; M, M-neuron; E, fast extensor tibiae motoneuron; F, flexor tibiae motoneurons; DCMD, descending contralateral movement detector neuron. In this and all subsequent figures, excitatory and inhibitory connections are shown as T-bars and filled circles respectively. Details in text. (From Pearson, 1983.)

The pattern of synchronous extensor and flexor activity evoked by the C-neurons is similar to that observed in electromyogram (EMG) recordings in behaving animals when they rapidly flex their tibia from a position close to full flexion (Pearson and Robertson, 1981). These rapid flexions in behaving animals can be evoked by visual, auditory, and tactile stimuli. Correspondingly, the C-neurons have been found to receive strong input from all these sensory sources. One of the most interesting of these inputs is from a pair of large visual-movement-detecting neurons, the descending contralateral movement detectors (DCMDs). The DCMDs fire in high-frequency bursts in response to movements of small objects in the contralateral visual field. Since movements of predators very likely cause the DCMDs to be activated, it is appropriate that these interneurons make powerful excitatory connections to the C-neurons so as to initiate the first phase of the jump.

The DCMDs also make strong excitatory connections to both M-neurons, which in turn inhibit hindleg flexor tibiae motoneurons. Normally activity in the DCMDs does not activate the M-neurons since the threshold for spike initiation in these neurons is high. The necessary condition for activation of the M-neurons by the DCMDs (and by other external stimuli) is concomitant input from leg proprioceptors (Steeves and Pearson, 1982). The latter occurs during the cocontraction phase of the jump. Discharge of the M-neurons during cocontraction inhibits flexor activity and thereby triggers the rapid extension of the hindleg tibia (Pearson *et al.*, 1980). Thus, the jump-trigger neurons, the M-neurons, can be activated by external stimuli only when the animal is prepared to jump, i.e., in the cocontraction phase. It is important that the activity in the M-neurons is regulated in this manner for it ensures that they will not exert their strong inhibitory action on flexor motoneurons during behaviors other than the jump.

The maintenance of the cocontraction phase is achieved by the combination of two mechanisms. The first is direct excitatory coupling of the fast extensor tibiae motoneuron to flexor motoneurons, and the second is positive feedback from leg receptors to both groups of motoneurons. Numerous afferents contribute to this feedback during cocontraction (Heitler and Burrows, 1977b), the most obvious being cuticular receptors activated by deformations of the cuticle. The advantage of having motoneurons excited by sensory feedback is that it reduces the complexity of integrative events in the central nervous system. If, for example, extensor activity was not maintained by sensory input, then tonic central inputs would have to persist during the cocontraction phase. As it is, a phasic central signal is required to initiate the cocontraction, but activity is maintained by positive feedback from peripheral receptors.

3. Locust Flight

There has been considerable interest in the nature of the mechanisms producing the motor pattern for flight in the locust. Weis-Fogh (1956) first proposed that flight movements depend on specific peripheral stimuli, but this was soon shown to be incorrect when Wilson (1961) demonstrated that the basic flight rhythm could be generated in deafferented preparations. Although it is now abundantly clear that a central rhythm generator can produce powerful oscillations in the activity of flight motoneurons and interneurons (Robertson and Pearson, 1982, 1983), it is equally clear that the properties of this central oscillator cannot fully account for the normal flight pattern. Wilson and his colleagues (Wilson and Gettrup, 1963; Wilson and Wyman, 1965) originally proposed that sensory input functioned in a tonic manner to elevate the frequency of the centrally generated rhythm. The first indications that this was incorrect came from Wendler's (1974) observation that the central rhythm could be entrained by phasic afferent input. More recently it has been shown that activity in the forewing stretch receptors can influence the central rhythm in a phase-dependent manner (Fig. 3) (Pearson *et al.*, 1983, and that phasic activity in wind-sensitive hairs and wing campaniform sensillae can reset the flight rhythm (Bacon and Möhl, 1983; Horsmann *et al.*, 1983; Horsmann, 1981, cited in Wendler, 1983b). Thus today it is generally believed that the generation of the flight motor pattern depends on the co-operative interaction of central and peripheral components (Wendler, 1983a,b).

Before considering the neural circuits involved in the central generation of the flight rhythm, it is useful to describe the characteristics of the central oscillator. First, the repetition rate of the rhythmic motor activity is about 12/sec (compared with about 20/sec in the intact animal). Second, the interval from depressor to elevator bursts remains largely independent of changes in the cycle period, whereas the interval from depessor to elevator activity changes markedly (Fig. 4A). Since depressor depolarizations never occur in the absence of a preceding elevator depolarization (Hedwig and Pearson, 1984), the central motor pattern is characterized by a tightly linked elevator–depressor depolarization sequence. Third, the duration of depressor depolarizations is less dependent on cycle time than the duration of the elevator depolarizations (Fig. 4B). And finally, there is no significant shift in the timing of depolarizations in forewing and hindwing elevator motoneurons (Fig.

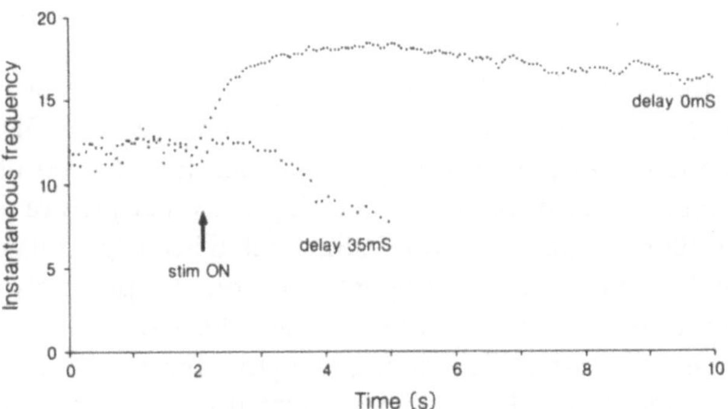

FIGURE 3. Phase-dependent influence of forewing stretch receptors on wingbeat frequency in deafferented preparation. Records of instantaneous frequency (reciprocal of cycle time) versus time for two trials have been superimposed. In both trials identical stimulus trains were delivered to both forewing stretch receptors beginning at about 2 sec following the onset of flight activity. In the first trial there was zero delay between each depressor spike (used to trigger each stimulus in the train) and the onset of the stimulus train. In the second trial the delay was 35 msec. With zero delay there was a marked elevation in wingbeat frequency and a prolongation in the duration of flight activity. These effects did not occur with a 35-msec delay. The profile of flight activity for a stimulus delay of 35 msec was similar to that produced by the preparation without stretch receptor stimulation (From Pearson *et al.*, 1982).

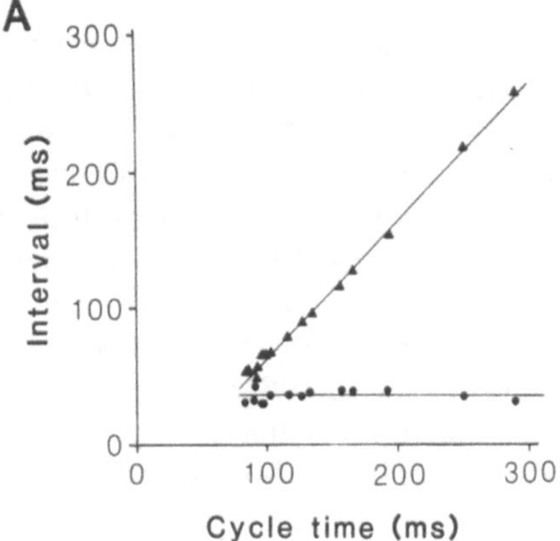

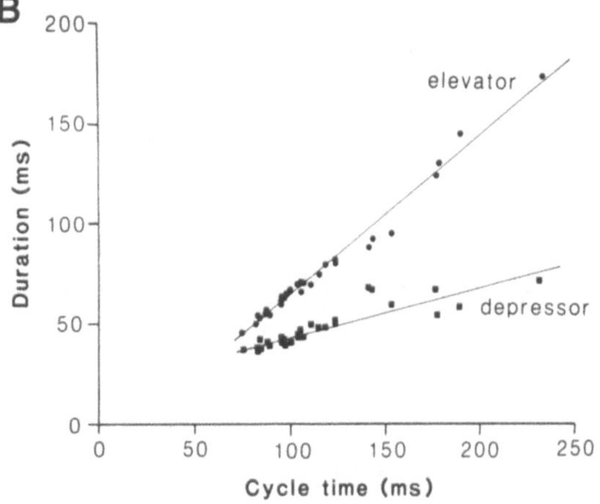

FIGURE 4. Characteristics of the central motor pattern for flight. (A) Graph showing that the interval between the last spike in an elevator burst and the last spike in the next depressor burst (●) remains constant as cycle time varies. Data from simultaneous recordings in a posterior tergocoxal and a first basalar in the metathoracic ganglion. (▲) are the intervals between the last spike in a depressor burst and the last spike in the next elevator burst. Lines fitted by eye. (B) Graph showing the relationship between the cycle time and the durations of the depolarizations in elevator (●) and depressor (■) motoneurons. Data from simultaneous recordings in a tergosternal and a first basalar motoneuron in the mesothoracic ganglion. Durations of the depolarizations were measured at points 5 mV above the minimum membrane potential immediately preceding each depolarization. The lines are best fits (r = 0.99 for elevator and 0.89 for depressor). The graph shows that the changes in the cycle time occur mainly by changes in the duration of the elevator depolarizations, whereas the duration of depressor depolarizations remain relatively constant. (From Hedwig and Pearson, 1984.)

5A), whereas depolarizations in hindwing depressor motoneurons usually precede those in forewing depressor motoneurons by 5 to 10 msec (Fig. 5B).

Recently we have developed a preparation that has allowed us to investigate the neuronal circuits responsible for generating these patterns of activity in flight motoneurons (Robertson and Pearson, 1982). Initial investigations using this preparation indicate that interactions among interneurons are primarily responsible for patterning motoneuronal activity (Robertson and Pearson, 1983, 1985). Over 50 interneurons phasically active with the flight rhythm have been identified morphologically and physiologically. These flight interneurons are distributed within the six neuromeres (three thoracic and first three abdominal) of the thoracic ganglia, an arrangement that provides support for the pleural appendage theory of the evolutionary origin of the insect wing (Robertson *et al.*, 1982; Kukalová-Peck, 1983). We originally proposed (Robertson and Pearson, 1983) that flight interneurons may be segregated into two functional categories, one group providing synaptic input to motoneurons and another group involved in the generation of the rhythm. However, more recently we have shown (Robertson and Pearson, 1985) that the functional organization of interneurons is more complex and that individual interneurons can both drive motoneurons and participate in pattern generation.

With the basic properties and organizational features of flight interneurons to some extent characterized (Robertson and Pearson, 1984), the issue becomes one of trying to understand how they interact to produce the pattern that is transmitted to motoneurons. There is no evidence to suggest that rhythmicity resides in the endogenous properties of a single interneuron or a single class of interneurons. Thus it appears that the network properties of the system are important determinants in establishing rhythmicity. Accordingly, our recent work has been concerned with establishing the circuit connections among flight interneurons (Robertson and Pearson, 1985).

The interneuron that has central importance in the neural circuit for flight is 301 (Fig. 6A). The cell body of 301 is located in the mesothoracic ganglion and its

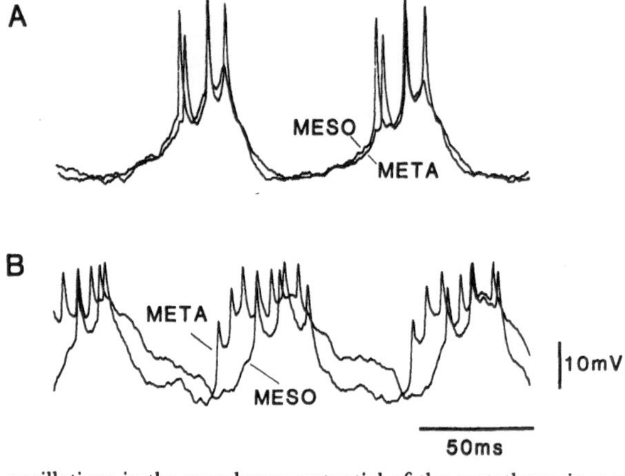

FIGURE 5. Characteristics of the central motor pattern for flight. (A) Superimposed simultaneous recordings in ipsilateral elevator motoneurons in the mesothoracic and metathoracic ganglia. Note that the oscillations in membrane potential were almost exactly in-phase and the spikes in the two neurons occurred almost synchronously. (B) Superimposed simultaneous recordings in ipsilateral depressor motoneurons (second basalars) in the mesothoracic and metathoracic ganglia. Note that oscillations in the membrane potential of the metathoracic motoneuron led those in the mesothoracic motoneuron by 4 to 7 msec. (From Hedwig and Pearson, 1984.)

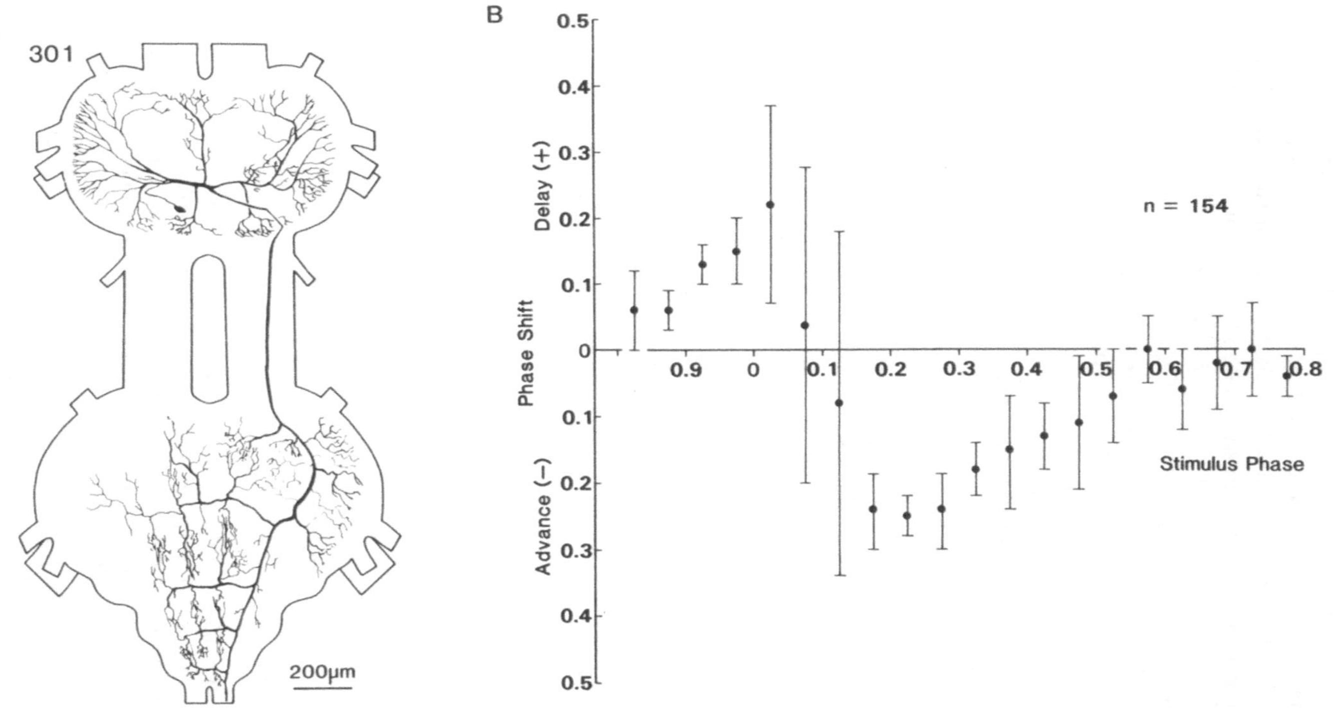

FIGURE 6. Characteristics of interneuron 301. (A) Drawing of the structure of interneuron 301 in the mesothoracic (top) and metathoracic (bottom) ganglia. (B) Phase-response curve summarizing the resetting effects of short-duration (30 msec) pulses of depolarizing current (about 10 nA) delivered to 301 during flight sequences. Cycle period was taken from the start of activity in successive bursts of dorsal longitudinal motoneuron activity. Stimulus phase was measured as the ratio of the latency of the stimulus (the interval from the start of the perturbed cycle to the start of the stimulus) to the average period of the two preceding unperturbed cycles. Phase shift was measured as the ratio of the difference between the period of the perturbed cycle and the average period, to the average period. To avoid biases introduced by irregular flight sequences and the tendency for the frequency to decrease during a single flight sequence, the only data points collected were those taken when the average period at the presentation of the stimulus was within one standard deviation of the mean cycle period for all flight sequences throughout that experiment. Stimuli occurring late in a cycle had a greater effect on the subsequent cycle than on the cycle in which they began. This was because the stimulus had a finite length and late stimuli overlapped the end of the cycle. So for the stimulus phases greater than 0.8, the phase shift was measured for the subsequent cycle and plotted to the left of stimulus phase zero. Note that pulses delivered to 301 can reliably delay and advance the occurrence of the subsequent cycle. (From Robertson and Pearson, 1985.)

axon descends in the contralateral connective and eventually branches extensively in the flight neuropil of the metathoracic ganglion. Short-duration depolarizing current pulses delivered to 301 reset the flight rhythm, either advancing or delaying the occurrence of the subsequent cycle depending on the time of application (Fig. 6B). No other interneuron we have studied has provided a similarly clear phase-response curve. Spikes in 301 result in short-latency inhibitory postsynaptic potentials in some flight interneurons (e.g., 511, Fig. 7A) and longer latency depolarizing potentials in other interneurons (e.g., 501, Fig. 7B). An interesting property of the delayed depolarizing potentials is that their polarity can be reversed by *hyperpolarizing* the postsynaptic neuron (Fig. 7C–E). This indicates that it is caused by an ionic conductance *decrease* across the postsynaptic membrane. Of the two possible mechanisms that could cause such a postsynaptic potential, a direct decreased conductance synapse or disinhibition with graded and tonic release of transmitter at the second synapse, we believe the latter to be the most reasonable explanation. This is consistent with the latency being about twice that of monosynaptic connections, a known disinhibitory pathway from 301 to 501 via 511, and the blocking of both the inhibitory postsynaptic potentials from 301 and the delayed excitatory potentials from 301 to 501 by the application of picrotoxin.

Due to the nature of the disinhibitory pathway from 301 to 501, a burst of

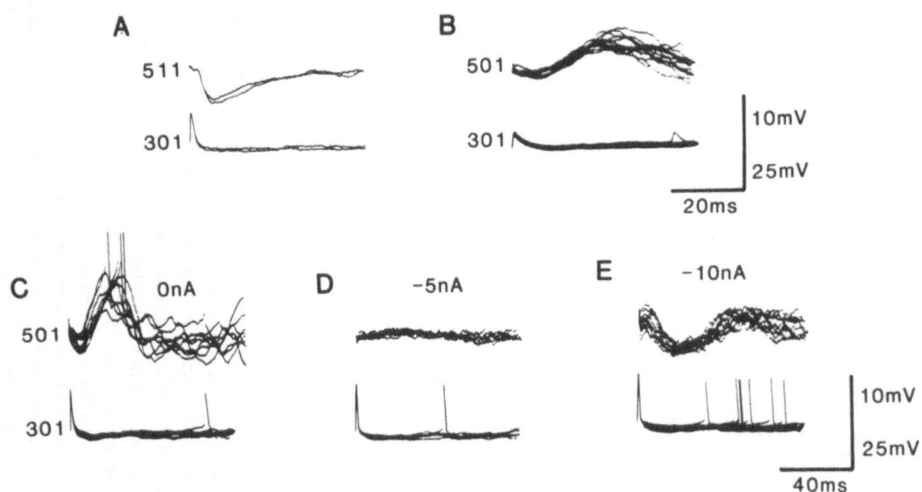

FIGURE 7. Types of connection from 301. (A) Inhibitory postsynaptic potential recorded in 511 a short and constant latency after each spike in 301 which was used to trigger successive oscilloscope sweeps. This is taken as indicative of a probable monosynaptic connection. (B) Depolarizing potential recorded in 501 a long and constant latency after each spike in 301. Note that the latency and duration of the postsynaptic potential shown here are approximately twice those of the postsynaptic potential shown in A. This type of connection underlies delayed excitation from 301 to other neurons. (C–E) Passage of hyperpolarizing current into 501 (the postsynaptic neuron) reverses the delayed excitatory potential recorded in 501 following each spike in 301. (C) No current is passed. (D) 5 nA of hyperpolarizing current is passed and the amplitude of the postsynaptic potential is negligible. (E) 10 nA of hyperpolarizing current is passed and the postsynaptic potential is reversed. Note that the time scale in C, D, and E is different to that in A and B. (From Robertson and Pearson, 1985, with modifications.)

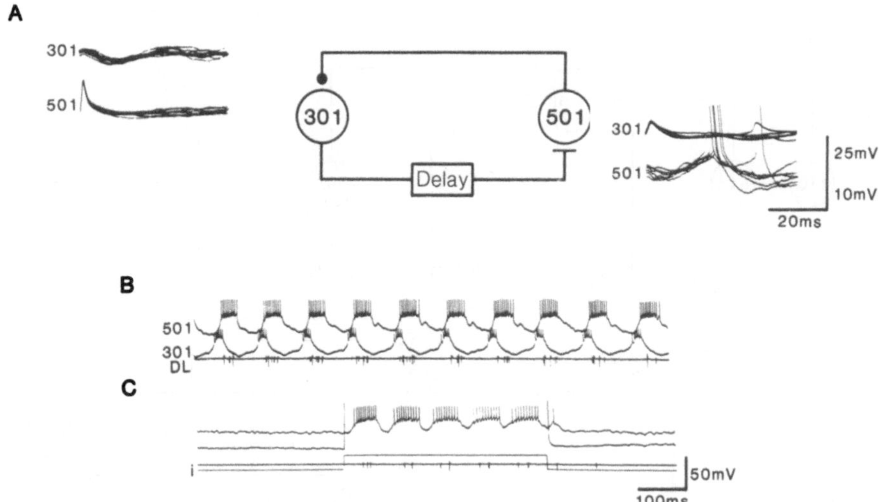

FIGURE 8. The circuit formed by 301 and 501. (A) Diagrammatic representation of delayed excitation from 301 to 501, and feedback inhibition from 501 to 301. Delayed excitation, here represented as an excitatory connection incorporating a delay box, probably results from a disinhibitory pathway with tonic and graded release of transmitter at the second synapse. The insets show examples of the inhibitory postsynaptic potential in 301 following each spike in 501 (left) and the delayed excitatory potential in 501 following each spike in 301 (right) as visualized using multiple oscilloscope sweeps triggered off the rising phase of the presynaptic spike. (B) Simultaneous intracellular recordings of 301 and 501 during a flight sequence to show their bursting activity in the central motor pattern. The time of depression is monitored extracellularly by recording *en passant* from the axons of dorsal longitudinal motoneurons (DL) in nerve 6 of the prothoracic ganglion (cut distally). (C) Stimulation of 301 with a long-duration (monitored with the current trace, i) pulse of constant depolarizing current (about 10 nA) induces rhythmical bursting activity in 501 and dorsal longitudinal motoneurons. Note that the cycle frequency and the phase relationships of the bursting neurons are similar to those observed in the central motor pattern (compare with B). The 301 trace disappears during the passage of current due to the use of high-resistance electrodes preventing the bridge from being balanced. B and C are from the same experiment. (From Robertson and Pearson, 1985, with modifications.)

spikes in 301 occurring in the elevator phase of flight activity causes 501 to generate a burst in the depressor phase (see Fig. 8B). This phase shift together with the existence of an inhibitory connection from 501 to 301 results in a network with properties appropriate for the generation of oscillatory activity (Fig. 8A). It is of significance, therefore, that constant current depolarization of 301 can induce rhythmical activity very similar to that expressed during flight sequences (Fig. 8D). Although we consider the 301–501 circuit to be an important element in the neural circuit for flight (Fig. 9), we recognize that other circuit elements may be equally or even more important since we know that rhythmical bursts can be recorded from flight interneurons after the actions of 301 have been suppressed by the application of picrotoxin (Robertson and Pearson, 1985). The circuit illustrated in Fig. 9 includes all the interneuronal connections discovered to date, as well as the known connections from the forewing stretch receptor. Clearly this circuit is incomplete, e.g., we have not identified any input connections to 401 or

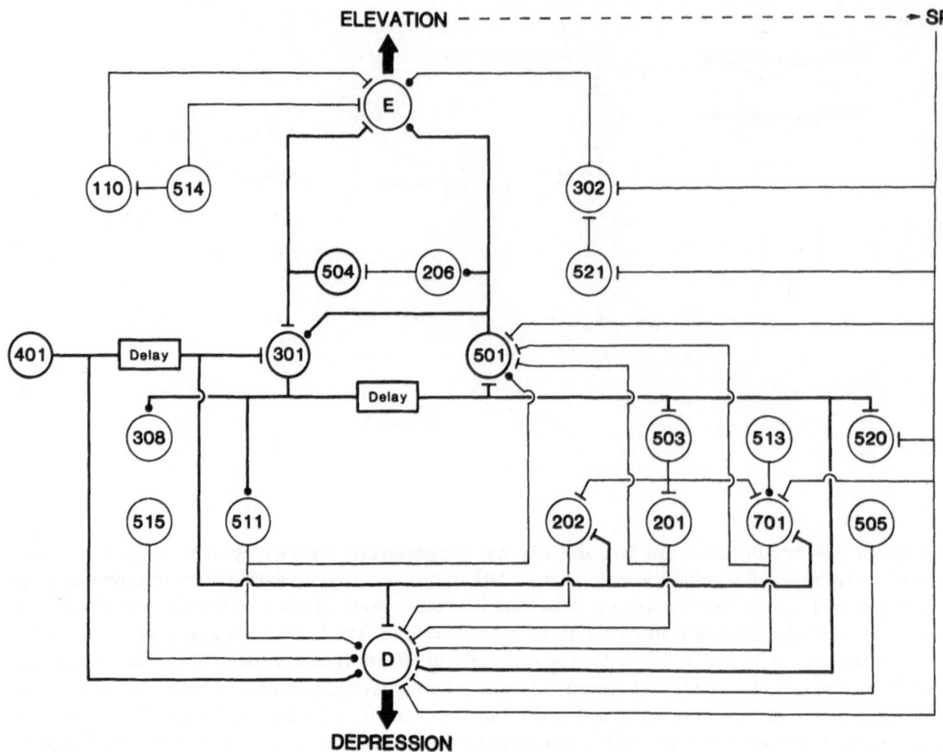

FIGURE 9. Circuitry in the locust flight pattern generator. Diagrammatic representation of all the connections found between flight neurons and the wing stretch receptor to date. Elevator motoneurons (E) cause elevation of the wing which mechanically stimulates (dashed line) the stretch receptor (SR) which feeds back into the network. Depressor motoneurons (D) cause depression of the wing. The delay box in an excitatory pathway indicates that the postsynaptic potential is of the delayed excitatory type shown in Fig. 7B. Bold lines belong to interneurons stimulation of which is known to be able to reset the rhythm. In order not to complicate the diagram, no distinction has been made between right and left or between fore and hind. Further explanation in text.

output connections of 308. Nevertheless, this circuit does suggest possible mechanisms for accounting for some characteristics of the central motor pattern.

1. The tight linkage of elevator and depressor depolarizations (depressor depolarizations following elevator depolarizations with a constant latency) may be due to 504 driving elevator motoneurons and 301. 301 causes a delayed burst of activity in 503 which in turn excites depressor motoneurons via 201. Therefore a burst in 504 will evoke a depolarization in elevator motoneurons followed, after a constant latency, by a depolarization in depressor motoneurons.
2. The relatively constant duration of depressor bursts may be because the generation of these bursts depends primarily on the 301–501 circuit. The outputs from this circuit are organized to promote depressor depolarizations (delayed excitation via 503 and 201, and removal of inhibition via

511) and to inhibit elevator activity (directly from 501). The 301–501 circuit exhibits characteristics that have been shown in another system to result in burst duration remaining independent of cycle period (Getting, 1983). These characteristics promote relatively constant duration bursts of activity in 301 and 501 and thus give relatively constant duration depolarizations in depressor motoneurons.

3. The shift in timing of depolarizations in forewing and hindwing depressor motoneurons (hind leading fore) and the lack of any shift in timing in fore and hindwing elevator motoneurons may be due to the organization of the interneurons providing the excitatory synaptic input to the motoneurons. Common inputs to hindwing and forewing elevator motoneurons are derived from interneurons 504 and 514, both of which are located in the metathoracic ganglion. These interneurons connect directly to elevator motoneurons and the difference in path length from source to forewing and hindwing motoneurons amounts to the meta-meso distance. This imparts a conduction delay of approximately 1 msec which would not be obvious in simultaneous recordings from hindwing and forewing elevator motoneurons. A common source of input to depressor motoneurons is 503 located in the metathoracic ganglion. In this case, however, 503 does not activate the motoneurons directly, but goes through premotor interneurons (primarily 201) which are located in the ganglion anterior to the motoneurons they drive. Thus for hindwing depressor motoneurons, the pathway is from metathoracic 503, through mesothoracic 201, and back to metathoracic motoneuron. For forewing motoneurons, the equivalent pathway is metathoracic 503, through prothoracic 201, and back to mesothoracic motoneuron. The difference in path length thus amounts to twice the meso-pro distance minus the meso-meta distance. In locusts the distance from the prothoracic to the mesothoracic ganglion is considerably longer than that from the mesothoracic to the metathoracic ganglion. Assuming conduction speed to be similar in ascending and descending interneurons, the conduction delay introduced by extra path length for the excitation of forewing depressors is from 5 to 7 msec, a value close to that observed (Hedwig and Pearson, 1984).

4. Conclusions

The rationale implicit in the work reviewed here is that by investigating the properties of individual neurons and by laboriously cataloguing interconnections between neurons, we may better understand how nervous systems control behavior. The circuit diagrams of Figs. 2 and 9 summarize the connections and pathways of neurons closely involved in the jumping and flight systems of the locust. Although in neither case has the validity of the circuit been tested by electronic or mathematical simulations (this issue is discussed with reference to the *Tritonia* swimming system in Chapter 1, this volume), in both cases it is clear that these circuits can account for some features for the observed motor patterns. Instead of concentrating on the circuitry *per se* and in view of the rationale stated above, we will conclude with some general observations about the organization of insect motor systems.

The concept of identifiability of neurons, particularly in arthropods, is really in no need of further substantiation (Hoyle, 1977; Bullock, 1980). However, it is particularly striking in our work on the locust just how readily identifiable neurons can be. The thoracic interneurons mentioned here have quite distinctive as well as esthetically pleasing morphologies. This together with their powerful physiological properties work to confer on some neurons prestige that perhaps they do not completely deserve given the depth of our ignorance of other interneurons. Nevertheless, for a few interneurons, their physiological properties do suggest a unique functional role. The C and M in the jump system each have a single important function and, given their properties, it is unlikely that they are involved in other behaviors. Moreover, searches in the thoracic ganglia have failed to reveal any other neuron that might be able to substitute for them. This bespeaks a high level of functional specialization of some locust interneurons. Such specialization is less obvious in the flight system. This might indicate merely a lack of understanding of the precise roles of interneurons. However, greater numbers and interactions of interneurons are involved in the flight system and the cooperation of interneurons in homologous sets becomes evident (Robertson and Pearson, 1983). In addition, for some other interneurons, more than one interneuron with apparently identical morphology and physiology can be found in the same hemiganglion (Robertson and Pearson, 1983). At present, such groups are treated as units, but this may not be possible in the future. This level of complexity in the flight system is challenging rather than daunting, for if we can adequately explain the mechanisms in this system, then we may gain an insight into the control of behavior by more complex nervous systems.

As more information on identified neurons has amassed, it has been possible to use this information to address evolutionary issues (Bacon, 1980; Dickinson, 1980; Robertson *et al.*, 1982; King and Valentino, 1983). Earlier we mentioned that data on the organization of flight interneurons were used to support the pleural appendage theory for the evolutionary origin of the insect wing (Robertson *et al.*, 1982; Kukalová-Peck, 1983). The particular organizational feature was that flight interneurons are serially repeated in abdominal neuromeres without any apparent functional necessity for such an organization. This illustrates a lack of optimal functional design (i.e., minimum numbers of neurons, connections, etc.) in the system and it follows that other features that add to the complexity of a system might exhibit characteristics either of poor functional design or which are inexplicable in such terms. This is evident to some extent in the jumping system also. The primary role of C-neuron in this system is to drive simultaneously flexor and extensor motoneurons of the *meta*thoracic tibia. Yet C-neuron is located in the *meso*thoracic ganglion. Similarly, M-neuron functions to inhibit metathoracic flexor tibiae motoneurons, yet it shows axonal branching in the mesothoracic ganglion as well. These morphological features are reminiscent of a system that may have been segmentally arranged before the overt specialization of the metathoracic legs for enhanced jumping ability. With these examples in the flight and jumping systems, we conclude that it may not always be possible to establish meaningful function in all observed features of a system.

It is of some interest that the role of sensory input in these motor systems is not confined simply to compensation for perturbations in the environment or for noise in the motor pattern generator. In both cases, afferent input is a fundamental necessity for production of the behavior. During a jump, cocontraction of extensor and flexor motoneurons is maintained partly via cuticular mechanoreceptors and rapid extension cannot be triggered by the M-neuron until its excitability has been increased by proprioceptive input. During flight, phasic afferent signals from the wing stretch receptors (Pearson *et al.*, 1982), the wind-sensitive head hairs (Bacon and Möhl, 1983; Horsmann *et al.*, 1983), and the campaniform sensillae (Horsmann, 1981, cited in Wendler, 1983b) interact with the central pattern generator to ensure that a behaviorally appropriate motor pattern in produced (see also Wendler, 1983a,b) It is worth stressing that this does not invalidate the concept of central pattern generation. It does, however, limit it and results in a need for greater specificity with the terminology. So, in the locust, the behavior of flight is produced by a flight pattern generator, an element of which is the central pattern generator which contains one or more oscillatory circuits (possibly separable, but not into homologous forewing and hindwing oscillators). In contrast with the jumping systems, at present there is no acceptable explanation of the cellular mechanisms by which sense organs help to generate flight motor activity. This can be done only with a knowledge of the circuitry of the central component. That such an approach is possible is attested to by the monosynaptic projections of the wing stretch receptor illustrated in Fig. 9 (see also Pearson *et al.*, 1982).

It could be argued that to a first approximation the circuitry underlying the motor pattern of a locust's jump has been worked out. It might also be claimed that a central circuit for locust flight is well under way to completion. One interesting point is that the motor patterns are produced by nearly universal neuronal processes (including, for example, coactivation of antagonists, proprioceptive gating, graded interactions, disinhibition). That is to say the overall circuits are unique only in the way that well-described components are assembled. This is heartening for those in search of general principles of neuronal organization. The next step is to look more closely at the control of behavior rather than motor patterns and also at the interactions between systems. For example, how is the jump and subsequent flight co-ordinated: purely peripherally as suggested by Weis-Fogh's (1956) sensory criteria for flight (i.e., wind excitation and lack of tarsal inhibition) or are there central interactions as well? The means to find out are at our disposal.

References

Adams, M. E., and O'Shea, M., 1983, Peptide cotransmitter at a neuromuscular junction, *Science* **221**:286–289.

Bacon, J. P., 1980, An homologous interneurone in a locust, a cricket and mantid, *Verh. Dtsch. Zool. Ges.* **73**:300.

Bacon, J. P., and Möhl, B., 1983, The tritocerebral commissure giant (TCG) wind-sensitive interneurone in the locust. I. Its activity in straight flight, *J. Comp. Physiol.* **150**:439–452.

Bullock, T. H., 1980, Reassessment of neural connectivity and its specification, in: *Information Processing in the Nervous System* (H. M. Pinsker and W. D. Willis, Jr., eds.), Raven Press, New York, pp. 199–220.

Burrows, M., 1981, Local interneurones in insects, in: *Neurones without Impulses* (A. Roberts and B. M. H. Bush, eds.), Cambridge University Press, Cambridge, pp. 199–121.

Burrows, M., 1982, Interneurones co-ordinating the ventilatory movements of the thoracic spiracles in the locust, *J. Exp. Biol.* **97:**385–400.

Dickinson, P., 1980, Neuronal control of gills in diverse *Aplysia* species: Conservative evolution, *J. Comp. Physiol.* **139:**17–23.

Evans, P. D., and O'Shea, M., 1978, The identification of an octopaminergic neurone and the modulation of a myogenic rhythm in the locust, *J. Exp. Biol.* **73:**235–260.

Getting, P. A., 1983, Mechanisms of pattern generation underlying swimming in *Tritonia.* II. Network reconstruction, *J. Neurophysiol.* **49:**1017–1035.

Hedwig, B., and Pearson, K. G., 1984, Patterns of synaptic input to identified flight motoneurons in the locust, *J. Comp. Physiol.* **154:**745–760.

Heitler, W. J., 1974, The locust jump. Specializations of the metathoracic femoral-tibial joint, *J. Comp. Physiol.* **89:**93–104.

Heitler, W. J., and Burrows, M., 1977a, The locust jump. I. The motor programme, *J. Exp. Biol.* **66:**203–219.

Heitler, W. J., and Burrows, M., 1977b, The locust jump. II. Neural circuits of the motor programme, *J. Exp. Biol.* **66:**221–241.

Horsmann, U., 1981, *Flugrelevante Afferenzen und ihre Verarbeitung bei Wanderheuschrecke (locusta migratoria* L.), Diplomarbeit, Köln.

Horsmann, U., Heinzel, H. G., and Wendler, G., 1983, The phasic influence of self-generated air current modulations on the locust flight motor, *J. Comp. Physiol.* **150:**427–438.

Hoyle, G. (ed.), 1977, *Identified Neurons and Behavior of Arthropods,* Plenum Press, New York.

King, D. G., and Valentino, K. L., 1983, On neuronal homology: A comparison of similar axons in *Musca, Sacrophaga* and *Drosophila* (Diptera: Schizophora), *J. Comp. Neurol.* **219:**1–9.

Kukalová-Peck, J., 1983, Origin of the insect wing and wing articulation from the arthropodan leg, *Can. J. Zool.* **61:**1618–1669.

Pearson, K. G., 1976, The control of walking, *Sci Am.* **235**(6)**:**72–86.

Pearson, K. G., 1983, Neural circuits for jumping in the locusts, *J. Physiol. (Paris)* **78:**765–771.

Pearson, K. G., and Robertson, R. M., 1981, Interneurons coactivating hindleg flexor and extensor motoneurons in the locust, *J. Comp. Physiol.* **144:**391–400.

Pearson, K. G., Heitler, W. J., and Steeves, J. D., 1980, Triggering of locust jump by multimodal inhibitory interneurons, *J. Neurophysiol.* **43:**257–278.

Pearson, K. G., Reye, D. N., and Robertson, R. M., 1983, Phase-dependent influences of wing stretch receptors on flight rhythm in the locust, *J. Neurophysiol.* **49:**1168–1181.

Robertson, R. M., and Pearson, K. G., 1982, A preparation for the intracellular analysis of neuronal activity during flight in the locust, *J. Comp. Physiol.* **146:**311–320.

Robertson, R. M., and Pearson, K. G., 1983, Interneurons in the flight system of the locust: Distribution, connections and resetting properties, *J. Comp. Neurol.* **215:**33–50.

Robertson, R. M., and Pearson, K. G., 1984, Interneuronal organization in the flight system of the locust, *J. Insect Physiol.* **30:**95–101.

Robertson, R. M., and Pearson, K. G., 1985, Neural circuits in the flight system of the locust. *J. Neurophysiol.* **53:**110–128.

Robertson, R. M., Pearson, K. G., and Reichert, H., 1982, Flight interneurons in the locust and the origin of insect wings, *Science* **217:**177–179.

Romer, H., Rheinlaender, J., and Dronse, R., 1981, Intracellular studies on auditory processing in the metathoracic ganglion of the locust, *J. Comp. Physiol.* **144:**305–312.

Siegler, M. V. S., and Burrows, M., 1983, Spiking local interneurons as primary integrators of mechanosensory information in the locust, *J. Neurophysiol.* **50:**1281–1295.

Steeves, J. D., and Pearson, K. G., 1982, Proprioceptive gating of inhibitory pathways to hindwing flexor motoneurons in the locust, *J. Comp. Physiol.* **146:**507–515.

Weis-Fogh, T., 1956, Biology and physics of locust flight. IV. Notes on sensory mechanisms in locust flight. *Philos. Trans. R. Soc. London Ser.* B **239:**553–584.

Wendler, G., 1974, The influence of proprioceptive feedback on locust flight coordination, *J. Comp. Physiol.* **88:**173–200.

Wendler, G., 1983a, The interaction of peripheral and central components in insect locomotion, in: *Neuroethology and Behavioral Physiology* (F. Huber and H. Markl, eds.), Springer-Verlag, Berlin, pp. 42–53.

Wendler, G., 1983b, The locust flight system: Functional aspects of sensory input and methods of investigation, in: *BIONA—Report 2* (W. Nachtigall, ed.), Gustav Fischer, Stuttgart, pp. 113–125.

Westin, J., and Ritzmann, R. E., 1982, The effect of single giant interneuron lesions on wind evoked motor responses in the cockroach, *Periplaneta americana, J. Neurobiol.* **13:**127–140.

Wilson, D. M., 1961, The central nervous control of locust flight, *J. Exp. Biol.* **38:**471–490.

Wilson, D. M., and Gettrup, E., 1963, A stretch reflex controlling wingbeat frequency in grasshoppers, *J. Exp. Biol.* **40:**171–185.

Wilson, D. M., and Weis-Fogh, T., 1962, Patterned activity of co-ordinated motor units studied in flying locusts, *J. Exp. Biol.* **40:**643–667.

Wilson, D. M., and Wyman, R. J., 1965, Motor output patterns during random and rhythmic stimulation of locust thoracic ganglia, *Biophys. J.* **5:**121–143.

Wohlers, D. W., and Huber, F., 1982, Processing of sound signals by six types of neurons in the prothoracic ganglion of the cricket, *Gryllus campestris* L, *J. Comp. Physiol.* **146:**161–174.

Weis-Fogh, T., 1956, Biology and physics of locust flight. IV. Notes on sensory mechanisms in locust flight, *Phil. Trans. R. Soc. London Ser. B* 239:553–584.
Wendler, G., 1974, The influence of proprioceptive feedback on locust flight coordination, *J. Comp. Physiol.* 88:173–200.
Wendler, G., 1983, The interaction of peripheral and central components in insect locomotion, in: *Neuroethology and Behavioral Physiology* (F. Huber and H. Markl, eds.), Springer-Verlag, Berlin, pp. 42–53.
Wendler, G., 1985, The locust flight system: Functional aspects of sensory input and methods of investigation, in: *Insect Locomotion* (M. Gewecke and G. Wendler, eds.), Paul Parey, Hamburg, pp. 113–125.
Westin, J., and Ritzmann, R. E., 1982, The effect of single giant interneuron lesions on wind-evoked motor responses in the cockroach, *Periplaneta americana*, *J. Neurobiol.* 13:127–139.
Wilson, D. M., 1961, The central nervous control of flight in a locust, *J. Exp. Biol.* 38:471–490.
Wilson, D. M., and Gettrup, E., 1963, A stretch reflex controlling wingbeat frequency in grasshoppers, *J. Exp. Biol.* 40:171–185.
Wilson, D. M., and Weis-Fogh, T., 1962, Patterned activity of co-ordinated motor units, studied in flying locusts, *J. Exp. Biol.* 39:643–667.
Wilson, D. M., and Wyman, R. J., 1965, Motor output patterns during random and rhythmic stimulation of locust thoracic ganglia, *Biophys. J.* 5:121–143.
Wohlers, D. W., and Huber, F., 1982, Processing of sound signals by six types of neurons in the prothoracic ganglion of the cricket, *Gryllus campestris* L., *J. Comp. Physiol.* 146:161–173.

3

Neural Mechanisms for the Production of the Lobster Pyloric Motor Pattern

JOHN P. MILLER AND ALLEN I. SELVERSTON

1. Introduction

In order to understand and explain the biological mechanisms underlying an animal's behavior, we must investigate the structure and physiology of its nervous system. One form of behavior that is particularly amenable to physiological analysis is the class of rhythmic behaviors such as locomotion, respiration, mastication, copulation, and circulation (Delcomyn, 1980; Kristan *et al.*, 1977). The behaviors are easily observed and quantified, and the components of the nervous systems involved in their production are usually straightforward to identify and manipulate experimentally. As a result of the work being done on several different rhythmic behaviors, general principles of nervous system organization and function are continuing to emerge (see Chapters 1, 4, 5 of this volume).

A system that is ideally suited to the study of rhythmic motor pattern generation is the stomatogastric system of decapod crustacea (see Selverston *et al.*, 1976). This system is distinguished by the following features: (1) the behavior of the stomach, including the neuromuscular physiology and the activity in each motor axon, can be observed and quantified; (2) most of the neurons responsible for generation of the motor activity controlling the behavior are located within the stomatogastric ganglion which contains only about 30 neurons; (3) all of these neurons are identified; (4) subthreshold integrative activity can be recorded from the somata of

JOHN P. MILLER • Department of Zoology, University of California, Berkeley, California, 94720.
ALLEN I. SELVERSTON • Department of Biology, University of California, San Diego, La Jolla, California 92093.

these identified neurons; and (5) central nervous system input to the ganglion comes through a single nerve from a group of relatively small ganglia, allowing detailed study of central control over the behavior. Two independent rhythmic behaviors are executed in the foregut—mastication of food by internal "teeth" in the gastric mill region and pumping/filtering of the resulting particles in the "pyloric" region. The muscles that drive these behaviors are innervated by axons from neurons in the stomatogastric ganglion. In the California spiny lobster, a subset of 13 out of the 30 neurons in the stomatogastric ganglion are the motoneurons that innervate the pyloric musculature. These 13 motoneurons, along with one interneuron with which they interconnect, are called the "pyloric network."

In this chapter, we review recent work that addresses three specific questions concerning the neural basis of the pyloric pattern: (1) Why do the pyloric neurons fire in bursts of activity rather than tonically? (2) How are these bursts of activity coordinated to occur with the observed phase relationships? (3) What determines the overall frequency of the pattern? Although we will *not* deal extensively with mechanisms of central control over the stomatogastric motor pattern, one other chapter in this volume examines those questions in considerable detail (see Chapter 4).

2. Structure and Operation of the Pyloris

The stomach of the lobster is ectodermally derived, is lined with chitin, and is operated by striated musculature. The pyloric region at the rear of the foregut is composed of a complex set of saclike "setae" distributed along a canal that connects the gastric mill region of the stomach to the hindgut and hepatopancreas duct (Maynard and Dando, 1974). Rhythmic contractions of the pyloric musculature act to pump food particles from the foregut into the hindgut and also act to filter large particles back into the gastric mill region for further mastication (Hartline and Maynard, 1975).

Our standard physiological experiments are performed on preparations in which the stomatogastric ganglion, the esophageal ganglion, the pair of commissural ganglia, and all of the related interconnecting nerves are dissected off the lobster stomach, as shown in Fig. 1A. When placed in a chamber containing cooled, oxygenated physiological saline solution containing antibiotics, the pyloric pattern persists for at least two days at a very regular pace, as shown in Fig. 1B. If the commissural and esophageal ganglia are removed—either by cutting the stomatogastric nerve or placing a sucrose block at the position indicated with an asterisk—the pyloric pattern slows down, becomes irregular, and ultimately stops completely. Thus, several cells in the higher ganglia must contribute to the generation and maintenance of the pyloric pattern in the intact preparation, even though all of the motoneurons are located in the stomatogastric ganglion.

The stomatogastric neurons that participate in the production of the pyloric pattern have been identified and their circuitry studied extensively (Maynard, 1972; Maynard and Selverston, 1975; Hartline and Gassie, 1979; Eisen and Mar-

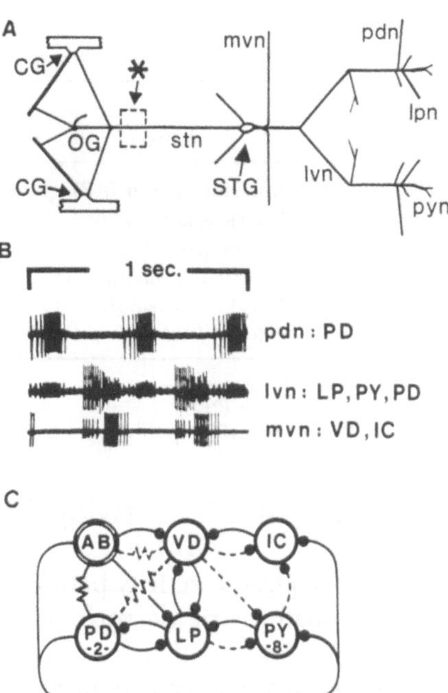

FIGURE 1. (A) Diagram of the stomatogastric ganglion preparation as it looks when pinned out in a Sylgard-lined chamber. The abbreviations of the nerves are given in Table I. Four are included: the two commissurals (R-CG and L-CG), the esophageal (OG) and the stomatogastric (STG). The asterisk indicates the position where sucrose blocks are applied. A full description is given in Selverston *el al.* (1976). (B) Burst pattern of the pyloric network. These extracellular recordings were obtained from preparations as shown in A. (C) Synaptic connectivity "circuit" diagram of the pyloric network. The inhibitory chemical synapses are indicated by black dots and electrotonic synapses by resistor symbols. Dashed lines indicate weak synapses, solid lines indicate strong synapses. All abbreviations are listed and defined in Tables I and II.

der, 1982). A circuit diagram representing the synaptic connectivities is shown in Fig. 1C. The neuron names corresponding to the abbreviations in these figures are given in Tables I and II. Except for the lumping of two classes of cells with slightly different properties (i.e., the "early" and "late" pyloric cells, or PE and PL cells) into one class (the PY cells), this diagram is considered to be complete and precise. Note that all but one of the pyloric neurons are motoneurons. However, as well as sending axons out to the stomach musculature, the motoneurons also make synaptic connections with one another. One neuron, the anterior burster (AB), is an interganglionic interneuron, making connections with neurons in the stomatogastric and commissural ganglia.

TABLE I
Nerves Containing Axons of Cells Involved in Pyloric Pattern

Nerve	Abbreviations
Inferior esophageal	ion
Superior esophageal	son
Inferior ventricular	ivn
Stomatogastic	stn
Median ventricular	mvn
Lateral ventricular	lvn
Lateral pyloric	lpn
Pyloric dilator	pdn
Pyloric	pyn

TABLE II
Pyloric Neurons in the Stomatogastric Ganglion

Neuron	Number	Axon location	Approximate phase
Anterior burster (AB)	1	stn	0
Pyloric dilator (PD)	2	pdn	0
Lateral pyloric (LP)	1	lpn	0.5
Ventricular dilator (VD)	1	mvn	0.75
Inferior cardiac (IC)	1	mvn	0.5
Pyloric (PY)[a]	8	pyn	0.65

[a]2 classes:
Early (PE)
Late (PL)

3. Mechanisms Underlying Burst Generation in Pyloric Cells

How might the activity pattern shown in Fig. 1B be produced by the neurons in the stomatogastric nervous system? Theoretical models for the generation of rhythmic neuronal activity can be subdivided into two general classes. In one class of models, rhythmic activity of neurons is attributed to *intrinsic* properties of the neuronal membrane itself: cyclic conductance changes lead to cyclic voltage changes called "bursting pacemaker potentials" (BPPs). If several such "endogenous burster" cells are synaptically interconnected, a stable rhythmic pattern could (theoretically) result. Cells capable of generating BPPs have been identified in several invertebrate species. In a second class of models, rhythmic activity of neurons is attributed to properties of the network of neuronal interconnections rather than to intrinsic neuronal properties. Theorists have demonstrated that model neurons that are not intrinsically capable of producing BPPs can be interconnected into "circuits" incorporating negative or postitive feedback loops that are, nevertheless, oscillatory. If model parameters are adjusted approximately, rhythmic activity patterns "emerge" from such networks of cells.

In the lobster stomatogastric system, rhythmic activity does, in fact, derive from both classes of mechanisms. Some of the pyloric cells can generate BPPs under certain conditions and reciprocally inhibitory synaptic connections (i.e., negative feedback loops) contribute to burst generation.

3.1 Generation of Bursting Pacemaker Potentials

Bursting pacemaker potentials result from the integration of several discrete ionic conductance mechanisms, including (1) conductances that slowly depolarize the resting membrane to a "plateau" threshold, (2) conductances that result in a rapid, regenerative depolarization to the "plateau" level above the threshold for action potential generation, and (3) conductances that regeneratively repolarize the membrane to the resting (inactive) state (for a good review, see Koester and Byrne, 1980). Any neuron that can produce these oscillatory potentials in the absence of any synaptic input can be considered a true "endogenous burster" (Alving, 1968).

We have tested individual pyloric neurons for their ability to generate BPPs by isolating each cell from synaptic input and observing the resulting activity (Miller and Selverston, 1982a) Isolation of a particular neuron was accomplished by a combination of the following techniques: (1) the inputs from different ganglia were blocked with a sucrose pool on the stomatogastric nerve (asterisk, Fig. 1A), (2) cells within the stomatogastric ganglion presynaptic to the "test" cell were killed using the dye-sensitized photoinactivation technique (Miller and Selverston, 1979), and (3) any remaining presynaptic cells were hyperpolarized below their threshold for transmitter release. Figure 2A shows the activity recorded from three pyloric motor neurons that were isolated from all other pyloric cells in one such experiment. The left part of the panel shows their bursting behavior before the stomatogastric nerve was blocked. Immediately after the nerve was blocked with a sucrose pool, however, all rhythmic activity ceased, as shown in the right part of the panel. Two neurons fired tonically at a low rate, and one became completely inactive. In fact, none of the pyloric motoneurons produced BPPs when totally isolated from all synaptic input in experiments such as these.

The AB cell, which is the only strict interneuron in the pyloric circuit, displayed a variable behavior. Bursting activity would sometimes persist after the motoneurons stopped bursting. This led to its tentative classification as an endogenous burster. Recent experiments utilizing more extensive sucrose blocks (i.e., having greater lengths of nerve in the sucrose pool, with the nerve desheathed for greater penetrance of sucrose) have demonstrated that the AB cell is also incapable of generating BPPs in total synaptic isolation (F. Nagy and J. P. Miller, unpublished data). Therefore, none of the pyloric cells can be classified as true endogenous bursters, according to the strict definition presented above.

Bursting is reinitiated in pyloric cells, however, by activity in the stomatogastric nerve. Figure 2B shows the response of three isolated pyloric cells to a 1.5-sec stimulus to the stomatogastric nerve. The cells began an episode of bursting that long outlasted the stimulus. In this case, the cells burst for approximately 30 sec before returning to the nonpatterned activity normally observed under sucrose block. Thus, these cells can be classified as "conditional bursters," with BPP generation being contingent on the presence of a "neuromodulatory substance." Such

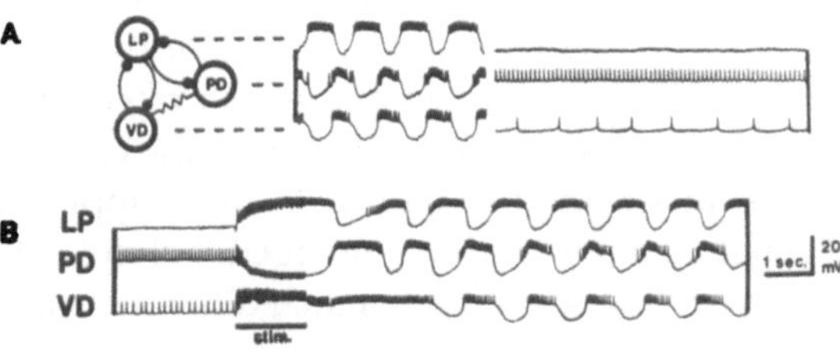

FIGURE 2. Modulation of bursting pacemaker potential generation by inputs in the stomatogastric nerve. For these records, all cells but the VD, LP, and PD were either photoinactivated or hyperpolarized. (A) When stomatogastric nerve (stn) inputs were active (left side of panel), these cells generated BPPs. When the stomatogastric nerve was blocked, the LP became inactive and the VD and PD cells fired tonically. (B) During sucrose block of the nerve, a short (1.5 sec) stimulus at 50 Hz to the stn, between the block point and the stomatogastric ganglion, reinitiated BPP generation for about 30 sec.

conditional bursting can be reliably evoked from AB, PD, LP, and VD cells. The other pyloric neurons—the IC and PY cells—can also be induced to generate regenerative potential changes when the stomatogastric nerve is active. However, only a subset of the conductance mechanisms necessary for BPPs is present. The IC and PY cells display the ability to jump up to a "plateau potential" equivalent to the depolarized phase of a BPP, but require a "trigger pulse" of current input in order to display this potential jump.

Several neuromodulatory substances capable of initiating BPP and plateau potential generation in these cells have recently been identified, and the results of those studies are summarized in other chapters in this volume (Chapters 4 and 17). These substances are undoubtedly released into the stomatogastric ganglion (STG) from other neurons with cell bodies in the esophageal and commissural ganglia that are involved in the initiation, maintenance, and modualtion of pyloric activity. Several such neurons have indeed been identified and characterized (Claiborne and Selverston, 1984; Dickinson and Nagy, 1983; Moulins and Cournil, 1982; Nagy and Dickinson, 1983; Russell and Hartline, 1981; Robertson and Moulins, 1981). The biophysical mechanisms underlying the intiation and generation of BPPs in pyloric cells are currently under investigation (Gola and Selverston, 1981; Russell and Hartline, 1982).

3.2 Generation of Bursts Due to Reciprocal Inhibition

Several sets of reciprocally inhibitory cell pairs are embedded within the pyloric circuitry. For example, the VD and LP cells each make an inhibitory chemical synapse upon the other, as shown in Fig. 1C. McDougall (1903) and Brown (1914) first hypothesized that such reciprocally inhibitory connections might result in alternating bursts of activity, even if the cells were incapable of generating BPPs. In order to test whether or not reciprocal inhibition might contribute to burst generation in the stomatogastric ganglion, individual cell pairs were isolated from other pyloric neurons and tested for their ability to generate alternating bursts of activity. These experiments were performed with the stomatogastric nerve blocked, so that none of the cells being tested were capable of generating BPPs. It was found that only one of the many pairs of reciprocally inhibitory cells—the LP and PD cells—could generate stable, alternating bursts of action potentials. A recording showing this activity is shown in Fig. 3. In this case, all pyloric cells other than the LP and PDs were either photoinactivated or hyperpolarized. By passing a low-amplitude, steady direct current into the cell bodies of LP and PD cells, their overall activities could be "adjusted" to an appropriate range within which a a stable alternating burst pattern "emerged." Note, however, that the amplitude and time course of these bursts were much different than those of the bursts produced when BPPs were being generated (compare with Fig. 2B). Thus, although reciprocal inhibition may be contributing to burst production, the generation of BPPs, when commissural inputs are active play a much more important role.

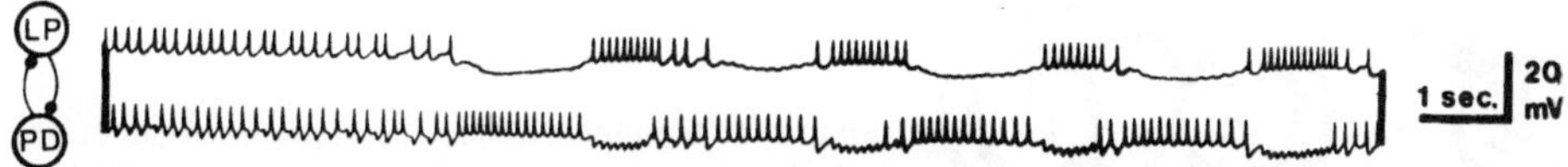

FIGURE 3. Generation of bursts due to reciprocal inhibition. For this recording, the stn was sucrose blocked as in Fig. 2 and all cells but the LP and PD neurons were killed or hyperpolarized. When the LP and PD neurons were slightly depolarized with steady current, they went into stable reciprocal oscillations.

4. Mechanisms Underlying Phase Coordination of the Bursts

The pyloric pattern shown in Fig. 1B is basically triphasic, with one phase of dilator activity being followed by two constriction phases. This characteristic pattern is essentially invariant from prepartion to preparation, and can be maintained for at least two days without significant variation if several experimental protocols are followed: (1) All nerves connecting the stomatogastric ganglion to the commissural ganglia must be undamaged. (2) Antibiotic and glucose must be added to the physiological saline solution. (3) Temperature of the saline must be accurately controlled. (4) The saline must be constantly perfused.

What factors determine the precise phase relationships of the bursts within this pattern? Although at least five of the pyloric cells can generate BPPs when commissural inputs are active, there must be some mechanisms that allow these bursts to be coordinated with one another. It is the synaptic circuitry that determines the basic sequence of the bursts. As discussed above, reciprocal inhibitory synaptic interactions can result in reciprocal bursting in one cell pair in the ganglion. In other words, the 180° phase shift in their bursts was due to their mutual inhibition of one another; when one cell fired, the other was shut off. When the inactive cell escaped from its inhibition (by synaptic fatigue, accommodation, postinhibitory rebound, or some other "restorative" property) its subsequent activity would, in turn, inhibit its partner.

Activity patterns of greater complexity emerge from circuits in which several or many such inhibitory loops are interwoven. In an earlier study (Miller and Selverston, 1982b), we showed how the phases of the bursts in the pyloric pattern could be explained in terms of the inhibitory synaptic circuitry and how the three-phase rhythm could be interpreted as a combination of two biphasic rhythms. This is simplified and summarized schematically in Fig. 4. The AB/PD group strongly inhibits all other cells in the pyloric circuit, shutting them off completely. When the AB/PD burst terminates, these other cells are released from inhibition. One of those cells (the LP cell) strongly inhibits the PD/AB group, and therefore plays the role of a reciprocal inhibitor to the PD/AB group. The cells that are released from inhibition when the AB/PD burst terminates (i.e., the LP, IC, VD, and PY cells) can themselves be subdivided into two groups that are reciprocally inhibitory. Within the period of time between PD/AB bursts, these two subgroups produce reciprocal bursts.

FIGURE 4. Simplified schematic representation of the pyloric network as interwoven reciprocally inhibitory loops.

Figure 4 is a very simplistic summary of our interpretation of earlier work and only explains the basic order of the bursts (i.e., PD/AB to LP/IC to VD/PY; see Fig. 1B). The precise timing of the bursts during that progression is dependent on several intrinsic cellular and synaptic properties. All pyloric cells express the ability to generate "plateau potentials" when inputs from the commissural and esophageal ganglia are active. These cells have two stable resting potential levels: one level below the threshold for spike initiation and one level above that threshold. The cells can be triggered to jump from one level to the other by relatively small currents. Inhibitory synaptic currents can initiate a jump from an active state to an inactive state and postinhibitory rebound can initiate a transition from an inactive to an active state. Therefore, the precise point in the cycle when a pyloric cell will make one of these transitions (i.e., the *phase* of the onset or offset of a burst) will depend very strongly on the amplitudes and time courses of the "trigger pulses" (i.e., the relative synaptic efficacies and the amplitudes of the post-inhibitory rebound) and on the sensitivity of the cell to those pulses (i.e., the voltage and time dependence of the regenerative currents underlying the plateau potentials). Changes in any of these biophysical parameters would be expected to have significant effects on the phase relationships of the bursts and therefore on the behavior. In a freely behaving animal, the phase relationships of the pattern are observed to vary significantly over time, and changes in the biophysical parameters listed above might easily account for the observed pattern variability. Mechanisms underlying changes in plateau potential properties and in relative synaptic efficacies and time courses are described in more detail in Chapter 4.

5. Mechanisms Underlying Frequency Control of the Pattern

Just as in the case for the *phases* of the pyloric pattern, the overall *frequency* of the pattern is essentially invariant if the dissection, perfusion, oxygenation, and temperature of the preparation are carefully controlled. What mechanisms control the overall frequency of the rhythm? Our experiments indicate that the AB interneuron plays the major role in setting the pyloric pattern frequency. When the AB cell is hyperpolarized, its burst frequency decreases and the pyloric pattern slows down. Depolarization of AB increases its burst frequency and also increases the frequency of the whole pattern. (Polarizations of the PD cells, which are strongly elecrically coupled to the AB, have similar effects.) When the AB cell is killed by photoinactivation, the pattern slows down to less than two thirds of its normal freequency as shown in Fig. 5. [Killing the PD cells does *not* decrease the frequency significantly (Miller and Selverston, 1982b).] If equivalent experimental manipulations of the VD, IC, or PY cells are performed, no significant frequency changes are observed. Polarization or photoinactivation of the LP cell does cause changes in the frequency, but to a much smaller degree than observed when the AB is polarized or killed (Selverston and Miller, 1980).

All of these data can be easily understood, considering that (1) the AB cell exhibits the strongest BPP currents, (2) those BPPs have the highest inherent frequency of any of the pyloric neurons, (3) the AB cell is the only cell that inhibits all other cells in the pyloric circuit (except for its synergists, the PD cells), and (4) those inhibitory synapses are extremely efficacious. Thus, from an engineering standpoint, the AB cell satisfies all the criteria for a frequency controller, or "clock," for the pyloric network. It is interesting that the AB cell is the only strict interneuron within the pyloric network and is also the only pyloric neuron to send an axon up into the commissural ganglia.

The frequencies of the pyloric patterns in other species of lobsters seem to be controlled by a substantially different mechanism. As discussed in Chapter 4, the clocks for the pyloric rhythm in species other than *Panulirus interruptus* are a pair of cells in the commissural ganglia called "commissural pyloric oscillators" (CPOs). The CPOs entrain the pyloric network to their own inherent frequency, and accelerate or slow down the pyloric cycle when they are depolarized or hyperpolarized. Polarization of the AB cell in these species does not affect the pyloric pattern frequency as it does in *Panulirus interruptus.* Furthermore, when the stomatogastric nerve is blocked with sucrose in *Homarus,* the pyloric rhythm stops, but the CPOs keep oscillating at their normal frequency.

Neurons fitting the criteria of CPOs have never been found in the commissural ganglia. However, the pair of "P cells" (Russell and Hartline, 1978) in *Panulirus* commissural ganglia display some of the properties of the *Homarus* CPOs. P cells make excitatory synapses with the same pyloric neurons in *Panulirus* as do the CPO cells in *Homarus.* In intact preparations, the P cells fire in bursts that are phase locked to the pyloric rhythm. However, this phasic nature of the P cell discharge is due to their inhibition by the AB cell. When the AB cell is killed, or when the AB axon in the stomatogastric nerve is blocked by focal photoinactivation, the

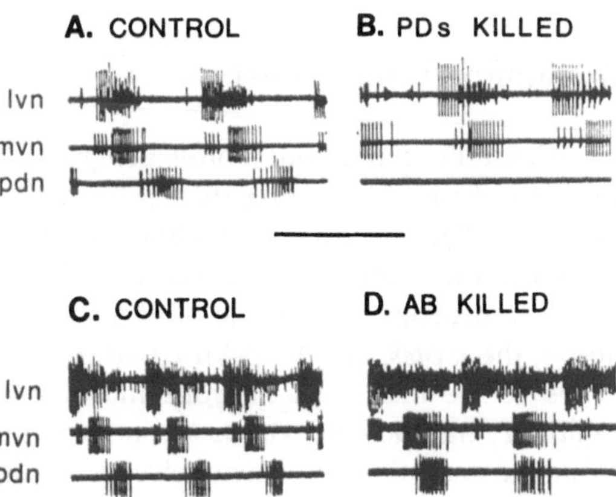

FIGURE 5. The AB is the primary frequency controller. (A) When the PD cells are killed, the frequency and phase relationships of the pattern are changed very little. (B) When the AB cell is killed, the frequency decreases significantly and phase relationships change noticeably. Scale bar equals 0.5 sec. for panels A and B, and 1.0 sec. for panels C and D. (Abbreviations from Table I.)

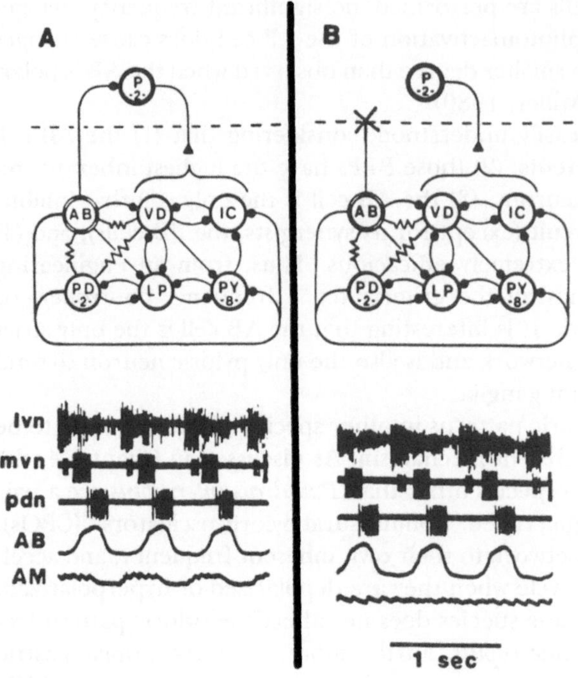

FIGURE 6. Photoinactivation of the AB axon has no effect on the frequency or phases of the pyloric pattern. (A) In the control situation, the commissural P cells were rhythmically inhibited by the AB cell and were constrained to fire in bursts. These bursts of P-cell excitatory postsynaptic potentials (EPSPs) were monitored by an intracellular recording in the AM cell—a member of the gastric circuit that also received P-cell EPSPs. Also shown are extracellular recordings from three nerves and an intracellular recording from the AB cell body. (B) The AB cell's axon was selectively blocked by filling the whole AB with Lucifer Yellow and illuminating its axon in the stn with a narrow beam of light. Only the AB axon was blocked and all other axons including those of the P cells were left intact. Since the P cells were released from their rhythmic inhibition from the AB cell, they fired tonically, as monitored by the "demodulation" of the EPSPs in the AM cell. There were no significant changes in the pyloric pattern. (Abbreviations from Table I.)

P cells fire tonically. Moreover, as shown in Fig. 6, the effect of the P cells on the pyloric rhythm is the same, whether or not they are firing tonically or in pyloric-phased bursts. It is an interesting possibility that the *Panulirus* P cells and the CPO cells in *Homarus* are evolutionary homologues with divergent properties. This would represent a fairly fundamental and significant difference between the physiological strategies of these two homologous systems which, in all other respects, appear to be remarkably similar.

6. Summary and Conclusions

Centrally generated motor programs such as the pyloric pattern can be thought of as representing the "information" that codes for the corresponding behavior. The information coding for the rhythmic movements of the pyloric region in the spiny lobster is created within the stomatogastric ganglion by only fourteen neurons, all but one of which are motoneurons. As we have summarized above, the *existence* of the pyloric pattern results from oscillatory membrane properties of the individual neurons in combination with the multiple reciprocally inhibitory interactions within the network. The precise *phase relationships* derive from the synaptic connectivity circuit and depend on relative synaptic efficacies, postinhibitory rebound properties, and the kinetics of the plateau potential and

BPP generation mechanisms. The overall *cycle frequency* is determined largely by the AB interneuron via its strong intrinsic oscillatory currents and its very strong synapses with the rest of the pyloric neurons.

It is clear that the pyloric pattern produced by an isolated nervous system represents the fundamental pattern. However, the *in vivo* behavior of the pylroic region is strongly influenced by sensory feedback and endogenous modulatory substances. The identification and physiological action of these factors on the well-characterized pyloric network promises to yield important insights into the problem of how centrally generated patterns can be modified by the environment.

References

Alving, B. O., 1968, Spontaneous activity in isolated somata of *Aplysia* pacemaker neurons, *J. Gen Physiol.* **51**:29–45.

Brown, T. A., 1914, On the nature of fundamental activity of the nervous centres, together with an analysis of the conditioning of the rhythmic activity in progression and a theory of the evolution of the function in the nervous system, *J. Physiol.* **48**:18–46.

Claiborne, B. J., and Selverston, A. I., 1984, Localization of stomatogastric IV neuron cell bodies in lobster brain, *J. Comp. Physiol.* **154**:27–32.

Delcomyn, F., 1980, Neural basis of rhythmic behavior in animals, *Science* **210**:492–498.

Dickinson, P. S., and Nagy, F., 1983, Control of a central pattern generator by an identified modulatory interneurone in Crustacea. II. Induction and modification of plateau properties in pyloric neurones, *J. Exp. Biol.* **105**:59–82.

Eisen, J. S., and Marder, E., 1982, Mechanisms underlying pattern generation in the lobster stomatogastric ganglion as determined by selective inactivation of identified neurons. III. Synaptic connections of electrically coupled pyloric neurons, *J. Neurophysiol.* **48**:1392–1415.

Gola, M., and Selverston, A. I., 1981, Ionic requirements for bursting activity in lobster stomatogastric neurons. *J. Comp. Physiol.* **145**:191–207.

Hartline, D. K., and Gassie, D. V., 1979, Pattern generation in the lobster *(Panulirus)* stomatogastric ganglion. I. Pyloric neuron kinetics and synaptic interconnections, *Biol. Cybern.* **33**:209–222.

Hartline, D. K., and Maynard, D. M., 1975, Motor patterns in the stomatogastric ganglion of the lobster *Panulirus argus, J. Exp. Biol.* **62**:409–420.

Koester, J., and Byrne, J. H., 1980, Molluscan nerve cells: From biophysics to behavior, *Cold Spring Harbor Rep.. Neurosci.* **1**:125–180.

Kristan, W. B., Burrows, M., Elsner, N., Grillner, S., Huber, F., Jankowska, E., Pearson, K., Sears, T., and Stent, G. S., 1977, Neural control of movement, in: *Function and Formation of Neural Systems* (G. S. Stent, ed.), *Berlin Dahlem Konferenzen*, pp. 329–354.

Maynard, D. M., 1972, Simpler networks, *Ann. N. Y. Acad. Sci.* **193**:59–72.

Maynard, D. M., and Dando, M. R., 1974, The structure of the stomatogastric neuromuscular system in *Callinectes sapidus, Homarus americanus* and *Panulirus argus (Decapoda crustacea), Philos. Trans. R. Soc. London Ser. B* **268**:161–220.

Maynard, D. M., and Selverston, A. I., 1975, Organization of the stomatogastric ganglion of the spiny lobster. IV. The pyloric system, *J. Comp. Physiol.* **100**:161–182.

McDougall, W., 1903, The nature of inhibitory processes within the nervous system, *Brain* **26**:153–191.

Miller, J. P., and Selverston, A. I., 1979, Rapid killing of single neurons by irradiation of intracellularly injected dye, *Science* **206**:702–704.

Miller, J. P., and Selverston, A. I., 1982a, Mechanisms underlying pattern generation in the lobster stomatogastric ganglion as determined by selective inactivation of identified neurons. II. Oscillatory properties of pyloric neurons, *J. Neurophysiol.* **48**:1378–1391.

Miller, J. P., and Selverston, A. I., 1982b, Mechanisms underlying pattern generation in the lobster stomatogastric ganglion as determined by selective inactivation of identified neurons. IV. Network properties of the pyloric system, *J. Neurophysiol.* **48**:1416–1432.

Moulins, M., and Cournil, I., 1982, All or none control of the bursting properties of the pacemaker neurons of the lobster pyloric pattern generator, *J. Neurobiol.* **13:**447–458.

Nagy, F., and Dickinson, P. S., 1983, Control of a central pattern generator by an identified modulatory interneurone in crustacea. I. Modulation of the pyloric motor output, *J. Exp. Biol.* **105:**33–58.

Robertson, R. M., and Moulins, M., 1981, Oscillatory command input to the motor pattern generators of the crustacean stomatogastric ganglion, *J. Comp. Physiol.* **143:**453–463.

Russell, D. F., and Hartline, D. K., 1978, Bursting neural networks: A reexamination, *Science* **200:**453–456.

Russell, D. F., and Hartline, D. K., 1981, A multiaction synapse evoking both EPSPs and enhancement of endogenous bursting, *Brain Res.* **223:**19–38.

Russell, D. F., and Hartline, D. K., 1982, Slow active potentials and bursting motor patterns in pyloric network of the lobster, *Panulirus interruptus*, *J. Neurophysiol.* **48:**914–937.

Selverston, A. I., and Miller, J. P., 1980, Mechanisms underlying pattern generation in the lobster stomatogastric ganglion as determined by selective inactivation of identified neurons. I. The pyloric system, *J. Neurophysiol.* **44:**1102–1121.

Selverston, A. I., Russell, D. F., Miller, J. P., and King, D., 1976, The stomatogastric nervous system: Structure and function of a small neural network, *Prog. Neurobiol.* **7:**215–290.

4

Extrinsic Inputs and Flexibility in the Motor Output of the Lobster Pyloric Neural Network

MAURICE MOULINS AND FRÉDÉRIC NAGY

1. Introduction

In recent years, our understanding of motor behavior in terms of single-cell activity has been primarily concerned with determining the functional structure of central pattern generators (CPGs) (Selverston, 1980). To analyze such a structure, i.e., to identify the neuronal components and determine the mutual interactions between these components, "naked" CPGs (i.e., completely deafferented CPGs) must be used. Nevertheless, the patterned activity that can be recorded from such an isolated CPG is relatively stereotyped, and until now, little was known about the mechanisms by which such a network could exhibit flexibility in its output in the intact animal. The goal of this chapter is to show how extrinsic inputs to a well-known CPG (the pyloric network of Crustacea, see Chapter 3, this volume) can continuously control the expression of the intrinsic properties of the neurons and thereby continuously "rewire" the network.

2. The Pyloric Neural Network and Its Behavioral Repertoire

In lobsters the rhythmic movements of the posterior part of the stomach (the pyloric chamber) are produced by three groups of muscles: the dilators (D), the anterior constrictors (C_1), and the posterior constrictors (C_2) (Fig. 1A). Electro-

MAURICE MOULINS AND FRÉDÉRIC NAGY • Laboratory of Comparative Neurobiology, CNRS and University of Bordeaux I, Arcachon, France.

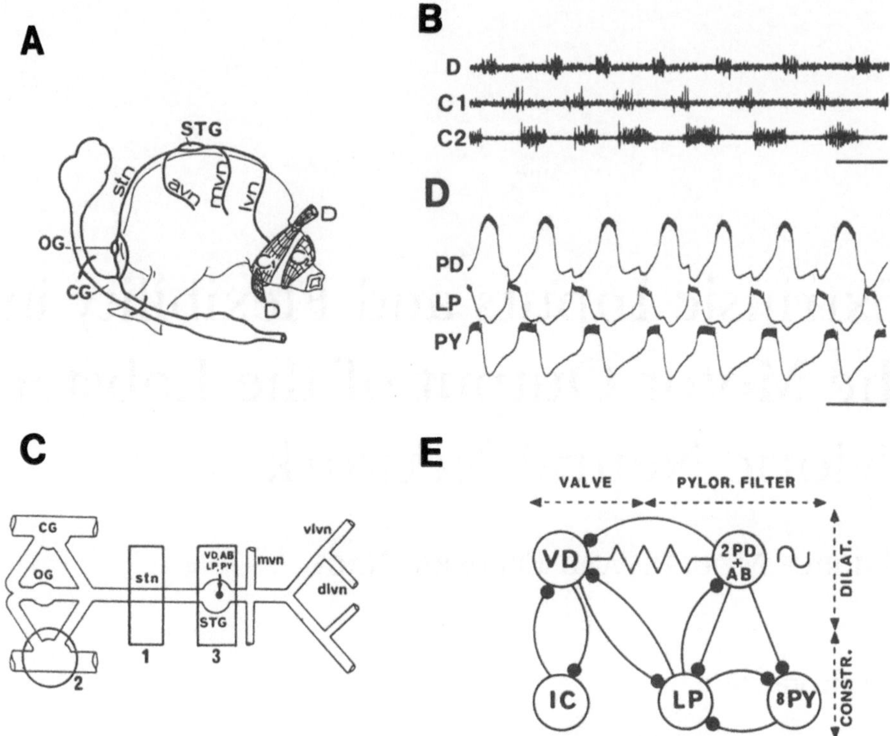

FIGURE 1. The lobster pyloric pattern generator and its motor output. (A) The *in situ* disposition of the stomatogastric nervous system on the foregut with the stomatogastric ganglion (STG), the esophageal ganglion (OG), and the commissural ganglia (CG). The stomatogastric nerve (stn) is the only input nerve to the STG. (B) Electromyographic recordings from the pyloric dilator muscles (D) and the anterior (C_1) and posterior (C_2) constrictor pyloric muscles (see A) in a freely moving intact animal. (C) Isolated stomatogastric nervous system as used to record extracellularly from the pyloric motor output nerves (avn, anterior ventricular nerve; mvn, medial ventricular nerve; dlvn, dorsal lateral ventricular nerve; vlvn, ventral lateral ventricular nerve) and intracellularly from the cell bodies of the pyloric neurons in the STG. 1, 2, and 3 denote vaseline pools used for perfusion of an isotonic sucrose solution on the stn (1) or modified salines on the CG(2) and STG(3). (D) Simultaneous intracellular recordings from PD, LP, and PY motor neurons innervating the D, C_1 and C_2 pyloric muscles, respectively. Note the similarity between the three-phased patterned activity obtained from an isolated stomatogastric nervous system with the EMG activity (B) obtained from the intact animal. (E) The pyloric network; the black dots represent chemical inhibitory synapses and the resistor is an electrotonic coupling. The symbol ~ denotes the endogenous oscillators (pacemakers). Horizontal bar, 1 sec; vertical bar, 20 mV.

myographic (EMG) recordings from the intact animal show that pyloric motor output is a triphasic pattern with repetitive cycles of contraction in the sequence D–C_1–C_2 (Fig. 1B). From the pioneering work of Maynard (1972) and subsequent works of Selverston and collaborators (Selverston *et al.*, 1976; see Chapter 3, this volume), it is known that in *Panulirus* this sequence is basically organized by a small neural circuit located within the stomatogastric nervous system (Fig. 1C). This system remains able to produce a replica of the normal motor pattern after isolation in a Petri dish (Fig. 1D), and for this reason it has been possible to identify all the pyloric neurons and determine their synaptic wiring diagram (Fig. 1E). Similar results have been obtained for *Homarus* (Robertson and Moulins, 1981; Moulins and Cournil, 1982) and for *Jasus* (Nagy and Dickinson, 1983).

The pyloric circuit consists of 14 neurons whose cell bodies lie in the stomatogastric ganglion (STG) (Fig 1A). Twelve of these cells drive the rhythmic movements of the pyloric chamber while the remaining two neurons, VD and IC, produce movements of the cardiopyloric valve. For the pyloric chamber alone, three classes of neurons can be distinguished in the circuit: three dilators (AB and 2 PD), the single anterior constrictor (LP), and the eight posterior constrictors (PY) (Fig. 1E). All are motor neurons (innervating muscles D, C_1, and C_2, respectively) except the interneuron AB which projects anteriorly to the commissural ganglia. Neurons within each class are electrically coupled and each class will be considered here as a functional unit. When the stomatogastric nervous system is isolated from the rest of the stomach (Fig. 1C), the output of the pyloric circuit consists of a sequential cyclic activity in the three functional groups of neurons (PD–LP–PY) (Fig. 1D). This rhythmic output results mainly from (1) the endogenous oscillatory properties of the dilator neurons that act as pacemakers for the circuit, (2) the ability of the constrictor neurons to generate "plateau" potentials (Russell and Hartline, 1978), a property that also promotes bursting, (3) the strong phasic inhibition imposed by the dilators on the constrictors, which in turn can fire only when the dilators are silent, and (4) the reciprocal inhibition between the anterior (LP) and posterior (PY) constrictors which partly explains their successive activation during the dilator interburst interval.

Since the pyloric circuit is composed almost entirely of motor neurons, its suprathreshold activity can be monitored in the intact animal by EMG recording from the D, C_1, and C_2 muscles (Rezer and Moulins, 1983). These recordings show that the pyloric output can manifest quite a high degree of flexibility, a feature that is not easily predicted from consideration of the organization of the pyloric network alone.

A summary of this flexibility is as follows:

1. In most experiments on the isolated stomatogastric nervous system, the pyloric circuit cycles continuously and on this basis it has been assumed that in the intact animal the pyloric circuit is also continuously active. However, in *Homarus* at least, the pyloric output can cease completely for 2 hr or more and then start again (E. Rezer and I. Cournil, unpublished data).
2. Examination of pyloric EMG sequences shows that the pattern can change considerably during the same experiment. First, the phase at which each muscle is activated in each cycle can vary. Second, the participation of every muscle in a given sequence is not obligatory; i.e., whereas most of the time pyloric output is a triphasic pattern, biphasic or monophasic sequences can also occur. Failure to participate during periods of rhythmicity can involve not only the two constrictor muscles (C_1 or C_2), but also dilator muscles (D) which are innervated by the pyloric pacemaker neurons PD (see Fig. 8A).
3. The period of the pyloric rhythm also displays variations. Most notably when the animal begins to feed, the pyloric output can switch abruptly

from a pattern with long and variable cycle period to one with a short and regular period that can continue for several hours after feeding.

Such variability in the pyloric output cannot be explained by simply considering the cellular and synaptic properties of the pyloric circuit itself. In recent years, considerable efforts have been made to understand how extrinsic inputs could be responsible for the observed flexibility in the output behavior of the circuit. We describe here an identified neuromodulatory input that controls the burstiness of the pyloric neurons and by this mechanism could be responsible for most aspects of the flexibility described above.

3. Triggering of the Pyloric Pattern: Induction of Burstiness in Neurons of the Pyloric Circuit

Rhythmic pyloric output is generated mainly by a group of three oscillatory neurons (2 PD and AB) that are acting as pacemakers in the circuit. This means that the rhythmic pattern of the circuit can stop only when these neurons are no longer oscillating (or are oscillating below their firing threshold). This could be achieved by a strong inhibitory tonic input maintaining the oscillatory neurons in a hyperpolarized state from which they cannot oscillate while, in turn, the pyloric behavior will recommence after release from this inhibition. Until now, however, such an inhibitory input has not been identified. Furthermore, we now have sufficient data to indicate that the pacemaker neurons oscillate (and rhythmic pyloric output produced) only when "permissive" inputs coming from more rostral ganglia are active.

All the inputs from the rostral ganglia to the pyloric circuit can be reversibly suppressed in *in vitro* experiments by sucrose block of axonal conduction in the stomatogastric nerve (see Fig. 1C). Under these conditions, in *Homarus* and *Jasus*, and also in *Panulirus* (F. Nagy and J. P. Miller, unpublished observations), the pyloric rhythm disappears completely with all the neurons becoming silent or tonically active (Fig. 2A, B). The time delay for this effect varies between preparations, but the pyloric rhythm usually stops within minutes after conduction block while pyloric cycling resumes either by unblocking the stomatogastric nerve or by delivering tonic electrical stimulation to this nerve, posteriorly to the sucrose block. The conduction block does not result in any hyperpolarizing effect that could explain the cessation of cycling in the pacemaker neurons, but rather it has now been demonstrated that it suppresses inputs that are necessary for the expression of the regenerative bursting properties of both the dilator and the constrictor neurons.

The oscillatory behavior of the membrane potential of *dilator neurons* consists of a two-phased trajectory during each cycle. The first phase is a relatively slow depolarization (pacemaker potential) which, if sufficient, reaches a threshold from which develops the second phase, a sudden depolarization giving rise to a short plateau (driver potential) underlying a burst of spikes. If the neuron is hyperpolarized, the pacemaker potential cannot reach the threshold for the driver potential and spontaneous oscillations do not appear. Nevertheless, it is always possible

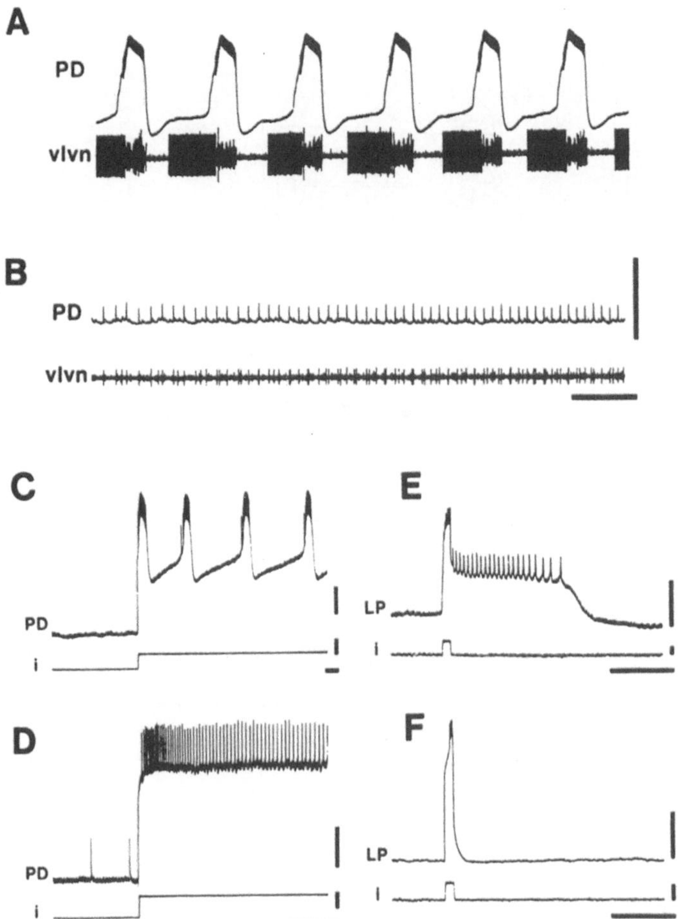

FIGURE 2. Suppression of burstiness in pyloric neurons by isolating the stomatogastric ganglion from rostral ganglia *(Homarus)*. (A,B) The pyloric rhythm as recorded intracellularly from a PD neuron and extracellularly from the vlvn (A) is abolished after blocking axonal conduction in the stn (B) (see Fig. 1). (C–F) The bursting properties of a dilator PD (C) and a constrictor LP (E) neuron are abolished (D and F, respectively), after conduction block in the stn. Horizontal bars, 1 sec; vertical bars, 20 mV in A, B, 10 mV and 1.5 nA in C–F.

to obtain oscillations by experimental depolarization of the neuron (Fig. 2C) providing the stomatogastric ganglion is connected to more rostral ganglia. As shown recently in *Homarus* (Moulins and Cournil, 1982), when the stomatogastric ganglion is deafferented, the ability for all the dilator neurons to produce regenerative depolarization disappears. In this situation, the neurons behave "passively"; experimental depolarization induces only tonic firing (Fig. 2D) and they do not exibit any bursting capability.

The *constrictor neurons* are not endogenous oscillators in that they do not have the facility to develop a pacemaker potential. Nevertheless, if some input drives their membrane potential above a given threshold, they develop a membrane

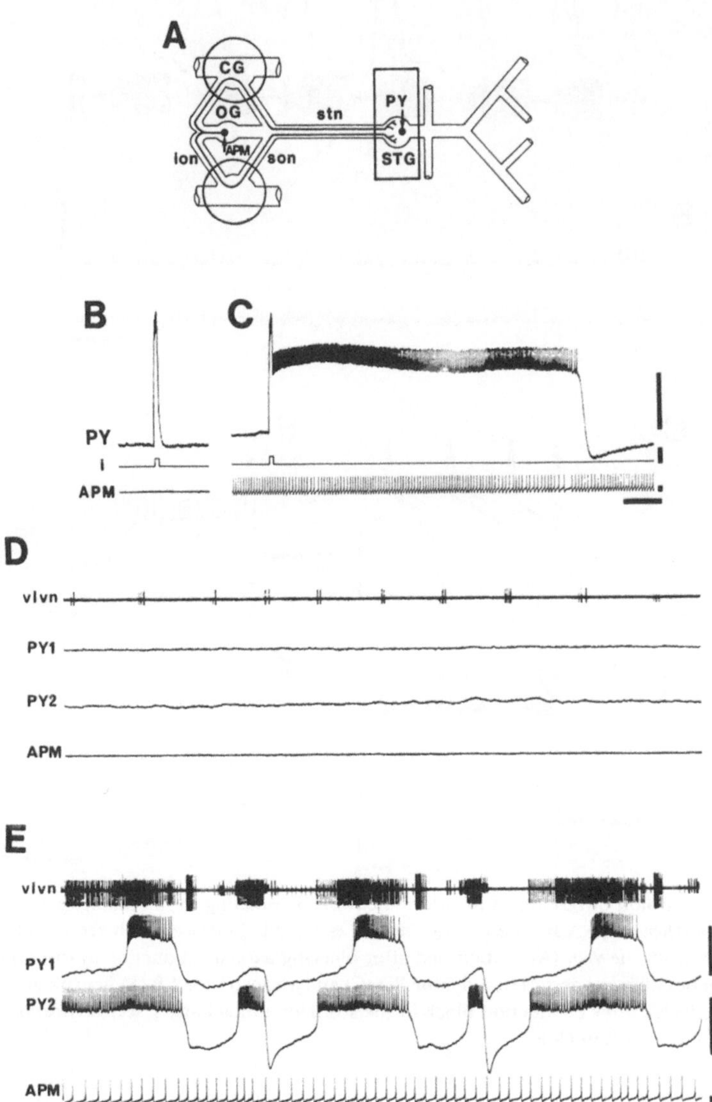

FIGURE 3. APM induces the burstiness of the pyloric neurons (A–C) and can switch on the pyloric pattern generator (D, E) *(Jasus)*. (A–C) When chemical synaptic activity is blocked in the CG by perfusion of a modified saline with OCa^{2+} + Co^2 (A), the constrictor neuron (PY) no longer exhibits any burstiness (B). In this experimental situation, firing of APM (see A) induces plateau properties in the constrictor neuron (C). Note that the synaptic inhibitions in the pyloric network had been weakened with saline containing 10^{-5}M picrotoxin. (D, E) During a period when the pyloric pattern generator is not cycling (D), APM firing induces rhythmic output (E). Horizontal bars, 2 sec in B, C, 1 sec in D, E; vertical bars, 20 mV and 2.5 nA.

potential trajectory directly comparable to the second phase of oscillation of the dilators. This again consists of a rapid depolarization followed by a relatively long plateau on which the cell fires, then termination by a sudden repolarization (Russell and Hartline, 1978). Here again the ability to generate plateau potentials can be demonstrated by the injection of a brief pulse of depolarizing current that induces a plateau response of considerably longer duration than the initial triggering pulse (Fig. 2E). As shown originally by Russell and Hartline (1978) for *Panulirus,* this plateauing property of the constrictors can be observed only when the pyloric circuit receives descending inputs. Following isolation of the stomatogastric ganglion, injection of current pulses never provokes a plateau potential in a penetrated cell (Fig. 2F). Thus the effect of suppressing descending inputs to constrictors, as for the dilators, causes them to behave as "passive" neurons without any apparent endogenous burstiness.

Considerable efforts have been made to characterize the extrinsic inputs that are able to turn on the pyloric pattern. Nagy *et al.* (1981) have identified in the esophageal ganglion of *Jasus* a neuron (the anterior pyloric modulator, APM) that projects via two axons to the stomatogastric ganglion (Fig. 3A) and that participates in the induction of the burstiness of the pyloric neurons. In *Jasus* it is possible, by blocking synaptic activity in the commissural ganglia (Fig. 3A), to obtain isolated preparations in which most of the pyloric neurons have lost their regenerative properties (Fig. 3B). In such preparations the specific discharge of APM (provoked by current injection) reinvokes the regenerative properties of pyloric neurons (Fig. 3C). This induction effect of APM discharge can restore rhythmicity in a spontaneously silent pyloric network (Fig. 3D, E). Furthermore, APM's effects are abolished when the stomatogastric ganglion is bathed with saline containing atropine, a cholinergic muscarinic antagonist (Nagy and Dickinson, 1983), whereas cholinergic agonists such oxotremorine or pilocarpine mimic the effects that APM exerts on the pyloric neurons (Nagy *et al.*, 1984). From these observations, it can be concluded that APM discharge is sufficient to induce the regenerative properties of pyloric neurons and that it does it probably via a cholinergic mechanism. However, APM is only one element of a group of modulatory inputs controlling the pyloric network (see Marder, 1984, for review, see also discussion), and among these inputs APM is probably not the only element able to induce burstiness in pyloric neurons, since in many experiments it remains silent although the pyloric circuit continues to cycle normally. In this situation, however, APM spike discharge can have additional complex modulatory effect on the ongoing pyloric motor pattern.

4. Control of the Pyloric Pattern: Efficacy of Synapses within the Pyloric Circuit

One obvious type of flexibility in the expression of rhythmic output of a pattern generator resides in an ability to shift the phase at which individual neurons fire in the cycle. Such phase shifts are actually observed in the pyloric pattern recorded by electromyography in the intact animal. The phase relationships between bursting of pyloric neurons are determined mainly by their synaptic connections within the pyloric network. Phase shifts could thereby result from modifications in the efficacy of these synaptic connections. Nagy and Dickinson (1983) have shown that a discharge of interneuron APM does indeed provoke modifications of the efficacy of synaptic relations among pyloric neurons, and, as a result, provokes changes in the phase of discharges of these neurons in the pyloric cycle. This occurs because APM firing is able not only to induce the regenerative properties of the pyloric neurons, but can also modify the expression of these properties once they are developed.

The effects of APM firing on an ongoing pyloric pattern are considered in the experiment of Figs. 4 and 5, conducted on isolated stomatogastric nervous systems. The pacemaker neurons (in this case PD) are known to strongly inhibit the constrictor neurons (LP, PY) (see Fig. 1E). This inhibition triggers the active repolarization of the plateau potentials in the constrictor neurons and, as a result, termi-

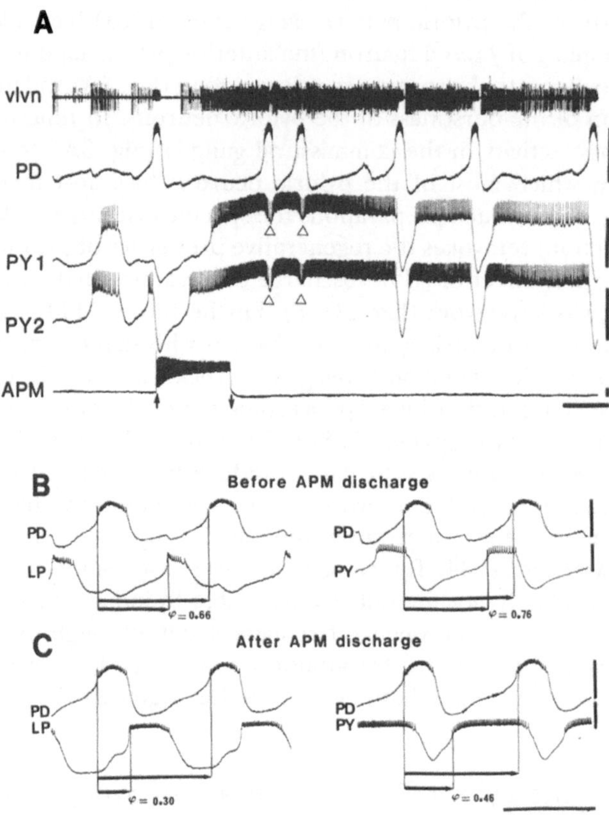

FIGURE 4. APM controls the phase relationships between the discharges of the pyloric neurons *(Jasus)*. (A) A short imposed burst in APM induces a long-term increase in the firing frequency of the constrictor neurons and considerably reduces the cyclic inhibitions of the constrictors (white arrow heads) by the dilators. (B,C) following a 6 sec discharge of APM at 5 Hz, the phase of LP bursting in the PD cycle shifts from 0.66 to 0.30 (B, C, left), while the phase of PY changes from 0.76 to 0.46 (B, C, right). (B) is a control recorded 10 sec before the onset of APM discharge; (C) was recorded 4 sec after the end of the APM discharge. Horizontal bars, 1 sec, vertical bars, 20 mV and 2.5 nA in A.

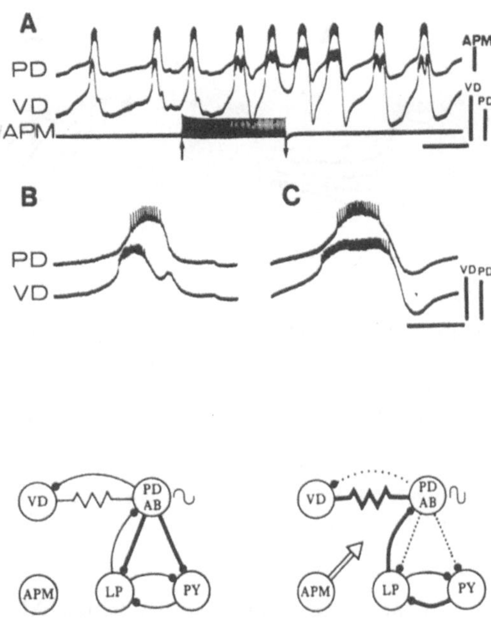

FIGURE 5. APM can produce a functional rewiring of the pyloric network *(Jasus)*. (A–C) An APM discharge alters the balance between the effects of the inhibitory chemical synapse between PD and VD and those of the electrical synapse between the two neurons. At the beginning of the recording in A, PD and VD display their usual phase relationships (see B). Following APM discharge, the efficacy of the chemical synapse is greatly reduced and the bursts of the two neurons become synchronous (see C). (D) Diagram of synaptic relationships within the pyloric network; heavy lines represent the synapses that are functionally the most important in producing the pyloric pattern. (E) After APM discharge, the network is functionally rewired; heavy lines indicate relationsips that have become the most important; dotted lines indicate synapses that have become less effective. Horizontal bars, 2 sec in A, 0.5 in B and C; vertical bars, 50 mV for APM, 20 mV for PD and VD.

nates each constrictor burst. It is due largely to this mechanism that the pacemaker neurons are able to entrain the rhythmic discharge of the constrictors. When APM fires, this cyclic inhibition is markedly reduced, and the pacemaker is no longer able to fully repolarize the constrictor neurons. This effect is particularly evident when the spontaneous pyloric rhythm is rather slow, where APM firing strongly increases the strength and duration of the plateau potentials of the constrictors (Fig. 4A). It also occurs during the normal pyloric rhythm when a burst invoked in APM discharge causes the constrictor neurons to be incompletely repolarized by the pacemaker activity and the former start to fire relatively earlier during the next cycle. That is, there is a strong positive phase shift of constrictor discharge in the pacemaker cycle (compare Fig. 4B and C). A similar observation can be made when considering the synaptic relationships that exist between the dilators (group PD-AB) and a neuron of the cardiopyloric valve (VD). The pyloric dilators inhibit the VD neuron which is also electrically coupled to the pyloric dilators. At the beginning of the recording in Fig. 5A, the VD burst starts before PD burst onset and stops near the middle of the PD burst (at the maximum of PD discharge frequency) (Fig. 5B). Following APM firing, the VD burst is considerably prolonged and the two neurons fire at almost the same phase (Fig. 5C). Here again, the inhibition from the dilators is unable to fully repolarize VD and appears to be reduced (or the electrical coupling the two neurons increased) by APM discharge. Indeed, when APM fires, the efficacy of most of the synaptic relations in the pyloric network is modified, as are the phase relationships between the discharge of most of the pyloric neurons. The anterior pyloric modulator thus has a strong modulating effect on the pyloric pattern and this functional rewiring of the pyloric network is summarized in Fig. 5D, E.

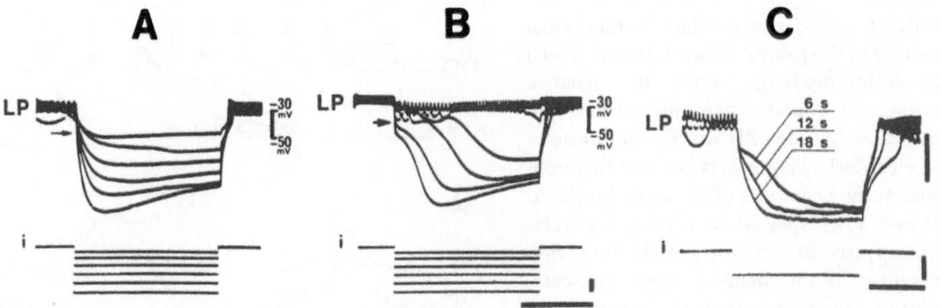

FIGURE 6. APM modulates the bursting characteristics of the pyloric neurons *(Jasus).* (A) Control; during spontaneous plateaus, hyperpolarizing pulses (injected via a second microelectrode) produce an initial passive repolarizing response (until the level of the arrow) followed by an active hyperpolarization. (B) Same tests when APM is firing; the active hyperpolarization is considerably slowed down. (C) APM effects are long term. Using the same experimental condition as in A and B, a hyperpolarizing pulse is delivered 6, 12, and 18 sec after the end of the firing of APM. The retardation of the active hyperpolarization is still obvious 18 sec after the end of APM firing. Horizontal bars, 0.5 sec; vertical bars, 20 mV and 2.5 nA.

What is the mechanism by which APM reduces the ability of inhibitory synapses to repolarize postsynaptic neurons in the pyloric network and as a result decreases the apparent efficacy of these synapses? There are sufficient data to suggest that this is achieved largely by modification, associated with APM firing, of the regenerative properties of the postsynaptic neurons (Dickinson and Nagy, 1983). Figures 6A and B show, for example, that when APM is firing, the time course of the active repolarization from a plateau in a constrictor neuron is considerably increased. As a result, this active repolarization becomes less easily triggered by the inhibitory synaptic inputs from the pacemaker neurons, and the inhibition from the pacemakers is apparently reduced.

Finally, as can be seen in Figs. 4A and 5A, the effect of APM firing on the synaptic efficacy within the pyloric circuit is of relatively long duration, outlasting considerably the duration of APM firing. It is interesting to note that the modifications caused by APM firing on the time course of the active phase of repolarization are also long lasting and decrease gradually with time after the end of APM activity (Fig. 6C).

5. Control of the Pyloric Cycle Frequency: Efficacy of Extrinsic Phasic Inputs to the Pyloric Circuit

Another feature of the flexibility in rhythmic output of a pattern generator resides in possible modifications of its cycle period (i.e., variations in the frequency of the rhythm). For *Jasus* and *Homarus,* EMG recordings in the intact animal show that the pyloric period can vary from about 1.25 to 5 sec or more. The frequency of oscillation of the pyloric pacemaker neurons, which governs the overall frequency of the pyloric output, is a function of their membrane potential. It follows that any tonic extrinsic inputs regulating the mean membrane potential of the pacemaker would be able to determine at any time the frequency of pyloric output. Until now, however, we have no experimental evidence suggesting that the pyloric frequency is mainly regulated in this way. However, recently it has been shown (Robertson and Moulins, 1981; Moulins and Nagy, 1983) that the frequency of the pyloric output is in fact under the control of an extrinsic *phasic* input arising from an independent pyloric oscillator located in the commissural ganglia (the commissural pyloric oscillator, CPO). On *in vitro* preparations, the CPO not only entrains the pyloric circuit, but it can also achieve this with different coupling ratios (i.e., 1:1; 1:2; 1:3) according to the instantaneous "sensitivity" of the pyloric pacemaker neurons. This "sensitivity" to the extrinsic phasic inputs from the CPO is dependent on the burstiness of the pyloric neurons, and so is most probably controlled by modulatory neurons such as APM.

A neuron belonging to the CPO and known as the CP neuron has been identified in the commissural ganglion of *Homarus* (Nagy, 1981). This neuron has an oscillatory behavior characterized by regular peaks of depolarization on which appear high-frequency bursts of spikes interspersed with lower frequency discharge (Fig. 7B). The CP neuron projects to the stomatogastric ganglion, probably monosynaptically onto the pacemaker neurons, the bursting activity of which is always coordinated to that of CP (as in Fig. 7B with a ratio 1:2). The oscillatory behavior of the CPO (and of CP) is not generated by the pyloric circuit in the STG because CPO frequency remains unmodified when the commissural ganglion is completely isolated (Robertson and Moulins, 1981). Moreover, the strong coordination that occurs between the rhythm of CP oscillations and the rhythm of the pacemaker neurons in the pyloric circuit (PD-AB) is due to a descending influence from CP onto the pyloric pacemakers. When the frequency of oscillation of the pyloric pacemakers is experimentally increased by depolarization (Fig. 7C) or decreased by hyperpolarization (Fig. 7D), there is no resultant modification in the frequency of bursting of CP. By contrast, experimental manipulation of the bursting frequency of CP by current injection results in a strong modification of the bursting frequency of the pyloric pacemakers (Fig. 7E). The CP neuron receives a strong inhibition from the pyloric dilators (via AB), but curiously this inhibition is without influence on the frequency of bursting of CP (compare, for example, Fig. 7C and D).

The period of CPO cycling remains fairly constant (about 1.25 sec) from one preparation to another. Given the variability observed in pyloric cycle frequency, it therefore appears that frequency of the rhythmic pattern produced by the pyloric network finally depends on the ratio of coupling between the CPO and the pyloric pacemakers. In the experiment of Fig. 7, this ratio is 1:2 (Fig. 7B), but can be changed to 1:1 by experimental depolarization of PD (Fig. 7C) or to 1:3 by experimental hyperpolarization of PD (Fig. 7D), and as a result the pyloric frequency is respectively half, equal to, and one third that of the CPO. Changes in coupling ratio between the CPO and the pyloric pacemaker neurons depend on the ability of these neurons to generate a driver potential (and a single oscillation) when they receive a synaptic excitation from the CPO. In the experiment of Fig. 7, this capability can be artificially increased and decreased by injection of depo-

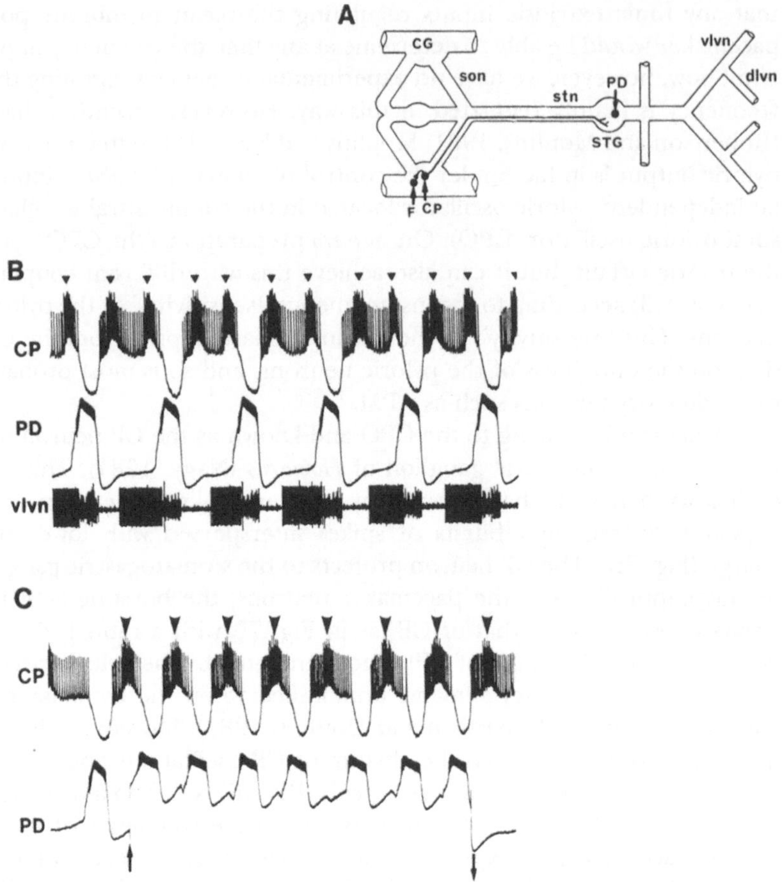

FIGURE 7. The commissural pyloric oscillator (CP) controls the frequency of the pyloric pattern generator *(Homarus)*. (A) The experimental preparation showing CP and a follower neuron of CP (F) in the CG (see Fig. 8C). (B) The oscillatory behavior of CP (each peak of depolarization is marked by a black arrow head) is coordinated (2:1) with the output of the pyloric pattern generator monitored intracellularly from PD and extracellularly from the vlvn. (C) Depolarization of PD increases its own bursting

larizing and hyperpolarizing current, respectively, but usually it depends on the endogenous burstiness of the pyloric pacemaker neurons. This suggests that any modulatory input to the pyloric network, which could control the burstiness of the pacemaker neurons, would also control the coupling between these neurons and the CPO, and in turn control the frequency of the pyloric output. Interneuron APM, which modulates the burstiness of all the pyloric neurons, could therefore exert such control. The evidence, albeit indirect, that supports this proposition is considered in the discussion.

The control of the coupling between the CPO and the pyloric neurons could also explain another aspect of the flexibility in pyloric output, namely, the partic-

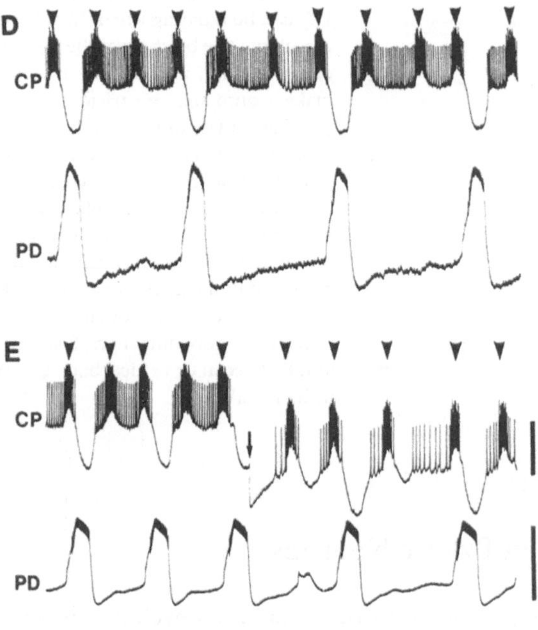

frequency and results in a coordination mode 1:1 with CP. (D) Hyperpolarization of PD decreases its bursting frequency and results in a coordination mode 3:1 with CP. Note that the period of CP oscillations remains the same in B, C, and D. (E) Conversely, hyperpolarization of CP reduces its bursting frequency and this results in an increase in PD cycle period. Horizontal bar, 2 sec; vertical bars, 20 mV.

ipation of a variable number of neurons in the pyloric sequence. Electromyographic recordings from the intact animal show that several neurons of the pyloric circuit, including the pacemakers, may remain inactive during pyloric rhythmicity (Fig. 8A). Similar results can be obtained from isolated preparations (Fig. 8B), and this phenomenon can be explained by the fact that the CPO also projects to the constrictor neurons and is able to entrain the constrictors as well as the dilators (Robertson and Moulins, 1981) (Fig. 8C). Thus any modulatory input neuron such as APM, which can selectively control the instantaneous burstiness of the pyloric neurons, would thereby effectively permit the latter to be entrained by the CPO with different coordination ratios.

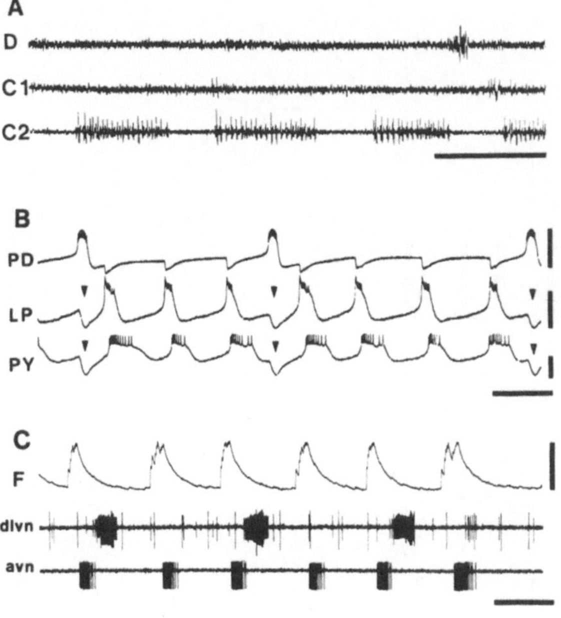

FIGURE 8. A variable number of neurons can participate in the pyloric sequence. (A) In the intact animal, the C_2 muscle (innervated by PY) can be active alone during several cycles (EMG recordings) *(Jasus)*. (B) Recordings from an isolated preparation of the stomatogastric nervous system showing that the constrictors (LP and PY) can be bursting when the dilators are silent. The black arrow heads indicate inhibition imposed by the pacemakers onto the constrictors *(Jasus)*. (C) The activity of CP, which is monitored here by recording from a postsynaptic F neuron in the CG (see Fig. 7A), can entrain the dilators (PD recorded in the dlvn) in a coordination mode 2:1 while a constrictor (I C recorded in the avn) is entrained with a different coordination mode (1:1) *(Homarus)*. Horizontal bars, 2 sec in A and B, 1 sec in C; vertical bars, 20 mV in B, 10 mV in C.

6. Discussion and Conclusions

6.1. Induction of Burstiness in Pyloric Neurons

All the neurons composing the pyloric network are able to develop regenerative depolarizations underlying their firing in bursts of action potentials. These regenerative events can be either endogenously rhythmic, like the bursting pacemaker potentials displayed by all the dilator neurons (Miller and Selverston, 1982), or triggered by synaptic inputs, like the plateau potentials of the constrictor neurons (Russell and Hartline, 1978). However, both bursting pacemaker potentials and plateau potentials are dependent on inputs from higher centers. Blocking conduction in the stomatogastric nerve suppresses intrinsic burstiness of the constrictor neurons (Russell and Hartline, 1978; Dickinson and Nagy, 1983), of the dilator neurons (Miller and Selverston, 1982; Moulins and Cournil, 1982), and of the interneuron AB (Moulins and Cournil, 1982; Nagy and Miller, in preparation). So even the pacemaker neurons of the network (PD-AB) are conditional bursters. As a result, when the stomatogastric ganglion is isolated from more rostral ganglia in *Homarus* (Moulins and Cournil, 1982), *Palinurus* and *Jasus* (Nagy *et al.*, 1984), and *Panulirus* (Nagy and Miller, in preparation), the pyloric network is no longer able to generate a rhythmic output. Under these conditions, electrical stimulation of the stomatogastric nerve can temporarily restore the regenerative properties in the pyloric neurons and trigger the pyloric rhythm. The pyloric network is therefore an interesting model of a rhythmic pattern generator, whose operation is mainly dependent on the intrinsic burstiness of all its component neurons, although for each neuron this bursting property is a potential capability only and requires conditioning inputs for its expression. As discussed hereafter, this allows a wide range of flexibility in the final motor pattern generated.

The induction of nonlinear membrane properties that can release the bursting capabilities of neurons has previously been demonstrated only for two neurons in mollusks. In the visceral ganglion of the snail *Otala*, a neurosecretory neuron (neuron 11) is a burster when the animal is active, but loses this ability when the animal estivates (Barker and Gainer, 1974). Similarly in *Lymnaea*, neurosecretory cells producing ovulating hormone are endogenous bursters during only one phase of the egg-laying cycle (Kits and Bos, 1981). In *Otala*, the burstiness of neuron 11 in an estivating animal can be experimentally reinduced with vasopressin or related peptides (Barker *et al.*, 1975; Ifshin *et al.*, 1975).

For all the neurons of the pyloric network, the regenerative properties underlying their firing in bursts can be reinduced by tonic firing of the single interneuron APM. This appears to be the first known example of an identified neuron capable of triggering the bursting ability of conditional oscillators in a CPG. The mechanisms by which APM exerts its action has still to be worked out in detail, but it appears to involve cholinergic muscarinic receptors of pyloric neurons (Nagy and Dickinson, 1983). The activation of these receptors could result in a Ca^{2+} inward current (Nagy *et al.*, 1984), which, in turn, triggers the sequence of conductances (Gola and Selverston, 1981) responsible for the bursting activity of the pyloric neurons. In the stomatogastric ganglion of *Panulirus*, dopamine, a putative transmitter of unidentified central afferents, was reported similarly to trigger rhythmic activity of a pyloric network treated with tetrodotoxin (Raper, 1979; Anderson and Barker, 1981). However, dopamine induces bursting pacemaker oscillations in the single interneuron AB (Anderson and Barker, 1981); it does not directly affect the constrictor neurons (Selverston and Miller, 1980; Eisen and Marder, 1984) and even depresses the bursting abilities of the dilator neurons PD (Marder and Eisen, 1984). In contrast, APM induces bursting in all of the pyloric neurons. Besides the fact that it is the only cell so far identified, APM is therefore the best candidate among the population of modulatory afferents to the stomatogastric ganglion (see Marder, 1984) for switching on the pyloric pattern. This does not imply, however, that APM is only a general activator. Indeed, APM can also modulate the pyloric motor pattern, once generated, and this is achieved by modulating the ongoing burstiness of individual pyloric neurons.

6.2. Modulation of Burstiness in Pyloric Neurons

Although discharge of APM alone is sufficient to induce burstiness in the pyloric neurons, its firing is not obligatory for pyloric cycling. In most isolated preparations of stomatogastric nervous system, APM does not fire spontaneously, but the pyloric network continues to display a normal rhythmicity, providing the stomatogastric ganglion remains connected to commissural ganglia. In these instances, however, APM firing strongly increases the burstiness of pyloric neurons and modifies its characteristics. The modulation exerted by APM on the spontaneous bursting mechanisms of the pyloric neurons can be compared with the control of endogenous burstiness in neuron R15 of the abdominal ganglion in *Aplysia* (Parnas *et al.*, 1974; Barker *et al.*, 1975; Mayeri *et al.*, 1979) or in neurons of the crustacean cardiac ganglion (Watanabe *et al.*, 1969; Cooke and Hartline, 1975; Lemos and Berlind, 1981). This modulation also bears similarities to the control by several putative neurotransmitters or neuromodulators of regenerative membrane properties in nonnervous tissues, like the plateau phase of the vertebrate cardiac action potential (Reuter, 1974; Giles and Noble, 1976; Ten Eick *et al.*, 1976) or the plateau phase of activity in pancreatic endocrine cells (Gagerman *et al.*, 1978).

However, the action of APM exhibits several peculiarities that result in a more complex control of the pyloric pattern generator. First, the increase in burstiness that APM provokes is different for each pyloric neuron, being maximal at least in *Jasus* for the constrictor neurons PY (Dickinson and Nagy, 1983). Depending on its firing frequency, therefore, APM can promote the bursting of a variable number of neurons in the network, including or not the dilator pacemaker neurons. As a result, APM firing can either provoke an increase in intensity of discharge of the constrictor neurons without altering the overall rhythm or provoke both an increase in the intensity of the discharge of all the pyloric neurons and an increase in the frequency of the rhythm, or even switch off momentarily the pyloric rhythm and induce strong tonic discharge in the constrictor neurons (Nagy and Dickinson, 1983).

Second, APM firing modifies the characteristics of burstiness in pyloric neurons. It modifies particularly the temporal characteristics of the active repolarization that provokes the transition between firing and nonfiring states of the pyloric neurons. This renders the pyloric neurons less sensitive to inhibitory synaptic inputs, which normally trigger their repolarization and terminate their discharges. Among the synaptic relations within the pyloric network, the inhibitory synapses from the dilator neurons (AB-PD) onto the other neurons are essential in determining the phase relationships of the neuronal discharges in the pyloric sequence (Eisen and Marder, 1984). Anterior pyloric modulator firing strongly decreases the efficacy of these inhibitions and, as a result, strongly modifies the phase relationships within the pyloric sequence. When an inhibitory synapse coexists with an electrical synapse, the synchronizing effect of the latter is favored and can lead to synchronous discharges of neurons that usually do not fire together (for instance, interneuron AB and neuron VD). In other words, the discharge of the modulatory interneuron APM results in a functional rewiring of the pyloric network. Because all but one of the neurons in the pyloric network are motor neurons, this functional rewiring is a powerful way of modulating the motor behavior of the pyloric filter. It must also be noted that for this functional rewiring of the pyloric network, it is not necessary to postulate any modification of the synaptic mechanisms themselves. The modification of synaptic efficacy within the pyloric network can simply be accounted for by changes, associated with APM firing, in the burstiness of the postsynaptic neurons.

Finally, promotion of burstiness in pyloric neurons increases the likelihood of depolarizing events triggering regenerative depolarizations and in turn bursts of spikes in these neurons. As a result, when APM fires, the pyloric neurons become more sensitive to excitatory inputs, and the pyloric network can be more easily entrained by a higher order oscillator.

6.3. Entrainment of Frequency of Pyloric Neuron Bursting

The pyloric network requires only tonic activation to operate and to produce a replica of the pyloric pattern recorded in the intact animal. It therefore fulfills the criteria for a CPG (Wilson and Wyman, 1965; Stein, 1977; Delcomyn, 1980; Selverston, 1980). In *Homarus,* however, a phasic activity associated with the pyloric rhythm is generated by an oscillator (CPO) located in the more rostral commissural ganglion (Robertson and Moulins, 1981). One of the neurons (CP) belonging to this oscillator has recently been identified (Moulins and Nagy, 1983). Neuron CP has endogenous bursting properties and its activity is phase locked to the motor output produced by the pyloric network. In addition, its axon projects to the stomatogastric ganglion, and manipulation of its bursting frequency entrains bursting of the pyloric pacemaker neurons. By contrast, manipulation of the pyloric pacemaker frequency has no effect on the bursting frequency of CP. The CP neuron therefore appears to be part of a master oscillator in the commissural ganglion which acts as a frequency controller of the pyloric pattern generator located in the stomatogastric ganglion. Such descending control of a pattern generator by a higher order oscillator has to be distinguished from mutual cooperation between equivalent oscillators, which occurs, for example, in the control of locomotion (Ikeda and Wiersma, 1964; Grillner, 1975; Kristan and Guthrie, 1977; Stein, 1978). Indeed, the situation reported here for the pyloric system challenges the commonly held idea that a pattern generator does not receive independent phasic inputs containing timing cues (Grillner, 1977). One of the characteristics of CPO activity is a great stability in cycling frequency. Whether determined indirectly by recording from its follower neurons (Robertson and Moulins, 1981) or directly by recording from CP itself, the CPO frequency always remains between 0.8 and 1 Hz. In contrast, the pyloric pattern generator shows no such inherent free-run frequency. When the pyloric network is pharmacologically activated in deafferented stomatogastric ganglion, the frequency of the resulting rhythm depends on the particular preparation and on the concentration of the activating agent (F. Nagy and J. A. Benson, unpublished observations). Thus the CPO could serve as a reference clock by which the potentially variable frequency of the pyloric pattern generator is more or less entrained. In the intact animal, EMG recordings show that when the animal begins to eat, the pyloric output pattern switches from a slow and irregular to a long-term fast and regular rhythm, the frequency of which is about 1 Hz (Rezer and Moulins, 1983). This transition during feeding has been interpreted as resulting from an increase in the degree of coupling between the pyloric network and the CPO.

However, the control exerted by the CPO appears to be more complex than simply that of a timer for the pyloric network as a whole. The CPO can entrain separately every neuron of the pyloric network. Moreover, it can entrain each pyloric neuron with a different coupling ratio, such as 1:1 for a constrictor neuron or 1:2 for a dilator neuron (Robertson and Moulins, 1981). This allows the constrictor neurons to burst during each CPO cycle, although the pacemaker neurons (PD-AB) are silent. Thus depending on how its neurons are coupled individually to the higher order oscillator, the pyloric network can produce those different motor patterns that have been recorded both in the intact animal (Rezer and Moulins, 1983) and in isolated preparations of *Homarus* and *Jasus* (Moulins and Nagy, 1983).

What determines the sensitivity of individual pyloric neurons to excitatory inputs from the CPO? Because of the intrinsic burstiness of the pyloric neurons, their sensitivity to CPO inputs could simply depend on their mean membrane potential. However, pyloric neurons are conditional bursters, so their sensitivity to excitatory inputs could be more accurately controlled by additional inputs that modulate their burstiness. Among such inputs, APM is a good candidate because, as already discussed, it promotes burstiness in all of the pyloric neurons and thereby favors the coupling of the whole network to the master oscillator. This possibility is supported by the following observations. In *Jasus,* APM firing can accelerate the pyloric rhythm until a maximum frequency of 1 Hz only (the inherent frequency of the CPO). If the spontaneous pyloric frequency is already about 1 Hz, an APM discharge does not cause it to accelerate. Moreover, an acceleration of the pyloric rhythm is always associated with recruitment of all of previously silent neurons in the network (Nagy and Dickinson, 1983). Finally, when rhythmic inputs from CPO are suppressed by experimentally blocking synaptic activity in the commissural ganglia, the pyloric rhythm, although very slow, cannot be accelerated by APM firing (Dickinson and Nagy, 1983). Therefore, it seems reasonable to propose that one important consequence of APM discharge is to increase the coupling of the pyloric pattern generator to the master oscillator CPO and to contribute to the triggering of the pyloric pattern associated with feeding.

Furthermore, the pyloric neurons can be entrained individually by the CPO, depending on their own state of burstiness. This interesting way to diversify the expression of the pyloric pattern could be dependent on the action of additional modulatory inputs more specific than APM and which would control the burstiness of single neurons in the pyloric network. Several putative neurotransmitters of afferent elements to the stomatogastric ganglion, such as dopamine, serotonin, and proctolin, have recently been reported to exert such a selective action (Marder, 1984; Marder and Eisen, 1984).

In conclusion, we have described several novel mechanisms that engender flexibility in the expression of the motor pattern produced by the pyloric pattern generator in the stomatogastric system. These mechanisms are based on the existence of intrinsic properties (burstiness) in the neurons composing the generator and on the control of these properties by modulatory inputs. The modulatory inputs can, on one hand, induce burstiness in the neurons composing the CPG, and as a result turn the pattern generator on. On the other hand, these inputs can modulate the pre-existing state of burstiness of the target neurons. This allows (1) control of the relative efficacy of synapses within the network, and so, control of the pattern produced by the generator and 2) control of the efficacy of phasic extrinsic inputs from a higher order oscillator, and as a result, control of the frequency of the rhythmic motor pattern.

ACKNOWLEDGMENTS. We thank Dr. A. J. Simmers for helpful criticisms of the manuscript and for correcting the final draft.

References

Anderson, W. W., and Barker, D. L., 1981, Synaptic mechanisms that generate network oscillations in the absence of discrete postsynaptic potentials, *J. Exp. Zool.* **216:**187–191.

Barker, J. L., and Gainer, H., 1974, Peptide regulation of bursting pacemaker activity in a molluscan neurosecretory cell, *Science* **184:**1371–1373.

Barker, J. L., Ifshin, M., and Gainer, H., 1975, Studies on bursting pacemaker potential activity in molluscan neurons. III. Effects of hormones, *Brain Res.* **84:**501–513.

Cooke, I. M., and Hartline, D. K., 1975, Neurohormonal alterations of integrative properties of the cardiac ganglion of the lobster *Homarus americanus*, *J. Exp. Biol.* **63:**33–52.

Delcomyn, F., 1980, Neural basis of rhythmic behavior in animals, *Science* **210:**492–498.

Dickinson, P. S., and Nagy F., 1983, Control of a central pattern generator by an identified modulatory interneurone in Crustacea. II. Induction and modification of plateau properties in pyloric neurones, *J. Exp. Biol.* **105:**59–82.

Eisen, J. S., and Marder, E., 1984, A mechanism for the production of phase shifts in a pattern generator, *J. Neurophysiol.* **51:**1375–1393.

Gagerman, E., Idahl, L. A., Meissner, H. P., and Taljedal, I. B., 1978, Insulin release, cGmP, cAMP and membrane potential in acetylcholine stimulated islets, *Am. J. Physiol.* **235:**493–500.

Giles, W., and Noble, S. J., 1976, Changes in membrane currents in bullfrog atrium produced by acetylcholine, *J. Physiol.* (London) **276:**103–123.

Gola, M., and Selverston, A. I., 1981, Ionic requirements for bursting activity in lobster stomatogastric neurones, *J. Comp. Physiol.* **145:**191–207.

Grillner, S., 1975, Locomotion in vertebrates, central mechanisms and reflex interaction, *Physiol. Rev.* **55:**247–304.

Grillner, S., 1977, On the neural control of movement. A comparison of different basic rhythmic behaviors, in: *Function and Formation of Neural Systems* (G. S. Stent, ed.), Dahlem Konferenzen, Berlin, pp. 197–224.

Ifshin, M., Gainer, H., and Barker, J. L., 1975, Peptide factor extracted from molluscan ganglia that modulates bursting pacemaker activity, *Nature (London)* **254:**72–73.

Ikeda, K., and Wiersma, C. A. G., 1964, Autogenic rhythmicity in the abdominal ganglia of the crayfish: The control of swimmeret movements, *Comp. Biochem. Physiol.* **12:**107–115.

Kits, K. S., and Bos, N. P. A., 1981, Pacemaking mechanisms of the after discharge of the ovulation hormone-producing caudo-dorsal cells in the gastropod mollusc, *Lymnaea stagnalis*, *J. Neurobiol.* **12:**425–439.

Kristan, W. B., and Guthrie, P. B., 1977, Acquisition of swimming behavior in chronically isolated single segment of the leech, *Brain Res.* **131:**191–195.

Lemos, J. R., and Berlind, A., 1981, Cyclic adenosine monophosphate mediation of peptide neurohormone effects on the lobster cardiac ganglion, *J. Exp. Biol.* **90:**307–326.

Marder, E., 1984, Mechanisms underlying neurotransmitter modulation of a neuronal circuit, *Trends Neurosci.* **7:**48–53.

Marder, E., and Eisen, J. S., 1984, Electrically coupled pacemaker neurons respond differently to the same physiological inputs and neurotransmitters, *J. Neurophysiol.* **51:**1362–1374.

Mayeri, E., Brownel, P., Branton, W. D., and Simon, S. B., 1979, Multiple, prolonged actions of neuroendocrine bag cells on neurons in *Aplysia*, I. Effects on bursting pacemaker neurons, *J. Neurophysiol.* **42:**1165–1184.

Maynard, D. M., 1972, Simpler networks, *Ann. N.Y. Acad. Sci.* **193:**59–72.

Miller, J. P., and Selverston, A. I., 1982, Mechanisms underlying pattern generation in lobster stomatogastric ganglion as determined by selective inactivation of identified neurons, II. Oscillatory properties of pyloric neurons, *J. Neurophysiol.* **48:**1378–1391.

Moulins, M., and Cournil, I., 1982, All-or-none control of the bursting properties of the pacemaker neurons of the lobster pyloric pattern generator, *J. Neurobiol.* **13:**447–458.

Moulins, M., and Nagy, F., 1983, Control of integration by exogenous inputs in crustacean neuronal circuits, *J. Physiol. (Paris)* **78:**755–764.

Nagy, F., 1981, Etude de l'expression d'activités motrices rythmiques organisées par des générateurs paucineuroniques du système nerveux stomatogastrique des Crustacés décapodes. Flexibilité intrinsèque aux réseaux moteurs; contrôle par les centres supérieurs; contrôle proprioceptif, Thèse d'Etat Université de Bordeaux I.

Nagy, F., and Dickinson, P. S., 1983, Control of a central pattern generator by an identified modulatory interneurone in Crustacea, I. Modulation of the pyloric motor output, *J. Exp. Biol.* **105:**33–58.

Nagy, F., Dickinson, P. S., and Moulins, M., 1981, Modulatory effects of a single neuron on the activity of the pyloric pattern generator in Crustacea, *Neurosci. Lett.* **23:**167–173.

Nagy, F., Benson, J. A., and Moulins, M., 1984, Cholinergic activation of burst generating oscillations mediated by opening of Ca^{++} channels in lobster pyloric neurons, *Soc. Neurosci. Abstr.* **10:**148.

Parnas, I., Armstrong, D., and Strumwasser, F., 1974, Prolonged excitatory and inhibitory synaptic modulation of a bursting pacemaker neuron, *J. Neurophysiol.* **37:**594–608.

Raper, J. A., 1979, Nonimpulse mediated synaptic transmission during the generation of a cyclic motor program, *Science* **205:**304–306.

Reuter, H., 1974, Localization of beta-adrenergie receptors and effects of noradrenaline and cyclic nucleotides on action potentials, ionic currents and tension in mammalian cardiac muscle, *J. Physiol. (London)* **242:**429–451.

Rezer, E., and Moulins, M., 1983, Expression of the crustacean pyloric pattern generator in the intact animal, *J. Comp. Physiol.* **153:**17–28.

Robertson, R. M., and Moulins, M., 1981, Oscillatory command input to the motor pattern generators of the crustacean stomatogastric ganglion, I. The pyloric rhythm, *J. Comp. Physiol.* **143:**453–463.

Russell, D. F., and Hartline, D. K., 1978, Bursting neural networks: A reexamination, *Science* **200:**453–456.

Selverston, A. I., 1980, Are central pattern generators understandable, *Behav. Brain Sci.* **3:**535–571.

Selverston, A. I., and Miller, J. P., 1980, Mechanisms underlying pattern generation in lobster stomatogastric ganglion as determined by selective inactivation of identified neurons, I. Pyloric system, *J. Neurophysiol.* **44:**1102–1121.

Selverston, A. I., Russell, D. F., Miller, J. P., and King, D. G., 1976, The stomatogastric nervous system: Structure and function of a small neural network, *Prog. Neurobiol.* **7:**215–290.

Stein, P. S. G., 1977, Application of the mathematics of coupled oscillator systems to the analysis of the neural control of locomotion, *Fed. Proc.* **36:**2056–2059.

Stein, P. S. G., 1978, Motor systems with specific reference to the control of locomotion, *Annu. Rev. Neurosci.* **1:**61–81.

Ten Eick, R., Nawrath, H., McDonald, T. F., and Trautwein, W., 1976, On the mechanism of the negative inotropic effect of acetylcholine, *Pflug. Arch. Eur. J. Physiol.* **361:**207–213.

Watanabe, A., Obara, S., and Akiyama, T., 1969, Acceleratory synapses on pacemaker neurons in the heart ganglion of a stomatopod, *Squilla oratoria, J. Gen. Physiol.* **54:**212–231.

Wilson, D. M., and Wyman, R. J., 1965, Motor output patterns during random and rhythmic stimulations of locust thoracic ganglia, *Biophysiol. J.* **5:**121–143.

5

Modulation of Central and Peripheral Rhythmicity in the Heartbeat System of the Leech

RONALD L. CALABRESE AND EDMUND A. ARBAS

1. Introduction

Rhythmic movements are programmed by peripheral myogenic oscillators or central neural oscillators called pattern generators (Delcomyn, 1980; Roberts and Roberts, 1983). For peripherally programed rhythms, such as heartbeat in vertebrates, the muscles involved usually produce an endogenous polarization rhythm that drives their contractions. Myogenic oscillators have been amenable to experimental analysis because muscle is readily accessible to cellular techniques (Noble, 1979; Jewell and Ruegg, 1966). For centrally programed rhythms, such as walking, breathing, and chewing, the basic pattern of motor discharge underlying the rhythm can be produced by a part of the CNS that is isolated from all phasic sensory input. Thus a central pattern generator—a network of central neurons that requires at most tonic input to produce an oscillatory output—drives the coordinated pattern of rhythmic motor outflow. Over the past twenty years, we have developed an experimental understanding of how central pattern generators work. Most of this progress has occurred in a few favorable invertebrate preparations (Selverston and Miller, 1984; Getting, 1983; Calabrese and Peterson, 1983) where, owing to the restricted number of neurons in the CNS and the ease with which these neurons are identified and experimentally manipulated, a rigorous cellular approach is possible.

Although rhythmic movements may seem superficially very stereotyped, they

RONALD L. CALABRESE AND EDMUND A. ARBAS • The Biological Laboratories, Harvard University, Cambridge, Massachusetts 02138.

are subject to considerable modulation in response to changes in the organism's internal and external environment. This modulation may be phasic, as when a walking animal alters its stepping pattern on encountering an obstacle, or it may be tonic, as when breathing and heartbeat are accelerated during exercise. Phasic and tonic alterations of rhythmic movements can be of two different types: changes in the fundamental timing of the rhythm (timing changes) or changes in the intensity of the rhythm (intensity changes). Myogenic rhythms are usually tonically modulated; these modulatory processes have in some cases been well studied (Noble, 1979; Evans and O'Shea, 1978) because the locus of modulation is limited to the muscle in question. Central rhythms have been less amenable to the study of modulatory processes because the site of modulation is not so limited. Although timing changes must, by definition, occur at the level of the central pattern generator, the exact site of action within the pattern generator is often difficult to determine. Moreover, intensity changes can occur at any level within a motor system, the pattern generator, the motor neurons, or the muscle. If we are to further our understanding of how rhythmic movements are controlled by central pattern generators, we must explore the modulatory processes that tune centrally programed rhythmic movements to environmental needs.

In this chapter, we will describe the types of modulation that occur within heartbeat system of the leech *Hirudo medicinalis* and our recent efforts to elucidate the underlying mechanisms. This system is particularly interesting for the study of the control of rhythmic movements because it is elaborated by a central pattern generator and a myogenic oscillator, both of which are subject to modulation.

1.1. The Hearts, Their Innervation, and Myogenic Properties

The leech is a segmented worm, and the segmental nature of the animal is reflected in the structure of its organ systems. The circulatory system is a closed network comprising four longitudinal tubelike vessels—one ventral, one dorsal, and two lateral—that extend the entire length of the animal, communicating directly at the front and rear of the animal and in every body segment by a series of branch vessels (Mann, 1962). Circulation is driven by the rhythmic constrictions of the lateral vessels, which we call hearts or heart tubes (Boroffka and Hamp,

FIGURE 1. (A) A schematic illustrating the neurons that innervate a heart in the third through eighth segments. Conventional excitatory neuromuscular junctions (■) and modulatory neuromuscular junctions (▲). In this and subsequent schematics, a neuron's soma, input region, and spike initiation site are indicated by a circle (unless otherwise noted), processes are indicated by lines, and neurons are indexed according to their ganglion of origin. (B) Simultaneous intracellular recordings and a tension recording (from the heart) illustrate the activity cycles of an HE motor neuron, an HA neuron, and a segment of heart that both neurons innervate. In this and subsequent records of heart activity, a section of heart was left innervated in only one segment, and the heart was immobilized for intracellular recording or connected to a tension transducer. See Maranto and Calabrese (1983b) for details of the methods employed. Both the HE and the HA neuron produce bursts of impulses interrupted by periodic barrages of inhibitory synaptic potentials. The heart beats in time with their bursts. (C) A schematic showing the inhibitory synaptic connections from the HN interneurons to the HE motor neurons and the HA neurons in the first eight segmental ganglia on one side. In this and subsequent schematics, inhibitory synapses are indicated by small filled circles. For clarity, the input region and spike initiation site of each

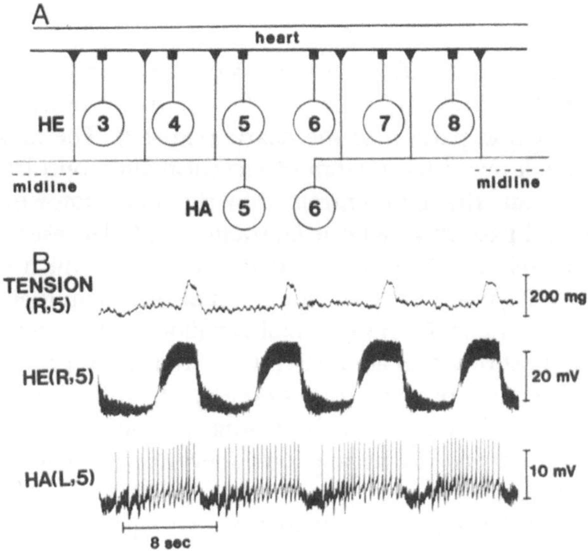

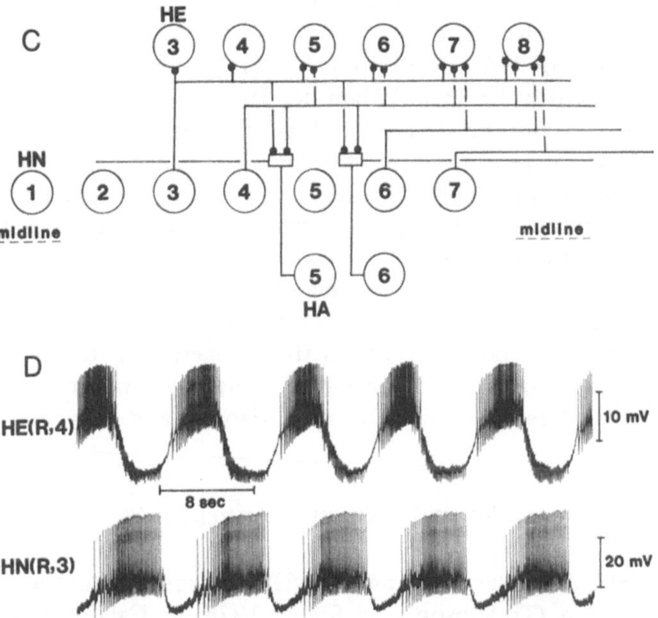

HA cell is indicated separately from its soma (circle) with a rectangle. The HE cells of the 8th through 15th ganglion all receive the same set of inputs from HN cells. Only four of the seven identified pairs of HN cells contact HE cells synaptically. (D) Simultaneous intracellular recordings from an HN interneuron and one of the HE motor neurons it inhibits in an isolated nerve cord preparation. Both cells produce bursts of impulses that are periodically interrupted by barrages of inhibitory synaptic potentials. Because the HN cell inhibits the HE cell, they burst in antiphase.

1969; Thompson and Stent, 1976a). The walls of the heart tubes are made up of spiral muscle cells (Maranto and Calabrese, 1984a), which contract rhythmically to produce the constriction pattern of the hearts (Fig. 1B). The muscle cells of the hearts can produce a myogenic rhythm of electrical and contractile activity, but, normally, this myogenic rhythm is entrained by rhythmic motor outflow driven by the heartbeat central pattern generator (Maranto and Calabrese, 1984b) (Fig 1B).

The CNS consists of a chain of 21 segmental ganglia, which communicate by paired connectives (Payton, 1981). At either end of this chain there are fused ganglionic masses called brains. Each segmental ganglion contains about 400 neurons (Macagno, 1980), most of which are bilaterally paired and repeated from ganglion to ganglion. The ganglia are sufficiently stereotyped, so that each neuron can be easily identified from animal to animal by its soma position and size and by its characteristic electrical activity. Each ganglion innervates its segment by bilaterally paired anterior and posterior roots.

The hearts of the leech are innervated by two types of neurons, HE motor neurons and modulatory HA neurons (Fig. 1A). The HE motor neurons occur as bilateral pairs in the 3rd through 18th segmental ganglia (Thompson and Stent, 1976a). [Neurons within the ganglia are indexed by ganglion number and body side; thus HE motor neurons are indexed from cell HE(L,3) to cell HE(R,18).] Each sends an axon out the ipsilateral anterior root of its ganglion and forms neuromuscular junctions on heart muscle cells (Maranto and Calabrese, 1984a). Each action potential in an HE motor neuron elicits an excitatory junctional potential in the heart muscle cells it innervates (Thompson and Stent, 1976a; Maranto and Calabrese, 1984a). The HA cells occur as bilateral pairs in the fifth and sixth segmental ganglia only (Calabrese and Maranto, 1984). Each HA neuron of the fifth ganglion sends axons out the contralateral roots of the fifth ganglion; it also sends axons out the contralateral roots of more anterior ganglia via an intersegmental axon in the contralateral anterior connective. The HA neurons of the sixth ganglion are similar, but their intersegmental projections are directed posteriorly. HA neurons form synaptic contacts on heart muscle cells that are ultrastructurally similar to those of the HE motor neurons, but their impulse activity does not give rise to recordable junctional potentials in the heart muscle (Maranto and Calabrese, 1984a). The HE motor neurons entrain the myogenic rhythm of the hearts by rhythmic exicitation (Maranto and Calabrese, 1984b), whereas the HA cells have a tonic modulatory influence over the hearts (Calabrese and Maranto, 1984).

1.2. The Heartbeat Central Pattern Generator

Both the HE motor neurons and the HA neurons are active in a precisely coordinated rhythm (Thompson and Stent, 1976a,b; Calabrese and Maranto, 1984) (Fig. 1B). These neurons are not inherently rhythmic, but rather tend to fire tonically (Calabrese, 1979). Their rhythmicity derives from cyclic inhibition that they receive from the heartbeat central pattern generator (Thompson and Stent, 1976b; Calabrese, 1979) (Fig. 1D). The pattern generator comprises seven bilateral pairs of identified HN interneurons that occur in the first seven segmental ganglia (Thompson and Stent, 1976b,c; Calabrese, 1977; Peterson and Calabrese, 1982)

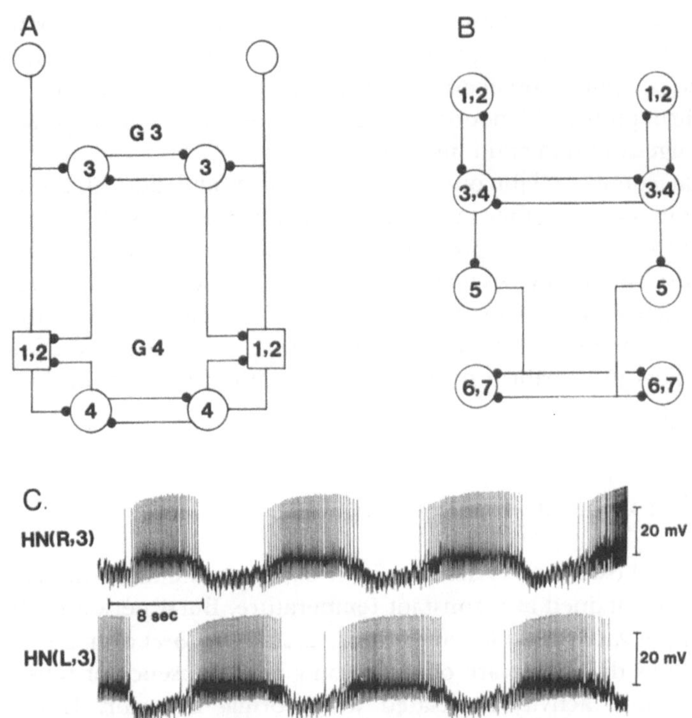

FIGURE 2. The central pattern generator for heartbeat. (A) Schematic showing the inhibitory synapses among the HN interneurons of the timing oscillator. Cells HN(1) and HN(2) are lumped together on either side because their properties and connections are indistinguishable. Their primary input regions and spike initiation sites are in the fourth ganglion (G4) and are indicated separately from their somata (circles) by open squares. The HN cells of the timing oscillator form a closed rectangle of reciprocal inhibitory connections that spans the third (G3) and fourth ganglion (G4). (B) Full schematic of the heartbeat central pattern generator showing all the inhibitory synapses among the identified HN interneurons. HN cells with similar inputs, outputs, and properties are lumped together for simplicity. The HN(5) interneurons link the timing oscillator to the more posterior premotor HN interneurons. (C) Simultaneous intracellular recordings showing antiphasic bursting in two reciprocally inhibitory HN interneurons of the timing oscillator. The reciprocal inhibitory interactions between HN cell pairs of the timing oscillator and the ability of those HN cells to recover from inhibition are instrumental in the production of the rhythmic bursting that times the heartbeat motor pattern. See Peterson (183a,b) for the detailed workings of the timing oscillator.

(Fig. 2A,B). The HN cells are connected to one another by inhibitory synapses and the pairs in the third, fourth, sixth, and seventh ganglia inhibit the HE motor neurons and the HA neurons (Thompson and Stent, 1976b,c; Calabrese, 1977) (Fig. 1C). The rhythmicity of the system is generated by the first four pairs of interneurons (Peterson and Calabrese, 1982) and derives from a blend of reciprocal inhibition and inherent membrane properties (Calabrese, 1979; Peterson, 1983a,b) (Fig. 2A,C). This timing oscillator imposes its activity rhythm on the other HN interneurons, which play a premotor coordinating role (Fig. 2B). The activity of the entire pattern generator (timing oscillator plus premotor interneurons) is imposed on the HE motor neurons and thus on the hearts. For a more complete review of the heartbeat motor system, see Calabrese and Peterson (1983).

There are two primary oscillators within the heartbeat motor system: the central timing oscillator and the peripheral myogenic oscillator. The central oscillator controls the peripheral oscillator phasically, through the HE motor neurons, and tonically, through the HA neurons. Both these oscillators undergo adaptive modulatory changes in timing and intensity.

In what context might one expect to find modulation of these oscillators? Heartbeat differs from other motor acts in that it does not stop or start. The leech depends on the continual expression of this motor act throughout life. The metabolic needs subserved by circulating blood vary, however, depending on whether the leech is active or inactive. Timing and intensity changes in heartbeat might be expected to compensate metabolic needs during vigorous activities. One might also predict changes in heartbeat to be associated with arousal of the animal through sensory input.

2. Modulation of the Central Oscillator

The central oscillator cycles at a fairly constant rate in unstimulated, isolated nerve cords maintained at a constant temperature. Burst period in the recording shown in Fig. 3 varies little (coefficient of variation = 5%) over 20 min; however, output rhythms this stable are obtained only in the absence of sensory feedback and other neural activity associated with normal behavior. If the rhythm is

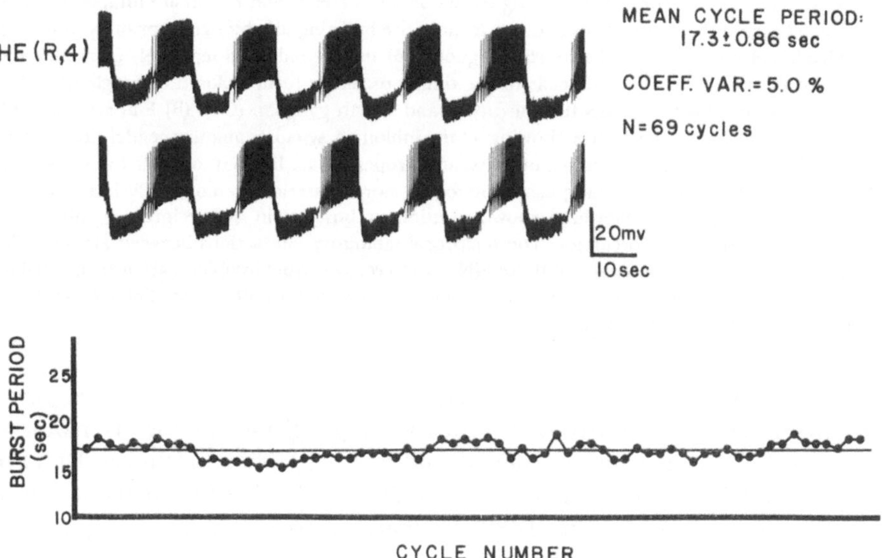

FIGURE 3. Cycling of the central oscillator in isolated, unstimulated nerve cords can be quite stable in rate. The upper panel shows two segments of a continuous recording of an HE (R,4) cell. Burst period is plotted against cycle number below for more cycles of the same recording. The solid line indicates the mean period of 17.3 sec.

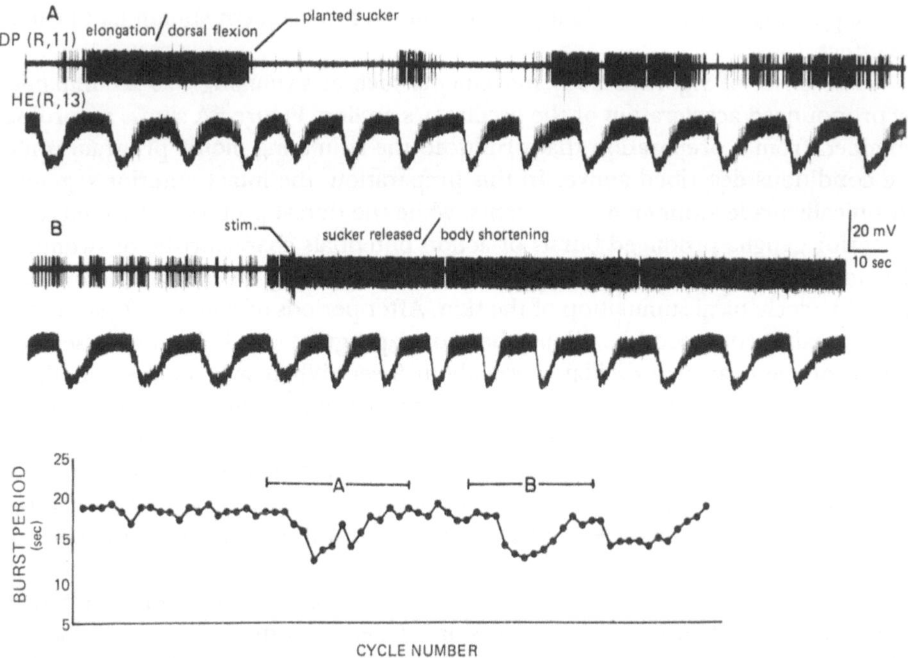

FIGURE 4. Fluctuations in period of the central oscillator occur with spontaneous movements and sensory stimulation in partially dissected preparations. Anterior 8 segments undissected; posterior ganglia 9 through 15 exposed for recording from HE cells (lower traces) and extracellular recording from DP nerve by suction electrode (upper traces). Anterior body movements were observed through a dissecting microscope and recorded on the voice channel of a tape recorder. Notation with upper traces reflects those observations. Burst period versus cycle number plotted below with bars marking times occupied by sequences A and B. (A) Acceleration of oscillator cycling occurred with spontaneous movements. (B) A tap on the anterior body (stim.) induced the animal to release its sucker and elicited shortening accompanied by acceleration of heartbeat oscillator cycling.

recorded under conditions that permit this type of activity, fluctuations in heartbeat period occur.

Changes in period of the central oscillator accompany the onset of body movements and follow certain types of sensory stimulation in partially dissected animals. Animals were prepared by dissecting posterior ganglia, while those body segments that contain HN cells were left undissected. Movements of these anterior body segments were partially restricted by a tether, but the preparations made movements similar to those associated with exploratory behavior and swimming in intact animals. The records of Fig. 4A show fluctuations in burst period that occurred with spontaneous movements of a partially dissected preparation. The cycling of the oscillator accelerated transiently when the animal elongated and planted its sucker on the floor of the recording chamber. A mechanical stimulus applied to the skin led to acceleration of the oscillator's cycling and shortening of the body wall (Fig. 4B). Similar acceleratory effects in other preparations saturated once the central oscillator was made to cycle very rapidly (with a period between 6 and 10 sec, depending on the individual preparation) and subsequent stimuli had little or no effect.

The onset of vigorous body movements such as swimming was accompanied by pronounced acceleration of the oscillator's cycling. Figure 5A shows recordings obtained from a preparation that produced the swimming motor program under the conditions described above. In this preparation, the intact anterior segments chronically made swimming movements, while the dorsal posterior (DP) nerves of posterior ganglia produced bursts of action potentials characteristic of swimming (Kristan *et al.*, 1974). The motor program for swimming was interrupted several times by mechanical stimulation of the skin. After periods of quiescence, swimming began again spontaneously. While the motor program was being expressed, the period of the heartbeat oscillator was about 9 sec. When swimming stopped, the burst period gradually lengthened over several cycles. When swimming resumed, the period again shortened to about 9 sec. Similar changes in the oscillator's period were observed repeatedly each time swimming was interrupted (Fig. 5). These and other observations indicate that the heartbeat rhythm undergoes timing changes associated with the activity level of the animal and with the degree of the animal's arousal through sensory input.

The changes in period of the central oscillator associated with motor activity or sensory stimulation also occur in isolated nerve cords. Figure 5B shows the effect of activating the swimming motor program in an isolated nerve cord by depolarizing a swim-initiating interneuron, cell 204 (Weeks and Kristan, 1978). When the motor program was initiated, the burst period of HN cell (R,4) shortened by about 40%, from a 23-sec period to a 14-sec period. When cell 204 was hyperpolarized and the bout of swim motor activity ceased, the period of the HN cell recovered gradually to prestimulus levels. We have observed similar accelerations of heartbeat oscillator in association with spontaneous bouts of swim motor activity that are often produced by isolated nerve cords (Willard, 1981). Thus, accelerations of heart rate associated with locomotory movements can be mediated through central interactions between motor circuits and do not require sensory feedback that might be provided by movement, or by build-up of an oxygen debt due to muscular activity.

Many neurons involved in swimming and other movements were individually tested for influence on the timing oscillator. Activity of any of the identified mechanosensory neurons, the "T," "P," or "N" cells (Nicholls and Baylor, 1968) can accelerate the period of the central oscillator. Motor neurons generally do not affect the oscillator. Other neurons vary in their ability to influence period of the oscillator's rhythm. Activity of the Retzius' cells—neurons thought to mediate their effects on motor systems through the release of serotonin—accelerates the rhythm of the central oscillator. Whether this influence acts directly on interneu-

→

FIGURE 5. Acceleration of heartbeat oscillator cycling is associated with expression of the motor program for swimming. (A) Changes in oscillator's period associated with swimming in a partially dissected preparation. (Recording conditions as in Fig. 4). Panels 1 and 2 show continuous recordings from DP nerve of segment 10 (upper traces) and an HE (R,11) cell (lower traces) in a chronically swimming preparation. Swimming was interrupted by applying a sharp mechanical stimulus to the undissected anterior body wall (A1: arrow) and eliciting shortening. Swimming resumed following an approximately

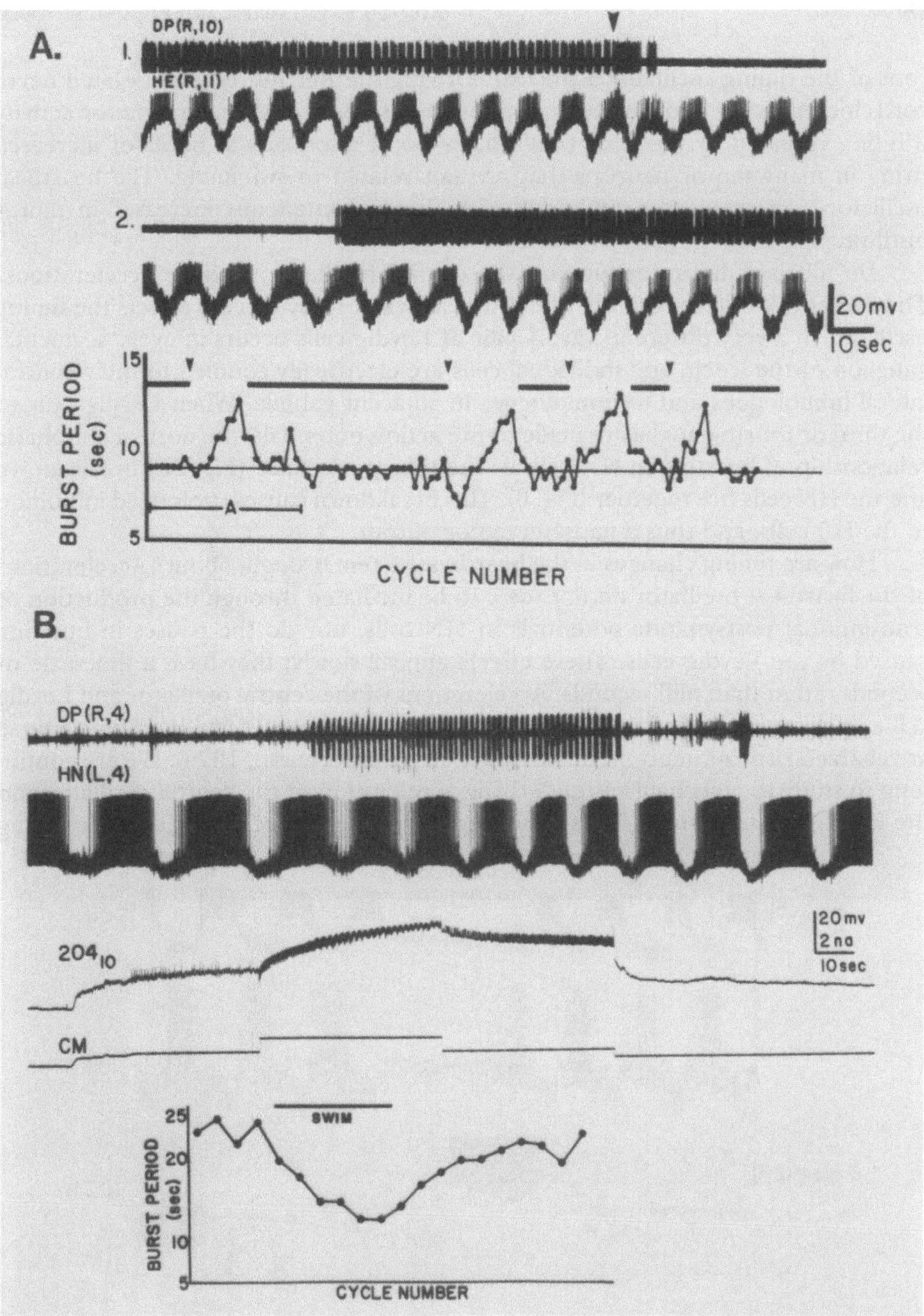

2-min period of quiescence. Burst period is plotted against cycle number below. The solid line indicates mean burst period (= 8.98 +/− 0.05 sec; n = 169 cycles) during expression of swimming motor program. The plot for the recording segment of A1, 2 is indicated by a bar labeled A. (B) Acceleration of the oscillator's cycling occurs with expression of motor program for swimming in isolated nerve cords. Swimming motor activity was monitored by an extracellular recording from DP nerve in ganglion 4 (upper trace); heartbeat oscillator activity by intracellular recording from an HN (L,4) cell (2nd trace). Swimming was induced by depolarizing swim-initiating interneuron 204 in ganglion 10 (3rd trace). The burst period of the HN cell is plotted against cycle number below, with the bout of swim motor program indicated by a bar. CM, current monitor for cell 204.

rons of the timing oscillator is unclear. Driving the Retzius' cells in isolated nerve cords increases the probability that spontaneous bouts of swimming motor activity will be expressed by the CNS (Willard, 1981); it also leads to bouts of increased firing in many motor neurons that are not related to swimming. The heartbeat oscillator accelerates concomitantly with these spontaneous increases in motor outflow.

Do all modulatory actions on the central oscillator produce accelerations? They do not. We have recently found that activity of Leydig cells affects the timing oscillator in a very different way. A pair of Leydig cells occurs in every segmental ganglion of the leech, and individual cells are electrically coupled to their contralateral homologues and to homologues in adjacent ganglia. When Leydig cells of the third or fourth ganglia are made to fire action potentials, the normal antiphasic relationship of bursting in HN cells of the timing oscillator (Fig. 2C) breaks down and the HN cells fire together (Fig. 6). This breakdown causes prolonged inhibition to the HE cells, and thus a pause in motor output.

How are timing changes in the heartbeat system brought about? Accelerations of the heartbeat oscillator do not seem to be mediated through the production of conventional postsynaptic potentials in HN cells, nor do the pauses in bursting caused by the Leydig cells. These effects appear slowly; they have a timescale of seconds rather than milliseconds. Accelerations of the central oscillator and Leydig cell effects outlast the stimuli that initiated them. Slow onset and long time course are characteristic of neuromodulatory actions (Kupfermann, 1979). We are continuing to study the mechanisms underlying acceleration of the central oscillator and the Leydig cell-mediated effects by testing the properties of HN cells in the timing

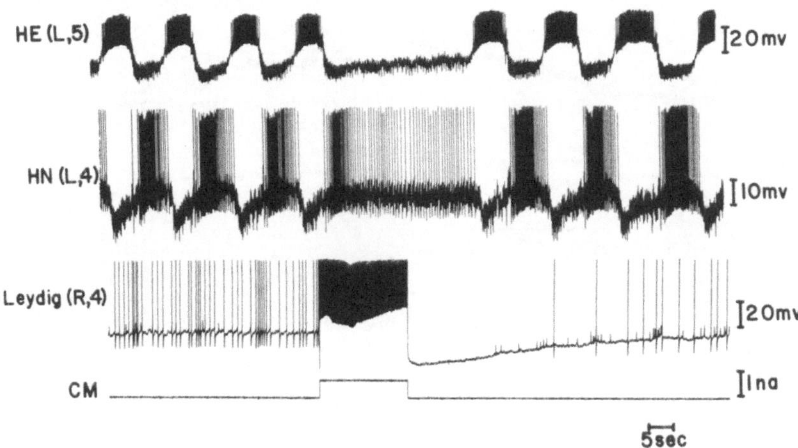

FIGURE 6. Leydig cell firing leads to transient pauses in motor output from the central oscillator. A Leydig cell of ganglion 4 (3rd trace) was made to fire at about 10 Hz by intracellular current injection. Cell HN (L,4) (2nd trace), which was cycling at a period of about 14 sec, went into a phase of prolonged firing after the onset of activity in the Leydig cell. This firing in turn caused prolonged inhibition of the HE cell (upper trace) and a pause in motor bursting. About 17 sec after Leydig cell's activity ceased, the normal activity of the HN cell was restored, and the heartbeat system continued to cycle without an apparent change in period. CM, current monitor for the Leydig cell.

oscillator before and after stimulation of acceleratory pathways and Leydig cells. We are most immediately interested in whether the modulatory actions are directed at altering the strengths of synaptic interactions within the circuit, the oscillatory properties of key HN cells, or both.

3. Modulation of the Peripheral Oscillator

As mentioned in the introduction, the hearts of the leech are themselves oscillators. They can produce a myogenic rhythm of electrical and contractile activity; this rhythm is entrained by the central pattern generator through the action of the HE motor neurons (Maranto and Calabrese, 1984b) (Fig. 1B). The myogenic properties of the hearts are not fixed: the HA neurons can modulate these properties. Figure 7A illustrates timing changes mediated by HA cells. These records were taken from a preparation where a section of a heart was left innervated in only one segment by a length of nerve cord that contained the full central pattern generator. The HA neuron was slightly hyperpolarized with injected current to slow its activity. At the beginning of the record, the HE motor neuron was firing normally, and the heart beat in time with its impulse bursts. When the HE cell was hyperpolarized with injected current to suppress its impulse activity, the heart ceased beating. When the HA neuron was twice depolarized and made to fire vigorously, however, a myogenic rhythm ensued. Termination of the depolarization in each case caused the myogenic rhythm to cease after one further beat. The latency from the beginning of the HA cell's depolarization to the start of the myogenic rhythm was much shorter for the second depolarization. This change in latency may reflect lingering effects from the first depolarization.

The results of this and other experiments indicate that the degree to which the myogenic rhythm is expressed by the heart may be controlled by the HA neurons. Phasic activity in the HA neurons does not seem to be essential in exerting this control: the latency from the beginning of the HA cells' activity to the start of the myogenic rhythm is long and variable, the effects of the HA cell seem to be long-lasting, and no fixed phase relation occurs between the rhythmic activity of an HA cell and myogenic activity of the heart when the HE motor neurons are experimentally silenced (Calabrese and Maranto, 1984). In about half of the preparations that were tested, the heart expressed a myogenic rhythm when the HA cells were firing normally. In those preparations where no myogenic rhythm was expressed, it could be induced by experimentally increasing the activity of an HA neuron. Since the HA cells fire weakly in isolated nerve cord preparations and since identified sensory cells can increase their firing (Calabrese and Maranto, 1984), we speculate that sensory input controls the firing frequency of HA cells in the intact animal, which in turn insures that the heart expresses a myogenic rhythm.

The HA cell is also capable of accelerating the ongoing myogenic rhythm of a heart when its firing rate is increased (Fig. 7B). Thus the HA neurons can tonically govern the period of the peripheral oscillator, but they do not instruct the phase of that oscillator. The tonic action of the HA neurons are contrasted to the phasic

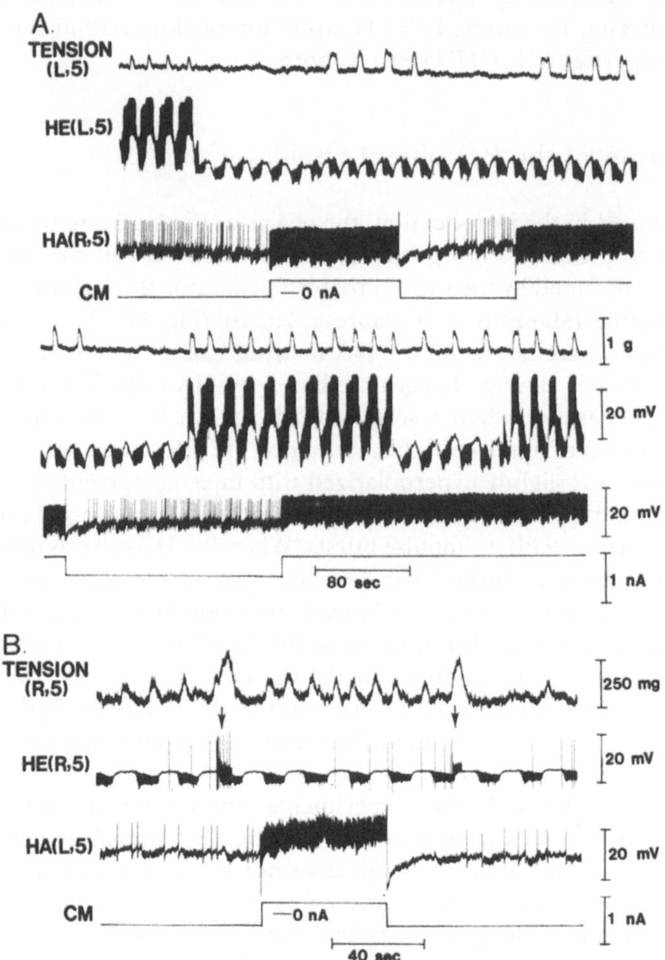

FIGURE 7. Effects of the HA neurons on the timing of the peripheral myogenic oscillator. (A) Simultaneous intracellular recordings from an HE cell and an HA cell, and a tension recording from a segment of heart that both neurons innervate. The two panels form a continuous record. The activity of the HE and of the HA cell was manipulated with injected current to demonstrate the ability of the HA cell to induce myogenic activity in the heart and of the HE cell to entrain that activity. (The HE cell was held hyperpolarized with a steady current to suppress its impulse activity and permitted to fire normally by terminating that current.) CM is the HA cell's current monitor. (B) Simultaneous intracellular recordings from an HE cell and an HA cell, and a tension recording from a segment of heart that both neurons innervate. The HE cell was hyperpolarized for the duration of the record with injected current to suppress its activity and permit the heart to express a myogenic rhythm. The sporadic bursts (arrows) that occurred in the HE cells were initiated at a peripheral spike initiation site (Maranto and Calabrese, 1984a). The activity of the HA cell was manipulated with injected current. When the HA cell was depolarized and made to fire vigorously, the myogenic rhythm sped up and then slowed down when the depolarization was terminated. CM, current monitor for the HA cell.

action of the HE motor neurons at the end of the records of Fig. 7A. If the HA cell was depolarized while the HE motor neuron expressed its normal activity rhythm, then there was no apparent effect of the HA neuron on the heart. Hyperpolarization of the HE motor neuron, however, revealed the covert action of the HA neuron on the heart's myogenic properties: the heart continued to beat rhythmically, but at a longer period than the HE motor neuron's activity period. When the HE motor neuron was then released from hyperpolarization, it quickly entrained the myogenic rhythm to its own activity period. Since the HA neurons can affect the myogenic period, they may match the period of peripheral heartbeat oscillator to that of the central oscillator and thus insure proper entrainment.

In addition to their timing effects, the HA neurons also influence the intensity of heartbeat as shown in Fig. 8. In this preparation, a section of heart was left innervated in only one segment by a length of nerve cord that contained the entire heartbeat central pattern generator. The innervating HE motor neuron was allowed to fire normally, and the heart was entrained to its activity rhythm. When the innervating HA neuron was depolarized with injected current so that it fired continuously, the tension produced during each beat increased in both amplitude and duration. The effect outlasted termination of the HA cell's depolarization by several cycles and did not require phasic activity in the HA neuron. This intensity change can often persist for an extended period. For example, the beat tension remained elevated for the duration of the records in Fig. 7A, once the HA cell was initially depolarized. The HA neurons can tonically modulate the beat tension of the hearts and are thus in a position to match cardiac output to the animals needs.

The mechanisms by which the HA neurons exert their timing and intensity effects on the heart remain elusive. It is unlikely, however, that conventional synaptic potentials mediate these effects. Figure 9 illustrates this point by contrasting the electrical effects of an HE motor neuron and an HA neuron on the same muscle cell. While each burst of impulses in the HE cell gives rise to summating excitatory junctional potentials that elicit plateau potentials in the muscle cell, there are

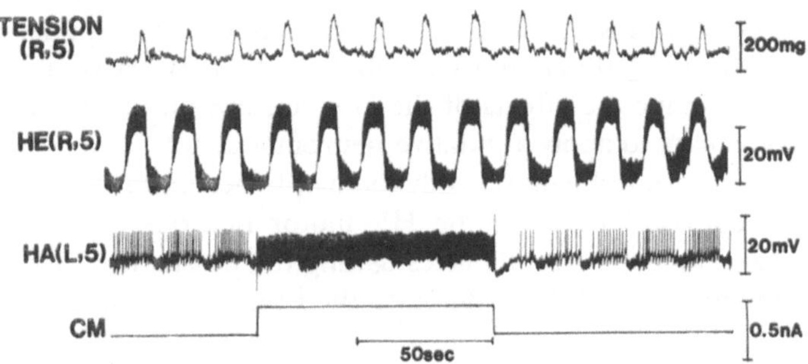

FIGURE 8. Effects of the HA neurons on intensity of the peripheral myogenic oscillator. The traces show simultaneous intracellular recordings from an HE cell and an HA cell, and a tension recording from a segment of heart that both neurons innervate. The contractile rhythm of the heart was entrained by the normal activity of the HE motor neuron, and the activity of the HA cell was manipulated with injected current to demonstrate its ability to increase both the amplitude and the duration of the tension produced by the heart during each beat. CM, current monitor for the HA cell.

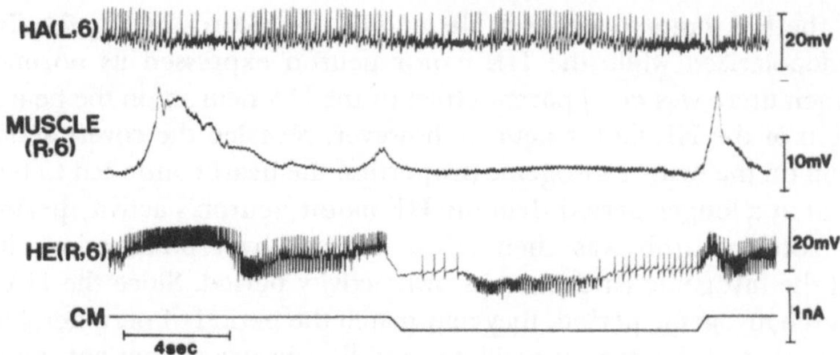

FIGURE 9. A comparison of the electrical effects of activity in the HE motor neuron and the HA neuron on a heart muscle cell. The traces show simultaneous intracellular recordings from an HE cell, an HA cell, and a muscle cell in a segment of heart that both neurons innervate. At the beginning of the records, a normal burst in the HE cell caused summating excitatory junctional potentials in the muscle cell which elicited a plateau potential. When the HE cell was hyperpolarized in midburst with injected current to slow its activity, the summating junctional potentials terminated before a plateau potential was elicited. (The HE cell's activity during the hyperpolarization caused small junctional potentials in the muscle that were not frequent enough to sum.) During the interval that the HE cell was hyperpolarized, the HA cell's normal activity did not cause any noticeable electrical events in the muscle cell. When the HE cell was released from hyperpolarization, it immediately evoked summating junction potentials in the muscle cell which elicted a plateau potential. CM, current monitor for the HE cell.

no obvious electrical events associated with the HA cell's impulses. At this point, we can only speculate that the HA cell's transmitter acts by modulating a voltage-dependent conductance to exert its timing effects, while it acts similarly (Nobel, 1979) or more directly on excitation–contraction coupling to exert its intensity effects (Weiss *et al.*, 1978; Benson *et al.*, 1981). In other systems, peptide neuromodulators are known to have both timing and intensity effects on muscle (Brown, 1967; Piek and Mantel, 1977; Sullivan, 1979; May *et al.*, 1979; Schwarz *et al.*, 1980; Benson *et al.*, 1981; Painter and Greenberg, 1982; Cottrell *et al.*, 1983; Adams and O'Shea, 1983). Recent experiments in our laboratory, using immunohistochemical techniques (Kuhlman *et al.*, 1984), have shown that the HA neurons can be labeled by antisera directed against the molluscan cardioacceleratory peptide, FMRFamide (Phe-Met-Arg-Phe-NH_2) (Price and Greenberg, 1977). We have also shown that FMRFamide can mimic the effects of the HA neurons on the heart (Calabrese *et al.*, 1984). The ability to apply a putative neuromodulator directly onto the heart muscle will allow us to probe more easily the biophysical mechanisms that underlie the HA cells' effects. Moreover, the HE motor neurons also label with anti-FMRFamide, and we are currently investigating the possibility that these neurons have a tonic modulatory role in addition to their phasic role described here.

4. Conclusion

Thus far, we have described how the central and peripheral oscillators that program heartbeat are separately modulated. Modulation at these two distinct sites must be linked if the two oscillators are to effect a coordinated change in circulation. Although we are only beginning to understand how parallel modulation of the two oscillators is brought about, it seems likely that the HA neurons will prove to be important in establishing the necessary coordination. Stimuli that lead to acceleration of the central oscillator can concomitantly increase the firing of HA cells (Fig. 10) (for effects of identified sensory cells see Calabrese and Maranto, 1984; Arbas and Calabrese, 1984). Changes in the period and strength of the peripheral oscillator might parallel the central acceleration.

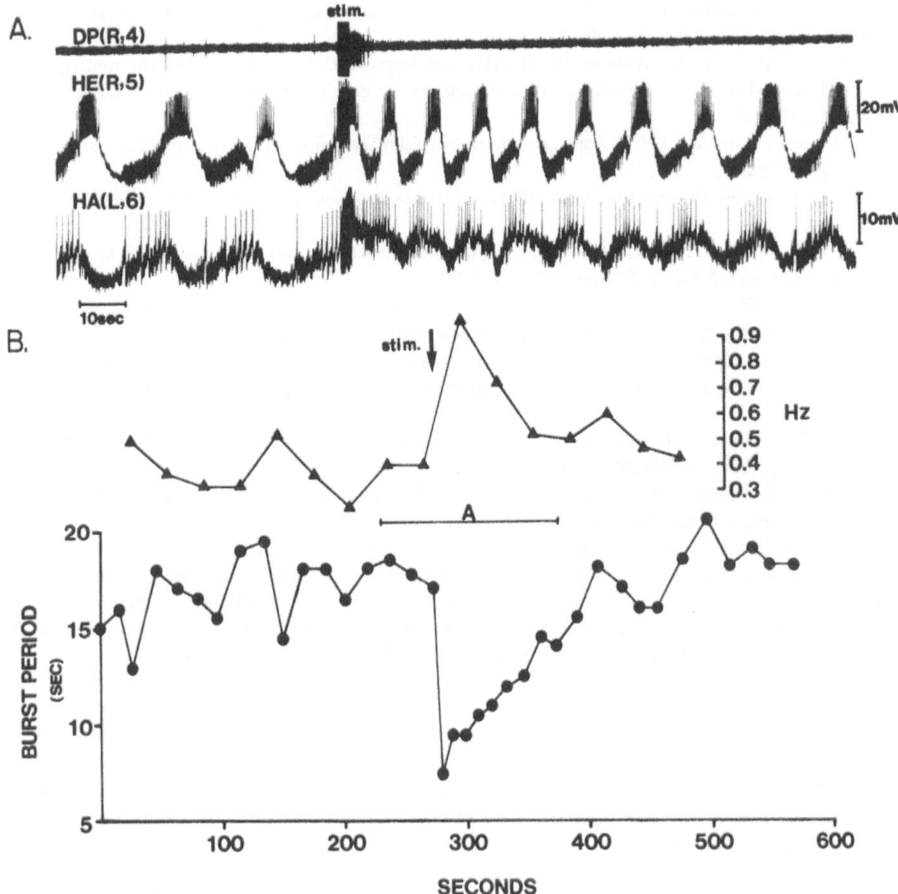

FIGURE 10. Concomitant change in period of the central oscillator and the firing frequency of an HA cell. (A) Simultaneous intracellular recordings from an HE cell and an HA cell in an isolated nerve cord (G2–G8) accompanied by an extracellular recording from a DP nerve. (B) Plot of HE cell burst period (●) and mean firing frequency of the HA cell (▲) over time from the same recording. HA cell firing frequency was averaged over 30-sec intervals, and each triangle is plotted in the center of its interval. The segment of recording shown in A is indicated in the graph by a bar. A 3-sec stimulus train of pulses at 5 Hz delivered to the DP nerve (stim.) accelerated the cycling of the central oscillator and increased the average firing frequency of the HA cell. Firing frequency of the HE cell was not altered significantly by the stimulus in this preparation.

ACKNOWLEDGMENTS. This research was supported by NSF grant BNS 81-21551 and an NIH Postdoctoral Fellowship to E. A. A. We wish to thank Christine S. Cozzens for valuable editorial assistance.

References

Adams, M. E., and O'Shea, M., 1983, Peptide co-transmitter at a neuromuscular junction, *Science* **221**:286–289.

Arbas, E. A., and Calabrese, R. L., 1984, Rate modification in the heartbeat central pattern generator of the leech, *J. Comp. Physiol.* **155**:783–794.

Benson, J. A., Sullivan, R. E., Watson, W. H., III, and Augustine, G. J., Jr., 1981, The neuropeptide proctolin acts directly on *Limulus* cardiac muscle to increase the amplitude of contraction, *Brain Res.* **213**:449–454.

Boroffka, I., and Hamp, R., 1969, Topographie des kreislaufsystems und zirkulation bei *Hirudo medicinalis, Z. Morph. Tiere* **64**:59–76.

Brown, B. E., 1967, Neuromuscular transmitter substance in insect visceral muscle, *Science* **155**:595–597.

Calabrese, R. L., 1977, The neural control of alternate heartbeat coordination states in the leech, *Hirudo medicinalis, J. Comp. Physiol.* **122**:111–143.

Calabrese, R. L., 1979, The roles of endogenous membrane properties and synaptic interaction in generating the heartbeat rhythm of the leech, *Hirudo medicinalis, J. Exp. Biol.* **82**:163–176.

Calabrese, R. L., and Maranto, A. R., 1984, Neural control of the hearts in the leech, *Hirudo medicinalis* III. Control of myogenicity and muscle tension by heart accessory neurons, *J. Comp. Physiol.* **154**:393–406.

Calabrese, R. L., and Peterson, E. L., 1984, Neural control of heartbeat in the leech, *Hirudo medicinalis, Symp. Soc. Exp. Biol.* **37**: **154**:393–406.

Calabrese, R. L., Kuhlman, J. R., and Li, C., 1984, FMRHamide-like substances in the leech: Bioactivity on the heartbeat system, *Neurosci. Abstr.* 10:46.2.

Cottrell, G. A., Schot, L. P. C., and Dockray, G. J., 1983, Identification and probable role of a single neuron containing the neuropeptide *Helix* FMRFamide, *Nature (London)* **304**:638–640.

Delcomyn, F., 1980, Neural basis of rhythmic behavior in animals, *Science* **210**:492–498.

Evans, P. D., and O'Shea, M., 1978, The identification of an octopaminergic neurone and the modulation of a myogenic rhythm in the locust, *J. Exp. Biol.* **73**:235–260.

Getting, P. A., 1983, Interaction of network, synaptic, and cellular properties in pattern generation, *Symp. Soc. Exp. Biol.* **37**:89–128.

Jewell, B. R., and Ruegg, J. C., 1966, Oscillatory contraction of an insect fibrillar muscle after glycerol extraction, *Proc. R. Soc. (London) Ser. B* **164**:428–459.

Kristan, W. B. Jr., Stent, G. S., and Ort, C. A., 1974, Neuronal control of swimming in the medicinal leech III. Impulse patterns of the motor neurons, *J. Comp. Physiol.* **94**:155–176.

Kuhlman, J. R., Calabrese, R. L., and Li, C., 1984, FMRFamide-like substances in the leech: Immunocytochemical localization, *Neurosci. Abstr.* **10**:46.1.

Kupfermann, I., 1979, Modulatory actions of neurotransmitters, *Annu. Rev. Neurosci.* **2**:447–465.

Macagno, E. R., 1980, The number and distribution of neurons in leech segmental ganglia, *J. Comp. Neurol.* **190**:283–302.

Mann, H., 1962, *Leeches (Hirudinea). Their Structure, Physiology, Ecology and Embryology,* Pergammon Press, New York.

Maranto, A. R., and Calabrese, R. L., 1984a, Neural control of the hearts in the leech, *Hirudo medicinalis* I. Anatomy, electrical coupling, and innervation of the hearts, *J. Comp. Physiol.* **154**:367–380.

Maranto, A. R., and Calabrese, R. L., 1984b, Neural control of the hearts in the leech, *Hirudo medicinalis* II. Myogenic activity and its control by heart motor neurons, *J. Comp. Physiol.* **154**:381–391.

May, T. E., Brown, B. E., and Clements, A. N., 1979, Experimental studies upon a bundle of tonic fibers in the locust extensor tibialis muscle, *J. Insect Physiol.* **25**:169–181.

Nicholls, J. G., and Baylor, D. A., 1968, Specific modalities and receptive fields of sensory neurons in the c.n.s. of the leech, *J. Neurophysiol.* **31**:740–756.

Noble, D., 1979, *The Initiation of the Heartbeat,* Clarendon Press, Oxford.

Painter, S. D., and Greenberg, M. J., 1982, A survey of the responses of bivalve hearts to the molluscan neuropeptide FMRFamide and 5-hydroxytryptamine, *Biol. Bull.* **162**:311–332.

Payton, B., 1981, Structure of the leech nervous system, in: *Neurobiology of the Leech* (K. J. Muller, J. G. Nicholls, and G. S. Stent, eds.), Cold Spring Harbor Laboratory, New York, pp. 27–34.

Peterson, E. L., 1983a, Generation and coordination of heartbeat timing oscillation in the medicinal leech. I. Oscillation in isolated ganglia, *J. Neurophysiol.* **49**:611–626.

Peterson, E. L., 1983b, Generation and coordination of heartbeat timing oscillation in the medicinal leech. II. Intersegmental coordination, *J. Neurophysiol.* **49**:627–638.

Peterson, E. L., and Calabrese, R. L., 1982, Dynamic analysis of a rhythmic neural circuit in the leech, *Hirudo medicinalis, J. Neurophysiol.* **47**:256–271.

Piek, T., and Mantel, P., 1977, Myogenic contractions in locust muscle induced by proctolin and by wasp, *Philanthus triangulum,* venom, *J. Insect Physiol.* **23**:321–325.

Price, D. A., and Greenberg, M. J., 1977, Structure of a molluscan cardioexcitatory neuropeptide, *Science* **197**:670–671.

Roberts, B. L., and Roberts, A. (eds.), 1983, Neural origin of rhythmic movements, *Symp. Soc. Exp. Biol.* **37**:195–221.

Schwarz, T. L., Harris-Warrick, R. M., Glusman, S., and Kravitz, E. A., 1980, A peptide action in a lobster neuromuscular preparation, *J. Neurobiol.* **11**:623–628.

Selverston, A. I., and Miller, J. P., 1984, Co-operative mechanisms for the production of rhythmic movements, *Symp. Soc. Exp. Biol.* **37**:55–87.

Sullivan, R. E., 1979, Proctolin-like peptide in crab pericardial organs, *J. Exp. Zool.* **210**:543–552.

Thompson, W. J., and Stent, G. S., 1976a, Neuronal control of the heartbeat in the medicinal leech. I. Generation of the vascular constriction rhythm by heart motor neurons, *J. Comp. Physiol.* **111**:261–279.

Thompson, W. J., and Stent, G. S., 1976b, Neuronal control of the heartbeat in the medicinal leech. II. Intersegmental coordination of heart motor neuron activity by heart interneurons, *J. Comp. Physiol.* **111**:281–307.

Thompson, W. J., and Stent, G. S., 1976c, Neuronal control of the heartbeat in the medicinal leech. III. Synaptic relations of heart interneurons, *J. Comp. Physiol.* **111**:309–333.

Weeks, J. C., and Kristan, W. B., Jr., 1978, Initiation, maintenance and modulation of swimming in the medicinal leech by the activity of a single neuron, *J. Exp. Biol.* **77**:71–88.

Weiss, K. R., Cohen, J. L., and Kupfermann, I., 1978, Modulatory control of buccal musculature by a serotonergic neuron (metacerebral cell) in *Aplysia, J. Neurophysiol.* **41**:181–203.

Willard, A. L., 1981, Effects of serotonin on the generation of the motor program for swimming by the medicinal leech, *J. Neurosci.* **1**:936–944.

6

Neural Network Analysis in the Snail Brain

PAUL R. BENJAMIN, CHRISTOPHER J. H. ELLIOTT, AND GRAHAM P. FERGUSON

1. Introduction

How can we hope to use the techniques of cellular neurobiology, which have been successfully used to analyze the properties of single neurons or small networks, to eventually understand a structure such as the snail brain which contains about 25,000 neurons? Our approach has been to carry out detailed analyses of specific neural networks underlying several types of function in the brain of the pond snail, *Lymnaea*, while collecting less specific information on more global aspects of brain organization and the larger scale of interactions within it. The concentration of the CNS into a compact brain and the distribution of cells of the same type over several ganglia (see, for instance, the neurosecretory cells of Fig. 1A) makes such a whole brain analysis almost inevitable even for simple mapping studies, but the global aspect of organization within the snail brain is also emphasized by our analysis of two interneurons that have follower cells in at least five of the ganglia of the CNS (Fig. 1C,D). Even neural systems underlying specific behavioral acts can be widely distributed and motoneurons responsible for whole body withdrawal responses occur in all nine ganglia of the central ganglionic ring (Fig. 1B).

Our intention in this review is to illustrate the usefulness of the *Lymnaea* CNS for neural network analysis by considering three different systems of neurons. Each of these three examples may be used to support the points made in the previous paragraph, but in addition, each network has its own features that are of more specific interest. The first system to be considered consists of two identified inter-

PAUL R. BENJAMIN, CHRISTOPHER J. H. ELLIOTT, AND GRAHAM P. FERGUSON • M.R.C. Neurophysiology Research Group, School of Biology, University of Sussex, Falmer, Brighton, Sussex, United Kingdom.

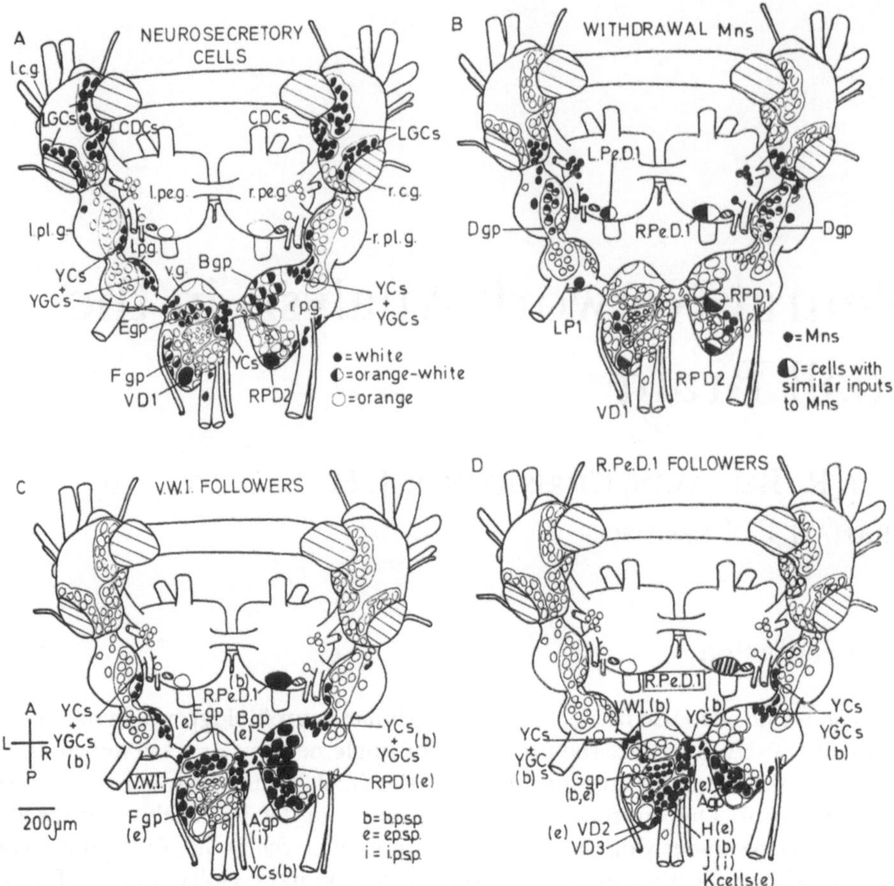

FIGURE 1. Maps of identified neurons on the dorsal surface of the CNS of *Lymnaea*. (A) Cell body colors. The white cells (shaded) are peptide-secreting neurons. These occur as uniquely identifiably giant neurons (RPD2, VD1), discrete bilaterally symmetrical clusters [the light green cells (LGCs) and caudodorsal cells (CDCs)], or in more scattered locations [yellow cells (YCs) and yellow-green cells (YGCs)]. The orange-white (half-shaded) Bgp (group), Egp, and Fgp probably contain a FMRF amidelike peptide (see text). Ganglion abbreviations: l.c.g., left cerebral ganglion; l.p.g., left parietal ganglion; l.pe.g., left pedal ganglion; l.pl.g., left pleural ganglion; r.c.g., right cerebral ganglion; r.p.g., right parietal ganglion; r.pe.g., right pedal ganglion; r.pl.g., right pleural ganglion; v.g., visceral ganglion. (From Benjamin *et al.*, 1980; Benjamin and Winlow, 1981; Benjamin, unpublished.) (B) Whole body withdrawal response motoneurons (Mns) (shaded) and other cells of unknown function (half-shaded) that receive similar types of synaptic input to the motoneurons either spontaneously or induced by sensory stimulation of the skin (From Ferguson and Benjamin, unpublished.) (C) Follower cells (shaded) of the visceral white interneuron (VWI, name inside box, cell body cross-hatched). Type of postsynaptic response shown in brackets. (From Benjamin, unpublished.) (D) Follower cells (shaded) of the giant dopamine-containing interneuron R.Pe.D.1 (name inside box, cell body cross-hatched). Type of postsynaptic response shown in brackets. (From Benjamin and Winlow, 1981; Winlow *et al.*, 1981; Benjamin, unpublished.) Ganglion outline used in A–D from Slade *et al.*, 1981.

neurons that control the electrical activity of three types of neurosecretory or peptide-secreting neurons, but in addition, both types of interneuron have many other nonneurosecretory follower cells. The second system is a network underlying whole body withdrawal responses in *Lymnaea.* This is a widely distributed electrotonically coupled motoneuron network whose activity is initiated by convergent sensory input (tactile and photic). Finally, the feeding system of *Lymnaea* will be reviewed as an example of a rhythmically active motor behavior produced by a central network of interneurons (Benjamin and Rose, 1980; Benjamin *et al.*, 1981; Benjamin, 1983b). Here some new features of the central pattern generator will be described.

2. General Features of Neuronal Organization in *Lymnaea*

Many neurons in the CNS of *Lymnaea stagnalis* are brightly pigmented, although colors vary from bright orange through white-orange to pure white (Fig. 1A). Neurons of various sizes can be mapped and this is shown in Fig. 1 for the dorsal surface of brain. A few large neurons can be recognized as unique individuals (eg., RPD1, RPD2, VD1, VD2, VD3, LP1, R.Pe.D.1, L.Pe.D.1), but most cells are identified as members of small clusters on the basis of color, position, and shared electrical properties, including synaptic connections. Marking cells with dye and staining them subsequently in histological sections with neurosecretory stains, such as Alcian Blue-Alcian Yellow (AB-AY), allows further classification of neurosecretory cells into several cell types (see Benjamin *et al.*, 1980).

Cell body color was a useful way of separating cells into neurosecretory and nonneurosecretory categories and all the white cells in Fig. 1A are classed as neurosecretory or peptide secretory cells on the basis of criteria reviewed elsewhere (Swindale and Benjamin, 1976; Benjamin *et al.*, 1980). Recent immunocytochemical studies have revealed many more peptide-containing neurons in the brain of *Lymnaea* (Schot *et al.*, 1981). A few of these correspond to the AB-AY–positive cells shown in Fig. 1A. Thus VD1 and RPD2, which stain yellow with AB-AY (Soffe and Benjamin, 1980), contain immunoreactive ACTH (Boer *et al.*, 1979) and the light green cells (growth hormone-producing cells) (Geraerts, 1976) stain with antisomatostatin (Schot *et al.*, 1981). Furthermore, recent work (P. R. Benjamin, D. Price, H. H. Boer, and M. J. Greenberg, unpublished data) shows that the Bgp, Fgp, and Egp cells of Fig. 1A all contain an FMRF amidelike peptide. However, many of the immunoreactive cells discovered by Schot *et al.* (1981) do not correspond with the "classical" neurosecretory cell types shown in Fig. 1A and these although at present unidentified, are of considerable future interest.

Identified interneurons of several types have been found in the feeding system (e.g., McCrohan and Benjamin, 1980a,b; Rose and Benjamin, 1981a,b), but outside this system only two interneurons have been found. These are the giant dopamine-containing interneuron, R.Pe.D.1 (Benjamin and Winlow, 1981; Winlow *et al.*, 1981) and the newly discovered visceral white interneuron (VWI). The location of these interneurons and their follower cells are shown in Fig. 1C,D and their synaptic effects will be considered in Section 3.

Electrotonic coupling between cells of the same type is common and occurs, for instance, between the giant cells VD1, RPD2 (Benjamin and Winlow, 1981), several types of neurosecretory cells (Benjamin *et al.*, 1976; Benjamin, 1983a; Benjamin and Rose, 1984), and withdrawal response motoneurons (see Section 4.4).

3. The Wide-acting Interneurons R.Pe.D.1 and the VWI

These two identifiable interneurons form part of a network controlling the activity of many cells in the visceral, parietal, and pleural ganglia. The distribution and type of synaptic effect on these follower cells are summarized in Fig. 1C,D, and this shows that the interneurons have many follower cells in common as well as reciprocal biphasic effects on each other. This discussion will concentrate on the synaptic effects that R.Pe.D.1 and VWI have on three types of peptide-secreting cells, the yellow cells, the yellow-green cells, and the Bgp, Egp, and Fgp cells. Two of these cell types produce peptides that affect the renopericardial system, whereas the third (the yellow-green cells) is also, like the yellow cells, probably involved in ion and water regulation. Some examples of follower cells that are non neurosecretory will also be given to emphasize the variety of follower cell responses.

The VWI has synaptic effects on all three types of peptidergic cells mentioned in the last paragraph. Fig. 2A shows that an evoked burst of spikes in the VWI excites cells of the Bgp and Egp. In fact, other data show that it excites all cells of these two clusters as well as cells of the Fgp (see Fig. 1A for the location of these cells). Recent results obtained in collaboration with the laboratories of M. J. Greenberg and H. Boer show that cells from all three clusters contain a cardioacceleratory FMRF amidelike peptide previously shown to exist in the *Lymnaea* by Geraerts *et al.* (1981) and Schot and Boer (1982). Furthermore, the VWI also has synaptic effects, in this case predominantly inhibitory, on two other types of neurosecretory cell, the yellow cells and yellow-green cells. An example of this inhibitory effect of the VWI on the yellow cell is shown in Fig. 2D where ongoing spike activity is prevented or reduced in frequency. Similar effects can be seen on the yellow-green cells. The VWI actually has biphasic (depolarizing followed by hyperpolarizing) effects on both cell types (see Fig. 2C for an example on a yellow-green cell), but the overall effect is to inhibit spike activity in both cell types because of the longer duration of the inhibitory phase of the biphasic postsynaptic potential (BPSP).

The yellow cells and yellow-green cells get their name from their staining reaction to the neurosecretory stain AB-AY and were first described by Wendelaar Bonga (1970). They have all the classical morphological features of neurosecretory cells with both central and peripheral release sites. The peripheral release sites are in the ureter of the kidney and there is morphological evidence that hormonal release into this structure is caused by hypo-osmotic environmental conditions (deionized water) (Wendelaar Bonga, 1972) which cause a drop in blood sodium and chloride levels (Soffe *et al.*, 1978). Preliminary data also suggest a role of the yellow-green cells in ion and water regulation (Soffe *et al.*, 1978). It thus appears that the VWI excites one type of neuron containing a cardio-excitatory peptide and inhibits two other types of cells involved in ion and water regulation, one of which releases peptides into the kidney.

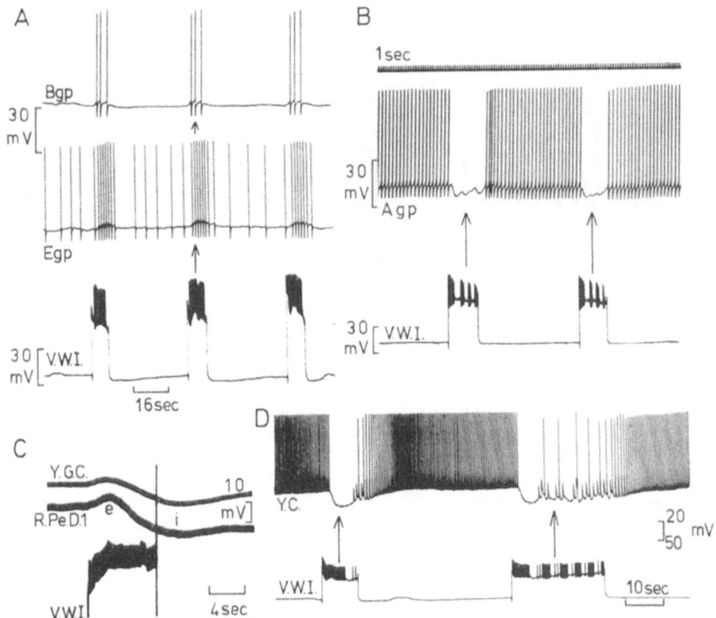

FIGURE 2. Follower cells of the VWI. In each record, activity in the VWI was evoked by current injection via the recording electrode. (A) Repeated bursts of VWI spikes excite the Bgp (group) and Egp cells. (B) Inhibition of Agp spike activity. (C) Biphasic responses in a yellow-green cell (YGC) and R.Pe.D.1 caused by a burst of spikes in the VWI. An initial depolarizing response, e, is followed by i, a longer duration hyperpolarizing wave. (D) Inhibition of a yellow cell (YC) caused by long periods of VWI spiking activity. (From Benjamin, unpublished data.)

These are not the only follower cells of the VWI, however; for instance, the VWI also inhibits the Agp cells of the right parietal ganglia (Fig. 2B) and the other major interneuron to be described in this section, R.Pe.D.1, although its postsynaptic effect is again nominally biphasic (Fig. 2C).

R.Pe.D.1 is the giant dopamine-containing neuron of *Lymnaea* previously described by Benjamin and Winlow (1981), which is known to have many monosynaptic connections with neurons in the right parietal and visceral ganglia (Winlow *et al.*, 1981). Some of these responses are shown in Fig. 3A. Here, R.Pe.D.1 was recorded with three types of follower cell in the visceral ganglion and this illustrates that the interneuron can produce three different types of postsynaptic potentials depending on the identity of the follower cell recorded. Each spike in R.Pe.D.1 is followed by a BPSP in the I cell, an inhibitory postsynaptic potential (IPSP) in the J cell, and an excitatory postsynaptic potential (EPSP) in the K cell. More recently, several new types of follower cell have been found (P. R. Benjamin, unpublished data). For instance, there are a few small cells in the G cluster (Fig. 3B) that have fast EPSP responses, a type of response not previously seen in *Lymnaea,* but described for the homologous dopamine cell in *Planorbis* (Berry and Cottrell, 1979).

Of more general interest for the present discussion are the new types of

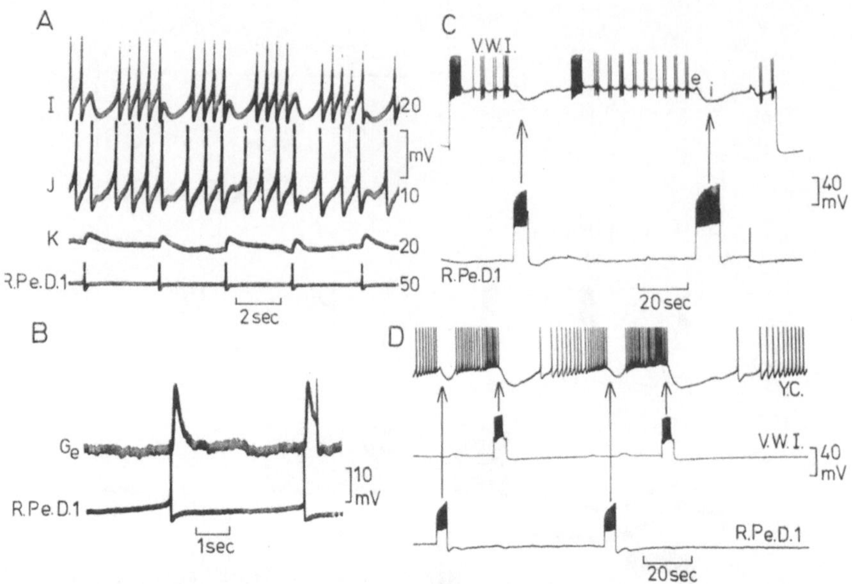

FIGURE 3. Follower cells of the giant dopamine-containing interneuron, R.Pe.D.1. (A) R.Pe.D.1 recorded with three visceral ganglion follower cells showing the different types of responses previously reported by Benjamin and Winlow (1981). The K cell has an EPSP response, the J cell a weak IPSP response, and the I cell a BPSP response (depolarization followed by hyperpolarization). Each spike in R.Pe.D.1 is accompanied by a PSP in the three follower cells. (B) A newly discovered fast EPSP response in a small Ggp cell designated, G_e. (C) Biphasic responses (e, depolarizing response followed by i, hyperpolarizing response) recorded on the VWI following bursts of spikes in R.Pe.D.1. The VWI was first activated by steady depolarizing current. The main effect of the R.Pe.D.1 BPSP is to inhibit VWI spiking. (D) Comparison of the effects of the two interneurons on yellow cell (YC) firing. R.Pe.D.1 and VWI were alternately activated. Both inhibited YC firing, but the VWI inhibition was larger in amplitude and longer in duration. (B–D from Benjamin unpublished data.)

R.Pe.D.1 follower cells shown in Fig. 3C,D. First, Fig. 3C shows that R.Pe.D.1 inhibits the VWI so that these two interneurons have reciprocal inhibitory connections. In fact, a burst of spikes in R.Pe.D.1 evokes a BPSP in the VWI (Fig. 3C), but the initial EPSP is ineffective in evoking spike activity. Second, R.Pe.D.1 inhibits the yellow cells and yellow-green cells, as does the VWI. Thus both interneurons have similar effects on these two cell types, although the strength of the response is different for the two cells. This is shown in Fig. 3D where both interneurons were alternately activated for short periods and the effects on a yellow-green cell spike activity examined. It can be seen that both interneurons inhibit the yellow-green cell, but the VWI is much more powerful in its effect than R.Pe.D.1. Note that the VWI inhibits the yellow cells and yellow-green cells in all five ganglia where they occur (Fig. 1C), whereas R.Pe.D.1 effects have a more limited distribution (Fig. 1D), having only about one half the total population as follower cells.

We can summarize the main points of this section by saying that there are two major wide-acting interneurons in the *Lymnaea* brain both of which inhibit the yellow cells and yellow-green cells, whereas only one cell type, the VWI, excites the

Bgp, Fgp, and Egp cells. Thus we have a central interneuronal system controlling different types of peptide-secreting cell, two of which at least are involved in control of the renopericardial system of the snail. The exact role of the two interneurons has yet to be ascertained, but their strong mutually inhibitory connections make it unlikely that they can both influence the peptide-secreting cells at the same time. Because they both have further synaptic effects on many other types of neuron in the *Lymnaea* CNS (Fig. 1C,D), other functions are likely.

4. The Whole Body Withdrawal Response System of *Lymnaea:* A Distributed Motor Network

4.1. Behavior and Underlying Muscular Actions

Whole body withdrawal responses are induced by photic (light-off or shadows) or tactile stimuli applied anywhere on the surface of the snail's body. The photically induced response is the more stereotyped of the two and consists mainly of a longitudinal shortening of the head-foot and a pulling forward of the shell to cover the body. Tactile stimuli produce responses that are dependent on both the site and the strength of the stimulus. Very strong tactile stimuli produce a sequence of responses that include longitudinal shortening as one part of the response, and it is this common component that has been investigated in the present study rather than the localized responses, such as tentacle withdrawal, which are produced on their own by weak stimuli and may partially be mediated by the peripheral nervous system.

The dorsal longitudinal muscle (DLM) and columellar muscle (CM) are the two main muscles involved in withdrawal behavior (Fig. 4). The DLM consists of three main bilaterally symmetrical bands of muscle on each side of the body (labeled 1, 2, and 3 in Fig. 4) that cover the entire dorsal surface. The anterior end of the muscle inserts into the diffuse muscle of the foot, while posteriorly the muscle ends under the first whorl of the shell (Fig. 4). Contractions of this muscle mainly cause a shortening of the head-foot. The bilaterally symmetrical CM muscle runs laterally on either side of the body inside the DLM (Fig. 4). At the anterior end of the body, the CM from each side crosses over the midline and inserts in the foot. Posteriorly,

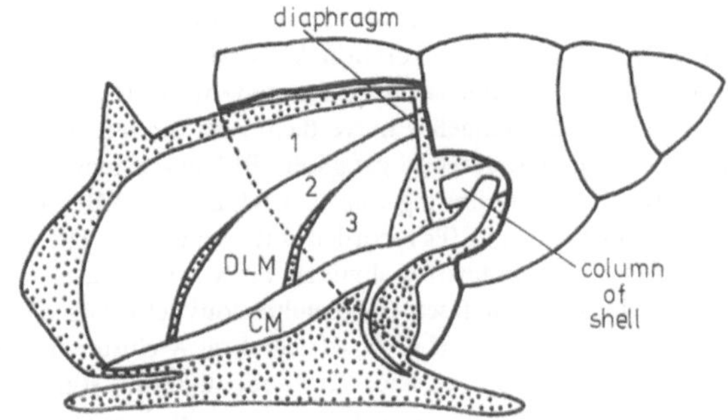

FIGURE 4. Arrangement of the muscle underlying the whole body withdrawal response. The snail is shown in longitudinal section with only muscles on the right side of the body displayed. Thus the CM is medial to the three strands (1,2,3) of the DLM. (From Ferguson and Benjamin, unpublished data.)

they fuse and join a third posterior branch of the muscle. All three branches, once fused, eventually insert on the column of the shell. Contraction of this complex muscle longitudinally shortens the ventral part of the anterior head-foot together with the posterior foot and also pulls the shell over the body. The actions of the DLM and CM underlie the main components of the observed behavior and electromyographic recordings show that they are activated simultaneously during withdrawal induced by sensory stimuli (Fig. 5C).

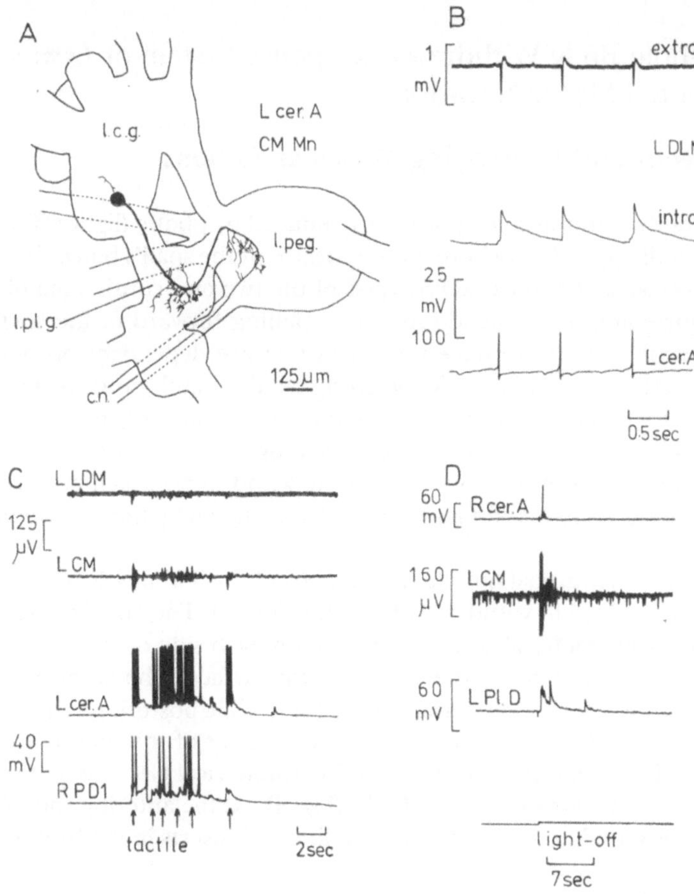

FIGURE 5. Motoneurons, muscles, and sensory input underlying the whole body withdrawal response. (A) Morphology of a left cerebral A cluster cell (L cer. A) revealed by Lucifer Yellow injection. This is one of the two columellar muscle motonerurons (CM Mn) occurring in this cluster. It eventually projects to the left columellar nerve (l.c.n.) via the left pleural ganglion (l.pl.g.) and left pedal ganglion (l.pe.g.). l.c.g., left cerebral ganglion. (B) Intracellular recording of a DLM motoneuron from the left cerebral A cluster showing that it produces 1:1 EJPs in the muscle. Large intracellular (intra.) and extracellular (extra.) EJPs recorded at the same time. The extracellular recording presumably represents the activity of a number of simultaneously active muscle fibers, whereas the intracellular recording is from a single muscle fiber. (C) Simultaneous activation of the left dorsal longitudinal muscle (LDLM), left columellar muscle (LCM), left cerebral A cluster neuron, and giant neuron, RPD1, by repeated tactile stimulation of the skin. (D) Photic (light-off) stimulation of the skin produces similar activation of a cerebral motoneuron (in this case a right, R cell) and the columellar muscle and EPSPs on a left pleural Dgp neuron (L Pl.D). (From Ferguson and Benjamin, unpublished data.)

4.2. Identification and Morphology of Motoneurons and Their Effects on DLM and CM

Tracing nerves from the CNS to the DLM and CM revealed a complex pattern of innervation involving nerves originating from most of the ganglia of the brain. Cobalt chloride back filling of nerves revealed the likely locations of putative motoneurons and Lucifer Yellow injections of these cells showed directly that neurons identified *in vivo* projected down the same nerves. In a semi-intact preparation, where longitudinal contraction of the body could be induced by sensory stimulations of the skin, it was possible to carry out extracellular and intracellular recording of muscle fibers together with intracellular motoneuron recording. This together with the anatomical evidence provided firm evidence for motoneuronal function. Motoneurons were found as single cells or in small clusters in all nine ganglia of the CNS (excluding the buccals) and Fig. 1B shows the location of their cell bodies. Detailed evidence for motoneuronal function will only be given for the cerebral A cluster cells as an example of the type of data obtained. These are the largest group of withdrawal motoneurons in the *Lymnaea* brain and each cluster contains from 12 to 15 cells. Bilaterally symmetrical clusters of cells occur on the posterior medial surface of the dorsal lobe of the cerebral ganglia (Fig. 1B). Cerebral A clusters contain motoneurons for either the CM or DLM muscles, although the majority are DLM motoneurons. The anatomy of a Lucifer-Yellow-filled left CM motoneuron is shown in Fig. 5A. Its axon projects through the left pleural and pedal ganglia and eventually leaves the latter via the left columellar nerve. Extensive dendritic branching occurs in the neuropil of both the pleural and pedal ganglia. Dorsal longitudinal muscle motoneurons have similar sorts of morphology except that they project to the ipsilateral superior and inferior cervical nerves.

Electrophysiological evidence for motoneuronal function, in this case for a cell acting on the DLM, is shown in Fig. 5B. Here, in a semi-intact preparation, intracellular recording of the motoneuron was carried out at the same time as both intracellular and extracellular recording of the DLM. A long depolarizing current pulse injected into the motoneuron evoked a series of action potentials that were followed by 1:1 excitatory junction potentials (EJPS) and corresponding extracellular potentials in the muscle. This excitatory action of motoneurons was a general feature of the withdrawal response system and so far motoneurons with inhibitory effects on muscle have not been found.

4.3. Sensory Input to Motoneurons

The same semi-intact preparation that allows simultaneous motoneuron and muscle recordings was also used to apply tactile and light-off stimulus to the skin. Both these stimuli caused activation of both motoneurons and withdrawal response muscles (Fig. 5C, D). In Fig. 5C, repeated tactile stimuli applied to the anterior foot caused EJPS in both the left DLM and CM and simultaneous spikes in a cerebral A cluster cell. It also excited the giant cell RPD1 which plays no obvious role in the withdrawal response. In fact, a number of identified neurons of unknown function (for instance, the pleural Dgp neuron of Fig. 5D) either receive sensory and/or common spontaneous synaptic input with withdrawal response motoneurons and their location is shown in Fig. 1B (half-shaded cells). Another recording (Fig. 5D) shows simultaneous muscle and cerebral A cluster activation, but in this case light-off stimulation (to similate shadows) was used. This dual sensitivity to tactile and light-off stimuli illustrated in Fig. 5 was the common feature of all the withdrawal response motoneurons. Furthermore, strong tactile stimuli applied anywhere on the skin of the semi-intact preparation produce similar coactivation of motoneurons in all their central locations as does light-off normally applied to the whole preparation. This implies a good deal of sensory convergence particularly in the case of tactile stimulation.

Cutting the optic nerve did not prevent light-off–induced contractions of the DLM and CM, but cutting the rest of the nerves to the skin did. This suggests that the light-off skin receptors described by Stoll (reviewed, 1976) are mediating the photically induced withdrawal response and that direct activation of central neurons is not involved.

4.4. Coordination of Motoneuron Activity

We can conclude from Fig. 5 that sensory activation of centrally located motoneurons is responsible for the main part of the withdrawal response involving whole body shortening. This does not occur in the absence of the central ganglia and it requires coactivation of all the motoneurons concerned, because they innervate distinct, although overlapping regions of the CM and DLM. We have mentioned the convergence of sensory input as an important coordinating factor for producing synchronized motoneuronal firing. However, another important factor in the synchronization of motoneuronal activity is the electrotonic coupling that probably occurs between the majority of the motoneurons occurring in the nine ganglia of the CNS, but is most obvious when ipsilateral cells are recorded together. An example of this coupling is shown in Fig. 6 where a left cerebral A cluster neuron was recorded with a left parietal motoneuron. Passing a depolariz-

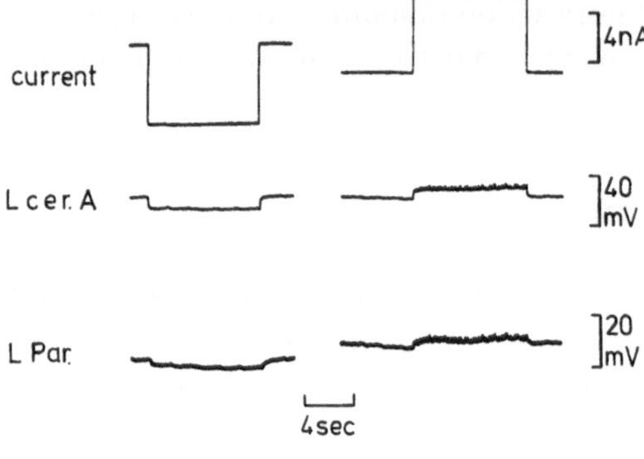

FIGURE 6. Electrotonic coupling of withdrawal response motoneurons. Strong coupling occurs between the cerebral A cluster cells (in this case a left one, L) and other withdrawal response motoneurons occurring in other ganglia on the same side of the brain, such as the left parietal (L Par.) cell shown here. Hyperpolarizing and depolarizing square current pulses of the same strength (top trace) were injected into the cerebral cell. The middle trace shows the voltage change in this cell recorded with a second electrode. Similar although attenuated voltage responses were recorded in the left parietal cell (bottom trace). Coupling ratios were 0.3 for both polarities of current indicating a nonrectifying junction. (From Ferguson and Benjamin, unpublished data.)

ing or hyperpolarizing current pulse of the same strength to the cerebral A cluster cell produced similar but attenuated responses in the left parietal cells, indicating a nonrectifying electrotonic junction. However, not all motoneurons have equally strong effects on motoneurons located in other ganglia and this is rather variable. However, one can make the general statement that the cerebral A cluster cells always have strong effects on all the other motoneurons and activation of these cells by strong depolarizing currents usually causes spikes in other cells of the network.

It can be concluded that coactivation of the widely distributed motoneuronal network necessary for whole body shortening to occur is mainly due to convergent sensory input from the skin whose effects are reinforced by electrotonic coupling between motoneurons.

5. The Interneuronal Network Underlying Rhythmic Feeding Movements

5.1. Behavior and Motoneuronal Organization

Lymnaea stagnalis is a browsing herbivore that feeds off algal films or floating pond weed. Ingestion of food is accomplished by a complex muscular structure called the buccal mass which carries out repeated cycles of sterotyped movement (Rose and Benjamin, 1979) during the consummatory phase of feeding behavior. The activity of muscles and motoneurons responsible for buccal mass movements have been analyzed in semi-intact preparations and cyclical movements occur spontaneously or can be initiated by addition of sugars such as sucrose to the preparation (Rose and Benjamin, 1979). The movements we have described thus serve ingestion, not egestion or other behaviors of the buccal mass (compare McClennan, 1982).

Within each feeding cycle, four different phases of buccal mass movements have been described. Three of these are protraction (P), retraction (R1, rasp or bite), retraction 2 (R2, swallow), which are separated by the fourth inactive phase (I), a variable interval between each feeding cycle. Examples of motoneurons firing during each of three active phases of the feeding cycle are shown in the three bottom traces of Fig. 7B (from Rose and Benjamin, 1981b). In Fig. 7B, motoneuronal firing was initiated by a higher order interneuron, the slow oscillator (SO), but in 15–20% of freshly dissected snails the activity occurs spontaneously in isolated preparations of buccal and cerebral ganglia.

5.2. The Interneuronal Network

The feeding rhythm in *Lymnaea* is generated by a network of interneurons that (like the motoneurons) have their cell bodies located in the buccal ganglia. In our previous study, three types of pattern-generating interneurons were identified,

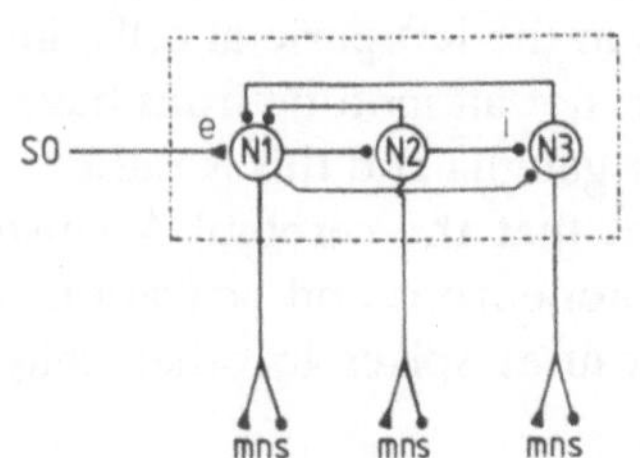

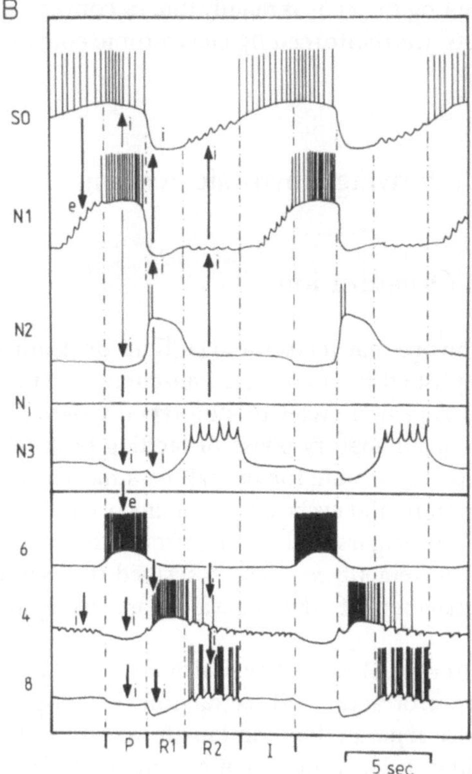

FIGURE 7. Interneuronal circuitry underlying feeding in *Lymnaea.* (A) Synaptic connections between CPG interneurons are shown inside the box. Connections of CPG interneurons with the slow oscillator (SO) and feeding motoneurons (mns) are shown outside the box. Inhibitory connections (i) are indicated by solid dots and excitatory ones (e) by solid triangles. (B) Firing patterns of the CPG interneurons (N1, N2, N3, N_i) and motoneurons (6, 4, and 8 cells below solid horizontal line) shown for two feeding cycles. The protraction (P), rasp (R1), swallowing (R2), and inactive (I) phases of the feeding cycle are indicated by vertical dashed lines. Thick lines on arrows indicate connections between cells shown by direct recording, thin lines are connections surmized from simultaneous recordings (based on data from Benjamin and Rose, 1980; Rose and Benjamin 1981a,b; Elliott and Benjamin, unpublished data).

called N1, N2, or N3 cells that fired in sequence to provide consecutive phases of synaptic inputs to the motoneurons (Fig. 7B). These interneurons had both excitatory or inhibitory effects on different motoneurons. Each motoneuron type received different patterns of inhibitory or excitatory inputs that made them fire in a particular phase of the feeding cycle (in Fig. 7B, the arrows indicate inputs to three examples of motoneurons). Intrinsic properties such as postinhibitory rebound and spike adaptation determined how the motoneuron responded to interneuronal input, but the rhythm was basically generated by a central pattern generator (CPG) of interneurons.

Sequences of activity of the type N1, N2, N3 occurred spontaneously in isolated preparations (Rose and Benjamin, 1981b), but they were irregular and of a much longer period than the intact feeding rhythm unless the system was activated by higher order interneuron. This can be the SO located in the buccal ganglia or CV1 located in the cerebral ganglia (Benjamin *et al.*, 1981).

5.3. The Role of the SO

Previously it had been assumed by Rose and Benjamin (1981a) that there were a pair of SO cells, one in each buccal ganglion, because in different preparations sometimes a cell was found on the left side and at other times on the right. However, a systematic search for left and right cells in the same preparation has failed in 65 preparations. Thirty-two were found on the left side and 33 on the right. The presence of a single SO cell makes the interpretation of the supression experiment reported in Rose and Benjamin (1981a) more straightforward. Three spikes in an already active SO cell were inhibited by hyperpolarization and activity in motoneurons continued with the usual synaptic inputs from the CPG interneurons. Given that there appears to be only one SO present in each snail, this experiment proves that activity in the SO is not essential for CPG oscillations and that it cannot be classed a command neuron on the strict necessity criterion of Kupfermann and Weiss (1978), although its activity is required to maintain the frequency of rhythm seen in the intact snail.

Injecting Lucifer yellow into the SO showed that its ability to activate feeding on both sides of the brain had an anatomical substrate in the axonal projection from the cell body to the opposite buccal ganglion (Fig. 8A). However, the most distinctive feature of SO anatomy was the U-bend arrangement of the axon. Thus the main axon of the cell first projected to the opposite side and then back again to the same side in a characteristic and consistent way in each preparation. The cell shown in Fig. 8A is located in the left buccal ganglion, but cells from the right side showed similar mirror image anatomy to the illustrated cell. A further feature of the SO is the extensive dendritic branching that occurs in the neuropil of both buccal ganglia and this is more extensive than that occurring for the N cells (Fig. 8B,C). The SO had no projections along peripheral nerves and so its axonal and dendritic processes were confined to the buccal ganglia.

The SO activates the CPG principally by exciting the N1 cells that come to threshold following a sequence of facilitating EPSPS, 1:1 with SO spikes (Fig. 7B). The connections between the N1, N2, and N3 cells are also summarized in Fig. 7A and the activity patterns of the cells in Fig. 7B. It should be noted that Fig. 7B is a modified version of Fig. 15 of Rose and Benjamin (1981b). The crucial difference between the figure here and the one from the earlier paper is that the solid arrows between cells in Fig. 7B indicate that the connections have now been demonstrated by direct recordings between the cells (C. J. H. Elliott and P. R. Benjamin, unpublished data), whereas previously many of the connections had been surmized from synaptic inputs to the CPG interneurons.

As well as more detailed recordings of the conections between the N cells, more details of the morphology and electrical properties of each of the CPG interneuron types have been obtained and will be described next.

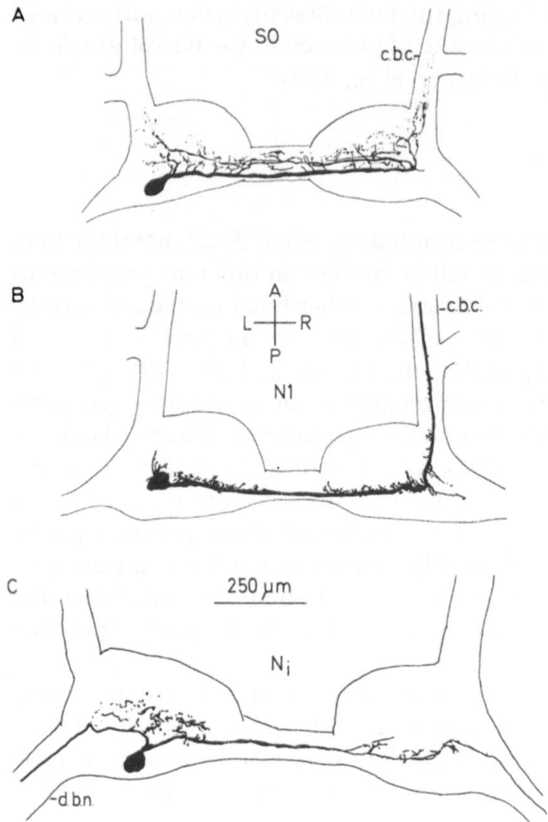

FIGURE 8. Drawings of Lucifer Yellow filled feeding interneurons reconstructed from photomicrographs of whole mounts. (A) Left slow oscillator (SO) showing the characteristic U-bend arrangement of the axon and the complex dendritic branching pattern occurring in the neuropil of both buccal ganglia. A few fine processes penetrate the proximal region of the right cerebrobuccal connective (cbc), but the cell has no peripheral nerve projections. (B) Left N1 CPG interneuron, showing a contralateral axonal projection to the right buccal ganglion which continues into the right cerebrobuccal connective, reaching all the way to the cerebral ganglion (only proximal part of the cell shown in this drawing). Dendrite processes occur in both buccal ganglia, but they are not as extensive as those of the SO. (C) Left N_i interneuron with ipsilateral and contralateral axonal projections to the dorsobuccal nerves (DBN). Dendritic branching is most extensive in the left buccal ganglion, but also occurs in the right buccal neuropil. (From Elliott and Benjamin, unpublished data.)

5.4. N1 Interneurons

Rose and Benjamin (1981b) showed the N1 cells are an electronically coupled subnetwork of the CPG with up to ten cells in each buccal ganglia. They had oscillatory properties so that injections of steady current into one member of the subnetwork caused it to burst along with other N1 cells recorded at the same time. This could occur in the absence of activity in N2 and N3 cells (see Fig. 5 of Rose and Benjamin, 1981b) implying a basic oscillatory capability of the N1 subnetwork, but more typically N1 bursts were terminated as N2 neurons became active and thus a regular rhythm (N1, N2, N3, N1 . . .) began. The ability of an N1 interneuron to drive the CPG in this pattern is shown in Fig. 9B and this provides direct evidence that N1 cells are capable of initiating the sequence of interneuronal bursts shown in Fig. 7B and that they form part of the feeding CPG. In Fig. 9B, the rhythm is driven to the highest frequency that could be obtained by injection of depolarizing current into N1. This ability of the N1 cell to activate the feeding system was compared with injecting steady current into the SO in the same preparation. Both cells were capable of initiating the pattern typical of bursting in the 4-cluster cell (a retractor motoneuron), but the SO was able to maintain a higher frequency and intensity of motoneuron bursting than N1. Note that in both Fig. 9A and B N2 input was prominent on both the SO and the 4-cluster cell (dashed lines).

Perhaps it is not surprising that the SO could maintain a higher frequency of motoneuron bursting than N1 because the SO strongly excites all the N1 cells (Rose and Benjamin, 1981b), whereas the injection of current into a single N1 cell would have been relatively ineffective in activating other N1 cells even though they are electrotonically coupled. The rise time of the prepotential feeding to firing in the N1 cell of Fig. 9B is rather slow compared with that occurring with SO activation (see Fig. 7 of Rose and Benjamin, 1981b), and the presence of presumed electrotonic EPSPs on the N1 cell (probably from other N1 cells) in Fig. 9B shows that the N1 cells are not firing exactly together as they do with SO activation of the system. This suggests that the SO is extremely effective in causing all the N1

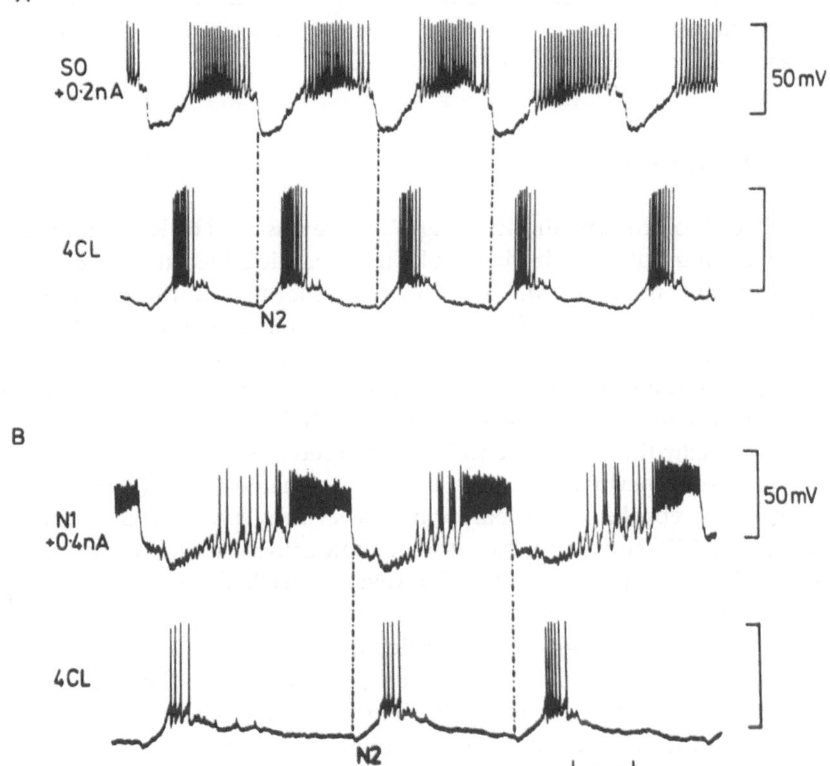

FIGURE 9. Activation of bursting activity in the 4-cluster cell (retractor motoneuron) by intracellular current injection of the SO (A) or an N1 interneuron (B), in the same preparation. Note that A and B are on the same time base. Both cells can activate the motoneuron into a feeding rhythm, but the SO is able to maintain a higher frequency rhythm than the N1 cell. Dashed lines indicate onset of the N2 phase of synaptic input. (From Elliott and Benjamin, unpublished data.)

cells to fire precisely together and it is the only buccal cell type capable of producing a regular and higher frequency rhythm typical of the intact snail.

Electrophysiology shows the N1 cells are a homogenous population of cells in both their patterns of activity and synaptic connections and this is supported by morphological studies with Lucifer Yellow. The axons of N1 cells (left or right) always projected to the opposite buccal ganglia which they traverse and eventually reach the cerebral ganglia through the cerebrobuccal connective. A typical example of a left-side cell is shown in Fig. 8B. This also shows the dendritic branching patterns of the N1 cells that are present in both left and right buccal ganglia.

The projection of the N1 cells to the cerebral ganglia is particularly interesting because previous studies by McCrohan (unpublished data) showed that motoneurons present in the cerebral ganglia, like their buccal ganglion counterparts, also received synaptic inputs during the N1 phase of the feeding cycle. These were assumed to come from the buccal ganglia because cutting the cerebrobuccal connective abolished the input. The morphological data presented here suggest that the N1 cells, as well as having a role in the CPG, may also have a further role in carrying the motor pattern to the other parts of the brain, so that corollary discharge neurons are themselves part of the CPG.

5.5. N2 Interneurons

The N2 cells of the type shown in Fig. 7B were classified by Rose and Benjamin (1981b) on the basis of two criteria: first by the synaptic effects they had on specific types of motoneurons and second by the fact that they fired during the appropriate (R1) phase of the feeding cycle. Our recent work confirms that these cells are members of a distinct class of cell. However, it has also revealed several new types of neuron that only fulfill one of the two criteria for N2 identification. At present it is convenient to consider them as subclasses of the N2 category of interneurons, although eventually they may have to be considered as separate classes.

One important new type of cell has been designated the N_i (N inert) because in the isolated cerebrobuccal preparation it never fires spontaneously when other members of the CPG are active. Nor does it show activity when the CPG rhythm is activated and maintained by the SO. However, N_i cells do have similar synaptic connections to the N2 cells described in the previous paragraph. Many N_i cells have been filled with Lucifer Yellow and a drawing of one is shown in Fig. 8C. The most striking feature of this cell is the peripheral projections that the cell has in both the left and right dorsobuccal nerves, which innervate the proesophagus (Benjamin *et al.*, 1979). All the other N_i cells have peripheral projections, but they are not all identical in that some have projections to the other buccal nerves. We do not know yet what normally activates the N_i cells, but it may be that they are activated by sensory inputs from the periphery.

Yet another class of putative N2 cells fires during the N2 phase of the feeding cycle, but has central connections that differ from those of the type described by Rose and Benjamin (1981b) and shown in Fig. 7B. This cell type, like the N_i, also has peripheral nerve projections and this seems a general feature of N2 cells.

5.6. N3 Interneurons

The firing pattern of the N3 cells described by Rose and Benjamin (1981b) is shown in Fig. 7B. This cell type has inhibitory connections with motoneurons, like the 8 cells (R2 retractor motoneurons), but they also inhibit the N1 cells and SO. The N3 cell shown in Fig. 7B only fires during the second phase of retraction, R2, and although it can account for some of the IPSPs seen in the 4 cell (R1 retractor motoneuron), it cannot be responsible for the EPSPs on other motoneurons, like the 3 cells, which occur during both the R2 and I phase of the feeding cycle. In fact, we have recently discovered a second type of N3 cell that fires for a longer period of time, during both R2 and I phases of the feeding cycle that produces the longer lasting EPSPs seen on the 3 cells. Morphological studies also show that, like the N2 cells, the N3s also have peripheral projections along buccal ganglion nerves projecting to the periphery.

5.7. Significance of Peripheral Projections of N2 and N3 Interneurons

The N2 and N3 interneurons have peripheral projections along nerves of the buccal ganglia, but the N1s do not and it is interesting to speculate why this might be. Peripheral axons could have sensory or motor functions, although the former seems more likely in view of the fact that sensory interneurons have previously been found in the buccal ganglia of *Navanax* (Spray *et al.*, 1980) and *Philine* (Dorsett and Sigger, 1981). Why then should the N2 and N3 cells have these supposed sensory projections and not the N1s? The N2 cells fire during the R1 or rasp phase of the feeding cycle when interference or "load" from unpredictable factors such as hardness of food or uneven substrate might be expected to occur. This leads to the suggestion that it would be more sensible for N2 interneurons to be influenced more by sensory feedback compared with the N1s which are responsible for the protraction phase of the feeding cycle where relatively unloaded movements of the buccal mass occur. A similar argument could be used for the N3 cells that are active during the R2 or swallowing phase of the feeding cycle. Swallowing again might have unpredictable load factors such as size and compressibility of the food item once it is in the cavity of the buccal mass. This suggested role for sensory feedback needs to be tested in semi-intact preparations when receptors would be able to provide their normal input to the CNS. This would be particularly significant for the N_i cells with receive no input from the CNS.

6. Discussion

Simultaneous investigations of neural networks underlying several types of behavior in *Lymnaea* give information on the firing patterns and synaptic connections of neurons in each system, although in-depth biophysical and pharmacological studies have not been carried out. We would argue that this broad comparative approach to different behavior systems in one animal is a necessary strategy to facilitate understanding the neural basis of behavior in a CNS as complex as the snail brain, containing about 25,000 neurons. This does not rule out more detailed biophysical studies, but these need to be carried out in a selective way. This comparative approach to neural network analysis could be justified further if it could be shown than the neural networks we describe in *Lymnaea* contain the main features necessary for their behavioral function. We will argue next that this is at least partly true for the feeding and withdrawal systems.

6.1. The Feeding System

In a complicated behavior such as feeding, the neuronal substrate of numerous behavioral features need to be described. These have been discussed in detail by Benjamin (1983b), and here we will restrict our discussion to the interneuronal network (CPG) responsible for rhythmic activity of the buccal mass. This interneuronal network is capable of oscillating in the absence of a food stimulus and indeed in the absence of any sensory feedback from the periphery (Benjamin and Rose, 1979). This allows us to consider the CPG as a rhythm-generating system separate from sensory feedback, although this could not be ignored in any complete analysis of the feeding system.

The CPG circuit in *Lymnaea* has to be capable of producing three phases of muscular activity per feeding cycle, and ten types of motoneurons, each having distinct firing patterns, have been found. This complex motoneuronal firing pattern is produced by three consecutive phases of synaptic input, each of which can be excitatory or inhibitory on different motoneurons. The activity of a particular motoneuron is determined by the unique combination of synaptic input it receives (Rose and Benjamin, 1981a) and by its own intrinsic membrane properties leading to spike adaptation or postinhibitory rebound (Benjamin and Rose, 1979). We searched for interneurons that fired during the different phases of the feeding cycle and that also had dual action effects of the appropriate sign on different motoneurons. Three types of interneurons were found in the buccal ganglia, N1–N3, which fired consecutively (N1→N2→N3) and apparently had monosynaptic connections with the feeding motoneurons (Rose and Benjamin, 1981b). These interneurons were able to provide all the inputs required to produce the motoneuron firing patterns.

The activity of the N cells could explain in a satisfactory way how the motoneurons fired, but two further questions arise: (1) how do the CPG interneurons fire in the correct sequence, and (2) what properties of the N cells allow them to oscillate in the absence of any outside input? Early clues to the first of these questions were given by the inhibitory inputs which the N cells received. These could account for the firing pattern of each cell type and Benjamin and Rose (1980) explained how the sequence of activity occurred if certain inhibitory connections between the interneurons were assumed. Their model is reproduced in Fig. 7. In this figure the activity in the N cells is initiated by the SO providing excitation to the N1 cells, but as Fig. 9B shows, steady depolarization of the N1 can also produce similar patterns of motoneuron activity. Consequently, activity in the SO is unnecessary for basic rhythmic activity in the CPG to occur (see also Fig. 6 of Rose and Benjamin, 1981b). Essentially, the N1→N2→N3 sequence occurs as a result of the inhibitory connections between the N cells themselves. Thus a burst of spikes in the N1 inhibits the N2 and N3 cells. On termination of the N1 burst, the N2 interneurons are released from inhibition and rapidly depolarize. The resulting excitation of N2 cells (which occurs more rapidly than in the N3 cells) re-established the inhibition of the N3 interneuron and the N3 cells fire erratically after the end of the N2 burst, following recovery from two phases of inhibition. Certain problems arise from this model, for instance, how the N1 cells stop firing, but this can partly be explained by the inhibitory effect of N2 on N1, once N2 starts firing, but may mostly be due to the endogenous properties of the N1 cells.

Direct evidence that the N cells are connected in this way was not obtained until the recent work outlined in this chapter was carried out, but now we can confirm much of this connectivity, which has been established by direct recording. So far, the model can reasonably be supported by the data.

There remains to be answered the second question raised at the beginning of the previous paragraph concerning the basic origins of the oscillatory activity. One possibility is that the inhibitory connections between the N cells provide this oscillating capability either on the basis of reciprocal inhibition between the N1 and N2 cells or using all three N1, N2, N3 subnetworks of CPG, as in the recurrent cyclic inhibition model of Kling and Székely (1968). Either of these network models are possible, although Benjamin and Rose (1980) have argued that the *Lymnaea* feeding system differs significantly from recurrent cyclic inhibition. It seems more likely that the properties of the cyberchron neurons originally described by Kater (1974) for *Helisoma* feeding provide a more satisfactory model for the CPG in *Lymnaea* system than either reciprocal inhibition or recurrent cyclic inhibition. Some of the cyberchrons are endogenous oscillators (Merickel and Gray, 1980) and they form an electrotonically coupled network that synchronizes their activity. They appear to be similar to the N1 cells of *Lymnaea* which also have some of the properties of endogenous oscillators in that steady current injected into one or more cells initiates burst activity in the injected cells as well as other electronically coupled N1s. Also, induced oscillatory activity in the N1s drives the N2 and N3 cells in the correct order. On this model, the N1 cells are seen as endogenous oscillators entrained to the rest of the system by the inhibitory connections previously described. What is required at present is a detailed study of the membrane properties of isolated N1 cells similar to that of Merickel and Gray (1980) on *Helisoma* to provide further evidence that this theory is correct.

So what are the main features of the *Lymnaea* CPG system we have been able to describe? First, we showed that to obtain multiphasic motoneuron firing patterns, each cell must receive several consecutive synaptic inputs in each feeding cycle. Second, we showed that these inputs are produced by interneurons firing at particular phases of the feeding cycle that excite or inhibit all the motoneurons receiving synaptic input during that phase. Third, we showed that inhibitory connections between the interneurons determine their sequence of activity, when they fire and in what order. What we have *not* explained adequately is how the basic oscillatory mechanism works in detail, although we may have a model for it in the cyberchron neurons of *Helisoma* (Kater, 1974) which appear to be homologous to the N1 cells of *Lymnaea.*

Finally it should be pointed out that higher order interneurons occur in the *Lymnaea* feeding system which can initiate and maintain feeding (the SO for instance) or affect the intensity of motoneuron bursting (the serotonergic cerebral giant cells) (McCrohan and Benjamin, 1980a), and so these also need to be considered in any complete model of the system.

6.2. The Whole Body Withdrawal System

Feeding in *Lymnaea* involves a centrally generated motor program influenced but not primarily generated by sensory input. In contrast, whole body withdrawal reflexes are initiated by sensory inputs that activate normally silent motoneurons. Motoneurons from different ganglia innervate separate parts of the two main muscles involved in withdrawal, and so the main problem was to understand how activity in this widely distributed motoneuronal system could be coordinated to produce the total response. Two types of stimuli were involved, photic (light-off was used to simulate the shadow stimulus) and tactile. Strong tactile stimuli applied locally to different parts of the body produced a behavioral response whose main component (longitudinal shortening) was similar to the more stereotyped response produced by photic stimulation. Both types of stimulus excited all the motoneurons innervating the DLM and CM, so one important coordination mechanism was the convergence of sensory inputs onto these cells. Synchronized excitatory input to these motoneurons causes cocontraction of all parts of both these muscles. A second mechanism coordinating motoneuron activity was the widespread occurrence of strong electrotonic coupling of ipsilateral motoneurons. This at first seems surprising given that the cell bodies of the withdrawal response motoneurons occur in all nine central ganglia, but there is a possible anatomical basis for this in that all the motoneurons from one side of the brain have at least one axon in the neuropil of the ipsilateral pleural or pedal ganglia (G. P. Ferguson and P. R. Benjamin, unpublished data).

Although sensory convergence plays an important role in coordination of motoneuronal activity, nothing is known of the central pathways involved. More is known about the tactile pathways involved in localized withdrawal responses (e.g., Lever *et al.*, 1978) and these presumably involve activation of motoneurons located in one ganglion as well as the peripheral nervous system rather than the whole network described here.

ACKNOWLEDGMENTS. We thank the M.R.C. and S.E.R.C. for financial support.

References

Benjamin, P. R., 1983a, Electrical properties of the Dark Green Cells, neurosecretory neurones in the brain of the pond snail, *Lymnaea stagnalis, Comp. Biochem. Physiol.* **75A:**549–556.

Benjamin, P. R., 1983b, Gastropod feeding: Behavioural and neural analysis of a complex multicomponent system, in: *Neural Control of Rhythmic Movements* (A. Roberts and B. L. Roberts, eds.), Cambridge University Press, Cambridge pp. 159–193.

Benjamin, P. R., and Rose, R. M., 1979, Central generation of bursting in the feeding system of the snail, *Lymnaea stagnalis, J. Exp. Biol.* **80:**93–118.

Benjamin, P. R., and Rose, R. M., 1980, Interneuronal circuitry underlying cyclical feeding in gastropod molluscs, *Trends Neurosci.* **3:**272–274.

Benjamin, P. R., and Rose, R. M., 1984, Electrotonic coupling and afterdischarges in the Light Green Cells: A comparison with two other cerebral ganglia neurosecretory cell types in the pond snail, *Lymnaea stagnalis, Comp. Biochem. Physiol.* **77A:**67–74.

Benjamin, P. R., and Winlow, W., 1981, The distribution of three wide-acting synaptic inputs to identified neurons in the isolated brain of *Lymnaea stagnalis* (L.), *Comp. Biochem. Physiol.* **70A:**293–307.

Benjamin, P. R., Swindale, N. V., and Slade, C. T., 1976, Electrophysiology of identified neurosecretory neurones in the pond snail, *Lymnaea stagnalis* L., in: *Neurobiology of Invertebrates,* III *Gastropoda Brain* (J. Sálanki, ed.), Akadémiai Kiadó, Budapest, pp. 123–138.

Benjamin, P. R., Rose, R. M., Slade, C. T., and Lacy, M. G., 1979, Morphology of identified neurones in the buccal ganglia of *Lymnaea stagnalis, J. Exp. Biol.* **80:**119–135.

Benjamin, P. R., Slade, C. T., and Soffe, S. R., 1980, The morphology of neurosecretory nerones in the pond snail, *Lymnaea stagnalis,* by the injection of Procion Yellow and Horseradish Peroxidase, *Philos. Trans. R. Soc. London Ser. B* **290:**449–478.

Benjamin, P. R., McCrohan, C. R., and Rose, R. M., 1981, Higher order interneurones which initiate and modulate feeding in the pond snail, *Lymnaea stagnalis,* in: *Invertebrate Neurobiology: Mechanisms of Integration* (J. Sálanki, ed.), Pergamon Press, Oxford, pp. 171–200.

Berry, M. S., and Cottrell, G. A., 1979, Ionic basis of different synaptic potentials mediated by an identified dopamine-containing neuron in *Planorbis, Proc. Soc. (London) Ser. B.* **203:**427–444.

Boer, H. H., Schot, L. P. C., Roubos, E. N., Maat, A. ter, Lodder, J. C., Reichelt, D., and Swabb, D. F., 1979, ACTH-like immunoreactivity in two electrotonically coupled giant neurons in the pond snail *Lymnaea stagnalis, Cell Tissue Res.* **203:**231–240.

Dorsett, D. A., and Sigger, J. N., 1981, Sensory fields and properties of the oesophageal propriocepters in the mollusc, *Philine, J. Exp. Biol.* **94:**77–93.

Geraerts, W. P. M., 1976, Control of growth by the neurosecretory hormone of the Light Green Cells in the freshwater snail *Lymnaea stagnalis, Gen. Comp. Endocrinol.* **29:**61–71.

Geraerts, W. P. M., Leeuwen, J. P. Th. M. van, Nuyt, K., and With, N. D. de, 1981, Cardioactive peptides of the CNS of the pulmonate snail *Lymnaea stagnalis, Experientia* **27:**1168–1169.

Kater, S. B., 1974, Feeding in *Helisoma trivolvis:* The morphological and physiological basis of a fixed action pattern. *Am. Zool.* **14:**1017–1036.

Kling, U., and Székely, G., 1968, Simulation of rhythmic nervous activities, I. Function of networks with cyclic inhibitions, *Kybernetik* **5:**89–103.

Kupfermann, I., and Weiss, K. R., 1978, The command neuron concept, *Behav. Brain Sci.* **1:**3–39.

Lever, A. J., Bruins, H. J., Geigenhuis, C., Everts, W. M., and Fokkema, B. S., 1978, The neural organization of the tentacle contraction reflex of the pond snail *Lymnaea stagnalis* (L.), *Proc. Kon. Akad. Wetensch. Ser. C* **81:**265–278.

McClennan, A. D., 1982, Movements and motor patterns of the buccal mass of *Pleurobranchaea* during feeding, regurgitation and rejection, *J. Exp. Biol.* **98:**195–211.

McCrohan, C. R., and Benjamin, P. R., 1980a, Patterns of activity and axonal projections of the cerebral giant cells of the snail, *Lymnaea stagnalis, J. Exp. Biol.* **85:**149–168.

McCrohan, C. R., and Benjamin, P. R., 1980b, Synaptic relationships of the cerebral giant cells with motoneurones in the feeding system of *Lymnaea stagnalis, J. Exp. Biol.* **85:**169–186.

Merickel, M., and Gray, R., 1980, Investigation of burst generation by the electrically coupled cyberchron network in the snail *Helisoma* using a single-electrode voltage clamp, *J. Neurobiol.* **11:**73–102.

Rose, R. M., and Benjamin, P. R., 1979, The relationship of the central motor pattern to the feeding cycle of *Lymnaea stagnalis, J. Exp. Biol.* **80:**137–163.

Rose, R. M., and Benjamin, P. R., 1981a, Interneuronal control of feeding in the pond snail *Lymnaea stagnalis.* I. Initiation of feeding cycles by a single buccal interneurone, *J. Exp. Biol.* **92:**187–201.

Rose, R. M., and Benjamin, P. R., 1981b, Interneuronal control of feeding in the pond snail *Lymnaea stagnalis.* II. The interneuronal mechanisms generating feeding cycles, *J. Exp. Biol.* **92:**203–228.

Schot, L. P. C., and Boer, H. H., 1982, Immunocytochemical demonstration of peptidergic cells in the pond snail *Lymnaea stagnalis* with an antiserum to the molluscan cardioactive tetrapeptide FMRF-amide, *Cell Tissue Res.* **225**:347–354.

Schot, L. P. C., Boer, H. H., Swaab, D. F., and Noorden, S. van, 1981, Immunocytochemical demonstration of peptidergic neurons in the central nervous system of the pond snail *Lymnaea stagnalis* with antisera raised to biologically active peptides of vertebrates, *Cell Tissue Res.* **216**:273–291.

Slade, C. T., Mills, J., and Winlow, W., 1981, The neuronal organization of the paired pedal ganglia of *Lymnaea stagnalis* (L.), *Comp. Biochem. Physiol.* **69A**:789–803.

Soffe, S. R., and Benjamin, P. R., 1980, Morphology of two electrotonically coupled giant neurosecretory neurons in the snail, *Lymnaea stagnalis, Comp. Biochem. Physiol.* **67A**:35–46.

Soffe, S. R., Benjamin, P. R., and Slade, C. T., 1978, Effects of environmental osmolarity on the blood composition and light microscope appearance of neurosecretory neurones in the snail *Lymnaea stagnalis, Comp. Biochem. Physiol.* **61A**:557–584.

Spray, D. C., Bennett, M. V. L., and Spira, M. E., 1980, Synaptic connections of buccal mechanosensory neurons in the opisthobranch mollusc, *Navanax inermis, Brain Res.* **182**:271–286.

Stoll, C. J., 1976, Extraocular photoreception in *Lymnaea stagnalis* L., in: *Neurobiology of Invertebrates: Gastropoda Brain* (J. Salanki, ed.), Akadémiai Kiadó, Budapest, pp. 487–495.

Swindale, N. V., and Benjamin, P. R., 1976, The anatomy of neurosecretory neurones in the pond snail *Lymnaea stagnalis* (L.), *Philos. Trans. R. Soc. London* (Ser. B) **274**:169–202.

Wendelaar Bonga, S. E., 1970, Ultrastructure and histochemistry of neurosecretory cells and neurohaemal areas in the pond snail *Lymnaea stagnalis* (L.), *Z. Zellforsch.* **108**:190–224.

Wendelaar Bonga, S. E., 1972, Neuroendocrine involvement in osmoregulation in a freshwater mollusc *Lymnaea stagnalis, Gen. Comp. Endocrinol.* (Suppl.) **3**:308–316.

Winlow, W., Haydon, P. G., and Benjamin, P. R., 1981, Multiple postsynaptic actions of the giant dopamine-containing neurone R.Re.D.1 of *Lymnaea stagnalis* (L.), *J. Exp. Biol.* **94**:137–148.

7

Nonspiking and Spiking Local Interneurons in the Locust

MALCOLM BURROWS

1. Introduction

Since the discovery of interneurons that normally do not produce action potentials (nonspiking interneurons) (for reviews, see Roberts and Bush, 1981), the question of why such neurons should exist has always been present. What advantages do they confer? Does their graded transmission of signals, an analogue process, offer greater flexibility than transmission effected by spikes, a digital process? Are these neurons specifically needed for local processing, because an individual neuron can perform many simultaneous but relatively independent computations at one time?

These questions have been thrown into even sharper focus by the discovery, within the same local circuits, of a class of interneurons with very similar shapes, but which normally spike, and use that spike to effect synaptic transmission (Burrows and Siegler, 1982). Both these nonspiking and spiking interneurons are local interneurons that lack an axon and have processes restricted to a small and often discrete region of the nervous system. The question of why there should be these two distinct types of local interneuron is not answerable by direct experiment, but an answer may emerge from an understanding of their roles in local integrative processes. The relevance of such an understanding can be judged by the abundance of local interneurons in a nervous system. For example, it has been estimated that about 60–70% of the cell bodies in a segmental ganglion of an arthropod may belong to local interneurons (Siegler and Burrows, 1979; Reichert *et al.*, 1982).

To tackle these problems, we need to study the nervous system of an animal that has both types of interneuron and uses them in the same behavior. The segmental ganglion of the locust that controls the movements of the hindlegs and

MALCOLM BURROWS • Department of Zoology, University of Cambridge, Cambridge, United Kingdom.

hindwings meets these requirements. This ganglion can be exposed to allow its neurons to be impaled with intracellular electrodes while many of the natural movements of the legs can continue. An essential correlation can therefore be made between the electrical activity of a neuron and the movement in which it is participating. In this way most of the leg motor neurons (Burrows and Hoyle, 1973), many of the projections of sensory neurons from the hindlegs (Pfluger *et al.*, 1981), some neurosecretory neurons (see review by Evans, 1980), and some intersegmental interneurons (Pearson *et al.*, 1980; Pearson and Robertson, 1981) have been identified. All indications to date are that the movements seen under experimental conditions are very similar to those in the normal locust. It is reasonable to expect, therefore, that the electrical activity recorded from the neurons will also be very similar to that which occurs in the normal animal. It must be remembered, however, that dissection may change some properties of the neurons through the action of any trophic factors or neuromodulators that may have been released. By building upon this considerable base of knowledge and by concentrating on the movements about the distal joints of a hindleg, it has been possible to reveal some

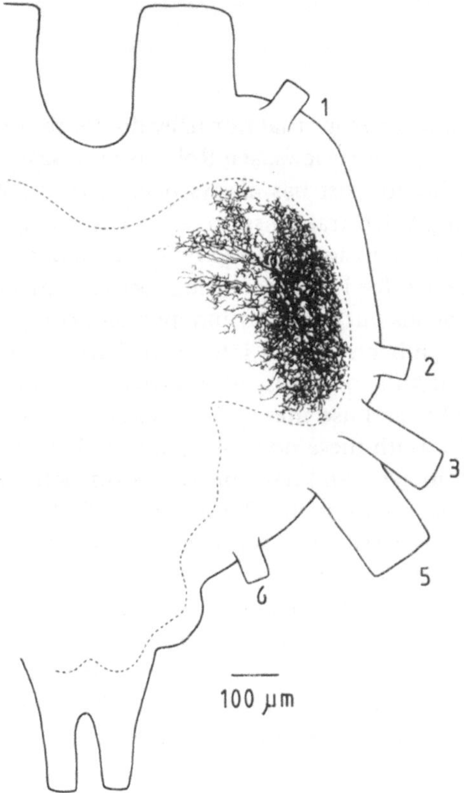

FIGURE 1. Drawing of a nonspiking local interneuron that was stained by injecting cobalt into one of its neuropilar processes and the stain subsequently intensified with silver. Half of the metathoracic ganglion is shown, with lateral nerves 1, 2, 3, 5, and 6 numbered, the others omitted. The dashed line indicates the boundary of the neuropil.

of the local integrative processes that are involved in their control. The task is to describe the actions of both nonspiking and spiking local interneurons and thereby elucidate their respective roles.

2. Morphology of the Local Interneurons

Superficially, both nonspiking and spiking local interneurons look similar, and there are certainly no obvious clues from their structure to indicate their physiological type (Figs. 1, 2). They both have small cell bodies, some 10–20 μm in diameter, in the cortex of the ganglion and enormous numbers of fine branches that suggest either that they make many connections or that each connection involves many branches. The branches are all restricted to the central neuropil of one ganglion and usually to one half of that ganglion. There are a few exceptions: some neurons of either type have branches in both halves of the ganglion and some spiking local interneurons (e.g., Fig. 2) have branches that extend to the next fused segmental ganglion. None, however, have an axon that emerges through any of the peripheral nerves. It is for these reasons that they are called local interneurons.

In general, the nonspiking interneurons appear to be a more diverse type;

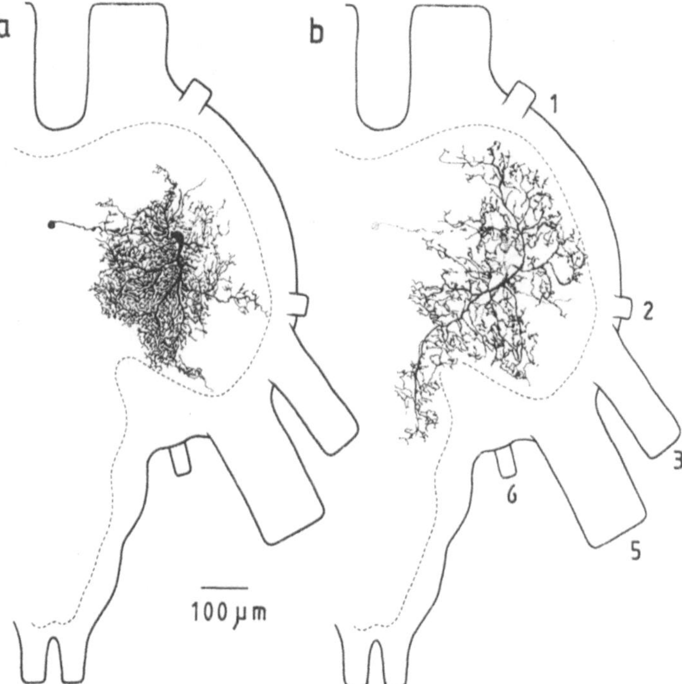

FIGURE 2. Drawings of a spiking local interneuron stained by injecting cobalt into its soma. Its branches are in two planes: (a) in the ventral regions of the neuropil, (b) in the more dorsal regions with the main ventral branches stippled. The cell body is at the ventral midline, making this interneuron a member of the midline group. This ganglion is from a smaller locust than the one shown in Fig. 1.

their shapes are more varied and their cell bodies scattered in both the ventral and dorsal cortex. By contrast, the spiking interneurons that have been revealed belong to one of two groups according to the position of their cell bodies; an anterior-lateral group and a midline group (Siegler and Burrows, 1984). Within a particular group, all the interneurons conform to a basic pattern of branching; neurons with differing physiological characteristics modify this basic plan in different ways (Burrows and Siegler, 1984). It is unlikely, however, that all the local interneurons within this ganglion have been described. The diversity of nonspiking interneurons,

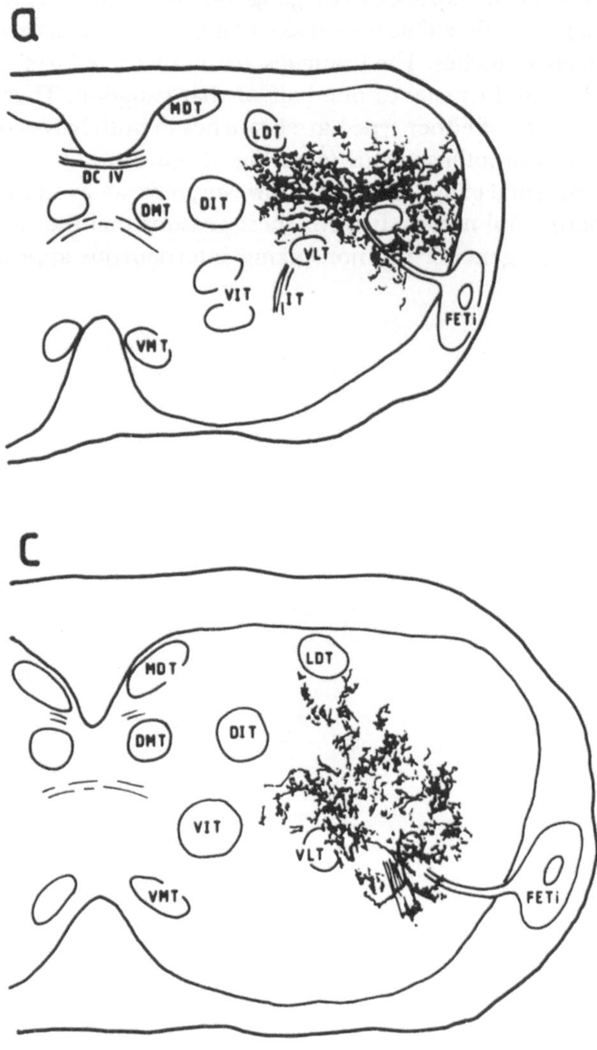

FIGURE 3. Transverse sections through metathoracic ganglia from locusts of different sizes at a plane that passes through the soma and initial part of the primary neurite of a large motor neuron, the fast extensor tibiae (FETi). The intent is to show the regions of neuropil occupied by four of the elements in local circuits. (a) A motor neuron, the slow extensor of the tibia. (b) Mechanosensory neurons, the projections of axons in the lateral nerve in the femur (N5b3). (c) A nonspiking local interneuron. (d) A

as compared with the spiking ones, may thus simply be a reflection of the methods used to reveal them. To find nonspiking interneurons, an electrode was introduced into the neuropil, whereas for the spiking interneurons the search was restricted to the ventral and medial region of cell bodies. Thus more diversity would be expected from the first strategy of search.

A more detailed examination of the morphology of individual local interneurons of each type reveals distinctive features. The branches of the nonspiking inter-

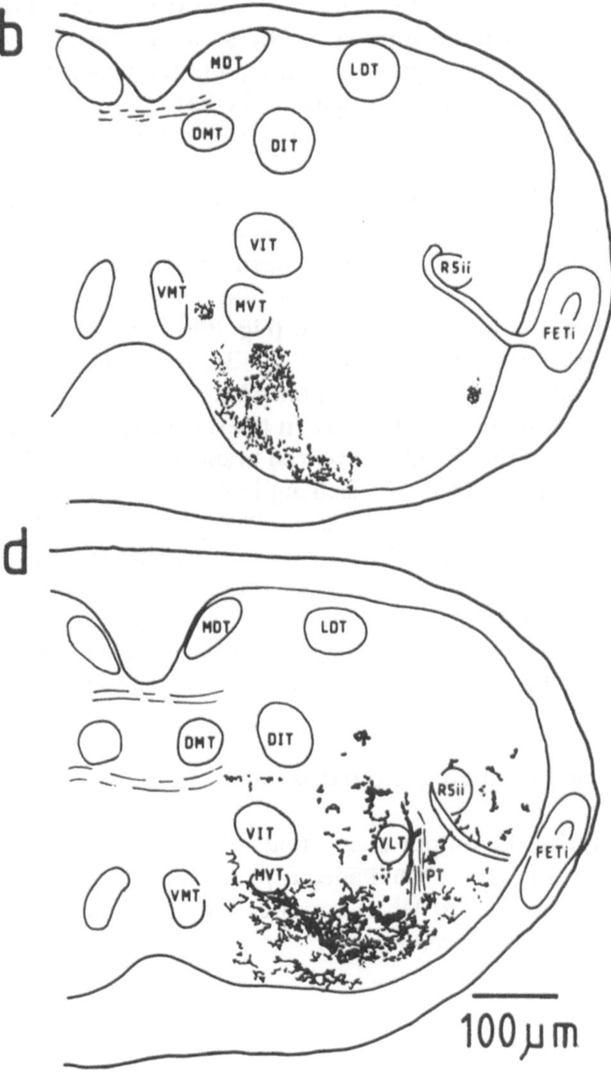

spiking local interneuron. Abbreviations (based on Tyrer and Gregory, 1982; Siegler and Burrows, 1984): DCIV, dorsal commissure IV; DIT, dorsal intermediate tract; DMT, dorsal median tract; FETi, fast extensor tibiae motor neuron; IT, I-tract; LDT, lateral dorsal tract; MDT, median dorsal tract; MVT, median ventral tract; PT, perpendicular tract; R5ii, root of nerve 5; VIT, ventral intermediate tract; VLT, ventral lateral tract; VMT, ventral median tract.

neurons are of similar texture throughout. Those of the spiking interneurons are divided into two distinct regions linked by a single fine process: a ventral region of stout primary and secondary branches that give rise to numerous fine branches (Fig. 2A) and a dorsal region of fewer and less densely packed branches that are typically of uneven diameter along their length (Fig. 2B).

The most important morphological difference is the area of neuropil to which the interneurons project and is dramatically revealed in cross sections of the ganglion (Fig. 3). Branches of the nonspiking interneurons are restricted to the more dorsal regions of neuropil. Some branches extend as far dorsal as the lateral dorsal tract (LDT) (one of a number of longitudinal tracts of fibers that are recognized and whose names are given in full in the caption to Fig. 3), and the majority are dorsal to the primary neurite of a large leg motor neuron, the fast extensor of the tibia (FETi) (Fig. 3C). Branches of the spiking interneurons are distributed in a quite different way. Their sparse dorsal branches are dorsal to the ventral lateral tract (VLT) and the primary neurite of FETi, but do not extend as far dorsal as those of the nonspiking interneurons (Fig. 3D). The numerous ventral branches extend along the ventral boundary of the neuropil beneath the median ventral tract (MVT) and as far medial as the ventral median tract (VMT).

By examining the branches of a leg motor neuron (Fig. 3A) and the central projections of some mechanoreceptors on the femur (Fig. 3B) in the same way, it can be seen that the majority of the branches of a nonspiking interneuron, but only the dorsal branches of a spiking interneuron, are in the same region of neuropil as the motor neuron. On the other hand, the ventral branches of the spiking interneuron, but none of the branches of the nonspiking interneuron, are in the same region of neuropil as those of the sensory neurons. The anatomy therefore allows the following predictions to be tested by physiological experiment. Nonspiking local interneurons may make synaptic connections with motor neurons, but are unlikely to do so with sensory neurons. Spiking local interneurons may receive direct synaptic inputs from sensory neurons and may also synapse upon motor neurons. Interconnections between the local interneurons may also be expected.

3. Output Connections of Local Interneurons

To reveal and to characterize the output connections of the local interneurons, intracellular recordings can be made simultaneously from one of them and from a putative postsynaptic neuron. First let us consider the connections of a nonspiking interneuron with a motor neuron (Fig. 4).

3.1. Nonspiking Interneurons

The membrane potential of one interneuron was altered selectively by injecting current through a recording microelectrode. At no time does this procedure elicit spikes in the interneuron and yet it effects a change in a postsynaptic neuron (Burrows and Siegler, 1978). A depolarization of an interneuron evokes either a

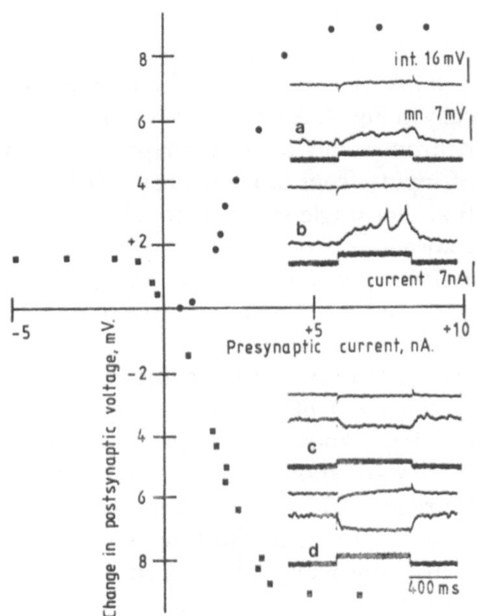

FIGURE 4. Graded effects of nonspiking interneurons upon a motor neuron, a flexor of the tibia. Four interneurons are shown, all recorded sequentially in one locust while a recording was maintained from the same motor neuron: two have excitatory effects and two inhibitory ones. The graph plots the graded relationship between the current injected into a neuropilar process of an interneuron against the resultant change in voltage of the motor neurone recorded in its soma. (a, b) Recordings of an interneuron with excitatory effects. With sufficient current the motor neuron is induced to spike. (c, d) Recordings of an interneuron with inhibitory effects. The greater the current, the larger the hyperpolarization in the motor neuron. Recordings were made from the soma of the motor neuron and from the neuropilar processes of the interneurons. (Based on data taken from Burrows, 1980.)

depolarization (Fig. 4A) or a hyperpolarization (Fig. 4C) of postsynaptic motor neurons. The more current that is injected, the more the interneuron is depolarized and the greater is the postsynaptic change in voltage (Fig. 4B,D). The graph of the presynaptic current against the change in postsynaptic voltage in Fig. 4 shows that transmission is continuously graded until a plateau is reached. The mechanism involves the graded release of chemical transmitter and not electrical transmission. For example, hyperpolarizing effects are associated with an increased conductance of the postsynaptic membrane and have a reversal potential (Burrows and Siegler, 1978).

The absence of a presynaptic spike makes it difficult to determine whether the connection between an interneuron and a motor neuron is direct, because there is no obvious point from which to measure the synaptic delay. A pulse of current that simulates a spike evokes a postsynaptic potential with a delay similar to that at a spiking synapse believed to be monosynaptic. But not all of the nonspiking synapses have such short delays, in insects or in crustaceans (Burrows and Siegler, 1978; Graubard, 1978). Some of the delay is undoubtedly caused by the distance of the stimulating and recording electrodes from the synaptic sites, but in the locust a more difficult assessment is involved because the interneurons are themselves interconnected (Burrows, 1979a). Thus an effect observed in a motor neuron could be the result of the current activating other local interneurons, which in turn make connections with this particular motor neuron. To be certain that a connection embedded deep in a central nervous system is actually direct requires electron microscopy of the neurons studied physiologically. For the present we must accept that the shorter and even some of the longer latencies indicate probable direct connections.

Only very small voltage changes in an interneuron are needed to evoke a measurable change in a postsynaptic motor neuron. For example, a depolarization of 1–2 mV is sufficient for the interneuron with an excitatory action shown graphically in Fig. 4. An electrode inserted into the larger branches of a nonspiking interneuron readily records synaptic potentials of this magnitude generated as a result of inputs from other neurons. It is not surprising, although it is challenging, to find that single synaptic potentials generated in this way can effect the release of transmitter from an interneuron and evoke a postsynaptic potential (Burrows, 1979b.) This finding raises the possibility that local processing may occur within one interneuron: an input to one region will have a strong effect on nearby output synapses, but perhaps no effect on more distant output synapses. For such processing to occur, three requisites must be met. First, there must be an appropriate arrangement of the input synapses on the nonspiking interneuron, but to date nothing is know of this. Second, input and output synapses must be intermingled as they are known to be on neurons in many animals (see reviews in Schmitt and Worden, 1979) including the locust (Watson and Burrows, 1982, 1983). Third, synaptic potentials must decrement rapidly with distance. Electrotonic models of locust nonspiking interneurons indicate that such decrement does occur for inputs on the fine branches (Rall, 1981). Thus it is a reasonable expectation, but an extremely difficult experimental task to prove that local processing does occur within one neuron.

In some interneurons, sufficient transmitter is released at the potential recorded upon penetration to exert measurable postsynaptic effects. An applied hyperpolarizing current first reduces and then abolishes release, while a depolarization increases release. For example, depolarizing the interneuron in Fig. 5a evoked a hyperpolarization of a postsynaptic motor neuron, whereas hyperpolarizing it allowed the motor neuron to depolarize and then spike. The explanation of these effects is that the interneuron releases transmitter with an inhibitory effect

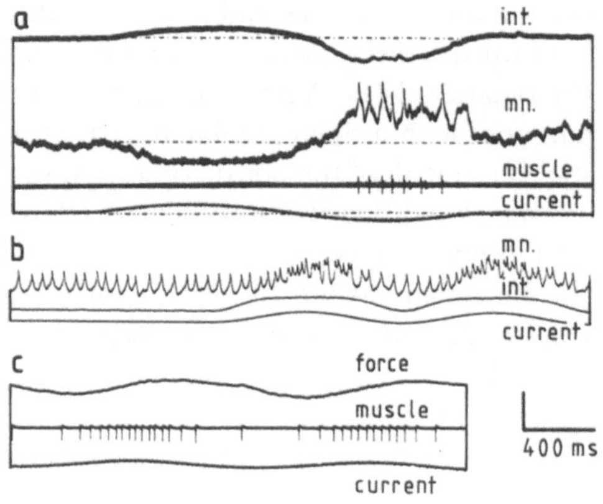

FIGURE 5. Graded and sustained effects of nonspiking interneurons upon the membrane potentials of motor neurons and on the force they cause to be generated at muscles. (a) A sine wave of current is injected into an interneuron. When depolarizing, the current evokes a hyperpolarization of a postsynaptic flexor tibiae motor neuron. When hyperpolarizing, the current evokes a depolarization and a series of spikes in the motor neuron. (b) Depolarizing sine waves of current injected into an interneuron evoke sinusoidal changes in the frequency of spikes in the slow extensor tibiae motor neuron. (c) In another locust, similar stimulation of an interneuron (not shown) leads to sinusoidal changes in the force of the extensor muscle caused by this motor neuron. Calibration: voltage interneuron (int), 20 mV, motor neuron (mn), (a) 4 mV, (b) 7 mV; force, 0.2 g; current, 15 nA.

on the motor neuron and at its "resting potential" keeps the motor neuron tonically hyperpolarized. For a neuron that is continually bombarded by inputs from other neurons, the term "resting potential" has no real meaning. It is probable that all, and not as has been suggested a distinct population (Wilson and Phillips, 1982), of nonspiking interneurons can act in this way. Only a small shift in potential will be necessary for most interneurons to be pushed into a new working range. The set of this working range is open to modification by changes in synaptic input from other neurons as a result of different sensory information or alterations in behavior and through the action of neurohormones or neuromodulators.

The ability to sustain the graded release of transmitter is a further important facet of the action of nonspiking interneurons. Its effectiveness can be shown by modulating with sine waves the membrane potential of an interneuron that excites the single slow motor neuron to the extensor tibiae muscle (Fig. 5b, c). The membrane potential of the interneuron follows the waveform of the injected current, and its effect on the motor neuron is converted into a sinusoidally varying frequency of motor spikes (Fig. 5b). At the muscle, this pattern of motor spikes is converted into a sinusoidally varying force that follows, with a delay, the form of the membrane potential in the interneuron (Fig. 5c). Thus delicate and smooth changes of force can be effected and then sustained.

3.2. Spiking Interneurons

In contrast to the many output connections known for nonspiking local interneurons, only a few have been shown so far for spiking local interneurons. Some connections are again made with motor neurons (Burrows and Siegler, 1982). Transmission is effected by the spikes that evoke conventional chemically mediated synaptic potentials in postsynaptic neurons. Tests have not been appropriate to reveal whether graded synaptic transmission also occurs. Synaptic delays are commensurate with the connections to the motor neurons being direct. The few connections so far revealed all involve interneurons of the midline group and have been inhibitory. Whether this is a significant finding about the organization of neurons within this ganglion awaits further tests. The connections form an essential part of local reflexes whereby the movement of a leg is altered as a result of mechanosensory information it receives.

3.3. Number of Connections

The number of output connections made by an individual local interneuron in a complex nervous system cannot in practice be easily determined. It is feasible only to define some of the output connections either by intracellular recording or by stimulating the interneuron directly and observing its effects. Both methods may give a misleading picture. The first may select only the larger neurons and of course will always reflect the effort expended in its pursuit, whereas the second, dependent as it is on the behavioral context, may give a different answer in different locusts. Using these methods and these caveats of interpretation, a general conclusion seems to be that nonspiking interneurons have a greater and more widespread effect on the output of the motor neurons than do the spiking local interneurons. Each nonspiking interneuron activates a set of motor neurons to produce a component of the normal movements of the leg (Burrows, 1980). Some have actions restricted to one joint, whereas others can affect the movement of at least three joints. Each movement of a leg therefore results from the concerted action of many nonspiking interneurons exerting their effects in parallel on overlapping sets of motor neurons.

By contrast, depolarization of a spiking local interneuron does not usually produce an overt movement of a leg. A possible explanation could be that all the connections are inhibitory. Even if this were true, some movement would be expected through the effects of disinhibition. The conclusion to be drawn is that the number of motor neurons affected by one spiking local interneuron is small or that the effects in the context of the experiment are weak.

4. Interneuron Action during Leg Movements

Not only are both types of local interneurons responsible for controlling movements of a leg, they are also themselves affected by those movements. During voluntary movements, it is difficult to disentangle the feedback effects from those

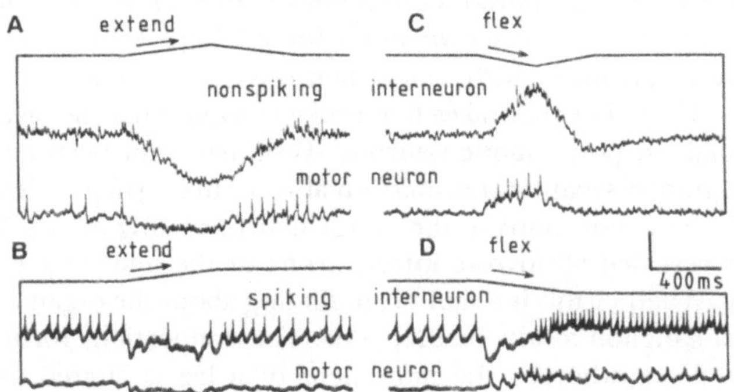

FIGURE 6. Both nonspiking (a, c) and spiking (b, d) local interneurons may respond to imposed movements of a hind leg. Simultaneous intracellular recordings were made from an interneuron and the slow extensor tibiae motor neuron. The nonspiking interneuron had an excitatory connection with the motor neuron, but the spiking interneuron made no connection. (a, b) Extension of the tibia through 40° evoked a large hyperpolarization in the nonspiking interneuron and a gradual repolarization during flexion. The waveform in the interneuron is a reflection of the imposed movement. In the spiking interneuron the tonic sequence of spikes was interrupted during the movement, only to resume at its previous level at the new extended position. (c, d) Flexion of the tibia evoked a depolarization of the nonspiking interneuron and a repolarization on extension that resulted in a more negative membrane potential than before the movement. The potential was slow to recover when the movement finished. In the spiking interneuron, there was an initial hyperpolarization followed by an increase in the frequency of spikes. Calibration: voltage interneuron (a, c) 7 mV, (b, d) 16 mV, motor neuron (a, c) 7 mV, (b, d) 16 mV; movement, 80°.

initiating the movements. Imposed movements are therefore easier to interpret. A nonspiking interneuron may signal such a movement by a graded change in its membrane potential and a spiking interneuron by a change in the frequency of its spikes. For example, during an imposed *extension* of the tibia, a nonspiking interneuron is gradually hyperpolarized by about 10 mV and then gradually repolarizes during *flexion* (Fig. 6A). In a spiking interneuron, the tonic sequence of spikes is interrupted during the extension, only to resume at its previous rate when the movement is complete (Fig. 6B). During an imposed *flexion* of the tibia, the same nonspiking interneuron is gradually depolarized by about 10 mV, but does not spike (Fig. 6C). When the movement is reversed, the interneuron hyperpolarizes, overshoots its original resting potential, and only gradually recovers. The spiking interneuron is again inhibited during the imposed movement, but at the new more flexed-position spikes at a much faster rate (Fig. 6D).

From these experiments it is clear that sensory feedback can have profound effects on local interneurons of both types. It becomes necessary to establish the sensory pathways exerting these feedbacks effects. Consideration will limited to be pathways from mechanosensory receptors on a leg.

5. Input Connections of Local Interneurons

Sensory neurons from a leg do not apparently make direct connections with the nonspiking local interneurons (Siegler and Burrows, 1983). Stimulation of individual sensory hairs produces sequences of synaptic potentials, usually inhibitory, but which are not correlated in a 1:1 fashion with sensory spikes. Thus, although these interneurons may be strongly influenced by sensory input, the pathway is not direct; a finding that accords well with the morphology (Fig. 3B,C).

Spiking local interneurons, however, appear to receive direct synaptic inputs from mechanoreceptors on the surface of a leg (Siegler and Burrows, 1983) (Fig. 7). Touching a particular hair excites an interneuron and usually evokes a spike or even a burst of spikes (Fig. 7A). The intensity of the stimulus is signaled by the frequency of spikes in the interneuron, but this simple relationship may not always pertain, for each interneuron works only in the context of its interactions with many other neurons. Hyperpolarizing the interneuron prevents the generation of spikes and at the same time accentuates the underlying excitatory synaptic potentials (EPSPs). When this is done, spikes from a particular sensory hair can be correlated with depolarizing potentials in an interneuron (Fig. 7B,C). Signal averaging shows even more clearly that the sensory spikes evoke excitatory synaptic potentials in the interneuron (Fig. 7D). Estimates of synaptic delay indicate that the sensory neurons synapse directly upon the interneurons. Other tests show that an increased conductance underlies the EPSP, suggesting that it is chemically mediated (Siegler and Burrows, 1983).

A spiking local interneuron may receive inputs from many hairs, and in turn each sensory axon from a hair synapses upon more than one interneuron. The effective combination of sensory inputs to an interneuron can be found only by experiment, and this has so far been achieved for 21 spiking local interneurons in

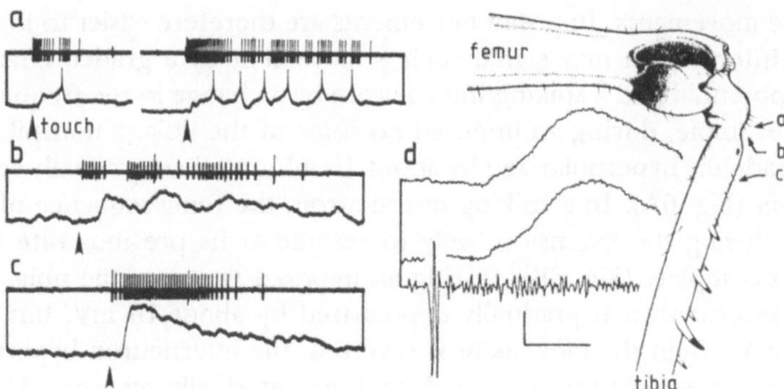

FIGURE 7. Sensory neurons from hairs on a leg synapse upon spiking local interneurons. Individual hairs (indicated in the diagram on the right) were stimulated mechanically (arrowheads), whereas the spikes of the single sensory neurons that innervate each of them were recorded extracellularly from a nerve in the femur and an intracellular electrode recorded from the soma of an interneuron. (a) The interneuron is at its normal membrane potential. Touching a hair evokes a spike or a burst of spikes. (b, c) The interneuron is hyperpolarized to reveal the excitatory synaptic potentials (EPSPs) that occur whenever there are spikes in the sensory neurons. (d) A signal-averaged record to show the relationship between a sensory spike and the EPSP in the interneuron. Each EPSP is averaged from 64 occurrences of the sensory spikes. Calibration, (a) 8 mV, (b, c) 4 mV, (d) 0.4 mV; time, (a) 200 msec, (b, c) 100 msec, (d) 22 msec.

the midline group (Burrows and Siegler, 1984). The receptive field of an interneuron is defined according to the area of a leg which provides its inputs. Implicit in such a method of description are the inherent difficulties of stimulating all the possible receptors and therefore of defining all the boundaries of the fields. Within these limitations of experiment, some interneurons have small and simple receptive fields with excitation limited to a single segment of a leg or to one axis of the entire leg. Other interneurons, however, have larger and more complex fields with inhibitory as well as excitatory areas. In the example shown in Fig. 8, hairs on the ventral

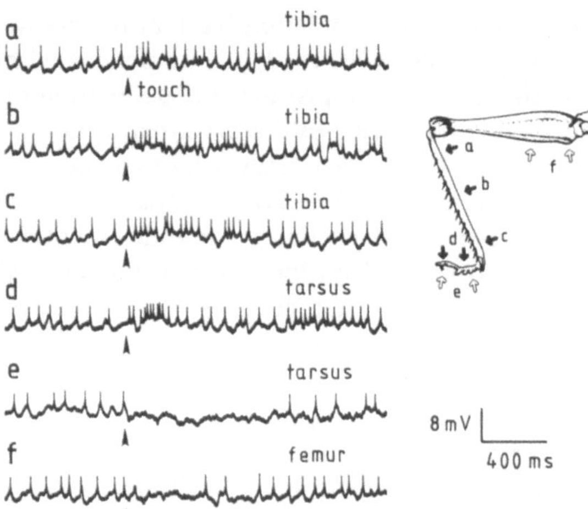

FIGURE 8. The receptive field of a spiking local interneuron. Intracellular recordings were made from the soma of an interneuron, while hairs on the regions of the leg indicated in the diagram were touched with a small brush. Four regions had an excitatory effect (solid arrows): on the ventral surface of the tibia (a–c) and on the dorsal surface of the tarsus (d). Two regions were inhibitory (open arrows); the proximal ventral region of the femur (f) and the ventral surface of the tarsus (e). Touching elsewhere on this leg was without effect.

surface of the tibia (Fig. 8A–C) and the dorsal surface of the tarsus (Fig. 8D) provide the excitatory input. Inhibitory input is provided by hairs on the proximal ventral region of the femur (Fig. 8F) and on the ventral surface of the tarsus (Fig. 8E). Hairs elsewhere have no effect. No direct inhibitory inputs are known from mechanosensory hairs to these spiking local interneurons, so that the inhibitory fields must result from interactions among the interneurons themselves. Interneurons with the same receptive fields and the same morphology can be recognized from locust to locust, so that it is probable that they are identifiable.

The organization of the receptive fields of these interneurons seems to reflect the organization of the local reflexes. Thus touching a particular array of hairs will activate a particular group of interneurons and evoke a particular movement of the leg. The reflexes elicited by the hairs adjust posture or movement of a leg if it encounters an external object. Other interneurons are excited only by movement of the joints, either as their loading changes in posture, or during voluntary movements. They underly resistance or interjoint reflexes of the leg.

6. Conclusions

The pathways involving the local interneurons can be summarized in the diagram in Fig. 9. A major pathway for mechanosensory neurons from the leg is to the spiking local interneurons, although some also synapse directly upon leg motor neurons (Pearson *et al.,* 1976) and probably upon intersegmental interneurons. None apparently synapse upon the nonspiking interneurons. A major pathway from the nonspiking interneurons is to the motor neurons, although they may also receive inputs from the spiking local interneurons and many other neurons. Inter-

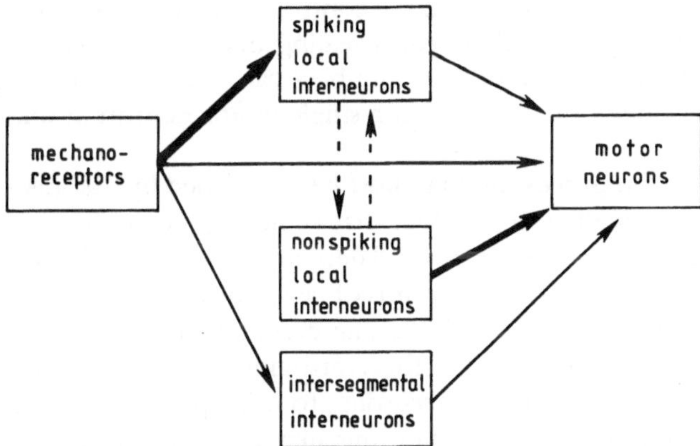

FIGURE 9. Diagram of some of the pathways revealed so far involving local interneurons. A major pathway for the sensory input is believed to be through the spiking local interneurons. Similarly, a major controlling effect on the motor neurons is exerted by the nonspiking local interneurons. The solid lines indicate known pathways, the dashed lines suspected pathways. It is certain that it will be necessary to incorporate more pathways in the future.

actions among the nonspiking interneurons are known (Burrows, 1979a), and interactions among the spiking interneurons are implicated from the finding of inhibitory regions in their receptive fields. The interactions among the spiking interneurons most likely take the form of a web of lateral inhibition, a common feature of many populations of second-order interneurons. Interactions between the nonspiking and spiking local interneurons are required to explain the patterns of synaptic potentials in the nonspiking interneurons when the hairs on a leg are stimulated. Clearly, what has been revealed are complex local circuits for the integration of sensory information and its conversion to appropriate motor patterns. From this complexity, can any generalizations be drawn about the role of the two types of local interneuron?

6.1. Can the Function of the Interneurons Be Defined?

The nonspiking interneurons may in large part be responsible for the shaping of the motor output, whereas the spiking local interneurons provide the first stage in the integration of the vast inflow of sensory signals. By virtue of their graded and sustained release of transmitter, modified by very small changes in voltage, the nonspiking interneurons can control the frequency of motor spikes in a precise and smoothly varying manner. This is exactly the type of control necessary for the smooth transitions of force that are required during posture and locomotion in an animal that controls muscular force by the frequency of spikes in a small number of motor neurons rather than by recruitment. The movements are also coordinated, in part by virtue of the pattern of connections with the motor neurons and in part by the pervasive influence of the prevailing behavioral context. No interneuron can, by itself, elicit a complete movement. Instead, it is always the concerted action of a group that brings about a voluntary movement. The number of nonspiking interneurons involved in the control of one leg cannot yet be accurately assessed, for there are no methods for revealing an entire population of local interneurons. An indication of the likely numbers can, however, be gauged from the fact that 12 have been found to affect a single motor neuron of one muscle (Burrows, 1980).

The scale of the processing that must be performed by the spiking local interneurons can be judged from the fact that some 10,000 sensory axons enter the metathoracic ganglion from a hindleg. There are probably no more than 140 spiking local interneurons in the two groups of cell bodies that have been examined in each half of the ganglion, although it seems certain that more remain to be discovered. Considerable convergence is therefore indicated. The surface of the leg is mapped onto these interneurons according to principles that are not yet obvious. The receptive fields of some interneurons may overlap, but it appears that each mechanosensory receptor on the leg has direct access to at least one of these spiking local interneurons.

These two divisions of function into motor control and sensory integration are oversimplifications. The separation of action becomes blurred when other aspects of both types of local interneuron are considered. The nonspiking interneurons, although not apparently receiving direct connections from sensory neurons, are strongly affected by them. Moreover, the effect of these interneurons on the motor neurons can change when the prevailing sensory inflow changes. These alterations in sensory inflow can be quite small, as would be caused by the change in angle of one joint of a leg (Siegler, 1981). The spiking interneurons also have a motor function in addition to their sensory one. Some synapse directly upon motor neurons thereby forming three neuron arcs that underly local reflexes. This dual action accords well with their morphology: their ventral branches project to the same region of the neuropil as the sensory neurons, while their dorsal branches project to the same region as the leg motor neurons.

6.2. Does an Individual Interneuron Have a Function?

This is the wrong type of question that the method of penetrating individual neurons with a microelectrode inevitably forces to the surface. Stimulating some neurons with such an electrode can elicit an obvious effect, but for the majority it does not. This does not mean that the function of the former has been defined and that of the latter not. For all neurons, the effectiveness of their output connections will depend on the behavioral context. Thus to observe an effect from some neurons will depend on the appropriate action of other neurons. In practice this may mean that different effects can be elicited from the same neuron in different animals, or in the same animal under different circumstances. Moreover, for a neuron embedded deeply in an interconnected web with many parallel pathways and feedback loops, the test necessary to reveal an action may be impossible to devise. It becomes obvious that an interneuron cannot be viewed in isolation, for to do so would result in ascribing it a single action. Instead, an interneuron must be viewed in the context of its interconnections with the other neurons in the nervous system. Then its actions are seen to be many and complex.

6.3. Future Strategies

In a nervous system of an alert animal, it is likely that the actions of the local interneurons will emerge, partly by chance and partly by knowing in which combinations with other interneurons they should be examined. It will also be necessary to describe their morphological and physiological features in as much detail as possible. The process should be repeated until the findings are consistent and can be linked together into larger and larger circuits associated with the behavior being studied. This strategy of describing how morphologically known interneurons are connected functionally and when they act as a way of defining causality is a long and irksome process. An intermediate strategy, and the one adopted here, is aimed at finding the principles underlying the operation of the nervous system. The different types of interneuron are distinguished as far as possible and their obvious interactions described to the extent that they are thought to be representative of the type as a whole. The strategy requires that the appropriate kind of explanation is selected for each of the observed responses. Alas, most of the explanations that spring to mind prove quickly to be incorrect. It is essential therefore to distinguish between the basic observations and the inferences that are based on them. It is precisely this reason that this article has concentrated more on observation than on inference.

ACKNOWLEDGMENTS. Much of the original experimental work reported here was carried out jointly with M. V. S. Siegler, and was supported by N.I.H. grant NS16058, and by a grant from the SERC (UK).

References

Burrows, M., 1979a, Synaptic potentials effect the release of transmitter from locust nonspiking interneurons, *Science* **204:**81–83.

Burrows, M., 1979b, Graded synaptic transmission between local pre-motor interneurons of the locust, *J. Neurophysiol.* **42:**1108–1123.

Burrows, M., 1980, The control of sets of motoneurons by local interneurones in the locust, *J. Physiol.* **298:**213–233.

Burrows, M., and Hoyle, G., 1973, Neural mechanisms underlying behavior in the locust *Schistocerca gregaria.* III. Topography of limb motorneurons in the metathoracic ganglion, *J. Neurobiol.* **4:**167–186.

Burrows, M. and Siegler, M. V. S., 1978, Graded synaptic transmission between local interneurones and motoneurones in the metathoracic ganglion of the locust, *J. Physiol.* **285:**231–255.

Burrows, M. and Siegler, M. V. S., 1982, Spiking local interneurons mediate local reflexes, *Science* **217:**650–652.

Burrows, M., and Siegler, M. V. S., 1984, The morphological diversity and receptive fields of spiking local interneurons in the locust metathoracic ganglion, *J. Comp. Neurol.* **224:** 483–508.

Evans, P. D., 1980, Biogenic amines in the Insect nervous system, *Adv. insect Physiol.* **15:**317–473.

Graubard, K., 1978, Synaptic transmission without action potentials: Input–output properties of a non-spiking presynaptic neuron, *J. Neurophysiol. 41:1014–1025.*

Pearson, K. G., and Robertson, R. M., 1981, Interneurons coactivating hindleg flexor and extensor motoneurons in the locust, *J. Comp. Physiol.* **144:**391–400.

Pearson, K. G., Wong, R. K. S. and Fourtner, C. R., 1976, Connexions between hair-plate afferents and motoneurones in the cockroach leg, *J. Exp. Biol.* **64:**251–266.

Pearson, K. G., Heitler, W. J., and Steeves, J. D., 1980, Triggering of locust jump by multimodal inhibitory interneurons, *J. Neurophysiol.* **43:**257–278.

Pfluger, H. J., Braunig, P., and Hustert, R., 1981, Distribution and specific central projections of mechanoreceptors in the thorax and proximal leg joints of locusts. II. The external mechanoreceptors: Hair plates and tactile hairs, *Cell Tissue Res.* **216:**79–96.

Rall, W., 1981, Functional aspects of neuronal geometry, in: *Neurones without Impulses* (A. Roberts and B. M. H. Bush, eds.), Cambridge University Press, Cambridge, pp. 223–254.

Reichert, H., Plummer, M. R., Hagiwara, G., Roth, R. L., and Wine, J. J., 1982, Local interneurons in the terminal abdominal ganglion of the crayfish, *J. Comp. Physiol.* **149:**145–162.

Roberts, A., and Bush, B. M. H., (eds.) 1981, in: *Neurones without Impulses* Cambridge University Press, Cambridge, Schmitt, F. O., and Worden, F. G., 1979, *The Neurosciences: Fourth Study Program,* MIT Press, Cambridge, Massachusetts.

Siegler, M. V. S., 1981, Posture and history of movement determine membrane potential and synaptic events in nonspiking interneurons and motor neurons of the locust, *J. Neurophysiol.* **46:**296–309.

Siegler, M. V. S., Burrows, M., 1979, The morphology of local non-spiking interneurones in the metathoracic ganglion of the locust, *J. Comp. Neurol.* **183:**121–148.

Siegler, M. V. S., and Burrows, M., 1983, Spiking local interneurons as primary integrators of mechanosensory information in the locust, *J. Neurophysiol.* **50:**1281–1295.

Siegler, M. V. S., and Burrows, M., 1984, The morphology of two groups of local spiking interneurons in the metathoracic ganglion of the locust, *J. Comp. Neurol.* **224:**463–482.
Tyrer, N. M., and Gregory, G. E., 1982, A guide to the neuroanatomy of locust suboesophageal and thoracic ganglia, *Philos. Trans. R. Soc. London Ser. B* **297:**91–123.
Watson, A. H. D. and Burrows, M., 1982, The ultrastructure of identified locust motor neurones and their synaptic relationships, *J. Comp. Neurol.* **205:**383–397.
Watson, A. H. D. and Burrows, M., 1983, The morphology, ultrastructure and distribution of synapses on an intersegmental interneurone of the locust, *J. Comp. Neurol.* **214:**154–169.
Wilson, J. A., and Phillips, C. E., 1982, Locust non-spiking interneurons which tonically drive antagonistic motor neurons: Physiology, morphology and ultrastructure, *J. Comp. Neurol.* **204:**21–31.

PART II
DEVELOPMENT

8

Metamorphosis of the Insect Nervous System

Influences of the Periphery on the Postembryonic Development of the Antennal Sensory Pathway in the Brain of *Manduca sexta*

JOHN G. HILDEBRAND

1. Introduction

1.1. Role of Intercellular Contacts in Neural Development

During the development of the nervous system, specific cell–cell contacts form among maturing neurons. The mechanisms by which those specific interactions take place and lead to the formation of appropriate synaptic connections and the establishment of normal neural pathways are of great interest in contemporary developmental neurobiology. Similarly intriguing is the question of the role of such cell–cell contacts in the survival and differentiation of the participating neurons.

In vertebrates, neurons often require contact with their targets to survive and mature normally (e.g., Cowan and Wenger, 1967; Black *et al.*, 1972; Landmesser and Pilar, 1972). Many excitable cells have been found also to require innervation for their normal development and maintenance of their normally differentiated characters (e.g., reviewed by Harris, 1974; Purves, 1976). This role of neuronal

JOHN G. HILDEBRAND • Department of Biological Sciences, Columbia University, New York, New York, 10027.

contact has been most fully documented in the case of motor innervation of developing muscle fibers (e.g., Dennis, 1981). Motor innervation influences the developing muscle fiber to localize its acetylcholine receptors in the postsynaptic membrane. In other cases, innervation has been found to influence strongly the development of target neurons (e.g., Levi-Montalcini, 1949; Guth, 1957; Black *et al.*, 1971).

In a variety of invertebrate systems, the development of appropriate and properly timed innervation is important for the survival and differentiation of target neurons. For example, the development of the first-order interneurons in the arthropod visual system depends on the ingrowth of their presynaptic inputs, the axons of the photoreceptor cells (Anderson, 1978; Macagno, 1977, 1979, 1981; Maxwell and Hildebrand, 1981; Meinertzhagen, 1977; Mouze, 1978; Meyerowitz and Kankel, 1978). The development of the photoreceptor cells, however, is largely independent of the target brain tissue (Kopéc, 1922; Gottschewski, 1960; Maxwell and Hildebrand, 1981; Eichenbaum and Goldsmith, 1968; Schneider, 1964, 1966). In no case is the mechanism of such influences of neurons over the development of their targets understood.

A central aim of our research has been to analyze and eventually to understand the mechanisms of influences exerted by ingrowing peripheral sensory axons over the development of their target neurons in the CNS. We pursue this goal through studies of an experimentally favorable, invertebrate animal, the hawk moth *Manduca sexta* (Lepidoptera, Sphingidae). Many developmental events that take place in the embryos of most species of animals (including much of the development of their adult nervous systems) are delayed in this large and easily reared insect until relatively late postembryonic stages of its life cycle.

2. Background

2.1. Insect Metamorphosis and the Transformation of the Nervous System

Holometabolous insects exhibit a pattern of postembryonic development in which the last larval instar is transformed to the adult by way of a pupal stage (for a survey and overview of insect metamorphosis, see Highnam, 1981). In the pupa, larval tissues are histolyzed to various degrees, and adult structures develop, again variously, from particular cells dedicated to this purpose and set aside—determined but largely undifferentiated—at an earlier stage in the animal's life. The orders Lepidoptera (moths and butterflies), Hymenoptera (ants, bees, wasps), and Diptera (flies) undergo a more or less spectacular version of this so-called complete metamorphosis. Following extensive breakdown of larval tissues in these insects, their adult appendages and certain other parts of the adult body develop from imaginal disks, which are clusters of cells that arise as invaginations of the embryonic epidermis and have no function in the larva, persisting only to evert and produce adult structures at metamorphosis.

These holometabolous insects "live two lives." Embryonic development produces a larva specialized structurally and behaviorally to feed. Later, during postembryonic life and under the control of known hormonal signals (reviewed by Granger and Bollenbacher, 1981; Coudron *et al.*, 1981), metamorphosis converts this assimilative stage into a reproductive adult, which has a dramatically different structure and is equipped with a transformed nervous system specialized for adult behaviors such as flight, courtship, mating, and oviposition.

As the exoskeleton and musculature are reshaped during metamorphosis, the nervous system is remodeled to accommodate the new sensory, motor, and integrative functions associated with the acquisition of the adult repertory of behavior. In the periphery, new sensory neurons arise from epidermal precursor cells as the imaginal disks develop into adult structures (Lawrence, 1966). In the CNS, neuroblasts and anlagen (primordia) of adult formations (such as the optic ganglia) generate adult structures by means of proliferation of neurons and glia, programed cell death, elaboration of neuropil, formation of synapses, and migration and fusion of ganglia. These postembryonic developmental events have been explored in a variety of insect species (reviewed by Edwards, 1969; Pipa, 1973, 1978) including *Manduca sexta* (e.g., Taylor and Truman, 1974; Truman and Reiss, 1976; Prescott *et al.*, 1977; Levine and Truman, 1982; Truman and Schwartz, 1982).

2.1.1. *Manduca sexta*

Embryonic development of *Manduca* culminates in the emergence of a first-instar larva four days after oviposition. A sequence of molts, controlled by juvenile hormone, ecdysteroids, and eclosion hormone (Riddiford and Truman, 1978; Truman *et al.*, 1981) and occurring over a period of nearly three weeks, carries the larva through five instars and finally yields the pupa. Adult development, triggered by ecdysteroid hormones acting in the absence of juvenile hormone (Riddiford and Truman, 1978; Bollenbacher *et al.*, 1981), commences soon after the larval-pupal molt. In *Manduca* reared in our laboratory under controlled environmental conditions and a long-day photoperiod regimen (Sanes and Hildebrand, 1976a; Prescott *et al.*, 1977; Tolbert *et al.*, 1983), adult development normally proceeds through 18 staged days and ends with the eclosion of the adult moth about 19 days after pupal ecdysis. The developmental stage of a metamorphosing insect can be determined in the laboratory on the basis of morphological changes in the cuticle, in structures visible through it, or in internal organs (Sanes and Hildebrand, 1976a).

Throughout the larval stages, the CNS of *Manduca* grows substantially but changes little in appearance (Prescott *et al.*, 1977). During metamorphosis, however, a dramatic upheaval transforms the CNS from the larval to the adult form (Prescott *et al.*, 1977). Most notably, the brain (supraesophageal ganglion) enlarges greatly and alters profoundly in form as the prominent optic and antennal lobes develop (illustrated in Hildebrand, 1980).

In the case of the optic lobes, the axons of myriad developing retinula (photoreceptor) cells, apparently contact, guided by remnants of the larval stemmatal nerves, grow from each developing compound eye into the anlage of the ipsilateral optic lobe (Maxwell and Hildebrand, 1981). As the sensory axons enter the brain, and during ensuing days of adult development, the optic ganglia (lamina, medulla, and lobula complex) grow dramatically. The lamina, which is the first-order optic ganglion and the synaptic target of the retinula-cell axons, increases in volume more than twentyfold by the time of adult eclosion (Maxwell and Hildebrand, 1981).

2.2. Antennal Sensory Pathway in Adult *Manduca sexta*

Both larval and adult *Manduca* have a pair of antennae, bilateral cephalic appendages that are specialized for certain chemosensory and mechanosensory functions and probably also serve thermoreceptive and hygroreceptive roles. The larval antenna is a simple structure (Dethier, 1941), which is apparently not directly developmentally related to the adult antenna that replaces it during metamorphosis (Sanes and Hildebrand, 1975, 1976a) and therefore will not be considered further here.

2.2.1. Functional Organization

As shown schematically in Fig. 1, the adult antenna in *Manduca* comprises two basal segments, the scape and pedicel, and a long flagellum divided into about 80 annuli (Sanes and Hildebrand, 1976a). The basal segments contain mechanosensory organs, including Johnston's organ in the pedicel and plates of Böhm bristles

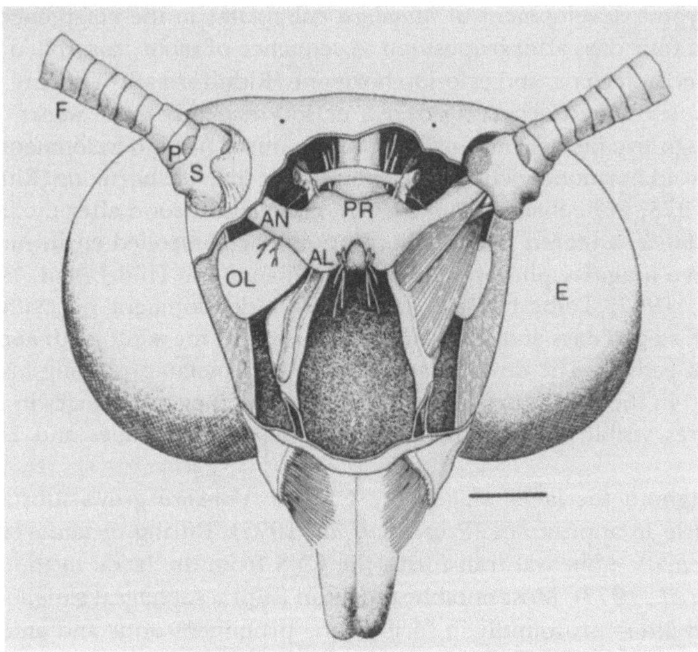

FIGURE 1. Drawing a cut-away view of the head of an adult *Manduca sexta* (dorsal aspect) showing the brain exposed by removal of cuticle from the top of the head and excision of underlying muscles and other structures. The extrinsic musculature of the left antenna (on the right side of the figure) is still in place. The antennal nerve (AN) enters the brain at the antennal lobe (AL), carrying axons from sensilla of the antennal flagellum (F) and basal segments, the scape (S) and pedicel (P). Also shown are the protocerebrum (PR), optic lobe (OL), and compound eye (E). Scale bar = 1 mm.

on the scape and pedicel (Camazine and Hildebrand, 1979). The flagellum is sexually dimorphic and bears large numbers of sensilla (small sensory organs) of several types, the great majority of which are olfactory. A male flagellum, for example, possesses about 2.6×10^5 sensory neurons associated with about 10^5 sensilla, of which approximately 40% are male-specific *sensilla trichodea* (Sanes and Hildebrand, 1976a). Each of these long hairlike sensilla is innervated by a pair of male-specific olfactory neurons specialized to detect components of the sex-pheromone mixture released by a sexually receptive, calling female moth (Sanes and Hildebrand, 1976b; Schweitzer *et al.*, 1976). One neuron in each of these sensilla responds very sensitively and selectively to bombykal, the hexadecadienal that is a prominent component of the pheromone mixture (Starratt *et al.*, 1979; J. G. Hildebrand and K.-E. Kaissling, unpublished observations). The female flagellum, which lacks the *sensilla trichodea* and their specialized olfactory neurons, is unresponsive to the female sex pheromone. Both male and female antennal flagella, however, carry large complements of sensilla of other types, many or most of which are olfactory sensilla that respond to volatile substances given off by plants, such as aldehydes, alcohols, and esters (Schweitzer *et al.*, 1976; I. D. Harrow, R. Kovelman, and J. G. Hildebrand, in preparation).

Sensory axons from receptor cells associated with antennal sensilla project through the antennal nerve, which enters the brain at the level of the ipsilateral antennal lobe (AL) in the deutocerebrum (Fig. 1). The sensory axons run to the brain without branching or forming synapses outside the CNS and project to targets primarily in the deutocerebrum (Sanes *et al.*, 1977; Hildebrand *et al.*, 1979; Camazine and Hildebrand, 1979; Matsumoto and Hildebrand, 1981). In addition to the AL, the deutocerebrum includes an "antennal mechanosensory and motor center" situated ventral to the AL and between the AL and the tritocerebrum and subesophageal ganglion (Fig. 2). Mechanosensory axons from the basal segments of the antenna project to this neuropil in the ipsilateral deutocerebrum and terminate there in characteristic patterns (Camazine and Hildebrand, 1979; S. M. Camazine, I. D. Harrow, K. S. Kent, and J. G. Hildebrand, in preparation).

The AL (schematized in Fig. 2) consists of a coarse, central region of neuropil (largely neurites of deutocerebral neurons) surrounded by an array of about 60 spheroidal glomeruli and bordered by three groups of neuronal somata (a large ventrolateral cluster, a smaller dorsomedial cluster, and a still smaller anteroventral group) totalling several hundred cells (Sanes *et al.*, 1977; Hildebrand *et al.*, 1979; Matsumoto and Hildebrand, 1981; Schneiderman *et al.*, 1983). Axons of olfactory neurons and possibly certain other receptor cells project to and terminate within the glomeruli of the ipsilateral AL (Camazine and Hildebrand, 1979; S. M. Camazine, I. D. Harrow, K. S. Kent, and J. G. Hildebrand, in preparation) where they form chemical synapses with dendrites of AL neurons (Matsumoto and Hildebrand, 1981; Tolbert and Hildebrand, 1981). The glomeruli are differentially staining, condensed knots of neuropil 50–100 μm in diameter, which are made up of terminals of sensory axons, dendritic processes of AL neurons, all of the chemical synapses that have been observed in the AL, and possibly other cellular elements, and are incompletely enveloped by an investment of glial processes (Hildebrand *et al.*, 1979; Matsumoto and Hildebrand, 1981; Tolbert and Hildebrand, 1981).

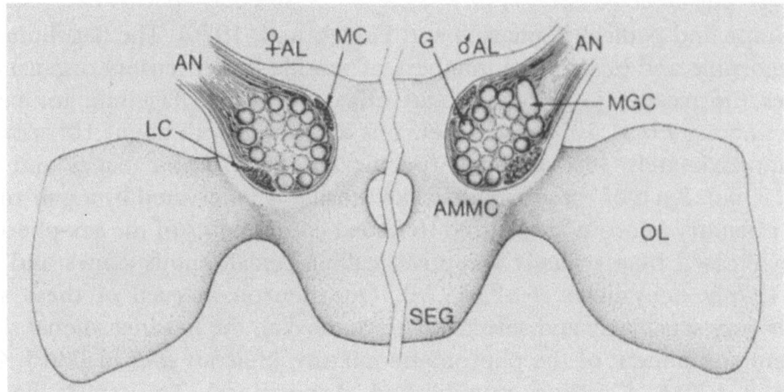

FIGURE 2. Frontal, cut-away views of male (on the right) and female (on the left) antennal lobes (AL). Antennal sensory axons enter the AL through the antennal nerve (AN) and many (largely olfactory) project to the "ordinary" glomeruli (G) in the AL, while others (mechanosensory) enter the antennal mechanosensory and motor center (AMMC) of the deutocerebrum lying ventral to the AL. The male AL contains the sexually dimorphic macroglomerular complex (MGC), which receives male-specific antennal sensory afferent input. Two of the three groups of AL neurons are shown: the lateral cell group (LC) and the medial cell group (MC). Also shown are the subesophageal ganglion (SEG) and optic lobes (OL). Scale bar = approximately 500 μm.

We have studied the neurons of the AL by intracellular recording and staining (Matsumoto and Hildebrand, 1981; Harrow and Hildebrand, 1982; Montague *et al.,* 1983; Christensen and Hildebrand, 1984; I. D. Harrow and J. G. Hildebrand, in preparation). Of the several thousand AL neurons we have examined, all have had their cell bodies in the AL and most, on the basis of the form and location of their dendritic arborizations, have clearly fallen into the categories outlined in Fig. 3. Many, and possibly a majority (Matsumoto and Hildebrand, 1981), of the AL neurons are amacrine, *local interneurons* with their dendritic aborizations confined to the AL and distributed in most or all of the glomeruli. Their somata are found in the lateral and medial cell groups. The *output neurons* are projection neurons with axons extending into the protocerebrum, dendritic arborizations confined to single glomeruli, and somata in all three AL cell groups. These output neurons can be further subclassified on the basis of their patterns of axonal projection in the protocerebrum (Montague *et al.,* 1983).

Like the antennae, the ALs are sexually dimorphic. In addition to the "ordinary glomeruli" (described above), which are present in ALs of both males and females, the male AL exclusively contains a large *macroglomerular complex* (MGC) (diagrammed in Figs. 2 and 3). Moreover, Matsumoto and Hildebrand (1981) recognized four sexually dimorphic types of AL neurons (schematized in Fig. 3: putatively female-specific type Ib local interneurons and male-specific types IIa and IIb local interneurons and type III output neurons). In the male AL, the MGC is the target of the axons of pheromone-detecting antennal sensory neurons (Camazine and Hildebrand, 1979; S. M. Camazine, I. D. Harrow, K. S. Kent, and J. G. Hildebrand, in preparation) and a site of dendritic arborization of all recognized male-specific AL neurons, including all AL neurons that respond postsynaptically to stimulation of the antenna with female sex pheromone (Matsumoto and Hildebrand, 1981). Thus the MGC is the primary or exclusive site for first-order synaptic processing of sensory information about the female sex pheromone. The ALs of several other species of insects have likewise been found to have sexually dimorphic glomeruli and interneurons implicated, on both anatomical and physiological grounds, in the processing of olfactory information about sex pheromone (e.g., Jawlowski, 1954; Boeckh *et al.*, 1977; Boeckh and Boeckh, 1979; Prillinger, 1981).

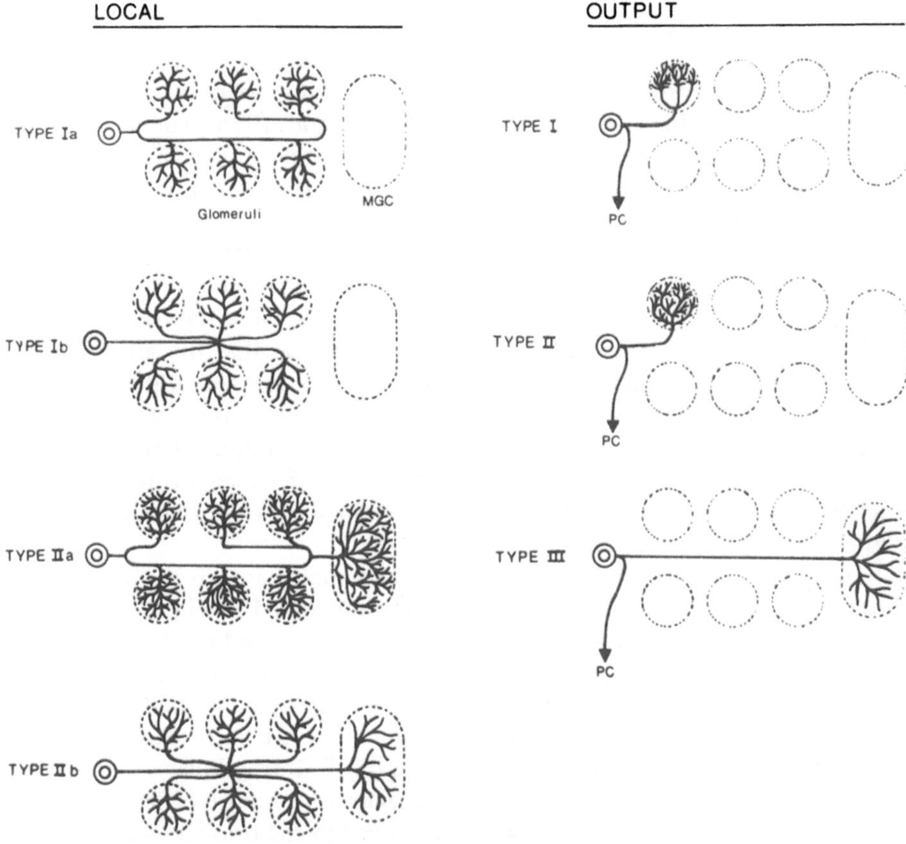

FIGURE 3. Schematic diagram of the morphological types of local interneurons and output neurons that have been recognized in the AL by means of intracellular staining and classified according to the nature of their arborizations in the AL neuropil. Type I local interneurons have arborizations in all or most of the ordinary glomeruli, but not in the MGC of male ALs. Types Ia and Ib differ mainly with respect to their branching pattern in the AL neuropil. Types IIa and IIb local interneurons ramify in most or all of the glomeruli, including the MGC, and thus are male-specific, sexually dimorphic cells. Type I output neurons have arborizations in a single ordinary glomerulus, as do type II output neurons, which differ from type I in having their somata in the anterior cell group of the AL and exhibiting a different pattern of arborization within "their" glomeruli. Type III output neurons are male-specific cells with their arborizations within the MGC. All of these types of AL output neurons send axons into the protocerebrum (PC), variously projecting to the calyces of the mushroom bodies, the lateral protocerebral lobe, and other regions of the protocerebrum (Montague *et al.*, 1983). (Reprinted with the publisher's permission, from Matsumoto and Hildebrand, 1981.)

2.2.2. Postembryonic Development

Within the first two of the 18 staged days of metamorphic adult development, mitotic divisions of precursor cells in the epidermis of the everted antennal imaginal disk generate new neurons, which immediately begin to differentiate morphologically and neurochemically into the characteristically bipolar, cholinergic sensory neurons of the antennal flagellum (Sanes and Hildebrand, 1976a,b,c). Only in the last five days of adult development do the sensory neurons develop the capacity to transduce natural stimuli into electrophysiological activity, as revealed by the electroantennogram method (Schweitzer *et al.*, 1976).

The nascent sensory neurons send their axons toward the brain along preformed nerve tracks, which provide contact guidance for the growing fibers, in the lumen of the developing antenna (Sanes and Hildebrand, 1975). The sensory fibers begin to reach the brain on day 3, and all have apparently arrived by day 10 (Sanes and Hildebrand, 1976b; Camazine and Hildebrand, 1979; S. M. Camazine, I. D. Harrow, K. S. Kent, and J. G. Hildebrand, in preparation). Based on preliminary [^{3}H]thymidine autoradiographic birthdating experiments (K. S. Kent and S. G. Hoskins, unpublished data) and findings in a related lepidopterous insect (Nordlander and Edwards, 1970), we believe that all or nearly all of the AL neurons are born prior to the arrival of antennal sensory axons, so that the afferent fibers encounter a full complement of potential target neurons. Figure 4 diagrams these events, through which two kinds of cellular lineages—one peripheral and the other central—generate postmitotic cells that differentiate into populations of neurons quite remote (up to 2–3 cm) from each other, but destined to interact as the peripheral sensory neurons send their axons to mingle and form synapses with the neurites of their central target neurons.

Our current understanding of the developmental events in the antennal flagellum and the AL during metamorphosis is summarized in Fig. 5. Before and during the ingrowth of the first antennal afferent fibers, and until day 5 of adult development, the AL neuropil is histologically homogeneous (Tolbert *et al.*, 1983). On day 5, glomeruli begin to be discernible in sectioned and stained ALs, but the sensory axons, in preparations in which they are stained intracellularly, do not exhibit a glomerular projection pattern before day 7. Intracellular staining of AL neurons has shown that their dendritic arborizations are more or less sparse and immature prior to day 9 and that they mature through elaboration and sculpturing between days 9 and 13.

Although some chemical synapses are evident in electron micrographs of thin sections of AL glomerular neuropil as early as day 5, the packing density of synapses increases abruptly at least fourfold (even as the glomerular volume is also increasing) to a mature level in the interval between days 8 and 12 (Tolbert *et al.*, 1983). Moreover, intracellular recordings begin to detect weak synaptic activity between antennal afferents and AL neurons on day 9 and demonstrate mature synaptic function by day 13. Thus by our criteria, synapses appear to develop in the AL largely during days 9 through 12 and to be structurally and physiologically mature by day 13, five days before the antenna is exposed (by eclosion of the adult) to the environment and before the electroantennogram indicates that the sensory

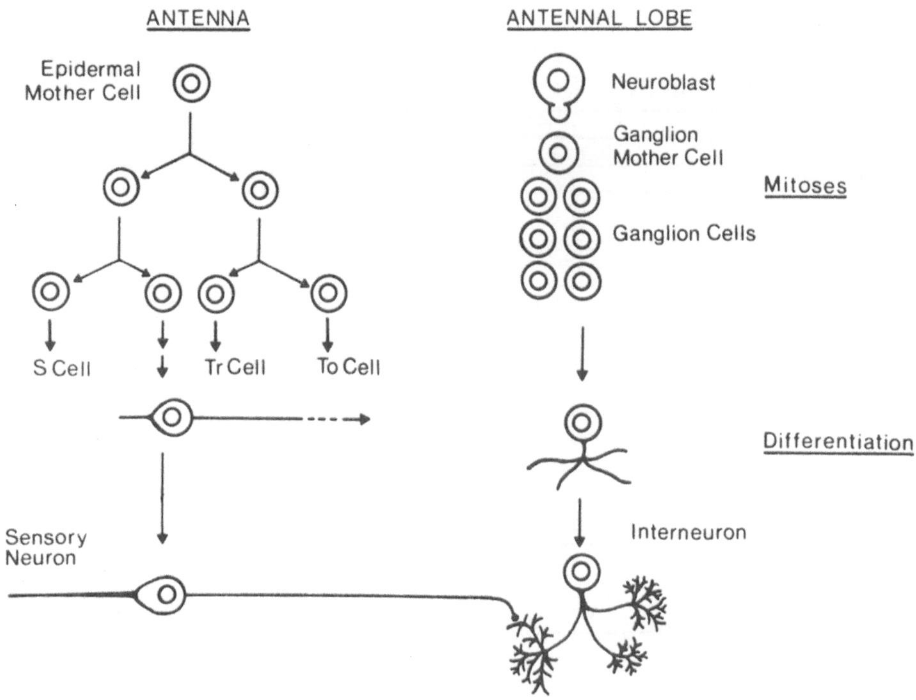

FIGURE 4. Schematic representation of the postembryonic development of the two principal populations of neurons associated with the AL: the sensory (largely olfactory) neurons of the antennal flagellum (left) and the central neurons, both local and output, of the AL (right). In the epidermis of the developing adult antennal flagellum, epidermal mother cells divide mitotically to produce a cell lineage that ultimately gives rise to the various types of cells that form a sensillum, including the tormogen cell (To), trichogen cell (Tr), glial-like supporting (S) cell, and sensory neuron(s). This developmental pattern has also been observed in other insects that have been studied (reviewed in Lawrence, 1966). The axons of the sensory neurons grow along pre-existing "guide" fibers in the antenna and enter the developing AL neuropil. There the sensory axons encounter neurites of differentiating AL neurons that have been produced mitotically in a lineage deriving from a large neuroblast by way of its daughter "ganglion mother cells" (Nordlander and Edwards, 1969). (Reprinted, with the publisher's permission, from Hildebrand *et al.*, 1982.)

transduction mechanism of the antennal sensory cells begins to function. Some evidence suggests, however, that synapses may continue to mature neurochemically after day 13. Thus we have found that the development of α-bungarotoxin-binding activity in the AL, which represents putative acetylcholine receptors, continues throughout nearly all of adult development and until eclosion of the moth (Sanes *et al.*, 1977).

Several of these observations are striking and provoke a search for developmental mechanisms. We would especially like to explore further the following findings: that most or all of the functional synapses between antennal sensory axons and their target AL neurons develop (or at least mature) suddenly, after most of the axons have entered the Al neuropil; that this burst of synaptic development

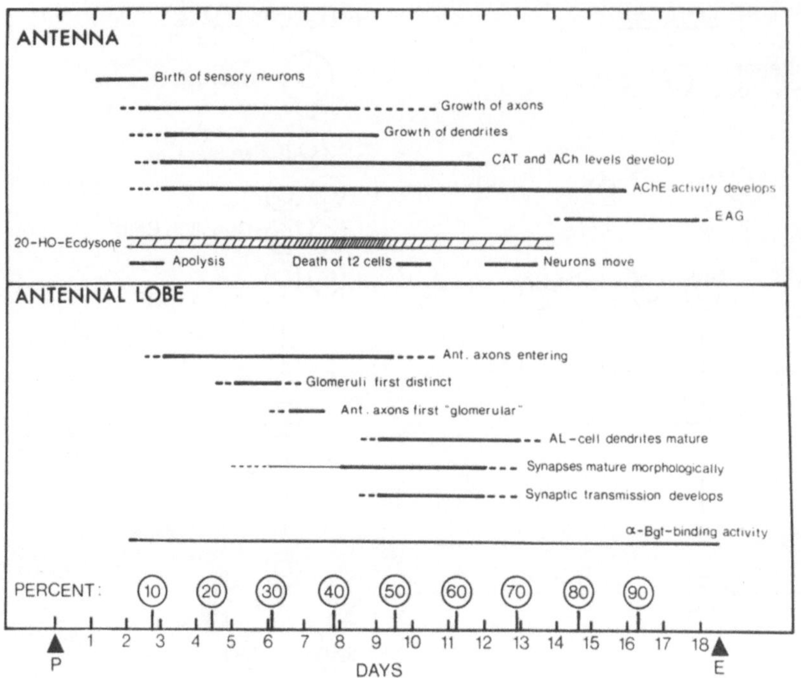

FIGURE 5. Summary of the development of the antennal flagellum (top) and AL (bottom) during post-pupal adult development. The events outlined in this timetable, with reference to days following pupation and to approximate percentage of adult development (which begins on day 1), are discussed in the text and in greater detail in the cited references. Also shown schematically is the sharp rise in the titer of circulating ecdysteroids beginning on day 2 after the larval-pupal molt, which initiates adult development, and the subsequent precipitous decline on day 9 (Bollenbacher *et al.*, 1981), which may control other developmental events (see text). Abbreviations: acetylcholine, ACh; acetylcholinesterase, AChE; antennal lobe, AL; choline acetyltransferase, CAT; eclosion, E; electroantennogram, EAG; pupation, P. (Reprinted, with the publisher's permission, from Hildebrand *et al.*, 1982.)

occurs after the participating axons and dendrites have intermingled and presumably interacted in the AL neuropil for several days; and that the "neural wiring" and synaptic connectivity of the AL develop and apparently become consolidated before the antennal neurons can report sensory information about the environment to the system.

The changes in titers of ecdysteroids diagrammed in Fig. 5 are important in the control of the developmental events outlined in that figure. Like other events in metamorphic adult development, the onset of which is signaled by apolysis, the birth of the antennal sensory neurons is apparently triggered by the rising titer of circulating ecdysteroids on days 2–3 after the larval-pupal molt. It is intriguing to note, moreover, that a dramatic episode of programmed cell death—of the cell t2 in each emerging trichoid sensillum in the developing male antennal flagellum (Sanes and Hildebrand, 1976b)—occurs on days 9–10, just when the titer of ecdysteroids has peaked and begun a precipitous decline. Perhaps this programmed cell death, as well as the other striking developmental events (outlined above) that also begin about the same time, are initiated or coordinated by the sharply declining ecdysteroid titer or an associated endocrine event.

Of particular interest to us recently has been the question: what factors influence or control the development of sexually dimorphic elements in the AL and the establishment of the specialized synaptic regions, the glomeruli, in its neuropil? Thus we have come to consider the role of the periphery in the postembryonic development of the ALs.

3. Influences of Sensory Afferent Inputs on the Development of the Antennal Lobes in *Manduca*

By means of surgical manipulations of *Manduca* larvae and pupae, we have explored the role of normal contacts between developing sensory afferents and their AL targets in the differentiation and maturation of both the peripheral (presynaptic) and central (postsynaptic) populations of neurons.

These studies began with a crude but decisive assessment of the dependency of the developing antennal sensory neurons on their CNS targets (Sanes *et al.*, 1976). We examined antennae from mature moths that had been surgically debrained prior to the mitotic birth of the sensory neurons in the everted antennal imaginal disk, so that those sensory neurons had no experience, intimate or remote, of their normal CNS target tissue. Despite this absolute deprivation of any influence from their normal targets, and indeed of any direct contact with the CNS, the antennal sensory neurons developed essentially normally with respect to morphological, electrophysiological, and neurochemical criteria. Like the first-order sensory neurons (retinula cells) of the optic pathway (Maxwell and Hildebrand, 1981), the primary sensory neurons of the neighboring antennal-sensory pathway exhibit no obvious dependency on their CNS targets for their survival and normal development.

Our more recent experiments have addressed the reciprocal developmental relationship—the dependency of the AL targets on their normal antennal afferent inputs. These studies have taken advantage of the fact that the antenna develops from an imaginal disk, which can be excised and discarded or transplanted in late-instar larvae, before the mitotic birth of the antennal sensory neurons.

3.1. Effects of Chronic Deafferentation

We first examined the effects of chronic deafferentation on development of the primary sensory center in the brain. This kind of experiment had been carried out previously in the visual systems of a number of species of arthropods (see Section 1 above), yielding results that shaped our expectations for our own studies. As in the other species that have been studied, removal of the primordium of the compound eye in *Manduca* (Maxwell and Hildebrand, 1981) results in the death of the lamina monopolar neurons with which many of the retinula cells would have made synaptic contacts.

By contrast, removal of the antennal disk leads to development of a stunted and somewhat displaced ipsilateral AL (Sanes *et al.*, 1977; Hildebrand *et al.*, 1979). This deprivation of antennal afferent inputs does not result in extensive death of CNS target neurons. On the contrary, many, and perhaps most or all, of the AL neurons survive and differentiate at least partially in the absence of their normal antennal afferent inputs. The resulting "unafferented" AL, examined in an adult moth, resembles a normal AL at about day 5–6 of adult development except that the orderly array of nascent glomeruli characteristic of the latter (Tolbert *et al.*, 1983) is replaced by a poorly organized patchwork of irregularly shaped "protoglomerular" knots of condensed neuropil in the former (Hildebrand *et al.*, 1979). The unafferented AL has the three groups of neuronal somata in approximately normal locations (lateral, medial, and anterior) and resides in a position characteristic of a developing AL prior to day 7 of adult development, ventromedial to the position of a normal adult AL. Intracellularly stained AL neurons in these unafferented ALs resemble the immature neurons of normal, developing ALs at early stages of adult development (K. S. Kent, S. G. Matsumoto, R. A. Montague, and A. M. Schneiderman, unpublished observations). The local interneurons have less extensive and sparser dendritic arborizations than their normal, mature counterparts. The output neurons have axons projecting into the protocerebrum, but generally exhibit a more or less diffuse dendritic arborization in the AL that contrasts sharply with the uniglomerular dendritic tuft of a normal, mature output neuron.

From these observations, it appears that antennal afferent innervation of a developing AL is needed not for the survival of at least a majority of the AL neurons, but to stimulate, organize, or sustain the normal elaboration of their dendritic arbors and the crystallization of the normal array of discrete glomeruli. Indeed, further experimentation has revealed that the antennal afferent axons exert even more dramatic influences over the developing AL.

3.2. Development of Sexually Dimorphic Components

In order to learn more about possible peripheral influences in AL development, we have focused on the male-specific MGC and its associated male-specific AL neurons. These components of the Al are experimentally advantageous in that they are readily recognizable and innervated by known male-specific antennal sensory fibers (Matsumoto and Hildebrand, 1981).

In histological studies of metamorphosing ALs, aided by computer-graphic techniques, we have found that the nascent MGC can be identified, distinct from developing "ordinary" glomeruli, as early as day 5 of adult development—the stage at which glomeruli are first discernible in the AL neuropil (Schneiderman *et al.*, 1983). Moreover, although the number of ordinary glomeruli in both male and female ALs appears to vary in the range 57–61, no male AL in our experience has had fewer than 57 ordinary glomeruli, the typical number for a female AL. Thus we believe that the MGC arises independently and not through modification or fusion of other glomeruli in the male AL. We have searched for an anlage of the

MGC in normal ALs of males younger than day 5 of adult development and for a "proto-MGC" in deantennated male ALs, but we have not yet identified such a structure in either case.

To test for a role of antennal sensory axons in the development of the male-specific components of the AL, we devised a surgical procedure for producing antennal gynandromorphs (Schneiderman *et al.*, 1982). Using the larval antenna, which overlies the antennal imaginal disk (Sanes and Hildebrand, 1976a) as a "handle," we transplant an antennal disk from a fifth-instar donor larva to the same site in the head of a recipient larva of the same developmental stage. In a large fraction of the animals that have received a grafted antennal disk, either from a donor of the same or opposite sex or from themselves (reimplantation), a normal antenna characteristic of the donor's sex develops (Fig. 6), and its sensory neurons send axons into the ipsilateral AL where they make synapses with neurites of AL neurons. In animals in which the grafted antenna develops from a reimplanted disk or from a disk obtained from a donor of the same sex, the mature AL is indistinguishable from that on the control (unoperated) side of the experimental animal's brain. When the grafted antenna develops from a disk of the opposite sex, however, the

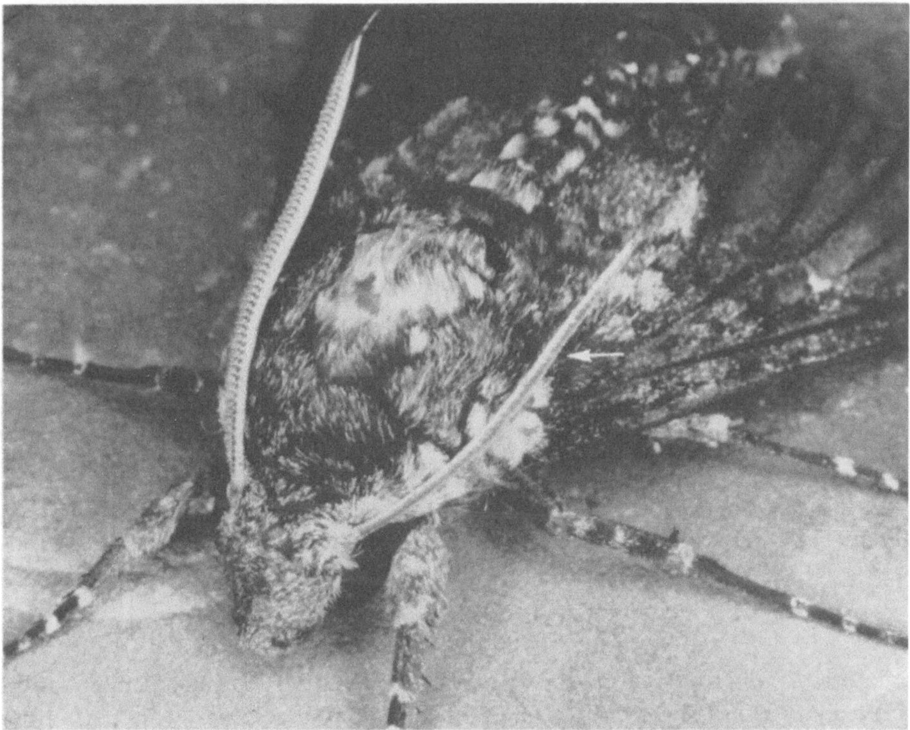

FIGURE 6. Photograph of a surgically gynandromorphic adult male *Manduca*, showing the grafted female antenna (arrow) and the normal male antenna on the contralateral side of the head.

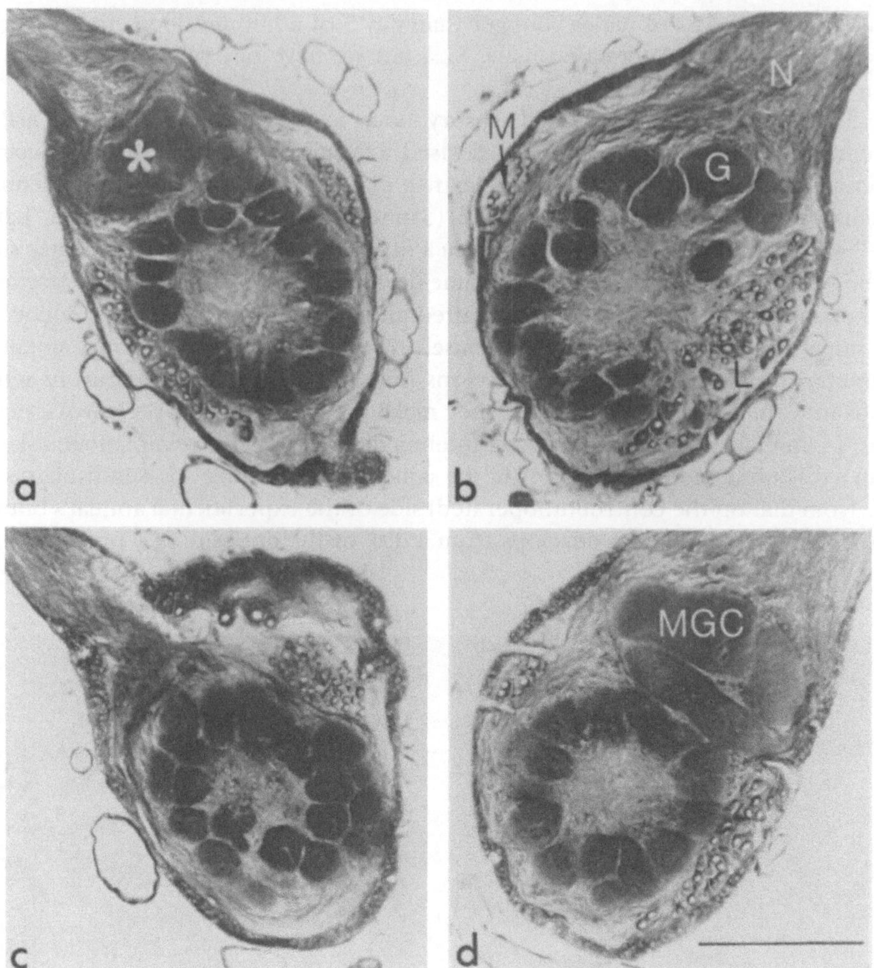

FIGURE 7. Frontal sections, prepared from paraffin-embedded tissue and stained with Cresylecht violet and Luxol fast blue, through the surgically gynandromorphic and normal, control ALs of female (a, b) and male (c, d) moths. (b) The antennal nerve (N) enters the AL of a normal female. The ordinary glomeruli (G) are arrayed around a central region of coarser neuropil and are bordered by the lateral (L) and medial (M) groups of AL neurons. The normal male AL (d) characteristically exhibits the sexually dimorphic MGC. In the gynandromorphic female AL (a), a structure resembling the MGC (*) can be seen, but the gynandromorphic male (AL) (c) lacks an MGC. Scale bar = 200 μm. (Reprinted, with the publisher's permission, from Schneiderman *et al.*, 1982.)

ipsilateral AL differs from the contralateral (unoperated) AL and resembles an AL of the donor's sex. Figure 7 illustrates these findings: a genetically male AL innervated by sensory axons from a grafted female antenna lacks a recognizable MGC and resembles a normal female AL; a genetically female AL innervated by male antennal axons exhibits an MGC and resembles a normal male AL.

Thus innervation by male antennal sensory axons apparently is required for the development of characteristically male histological features in male ALs and can masculinize a genetically female AL that receives sensory innervation from a grafted male antenna. This influence of the male antennal axons also operates at the cellular level, clearly affecting the development of AL neurons (Schneiderman *et al.*, 1982). By means of intracellular recording and staining, we have found that numerous malelike interneurons develop in a female AL innervated by sensory axons from a grafted male antenna. These cells have somata in the lateral and medial cell groups of the AL and dendritic arborizations in the MGC-like structure in the surgically gynandromorphic AL and thus resemble the male-specific AL neurons schematized in Fig. 3. Although we have visualized intracellularly stained examples of apparently normal, male-specific type III output neurons in female ALs innervated by male antennal axons (A. M. Schneiderman and R. A. Montague, unpublished observations), we have not yet succeeded in recording from such cells. Many more AL cells resembling the male-specific type II local interneurons have been impaled and stained intracellularly (Fig. 8), and in most cases, these cells have been shown to receive characteristic excitatory synaptic input when the grafted male antenna is stimulated with the female sex pheromone component bombykal (Schneiderman *et al.*, 1982). Figure 8 shows two examples of such malelike interneurons in female-gynandromorphic ALs as well as a male-specific type II local interneuron in a normal male AL. Such cells are of course never detected in normal female ALs.

Obvious questions arising from these findings are: how extensively is the development of the CNS ultimately affected by male antennal-afferent innervation and is the behavior of a female gynandromorph (with grafted male antennae) altered? We have begun to explore these issues experimentally. In our studies of the brain and rostral ganglia, we have not yet recognized anatomical differences, other than

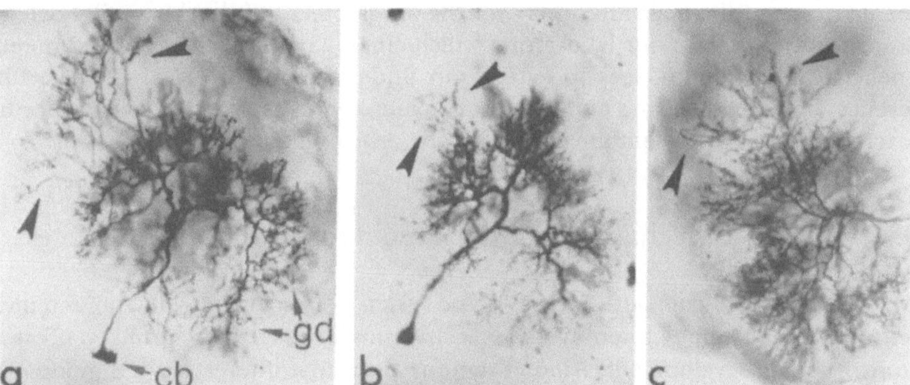

FIGURE 8. AL neurons stained intracellularly with cobalt sulfide. (a) Normal male-specific local interneuron; (b, c) similar interneurons from ALs of female gynandromorphs. Each cell had arborizations in the region of the MGC (marked with arrowheads) and received excitatory synaptic input when the ipsilateral male antenna was stimulated with the female sex pheromone bombykal. Also labeled for cell (a) are the cell body (cb) and dendritic arborizations in ordinary glomeruli (gd). Scale bar = 200 μm. (Photomicrographs (a) and (b) are reprinted, with the publisher's permission, from Schneiderman *et al.*, 1982.)

those in the ALs, between CNS structures in males and females or in antennal gynandromorphs. But the search continues. At the same time, we have begun to ask whether adult female gynandromorphs express altered, malelike behaviors in response to female sex pheromone in a wind tunnel (A. M. Schneiderman, J. G. Hildebrand, P. Brennan, and J. H. Tumlinson, in progress). Normal males respond characteristically to female sex pheromone with arousal, patterned flight, positive anemotaxis, and approach to and interaction with the pheromone source in a wind tunnel. Females, whose antennae cannot even detect the pheromone, exhibit none of these behaviors when exposed to the sex pheromone. If our studies, which are as yet inconclusive, establish that female antennal gynandromorphs do indeed express some or all of the male-specific behavioral responses to pheromone, then it will be evident that the innervation of the genetically female CNS by male antennal sensory neurons has consequences extending far beyond the ALs.

The striking effects of male sensory innervation require physical contact between the afferent fibers and the target AL; if the axons fail to reach the AL, then it is identical to other unafferented ALs. From this finding we infer that cell–cell contacts between male antennal sensory axons and target neurons in the AL are necessary for the afferent fibers to influence the development of the AL neurons. On this basis we have hypothesized that the male-specific axons (the fibers of pheromone-detecting sensory neurons) possess molecular "factor(s)"—possibly polypeptides—that are absent from female afferents and mediate the male axons' influence over developing AL target elements. By means of two-dimensional polyacrylamide gel electrophoresis of polypeptides extracted from antennal nerves and ALs, we have observed certain subtle but reproducible differences in the patterns of polypeptides revealed by silver staining and by autoradiography following labeling *in vivo* with [^{35}S]methionine (Kingan and Hildebrand, 1983). Whether any of these molecular differences between male and female antennal nerves is related to the influence of the male afferents over the development of the AL is still unclear. Studies in progress in our laboratory, which employ both biochemical and immunochemical techniques, seek to extend our understanding of such molecular differences between male and female sensory axons and to explore the role of specific macromolecules of the antennal nerve in the development of the AL.

3.3. Toward a Mechanistic Understanding of Afferent Influences

Whatever the chemical nature of the "signals" associated with the antennal sensory axons, we hope to unravel the mechanisms involved in the primary-afferent control of AL development. Among the questions that interest us and guide our experimentation are the following: Are hormonal changes, such as the last decline in the titer of circulating ecdysteroids, responsible for the coordinated developmental events in the antennal pathway on days 9–12 of metamorphic adult development? Is electrophysiological activity in the developing antennal sensory axons necessary for the formation of normal glomeruli and synaptic connections in the AL? During metamorphosis, is there a "critical period" for antennal-afferent influences to be exerted over AL development? Do the male-specific antennal axons induce a malelike course of development in their target neurons in the AL, or does contact from the male afferent fibers rescue a subpopulation of AL neurons that is ordinarily destined to die in the absence of that contact, as in a normal female AL? Are the components or actions of antennal sensory fibers that influence AL development species-specific? Can antennal sensory axons precipitate development of a glomerular neuropil in regions of the CNS other than the ALs? Answers to these and other questions should be attainable by virtue of the extraordinary experimental advantages of the metamorphosing *Manduca* antennal system and are goals of our current research.

4. Conclusions

Our studies of the postembryonic, metamorphic development of the antennal and optic pathways in *Manduca* have revealed that ingrowing afferent axons from the periphery (i.e., from the antenna and the compound eye) can strongly influence the development of their targets in the CNS. In the optic pathway, the role of the sensory input is most decisive. Most or all of the lamina monopolar neurons depend absolutely on normal afferent innervation for their survival and differentiation, and all but a small portion of the lamina fails to develop if the primary sensory axons (of retinula cells) from the compound eye have been prevented from reaching the developing optic lobe of the brain. The picture is different in the antennal pathway. The AL neurons depend on innervation by antennal sensory fibers in order to organize discrete glomeruli, elaborate mature dendritic arborizations, and develop sexually dimorphic characteristics. Many if not all of the AL neurons, however, survive and partially differentiate in the absence of their normal afferent inputs.

Thus the first-order interneurons in the visual and olfactory pathways apparently depend to very different degrees on their inputs from the periphery for their normal development. Some central neurons require normal afferent contacts in order to survive; others survive and at least partially differentiate in the absence of their normal afferent contacts; and still others, although they may survive when deprived of their normal afferent inputs, appear to develop in ways dictated by the character (e.g., the sex) of their inputs from the periphery.

On the basis of our findings and those of other investigators studying a variety of species, we suggest that diverse classes of neurons in any species may follow different strategies or "rules" of development. In particular, appropriate synaptic inputs or contact with normal target cells may be more or less important in permitting or shaping the development of a certain nerve cell. We believe that factors such as the ultimate function and position of a neuron in a nervous system, and not its phyletic identity, account for those differences of cellular behavior during development. A challenging problem now is to discover *why* different types of neurons follow different rules of development.

ACKNOWLEDGMENTS. The research from the author's laboratory described in this chapter has been supported for many years by grants from the federal government (currently NSF grant BNS 83-12769, NIH grant AI-17711-04, and Army Research Office contract DAAG29-81-K-0091). The author is very grateful to R. A. Montague for photographic assistance and to the many co-workers whose research findings are summarized here.

References

Anderson, H., 1978, Postembryonic development of the visual system of the locust, *Schistocerca gregaria,* II. An experimental investigation of the formation of the retina-lamina projection, *J. Embryol. Exp. Morphol.* **46:**147–170.

Black, I. B., Hendry, I. A., and Iversen, L. L., 1971, Transsynaptic regulation of growth and development of adrenergic neurons in the mouse sympathetic ganglion, *Brain Res.* **34:**229–240.

Black, I. B., Hendry, I. A., and Iversen, L. L., 1972, Role of postsynaptic neurons in the biochemical maturation of presynaptic cholinergic nerve terminals in a mouse sympathetic ganglion, *J. Physiol. (London)* **221:**149–159.

Boeckh, J., and Boeckh, V., 1979, Threshold and odor specificity of pheromone-sensitive neurons in the deutocerebrum of *Antheraea pernyi* and *A. polyphemus* (Saturnidae), *J. Comp. Physiol.* **132:**235–242.

Boeckh, J., Boeckh, V., and Kühn, A., 1977, Further data on the topography and physiology of central olfactory neurons in insects, in: *Olfaction and Taste,* Volume 6 (J. Le Magnen, and P. MacLeod, eds.), Information Retrieval, London, pp. 315–321.

Bollenbacher, W. E., Smith, S. L., Goodman, W., and Gilbert, L. I., 1981, Ecdysteroid titer during larval–pupal–adult development of the tobacco hornworm, *Manduca sexta, Gen. Comp. Endocrinol.* **44:**302–306.

Camazine, S. M., and Hildebrand, J. G., 1979, Central projections of antennal sensory neurons in mature and developing *Manduca sexta, Soc. Neurosci. Abstr.* **5:**155.

Christensen, T. A., and Hildebrand, J. G., 1984, Functional anatomy and physiology of male-specific pheromone-processing interneurons in the brain of *Manduca sexta, Soc. Neurosci. Abstr.* **10:**862.

Coudron, T. A., Law, J. H., and Koeppe, J. K., 1981, Insect hormones, *Trends Biochem. Sci.* **6:**248–251.

Cowan, W. M., and Wenger, E., 1967, Cell loss in trochlear nucleus of chick during normal development and after radical extirpation of the optic vesicle, *J. Exp. Zool.* **164:**267–280.

Dennis, M. J., 1981, Development of the neuromuscular junction: Inductive interactions between cells, *Annu. Rev. Neurosci.* **4:**43–68.

Dethier, V. G., 1941, The antennae of lepidopterous larvae, *Bull. Mus. Comp. Zool. Harvard Univ.* **87:**455–507.

Edwards, J. S., 1969, Postembryonic development and regeneration of the insect nervous system, *Adv. Insect Physiol.* **6:**97–137.

Eichenbaum, D. M., and Goldsmith, T. H., 1968, Properties of intact photoreceptor cells lacking synapses, *J. Exp. Zool.* **169:**15–32.

Gottschewski, G. M. H., 1960, Morphogenetische Untersuchungen an *in vitro* wachsenden Augenanlagen von *Drosophila melanogaster, Wilhelm Roux' Arch.* **152:**204–229.

Granger, N. A., and Bollenbacher, W. E., 1981, Hormonal control of insect metamorphosis, in: *Metamorphosis—A Problem in Developmental Biology,* 2nd ed. (L. I. Gilbert and E. Frieden, eds.), Plenum Press, New York, pp. 105–137.

Guth, L., 1957, Effects of glossopharyngeal nerve transsection on the circumvallate papilla of the rat, *Anat. Rec.* **28:**715–731.

Harris, A. J., 1974, Inductive functions of the nervous system, *Annu. Rev. Physiol.* **36:**251–305.

Harrow, I. D., and Hildebrand, J. G., 1982, Synaptic interactions in the olfactory lobe of the moth *Manduca sexta, Soc. Neurosci. Abstr.* **8:**528.

Highnam, K. C., 1981, A survey of invertebrate metamorphosis, in: *Metamorphosis—A Problem in Developmental Biology,* 2nd ed., (L. I. Gilbert and E. Frieden, eds.), Plenum Press, New York, pp. 43–73.

Hildebrand, J. G., 1980, Development of putative acetylcholine receptors in normal and deafferented antennal lobes during metamorphosis of *Manduca sexta*, in: *Receptors for Neurotransmitters, Hormones, and Pheromones in Insects* (D. B. Sattelle, L. M. Hall, and J. G. Hildebrand, eds.), Elsevier/North Holland, Amsterdam, pp. 209–220.

Hildebrand, J. G., Hall, L. M., and Osmond, B. C., 1979, Distribution of binding sites for ^{125}I-labeled α-bungarotoxin in normal and deafferented antennal lobes of *Manduca sexta*, *Proc Natl. Acad. Sci. U.S.A.* **76**:499–503.

Hildebrand, J. G., Matsumoto, S. G., Tolbert, L. P., Schneiderman, A. M., and Camazine, S. M., 1982, Postembryonic development of the antennal lobes in the moth *Manduca sexta*, *Neurosci. Res. Prog. Bull.* **20**:891–900.

Jawlowski, H., 1954, Über die Struktur des Gehirnes bei *Saltatoria*, *Ann. Univ. Mariae Curie-Sklodowska Sect.* C **11**:403–434.

Kingan, T. G., and Hildebrand, J. G., 1983, Sexually dimorphic expression of polypeptides in antennal sensory neurons in the brain of *Manduca sexta*, *Soc. Neurosci. Abstr.* **9**:835.

Kopéc, S., 1922, Mutual relationship in the development of the brain and eyes of Lepidoptera, *J. Exp. Zool.* **36**:459–468.

Landmesser, L., and Pilar, G., 1972, The onset and development of transmission in the chick ciliary ganglion, *J. Physiol. (London)* **222**:691–713.

Lawrence, P. A., 1966, Development and determination of hairs and bristles in the milkweed bug, *Oncopeltus fasciatus* (Lygaeidae, Hemiptera), *J. Cell Sci.* **1**:475–498.

Levi-Montalcini, R., 1949, The development of the acoustico-vestibular centers in the chick embryo in the absence of the afferent root fibers and of descending fiber tracts, *J. Comp. Neurol.* **91**:209–242.

Levine, R. B., and Truman, J. W., 1982, Metamorphosis of the insect nervous system: Changes in morphology and synaptic interactions of identified neurones, *Nature (London)* **299**:250–252.

Macagno, E. R., 1977, Abnormal synaptic connectivity following UV-induced cell death during *Daphnia* development, in: *Cell and Tissue Interactions* (J. W. Lash and M. M. Burger, eds.), Raven Press, New York, pp. 293–309.

Macagno, E. R., 1979, Cellular interactions and pattern formation in the development of the visual system of *Daphnia magna* (Crustacea, Branchiopoda). I. Interactions between embryonic retinular fibers and laminar neurons, *Dev. Biol.* **73**:206–238.

Macagno, E. R., 1981, Cellular interactions and pattern formation in the development of the visual system of *Daphnia magna* (Crustacea, Branchiopoda). II. Induced retardation of optic axon ingrowth results in a delay in laminar neuron differentiation, *J. Neurosci.* **1**:945–955.

Matsumoto, S. G., and Hildebrand, J. G., 1981, Olfactory mechanisms in the moth *Manduca sexta:* Response characteristics and morphology of central neurons in the antennal lobes, *Proc. R. Soc. (London)* Ser. B **213**:249–277.

Maxwell, G. D., and Hildebrand, J. G., 1981, Anatomical and neurochemical consequences of deafferentation in the development of the visual system of the moth *Manduca sexta*, *J. Comp. Neurol.* **195**:667–680.

Meinertzhagen, I. A., 1977, Development of neuronal circuitry in the insect optic lobe, in: *Society for Neuroscience Symposia*, Volume II, Approaches to the Cell Biology of Neurons (W. M. Cowan and J. A. Ferrendelli, eds.) Society for Neuroscience, Bethesda, Maryland, pp. 92–119.

Meyerowitz, E. M., and Kankel, D. F., 1978, A genetic analysis of visual system development in *Drosophila melanogaster*, *Dev. Biol.* **62**:112–142.

Montague, R. A., Kent, K. S., Imperato, M. T., and Hildebrand, J. G., 1983, Projections of antennal-lobe output neurons in the brain of *Manduca sexta*, *Soc. Neurosci. Abstr.* **9**:216.

Mouze, M., 1978, Role des fibres post-rétiniennes dans la croissance du lobe optique de la larve d'*Aeshna cynea* Müll. (Insecte Odonate), *Wilhelm Roux' Arch.* **184**:325–350.

Nordlander, R. H., and Edwards, J. S., 1969, Postembryonic brain development in the Monarch butterfly, *Danaus plexippus plexippus*, L. I. Cellular events during brain morphogenesis, *Wilhelm Roux' Arch.* **162**:197–217.

Nordlander, R. H., and Edwards, J. S., 1970, Postembryonic brain development in the Monarch butterfly, *Danaus plexippus plexippus* L. III. Morphogenesis of centers other than the optic lobes, *Wilhelm Roux' Arch.* **164**:247–260.

Pipa, R. L., 1973, Proliferation, movement, and regression of neurons during the postembryonic development of insects, in: *Developmental Neurobiology of Arthropods* (D. Young, ed.), Cambridge University Press, United Kingdom, pp. 105–129.

Pipa, R. L., 1978, Patterns of neural reorganization during the postembryonic development of insects, *Int. Rev. Cytol.* (suppl.) **7**:403–438.

Prescott, D. J., Hildebrand, J. G., Sanes, J. R., and Jewett, S., 1977, Biochemical and developmental studies of acetylcholine metabolism in the central nervous system of the moth, *Manduca sexta, Comp. Biochem. Physiol.* **56C**:77–84.

Prillinger, L., 1981, Postembryonic development of the antennal lobes in *Periplaneta americana* L., *Cell Tissue Res.* **215**:563–575.

Purves, D., 1976, Long-term regulation in the vertebrate peripheral nervous system, *Int. Rev. Physiol. Neurophysiol. II* **10**:125–177.

Riddiford, L. M., and Truman, J. W., 1978, Biochemistry of insect hormones and insect growth regulators, in: *Biochemistry of Insects* (M. Rockstein, ed.), Academic Press, New York, pp. 307–357.

Sanes, J. R., and Hildebrand, J. G., 1975, Nerves in the antennae of pupal *Manduca sexta* (Lepidoptera: Sphingidae), *Wilhelm Roux' Arch.* **178**:71–78.

Sanes, J. R., and Hildebrand, J. G., 1976a, Structure and development of antennae in a moth, *Manduca sexta, Dev. Biol.* **51**:282–299.

Sanes, J. R., and Hildebrand, J. G., 1976b, Origin and morphogenesis of sensory neurons in an insect antenna, Dev. Biol. **51**:300–319.

Sanes, J. R., and Hildebrand, J. G., 1976c, Acetylcholine and its metabolic enzymes in developing antennae of the moth, *Manduca sexta, Dev. Biol.* **52**:105–120.

Sanes, J. R., Hildebrand, J. G., and Prescott, D. J., 1976, Differentiation of insect sensory neurons in the absence of their normal synaptic targets, *Dev. Biol.* **52**:121–127.

Sanes, J. R., Prescott, D. J., and Hildebrand, J. G., 1977, Cholinergic neurochemical development of normal and deafferented antennal lobes in the brain of the moth, *Manduca sexta, Brain Res.* **119**:389–402.

Schneider, I., 1964, Differentiation of larval *Drosophila* eye-antennal discs *in vitro, J. Exp. Zool.* **156**:91–104.

Schneider, I., 1966, Histology of larval eye-antennal discs and cephalic ganglia of *Drosophila* cultured *in vitro, J. Embryol. Exp. Morphol.* **15**:271–279.

Schneiderman, A. M., Matsumoto, S. G., and Hildebrand, J. G., 1982, Trans-sexually grafted antennae influence development of sexually dimorphic neurones in moth brain, *Nature (London)* **298**:844–846.

Schneiderman, A. M., Hildebrand, J. G., and Jacobs, J. J., 1983, Computer-aided morphometry of developing and mature antennal lobes in the moth *Manduca sexta, Soc. Neurosci. Abstr.* **9**:834.

Schweitzer, E. S., Sanes, J. R., and Hildebrand, J. G., 1976, Ontogeny of electroantennogram responses in the moth, *Manduca sexta, J. Insect Physiol.* **22**:955–960.

Starratt, A. N., Dahm, K. H., Allen, N., Hildebrand, J. G., Payne, T. L., and Röller, H., 1979, Bombykal, a sex pheromone of the sphinx moth *Manduca sexta, Z. Naturforsch.* **34c**:9–12.

Taylor, H. M., and Truman, J. W., 1974, Metamorphosis of the abdominal ganglia of the tobacco hornworm, *Manduca sexta, J. Comp. Physiol.* **90**:367–388.

Tolbert, L. P., and Hildebrand, J. G., 1981, Organization and synaptic ultrastructure of glomeruli in the antennal lobes of the moth *Manduca sexta:* A study using thin sections and freeze-fracture, *Proc. R. Soc. (London)* Ser. B. **213**:279–301.

Tolbert, L. P., Matsumoto, S. G., and Hildebrand, J. G., 1983, Development of synapses in the antennal lobes of the moth *Manduca sexta* during metamorphosis, *J. Neurosci.* **3**:1158–1175.

Truman, J. W., and Reiss, S. E., 1976, Dendritic reorganization of an identified motoneuron during metamorphosis of the tobacco hornworm moth, *Science* **192**:477–479.

Truman, J. W., and Schwartz, L. M., 1982, Programmed death in the nervous system of a moth, *Trends Nuerosci.* **5**:270–273.

Truman, J. W., Taghert, P. H., Copenhaver, P. F., Tublitz, N. J., and Schwartz, L. M., 1981, Eclosion hormone may control all ecdyses in insects, *Nature (London)* **291**:70–71.

9

The First Neuronal Growth Cones in Insect Embryos

Model System for Studying the Development of Neuronal Specificity

MICHAEL J. BASTIANI, SASCHA DU LAC, AND COREY S. GOODMAN

1. Introduction

Little is known about the mechanisms that generate neuronal specificity during development. Unraveling these mechanisms will require an understanding at the cellular and molecular level of how individual neurons recognize and interact with one another during development. Whereas mammalian nervous systems have an enormous number of neurons, the simpler nervous systems of invertebrates with their relatively small number of identified neurons can provide excellent model systems for studying neuronal specificity.

Ideally, the best model system would contain only a few identified neurons, each large and identifiable from birth, each connected with the others in a specific and unique way, and each accessible to the most advanced cellular, immunological, and molecular genetic techniques throughout their embryonic development. Although no such organism exists, a combination of two insects, one with accessible cells (grasshopper) and the other with accessible genes *(Drosophila)*, at their earliest stages of neuronal development, comes quite close.

Insects have relatively simple nervous systems which, in addition to their more

MICHAEL J. BASTIANI AND COREY S. GOODMAN • Department of Biological Sciences, Stanford University, Stanford, California 94305. SASCHA DU LAC • Neuroscience Program, Stanford University, Stanford, California 94305.

complex brains, include a chain of 15 or so relatively simple segmental ganglia, each containing about 1000 pairs of neurons. Unfortunately, if we wait until all of these neurons have extended processes and begun forming specific synaptic connections, then even this "simple" system appears hopelessly complex. However, if we look early enough, when only a handful of neurons have sent out processes, then the system is indeed simple enough to study neuronal specificity in its earliest forms: the specific affinities of growth cones for particular neuronal surfaces.

Neuronal specificity is largely achieved during embryonic development by a sequential series of cell recognition events. The events of cell recognition begin with the specific choices made by growth cones in terms of their selective affinity for particular axonal surfaces (selective fasciculation) and end with their specific choices in terms of their selective connectivity with particular neurons in their local neighborhood (synaptic specificity). At the early stages of neuronal development, cell recognition occurs most dramatically at the tips of growing axons, called growth cones, and at their fingerlike extensions, called filopodia.

Growth cones radiate many filopodia (approximately 0.1 μm in diameter, up to 50 μm in length) that transiently explore their environment in a cycle of extension and retraction. Many of these filopodia contact other cell surfaces. To some of these surfaces they strongly adhere and to others their adhesion is much weaker. If adhesion is weak, during the contractile phase, the filopodium is retracted; if, however, its adhesion is strong, then tension in that direction is increased during the contractile cycle and the leading tip of the growth cone advances toward the point of attachment (Bray, 1982; Letourneau, 1982).

Our results in the grasshopper embryo suggest that differential filopodial adhesion guides neuronal growth cones. The selective affinity that growth cones and their filopodia show for particular neuronal surfaces gives rise to the stereotyped patterns of selective fasciculation underlying the orthogonal scaffold of axon bundles (Figs. 1, 2). It is within the neighborhoods dictated by these axonal associations that, later in development, specific synaptic connections are made with subsets of neighboring neurons.

A detailed analysis of these early recognition events at the cellular level has been possible in the grasshopper embryo because the neurons in its developing CNS are large and highly accessible from their birth to maturation. Over the past decade, these attributes of the grasshopper embryo have permitted rapid advances in our understanding of the cellular events of neuronal development (e.g., Bate, 1976; Goodman and Spitzer, 1979; Goodman *et al.*, 1979, 1981, 1982; Bate and Grunewald, 1981; Goodman and Bate, 1981; Goodman, 1982; Goodman and Pearson, 1981; Shankland *et al.*, 1982; Taghert *et al.*, 1984; Taghert and Goodman, 1984).

1.1. The G Neuron as a Model System

Many of our previous studies have focused on the analysis and manipulation of a single identified neuron, called "G" (from the 2nd division of NB 7-4), at a single choice point in the 40% grasshopper embryo (e.g., Raper *et al.*, 1983a–c,

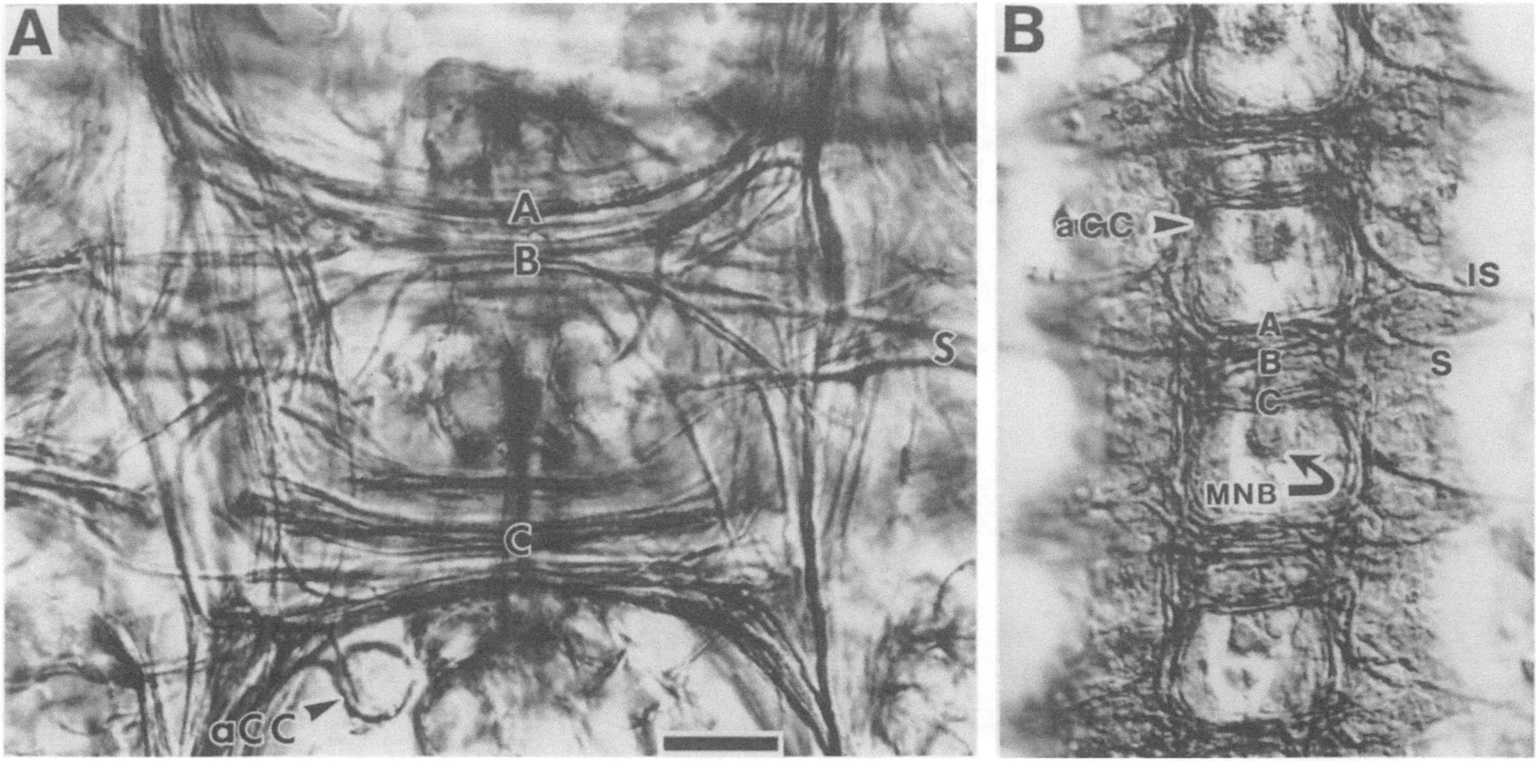

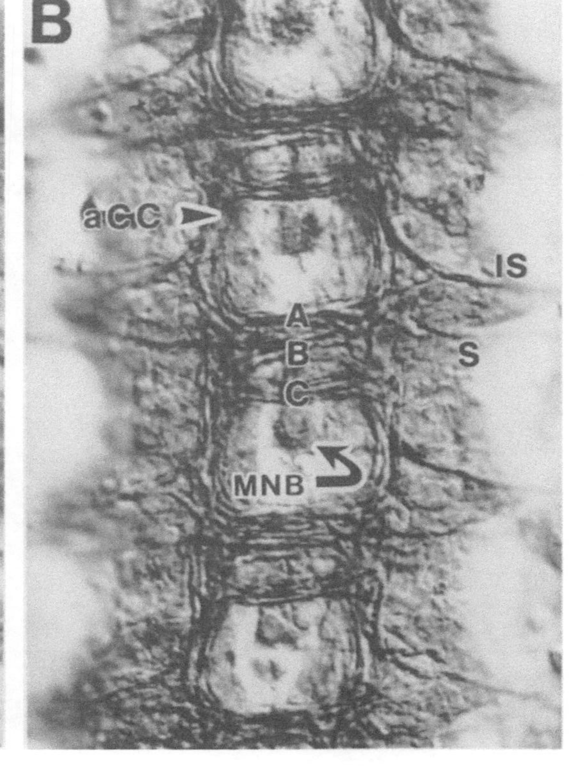

FIGURE 1. Embryonic axonal scaffold of grasshopper (A) at 40% development and *Drosophila* (B) at 12 hr stained, respectively, with the I-5 monoclonal antibody and an anti-tubulin monoclonal antibody (made by S. Lose, provided by T. Carr). A single segment which is the basic repeated unit of the scaffold is shown for grasshopper; three contiguous segments are shown for *Drosophila*. The scaffold consists of bundles of axons arranged as two bilaterally symmetrical longitudinal connectives running the length of the embryo, three lateral commissures per segment joining the two sides of the ganglion (one posterior, C, and two fused anterior, A and B), and two peripheral pathways per segment, the intersegmental nerve (IS) at the segment boundary and the segmental nerve (S) at the segment midline (IS not included here in the grasshopper). In grasshopper there are ~25 axon fascicles in each of the connectives and ~20 axon fascicles in each of the three commissures. A similar pattern is present in *Drosophila*. Later-developing neurons grow along this framework, each choosing particular axon bundles upon which to selectively fasciculate. In addition to the scaffold, a number of homologous cells can be identified in the photographs: i.e., the anterior corner cell (aCC) and the median neuroblast (MNB). Calibration bar: 20 μm for grasshopper (A) and 15 μm for *Drosophila* (B).

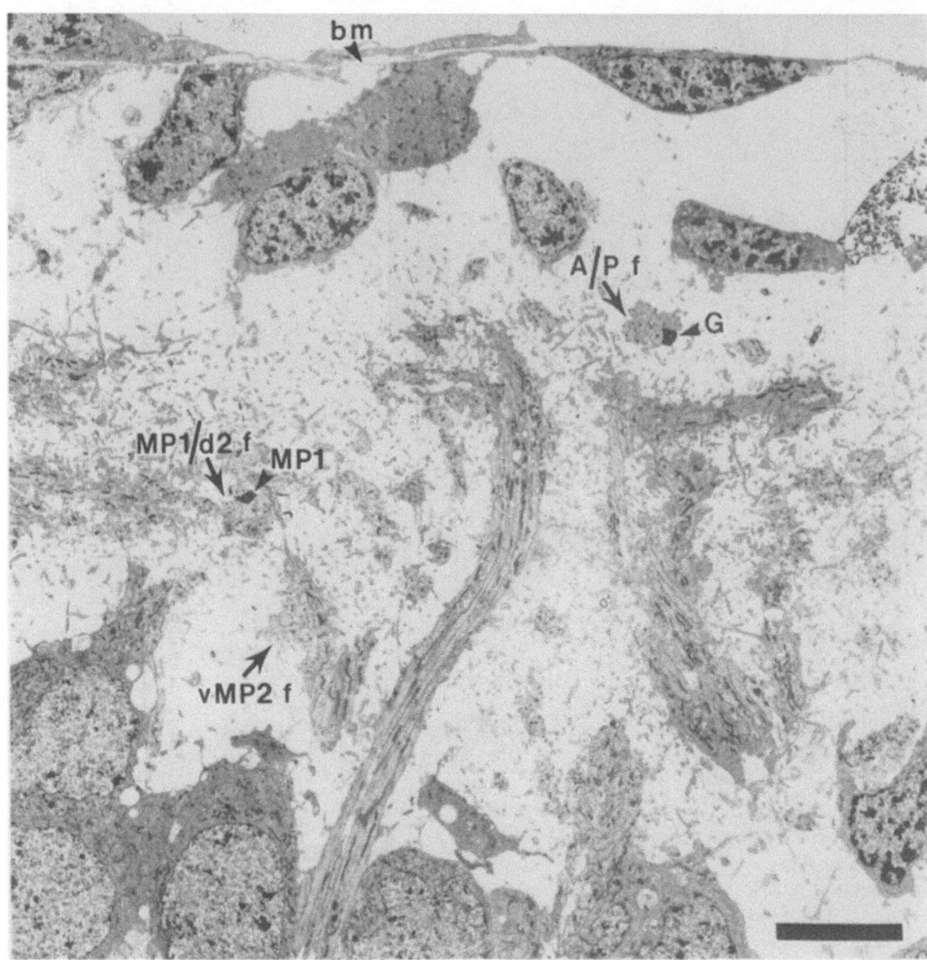

FIGURE 2. Electron micrograph montage of the embryonic neuropil at 42% of development in the mesothoracic (T2) segment. Cross sectional view taken at the level of the C commissure (as shown in Fig. 1A) showing the axon fascicles in the neuropil of the right hemisegment at the same age as the axon scaffold shown in Fig. 1A. The large profiles are axons of about 1–2 μm diameter; the abundant small profiles are filopodia of about 0.1–0.2 μm diameter that fill up much of the space around the axon bundles. There are about 25 longitudinal axon fascicles containing about 100 axons. The MP1 neuron within the MP1/dMP2 fascicle (MP1/d2 f) and the G neuron within the A/P fascicle (A/P f) were filled with HRP. Midline precursor 1 and dMP2 pioneer one of the first longitudinal axonal fascicles (MP1/d2 f) and vMP2 pioneers another (vMP2 f). The axon bundles to the left (toward the midline) of MP1/d2 fascicle represent commissural fascicles interconnecting the two hemisegments. Basement membrane, bm; neuron, N. Dorsal is at top; lateral is to the right. Scale bar, 10 μm.

1984; Bastiani *et al.,* 1984a,b; Goodman *et al.,* 1984). At this stage, the G growth cone is within filopodial grasp of ~25 different longitudinal axon fascicles (made up of ~100 different axons), and yet invariably chooses to fasciculate on the A/P fascicle (initially consisting of the A1, A2, P1, and P2 axons) (Fig. 2). Extensive ultrastructural and experimental analysis demonstrates that G is able to distinguish the two P axons from the two A axons within the A/P fascicle. Furthermore, when only the P axons are ablated, the G growth cone behaves abnormally, whereas ablation of the A axons has little affect. Thus, the degree of neuronal specificity at these early stages is exquisite: the G neuron can specifically recognize two axon surfaces (the Ps) as compared with the other ~100 axons within filopodial grasp.

Recent studies reveal that the *Drosophila* embryo has a neuron homologous to the grasshopper G neuron (Thomas *et al.*, 1984); just as in the grasshopper, the tip of the fly G growth cone associates with the P axons in the A/P fascicle. These and other results indicate that the early *Drosophila* embryo is a miniature replica of the grasshopper embryo in terms of its identified neurons, their growth cones, and their selective fasciculation choices (Fig. 1B).

1.2. A Simpler Model System

These studies bridge the gap between the accessible cells of large insects and the accessible genes of small insects, and open the way for a combined cellular and molecular genetic analysis of cell recognition during neuronal development in insect embryos. However, at this juncture, it is wise to pause and ask: which neurons compose the best model system for future studies? We began several years ago by studying the selective affinity of the G growth cone at 40% of development when only 100, rather than all 1000, neurons had extended processes. And although much has been learned by studying the G neuron, in many respects the embryonic neuropil at this stage is still very complex, and thus 100 neurons is perhaps too many for the precise cellular and molecular analysis we envision.

In this paper we describe our recent studies on an even simpler model system within this "simple system." We shift our attention to the earliest recognition events in the CNS involving less than ten neurons (Bate and Brunewald, 1981; Goodman *et al.*, 1982; Taghert *et al.*, 1982; Bastiani and Goodman, 1984a, b). The first seven growth cones in the longitudinal axon fascicles of the CNS (from the MP1, dMP2, vMP2, pCC, aCC, U1, and U2 neurons) are able to distinguish one another's surfaces, and by their specific interactions, selectively fasciculate with one another to form the first three longitudinal axon bundles (Fig. 3A,C). Many of these same cellular events occur in the *Drosophila* embryo (Thomas *et al.*, 1984). Uncovering the mechanisms underlying the specific recognition events by these $\sim 10^1$ neurons will serve as a model for the $\sim 10^3$ neurons that develop later in each insect segmental ganglion and perhaps even for the 10^{12} or so neurons that develop in more complex nervous systems.

1.3. Neuronal Precursor Cells

The ~1000 pairs of neurons in each thoracic ganglion are generated during embryogenesis from a repeated segmental pattern of neuronal precursor cells. With minor differences, each segment contains two bilaterally symmetric plates of 30 neuroblasts (NBs) (Bate, 1976), an unpaired median NB (MNB), and seven midline precursors (MPs) (Bate and Grunewald, 1981). Each NB is a stem cell, maintaining its large size as it divides repeatedly to produce a chain of smaller ganglion mother cells. Each ganglion mother cell in the chain divides once more, thus pro-

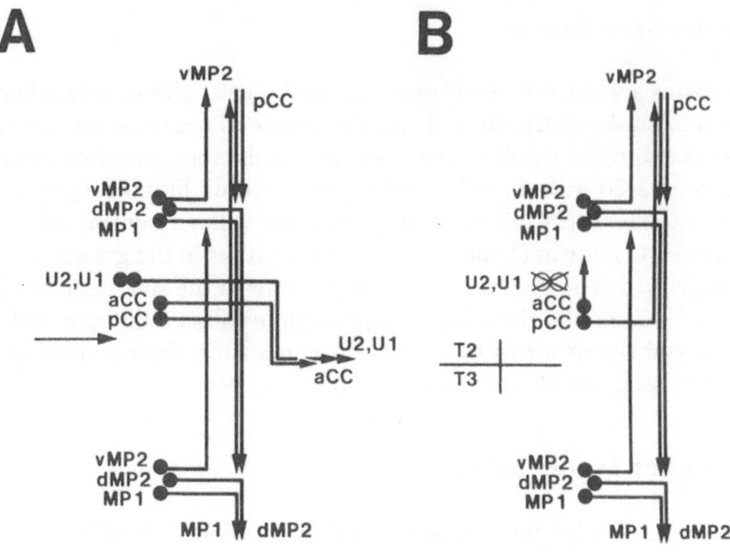

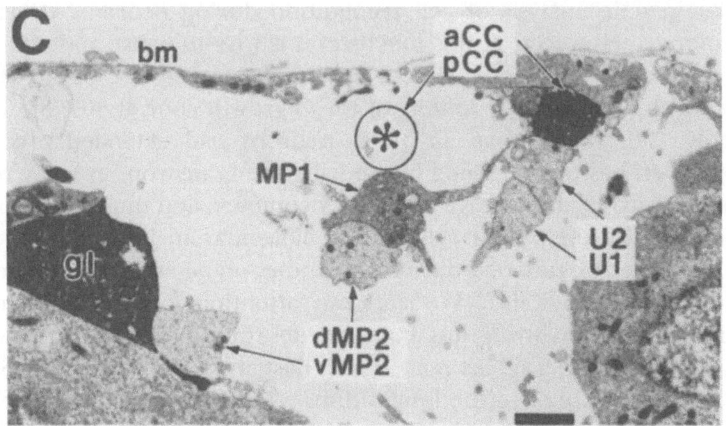

FIGURE 3. Schematic diagram and electron micrograph of the first three longitudinal axon fascicles in each hemisegment of the grasshopper embryo and the seven identified neurons whose axons fasciculate to form these three bundles. These neurons are: MP1, dMP2 and vMP2 (siblings), aCC and pCC (siblings), U1 and U2 (siblings). (A) The first three longitudinal axon fascicles (and the axons they contain) are the vMP2 fascicle (vMP2), the MP1/dMP2 fascicle (MP1, dMP2, pCC), and the U fascicle (U1, U2, aCC). The MP1 and dMP2 fasciculate as they extend posteriorly, whereas the vMP2 extends anteriorly and pioneers a more medial (and ventral) pathway; the vMP2 does not fasciculate with the MP1 and dMP2 growth cones when they meet near the segment border. The third longitudinal fascicle is pioneered by the U1 and U2 growth cones. The pCC growth cone extends anteriorly along the MP1/dMP2 fascicle; the aCC growth cone extends posteriorly along the U fascicle. The arrow in A marks the approximate location of the electron micrograph shown in C. (B) Selectively ablating the U1 and U2 cells prevents the posterior extension of the aCC without affecting the formation of the vMP2 or MP1/dMP2 fascicles nor the normal choice of the pCC to extend anteriorly along the MP1/dMP2 fascicle. (C) Electron micrograph showing the first three longitudinal fascicles (the vMP2 fascicle, the MP1/dMP2 fascicle, and the U fascicle) at 35% of development. They contain the vMP2, MP1 and dMP2, and the U1, U2, and aCC axons, respectively. The pCC axon extends anteriorly along the MP1/dMP2 fascicle and is not present in this plane of section (asterisk). The vMP2 growth cone extends anteriorly in a more ventral and medial position along a glial cell (gl) at the same time that the MP1 and dMP2 growth cones extend posteriorly in a more dorsal and lateral position initially along the dorsal basement membrane. The Us (U1 and U2) and aCC take an even more lateral and dorsal position. In this preparation, MP1, the aCC, and the glial cell (gl) have been filled with HRP. Scale bar = 1 μm.

ducing a chain of paired ganglion cells that subsequently differentiate into neurons (each NB produces from 6 to 100 progeny and then dies). The MPs divide only once to produce two neuronal progeny. In this paper, we examine the interactions of seven embryonic neurons that include descendants from MP1 (dMP1), MP2 (dMP2, vMP2), NB 1-1 (aCC, pCC), and another as yet unidentified NB (U1, U2).

2. The First Three Axon Fascicles in the Grasshopper Embryo

The first three longitudinal axon fascicles in the grasshopper embryo initially contain the axons of seven identified neurons: MP1 (one of the two progeny of MP1), dMP2 and vMP2 (the two progeny of MP2), aCC and pCC (siblings from the first division of NB 1-1), and U1 and U2 (siblings from an unidentified NB). The vMP2 fascicle contains the vMP2 axon, the MP1/dMP2 fascicle contains the MP1, dMP2, and pCC axons, and the U fascicle contains the U1, U2, and aCC axons (Fig. 3A,C; although the pCC axon is in the MP1/dMP2 fascicle, it is not present in the plane of section shown in Fig. 3C). Three pairs of siblings are included among these first seven neurons to extend growth cones in the longitudinal fascicles. One pair fasciculates together (U1 and U2), whereas the other two pairs (dMP2 and vMP2, aCC and pCC) make divergent choices in axon fascicle when confronted with this small number of embryonic neurons (Fig. 3C).

The dMP2 and vMP2 make the first divergent choice: the dMP2 (and MP1) growth cones turn posteriorly as their filopodia show a high affinity for the surface of the pCC (as compared with the aCC), while the vMP2 turns anteriorly as its filopodia show no affinity for the pCC, but rather selectively contact other neurons. This suggests that at this early stage of development, these two pairs of sibling neurons already express differential surface labels used for cell recognition. The dMP2 is different from the vMP2 as shown by its specific affinity for the pCC, and the pCC is different from the aCC as shown by its specific recognition by the MP1 and dMP2.

The pCC and aCC make the second divergent choice: the pCC extends anteriorly and fasciculates on the MP1/dMP2 axons (Fig. 3A,C), whereas the aCC growth cone waits and then extends posteriorly and laterally as it fasciculates on the later arriving U1 and U2 axons that pioneer the U fascicle (Fig. 3A,C). These two divergent choices by pairs of sibling neurons at these early stages of embryonic development offer the rare opportunity to study cell recognition during neuronal development in a truly simple system, as described below.

3. MP1 Distinguishes the Surface of the pCC from the aCC

The MP1, dMP2, and vMP2 neurons arise from midline precursors 1 (MP1) and 2 (MP2). There are two MP2s (one on each side) and one MP1 (at the midline) in each segment.) Each MP2 divides once to give rise to a ventral (vMP2) and dorsal (dMP2) daughter. The single MP1 gives rise to a pair of bilaterally symmetric daughters, each of which comes to lie dorsal to the two MP2 progeny, thus forming a trio of cells on each side of each segment (Figs. 4B, D, 5A). All three cells send

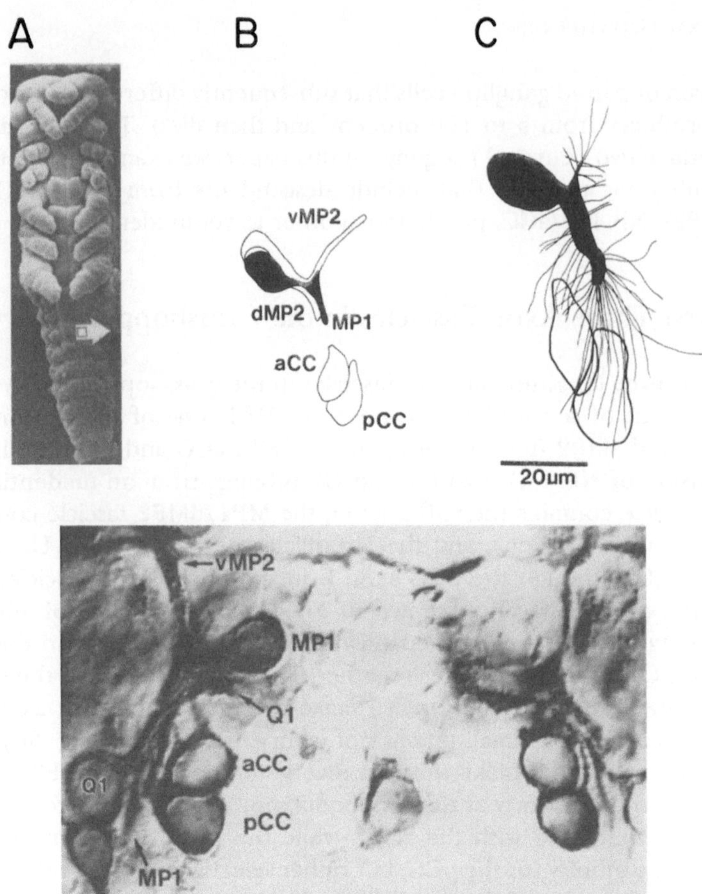

FIGURE 4. The location of the MP1, aCC, and pCC neurons in a single segment of the grasshopper embryo as shown in a scanning electron micrograph of the whole embryo (A), camera lucida drawings (B), and an I-5 MAb-stained preparation (D). (A) Scanning electron micrograph of a 32% grasshopper embryo, showing metameric arrangement of cephalic, three thoracic, and eleven abdominal segments. (B) Drawing of the cell bodies and growth cones of the identified neurons involved in pioneering the first two longitudinal axonal pathways. The vMP2 growth cone turns anteriorly while the MP1 and dMP2 growth cones turn posteriorly, pointing toward and contacting the pCC neuron. For details, see text. (C) Apparent selective filopodial adhesion from the MP1 growth cone to the pCC growth cone and cell body. Camera lucida drawing was made from cell filled with Lucifer Yellow dye, reacted with a rabbit anti-Lucifer Yellow antibody, and then a goat anti-rabbit HRP-conjugated second antibody. (D) Photograph of a whole-mount preparation of a grasshopper embryo stained with the I-5 MAb followed by HRP immunocytochemistry, showing the bilateral symmetry and spatial relationship of the cell bodies and growth cones described in this chapter. The MP1 and dMP2 growth cones are extending posterior between the aCC and pCC, and Q1. The MP1 cell body is darkly stained and in the same focal plane as its growth cone, while the dMP2 and vMP2 cell bodies are more ventral to the MP1 cell body and thus not shown in this photograph. The growth cone of the vMP2 extends anteriorly on both the left and right sides of the segment. The growth cone of the pCC can be seen extending anterior along the MP1 and dMP2 axons. Scale bar (20 μm) refers to C and D.

FIGURE 5. Selective affinity of the filopodia from the MP1 growth cone for the pCC neuron. Examples of electron micrographs taken from serial sections of the MP1 and dMP2 growth cones in the CNS of the grasshopper embryo. The growth cones of MP1 and dMP2 extend up to the dorsal basement mem-

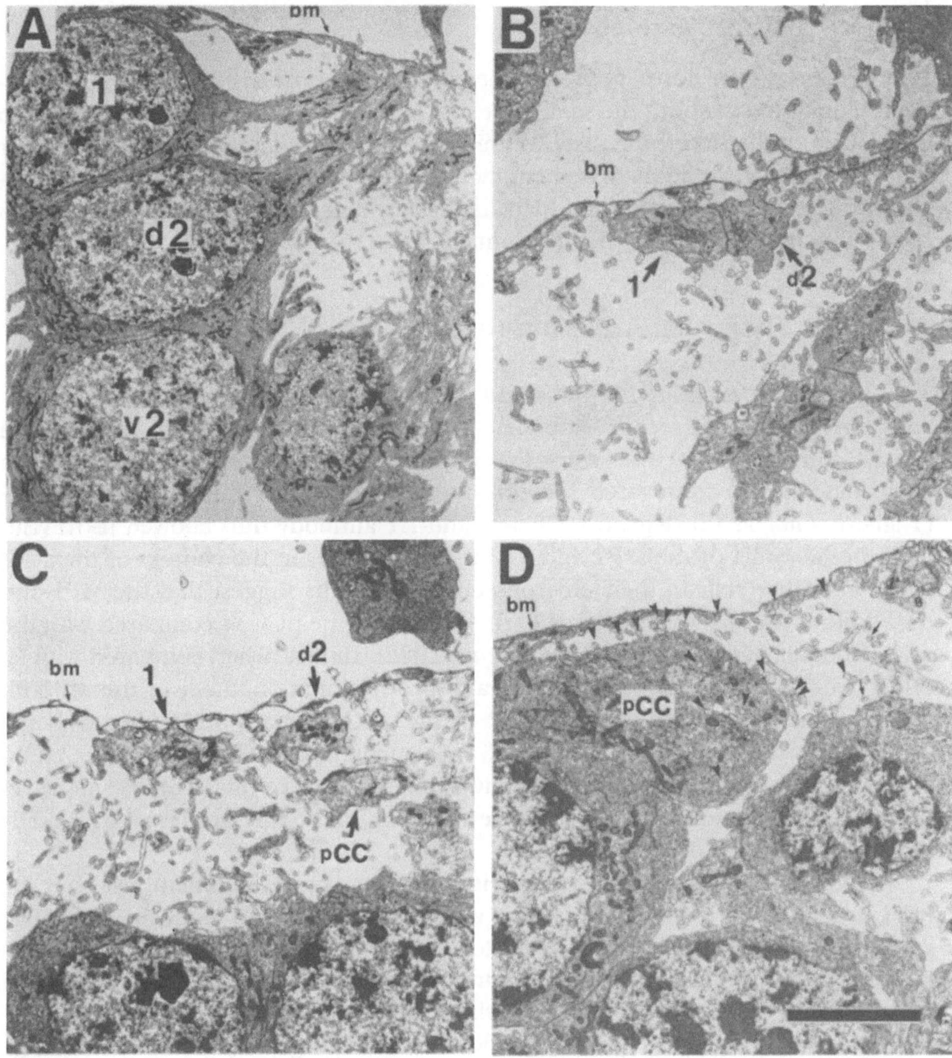

brane, turn, and grow posteriorly toward the pCC, suspended by their lamellipodia and filopodia from the basement membrane. (A) The cell bodies of the MP1 (1), dMP2 (d2), and vMP2 (v2); their axons extend to the dorsal basement membrane. Note that the axons of MP1 and dMP2 are in close apposition to each other near the basement membrane (bm). The vMP2 axon extends dorsally at a slightly more anterior position, so that only part of its axon is shown in this cross section. (B) This section is taken approximately 10 μm posterior to (A). The axons of MP1 (1) and dMP2 (d2) are closely adherent to each other. Note the lamellipodium extending dorsally to the basement membrane (bm) from dMP2 (d2). (C) This section, approximately 20 μm posterior to (A), shows the growth cones of MP1 (1), dMP2 (d2), and the pCC. The growth cones of MP1 (1) and dMP2 (d2) are not in extensive contact with each other, in contrast to (A) and (B), or with any other structure in their environment. The growth cones are suspended in this space by their filopodial contacts with other cells and the basement membrane. (D) Approximately 40 μm posterior to (A), the growth cone of the pCC is merging into its cell body. The growth cones of the MP1 and dMP2 have broken up into an array of filopodia. The filopodia from MP1 (arrowheads) are contacting the basement membrane (bm) (two), endfeet of epidermal cells (two), and the pCC (seven). Notice especially the four filopodia that are embedded within the pCC. These are filopodia that first contacted the pCC at its growth cone and have inserted deep into the cell. The dMP2 growth cone ends several microns before the MP1 growth cone and extends fewer filopodia this far posteriorly (arrows). Scale bar: 2 μm.

growth cones to the dorsal basement membrane. The growth cones reach the basement membrane at about the same time and in a variety of orientations, relative to each other and to the body axis. Within several hours, and irrespective of their initial orientation along the basement membrane, the growth cones make divergent choices: the vMP2 growth cone turns anteriorly and the dMP2 and MP1 growth cones turn posteriorly (Fig. 5B) (Goodman *et al.*, 1982).

3.1. MP1 Filopodial Adhesion to the pCC

In order to examine filopodia and their contacts with other embryonic neurons, we made rabbit serum antibodies to Lucifer Yellow (anti-LY) and visualized the dye-containing processes using this antibody in conjunction with a horseradish peroxidase (HRP)-conjugated second antibody and HRP immunocytochemistry (Taghert *et al.*, 1982) (Figs. 6A, 7). The anti-LY antibody thus allowed us to visualize the filopodia of the MP1 and dMP2 growth cones in the context of the aCC, pCC, and other cells in their environment. The results suggest that the MP1 and dMP2 filopodia selectively adhere to the surface of the pCC as compared with the aCC (Goodman *et al.*, 1982; Taghert *et al.*, 1982). In fact, when compared with all other neurons within filopodial grasp, a disproportionate number of the MP1 filopodia appear to adhere to the surface of the pCC cell body and its growth cone (which has just been initiated) (Figs. 4C, 7B). Furthermore, those filopodia that do contact the pCC tend to continue running along its surface (Fig. 7B). We interpret this selective contact to imply selective adhesion, although we have not directly measured adhesive forces.

When we examined 11 preparations at the stage of development just after the MP1 growth cone turned posteriorly, we consistently found extensive filopodial contacts from the MP1 growth cone to the pCC, suggesting selective filopodial adhesions to the surface of this cell. On average, 45% of the filopodia contacted the surface of the pCC, although this cell provides much less than 10% of the available cell surfaces. Thus, at the light level, the MP1 growth cone at this stage appears to demonstrate a strong tendency for filopodial contact with the pCC, less of a tendency with the aCC (even though this cell is closer to the MP1), and very little tendency with any of the other cells in its immediate environment. Within a short period, however, some of the MP1 filopodia begin to selectively adhere to the Q1 cell body, and later to other cell bodies along its pathway (Fig. 7C,D).

→

FIGURE 6. Whole-mount staining of the grasshopper embryo (33%) with three different antibodies (A, anti-LY; B, 1-5 MAb; C, D, Mes-3 MAb) to show the relationships of the neurons and longitudinal axon fascicles described in this chapter. (A) The MP1 neuron was filled with the fluorescent dye Lucifer Yellow. The embryo was then fixed and processed for HRP immunocytochemistry using the anti-LY antibody. The dye spread to the dMP2 and vMP2 neurons and then via the MP1 and dMP2 growth cones to the pCC. The secondary dye spread to the sibling aCC is thought to be through the pCC and not directly from the MPs. Notice that the growth cones are separated as shown in Fig. 5C. (B) 35% grasshopper embryo stained with the I-5 MAb. The pCC growth cone extends anteriorly on the MP1/dMP2 axons. The growth cone of the vMP2 neuron from the next posterior segment is extending anteriorly medial to the MP1/dMP2 fascicle and lateral to the pCC cell body. The aCC growth cone is still pointing anterior. The U growth cones have just reached the dorsal basement membrane, lateral to the

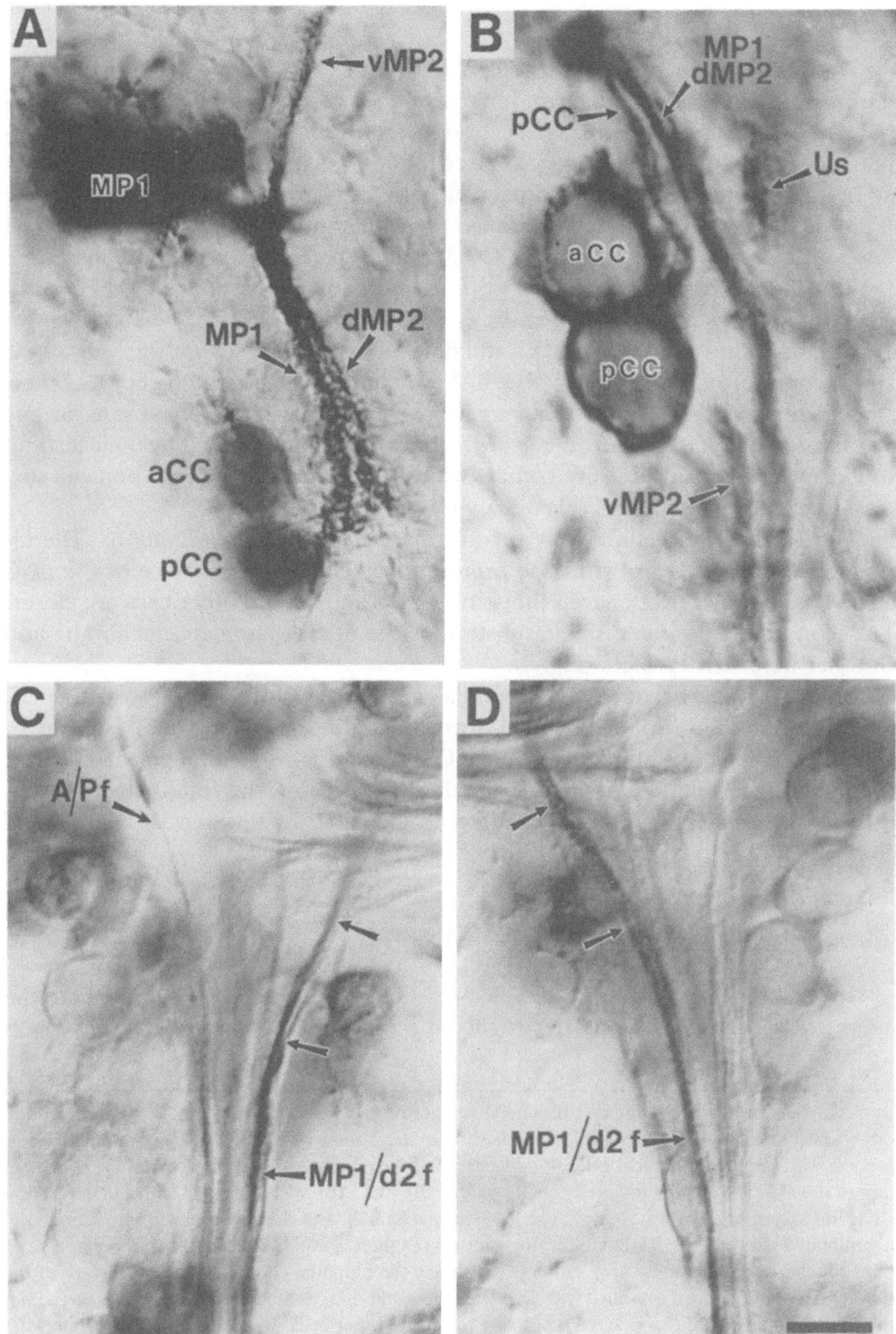

MP1/dMP2 pathway, and are about to turn posterior. (C, D) 38% grasshopper embryo stained with the Mes-3 MAb which recognizes a cell surface antigen on a small subset of neurons whose axons fasciculate. Many longitudinal axon bundles are present by this stage (e.g., A/P fascicle has formed in its lateral location), but only the MP1/dMP2 fascicle (MP1/d2 f) is stained by the Mes-3 MAb. C shows the left connective and D the right connective from the same segment showing the symmetry and specificity of the Mews-3 MAb staining. Scale bar: A, C, D, 15 μm; B, 10 μm.

We next examined the specific contacts of the MP1 filopodia with the pCC neuron by a transmission electron microscope serial section reconstruction (Bastiani and Goodman, 1984a,b). We reconstructed the MP1 growth cone and its filopodia at a developmental stage just after the MP1 growth cone turns posteriorly and before it comes into direct contact with the pCC growth cone. Approximately 60 μm of serial thin sections were taken from the anterior edge of the MP1 cell body to the posterior edge of the pCC cell body. The cell bodies and axons of the MP1, dMP2, vMP2, aCC, and pCC were identified in thin sections by their characteristic positions and shapes (Fig. 5).

Although the axons of the MP1 and dMP2 fasciculate together (Fig. 5B), their growth cones often do not (Figs. 5C, 6A). Furthermore, the leading edge of their growth cones need not be in contact with the dorsal basement membrane or any cell surface (Fig. 5C). Rather, the growth cones appear suspended by their leading filopodia which are in extensive contact with the basement membrane and cell surfaces, in particular, the pCC (Figs. 5D, 8).

The TEM serial section analysis confirms the light level observations. The filopodia from the MP1 growth cone preferentially contact the surface of the pCC cell body and growth cone, even though the aCC and several other cells are closer. The second most desirable single substrate is the dorsal basement membrane and not other neuronal surfaces (Fig. 8).

Thus, it appears that MP1's filopodia can distinguish the surface of the pCC from all other cells within filopodial grasp (including its sibling aCC), and in so doing show a selective affinity for the pCC surface (Fig. 8). Although the electron micrographs present a static picture of a dynamic process, the results strongly suggest that this selective affinity is likely to be mediated by differential adhesion of the MP1's filopodia to the pCC.

3.2. MP1 Filopodial Insertion Into the pCC

In the process of studying the selective adhesion of the MP1's filopodia to the pCC cell, we discovered another highly specific interaction between these cells

FIGURE 7. Selective dye coupling and filopodial contact revealed with Lucifer Yellow (A) and the antibody to Lucifer Yellow (B–D). (A) Filling MP1 with Lucifer Yellow reveals selective dye coupling via its filopodia with a small number of cells along its route; the coupling is direct to pCC and Q1. The coupling to the aCC is probably indirectly via its sibling, the pCC. (B) MP1 was filled with Lucifer Yellow just as in (A), but at a slightly younger age (33% versus 33.5%), and then treated with the rabbit anti-LY antibody followed by HRP immunocytochemistry (Taghert *et al.*, 1982). At this younger age, MP1 is dye coupled via its filopodia only to the pCC. Notice the filopodia coursing over the surface of the pCC. (C) The MP1 neuron was filled with Lucifer Yellow and then treated as in (B). Dye has spread not only to dMP2 and vMP2 (open arrow), but also to several other cells. In the posterior direction MP1 and dMP2 remained coupled to the pCC, Q1, and Q2 (arrow next to Q1; coupling to Q2 may be indirect via its sibling Q1). Nearer the MP1 growth cone at the segment border can be seen another cell weakly dye coupled (arrow in D). In the anterior direction, the vMP2 is coupled to the lateral landmark cell (LLC) even though its growth cone has entered the next anterior segment. (D) A higher magnification of the MP1 growth cone and the weakly dye coupled cell shown in (C). Notice that there seems to be a parallel array of filopodia extending posteriorly from the tip of the growth cone along the pathway that the growth cone will take. Scale bar: A, B, 10 μm; C, 50 μm; D, 15 μm.

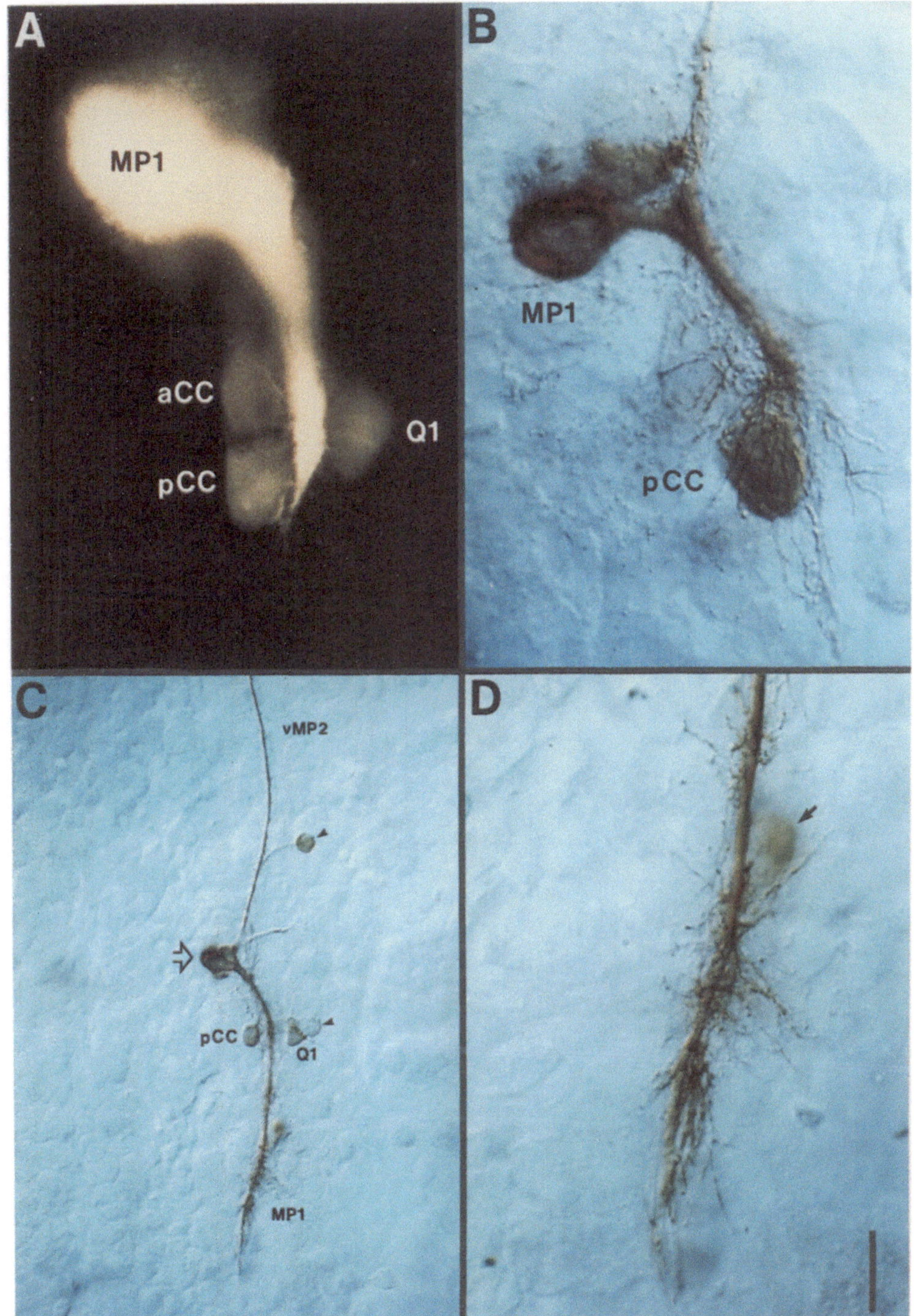
A
MP1
aCC
Q1
pCC
B
MP1
pCC
C
vMP2
pCC
Q1
MP1
D

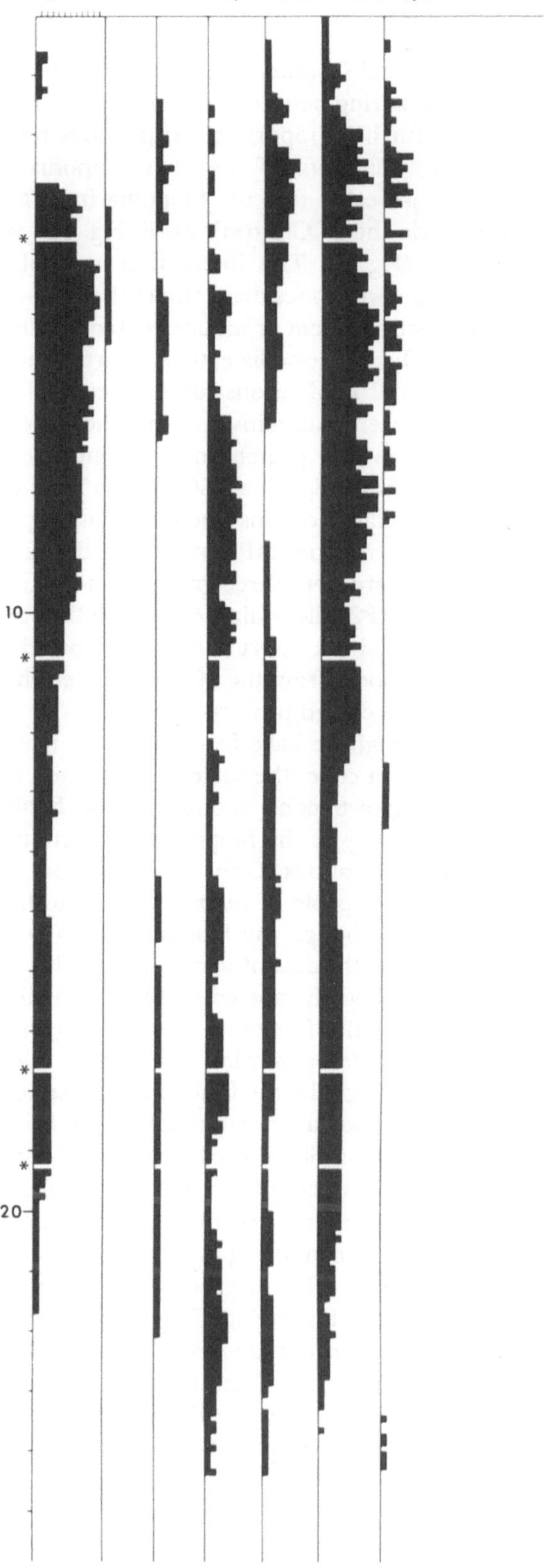

FIGURE 8. Selective affinity of the MP1 filopodia for the pCC neuron. Schematic diagram of a serial section reconstruction of all the contacts of the MP1 filopodia extending from the anterior 4 μm of its growth cone. Serial sections were taken from the anterior edge of the MP1 cell body to the posterior edge of the pCC cell body, about 60 μm, at a stage of development just after the MP1 growth cone had turned posterior, but before its growth cone had directly contacted the pCC (examples of sections shown in Fig. 5). All the filopodia, 28, from the distal 4 μm of the MP1 growth cone were reconstructed and their contacts with the environment identified. The most significant comparison is between the filopodial contacts with the aCC and the pCC. Even though the MP1 growth cone has equal or better access to the aCC, many fewer filopodia contact the aCC than the pCC. Few filopodial contacts are made with the dMP2, but this is not surprising because the dMP2 growth cone ended more anteriorly and is following the MP1 growth cone. The MP1 filopodia also contacted many different epidermal cells (epi) and unidentified filopodia (X filo) from other growth cones and epidermal cells. However, the next preferred single substrate of the filopodia after the pCC was the dorsal basement membrane (BM). There were relatively few places along the filopodia where they were not in contact with any substrate (Space). The vertical axis is distance in microns and the horizontal axis is number of filopodia. The smallest division is the contact of one filopodium for a distance of 0.1 μm. The asterisks represent sections that were lost.

(Bastiani and Goodman, 1984a,b). We have recently reported on similar interactions occurring between other identified neurons (Bastiani *et al.*, 1984b). In addition to further supporting our previous results on the extensive contact (and presumably adhesion) of the MP1 filopodia with the pCC cell surface, the TEM analysis revealed that the filopodia from the MP1 growth cone selectively insert deep into the pCC growth cone (Fig. 5D, 9) and induce the formation of coated vesicles (Fig. 9). This interaction is highly specific, since filopodia from other nearby growth cones that contact the surface of the MP1 and pCC growth cone do not penetrate them or induce coated vesicles.

Of the 28 filopodia extending from the leading 4 μm of the MP1 growth cone in our serial TEM reconstruction, seven filopodia contacted the pCC at its growth cone and were found inserted anywhere from 0.1 to 7 μm into the cell. Coated pits and vesicles were present in the membrane of the pCC just adjacent to the tips of the MP1 filopodia. The pCC growth cone in this reconstruction was riddled with ten filopodial insertions (the section in Fig. 9 shows five of these insertions): seven of these were from MP1, two from dMP2, and one could not be followed to its source. There were three growth cones closer to the pCC than those of either the MP1 or dMP2. Filopodia from these other growth cones contacted the pCC growth cone, yet did not insert and induce coated pits. The aCC lies just anterior to the pCC. Filopodia from the MP1 contacted the aCC growth cone, but did not insert and induce coated pits.

Although we have focused on the interaction of the MP1 filopodia with the pCC growth cone, the same phenomenon occurs between the pCC filopodia and the MP1 growth cone, and reciprocally between the MP1 and dMP2 growth cones. The specificity of the filopodial interactions and induction of coated vesicles has several implications. Only filopodia from particular cells (in this case MP1 and dMP2) are capable of inserting into another identified growth cone (in this case pCC), even though the filopodia of many other growth cones contact their surfaces. Just as the data of selective filopodial adhesion suggest that different neurons have differentially labeled surfaces that filopodia can distinguish among, so the data on selective filopodial insertion suggest that the filopodia from different neurons have differentially labeled surfaces that growth cones can distinguish among.

What might be the function of these selective filopodial insertions? This specificity of filopodial insertion appears to be identical to the specificity of filopodial adhesion that is likely to mediate growth cone guidance. One possible function of these insertions may be that the numerous adhesive junctions around the filopodial shank allow the inserted filopodium to mediate increased tension relative to surface filopodia contacting the same cell.

However, these insertions may function for cell communication rather than simply for cell anchoring. Growth cones typically make a series of choices on the way to their appropriate target. The selective adhesion of filopodia is likely to be a dynamic process such that the adhesive properties of a growth cone change during the course of these navigations. Interactions involved in one choice point might induce the cell to change its expression of cell surface molecules involved in (1) the filopodial adhesion at that choice point, or (2) the filopodial adhesion at a subsequent choice point. The specific events of insertion and induction of coated vesicles might mediate changes in a cell's expression by signaling to the cell's biochemical machinery via receptor-mediated endocytosis.

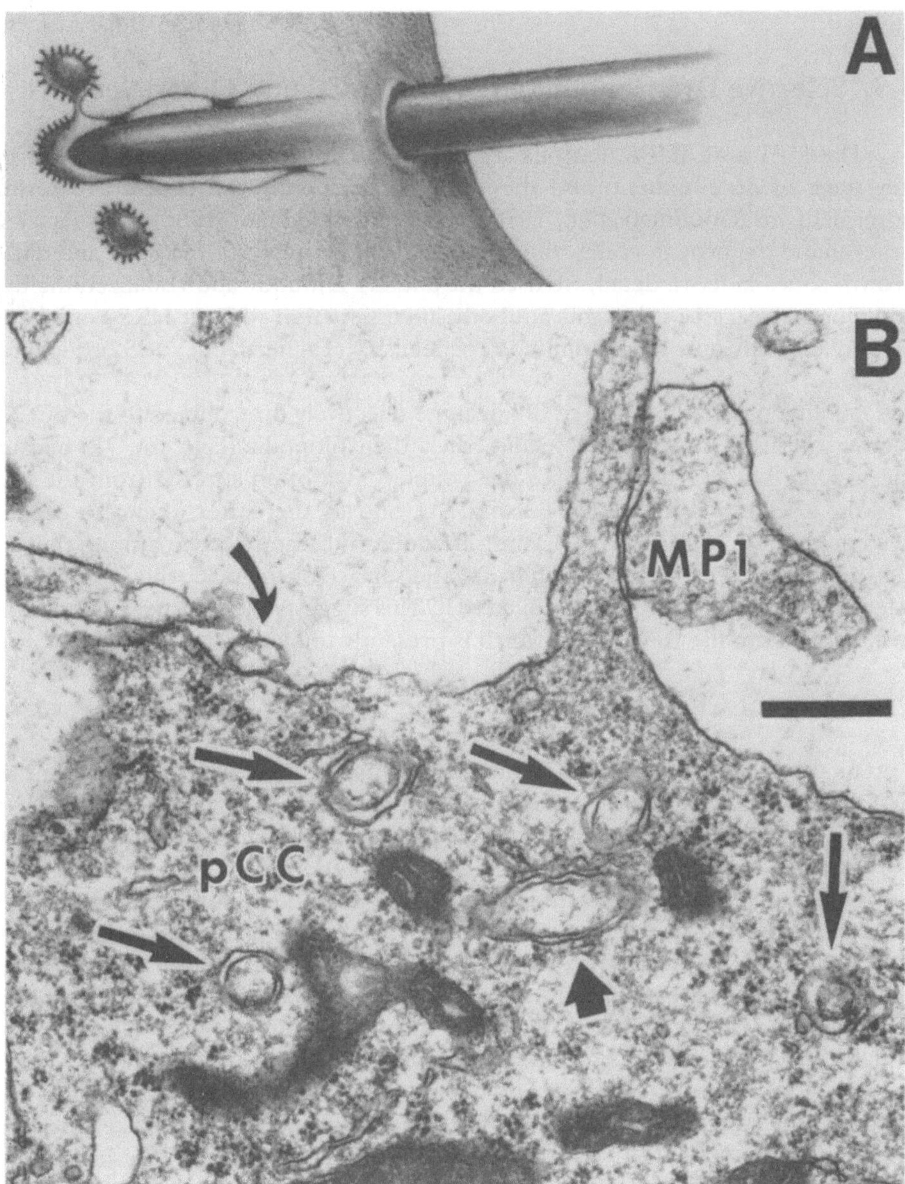

FIGURE 9. Selective filpodial insertion. (A) Artists drawing illustrating many of the important features of filopodial insertion and the induction of coated vesicles. The induction of coated vesicle formation is localized to the membrane just opposite the tip of the filopodium. The tip of the filopodium is morphologically distinct from its sides, with a denser matrix on both the extracellular and cytoplasmic side of its membrane. Punctate junctions are seen between the filopodium and growth cone along the sides of the insertion. The insertion process is probably active; the filopodium grows forward at the same time that junctions are forming along the sides and the growth cone is advancing. Filopodia from other cells contact the growth cone, but do not insert, again suggesting that insertion is not simply a passive engulfment of filopodia by the advancing growth cone. (B) Five filopodial insertions are seen in this cross section of a portion of the pCC growth cone. Four of the insertions are from the MP1 (small arrows) and one from dMP2 (the large arrow). The filopodium on the surface of the growth cone (curved arrow) is from MP1. This filopodium is in very close apposition to the growth cone membrane and a few microns distally inserts into the pCC growth cone and induces coated pits and vesicles. The larger profile (MP1) is the terminal filum of the MP1 growth cone. Scale bar: 0.5 μm.

3.3. Selective Dye Coupling by MP1 to pCC and Other Neurons

The MP1 and dMP2 neurons are highly dye coupled to one another during this stage of development, and they become transiently dye coupled to several other neurons (Goodman *et al.*, 1982; Raper and Goodman, 1982) (e.g., Fig. 7A). To examine the growth cones, filopodia, and dye coupling of the MP1 and dMP2 growth cones in more detail, we used the anti-LY antibody and visualized the dye-containing processes using this antibody in conjunction with an HRP-conjugated second antibody and HRP immunocytochemistry (Taghert *et al.*, 1982) (Figs. 6A, 7B–D).

Midline precursor 1 and dMP2 become selectively dye coupled to the pCC; at the outset this dye coupling is mediated via their filopodia (Figs. 6A, 7B; because the pCC and aCC are themselves dye coupled, dye often spreads from the pCC into the aCC as shown in Figs. 6A and 7A). Shortly thereafter, they also become dye coupled to Q1 (Fig. 7A,C). Thus, in addition to the selective affinity that the MP1 (and dMP2) growth cones show for the surface of the pCC, they also become selectively and transiently dye coupled to it. This suggests some form of cytoplasmic communication, the function of which is presently unknown.

4. Divergent Choices by the pCC and aCC Growth Cones

As described above, the pCC and aCC neurons are siblings arising from the first division of NB 1-1 in the next posterior segment; they migrate anteriorly to the location where they finally reside. By the time they migrate across the segment border, they already begin displaying cell-specific behavior and presumably cell-specific surface labels. The leading anterior edge of the pCC extends around the lateral edge of the aCC and points directly toward the MP1 and dMP2 growth cones. That the two cells have differentially labeled cell surfaces by this time is suggested by the differential contact of the MP1 filopodia with the pCC cell body as compared with the aCC cell body (Fig. 8), even though the MP1 filopodia have equal or even better access to the aCC surface (Fig. 4).

As the two corner cells cease their migration and take up residence just posterior to the MP1 and MP2 progeny, their growth cones then make divergent choices when confronted with the same environment, ultimately fasciculating in different axon bundles. The pCC growth cone extends anteriorly as it fasciculates with the MP1 and dMP2 axons. In contrast, the aCC growth cone continues to point anteriorly, and only after 10–15 hr does it extend posteriorly and then laterally as it fasciculates with the U1 and U2 axons, as described below.

4.1. pCC Selective Fasciculation onto MP1/dMP2 Fascicle

While the corner cells are migrating anteriorly, both cells have anteriorly directed leading edges that resemble growth cones. About the time their migration comes to an end, the leading edge of the pCC continues to extend anteriorly as a growth cone around the lateral edge of the aCC. This pCC growth cone fasciculates with the MP1 and dMP2 growth cones and extends anteriorly along their surface (Figs. 3A, 6B, 10A, 11A). The aCC growth cone, in contrast, remains relatively

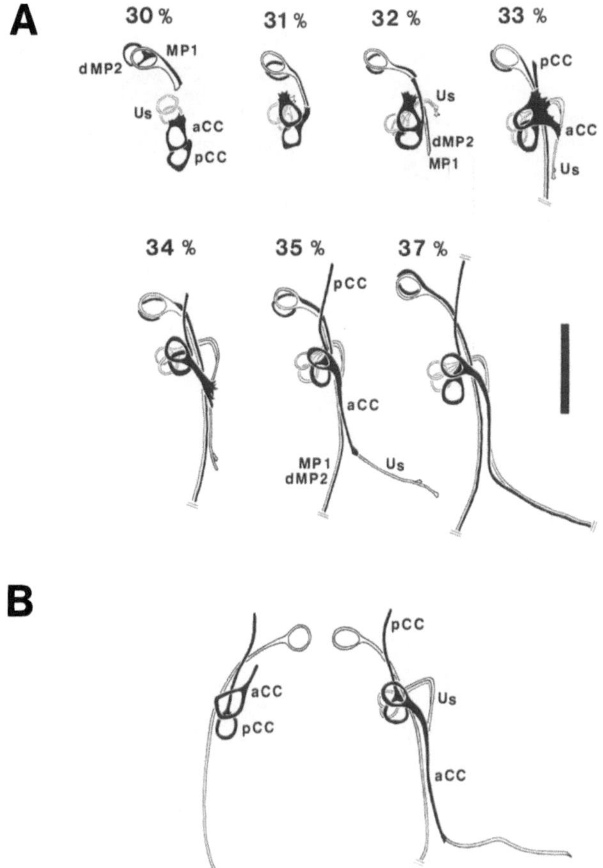

FIGURE 10. Development of the pCC and aCC neurons in normal and experimental embryos, in relationship to the MP1, dMP2, U1, and U2 axons. (A) Camera lucida drawings of MP1, dMP2, pCC, aCC, U1, and U2 neurons stained with the I-5 MAb between 30% and 37% of development, as they fasciculate in two longitudinal bundles (MP1/dMP2 and U fascicles). The MP1 and dMP2 axons elongate posteriorly, establishing the first longitudinal axon fascicle in this region of the neuropil. By about 31%, the tips of the MP1 and dMP2 growth cones reach the anteriorly migrating cell bodies of the aCC and pCC, in particular pointing at the anterior tip of the pCC growth cone. As the aCC and pCC neurons reach their final cell body position at 32%, the MP1/dMP2 axons elongate posteriorly past them. At this stage, the pCC growth cone extends anteriorly, fasciculating with the MP1/dMP2 axons. Its sibling, the aCC, does not extend along the axons taken by the pCC, but rather waits, its growth cone pointing anteriorly and remaining relatively stationary. However, at about 33%, the axons of the U1 and U2 neurons, whose cell bodies are located ventrally, arrive lateral to the MP1/dMP2 fascicle. As the U1 and U2 growth cones extend posteriorly and establish the U fascicle, the aCC growth cone swings laterally, crossing over the MP1/dMP2 fascicle to follow the U axons. At 35% the U axons turn laterally, pioneering the intersegmental nerve. The aCC growth cone follows the same pathway laterally. (B) Experimental test of the labeled pathways hypothesis. Camera lucida drawing of I-5 MAb–stained embryo showing an experiment in which the U cell bodies in the left hemisegment were ablated with a laser microbeam at 30% and the embryo cultured for 48 hr. The aCC in the control segment extended posteriorly upon the U fascicle as in normal, uncultured embryos. In the absence of the U axons, the aCC growth cone did not fasciculate with other axons, but rather remained pointing anteriorly. Scale bar: 50 μm.

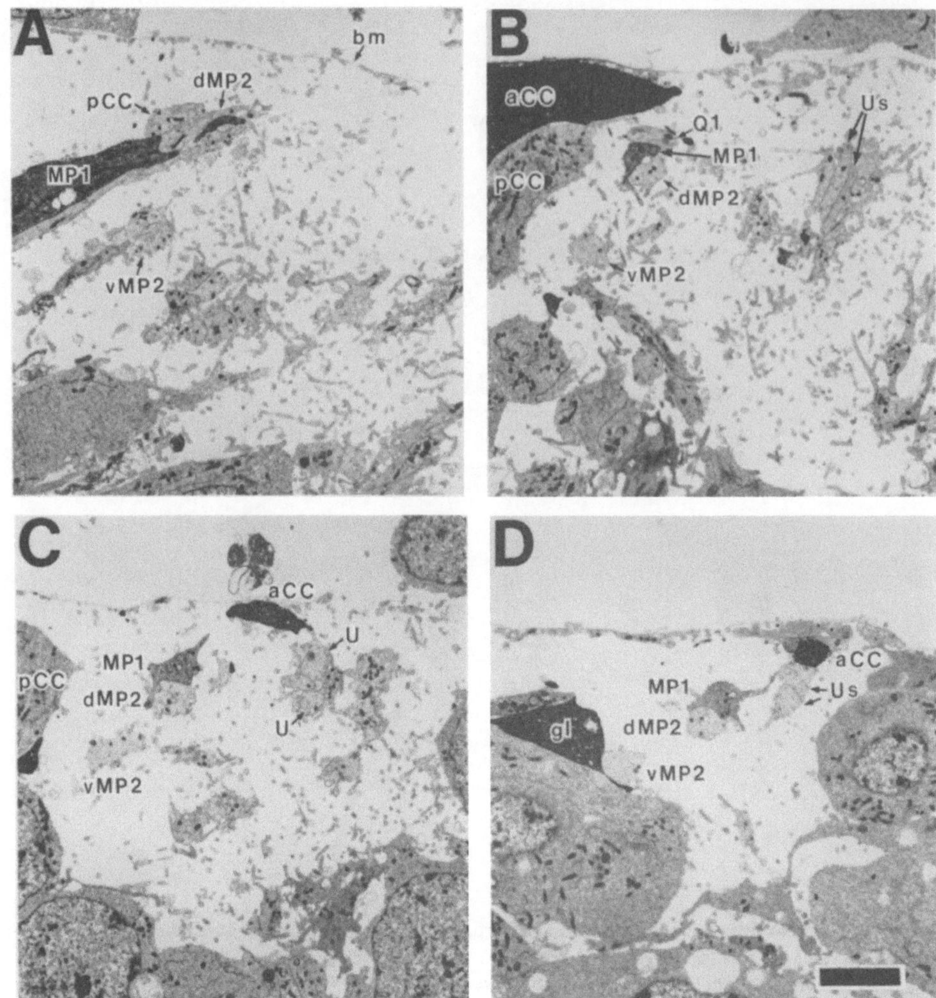

FIGURE 11. Selective affinity of the pCC growth cone for the MP1/dMP2 axons and of the aCC growth cone for the U axons, as revealed by HRP injections and semiserial section electron microscopy. An identified glial cell (gl), MP1, and the aCC have been filled with HRP. The light micrograph (E) shows the relationship of these cells in the whole-mount preparation from which these sections were cut (B, C, D show level of sections). (A) Section at the level of the MP1 cell body shows the axon of the filled MP1 extending dorsolaterally from its cell body with the axon of the dMP2 to form the MP1/dMP2 fascicle. At this level the pCC has just joined the fascicle. The vMP2 fascicle can be seen just ventral and medial. At this stage and level there are only two axons in this bundle; the vMP2 from the next posterior segment (labeled) and the vMP2 from the same segment (unlabeled). (B) Section through the cell body of the filled aCC shows its axon extending laterally over the MP1/dMP2 fascicle toward the two U cells. the pCC extends ventrolaterally around the aCC to join the MP1/dMP2 fascicle as shown in (A). The axon of Q1 contacts MP1 as it crosses to form one of the first fascicles in the C commissure. The vMP2 fascicle contains only the vMP2 axon at this level. (C) At the level of the pCC cell body, the aCC is still extending along the dorsal basement membrane (bm in A) toward the U fascicle. The ventromedial vMP2 fascicle contains only the vMP2 axon. The centrally located MP1/dMP2 fascicle contains only the filled axon of the MP1 and the unfilled axon of the dMP2. Note the densely filled profile of the glial cell just ventral to the pCC cell. (D) Section through the longitudinal connective shows only

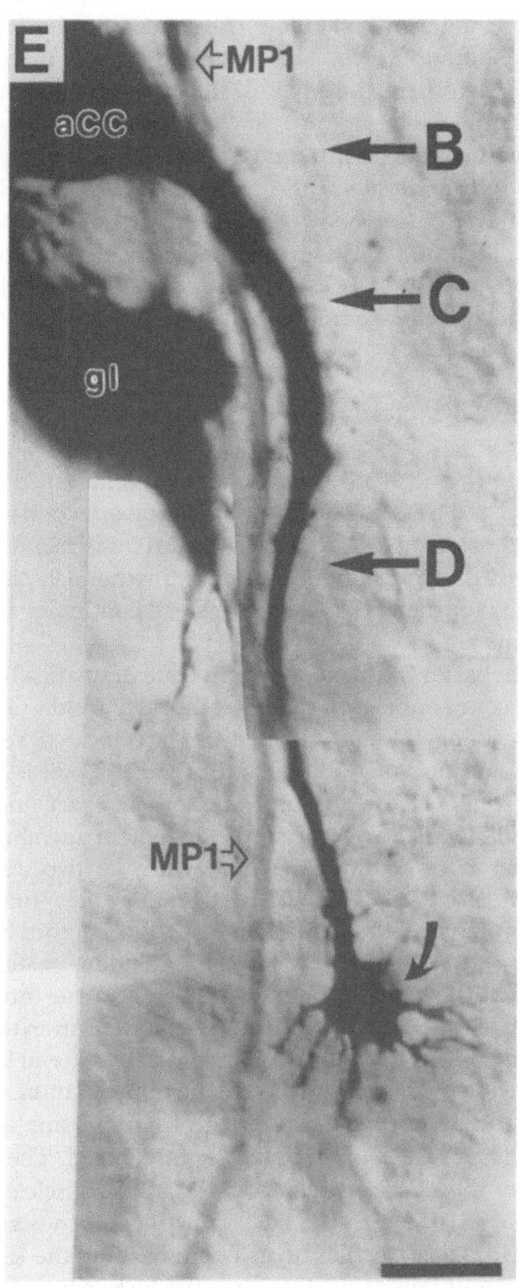

the three longitudinal fascicles (the vMP2, MP1/dMP2, and the U fascicles). The aCC fasciculates with the Us. The vMP2 is in close apposition to the filled profile of the glial cell. (E) Dorsal view of the whole-mount preparation from which the sections in A–D were taken, showing the MP1, aCC, and glial cell filled with HRP (B–D note locations of sections shown in those panels). At this stage (35%), the aCC growth cone has just reached the segment border where it will turn laterally to form the intersegmental nerve with the U axons. Anterior at top, lateral at right. Scale bar: A–D, 2 μm; E, 10 μm.

stationary and points anteriorly, its filopodia contacting the MP1/dMP2 fascicle, but showing no high affinity for it. Thus, confronted with the same environment, the MP1 and dMP2 axons in the MP1/dMP2 fascicle, the two sibling neurons, behave quite differently.

The pCC growth cone extends about 20 μm anteriorly along the MP1 axon and then stops abruptly and waits at the anterior edge of the MP1 axon, just anterior and lateral to the MP1 cell body (Fig. 10A). Within a few percent of development, however, the MP1 and dMP2 growth cones from the next anterior segment come within filopodial grasp of pCC growth cone. Only then does the pCC once again extend anteriorly as it fasciculates with the MP1 and dMP2 axons (Fig. 3A).

4.2. aCC Selective Fasciculation onto U Fascicle

The vMP2 from the next posterior segment extends anteriorly along the lateral side of the pCC and aCC cell bodies, pioneering a second axon fascicle that is located ventral, and often medial, to the MP1/dMP2 fascicle (Figs. 3A,C, 6B, 11). Just as the aCC growth cone shows no affinity for the MP1/dMP2 fascicle, so it also shows no affinity for the vMP2 fascicle, instead continuing to remain relatively stationary as it points anterior.

However, the behavior of the aCC growth cone dramatically changes when the U1 and U2 growth cones appear on the dorsal surface within filopodial grasp. The cell bodies of the sibling U1 and U2 neurons are located ventral, and slightly medial, to the final location of the aCC and pCC cell bodies. Although we know that the Us arise from a NB, we do not know their exact lineage. They extend growth cones dorsolaterally, arriving at the basement membrane lateral to the MP1/dMP2 fascicle. They do not fasciculate with the two existing longitudinal axon fascicles (vMP2 and MP1/dMP2), but rather they turn posteriorly and pioneer a third axon fascicle, the U fascicle, just a few microns lateral to the MP1/dMP2 fascicle (Figs. 3A,C, 6B, 10A, 11). After extending posteriorly, the Us turn laterally once again at the segment boundary to pioneer the intersegmental nerve.

Once the Us reach the basement membrane and begin extending posteriorly, the aCC growth cone begins to change its direction and extend laterally toward the U axons. This behavior of the aCC growth cone appears to be directly correlated with the arrival of the U axons, since although there is some temporal variability in when the Us arrive, these two events are tightly linked. The aCC growth cone extends directly over the vMP2 fascicle and MP1/dMP2 fascicle to reach the more lateral U fascicle (Figs. 10A, 11). The aCC growth cone continues to extend posteriorly along the U fascicle; it then turns laterally along the same pathway taken by the Us at the segment border as they form the intersegmental nerve.

This completes the initial cell recognition events by the first seven growth cones as they fasciculate to form the first three longitudinal axon fascicles as shown in Fig. 3. The sibling vMP2 and dMP2 growth cones choose different fascicles as do the sibling pCC and aCC growth cones; in contrast, the sibling U1 and U2 growth cones remain together (Fig. 3C).

4.3. Experimental Test of Labeled Pathways Hypothesis

The previous results on the G growth cone and the A/P fascicle as described in the introduction led us to propose the labeled pathways hypothesis (Goodman *et al.*, 1982; Raper *et al.*, 1983b; Bastiani *et al.*, 1984b). We experimentally tested the hypothesis by specifically ablating the axons within the A/P fascicle and examining the effects of these ablations on the behavior of the G growth cone (Raper *et al.*, 1984b). These experimental results confirmed the hypothesis and suggested that the neuropil was subdivided into many labeled axonal pathways (Goodman *et al.*, 1984).

This same hypothesis applies to the results we have presented here involving the development of the first three longitudinal axon fascicles by the first seven growth cones. We propose that these first three fascicles are differentially labeled on their cell surfaces, and that later growth cones are differentially determined in their ability to make specific choices of which labeled pathway to follow. In particular, the results suggest that the pCC and aCC neurons are differentially determined to follow different labeled pathways, the MP1/dMP2 fascicle and the U fascicle, respectively.

To test this hypothesis, we ablated the U1 and U2 neurons with a laser microbeam before they extended growth cones and before the corner cells had migrated to their final position. If the aCC growth cone turns laterally and then posteriorly along a stereotyped pathway because of its specific contact with the U axons, then the ablation of the Us should affect its behavior. The results, as described below, support this hypothesis.

For these experiments, the embryos were cultured outside of their eggshell, embryonic membranes, and yolk in the RPMI 1640 (Gibco)-based culture medium supplemented with insect hormones (Raper *et al.*, 1984a,b; Taghert *et al.*, 1984). Each manipulated embryo had its own internal control, since the aCC neuron on one side faced a perturbed environment in which the Us had been ablated, while the contralateral aCC in the same segment faced a control environment. The results of sham ablations were axon growth equivalent to controls.

Figure 10B shows the morphology of the control and experimental aCC after ~5% of development in culture following ablation of the Us. The control aCC extended along its normal posterior pathway. However, the growth cone of the experimental aCC continued to point anteriorly without choosing any particular pathway. Another internal control is provided by the pCC neuron. On both the control and experimental sides, the pCC extended along its normal anterior pathway.

These results thus provide further support for the labeled pathways hypothesis, and suggest that early in development the first three longitudinal axon fascicles are differentially labeled on their surfaces. In the absence of the Us, the aCC growth cone appears uninterested in the other four axons (vMP2, MP1, dMP2, pCC) and instead continues to point anteriorly (Fig. 3B). This example of exquisite specificity, in addition to the example of the G growth cone and A/P fascicle, has convinced us that many different molecules are differentially expressed on the surfaces of these early embryonic axon fascicles, or subsets of axons within them, and that they guide growing neurons to their appropriate targets by selective adhesion of their filopodia to these labeled axonal pathways.

4.4. Monoclonal Antibody Correlate of the Hypothesis

Given the specificity that the aCC growth cone shows for the U fascicle and that the G growth cone shows for the A/P fascicle, it seems reasonable to hypothesize that each of the ~25 longitudinal axon fascicles at 40%, and perhaps subsets of axons within the fascicles, are differentially labeled. Cell recognition during neuronal development is thus likely to involve the temporal and spatial expression of many different molecules.

Monoclonal antibodies (MAbs) have been generated which reveal cell surface antigens whose temporal and spatial expression correlate with these predictions, namely, neurons whose axons fasciculate together share common surface antigens (Kotrla and Goodman, 1984). The Mes-3 and Mes-4 MAbs recognize surface antigens that distinguish a single longitudinal axon fascicle from all other fascicles in the 40% axonal scaffold (Fig. 6C,D). Both MAbs recognize only the MP1/dMP2 fascicle, and thus demonstrate that an individual fascicle is antigenically distinct from all other fascicles within filopodial grasp while growth cones are choosing among them.

Although it would be interesting to isolate the surface molecule(s) recognized by the Mes-3 and Mes-4 MAbs, such procedures are difficult with such rare surface antigens. Furthermore, it is difficult to test molecular function in an embryo such as a grasshopper that does not have advanced genetics. Such molecular genetic approaches are much easier in the *Drosophila* embryo.

5. The First Growth Cones in the *Drosophila* Embryo

Whereas the grasshopper embryo has been an ideal system for the types of cellular studies described above in this paper, the *Drosophila* embryo has obvious attributes for a molecular genetic approach. The problem with studying cell recognition and neuronal specificity in the CNS of the *Drosophila* embryo has always been the small size of its seemingly inaccessible embryonic neurons. Until recently, nothing was known of the detailed cellular events underlying neurogenesis in *Drosophila*. However, we recently bridged this gap by showing that the early fly embryo CNS is a miniature replica of the grasshopper embryo in terms of its identified neurons, their growth cones, and their selective fasciculation choices when confronted with a common embryonic axonal scaffold (Thomas *et al.*, 1984).

In the present chapter we have described a simple model system in the grasshopper embryo: the cell recognition events involving the divergent choices of the growth cones of MP1 and dMP2 versus vMP2, and aCC versus pCC. These identified embryonic neurons extend their growth cones along the same stereotyped pathways in the *Drosophila* embryo.

The development in the fly embryo of these and several other highly accessible embryonic neurons was followed by intracellular dye injections (Fig. 12). Embryonic development in *Drosophila* takes 22 hr (at 25°C) compared with 20 days (at 33°C) in the grasshopper. The cellular events of neurogenesis that occur over a three-day period between 30% and 45% of development in the grasshopper, occur over a 3 hr period between hours 10–13 in the fly. In general, 1 hr of fly development equals one day of grasshopper development. In the fly embryo, NBs have begun to delaminate from the neuroepithelium by hour 6; 2 hr later, nearly the full pattern is evident and some NBs have already begun to divide and send their progeny dorsally. By hour 10, the first neurons (MP1, dMP2, vMP2, pCC, aCC, RP1, and RP2) have begun differentiating (the first growth cones appear at 9:45), extending axons that establish the first longitudinal, commissural, and peripheral axonal pathways.

Figure 12 shows the development of these identified neurons in *Drosophila* at hours 10, 11, and 12 of embryonic development (Thomas *et al.*, 1984). As in the grasshopper, the growth cones of MP1 and dMP2 extend posteriorly, whereas the growth cone of the vMP2 extends anteriorly, pioneering a second longitudinal axon fascicle. By hour 10.5, the MP1 and dMP2 growth cones have fasciculated with the pCC growth cone extending anteriorly from the next posterior segment. By 11 hrs, the MP1 growth cone has reached its homologue in the next posterior

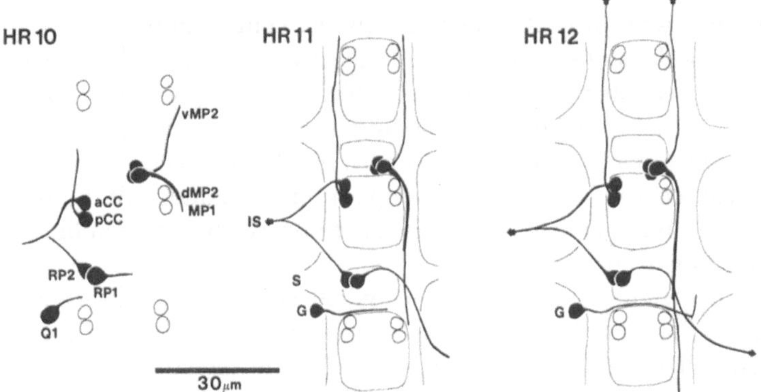

FIGURE 12. Timecourse of development of identified neurons in the *Drosophila* embryo. Wild-type O-R eggs were collected on yeasted agar plates and allowed to develop at 25°C. Since there is some variation in developmental time between eggs within a single collection, each egg was staged on the basis of external embryo morphology after removal of the chorion. Embryos were dissected from the vitelline membrane under saline, filleted on a microscope slide, and lightly fixed for 8 min in 2% paraformaldehyde. Neurons were filled with Lucifer Yellow via their cell bodies. Each developmental stage shown is a compilation of camera lucida drawings of LY fills from several embryos. Three contiguous segments are shown, marked by aCC and pCC cell bodies (open cell bodies). The first two axon fascicles in each of the longitudinal connectives are established by MP1, dMP2, pCC, and vMP2 (see text). Cell Q1 (at hour 10) crosses the segment in what will become the posterior commissure (cc); Q1's homologue in grasshopper establishes the fascicle followed by its later developing sibling, the G neuron. The G neuron in *Drosophila* crosses the segment between hours 11 and 12, grows past both the vMP2 fascicle and the MP1/dMP2 fascicle, and selectively fasciculates with a more lateral bundle (the A/P fascicle) running within the connective. Intersegmental nerve, IS; segmental nerve S.

segment. The patterns of growth parallel those in the grasshopper with minor modifications of temporal order. For example, in *Drosophila* the aCC extends its growth cone posteriorly and laterally to establish the intersegmental nerve simultaneous to the extension of the MP1, dMP2, and vMP2 growth cones to pioneer the first longitudinal fascicles. In grasshopper, the aCC growth cone does not extend posteriorly and laterally until some time after the MP1 and dMP2 growth cones have extended past it.

The early differentiating neurons in the grasshopper embryo give rise to a stereotyped orthogonal pattern of axon fascicles that is repeated in each segment of the neuroepithelium (Fig. 1A). This orthogonal scaffold of axon fascicles is common to the *Drosophila* embryo as well, the pattern of axonal fascicles in the fly simply representing a miniature version of the scaffold in the grasshopper embryo (Fig. 1B). For example, in the grasshopper embryo, the first two longitudinal axon bundles, the vMP2 fascicle and the MP1/dMP2 fascicle, are pioneered by the interactions of the MP1, dMP2, vMP2, and pCC neurons. The same is true in the fly.

6. Discussion

We began this chapter by asking how neuronal specificity was generated during embryonic development. Because unraveling these mechanisms is likely to require a detailed cellular and molecular analysis of individual recognition events, we turned our attention to the relatively simple central nervous system of insects in search of a model system. We focused our attention on a model system of seven identified neurons, each large and accessible, and each of which specifically recognizes and fasciculates with the others in a stereotyped pattern giving rise to the first three axon fascicles. These seven cells are highly accessible to cellular and immunological techniques in the grasshopper embryo, and with the recent studies of the same cells in the *Drosophila* embryo, are now ready for a combined molecular genetic and cellular analysis.

Neuronal specificity in this simple model system is largely achieved during embryonic development by a sequential series of cell recognition events. At these early stages, cell recognition occurs largely by the selective affinity that growth cones and their filopodia show for particular neuronal surfaces. These selective patterns of cell recognition give rise to the stereotyped patterns of selective fasciculation underlying the orthogonal scaffold of axon bundles. It is within the neighborhoods dictated by these axonal associations that, later in development, specific synaptic connections are made with subsets of neighboring neurons.

Many of our previous studies have focused on the analysis and manipulation of the G neuron as a model system, as it chooses the P axons within the A/P fascicle as compared with ~100 other axons within filopodial grasp (e.g., Raper *et al.*, 1983a,b, 1984; Bastiani *et al.*, 1984a,b; Goodman *et al.*, 1984). In the present chapter, we describe our recent studies on an even simpler model system, the earliest recognition events in the CNS involving only seven neurons: the MP1, dMP2, vMP2, pCC, aCC, U1, and U2 neurons. In the absence of the Us, the aCC appears completely uninterested in the other two sets of neurons, suggesting that the recognition molecules are highly specific. The aCC growth cone does not simply adhere to any available axons (e.g., vMP2, dMP2, MP1), even though other growth cones (e.g., its sibling pCC) may find some of those axons (e.g., MP1, dMP2) desirable. The fact that the aCC growth cone waits so long before the Us arrive, even though the vMP2, MP1, and dMP2 axons are available, further suggests that these other axons simply will not do.

These results are most easily explained by postulating the existence of at least three different surface-recognition labels at about 35% of development on these seven neurons: one on the vMP2 neuron; the second on the MP1, dMP2, and pCC neurons; and the third on the U1, U2, and aCC neurons. The Mes-3 and Mes-4 Mabs demonstrate the existence of surface antigens with just such specificities. Many more surface labels are required to explain the specific patterns of axonal affinities in later embryos. For example, at about 40% of development, the axons of ~100 different neurons fasciculate into ~25 different longitudinal bundles, one of which is the A/P fascicle. Ablation studies at this stage suggest the existence of many (probably more then 25) different surface labels (e.g., Goodman *et al.*, 1984).

A promising discovery is that by studying the large and accessible neurons of the grasshopper embryo for the past decade, we have in effect been studying the less accessible neurons of the *Drosophila* embryo. The fact that we can now study with these same identified neurons, their growth cones, and their patterns of selective fasciculation in the *Drosophila* embryo means that genetic and molecular approaches can be applied to unraveling the mechanisms of cell recognition and neuronal specificity. Thus with this simple model system, and with the attributes of the grasshopper and *Drosophila* embryos, we hope someday to understand how cells selectively recognize one another during neuronal development.

ACKNOWLEDGMENTS. We thank our colleagues Michael Bate, Robert Ho, Kathryn Kotrla, Jonathan Raper, Paul Taghert, and John Thomas for participation in some of the experiments reported here, and Francis Thomas for valuable assistance with the electron microscopy. Supported by NICHHD Developmental Biology Training Grant to M.J.B., NIMH Neuroscience Training Grant to S.d.L., and grants from the NSF and NIH to C.S.G.

References

Bastiani, M. J., and Goodman, C. S., 1984a, Neuronal growth cones: Specific interactions mediated by filopodial insertion and induction of coated vesicles, *P.N.A.S.* **81**:1849–1853.

Bastiani, M. J., and Goodman, C. S., 1984b, The first growth cones in the central nervous system of the grasshopper embryo, in: *Cellular and Molecular Approaches to Neuronal Development* (I. Black, ed.) Plenum Press, New York, pp. 63–84.

Bastiani, M. J., Pearson, K. G., and Goodman, C. S., 1984a, From embryonic fascicles to adult tracts: Organization of neuropil from a developmental perspective, *J. Exp. Biol.* **112**:45–64.

Bastiani, M. J., Raper, J. A., and Goodman, C. S., 1984b, Pathfinding by neuronal growth cones in grasshopper embryos. III. Selective affinity of the G growth cone for the P cells within the A/P fascicle, *J. Neurosci.* **4**:2311–2328.

Bate, C. M., 1976, Embryogenesis of an insect nervous system. I. A map of the thoracic and abdominal neuroblasts in *Locusta migratoria, J. Embryol. Exp. Morphol.* **35:**107–123.

Bate, C. M., and Grunewald, E. B., 1981, Embryogenesis of an insect nervous system. II. A second class of precursor cells and the origin of the intersegmental connectives, *J. Embryol. Exp. Morphol.* **61:**317–330.

Bray, D., 1982, Filopodial contraction and growth cone guidance, in: *Cell Behavior* (R. Bellairs, A. Curtis, and G. Dunn, eds.) Cambridge University Press, Cambridge, England, pp. 299–317.

Goodman, C. S., 1982, Embryonic development of identified neurons in the grasshopper, in: *Neuronal Development* (N. C. Spitzer, ed.), Plenum Press, New York, pp. 171–212.

Goodman, C. S., and Bate, M., 1981, Neuronal development in the grasshopper, *Trends Neurosci.* **4:**163–169.

Goodman, C. S., and Pearson, K. G., 1982, *Neuronal Development: Cellular Approaches in Invertebrates* (C. S. Goodman and K. G. Pearson, eds.) NRP Bulletin, MIT Press, Cambridge, Massachusetts.

Goodman, C. S., and Spitzer, N. C., 1979, Embryonic development of identified neurones: Differentiation from neuroblast to neurone, *Nature (London)* **280:**208–214.

Goodman, C. S., O'Shea, M., McCaman, R. E., and Spitzer, N. C., 1979, Embryonic development of identified neurons: Temporal pattern of morphological and biochemical differentiation, *Science* **204:** 219–222.

Goodman, C. S., Bate, C. M., and Spitzer, N. C., 1981, Embryonic development of identified neurons: Origins and transformation of the H cell, *J. Neurosci.* **1:**94–102.

Goodman, C. S., Raper, J. A., Ho, R., and Chang, S., 1982, Pathfinding by neuronal growth cones in grasshopper embryos, *Symp. Soc. Dev. Biol.* **40:**275–316.

Goodman, C. S., Bastiani, M. J., Raper, J. A., and Thomas, J. B., 1984, Cell recognition during neuronal development in grasshopper and *Drosophila,* in: *Molecular Bases of Neural Development* (W. M. Cowan, ed.), NRP Press, (in press).

Kotrla, K. J., and Goodman, C. S., 1984, Transient expression of a surface antigen on a small subset of neurons during embryonic development, *Nature (London)* **311:**151–153.

Letourneau, P. C., 1982, Nerve fiber growth and its regulation by extrinsic factors, in: *Neuronal Development* (N. C. Spitzer, ed.), Plenum Press, New York, pp. 213–254.

Raper, J. A., and Goodman, C. S., 1982, Transient dye coupling between developing neurons reveals patterns of intercellular communication during embryogenesis, in: *Cellular Communication during Ocular Development* (J. B. Sheffield and S. R. Hilfer, eds.), Springer-Verlag, New York, pp. 85–96.

Raper, J. A., Bastiani, M., and Goodman, C. S., 1983a, Pathfinding by neuronal growth cones in grasshopper embryos: I. Divergent choices made by the growth cones of sibling neurons, *J. Neurosci.* **3:**20–30

Raper, J. A., Bastiani, M., and Goodman, C. S., 1983b, Pathfinding by neuronal growth cones in grasshopper embryos: II. Selective fasciculation onto specific axonal pathways, *J. Neurosci.* **3:**31–41.

Raper, J. A., Bastiani, M. J., and Goodman, C. S., 1984a, Guidance of neuronal growth cones: Selective fasciculation in the grasshopper embryo, *Cold Spring Harbor Symp. Quant. Biol. Mol. Neurobiol.* **48:**587–598.

Raper, J. A., Bastiani, M. J., and Goodman, C. S., 1984b, Pathfinding by neuronal growth cones in grasshopper embryos. IV. The effects of ablating the A and P axons upon the behavior of the G growth cone, *J. Neurosci.* **4:**2239–2345.

Shankland, M., Bentley, D., and Goodman, C. S., 1982, Afferent innervation shapes the dendritic branching pattern of the medial giant interneuron in grasshopper embryos raised in culture, *Dev. Biol.* **92:** 507–520.

Taghert, P. H., and Goodman, C. S., 1984, Cell determination and differentiation of identified serotonin-containing neurons in the grasshopper embryo, *J. Neurosci.* **4:**989–1000.

Taghert, P., Bastiani, M., Ho, R. K., and Goodman, C. S., 1982, Guidance of pioneer growth cones: Filopodial contacts and coupling revealed with an antibody to Lucifer Yellow, *Dev. Biol.* **94:**391–399.

Taghert, P. H., Doe, C. Q., and Goodman, C. S., 1984, Cell determination and regulation during development of neuroblasts and neurons in the grasshopper embryo, *Nature (London)* **307:**163–165.

Thomas, J. B., Bastiani, M. J., Bate, C. M., and Goodman, C. S., 1984, From grasshopper to *Drosophila:* A common plan for neuronal development, *Nature (London)* **310:**203–207.

10

The Development of Serotonin-containing Neurons in the Leech

DAVID A. WEISBLAT AND WILLIAM B. KRISTAN JR.

1. Introduction

Given the rapid deployment of new experimental techniques in the last decade, many of the important questions in developmental neurobiology appear to be nearing their resolution. One of these questions is that of how each nerve cell acquires its differentiated characteristics: the appropriate membrane channels and receptors, the capacity to produce and store the appropriate transmitter(s), the correct pattern of process outgrowth, and, finally, the establishment of connections of just the right type and strength with just the right constellation of neurons to produce proper behaviors.

In focusing on any of these issues, it is necessary to determine the extent to which developmental events are controlled intrinsically (by factors within the cell) as opposed to extrinsically (by other cells or extracellular factors). Such studies require a well-characterized system, ideally a relatively simple one, in which cell types can be identified at many, if not all, developmental stages. Invertebrate nervous systems have been used extensively for such studies, particularly those of insects, nematodes, and leeches, (Goodman and Pearson, 1982). Each has its particular advantages. In the leech, the relatively simple nervous system is made even more simple by virtue of the fact that it is composed of segmentally repeating units. Many, perhaps all, of the neurons are uniquely identifiable and are accessible to microelectrodes for electrical recording and microinjection of dyes and other substances. Homologous cells can be individually identified from segment to segment, animal to animal, and even between different leech species. A number of synaptic

DAVID A. WEISBLAT • Department of Zoology, University of California, Berkeley, California 94720. WILLIAM B. KRISTAN, JR. • Department of Biology, University of California, San Diego, California 92093.

connections have been described and at least partial neuronal circuits for particular behaviors have been elucidated (Muller *et al.*, 1981). Thus, an endpoint toward which neurogenesis proceeds has been well characterized, allowing developmental questions to be clearly defined.

In addition, leech development is direct and proceeds from the first by stereotyped, holoblastic cleavages that give identifiable blastomeres. Embryos of the glossiphoniid leeches are quite large, moreover, and can develop to maturity isolated in simple media. Therefore, it is possible to microinject embryonic cells with any of various cell lineage tracers. By this means, the development of an otherwise nondescript blast cell can be followed until its offspring become unique, differentiated neurons.

Thus, in the leech, phenomena of interest in neurodevelopment can be examined in individual cells from the uncleaved egg to the time of postmitotic differentiation. We hope ultimately to understand the relationships between mechanisms operating at different phases of neurodevelopment, including cell cleavage, proliferation of neuroblasts, differentiation and axonogenesis, and the maturation of specific behaviors. In this account, we will relate the progress that is being made along these lines with respect to the development of the set of neurons in the segmental ganglia of the leech that contain the neurotransmitter serotonin (5-hydroxytryptamine). (A review of leech neurobiology, including neurodevelopment, is available in Muller *et al.*, 1981.)

2. Leech Neuroanatomy

2.1. The Ganglion

Each of the 32 homologous segmental ganglia of the leech ventral nerve cord (Fig. 1A) contain about 200 bilateral pairs of neurons and a few unpaired neurons as well (Fig. 1B). Many of these neurons have been characterized in adult leeches by a variety of physiological, anatomical, and biochemical criteria and their roles in controlling various behaviors are understood, especially in the 21 midbody ganglia between the head and tail (Muller *et al.*, 1981). In all leeches, the four segmental ganglia anterior and the seven segmental ganglia posterior to these 21 midbody ganglia are fused into a subesophageal ganglion and a tail brain, respectively, and will not be treated further in this discussion.

2.2. Serotonin-containing Neurons

A segmental ganglion in the midbody of an adult leech contains up to nine identifiable serotonin-containing neurons (Fig. 1C). These include a bilateral pair of giant neurons, the Retzius neurons, near the midline on the anteroventral aspect of the ganglion; a smaller pair of neurons, the anteromedial (AM) neurons, just anterior to the Retzius neurons; two lateral pairs of neurons, neuron pairs 21 and 61, on the dorsal and ventral aspects of the ganglion, respectively; and an unpaired

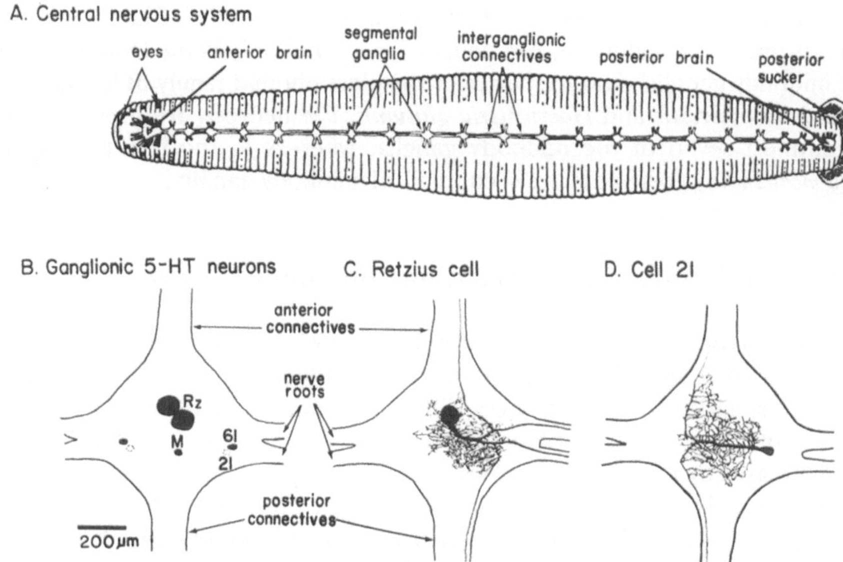

FIGURE 1. Anatomy of the central nervous system of the medicinal leech, showing the serotonin-containing neurons. (A) Diagram of the central nervous system superimposed on a sketch of the body. Connectives run between successive ganglia and two nerve roots exit the ganglion on either side. There are 21 midbody ganglia, as well as fused ganglia at each end forming the anterior and posterior brains. (B) A drawing of the ventral surface of a midbody segment indicating the location of the serotonin-containing cells. Retzius cell, Rz; medial cell, M. Cell 61 has also been called the VL (ventrolateral) cell, cell 21 the DL (dorsolateral) cell, and M the PM (posteromedial) cell (there is also an anteromedial cell in midbody ganglia 1-3). Cells 21 are shown in dotted outline because they are found on the opposite, i.e., dorsal, surface. Rz, 21, and 61 are in every midbody ganglion; there is a pair of M cells only in ganglion 1, with only a single M in all other midbody ganglia. (C, D) Camera lucida drawings of a RZ cell (C) and cell 21 (D) on the same scale as in B, from whole mounts of HRP-filled neurons (kindly supplied by Dr. Adrian Mason).

neuron, the posteromedial (PM) neuron, at the midline on the posteroventral aspect of the ganglion.

The presence of these neurons can be demonstrated by their selective accumulation of the dye neutral red *in vivo* (Stuart *et al.*, 1974), by a characteristic yellow-green fluorescence induced in these neurons on exposure to formaldehyde or glyoxylic acid (Lent, 1982), by treatment with commercially available serum antibodies raised against serotonin coupled to carrier protein (Stuart *et al.*, 1983), and by autoradiography after their selective accumulation of tritiated serotonin (Glover and Stuart, 1983). Except for the Retzius neurons, all the serotonin-containing cells appear to be interneurons, i.e., their processes are confined to the segmental ganglia and interganglionic connectives that make up the central nervous system of the leech.

Not all of these serotonin-containing neurons are present in every segmental ganglion. Systematic differences in their occurrence are observed according to segment number and species. For example, in the ganglia of the fifth and sixth midbody segments, which contain the sexual pores for these hermaphroditic worms, the Retzius neurons are reduced in size and only a single pair of lateral neurons, pair 61, is present. In addition, AM neurons are found only in the frontmost three of the midbody ganglia. And in species of the glossiphoniid family of leeches, such as *Helobdella triserialis* and *Haementeria ghilianii,* PM neurons are present only in the frontmost seven of the midbody ganglia, whereas in the hirudinid species *Hirudo medicinalis,* these cells are seen in all 21 midbody ganglia.

3. Behavioral Function of Serotonin-containing Neurons

Some behavioral roles for the serotonin-containing neurons in the leech have been elucidated. The activity level of an individual leech (e.g., the fraction of the time that it spends spontaneously crawling or swimming) increases with the concentration of serotonin in the bloodstream, and the level of serotonin in the blood can be increased by release from serotonin-containing neurons into the circulation (Willard, 1981). Together, these observations suggest that one role for serotonin-containing neurons in the leech is as neurohormonal modulators of locomotor activity. Recent behavioral and physiological experiments suggest that these effects of serotonin on swimming may be part of a more general control of the appetitive behavior of leeches (Lent *et al.*, 1983).

But there is good evidence for more specific effects of serotonin-containing neurons as well. For instance, spike activity in cell 61 initiates swimming activity quickly, selectively, and in isolated nerve cords, from which serotonin is released into a large volume and thus cannot accumulate to the levels needed to elicit swimming by the neurohormonal mechanism (Kristan and Nusbaum, 1983). Moreover, spikes in cell 61 are followed one-for-one with short, fixed latency by excitatory postsynaptic potentials (EPSPs) in cell 208, a previously identified swim pattern-generating interneuron (Fig. 2). Thus, it appears that serotonin-containing neurons can act by synaptic and neurohormonal mechanisms to affect leech behavior. Additional functions can be assigned to the Retzius cells, which, in addition to their central projections, also send processes out the segmental nerves to the body wall. High rates of impulse activity in the Retzius cells have been shown to increase mucus secretion from the skin (Lent, 1973). Retzius cell impulses also cause large inhibitory postsynaptic potentials (IPSPs) in the longitudinal muscles of *Haementeria* that are of variable latency and do not follow spikes in the Retzius cells one-for-one (A. P. Kramer, unpublished data). Retzius cell activity causes a relaxation of resting muscles and an increased rate of relaxation of contracted muscles in

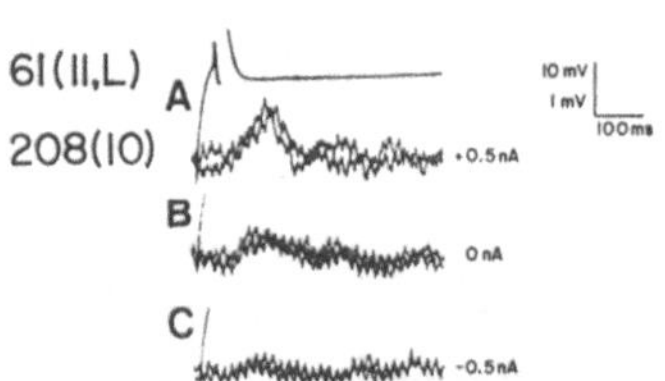

FIGURE 2. Synaptic effects of cell 61 on a pattern-generating interneuron, cell 208 (located between the Rz cells). The top trace is composed of several superimposed traces of single impulses in cell 61 (the second upward deflection is the artifact at stimulus offset). The bottom three traces are superimposed traces of the EPSP recorded in cell 208, at short and constant latency, with 0.5 nA of (A), depolarizing current, (B) no current, and (C) 0.5 nA of hyperpolarizing current passed into cell 208.

Hirudo, apparently by direct action of serotonin on the longitudinal muscles (Mason *et al.,* 1979; Mason and Kristan, 1982).

Detailed anatomical knowledge of the serotonin-containing neurons is available by the examination of horseradish peroxidase (HRP) and Lucifer Yellow-injected neurons by light and electron microscopy. Another promising technique is that of removing these neurons from the ganglion and growing them in culture for detailed studies of single cells and identified cell pairs in isolation (Ready and Nicholls, 1979; Fuchs *et al.,* 1982; Henderson, 1983).

Thus, it would seem that any characteristic of the differentiated serotonin-containing neurons that is of interest can be examined in detail. The knowledge available already about this set of neurons in adult leeches leads to specific questions about their development. For instance, regarding the nature of cell identity, what makes these cells produce, take up, and store the same transmitter, serotonin? And what makes them differ from each other in their position, arborizations, and synaptic connections? How is it that most of these cells occur as bilateral pairs and yet one is unpaired? How do the segment-specific and species-specific differences in their occurrence arise? Unfortunately, none of these questions can yet be answered in sufficiently satisfying detail. Still, a beginning has been made, or more precisely, a continuing, for the first studies of leech development were published over 100 years ago when C. O. Whitman (1878, 1892) studied the cleavage patterns and general developmental processes in glossiphoniid leech embryos.

4. Embryogenesis of the Leech

A modern summary of Whitman's observations is presented in Fig. 3, including depictions of *Helobdella* embryos at various developmental stages (Fernandez, 1980; Weisblat *et al.,* 1980a, 1984). The leech egg undergoes a series of stereotyped cleavages in early development (stages 0–6) that yield a number of identifiable blastomeres, including five bilateral pairs of cells called teloblasts. On each side there is an M, N, Q and two sister O/P teloblasts. The teloblasts can be quite large cells, over 100 μm across in various glossiphoniid leeches, which lay large, yolk-filled eggs. Each teloblast undergoes a series of several dozen highly asymmetric cleavages (stages 6–8) that generate a column of much smaller blast cells, referred to collectively as a germinal bandlet. The blast cells born first are at the distal end of each bandlet and are pushed further from the parent teloblast as each successive blast cell is formed; they will contribute to the more anterior segments of the embryo. On either side of the midline, the five germinal bandlets, designated by the lower case letter corresponding to their teloblast of origin, grow and merge to form left and right germinal bands lying on the dorsal surface of the embryo. In each germinal band, the m (mesodermal) bandlet lies underneath the other four bandlets, which are thus designated as ectodermal. Among these four, the n bandlet is most lateral and the q bandlet is most medial. The bandlets derived from the sister O/P teloblasts lie between them and can thereby be distinguished as distinct o (n-proximal) and p (q-proximal) bandlets (Weisblat and Blair, 1984).

As more blast cells are produced by the teloblasts, the germinal bands

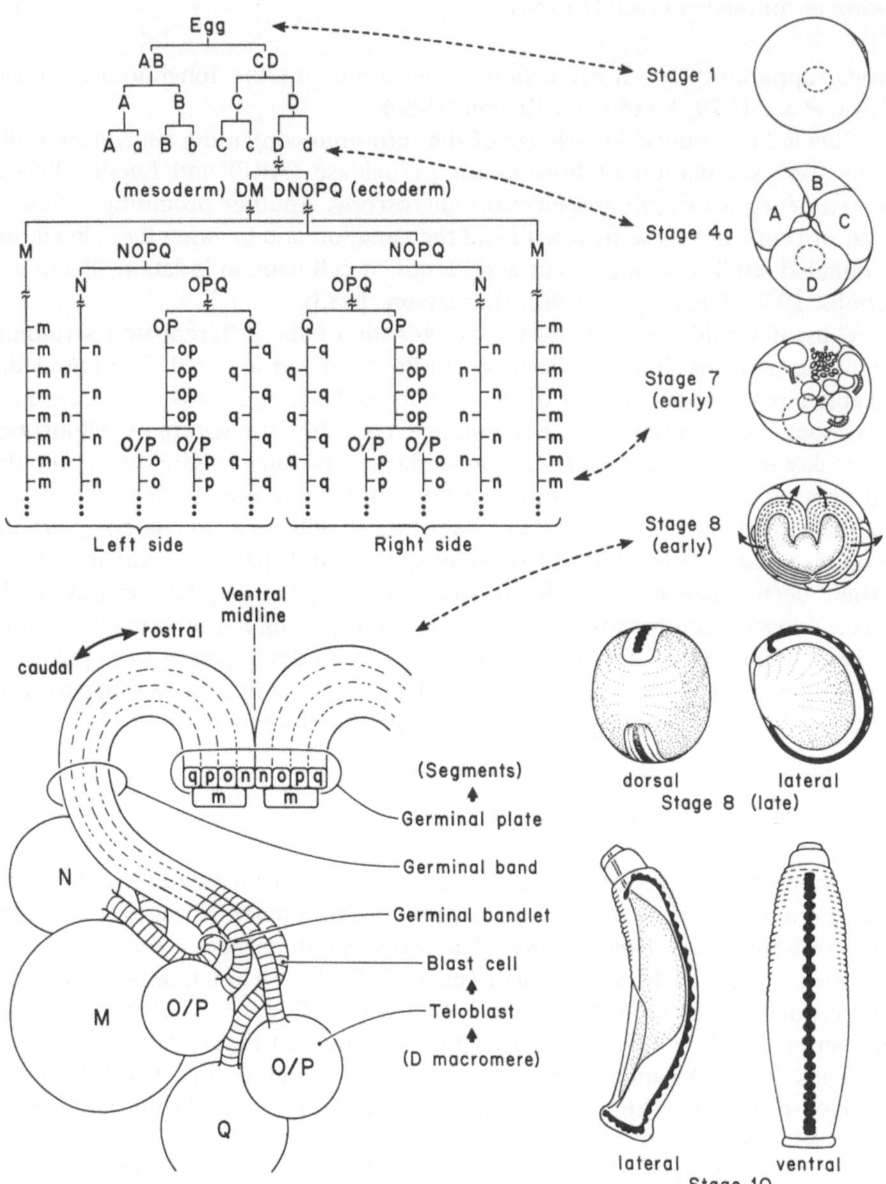

FIGURE 3. A summary of glossiphoniid leech development. Several key stages of embryogenesis are illustrated schematically along the right-hand margin of the figure, beginning with the uncleaved egg (stage 1). The egg contains a domain of special, yolk-free cytoplasm (indicated by the dashed circle) that is passed on selectively to the D macromere of the 8-cell embryo (stage 4a) and then to the M, N, O/P, and Q teloblasts (medium-sized circles in the stage-7 embryo). The family tree diagram at the upper left of the figure denotes the stereotyped cleavage pattern by which the teloblasts (M, N, O/P, Q) arise from the fertilized egg, and how each teloblast in turn gives rise to the many smaller blast cells. Note that, unlike N and Q, the O/P teloblasts on each side arise by the final cleavage of the NOPQ precursor blastomere and are thus distinguishable only by their relative position in the embryo. Dashed, double-headed arrows connect equivalent points in the family tree diagram and the schematic illustration. In the lower left-hand corner of the figure is an enlarged representation of part of an early stage-8 embryo, showing how the teloblasts give rise to columns of blast cells (germinal bandlets) that merge to form left

lengthen and arc so that their central portions move circumferentially across the surface of the embryo, gradually entering its ventral hemisphere. Eventually (stage 8), the left and right germinal bands meet and coalesce at the ventral midline. The coalescence begins at the future head of the animal and progresses rearward like a zipper; at completion of coalescence, the fused germinal bands constitute a structure known as the germinal plate. All the segmental tissues of the leech, including the segmental ganglia in the ventral nerve cord, are formed by the proliferation of the blast cells and their progeny within the germinal bands and germinal plate (stages 8–11). [As are the cleavages leading from the egg to the identifiable blastomeres, the proliferative divisions of the blast cells appear to be stereotyped as to timing, spindle orientation, and relative size of the daughter cells, according to the bandlet in which they lie (Zackson, 1982, 1984).] This proliferation causes the germinal plate to thicken and expand laterally, back into the dorsal half of the embryo. Eventually the lateral edges of the expanding germinal plate meet at the dorsal midline, closing the body tube of the developing leech (stage 10). In accord with the position of the first-born blast cells at the anterior ends of the germinal bands and the head-to-tail progression of germinal plate formation, subsequent tissue differentiation, including the maturation of the segmental nervous system, exhibits a general rostrocaudal gradient within the 21 midbody segments. Simultaneously, head and tail suckers form at the two ends of the animal. Also during this time, the gut forms around the yolk-filled remnants of the teloblasts and other embryonic blastomeres. The ensuing digestion of these cells provides nutrition for the embryo during the latter stages of embryogenesis. When its yolk is exhausted (end of stage 11), the juvenile leech is ready to take its first meal.

Within each germinal band, the bandlets from the five teloblasts maintain fixed positions relative to one another; when the germinal bands meet to form the germinal plate, the left and right n bandlets are in apposition at the ventral midline, where the nerve cord forms. For this reason, and because he believed that each teloblast was fated to give rise to a distinct tissue type, Whitman regarded the pair of N teloblasts as the sole progenitors of the leech nervous system. But this idea could not be tested because the cells within the germinal plate are too small in size and large in number for their origins and fates to be followed precisely with the techniques then available. In recent years, this question has been examined using HRP or fluorescent dyes conjugated to small peptides or dextrans as cell lineage tracers that can be injected into an identified cell in the living embryo. In subsequent development, these tracers are passed on exclusively to the progeny of the injected cell, which can thus be identified as such histochemically or by fluores-

---

and right germinal bands. The germinal bands migrate over the surface of the embryo and coalesce, zipperlike, along the ventral midline of the embryo into a structure called the germinal plate that wraps most of the way around the embryo by late stage 8 of development. Subsequently, as cells in the germinal plate proliferate, it expands circumferentially until its edges meet along the dorsal midline of the embryo in stage 10, thus forming the body tube of the leech. Segmental tissues of the adult leech, including the chain of ventral nerve cord ganglia indicated in solid black in the stage-10 embryo, arise from the germinal plate. (Reprinted from Weisblat *et al.*, 1984.)

cence microscopy (Weisblat *et al.*, 1978, 1980b; Braun and Gimlich, in preparation).

Using this technique, it was found that all teloblasts, not just the N teloblasts, generate neurons of the segmental ganglia, and that each teloblast generates non-neural cells as well. But even though the teloblasts are not fated to produce unique tissue types, there is a high degree of specificity in the spatial distribution of their progeny. In normal development, each teloblast gives rise to ipsilateral progeny exclusively. Moreover, the pattern formed by the labeled progeny of a given bandlet is much the same from segment to segment and animal to animal, is mirror symmetric to the pattern formed by the progeny of the contralaterally homologous bandlet, and is clearly distinct from the segmental pattern formed by the progeny of all the ipsilateral bandlets. Within each half-segmental ganglion, therefore, the cells fall into five *kinship groups,* according to their bandlet of origin. Each kinship group has a characteristic cell number and distribution (Fig. 4). Further experiments revealed that from each N teloblast two blast cells are needed to generate the N kinship group in each ipsilateral half ganglion, one for the anterior and the second for the posterior cells. Two q blast cells also are needed to generate one segmental complement of progeny, whereas a single m, o, or p blast cell itself generates an entire segmental complement of cells.

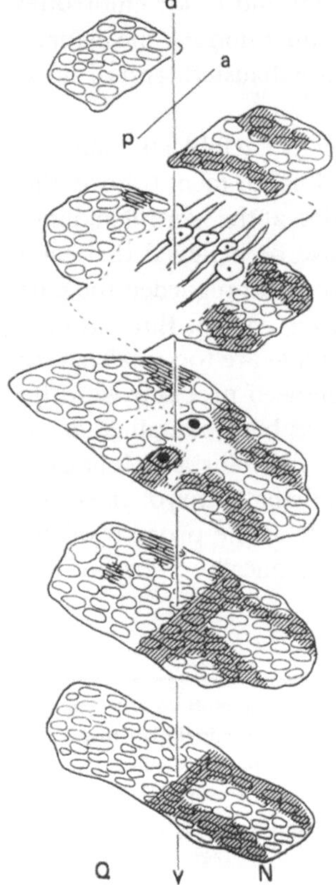

FIGURE 4. A schematic drawing of five horizontal sections through a midbody segment of a glossiphoniid leech, with the dorsal aspect at the top and the front edge facing away from the viewer. The two pairs of dark contours in the center of the second section from the top represent muscle cells in the longitudinal nerve track. The two dark contours in the middle section represent the neuropil glia, each descended from one N teloblast. The faint contours do not correspond to actual cells, but indicate approximate size and number of neurons in the ganglion. In the right-half ganglion, domains of descendants of the right N teloblast (i.e., N cell kinship groups) are crosshatched. Domains of descendants of the left Q teloblast are crosshatched in the left-half ganglion.

5. Embryonic Origins of the Serotonin-containing Neurons

Given the stereotyped locations of identified neurons in the adult leech ganglion, the results described above suggested that each identified neuron invariably arises from a particular teloblast. To test this hypothesis, it was necessary to be able to identify individual neurons and in the same preparation assay for the presence or absence of lineage tracer with single-cell resolution. For this purpose, the serotonin-containing neurons are well-suited, because they can be identified at times when the lineage tracers can also be detected (Stuart *et al.*, 1983). In these experiments, a rhodamine-conjugated (red-fluorescing) cell lineage tracer was injected into a teloblast early in development (stage 6). After about a week of further development, the embryos were fixed, counterstained, and the serotonin-containing cells identified immunohistochemically, using a fluorescein-conjugated (green-fluorescing) second antibody. In all cases, when an N teloblast was injected with lineage tracer, but not when any other teloblast was injected, the ipsilateral member of every bilateral pair of serotonin-containing neurons was labeled.

These results support the hypothesis that the ability of this set of neurons to accumulate serotonin is determined by their line of descent; but other alternatives exist. For instance, it is possible that the ability to accumulate serotonin is conferred on a neuron by its position within the nascent ganglion at the time of differentiation. Since the N teloblasts contribute a large proportion of the total number of neurons and to the same regions of each ganglion, it may just be a matter of chance that it is always an N-derived neuroblast that receives the postulated positional cue to accumulate serotonin. One way to test this hypothesis would be to change the environment around the differentiating neurons, by translocating them, or by killing some of the neurons in their vicinity. This aim was accomplished by experiments in which one N teloblast was labeled with a lineage tracer and the other one was ablated by injecting it with a toxic enzyme, deoxyribonuclease (DNase) (Blair, 1982; Blair and Weisblat, 1982; Stuart *et al.*, 1983). Such embryos develop more or less normally, but their segmental ganglia are deficient in cell number relative to unablated controls. Moreover, the progeny of the surviving N teloblast often cross over the ventral midline, occupying ectopic sites in the ganglion, so that the normal neighbor relations of cells are disrupted. But when such embryos were examined immunohistochemically, it was found that exactly one member of each of the pairs of serotonin-containing neurons was missing and that the surviving cells lay on both sides of the ganglia, in accord with the lineage-dependent model of neurogenesis.

Ablation experiments also provided information regarding the process by which unpaired cells arise from bilaterally paired precursor teloblasts. When the unpaired, PM neurons were examined in unablated, lineage tracer-labeled preparations, it was found that they derived from the injected N teloblast in only about half of the ganglia examined. In the N-ablated embryos, this unpaired serotonin-containing neuron was still present in *every* ganglion, but now each PM neuron contained the lineage tracer. One interpretation of these results is that two *potential* PM neurons arise, one from each N teloblast, but that only one develops into an adult serotonin-containing neuron, thanks to a competitive interaction that occurs independently within each segmental ganglion between the differentiating PM neurons themselves or between some precursors to them earlier in gangliogenesis.

This interpretation is strenghthened by the observation that just as the serotonin-containing neurons are beginning to differentiate (stage 9), serotonin antigenicity appears in a *pair* of neurons at the posteromedial site on the ventral aspect of the ganglion. About a day later, only one member of this pair is detectable immunohistochemically, as the unpaired PM neuron. It is believed that the cell losing its antigenicity actually dies, rather than simply losing its serotonin, because there is a slight loss of ganglionic neurons in maturation (Stewart and Macagno, 1984) and because sections through ganglia at this stage of development reveal vacuolated cells with pyknotic nuclei (our unpublished observations). The genesis of a similarly unpaired neuron arising in the P teloblast kinship group has been shown to involve a cell death (Kramer and Weisblat, 1985; Shankland and Weisblat, 1984). However, in the case of the PM neuron, the possibility that the cell losing antigenicity actually redifferentiates into a phenotypically distinct neuron has yet to be ruled out. Thus, unpaired neurons in the leech result from the death or redifferentiation of one member of a bilaterally derived pair of cells, in contrast to the grasshopper, where at least some of the unpaired neurons arise from a distinct, unpaired neuroblast (Bate, 1976).

A similarly transient expression of serotonin antigenicity, albeit with a slower time course, is observed for the lateral pair of serotonin-containing neurons in the maturation of the fifth and sixth midbody ganglia. Here, although only cell 61 is found in the ganglia of the adult leech, both the lateral serotonin-containing neurons 21 and 61 can be detected immunohistochemically in the late embryo. The staining in cell 21 fades only gradually over a period of several weeks (Stuart *et al.*, 1983).

6. Axonogenesis

An important aspect of the identifiability of individual neurons in the segmental ganglia of the adult leech is the unique branching pattern associated with each kind of cell. For example, each Retzius cell sends processes out all the ipsilateral segmental nerves in its ganglion and into both the anterior and posterior ipsilateral connectives as well. But even though the cell bodies of these giant serotonin-containing cells lie right beside each other at the midline of the ganglion, their processes normally never project into the contralateral nerves. Is this specificity generated during the initial outgrowth of Retzius cell processes or is it established by a selective retraction of an initially exuberant outgrowth? Two series of experiments have provided information on the process of axonogenesis in the serotonin-containing neurons. The first of these took advantage of the observation that the serotonin antigenicity of these neurons arises when they have only just begun to extend processes, and that the staining is distributed uniformly over the entire cell, so that the axons and growth cones can be observed readily in immunohistochemically processed preparations (D. K. Stuart and J. G. Glover, in preparation). In addition, since there is a rostrocaudal gradient of development for each neuronal type, a number of developmental time points can effectively be obtained from a single nerve cord. From such experiments it was clear that the Retzius cell processes were able to detect the correct pathways from the beginning and did not grow down inappropriate (e.g., contralateral) paths. In the second series of experiments, the morphology of Retzius neurons was examined in adult leeches raised from embryos in which one N teloblast had been ablated (Kramer and Blair, unpublished data). In these experiments, the one surviving Retzius neuron in the ganglion was identified by the position and relatively large size of its cell body and by its characteristically large and slow-action potentials, then filled with Lucifer Yellow via the same intracellular microelectrode. In nine of ten Retzius cells examined, all the processes were confined to the same side of the nerve cord, and none were extended into the inappropriate contralateral pathways. These results would suggest that, in contrast to the mechanism whereby the fate of the two potential unpaired neurons is decided, the restriction of Retzius cell processes to the ipsilateral half nerve cord does not depend on a competitive interaction with the contralateral Retzius cell, but rather on some cue intrinsic to the nerve cord. In one of the cells examined, however, there were processes on both sides of the nerve cord, in the right-hand anterior root and anterior connective, and in the left-hand posterior root and posterior connective. Thus, it may be that the normal barrier that restricts the Retzius cell processes to one side or the other is not insurmountable, and that competition between paired Retzius neurons cannot be totally eliminated as a factor contributing to their final arborization patterns. Hence, the distinction between cue-dependent and competition-dependent mechanisms in the elaboration of neuronal processes may be quantitative rather than qualitative.

7. Serotonin and Behavioral Development

Perhaps the most complex question regarding the development of the serotonin-containing neurons is their role in the maturation of the synaptic networks they serve to modulate in the adult leech. This problem has been addressed in experiments using a toxic serotonin analog, 5,7-dihydroxytryptamine (5,7-DHT) (Glover and Kramer, 1982). This compound selectively kills serotonin-containing neurons without demonstrable effect on any other neurons in the ganglion. Serotonin-containing neurons become sensitive to 5,7-DHT at the time they first express the serotonin antigenicity; thus they would be killed well before they have innervated their targets. To assess the behavioral effects of serotonectomy, 5,7-DHT was injected into the coelom of embryos of the leech species *Haementeria ghilianii,* causing the selective loss of serotonin-containing neurons. The injected embryos continued for quite some time along the usual pathway of morphogenetic and behavioral stages. Embryonic behavior begins with a period of myogenic peristaltic contractions that start soon after the germinal plate has formed and serve to promote the emergence of the embryo from the vitelline membrane, then progresses through a series of increasingly complex neurogenic contractions and body bends that normally culminate in walking and finally swimming. But the 5,7-DHT-injected embryos were somewhat sluggish in their overall behavior and never swam normally, in contrast to their sham-injected cohorts. Was this due to a failure of the swimming circuits to develop? Definitely not, because, on injecting these leeches with serotonin, or simply putting them into pond water to which serotonin had been added, they swam as well as their untreated cohorts at all stages of development. It cannot be concluded from these experiments, however, that the neural circuits responsible for generating the swimming behavior can develop completely in the absence of serotonin, because at least some of the serotonin-containing neurons had differentiated before they were killed. It is clear, though, that the circuit can develop and be moderately well expressed without practice.

8. Summary

What answers can we offer to the questions posed earlier regarding the development of the serotonin-containing neurons? Only the sort that lead to more questions, unfortunately, but the new questions are yet more clearly defined and still approachable within the experimental system we have chosen.

8.1. Cell Identity

From the studies of cell lineage, it seems that some essential factor in the genesis of serotonin-containing neurons is segregated into the N teloblasts during the early cleavages of the egg and thence into each n blast cell. Thus, the capacity for cells to make this particular transmitter segregates even before the point in development where neural and aneural cell lines separate. Is the postulated factor contained within the cytoplasm or the nucleus of the N teloblasts? Is it present in the uncleaved egg? These questions should be answerable by applying techniques used to study the nature and distribution of developmental determinants in other organisms such as ascidians (Whittaker, 1982; Jeffery *et al.*, 1983). One place to look for such a factor would be in the specialized, asymmetrically distributed cytoplasm of the leech egg referred to as the teloplasm (Fernandez and Olea, 1982).

The results of teloblast ablation experiments support the conclusion that lineage is critical in this aspect of neurogenesis. But there are also competition-dependent aspects to the genesis of the serotonin-containing neurons as well, as in the selection of a single PM neuron in each ganglion from two apparently equipotent contenders. Are the precursors to the various serotonin-containing neurons interchangeable with one another or are there critical differences between them, such as the presence or identity of second transmitters, for example, that are also lineage dependent? More detailed analyses of the lineages of the various cells and the advent of techniques for selectively killing even individual blast cells and their progeny within the germinal plate (Shankland, 1984; J. Braun, unpublished data), combined with immunohistochemical localization of putative peptide transmitters (Li and Calabrese, 1983) and other markers of neurodifferentiation (Zipser *et al.*,

1984; Hockfield and McKay, 1983), should allow further progress on this front as well.

8.2. Neuronal Circuitry and Behavior

At least one circuit, the one for swimming, can be assembled normally with a significant part of it, most of the serotonin-containing cells, missing. Do the surviving cells within the swim circuit make functional adaptations to this deficit, such as new or stronger synapses, or higher transmitter sensitivities? What is the result of this deficit on other serotonin-dependent behaviors, such as feeding? Because the 5,7-DHT-treated leeches can be raised to the size where electrophysiological experiments are practical and because the detailed knowledge of the swim circuitry in the hirudinid leeches provides an excellent baseline for comparison, experiments now in progress should soon generate information in this area. Comparable results for the glossiphoniid leeches await the elucidation of the appropriate neuronal circuitry for this group. But the effort should prove worthwhile, because the fact that embryos of glossiphoniid leeches can be raised to maturity following the ablation of one or more teloblasts (with the concomitant deletion of the corresponding neuronal kinship groups) should permit the analysis of the behavioral effects of selectively deleting other selected cell types.

It has been possible to identify neurons other than the serotonin-containing cells on the basis of cell body size and location, as they start to grow out their processes. By this means, the stereotyped development of the structure and electrical properties of mechanosensory neurons (Kuwada and Kramer, 1983; Kramer and Kuwada, 1983) and motor neurons (Kuwada, 1984) has been described in detail. These neurons are connected both monosynaptically and via identified interneurons to produce segmentally localized and whole-body behaviors. It is now possible to study this simpler neural network, to correlate morphogenetic and electrogenetic events with the development of behavior, and to manipulate the system to determine the mechanisms for the establishment of reflex circuits of neurons.

8.3. Evolution and Developmental Mechanism

The result that the character of the serotonin-containing neurons is determined by their line of descent is in agreement with the intuitively attractive notion that developmental potential is progressively restricted in successive cell divisions, leading ultimately to a uniquely determined differentiated cell or cell type. The general validity of this notion is best demonstrated by studies of the nematode *Caenorhabditis elegans,* in which, with few exceptions, the cell lineages are largely invariant both in normal development and after ablation of various cells (compare Sulston *et al.,* 1983). But no general rules by which such a stepwise restriction of developmental potential might operate have emerged in any system that has been examined. In the leech for instance, how is it that the neurons derived from the N teloblast are more closely related to certain (N-derived) epidermal cells than to other (M-, O-, P-, or Q-derived) neurons? And although it seems appropriately economical that all the serotonin-containing neurons are descended from the N teloblasts [analogous to the case in the grasshopper, where all octopamine-containing neurons seem to arise from the dorsal unpaired medial neuroblasts (Goodman *et al.*, 1979)], other experiments have shown that the three distinct dopamine-containing neurons derive independently from the O, P, and Q kinship groups (Blair, 1982; Stuart *et al.*, in preparation). Thus, even phenotypic traits that would seem to be of equivalent status in any differentiation hierarchy, such as production of serotonin or dopamine as neurotransmitter, exhibit differences in the points at which they undergo lineage restriction in development. In the absence of any valid generalizations about a rationale by which the postulated stepwise restriction of cell fate may operate, we should consider two possibilities. One is that this progressive restriction of developmental potential does operate systematically, but on as yet unknown cellular characteristics, to which the observed phenotypes are not directly linked. But a second alternative is that the observed lineages, resulting as they do from an evolutionary, rather than an engineering process, are innately *unsystematic* and therefore operate in innocence of the logical strictures which we seek to impose on them. If a definitive answer to this question is ever to be obtained, it will probably not be until molecular mechanisms for the developmental restrictions are known. But for those with patience, this is good news, for, whether or not there is a logic to the system, its reproducibility from generation to generation strongly suggests that it operates with clear-cut mechanisms whose workings *can* eventually be elucidated.

References

Bate, C. M., 1976, Embryogenesis of an insect nervous system. I. A map of the thoracic and abdominal neuroblasts in *Locusta migratoria, J. Embryol. Exp. Morphol.* **35:**107–123.

Blair, S. S., 1982, Blastomere ablation and the developmental origins of identified monoamine-containing neurons in the leech, *Dev. Biol.* **95:**65–72.

Blair, S. S., and Weisblat, D. A., 1982, Ectodermal interactions during neurogenesis in the glossiphoniid leech *Helobdella triserialis, Dev. Biol.* **76:**245–262.

Fernandez, J. H., 1980, Embryonic development of the glossiphoniid leech *Theromyzon rude:* Characterization of developmental stages, *Dev. Biol.* **76:**245–262.

Fernandez, J. H., and Olea, N., 1982, Embryonic development of glossiphoniid leeches, in: *Developmental Biology of Freshwater Invertebrates,* Alan R. Liss, New York, pp. 317–361.

Fuchs, P. A., Henderson, L. P., and Nicholls, J. G., 1982, Chemical transmission between individual Retzius and sensory neurones of the leech in culture, *J. Physiol. (London)* **323:**195–210.

Glover, J. C., and Kramer, A. P., 1982, Serotonin analog selectively ablates identified neurons in the leech embryo, *Science* **216:**317–319.

Glover, J. C., and Stuart, D. K., 1983, Differentiation of serotonin-containing neurons in the leech, *Soc. Neurosci. Abstr.* **9:**606.

Goodman, C. S., and Pearson, K. G. (eds.), 1982, Neuronal development: Cellular approaches in invertebrates, *NRP Bull.* **20:**777–942.

Goodman, C. S., O'Shea, M., McCaman, R. E., and Spitzer, N. C., 1979, Embryonic development of identified neurons: Temporal pattern of morphological and biochemical differentiation, *Science* **204:**1219–1222.

Henderson, L. P., 1983, The role of 5-hydroxytryptamine as a transmitter between identified leech neurones in culture, *J. Physiol. (London)* **339:**309–324.

Hockfield, S., and McKay, R., 1983, Monoclonal antibodies demonstrate the organization of axons in the leech, *J. Neurosci.* **3:**369–375.

Jeffery, W. R., Tomlinson, C. R., and Brodeur, R. D., 1983, Localization of actin messenger RNA during early ascidian development, *Dev. Biol.* **99:**408–417.

Kramer, A. P., and Kuwada, J. Y., 1983, Formation of the receptive fields of leech sensory neurons during embryonic development, *J. Neurosci.* **3:**2474–2486.

Kramer, A. P., and Weisblat, D. A., 1985, Developmental neural kinship groups in the leech, *J. Neurosci.* **5:**(in press).

Kristan, W. B., Jr., and Nusbaum, M. P., 1983, The dual role of serotonin in leech swimming, *J. Physiol. (Paris)* **78:**743–747.

Kuwada, J. Y., 1984, Normal and abnormal development of an identified leech motor neuron, *J. Embryol. Exp. Morphol.* **79:**125–137.

Kuwada, J. Y., and Kramer, A. P., 1983, Early embryonic development of identified sensory neurons in the leech CNS, *J. Neurosci.* **3:**2098–3111.

Lent, C. M., 1973, Retzius cells: Neuroeffectors controlling mucus release by the leech, *Science* **179:**693–696.

Lent, C. M., 1982, Fluorescent properties of monoamine neurons following glyoxylic acid treatment of intact leech ganglia, *Histochemistry* **75:**77–89.

Lent, C. M., Dickinson, M. H., and Marshall, C. G., 1983, Serotonin controls feeding behavior in the medicinal leech, *Soc. Neurosci. Abstr.* **9:**913.

Li, C., and Calabrese, R. L., 1983, Evidence for proctolin-like substances in the central nervous system of the leech, *Soc. Neurosci. Abstr.* **9:**76.

Mason, A., and Kristan, W. B., Jr., 1982, Neuronal excitation, inhibition and modulation of leech longitudinal muscle, *J. Comp. Physiol.* **146:**527–536.

Mason, A., Sunderland, A. J., and Leake, L. D., 1979, Effects of leech Retzius cells on body wall muscles, *Comp. Biochem. Physiol Comp. Pharmacol.* **63:**359–361.

Muller, K. J., Nicholls, J. G., and Stent, G. S. (eds.), 1981, *The Neurobiology of the Leech,* Cold Spring Harbor Press, New York.

Ready, D., and Nicholls, J. G., 1979, Identified neurones from leech CNS make selective connections in culture, *Nature (London)* **281:**67–69.

Shankland, M., 1984, Positional determination of supernumerary blast cell death in the leech embryo, *Nature (London)* **307:**9–15.

Shankland, M., and Weisblat, D. A., 1984, Stepwise commitment of blast cell fates during the positional specification of the O and P cell lines in the leech embroy, *Dev. Biol.* **106:**326–342.

Stewart, R. R., and Macagno, E. R., 1984, The development of segmental differences in cell number in the CNS of the leech, *Soc. Neurosci. Abstr.* **10:**512.

Stuart, A. E., Hudspeth, A. J., and Hall, Z. W., 1974, Vital staining of specific monoamine-containing cells in the leech nervous system, *Cell Tissue Res.* **153:**55–61.

Stuart, D. K., Glover, J. C., Blair, S. S., and Weisblat, D. A., 1983, Development of leech serotonin neurons examined with serotonin antibody, cell lineage tracer and cell killing, *Soc. Neurosci. Abstr.* **9:**604.

Sulston, J. E., Schierenberg, E., White, J. G., and Thomson, J. N., 1983, The embryonic development of the nematode *Caenorhabditis elegans, Dev. Biol.* **100:**64–119.

Weisblat, D. A., and Blair, S. S., 1984, Developmental indeterminacy in embryos of the leech *Helobdella triserialis, Dev. Biol.* **101:**326–335.

Weisblat, D. A., Sawyer, R. T., and Stent, G. S., 1978, Cell lineage analysis by intracellular injection of a tracer enzyme, *Science* **202:**1295–1298.

Weisblat, D. A., Harper, G., Stent, G. S., and Sawyer, R. T., 1980a, Embryonic cell lineages in the nervous system of the glossiphoniid leech *Helobdella triserialis, Dev. Biol.* **76:**58–78.

Weisblat, D. A., Zackson, S. L., Blair, S. S., and Young, J. D., 1980b, Cell lineage analysis by intracellular injection of fluorescent tracers, *Science* **209:**1538–1541.

Weisblat, D. A., Kim, S. Y., and Stent, G. S., 1984, Cell lineage in the development of the leech *Helobdella triserialis, Dev. Biol.* **104:**65–85.

Whitman, C. O., 1878, The embryology of *Clepsine*, *Q. J. Microsc. Sci. (N.S.)* **18:**215–315.

Whitman, C. O., 1892, The metamerism of *Clepsine*, in: *Festschrift zum 70. Geburtstage R. Leuckarts*, Engelman, Leipzig, pp. 385–395.

Whittaker, J. R., 1982, Muscle lineage cytoplasm can change the developmental expression in epidermal lineage cells of ascidian embryos, *Dev. Biol.* **93:**463–470.

Willard, A. L., 1981, Effects of serotonin on the generation of the motor program for swimming by the medicinal leech, *J. Neurosci.* **1:**936–944.

Zackson, S. L., 1982, Cell clones and segmentation in leech development, *Cell* **31:**761–770.

Zackson, S. L., 1984, Cell lineage, cell interactions and segment formation in the embryo of a glossiphoniid leech, *Dev. Biol.* **104:**143–160.

Zipser, B., Stewart, R., Flanagan, T., Flaster, M., and Macagno, E., 1984, Do monoclonal antibodies stain sets of functionally related leech neurons? *Cold Spring Harbor Symp. Quant. Biol.* **48:**551–556.

11

Dynamic Regulators of Neuronal Form and Connectivity in the Adult Snail *Helisoma*

S. B. KATER

1. Introduction

Among the most striking features to emerge from 15 years of investigation of *Helisoma* are the tremendous adaptability and resiliency of its nervous system and the precision with which remodeling can occur. Somewhat ironically, *Helisoma* was initally selected for investigation because of the existence of definable behavior, especially fixed action patterns, and a nervous system that was approachable because of its relative simplicity and assumed stability. However, as the analytical precision of investigations on this system improved, the view of a nervous system of immutable circuits has been refined to encompass a much more dynamic set of capabilities. It is apparent now that these more plastic attributes may be a general property of nervous systems. One might also speculate that the responses of perturbation-induced neural plasticity are actually an amplification of ongoing events normally present in all neuronal systems.

While it has been clear for many years that neuronal plasticity exists, the degree of precision and the mechanisms underlying such remodeling have, in large part, remained unknown. It is almost axiomatic that the precision of any analysis of neuronal change will be limited by the degree to which a steady-state system can be defined under stable conditions. With this rationale, we have focused on *Helisoma* because of its relative simplicity and its ability to act as a model system to study a range of topics in neuronal plasticity. The resilience and adaptability of the *Helisoma* nervous system has proved surprising. Even more so, the precision with which entirely new neuronal circuits can be routinely designed starkly emphasizes the inappropriateness of earlier views of fixed circuitry. A view which someday may

S. B. KATER • Department of Biology, University of Iowa, Iowa City, Iowa 52242.

prove most appropriate might consider that a given circuit or configuration of connections within a specific neuronal ensemble really represented a best fit to the environmental conditions that confront the nervous system. Perhaps ultimately, neurosciences will have a concept of neuronal connections more like that which physicists have for phenomena such as subatomic architecture. Rather than certain connections being rigidly specified, we may someday demonstrate that connections (as electrons in space) occur in a probabilistic fashion that can be biased by the environmental conditions in which the system presently exists.

This chapter provides some examples of neuronal plasticity in *Helisoma* and the rather simple rules that can govern the formation of connections. Highly specific and entirely novel sets of electrical connections can be routinely evoked experimentally in *Helisoma.* The analysis of these changes is based on the stable characteristics of a neuronal circuit that underlies the feeding behavior of this animal. Both the stable and plastic properties of these neurons were initially studied within the buccal ganglion environment. However, more recent experiments have taken advantage of cell-culture techniques to study more critically the mechanisms involved in these changes. The discussion will conclude with new experimental data that demonstrate that molecules released by neurons themselves can regulate neurite outgrowth. These molecules may not only control neuronal morphology, but they may also regulate the formation of connections in a nervous system.

2. Buccal Ganglion Control of Feeding *Helisoma*: A System for Studying Neuronal Plasticity

Our early work was concerned with the feeding behavior of this snail. The strategy was to make a series of progressively more reductionistic descriptions ranging from the behavior itself through the muscles, the motor neurons, and eventually to the neurons that ultimately decided whether or not motor neurons would generate a feeding motor program. In its simplest form, feeding involves the reciprocal movements of the rasplike radula against food substratum (Fig. 1). These movements are mediated by the highly specific phase-coupled contractions of the 54 muscles forming the buccal mass. In the absence of innervation, local contractions and uncoordinated movement of the buccal mass and radula can occur, but it is the activity of the buccal ganglia imposed upon the muscles of the buccal mass that patterns the feeding behavior. Because of their relatively large size, discrete locations, and contrasting coloration, many of the individual neurons within the buccal ganglia can be mapped to precise locations. The activity of identified motor neurons during feeding is derived from intrinsic membrane properties and inputs from extrinsic sources. The feeding behavior is a composite in which a central motor program (Kater, 1974; Kaneko *et al.,* 1978) within the buccal ganglia is integrated with receptor feedback from the buccal mass itself (Kater and Rowell, 1973). The central program determines the timing of the feeding activity, whereas the sensory feedback appears to regulate the strength of motor bursts.

One of our final considerations in analyzing the feeding behavior was to understand how buccal ganglion circuitry was "turned on." Two prominent, higher-order neurons in the cerebral ganglia with axons extending to the buccal ganglia were identified. These neurons (C_1R and C_1L), which are undoubtedly

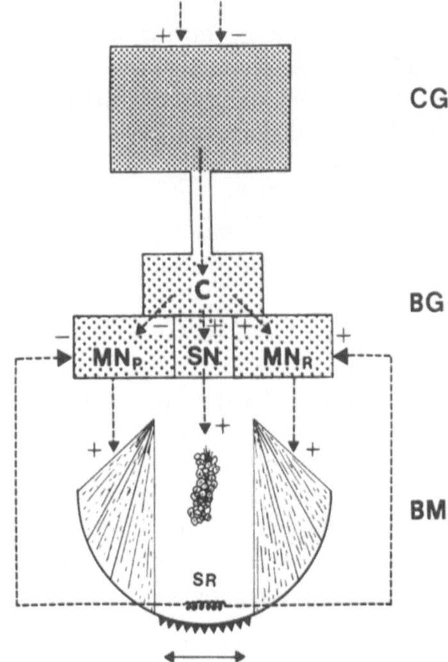

FIGURE 1. A schematic representation of the neural control of feeding in *Helisoma*. The filelike radula rasps back and forth against the substratum as a result of opposing contractions of muscles of the buccal mass (BM). The central motor program neurons (C) intrinsic to the buccal ganglia (BG) imposed the feeding rhythm on protractor (MN_P) and retractor (MN_R) motoneurons as well as on to the salivary effector neuron (SN). The output of the buccal ganglia is modulated by stretch receptors (SR) within the buccal mass and descending, higher-order input from the cerebral ganglia (CG).

homologous to a wide range of other giant serotonergic cerebral cells in other mollusks (see Granzow and Kater, 1977), have a profound effect on the feeding motor program. In many buccal ganglia that display an ongoing feeding motor program, one finds concurrent activity in the two bilaterally symmetrical cerebral neurons C_1. Frequently the experimental silencing of C_1, by passage of hyperpolarizing current, also results in cessation of the motor program. Conversely, this higher-order neuron can activate the feeding motor program in quiescent buccal ganglia. The complete feeding motor pattern can be activated by action potentials initiated experimentally in a previously quiescent C_1. Nueron C_1 exerts this "commandlike" function on the feeding motor program by what appears to be the humoral release of serotonin within the buccal ganglion (Granzow and Kater, 1977). Investigations such as those just described on the neuronal components of the feeding system have helped to define the cellular characteristics of an analyzable neural circuit and provided the basis for our subsequent investigations on neuronal plasticity.

From our studies on the neuronal basis of feeding, three particular buccal ganglion neurons have been selected for detailed observation and manipulation in studying neuronal plasticity: neuron 4 which provides excitatory chemical synaptic inputs to the salivary gland (Kater *et al.*, 1978a,b); neuron 5 whose axons branch extensively over the foregut; and neuron 19 which provides chemical excitatory input to the largest intrinsic feeding muscle of the buccal mass (Kater, 1974). Additionally, we have examined two serotonergic neurons: neuron C_1 of the cerebral ganglia and neuron P_5 of the pedal ganglia. For each of these neurons, we have extensive knowledge of their stereotyped morphology, biophysical properties, and interconnections with other elements. Our studies on the ability of these identified neurons to change during neuronal plastic events is based on the stereotypic nature of these characteristics. These studies have emphasized questions about the ability of individual, rather different, neurons to change selectively in response to environmental perturbations.

3. Neuronal Plasticity: The Adaptability and Resiliency of Buccal Ganglion Neurons

3.1. From CNS to Cell Culture

A variety of experimental configurations can be employed to examine neuronal plasticity in *Helisoma*. In this way, one can first observe and document the capabilities of component neurons in a normal environment and then examine these neurons under conditions that favor the study of the mechanisms of neuronal plasticity. Figure 2 shows three examples of identified neuron 5 examined in its normal form in the ganglion, after growth in response to axotomy, and after removal from the ganglion. Neuron 5 normally has a simple "balloon on a string"

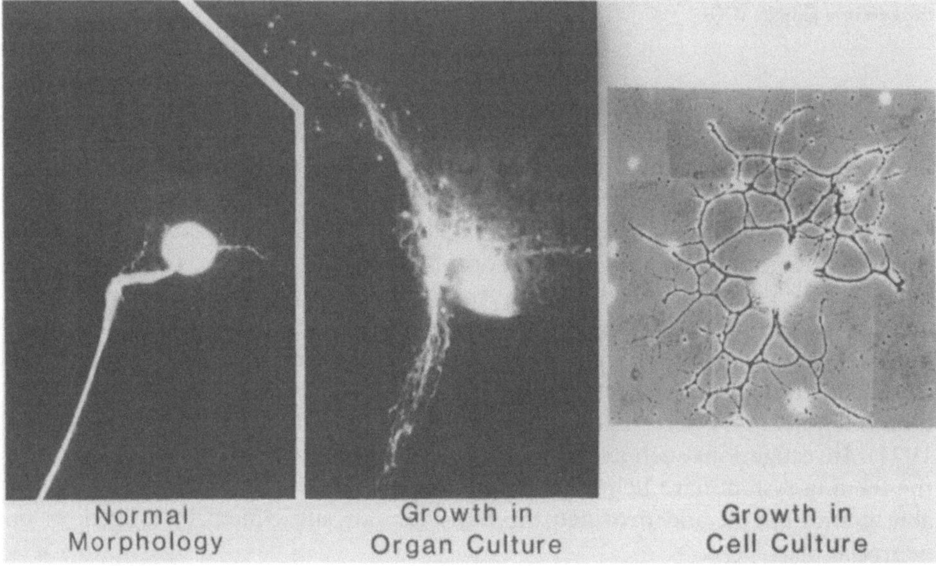

FIGURE 2. Examples of three different environmental conditions in which a given identified neuron (in this case buccal ganglion neuron 5) can be experimentally examined and manipulated. Within normal ganglia one can readily examine both physiological and morphological aspects of a normal neuron by standard intracelllular microelectrode recording and dye injection techniques. Similar methods, using Lucifer Yellow staining, can be used in axotomized neurons within normal buccal ganglia either *in situ* or in organ culture. Such neurons grow profusely and form definable new connections (middle panel). Any given identified neuronal cell body can be removed from a normal ganglian and placed in cell culture (right panel) where again both physiological and morphological aspects of this neuron can be examined, but in this case, in the most defined of environmental conditions.

morphology, a sparse dendritic tree, and a single prominent axon (Fig. 2, left panel). This basic form is very similar from animal to animal. When its major axon is severed by nerve crush, neuron 5 exhibits a characteristic growth pattern (Bulloch and Kater, 1981, 1982; Murphy and Kater, 1980a,b) in which multiple growth cones produce profuse neuritic sprouting that extends out major nerve trunks as well as across the central commissure (Fig. 2, center). Clearly, the pattern of this growth must be the product of intrinsic features of this neuron interacting with aspects of the particular terrain over which it grows. But one cannot readily manipulate either the intrinsic features of the neuron or its terrain within the buccal ganglion. To this end we have turned to cell culture (Wong *et al.*, 1981). The isolated neuron 5 shown at the right of Fig. 2 was a sphere when placed in culture and subsequently elaborated, through multiple growth cones, a complex pattern of neurite outgrowth. As might be expected, the morphology of this neuron is clearly different from that within the ganglion either when compared with normal development (left) or regeneration (center). This is the case for all neurons that we have grown in cell culture.

With this system we can partition the nervous system of *Helisoma* and study its neuronal form and connectivity at multiple levels of organization. Thus, our studies of neuroplasticity have interrelated observations made from the same identified neuron not only in the intact ganglion, *in situ,* but also in isolated cell culture. The technique of cell culture of individual identified neurons has added immeasurably to the precision with which we can address questions of the mechanisms underlying specific neuronal responses. In addition, it is now possible with this system to identify specific molecular components that may regulate neuronal growth and connectivity. Figure 3 illustrates the range of neuroplastic phenomena that we have investigated. All of these begin with anatomy-induced growth (Stamler *et al.*, 1982) and involve phenomena such as the reformation of broken connections (Murphy and Kater, 1978, 1980a), modulation of existing connections (Bulloch *et al.*, 1980), and formation of entirely new connections (Bulloch and Kater, 1981, 1982).

3.2. Neuron-specific Neuronal Plasticity

A basic question of long-standing in the neurosciences concerns the degree of individuality of the component neurons of a system. Our initial observation of cultured buccal ganglia in which all nerve trunks were severed gave the impression of homogeneity of neuritic sprouting (Stamler *et al.*, 1982). However, careful examination of individual neurons has revealed differences between specific neurons that might erroneously be "averaged" out by population studies. For example, when the esophageal nerve (which contains the axon of both neuron 5 and neuron 4) is crushed, neuron 5 displays a high-fidelity pathfinding in this nerve trunk during regeneration, essentially retracing the same pathways it occupied before axotomy. In marked contrast, neuron 4 grows indiscriminately and may send neurites into several "foreign" nerve trunks that its growth cones encounter (Murphy and Kater, 1980a,b). By treating neurons as individuals, one rapidly finds that they are encoded with sets of highly individualized neuroplastic responses. As will be seen

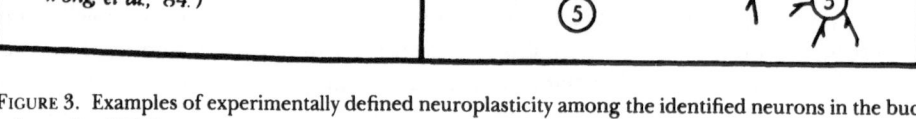

DEFINED NEURONAL CIRCUITRY AND REFERENCES	
Known electrical, chemical excitatory and chemical inhibitory synapses *(Kater and Rowell, '73; Kater, '74; Kaneko, et al., '78.)*	C′ C″ 5 19 4

KNOWN CLASSES OF CHANGE	
Adult neurons grow rapidly after axotomy *(Murphy and Kater, '80A; Stamler, et al., '82.)*	
Differential pathfinding during regeneration (cell 5 is accurate, cell 4 is indiscriminant) *(Murphy and Kater, '80B.)*	* ? *
Precise reformation of broken connections *(Murphy and Kater, '80A; Bulloch and Kater, '82.)*	
Breaking extant chemical connections Strengthening extant electrical connections *(Bulloch, et al., '80)*	
Formation of new chemical and electrical connections *(Bulloch, et al., '80; Bulloch and Kater, '81.)*	
Growth of identified neurons in cell culture *(Wong, et al., '81; Wong et al., '84.)*	19 4 5 → 19 4 5

FIGURE 3. Examples of experimentally defined neuroplasticity among the identified neurons in the buccal ganglia of *Helisoma*.

subsequently, growth cone behavior, growth rates, and responses to growth-regulating factors may all be neuron specific.

3.3. The Dynamic Nature of Connections within Adult Buccal Ganglia

It has been known for some time that synaptic inputs to mammalian spinal motor neurons are broken subsequent to axotomy of those motor neurons (Kuno and Llinas, 1970). Axotomy of *Helisoma* neurons also results in a major remodeling of connections. As with mammalian spinal motor neurons, chemical synapses can be broken. Additionally, existing electrical synapses can be modified and the strength of the connections can be increased (Bulloch *et al.*, 1980). Most striking, however, is the formation of entirely new chemical (Hadley, 1983) and electrical synaptic (Bulloch and Kater, 1981) connections. The remainder of this chapter reviews our present knowledge of the rules underlying the formation of new electrical synapses and the agents known to regulate their formation.

The events underlying the formation and breaking of neuronal connections are fundamental to the establishment of functional neuronal circuitry. Adult *Helisoma* neurons provide an opportunity to investigate these processes among highly stereotyped central neuronal connections. For example, neuron 5 is not coupled to its contralateral homologue (Fig. 4A) nor to neuron 4 in the normal adult buccal ganglia. However, axotomy-induced growth of neuron 5 inevitably results in the establishment of sets of new electrical synapses including connections with the contralateral neuron 5 (Fig. 4B), both neurons 4, and population of retractor motorneurons (Fig. 5). Although formation of new connections initially appears random, it rapidly becomes highly specific when all connections, except those between 5R and 5L, are broken. Thus, there appears to be a process of sampling, testing, making and breaking of connections which may indicate an important organizational scheme underlying what previously has seemed a rigid and stereotyped set of adult circuitry in *Helisoma*. Rather than specify neurons for certain connections, the underlying mechanism may provide opportunities for remodeling in order to achieve a "best fit" under a given set of circumstances. One could envision a group of equipotential connections in which the priority connection that was eventually selected and stabilized was different in response to different environmental behavioral inputs.

4. The Regulation of Electrical Synapse Formation by Growing Neurites

In examining the initial set of data on the electrical connections formed by neuron 5, it became clear that there was a subset of neurons that never connected under the conditions of these experiments (Fig. 5). This raised the question as to whether these neurons were intrinsically different from the neurons that con-

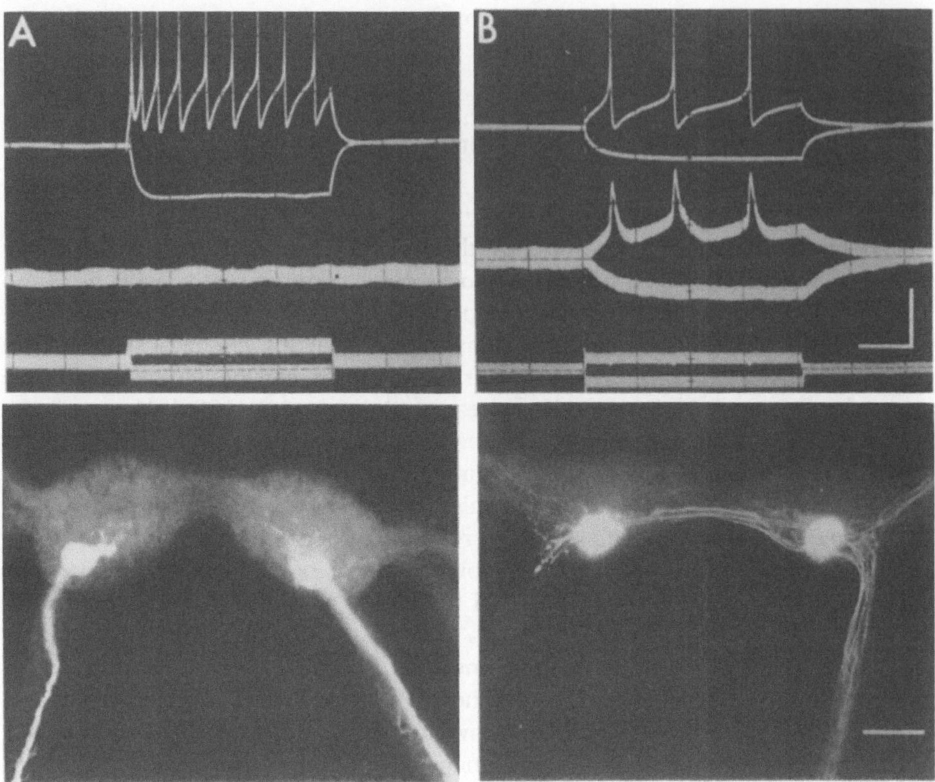

FIGURE 4. An example of both physiological and morphological changes between a pair of identified neurons in *Helisoma.* (A) A demonstration of the lack of electrical couplings between homologues of neuron 5 in normal buccal ganglia as well as a demonstration of the normal morphology of these neurons as shown by intracellular staining with Lucifer Yellow. (B) A demonstration of the new neurite outgrowth and subsequent formation of electrical couplings between the same identified neuron after seven days in organ culture. Calibration with electrophysiological records is 200 msec and 25 mV upper traces, 5 mV middle traces, and 2 nA lower traces. Calibration for fluorescent photomicrographs is 100 μm. (Modified from Bulloch and Kater, 1981.)

nected with neuron 5 or whether they were in a different "state" from those which connected with neuron 5. A series of investigations with R. D. Hadley and C. S. Cohan (Hadley *et al.*, 1983) has addressed the idea that in order to form electrical synapses, *both* potentially interconnecting neurons must have overlapping, newly generated neurite outgrowth. Figure 6 shows one test of this hypothesis in which we axotomized only a single neuron 5 in the buccal ganglia. Under these conditions, essentially no connections were made by these neurons, even though the axotomized partner sent extensive neurite outgrowth that overlapped the stable neuron 5. This contrasts with *bilateral* axotomy experiments that inevitably resulted in formation of electrical synapses. Thus, only when both potential partners had mutually overlapping new neurites did electrical synapses form. The reciprocal of this experiment was to determine conditions where a neuron that previously never

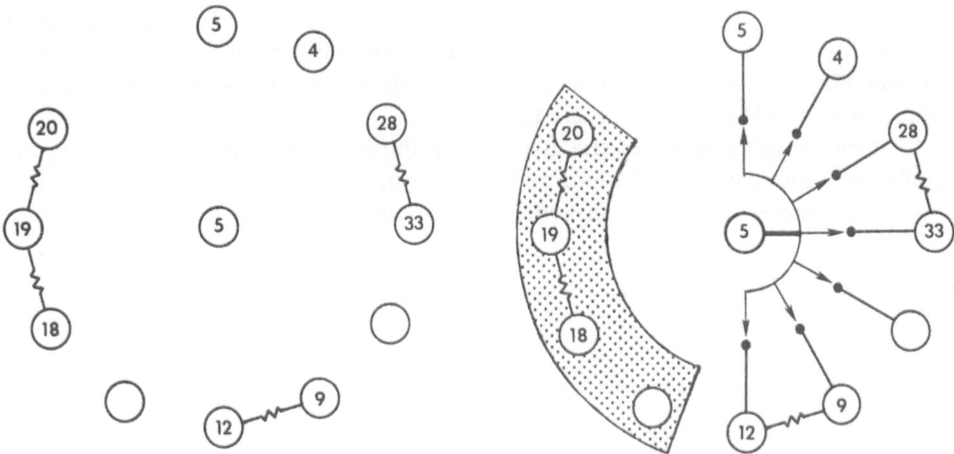

Figure 5. Neuronal plasticity as seen from the vantage point of a single identified neuron. The normal connections of neuron 5 (left) and connections made after isolation of ganglia and axotomy of many neurons (right). Many new connections are formed (arrows). The thatched area represents neurons not usually connecting with neuron 5 in these experiments.

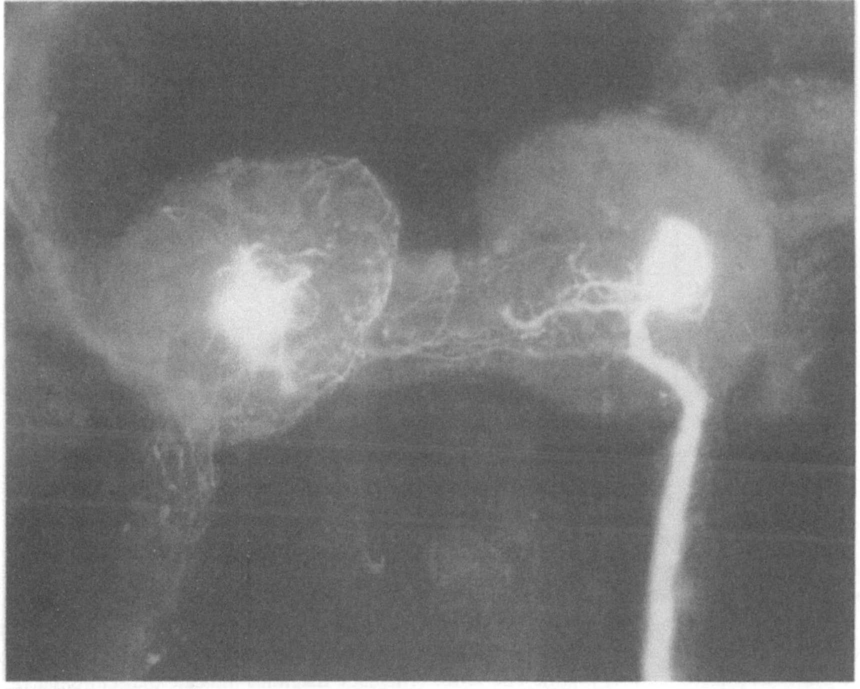

Figure 6. A more selective axotomy (zap axotomy) (Cohan *et al.*, 1983) of only a single neuron 5 alters connectivity patterns. These Lucifer Yellow-filled neurons demonstrate significant neurite overlap onto the stable neuron (right). Under these conditions no connections form between these elements.

connected, such as neuron 19, could connect with neuron 5. Indeed, when axotomies are performed appropriately so as to allow overlapping neurite outgrowth between both neuron 19 and neuron 5, strong electrical connections form between these neurons (Hadley and Kater, 1983).

A more rigorous test of the growth hypothesis has been performed using the ability to culture individual identified neurons in cell culture where temporal and spatial characteristics of outgrowth can be monitored more precisely. *Helisoma*

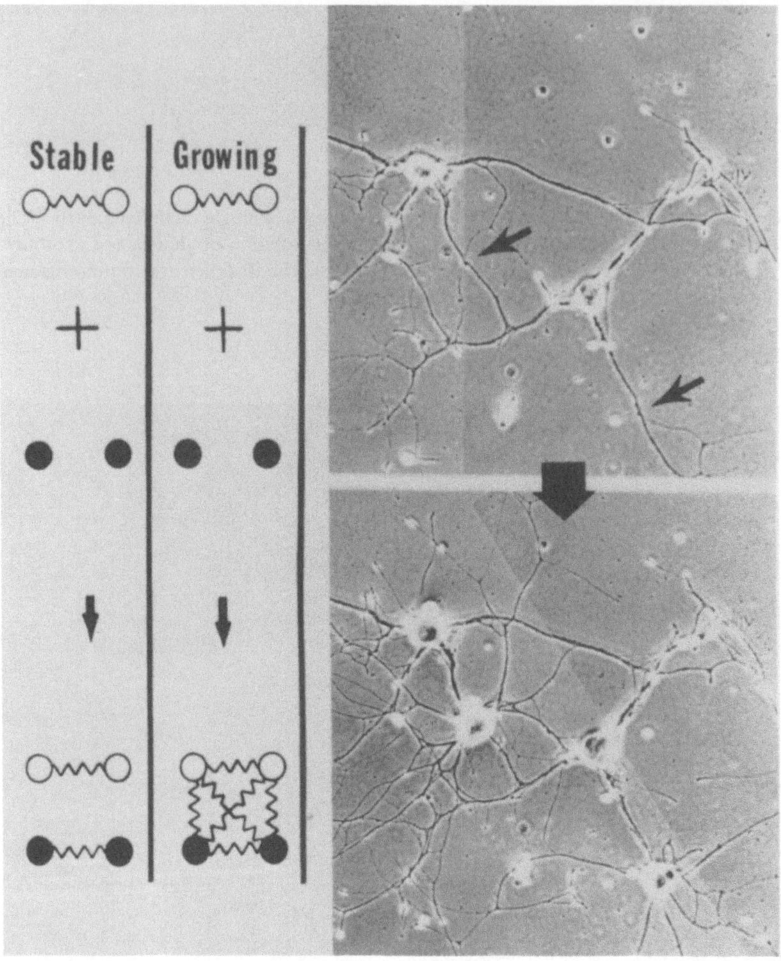

FIGURE 7. Strong electrical connections resulting from temporal coordination of growth. The upper phase-contrast photomicrograph are of established cultures of neurons (left) which can have secondary neurons subsequently added (arrows); one day later, growth from secondary neurons produces networks of asynchronous plated neurons (right). The lower schematic diagrams indicate that coupling results differ according to the growth status of the original neurons (open circles) at the time of plating of secondary neurons. If growing, the primary neurons couple with the growing secondary neurons. If the primary neurons are morphologically stable when the secondary neurons are added, then coupling with the growing secondary neurons fails to occur or is very weak. (Modified from Hadley *et al.*, 1983.)

neurons in cell culture undergo a characteristic sequence of sprouting, active growth, and cessation of outgrowth. It is possible to plate neurons at different temporal intervals which allows neurons at different stages of outgrowth to interact. Pairs of neurons can be plated one after another allowing either simultaneous or asynchronous outgrowth (Fig. 7). In cell culture, as was seen *in situ,* we have found that the primary rule for determining whether or not two neurons will form an electrical synapse is simply whether or not both neurons have overlapping, newly generated neurite outgrowth (Fig. 7). Even when one neuron extensively overlapped the processes of a "stable," preplated neuron, functional connections never formed.

With a clear dependence on growth for the formation of new electrical synapses, it now becomes important to examine those factors that might have the capability of regulating the growth of individual neurons. These factors that regulate growth may also exert regulatory effects on synaptogenesis.

5. Regulators of Neuronal Outgrowth

5.1. Brain-derived Growth Factor(s)

The ability to interrelate data from intact buccal ganglia with data from isolated buccal ganglion neurons in cell culture has provided several indications that the CNS itself is capable of regulating the outgrowth of its constituent neurons. R. G. Wong has defined a proteinaceous factor(s) released by neural tissue that promotes neuritic growth (Wong *et al.,* 1981). When cultured in a defined medium, both isolated neurons and neurons within intact ganglia initiate, at least, only very limited extensions. Under identical circumstances, but with the addition of this brain-derived factor(s), these neurons, both in isolation and in ganglia, display substantial outgrowth and form extensive networks of neurites. The full significance and nature of the molecular control of growth in normal-functioning *Helisoma* nervous system has not yet been established. It is noteworthy, however, that these observations are not just limited to isolated neurons in cell culture, but that intact buccal ganglia cultured *in vitro* also seem to require the same conditions for neurite outgrowth. This observation on intact ganglia and the fact that the factor is synthesized specifically by ganglia (Wong *et al.,* 1981, 1984) indicate that such substances may normally play key roles in the general regulation of neuronal growth in this system.

5.2. Neural Transmitters as Regulators of Growth Cone Behavior

We have considered the possibility that individual neurons might have the capacity for selectively regulating the neurite outgrowth of other specific neurons. We have observed a wide range of growth patterns of individual neurons in cell culture. On occasions some neurons seem to be attracted to other neurons, whereas on other occasions we see what appears to be repulsion (Fig. 8). Although many factors could affect such growth patterns, the example shown in Fig. 8 was in fact the impetus for us to explore whether rather simple molecules that may regulate neurite outgrowth might be released by neurons. One possible candidate molecule is the classical neurotransmitter.

In order to determine the effect of neural transmitters on outgrowth, we have developed both a microphotographic and low-light-level video system for precisely analyzing growth cone behavior (Kater and Hadley, 1982). Figure 9A illustrates a reliable method for quantitatively analyzing the behavior of individual growth cones from an identified *Helisoma* neuron. We have found that growth cones move across culture substrata at a linear rate. The motile behavior of growth cones of an individual identified neuron has highly stereotyped components. Individual neurons grow at rates that are characteristic of a particular identified neuron and respond to extrinsic agents in a neuron-specific fashion. Active *Helisoma* growth cones display numerous long filopodia which when viewed with time-lapsed video

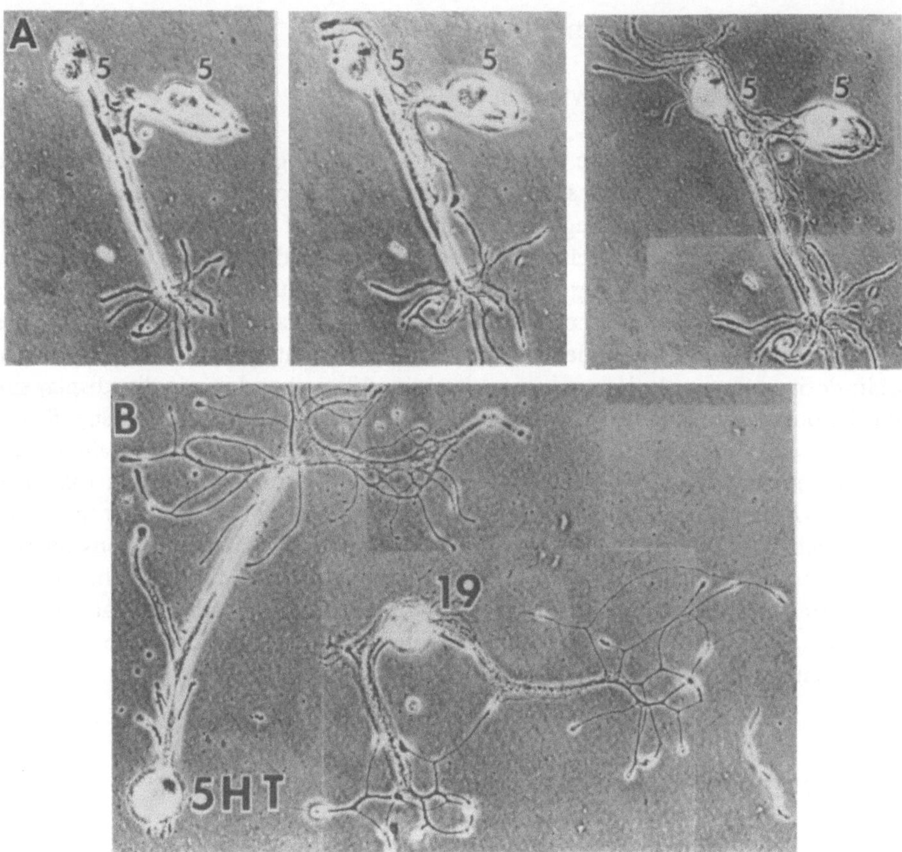

FIGURE 8. Selected patterns of outgrowth as can be seen between pairs of neurons in cell culture. Sometimes growth patterns are suggestive of neuron–neuron attraction (A), whereas other patterns might result from neuron–neuron repulsion (B).

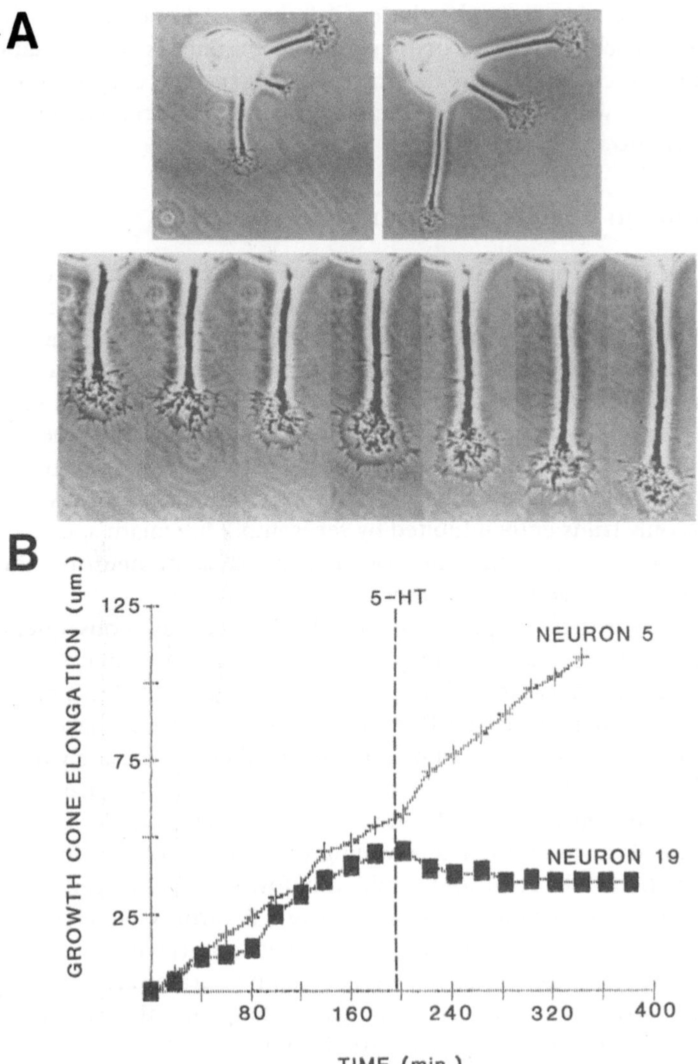

FIGURE 9. Photomicrographic measurements of growth cones and associated neurites allow quantitative analysis of growth cone motility. (A) The sequential photomicrographs shown in the lower portion were taken at 40-min intervals. The upper two micrographs indicate the overall morphology of neuron 5 at the onset and completion of these measurements. (B) The effect of serotonin (5-HT) on the growth rates of neuron 5 and neuron 19. Application of as little as 10^{-7} M 5-HT to a neuron 19 can abruptly halt growth cone elongation (■). Application of serotonin in excess of 5×10^{-5} M serotonin has no effect on neuron 5 growth cone elongation. (From Haydon, McCobb, and Kater unpublished data.)

techniques can be seen to probe the environment. Additionally, broad flattened lamellapodia are characteristic of these motile organelles. On cessation of outgrowth, dramatic changes occur in growth cone structure consisting of filopodial and lamellapodial withdrawal. These stable growth cones become phase bright and are easily recognizable.

5.2.1. Serotonin Regulates Growth Cones

Using the above quantitative methods, we have found that the neural transmitter serotonin has a neuron-specific effect on the behavior of growth cones in *Helisoma.* Bath application of as little as 10^{-7} M serotonin (5-HT) can abruptly halt the growth cones of neuron 19 (Fig. 9B). This is characteristically accompanied by a retraction of filopodia and lamellapodia and an abrupt cessation of elongation. In contrast, neuron 5 is never affected by serotonin even in concentrations 500 times greater. These two neurons represent the extreme responses to serotonin we have seen thus far. Other neurons can have intermediate responses. For example, neuron P_5 is only transiently inhibited by serotonin. One might speculate that continued trials on other identified neurons may reveal some neurons whose growth rates are even enhanced by serotonin.

Effects such as these on growth cone behavior may have consequences beyond that of simply altering neuronal morphology. As discussed earlier, for example, the formation of electrical synapses is dependent on the growth status of a neuron. In fact, experiments performed by P. Haydon and D. McCobb have demonstrated that specificity of connections can be obtained by the appropriately timed addition of serotonin to the medium. Neuronal ensembles in cell culture composed, for instance, of two neurons 5 and neuron 19 completely interconnect with one another in the absence of other intervening variables. However, in parallel experiments in which 10^{-6} M serotonin is added before neurites of cells 5 and 19 have overlapped, only connections between neurons 5 form. Connections between 5 and 19 do not form because neuron 19 ceases its outgrowth in response to serotonin. Thus, the presence or absence of a common neural transmitter at propitious times during synaptogenesis can specifically designate the resultant circuitry. Again, one might speculate as to the broader ramifications of such a finding. Is it possible that during neuronal plasticity (as well may be the case during development) the specific presence of particular neural transmitters could regulate the final pattern of connectivity (e.g., Kasamatsu and Pettigrew 1979)?

There are essentially three further directions that this work might take: (1) additional neural transmitters must be examined, in addition to a wider range of identified neurons; (2) we must determine whether such effects as observed in cell culture can be obtained *in situ;* and finally, (3) we must seek out the specific locus of action for these effects.

A. D. Murphy has obtained a set of data which demonstrate that, in fact, serotonin can have highly neuron-specific effects *in situ.* Addition of serotonin to medium in which buccal ganglia have been cultured selectively reduces the amount of neurite outgrowth produced by specific neurons. Neuron 19, for instance, produces significantly less outgrowth in the presence of serotonin, whereas neuron 5 is apparently unaffected (Fig. 10) just as was seen in cell culture. These results

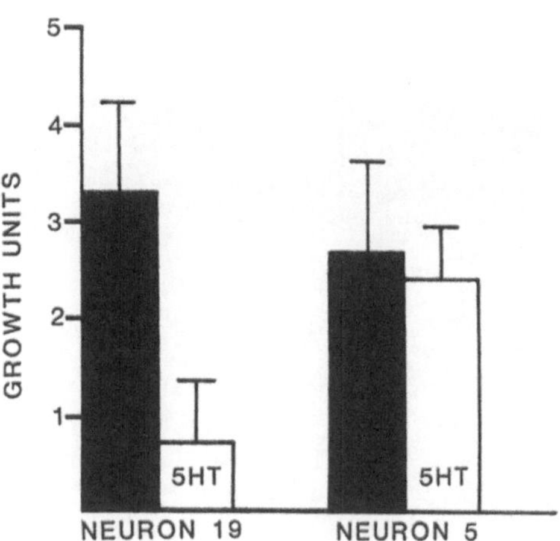

FIGURE 10. The response of identified neurons within intact ganglia to application of serotonin during organ culture of whole ganglia. Each bar represents the mean growth unit response of 12 individual neurons. Under these conditions, serotonin significantly retards the outgrowth of neurons 19 but not neurons 5, precisely as seen for isolated neurons in cell culture. (From Murphy and Kater, unpublished data.)

strongly suggest a most intriguing possibility. Namely, since it has already been established that cerebral neuron C_1 exerts its "commandlike" effect on the buccal ganglia feeding motor program by releasing serotonin (Granzow and Kater, 1977), one wonders whether such a release process could also alter the growth properties of specific neurons in this ganglion.

5.2.2. The Locus of Action of Serotonin

By focal application of serotonin from a micropipette, it has been possible to demonstrate that application of serotonin to the growth cones of neuron 19 always evokes the inhibition of motile activity. Similar treatments to cell bodies or neurites are without effect. A more dramatic demonstration of this local effect can be made by virtue of the fact that individual growth cones can be physically isolated from the rest of the neuron (Fig. 11A). Such isolated growth cones survive and grow for several hours. Under these conditions, the motility of isolated growth cones of neurons 19 can be inhibited by pipette application of serotonin (Fig. 11B). This effect is reversible since withdrawal of the serotonin pipette results in the resumption of motility (Fig. 11C). Again, it is worth considering what analogous processes might occur *in situ*. If we conceptually replace the micropipette shown in Fig. 11 with a serotonin-releasing terminal from C_1, we can see the plausibility for neurite-specific inhibition of outgrowth. This mechanism would allow for temporal and spatial inhibition of neurite outgrowth in a neuron-specific fashion. Perhaps it is precisely such situations that account for the pruning and trimming of the dendritic trees and axons to produce the characteristic morphologies of these identified neurons *in situ*.

Ultimately we would like to tie together earlier observations made *in situ* with molecular changes responsible for these selective modifications. As one avenue in this regard, C. S. Cohan has begun an analysis of the growth cone ionic current of *Helisoma* neurons. Are there ionic currents of interest in growth cones? Figure 12 demonstrates a patch-clamp record from the growth cone membrane of a *Helisoma*

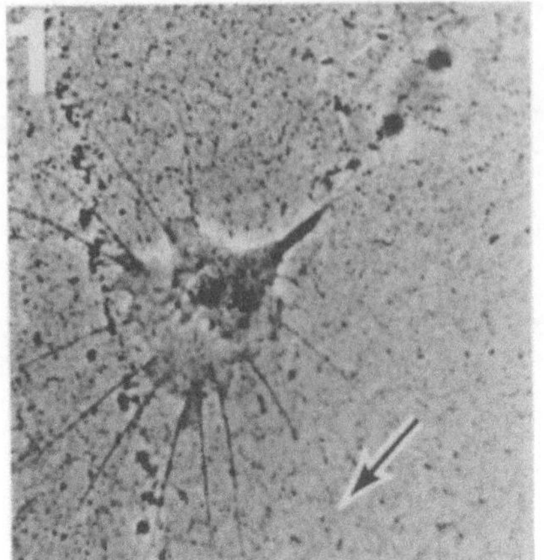

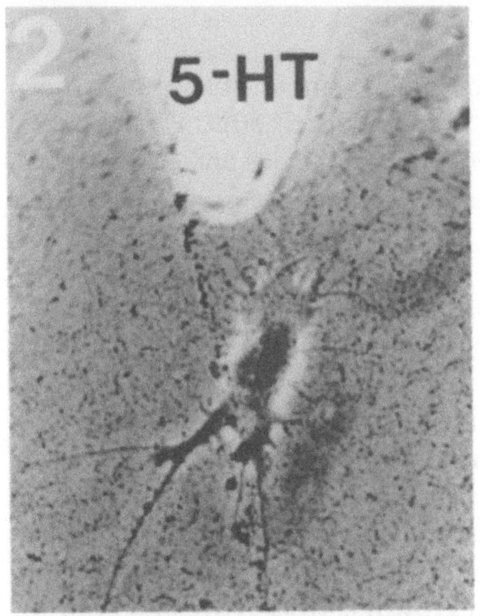

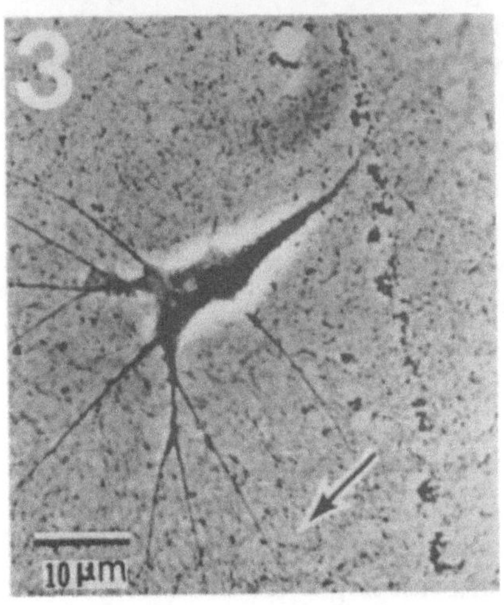

Figure 11. The behavior of isolated growth cones from a neuron 19 of the buccal ganglia of *Helisoma*. Growth cones can be isolated by microelectrode cuts of the neurites. (A) Shows such a growth cone in an active state of motility. (B) Shows the same cone shortly after the addition of a 10^{-6} M 5-HT-containing micropipette. Motility is abruptly halted. (C) Shows the restoration of active motility of this growth cone about 20 min after removal of the 5-HT-containing micropipette. (From Haydon and Kater, unpublished data.)

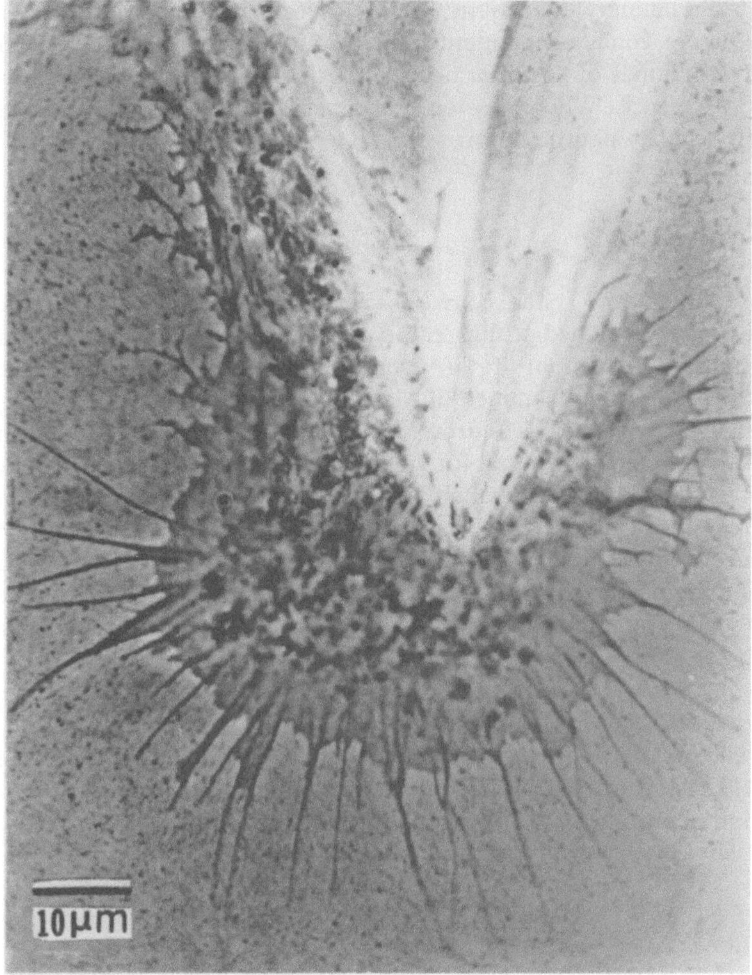

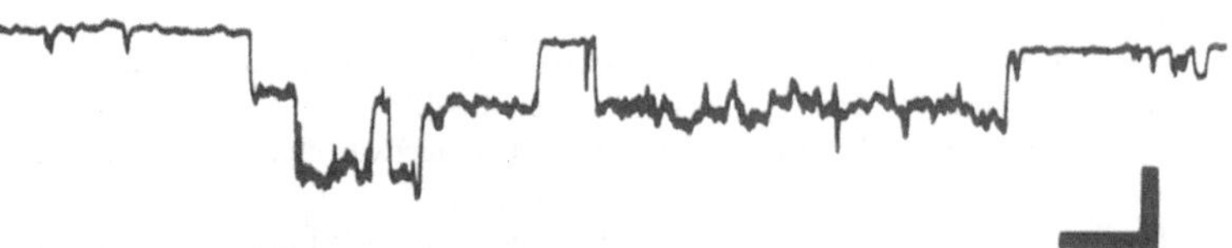

FIGURE 12. Patch-clamp recording from a growth cone of *Helisoma*. The upper panel is a photomicrograph of the growth cone of a neuron 5 with a patch-clamp pipette in place. The lower trace shows exemplary records of channel openings and closings as seen in a cell-attached patch recording configuration. Calibration, 400 msec, 4 pA. (From Cohan, Haydon, and Kater unpublished data.)

neuron 5. Preliminary indications are that there may be significant differences in whether or not voltage-dependent ionic channels are present in growing versus stable growth cones of *Helisoma* neurons. This should provide the future opportunity to ask precisely how serotonin and perhaps other neural modulators can selectively regulate neurite outgrowth.

6. Conclusions

The adult nervous system of *Helisoma* is capable of significant change. Its neurons are not constrained within embryonically determined circuits, but rather remain capable throughout life of reconnecting into new, quite predictable, patterns. This chapter has examined only a few of the many alternate configurations that a single neuron, such as neuron 5 or neuron 19, may take. The lesson of even this highly restricted view is that for any given neuron there is an array of different possibilities. Even simple environmental conditions, such as the presence of a particular neural transmitter, can bias what role a given neuron assumes. Perturbation-induced neural plasticity has revealed an enormous range of neuron-specific potential. One now must consider more seriously new possibilities for the growth and remodeling of connections that might occur during the normal course of animal behavior.

ACKNOWLEDGMENTS. This work was supported by PHS grants H.D. 18577, NS 15350, NS 18819 and NS 21217. I thank Drs. C. Cohan and P. Haydon for allowing citation of our unpublished work and for helpful comments on this manuscript and D. Dehnbostel for photograph assistance.

References

Barker, D. L., Wong, R. G., Kater, S. B., 1982, Separate factors produced by the CNS of the snail *Helisoma* stimulate neurite outgrowth and choline metabolism in cultured neurons, *J. Neurosci. Res.* **8**:419–432.

Bulloch, A. G. M., and Kater, S. B., 1981, Selection of a novel connection by adult molluscan neurons, *Science* **212**:79–81.

Bulloch, A. G. M., and Kater, S. B., 1982, Neurite outgrowth and the selection of novel electrical connections by adult *Helisoma* neurons, *J. Neurophys.* **48**:569–583.

Bulloch, A. G. M., Kater, S. B., and Murphy, A. D., 1980, Connectivity changes in an isolated molluscan ganglion during *in vitro* culture, *J. Neurobiol.* **11**:531–546.

Cohan, C. S., Hadley, R. D., and Kater, S. B., 1983, "Zap axotomy": Localized fluorescent excitation of single dye-filled neurons induces growth by selective axotomy, *Brain Res.* **270**:93–101.

Granzow, B., and Kater, S. B., 1977, Identified higher-order neurons controlling the feeding motor program of *Helisoma, Neuroscience* **2**:1049–1063.

Hadley, R. D., 1983, Neuronal growth and the formation of electrical synapses between identified neurons of adult *Helisoma* buccal ganglia, Ph.D. Thesis, University of Iowa.

Hadley, R. D., and Kater, S. B., 1983, Competence to form electrical connections is restricted to growing neurites in the snail, *Helisoma, J. Neurosci.* **3**:924–932.

Hadley, R. D., Kater, S. B., and Cohan, C. S., 1983, Electrical synapse formation depends upon interaction of mutually growing neurites, *Science* **221:**466–468.

Kaneko, C. R. S., Merickel, M., and Kater, S. B., 1978, Centrally programmed feeding in *Helisoma:* Identification and characteristics of an electrically coupled premotor neuron network, *Brain Res.* **146:**1–21.

Kasamatsu, T., and Pettigrew, J. D., 1979, Preservation of binocularity after monocular deprivation in the striate cortex of kittens treated with 6-hydroxydopamine, *J. Comp. Neurol.* **185:**130.

Kater, S. B., 1974, Feeding in *Helisoma trivolvis:* The morphological and physiological bases of a fixed action pattern, *Am. Zool.* **14:**1017–1036.

Kater, S. B., and Hadley, R. D., 1982, Video monitoring of neuronal plasticity, *Trends Nuerosci.* **5:**80–82.

Kater, S. B., and Rowell, C. H. Fraser, 1973, Integration of sensory and centrally programmed components in the generation of cyclical feeding activity of *Helisoma trivolvis, J. Neurophysiol.* **26:**142–155.

Kater, S. B., Rued, J. R., and Murphy, A. D., 1978a, Propagation of action potentials through electrotonic junctions in the salivary glands of the pulmonate mollusc *Helisoma trivolvis, J. Exp. Biol.* **72:**77–90.

Kater, S. B., Murphy, A. D., and Rued, J. R., 1978b, Control of the salivary glands of *Helisoma* by identified neurons, *J. Exp. Biol.* **72:**91–106.

Kuno, M., and Llinas, R., 1970, Alterations of synaptic actions in chromatolysed motorneurons of the cat, *J. Physiol.* **210:**823–838.

Murphy, A. D., and Kater, S. B., 1978, Specific reinnervation of a target organ by a pair of identified molluscan neurons, *Brain Res.* **156:**322–328.

Murphy, A. D., and Kater, S. B., 1980a, Sprouting and functional regeneration of identified neurons in *Helisoma, Brain Res.* **186:**251–272.

Murphy, A. D., and Kater, S. B., 1980b, Differential discrimination of appropriate pathways regenerating identified neurons in *Helisoma, J. Comp. Neurol.* **190:**395–403.

Stamler, J. F., Kater, S. B., Murphy, A. D., and Bulloch, A. G. M., 1982, Neuritic growth and neuroma formation by an isolated molluscan ganglion, *J. Neurobiol.* **13:**85–89.

Wong, R. G., Hadley, R. D., Kater, S. B., and Hauser, G., 1981, Neurite outgrowth in molluscan organ and cell cultures: The role of conditioning factor(s), *J. Neurosci.* **1:**1008–1021.

Wong, R. G., Barker, D. L., Kater, S. B., and Bodnar, D. A., 1984, Nerve growth-promoting factor produced in culture media conditioned by specific CNS tissues of the snail *Helisoma, Brain Res.* **292:**81–91.

Part III
Learning and Plasticity

12

Cellular Studies of an Associative Mechanism for Classical Conditioning in *Aplysia*

Activity-dependent Presynaptic Facilitation

THOMAS W. ABRAMS

1. Introduction

For much of this century, a central goal for both psychologists and neurobiologists has been to understand the mechanisms underlying associative learning. Of the various forms of associative learning, classical conditioning is perhaps the simplest and is thought to be the prototypical way in which an animal learns a predictive relationship (Pavlov, 1927; Kamin, 1969; Rescorla and Wagner, 1972). In classical conditioning, an animal alters its response to one stimulus as a result of the temporal pairing of this stimulus with a second event. During training, the animal comes to know that the first stimulus, called the conditioned stimulus, signals the occurrence of the second stimulus, the unconditioned stimulus. Because of its simplicity, many of the efforts to analyze cellular mechanisms of associative learning have focused on classical conditioning.

Classical conditioning was first demonstrated and has been most extensively studied in higher vertebrates. However, given the complexity of the neural circuits involved in the control of even the simplest behaviors in vertebrates, the analysis of learning has been difficult. Although recently, specific loci have been identified at which critical changes occur during learning (Gold and Cohen, 1981; Brons and

THOMAS W. ABRAMS • Center for Neurobiology and Behavior, College of Physicians and Surgeons, Columbia University, and Howard Hughes Medical Institute, New York, New York 10032.

Woody, 1980; Clark *et al.*, 1984), it has still not been possible to specify the cellular bases of these changes nor even to determine whether these changes are primary or merely consequences of changes at other sites. As a result, during the last two decades a number of biologists have turned to invertebrates to study conditioning. Initially it was questionable whether simple invertebrates actually show associative leaning. However, a number of studies have now demonstrated conditioning in both gastropod mollusks and insects (Horridge, 1962; Quinn *et al.*, 1974; Gelperin, 1975; Mpitsos and Collins, 1975; Crow and Alkow, 1978; Davis *et al.*, 1980; Hoyle, 1980; Lukowiak and Sahley, 1981; Chapters 13 and 14, this volume). These studies have also shown that many of the features of clasical conditioning seen in mammals are present in invertebrates, such as *Limax* and *Aplysia* (Sahley *et al.*, 1981; Hawkins *et al.*, 1938b).

These studies on invertebrates were initiated with two expectations: (1) that the mechanisms underlying associative learning might be more readily analyzable in invertebrate species, given their comparatively simple nervous systems; and (2) that these mechanisms might prove to be quite general. As illustrated by the study reviewed in this chapter, both of these expectations are beginning to be fulfilled. The cellular studies I describe here have been carried out in collaboration with Robert Hawkins, Tom Carew, Eric Kandel, and Lise Bernier. Before considering these, I begin by reviewing findings of behavioral studies of conditioning in *Aplysia*.

2. Classical Conditioning of the Defensive Withdrawal Reflex of *Aplysia*

2.1. *Aplysia* Shows Simple Associative Learning

Over the past several years ago, Walters *et al.* (1979), Carew *et al.* (1981), Lukowiak and Sahley (1981), and Ingram and Walters (1984) demonstrated that several behaviors of the marine snail *Aplysia californica* could be classically conditioned. One of the conditioning paradigms that was developed involved a simple protective reflex, the gill and siphon withdrawal reflex (Carew *et al.*, 1981; Lukowiak and Sahley, 1981). The external respiratory organ of *Aplysia*, the gill, is covered by a fold of skin called the mantle, which ends posteriorly in a fleshy excurrent siphon. Tactile stimulation of either the mantle or the siphon results in the withdrawal of the gill and siphon within the protective folds of the parapodia. This reflex seemed promising for a cellular analysis of conditioning since the neural circuitry had been worked out (Byrne *et al.*, 1974, 1978; Perlman, 1979; Hawkins *et al.*, 1981a) and two nonassociative forms of behavioral plasticity that this reflex undergoes, habituation and sensitization, had been extensively studied (Kandel and Schwartz, 1982).

The conditioning paradigm relies on the fact that the amplitude and duration of the reflex vary with the strength of the tactile stimulus. A very light touch to the siphon, which elicits a brief, small, gill and siphon withdrawal, was used as a conditioned stimulus. If during training this weak siphon stimulus is paired with an unconditioned stimulus (US), a moderate tail shock, which elicits a strong siphon withdrawal response (Fig. 1A,B), then after a series of pairing trials, the response to that same weak conditioned stimulus is markedly enhanced.

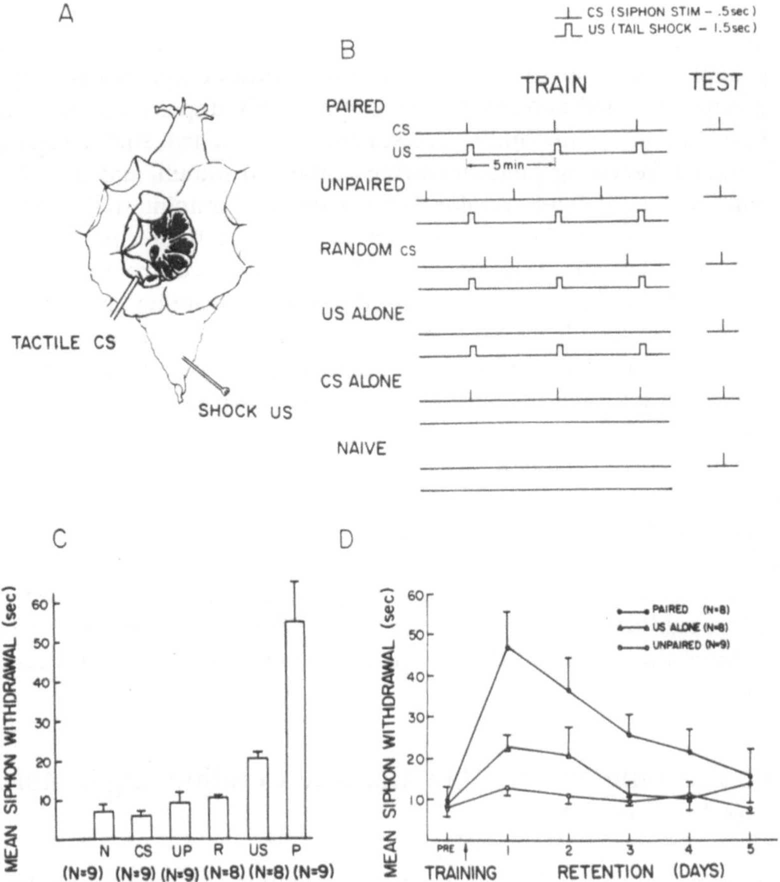

FIGURE 1. Classical conditioning of the defensive withdrawal reflex of *Aplysia.* (A) Dorsal view of *Aplysia* showing the sites used to deliver the conditioned and unconditioned stimuli. For illustrative purposes the parapodia and mantle shelf are shown retracted to reveal the gill and siphon. However, the behavioral studies were carried out with freely moving animals whose parapodia were surgically removed. The conditioned stimulus (CS) was a weak tactile stimulus to the siphon delivered with a nylon bristle that was inserted into the siphon and then briskly moved upward stimulating the inner surface of the siphon. The unconditioned stimulus (US) was a strong electric shock to the tail (50 mA for 1.5 sec) delivered with hand-held spanning electrodes. (B) Diagrammatic representation of the classical conditioning procedure and five control procedures. Training consisted of a series of stimulus presentations at 5-min intervals. The US alone group received pure sensitization training. After training, during the posttest, the duration of the siphon withdrawal was measured in response to the conditioned stimulus alone. In some experiments, a pretraining test with the conditioned stimulus alone was given to match groups and permit within-group comparisons. (C) Siphon withdrawal response 24 hr after training. The siphon withdrawal responses 24 hr after 31 training trials are shown for the six groups in B. Paired animals showed significantly greater reflex responsiveness than the control animals. The group that received US alone training had significantly greater responsiveness than the other control groups. Data in this and all other figures are expressed as means ± standard error of the mean. (D) Time course of retention of the conditioned response. After pretraining tests (PRE), three groups were given 30 trials of paired, US alone, or unpaired training and then were tested with the conditioned stimulus once every 24 hr. Within-group comparisons showed that the paired group exhibited significant retention of the learned response up to and including day 3; this group's response was significantly greater than either control group ($P < 0.005$). The US alone group, while significantly lower in retention than the paired group ($P < 0.01$), showed significant sensitization up to and including day 2. (From Carew *et al.*, 1981.)

To demonstrate an associative effect, it was necessary to examine a number of control groups of animals that received training without stimulus pairing. These control groups included animals receiving specifically unpaired presentations of the conditioned and the unconditioned stumuli (with the two stimuli separated by 2.5 min), animals receiving random pairing of the two stimuli, and animals receiving only tail shock (US) alone training. After a series of training trials, animals that received paired training showed a greater increase in their siphon withdrawal response to the conditioned stimulus than did any of the control groups (Fig. 1C) (Carew *et al.*, 1981). Animals that received US alone training showed a smaller though statistically significant increase in the siphon withdrawal response. This nonassociative enhancement of the reflex that results from a noxious stimulus in the absence of stimulus pairing is an example of sensitization (Pinsker *et al.*, 1970). Animals trained with unpaired or random stimulus presentations showed little or no increase in the reflex. Thus after training, the US alone group had a significantly larger response than the unpaired or random groups despite the fact that both groups were exposed to the sensitizing unconditioned stimulus an equal number of times (Fig. 1C). This probably occurs because the unpaired and random control groups received the conditioned stimulus repeatedly in an unreinforced manner and therefore habituated to it. Thus whereas the US alone group experienced sensitization alone, the other two control groups experienced both sensitization and habituation.

2.2. The Learning Produced by Classical Conditioning in *Aplysia* Lasts for Several Days

Experiments in which the time course of retention was investigated by testing the response once each day (Carew *et al.*, 1981) revealed that the paired effect increased immediately after training and peaked one day later (Fig. 1D). The responses of the paired animals than declined over the next several days, but were still significantly facilitated as compared with their pretraining responses on the second and third days after training. The sensitized group showed a similar time course of retention as the paired group, but the magnitude of the facilitation was significantly less. Thus the associative effect persists for at least several days. It is possible that this is an underestimate of the actual time course of retention since the repeated testing of the same animals after training habituates the reflex response. Although the paired effect increases gradually after training, associative effects of classical conditioning can be observed within minutes after training. This has allowed us in physiological experiments to study cellular mechanisms of conditioning within the first minutes and hours after training.

2.3. The Defensive Withdrawal Reflex Can Be Differentially Conditioned with Two Discriminative Stimuli

In the initial experiments, comparisons were made between the groups of animals having received conditioning training and various control groups of animals.

In physiological experiments to reduce variability in dissected preparations, it was desirable to have a within-animal control. Carew *et al.* (1983) therefore explored the possibility that responses to weak stimuli delivered to two different regions of the body could in fact be differentially conditioned within a single animal. The two regions that received the conditioned stimuli in these experiments were the anterior mantle skin and the siphon (Fig. 2A). Half the animals received siphon-paired training; in these, each weak siphon stimulus occurred immediately before the unconditioned stimulus, the tail shock (Fig. 2B). The second site on the mantle was stimulated in an unpaired fashion, 2.5 min after the unconditioned stimulus. The other half of the animals received mantle-paired training with a reciprocal procedure (Fig. 2B). In both groups, training produced a larger increase in the response to the paired tactile stimulus than to the unpaired stimulus (Fig. 2D). There also tended to be a small amount of sensitization as indicated by the slight increase in the siphon withdrawal response to the unpaired stimulus. This enhancement of the reflex produced by the tail shock in the absence of pairing is an example of sensitization. The power of these differential conditioning experiments is that each ani-

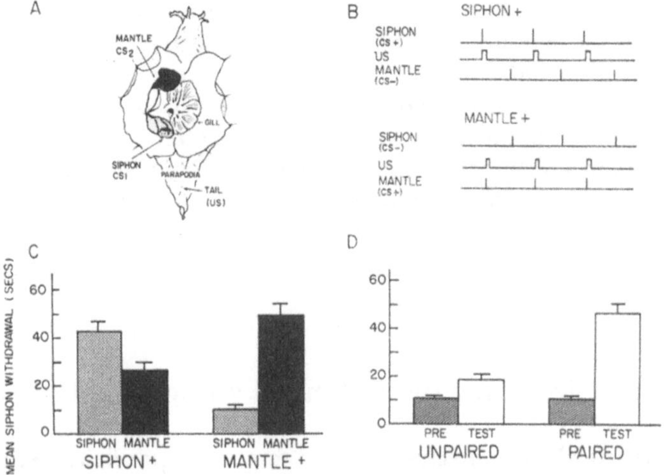

FIGURE 2. Differential conditioning of the siphon withdrawal reflex. (A) Dorsal view of *Aplysia* showing the two sites used to deliver conditioned stimuli: the siphon and the mantle shelf. The conditioned stimulus to the siphon (CS_1) was a weak tactile stimulus as described in Fig. 1. The conditioned stimulus to the mantle shelf (CS_2) was a weak electric shock. The US was an electric shock delivered to the tail. (B) Paradigm for differential conditioning. One group of animals (Siphon+) received the siphon stimulus (CS+) paired with the unconditioned stimulus and the mantle stimulus (CS−) specifically unpaired with the unconditioned stimulus. The other group (Mantle+) received the mantle stimulus paired and the siphon stimulus unpaired. Before training, animals were matched on the basis of pretraining responses and then assigned to the two groups. The intertrial interval was 5 min. (C) Results of an experiment using the paradigm illustrated in B. Testing was carried out 30 min after 15 training trials. Responses in the Siphon+ group (n = 12) were significantly more prolonged ($P < 0.05$) to the siphon stimulus than to the mantle stimulus, whereas responses in the Mantle+ group (n = 12) were significantly more prolonged ($P < 0.01$) to the mantle stimulus than to the siphon stimulus. (D) Pooled data from the experiment in C. Responses to the unpaired and paired stimuli after training are compared with the pretest scores for the same stimuli. The increase in the siphon withdrawal response to the paired stimulus was significantly greater than the change in the response to the unpaired stimulus ($P <$ 0.005). From Carew *et al.*, 1983).

mal serves as its own control for any such nonassociative changes. The differential effect of training on the responses to the paired and unpaired stimuli represents the associative effect. This differential paradigm has been useful in further studies of the conditioning. Using differential conditioning, Carew *et al.* (1983) were able to demonstrate that a single paired trial can produce associative learning in the siphon withdrawal reflex that lasts at least 24 hr.

2.4. Classical Conditioning Has a Narrow Temporal Window

One salient feature of the learning that must be accounted for by any cellular model is the temporal constraints on conditioning. In most types of classical conditioning, the animal only learns the predictive relationship if during training the conditioned stimulus is presented immediately before the unconditioned stimulus. Hawkins *et al.* (1983b) found that in *Aplysia* optimal conditioning occurs if the unconditioned stimulus begins about 0.5 sec after the onset of the conditioned stimulus. If the delay before the unconditioned stimulus is increased, the pairing specific effect is reduced. For example, if the onset of the unconditioned stimulus lags the onset of the conditioned stimulus by more than about 2 sec, it is as if the stimuli were unpaired and associative learning does not occur. Perhaps most interesting, if the order is reversed and the tail shock precedes the weak siphon stimulus (backward conditioning), there is also no associative effect. Thus there is a narrow temporal window in which the unconditioned stimulus must follow the conditioned stimulus for effective conditioning to occur.

What is the cellular basis of this window of temporal specificity? How does the nervous system determine that the conditioned and unconditioned stimuli occur in a temporally contiguous manner?

3. The Neural Circuit for the Defensive Withdrawal Reflex

Our efforts to identify mechanisms for temporal specificity were aided substantially by the fact that the neuronal organization of the reflex had already been delineated (see, for example, Byrne *et al.*, 1974; Perlman, 1979; and Hawkins *et al.*, 1981a). The reflex is mediated by a population of about 20 to 25 sensory neurons in the abdominal ganglion that innervate the siphon and that make direct synaptic connections with siphon motoneurons. A substantial part, roughly half, of the siphon withdrawal reflex is due to these monosynaptic excitatory connections between the siphon sensory neurons and siphon motoneurons (Byrne *et al.*, 1978); the remainder of the reflex is due to polysynaptic indirect input from the sensory neurons that reaches the motoneurons via interneurons. In our initial studies, we have focused on changes in the monosynaptic portion of this circuit that contribute to the conditioning.

As mentioned above, enhancement of the siphon withdrawal reflex occurs during sensitization, as well as during classical conditioning. Sensitization is produced by noxious stimuli, such as strong shocks to the head or tail, in the absence of stimulus pairing. If the sensitizing stimuli are sufficiently intense or prolonged,

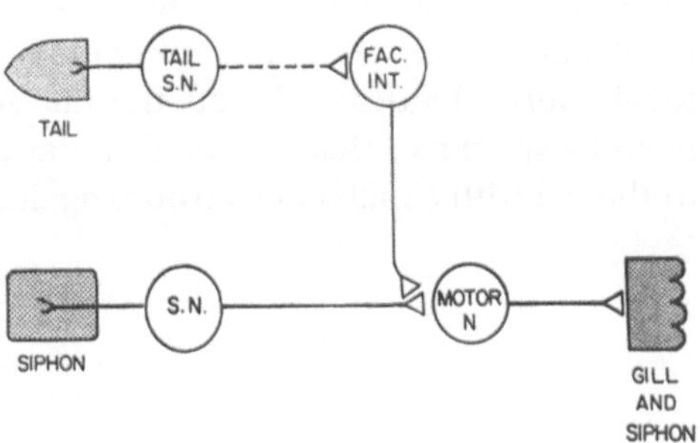

FIGURE 3. Neural circuit for the gill and siphon withdrawal reflex in response to siphon stimulation and for modulation of the reflex during sensitization. The sensory neurons (S.N.) that innervate the siphon make direct excitatory synaptic connections with a number of identified motoneurons (MOTOR N) that project to either the gill or the siphon. The siphon sensory neurons receive modulatory input from facilitator interneurons (FAC. INT.) that are excited by noxious stimuli to the tail such as the unconditioned stimulus, tail shock.

the siphon withdrawal response can be increased dramatically (Pinsker *et al.*, 1973). Because sensitization has some resemblance to classical conditioning, it is appealing to think that both types of learning might use similar mechanisms for enhancing the withdrawal response. The cellular mechanism underlying sensitization was understood in some detail. As shown in Fig. 3, the noxious stimulus excites a population of facilitator interneurons that act to presynaptically facilitate synaptic transmission from siphon and mantle sensory neurons to motoneurons, thus increasing the gain of the defensive withdrawal reflex (Castellucci and Kandel, 1976; Hawkins *et al.*, 1981b).

During sensitization, this synaptic facilitation occurs indiscriminantly among the entire population of these sensory neurons (Castellucci *et al.*, 1983). How does classical conditioning selectively increase the response elicited by just those sensory neurons that are excited by a tactile stimulus immediately prior to the delivery of the tail shock US?

4. Activity-dependent Enhancement of Presynaptic Facilitation as an Associative Mechanism

We hypothesized that the recent occurrence of action potentials in a sensory neuron increases the cell's ability to respond to the facilitatory input. It seemed probable to us that during conditioning, the aversive unconditioned stimulus acts via the same population of facilitator neurons that are involved in sensitization to facilitate synaptic transmission from sensory neurons in both paired and unpaired sensory pathways. However, according to this model, which we called *"activity-dependent enhancement of presynaptic facilitation,"* the facilitation would occur more effectively when a sensory neuron has fired action potentials just prior to the presentation of the unconditioned stimulus.

4.1. Cellular Tests of the Activity-dependent Facilitation Model

To directly test the activity-dependent facilitation model, we attemped to differentially facilitate the postsynaptic potentials from two sensory neurons to a common postsynaptic neuron by pairing activity in one of the two sensory neurons with the unconditioned stimulus. The paradigm we used was analogous to that used in the behavioral differential conditioning experiments, and where posssible the protocols were identical. We left the tail attached to the otherwise isolated central nervous system so that we could use as an unconditioned stimulus either a shock to the tail with capillary electrodes against the skin or direct stimulation of the tail

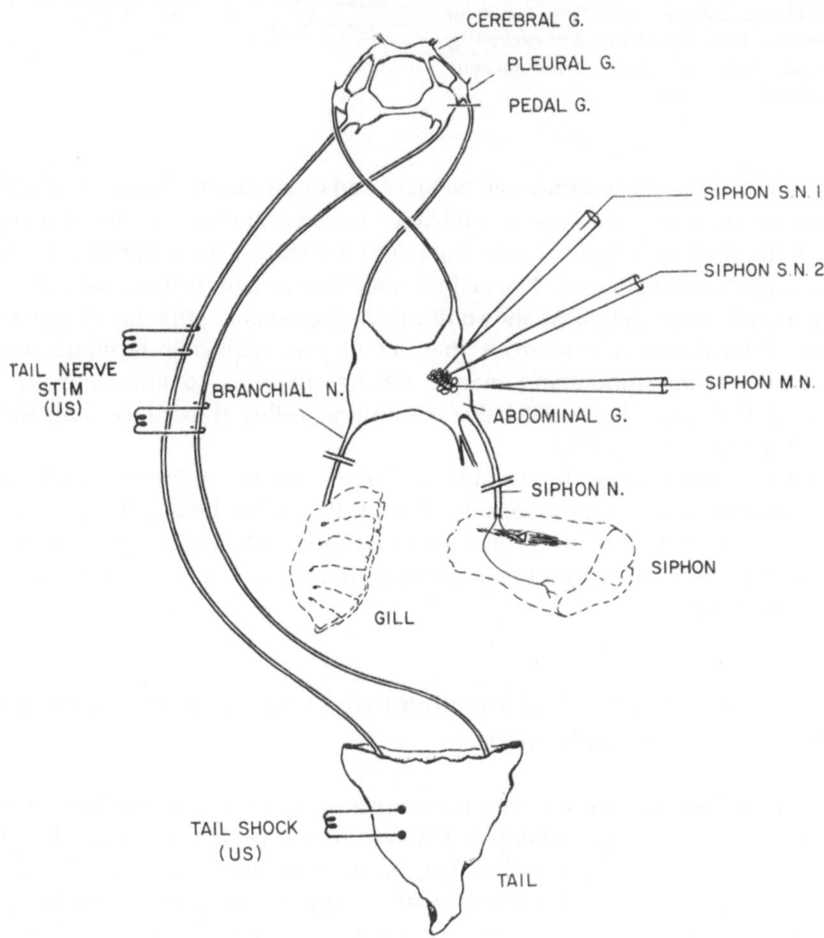

FIGURE 4. Neurophysiological preparation for studying the changes in the neural circuit for the reflex that occur during classical conditioning. The central ganglia (G) were removed from the animal and the pedal ganglia were left attached to the isolated tail via the posterior pedal nerves. The siphon sensory neurons have their somata clumped together in a cluster in the abdominal ganglion. Two presynaptic siphon sensory neurons (S.N.) were penetrated with micropipettes and stimulated with intracellular current pulses to fire action potentials. A postsynaptic siphon motoneuron (M.N.) was also penetrated and the EPSPs from the two sensory neurons were recorded. The postsynaptic cell was one of a handful of recently identified small siphon motoneurons located adjacent to the sensory neuron cluster. The unconditioned stimulus was the same as in the behavioral experiments: an electrical shock (50 mA for 1.5 sec) to the tail skin. The gill and siphon are shown for illustrative purposes (with broken lines) and were removed from the preparation in actual experiments.

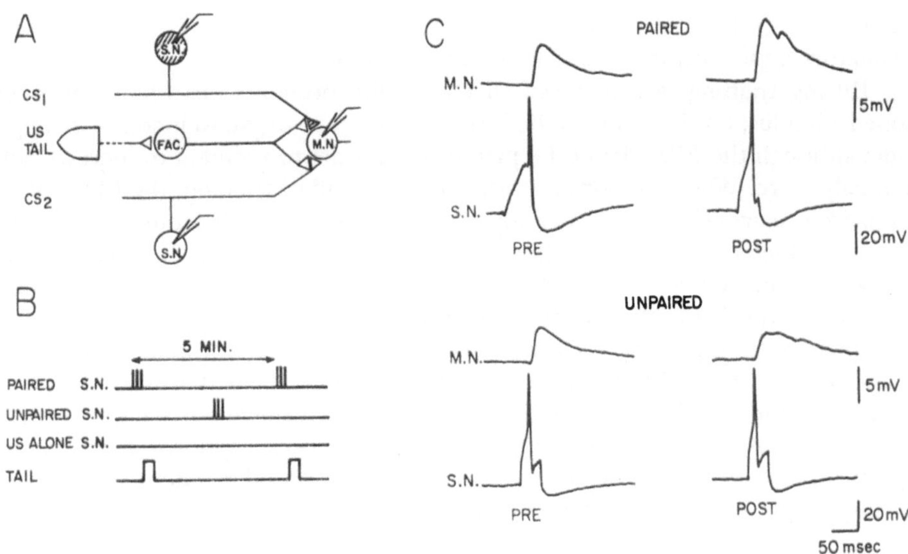

FIGURE 5. Activity-dependent facilitation of monosynaptic EPSPs in the neuronal circuit for the withdrawal reflex. (A) Neural circuit for testing activity-dependent facilitation hypothesis. (B) Protocol for cellular experiments. One sensory neuron (PAIRED S.N.) was caused to fire a 0.5-sec 10-Hz train of action potentials immediately prior to the tail stimulus. The control sensory neuron (UNPAIRED S.N.) was similarly caused to fire a train of action potentials, but 2.5 min after the unconditioned stimulus. In other experiments (shown in Fig. 7), the control sensory neuron was not stimulated during training, and served as a US alone, or sensitization, control. (C) Examples of the EPSPs produced in a common postsynaptic siphon motoneuron (M.N.) by action potentials in a paired and an unpaired sensory neuron (S.N.) before (PRE) and 1 hr after training (POST). At the time of this posttest, whereas the EPSP from the paired sensory neuron had increased to nearly 200% of its pretraining amplitude, the EPSP from the control cell in this same experiment showed no facilitation. Notice the late polysynaptic EPSP following the action potential in the paired sensory neuron after training; in addition to producing a larger monosynaptic EPSP in the motoneuron, the facilitated synaptic transmission from the paired sensory neuron is more effective in recruiting interneurons. (From Hawkins *et al.*, 1983a.)

nerve afferents that innervate the tail skin with extracellular electrodes on the posterior pedal nerves (Fig. 4). In these neurophysiological experiments, instead of stimulating two sites on the skin, we stimulated two individual siphon sensory neurons with intracellular current. Instead of measuring the duration of the siphon withdrawal, we studied the effects of training on the amplitudes of the monosynaptic excitatory postsynaptic potentials (EPSPs) that are produced by the two sensory neurons in a single postsynaptic siphon motoneuron (Figs. 4, 5A).

Training consisted of a series of five training trials at 5-min intervals (Fig. 5B). During each training trial, we fired a 0.5-sec train of six action potentials in one sensory neuron, the paired sensory neuron, just prior to the delivery of the unconditioned stimulus to the tail. The other sensory neuron, the unpaired control, was similarly caused to fire a train of six spikes, but 2.5 min after the unconditioned stimulus. In each experiment the two sensory neurons were arbitrarily selected from among the homogeneous population of cells in the siphon sensory neuron cluster (Fig. 4). On the average, they should both therefore have received similar facilitatory input during the unconditioned stimulus.

During training, we observed that after the presentations of the unconditioned stimulus, the EPSPs from both sensory neurons began to increase in amplitude, although the EPSP from the paired sensory neuron tended to increase substantially more. When measured shortly after the end of training, the EPSPs from the paired neurons were significantly greater than the EPSPs from the unpaired neurons, which on the average showed no facilitation (Figs. 5C, 7). This is exactly the type of differential effect that would be expected given the activity-dependent facilitation hypothesis. We have now seen similar differential results in more than 40 experiments. Moreover, the differences were maintained for more than 45 min in those experiments in which all three neurons were held that long (Fig. 5C). Notice that the fact that both EPSPs are recorded in a single postsynaptic neuron suggests that these differential changes are occurring presynaptically. Similar activity-dependent presynaptic facilitation has been observed independently in a second sensory system in *Aplysia* by Walters and Byrne (1983).

4.2. Activity-dependent Facilitation Differs from the Activity-dependence Hypothesis of Hebb

This activity-dependent facilitation mechanism bears some resemblance to widely known model proposed by Hebb in 1949 in which paired spike activity also played a critical role in synaptic plasticity. The hypothetical Hebb synapse differs in fundamental respects, however, from the mechanism described here. Hebb speculated that if spike activity in one neuron repeatedly coincided with spike activity in a postsynaptic cell, then the connection between these two cells might be strengthened (Hebb, 1949). In activity-dependent facilitation, on the other hand, it is spike activity in a facilitator neuron, rather than postsynaptic activity, that must be paired with activity in the presynaptic neuron for the optimal enhancement of synaptic transmission to occur (Fig. 6).

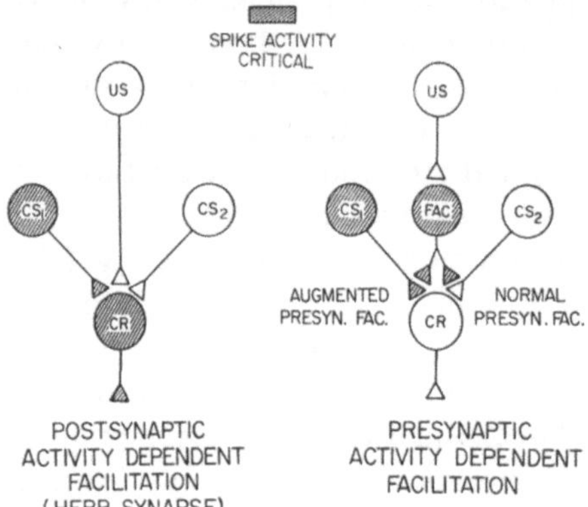

FIGURE 6. Comparison of two models to account for temporal specificity in classical conditioning: a Hebb-type mechanism and activity-dependent facilitation. In order to strengthen the synaptic connection, the Hebb synapse requires the pairing of presynaptic and postsynaptic spike activity, whereas activity–dependent facilitation requires the pairing of presynaptic activity with activity in facilitator neurons (FAC). CS_1, presynaptic neuron of the paired stimulus pathway; CS_2, presynaptic neuron of the unpaired stimulus pathway; CR, postsynaptic neuron producing the conditioned response. Input from the US excites (1) the postsynaptic neurons in the Hebb model and (2) the facilitator neurons in the activity-dependent facilitation model.

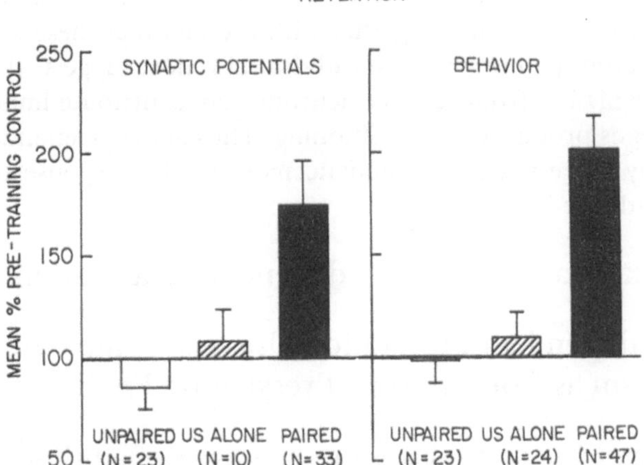

FIGURE 7. Comparison of cellcular data showing differential facilitation of EPSPs and behavioral data showing differential conditioning of the withdrawal reflex. Data are pooled from two types of experiments: paired versus unpaired experiments and paired versus US alone experiments. In both sets of results, we have plotted the response measured 5 to 15 min after a series of five training trials as a percentage of the response prior to training. (From Hawkins *et al.*, 1983a.)

In our conditioning paradigm, the postsynaptic motoneuron fires a burst of spikes in response to the unconditioned stimulus to the tail, so it seemed possible that a Hebb-type mechanism might contribute to changes in synaptic transmission during conditioning in *Aplysia.* However, in experiments in which we held the postsynaptic cell hyperpolarized, we found that it was not necessary for the postsynaptic motoneuron to fire during training for activity-dependent facilitation to occur. Furthermore, results of experiments in which we paired intracellularly stimulated spike trains in the motoneuron with spike trains in the sensory neurons indicated that pairing presynaptic and postsynaptic activity is not effective in facilitating the synaptic connection between the two cells (Carew *et al.*, 1984). Thus, at least in this system, it is activity-dependent facilitation, rather than a Hebb-type synaptic mechanism, that is responsible for the temporal specificity.

4.3. Comparison with Behavioral Results

Clearly this dependence of facilitation on the recent history of spike activity in a sensory neuron could provide a potential associative mechanism during learning. However, could the observed activity-dependence of presynaptic facilitation in the siphon sensory neuron account for the temporal specificity in the conditioning of the reflex? In Fig. 7 I compare the results of these experiments on synaptic transmission with behavioral data on siphon withdrawal from intact animals. Since in the physiology experiments, it was the amplitude of the synaptic potentials that was measured and in the behavior it was the duration of the withdrawal response, it is probably somewhat fortuitous that there is such a good quantitative match between the physiology and the behavior. However, the qualitative similarity between the two sets of results, together with the fact that these sensory neurons compose the afferent pathway for the withdrawal reflex, suggests that the changes in synaptic transmission from sensory neurons may contribute importantly to the behavioral changes produced by conditioning. Thus activity-dependent facilitation may explain why there is selective enhancement of the response to the stimulus that is paired with the US.

5. The Cellular Basis of Activity-dependent Facilitation

5.1. Activity-dependent Facilitation Uses the Same Cellular Mechanism as Conventional Presynaptic Facilitation

We next wanted to know what is the mechanism of activity-dependent facilitation and how activity enhances the facilitation response. Since similar synaptic facilitation occurs in these same siphon sensory neurons during sensitization, we have asked whether activity-dependent facilitation might employ the same cellular mechanism as is involved in conventional facilitation (Fig. 8).

In sensitization, a noxious stimulus excites facilitator neurons. Bernier *et al.* (1982) had shown that the input from the facilitator neurons activate the adenylate cyclase in the sensory neurons, producing a rise in cyclic AMP within these cells. Siegelbaum *et al.* (1982) and Castellucci *et al.* (1980) had demonstrated that increased cyclic AMP-dependent protein phosphorylation closes a K^+ channel called the S channel. Because outward current flowing through K^+ channels is responsible for the repolarization of the action potential, when the S-K^+ channel is blocked the action potentials in the sensory neurons repolarize more slowly. Since the sensory neuron action potentials are prolonged, more Ca^{2+} enters the presynaptic terminals during each spike and transmitter release is increased (Klein and Kandel, 1980). Klein *et al.* (1981) have demonstrated that even a small increase in the duration of the action potential in the sensory neuron can have a profound effect on synaptic transmission; for example, a 20% broadening of the action potential can double the EPSP amplitude.

This cascade shown in Fig. 8 represents only those processes known to be involved in short-term facilitation. Following a strong sensitizing stimulus, cyclic

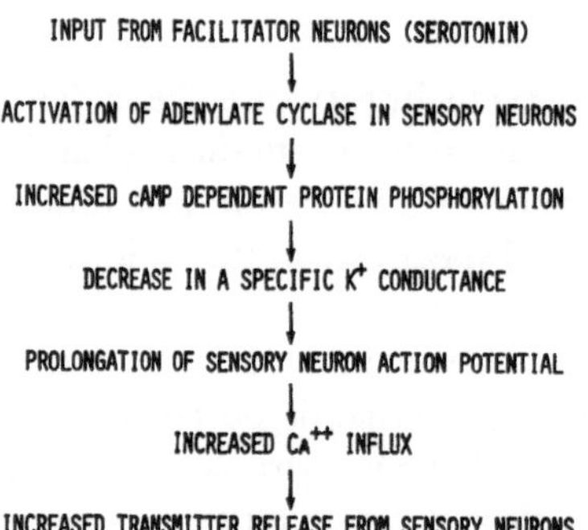

FIGURE 8. Flow diagram of the cellular mechanism underlying presynaptic facilitation in the siphon sensory neurons.

AMP levels rise and then typically decay in approximately 20 min (Bernier *et al.*, 1982). During the short term, the maintained closure of the S-K^+ channel is dependent on the ongoing synthesis of cyclic AMP (Castellucci *et al.*, 1983). However, both sensitization (Pinsker *et al.*, 1973) and facilitation of the sensory neuron synapses can last hours and days (W. Frost, V. C. Castellucci, and E. R. Kandel, in preparation). We therefore expect that on a longer time scale, more persistent changes, perhaps triggered by the transient rises in cyclic AMP, contribute to the enhancement of synaptic transmission; such long-term changes include long-term spike broadening (see below) and changes in synapse morphology (Bailey and Chen, 1983).

To investigate whether the same cellular mechanism involving presynaptic spike broadening might play a role in activity-dependent facilitation, we studied the effects of our conditioning paradigm on the shape of the presynaptic action potential. A series of experiments similar to those on synaptic transmission were carried out in which spike activity in one sensory neuron was paired with the unconditioned stimulus to the tail; the second sensory neuron was fired in a specifically unpaired manner 2.5 min after the unconditioned stimulus. To help us visualize changes in spike duration, we used a technique first developed by Klein and Kandel (1978) in studying conventional facilitation: bathing the ganglia in tetraethylammonium (TEA) saline. Fifty millimolar TEA blocks a substantial portion of the sensory neurons, K^+ channels that are not modulated by cyclic AMP, and thus slows the repolarization of the action potential. The cyclic AMP-modulated K^+ channels are relatively unaffected. Since the burden of repolarizing the action potential now rests principally on the serotonin-sensitive channels, the spike duration is specially sensitive to any changes in these channels; thus the broadening of the action potential during facilitation is magnified.

Our training protocol resulted in spike broadening in both the paired and unpaired sensory neurons, but this increase in spike width was significantly greater in the cell whose activity has been paired (Fig. 9A). These experiments provide several important insights:

1. Since the inward current during the late shoulder of the spike is carried by Ca^{2+} ions, this differential increase in the duration of the action potential indicates that there is a greater increase in the Ca^{2+} influx in the paired sensory neuron; it is this differential change in Ca^{2+} influx that probably underlies the increase in the EPSP amplitude during conditioning.
2. This differential change in the sensory neuron spike shape provides direct evidence that the pairing specific change responsible for enhanced facilitation of the paired EPSP occurs within the presynaptic neuron itself.
3. The differential effect persisted for at least 3 hr after training (Fig. 9B), suggesting it might contribute to a long-lasting change in the siphon withdrawal reflex.

Although this prolongation of the action potential could be due to changes in any of a number of ionic conductances, recent voltage clamp experiments indicate that the current which is modulated in an activity-dependent manner during training resembles the S-K^+ current both in its voltage dependence and its kinetics

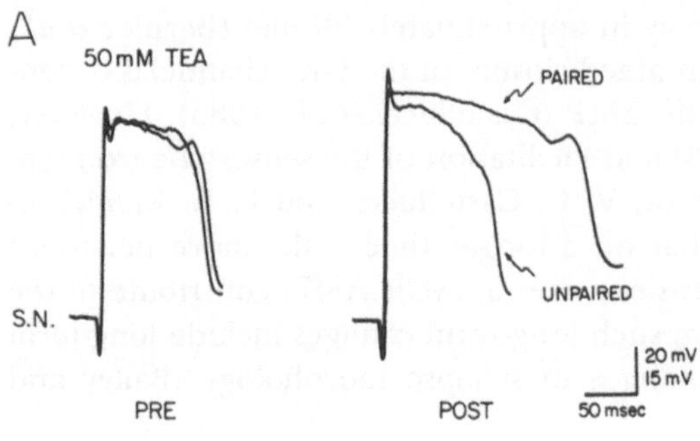

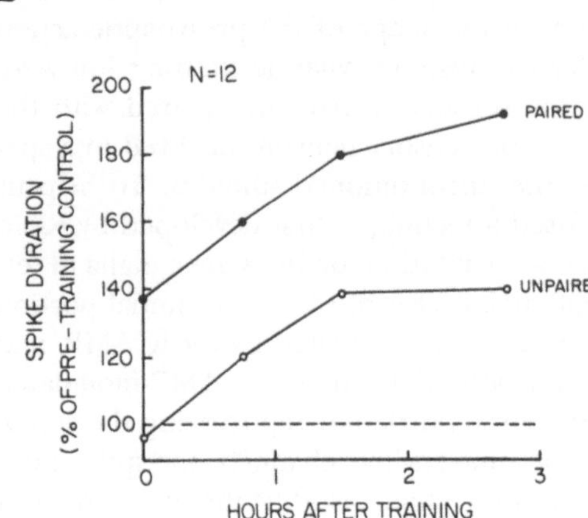

FIGURE 9. Activity-dependent broadening of the action potentials in the two presynaptic sensory neurons. (A) Examples of the action potentials in a paired and an unpaired sensory neuron before (PRE) and 3 hr after (POST) training. Training consisted of 15 training trials with a protocol similar to that illustrated in Fig. 5A (intertrial interval = 5 min). The two action potentials in each neuron have been superimposed. Action potentials were recorded in saline containing 50 mM tetraethylammonium (TEA). (B) Retention of differential spike broadening after training. Average spike duration after training expressed as a percentage of the spike duration prior to training in each sensory neuron. The action potentials in the paired neurons broadened significantly more than the action potentials in the unpaired neurons ($P < 0.05$).

(Hawkins and Abrams, 1984). These findings suggested that activity-dependent facilitation might use the same cellular machinery that is involved in conventional facilitation (Fig. 8), but that recent spike activity might allow the biochemical cascade to be activated more powerfully.

5.2. The Interaction between Spike Activity and the Facilitation Response

We next wanted to ask how spike activity in the sensory neurons enhances their ability to respond to facilitatory input. In investigating the basis of activity-dependent facilitation we have posed two questions:

1. If activity-dependent facilitation uses this same cyclic AMP-dependent cascade, what step in the cascade is modulated by spike activity in the sensory neuron?
2. Which component of the action potential interacts with the cascade and increases the facilitatory response?

In this analysis we wanted to study activity-dependent enhancement while we manipulated the ionic environment and also to carry out some experiments in the absence of synaptic transmission. Because this would interfere with the polysynaptic facilitatory input from the tail, we decided to bypass the facilitator interneurons completely by substituting for the unconditioned stimulus a brief puff of facilitatory transmitter onto sensory neuron somata. Our evidence suggests that the facilitatory system may use several transmitters. Three known endogenous transmitters, which are found in the region where the sensory neurons arborize, are capable of producing presynaptic facilitation in these cells by increasing cAMP levels and broadening the action potential: serotonin and the two small cardioactive peptides, SCP_A and SCP_B (Brunelli *et al.*, 1976; Klein and Kandel, 1978; Kistler *et al.*, 1985; Abrams *et al.*, 1984b). In addition, a yet unidentified, facilitatory transmitter is released by one group of identified facilitator interneurons, the L_{29} cells (Hawkins *et al.*, 1981a,b; Glanzman *et al.*, 1984). Since serotonin mimics the input from facilitator neurons, we hoped it could be used in studying the mechanism of activity dependence.

5.3. Paired Spike Activity Enhances the Response to Serotonin

As a first step, it was necessary to determine whether the response to the serotonin puff was enhanced in a pairing specific manner. A pressure ejection micropipette was used to apply brief puffs of serotonin (approximately 1 sec in duration) to the cell body of a siphon sensory neuron (Fig. 10). We compared the responses to serotonin puffs that were delivered alone with the responses to puffs delivered immediately after a 0.5 sec train of action potentials were stimulated in the sensory neuron. Since these brief puffs of 5-HT reached only receptors on or near the cell bodies of the sensory neurons, we used the increase in spike duration measured in the cell body as an index of the facilitation response. Sufficient time was allowed between individual puffs for the facilitation response to decay completely. We found that a train of 6 action potentials immediately prior to the puff was able to enhance the serotonin response significantly, much as was observed when tail shock was used as the unconditioned stimulus. Because the differential enhancement of the facilitation response to the serotonin puff was quite consistent, the puff provided a good model system in which to further analyze the basis of the activity dependence.

5.4. Spike Activity Enhances the Cyclic AMP Cascade

Using the puff technique, we are asking what step in the facilitation response is modulated by recent spike activity. Activity could affect the activation of adenylate cyclase or any of the subsequent steps that are interposed between cyclic AMP synthesis and the facilitation of synaptic transmission (Fig. 8), including protein phosphorylation and the closing of the S-K^+ channel. We chose to first test the possibility that cyclic AMP synthesis in response to a serotonin puff will be greater in cells that have recently fired action potentials. We compared the levels of cyclic AMP in clusters of sensory neurons that receive a puff of serotonin alone with the cyclic AMP responses of sensory neuron clusters that receive a puff immediately after a 0.5-sec train of action potentials has been generated in them. To prevent

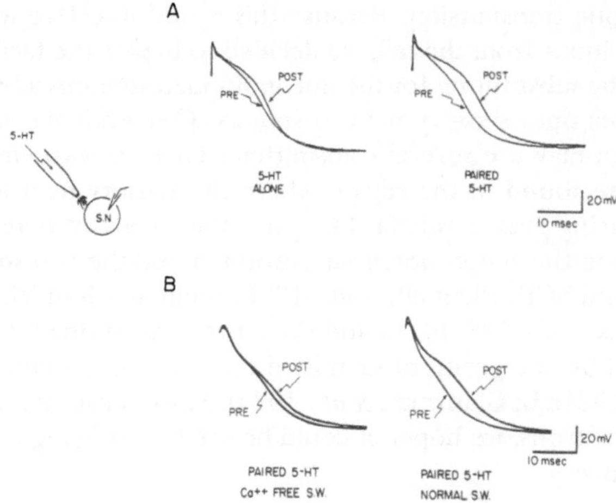

FIGURE 10. Activity-dependent enhancement of siphon sensory neuron responses to brief applications of serotonin. Brief (1 sec) puffs of serotonin (5-HT) were pressure ejected from an extracellular micropipette onto the soma of a single sensory neuron (shown on left). (A) Influence of paired spike activity on the serotonin response measured as the increase in action potential duration. In alternate trials the serotonin puff was delivered alone or immediately following a 0.5–sec train of six action potentials. The effect on the action potential duration was then measured 60 sec after the puff. Differential spike broadening from one typical experiment is shown here. Action potentials before (PRE) and after (POST) the puff alone and the paired puff are superimposed. Recordings were made in normal saline (without TEA). To allow more accurate measurement of the change in the action potential during pre- and post-tests the neuron was fired repetitively at 20 Hz for 2.5 sec and the action potentials measured at the end of the train; because two other K^+ currents (I_A and I_K) are inactivated during high frequency repetitive firing, the action potentials at the end of these spike trains are slower and their duration is particularly sensitive to any changes in the serotonin-sensitive current (I_S). Note that although the duration of the action potential increased only slightly after the serotonin puff alone, there was substantially more spike broadening after the paired serotonin puff. (B) Role of Ca^{2+} influx in activity-dependent enhancement of the serotonin response. In single sensory neurons, 0.5 sec trains of action potentials were paired with brief serotonin puffs alternately in normal (10^{-2}M Ca^{2+}) saline (S.W.) and in Ca^{2+}-free (10^{-7}M Ca^{2+}) saline. Action potential duration was measured in normal saline both before and 90 sec after pairing in each solution. In Ca^{2+} free saline trials, the bath was rapidly switched from normal Ca^{2+} to low Ca^{2+} for the paired serotonin puff and back to normal saline for the post-test 90 sec after the puff. Note that after action potentials were paired with serotonin in normal saline, the increase in action potential duration was substantially greater than after pairing in the absence of Ca^{2+} influx. Pre- and post-tests were conducted using 2.5 sec of high–frequency repetitive firing as in A.

this firing of the whole population of sensory neurons from exciting facilitator neurons, these experiments were done under conditions in which synaptic transmission is blocked; the spike train in the sensory neurons produced no measurable facilitation when given by itself. Initial data from these experiments indicate that sensory clusters that received puffs paired with activity had fourfold higher levels of cyclic AMP than clusters that received serotonin puffs alone (Kandel *et al.*, 1983; Abrams *et al.*, 1984a). Ocorr *et al.* (1983) have similarly found that sensory neurons in the pleural ganglion of *Aplysia* show increased levels of cyclic AMP in response to serotonin applications if the cells are depolarized just prior to the application. While still preliminary, these results suggest two things: (1) that activity-dependent facilitation does involve the same cyclic AMP-dependent mechanism as conventional facilitation; and (2) that paired spike activity may be able to increase the activation of the adenylate cyclase by facilitatory transmitter.

5.5. Calcium Influx May Act as the Signal Indicating That the Sensory Neurons of the Conditioned Stimulus Pathway Have Recently Been Active

The second question we have asked about the activity dependence is which element of the action potential is responsible for the increased response to facilitatory input. There are four possible candidates: Na^+ or Ca^{2+} influx, K^+ efflux, or membrane depolarization. Given the diverse roles of Ca^{2+} as an intracellular messenger, we chose Ca^{2+} as the first candidate to test. Ca^{2+} also seemed a good first possibility since a number of studies have shown that Ca^{2+} acting via calmodulin can activate adenylate cyclase in the mammalian brain (e.g., Brostrom *et al.*, 1978).

We hypothesized that the brief elevation of intracellular Ca^{2+} that occurs during spike activity might transiently increase the sensory neurons' response to serotonin. As a first test of this Ca^{2+} hypothesis, we asked whether paired spike trains in the absence of Ca^{2+} influx would still enhance the serotonin response. In these experiments, we paired spike trains with serotonin puffs, but eliminated Ca^{2+} influx by greatly reducing extracellular Ca^{2+} during the pairing. Since Ca^{2+} influx is responsible for the late depolarizing shoulder of the action potential, it was essential to measure the spike-broadening response in the presence of external Ca^{2+}; all pretests and posttests were conducted in normal saline. In alternate trials, we compared spike-broadening responses in single sensory neurons, after puffs paired with spike trains in Ca^{2+}-free saline (10^{-7} M Ca^{2+}), and after paired puffs in normal saline (10^{-2} M Ca^{2+}). The paired serotonin puff in normal saline resulted in a substantial increase in the duration of the action potential, whereas there was little or no spike broadening when the serotonin puff was paired with spike activity in Ca^{2+}-free saline (Fig. 10B). Since we have not directly altered the Na^+ or K^+ fluxes or the membrane potential changes that occur during the action potential, these results suggest that paired action potentials in the absence of Ca^{2+} influx are unable to enhance the facilitation response (Abrams *et al.*, 1983).

We have begun to test the Ca^{2+} hypothesis using several other approaches. For example, we are presently studying the effects of injecting the Ca^{2+} chelator EGTA and have found in preliminary experiments that EGTA will block the activity-dependent enhancement. Thus these results are consistent with our hypothesis that Ca^{2+} influx is critical for the enhancement of the facilitation response by paired spike activity.

6. Summary of Model

The cartoon shown in Fig. 11 sumarizes our hypothesis that Ca^{2+} influx in the sensory neurons produces activity-dependent enhancement of the facilitation

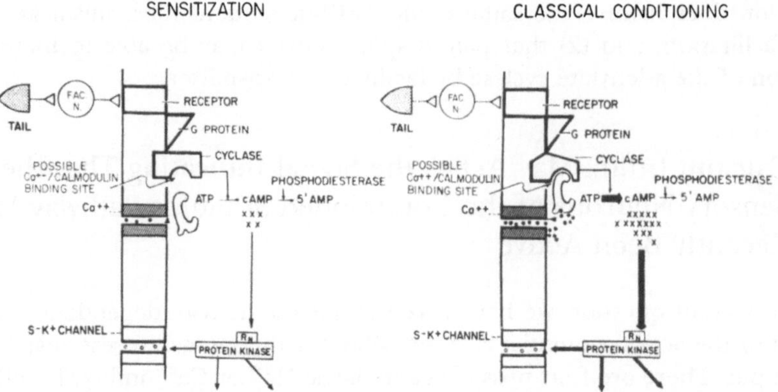

FIGURE 11. Cartoon summarizing hypothesis that Ca^{2+} provides the temporal specificity for conditioning by transiently modulating the adenylate cyclase. The details of the hypothesis diagrammed here are quite speculative. Based on studies in other species, the adenylate cyclase system is shown with the receptor to the facilitatory transmitter coupled to the catalytic unit of the cyclase via the interposed regulatory G protein (see, for example, Rodbell, 1981). Ca^{2+} is illustrated as binding via calmodulin to the catalytic unit since CA^{2+}/calmodulin has been found to interact with mammalian brain adenylate cyclase (Salter *et al.*, 1981). On the left, an inactive sensory neuron receives facilitatory input during sensitization. The adenylate cyclase is stimulated to synthesize cyclic AMP (XX) and cyclic AMP–dependent phosphorylation results in the closing of some of the S-K^+ channels. On the right, the sensory neuron is excited by the conditioned stimulus to fire a burst of action potentials that are accompanied by an influx of Ca^{2+}. The Ca^{2+} ions bind (. . .) to calmodulin which interacts with the cyclase potentiating its activation by facilitatory input. Multiple arrows extend from the cyclic AMP–dependent protein kinase to suggest that multiple substrates must be phosphorylated when cyclic AMP is elevated. Although there is evidence that one action of the kinase is to close the S-K^+ channel, it is not known whether the channel is directly phosphorylated (Schuster *et al.*, in preparation).

response by modulating the synthesis of cyclic AMP. When a modest amount of facilitatory input activates the cyclase system in the sensory neuron in the absence of spike activity, the cyclase synthesizes a small amount of cyclic AMP. This produces a limited activation of the protein kinase and a limited closing of the S-K^+ channel. Thus there is only a small amount of presynaptic facilitation. On the other hand, if this modest facilitatory input arrives shortly after the siphon sensory neuron has been active—such as occurs when the tail shock follows the conditioned stimulus to the siphon—it comes during a period of elevated intracellular Ca^{2+}. Under these conditions, the cyclase is activated more effectively or for a longer period, and greater cyclic AMP production results. The higher cyclic AMP levels cause greater activation of the protein kinase and increased presynaptic facilitation. Thus, if this model is correct, Ca^{2+} influx would provide the temporal specificity for conditioning by transiently modulating the adenylate cyclase system. We are beginning to explore whether the dual activation of the calmodulin-dependent cyclase by Ca^{2+} and transmitter gives this enzyme a critical associative role in learning.

Increased cyclic AMP dependent phosphorylation undoubtedly has other effects in addition to broadening the action potential. One attractive aspect of this model is that if spike activity enhances the early part of the cascade, conditioning would smplify all the facilitatory effects of a rise in cyclic AMP—not only spike broadening but also the triggering of any relatively persistent changes that may strengthen synaptic connections on the longer time scale.

A number of questions about this model still need to be answered including the following:

1. Is Ca^{2+} the only signal for modulating the facilitation response, or does its action require an additional component of the action potential, such as Na^+ influx, K^+ efflux, or membrane depolarization?
2. Which subunit of the cyclase system does Ca^{2+} or Ca^{2+}/calmodulin interact with to enhance the synthesis of cyclic AMP; does it act on the catalytic unit, the regulatory G unit, or on the receptor?
3. Are other steps in the facilitation cascade downstream from cyclic AMP synthesis (Fig. 8) also affected by paired spike activity?

Hopefully the answers to these questions will emerge as we characterize the molecular basis of activity-dependent facilitation in more detail. It is exciting to think that we are beginning to understand the biochemical basis of the temporal constraints on conditioning observed behaviorally. As mentioned earlier, effective learning does not occur if the tail shock begins more than about a second after the siphon stimulus. If the model in Figure 11 is correct, and Ca^{2+} coupled to calmodulin binds to the cyclase, then we would predict that following a train of action potentials, that binding should decay with a time course that matches the temporal window of conditioning. Less obvious is what prevents backwards conditioning from occurring successfully. Why doesn't the reciprocal sequence, in which spike activity and Ca^{2+} influx in the sensory neurons immediately follows the arrival of facilitatory input, result in equally effective activity-dependent enhancement of cyclic AMP synthesis? We expect that the basis of the sequential requirement that the conditioned stimulus precede the unconditioned stimulus will also be found in the dynamics of the interactions between Ca^{2+} and the cyclase.

7. Perspective

7.1. There Are a Number of Attractive Features about Activity-dependent Facilitation as an Associative Mechanism

7.1.1. An Activity-dependent Decrease in a K^+ Conductance Can Affect Neuronal Excitability

In addition to regulating the repolarizing phase of the action potential, outward current flowing through K^+ channels also influences the threshold for initiating an action potential. During presynaptic facilitation, the sensory neurons show both a decrease in spike threshold and an increase in their firing rate during a prolonged depolarizing current pulse (M. Klein, unpublished observations; Walters *et al.*, 1983). If a cyclic AMP-mediated closing of the S-K^+ channel or a similar K^+ channel occurred in an interneuron or a motoneuron, this would increase that neuron's responsiveness to excitatory synaptic input. Should this neuron's synthesis of cyclic AMP in response to facilitatory transmitter be enhanced by paired spike activity, then this "postsynaptic" activity-dependent facilitation would constitute a postsynaptic associative mechanism for learning using the identical cellular mechanisms as I have described here. Such changes in neuronal excitability due to decreases in K^+ conductance have been implicated in conditioning both in the grasshopper (Woolacott and Hoyle, 1977) and in another marine mollusk, *Hermissenda* (Alkon *et al.*, 1982).

7.1.2. Activity-dependent Facilitation Requires No Special Circuitry

Notice that activity-dependent facilitation is a fairly novel neuronal mechanism in that the summation of two temporally contiguous events is occurring on the biochemical level rather than through the electrical summation of excitatory synaptic inputs on the membrane level. Because this summation occurs within a single cell, there are no requirements for special circuitry. The facilitatory input could actually be quite widespread and still act in a functionally quite specific way if it modulates only those neurons that are active at the time of the input.

7.1.3. The Changes Produced by Learning Do Not Require New Neural Circuitry

As in sensitization which employs presynaptic facilitation, the learning can occur without the formation of new synaptic connections. This is not to rule out morphological changes in synapses; in the long term in classical conditioning of the siphon withdrawal reflex, as in sensitization, there is likely to be an increase in the density of presynaptic active zones for transmitter release (Bailey and Chen, 1983).

7.1.4. Associative Learning May Have Evolved from Simpler Forms of Learning

Activity-dependent facilitation can be viewed as a specific elaboration of conventional facilitation, providing it with associative specificity. It is appealing to think that just such an elaboration occurred in evolution, where a nonassociative form of behavioral plasticity may have evolved into a simple form of associative learning.

7.2. Activity-dependent Facilitation May Be a Phylogenetically Widespread Mechanism of Learning

All of the cellular machinery involved in activity-dependent facilitation including transmitter-activated decreases in K^+ conductances and modulation of the adenylate cyclase system by Ca^{2+} via calmodulin are phylogenetically widespread. Together with the lack of requirements for special circuitry, this suggests that this mechanism for temporal specificity could be involved in associative learning in phylogenitically diverse groups as well as in gastropod mollusks. There is now a suggestion that the interaction between intracellular Ca^{2+} and the adenylate cyclase may also be critical for conditioning in *Drosophila.* Dudai and Zvi (1984) and Livingstone *et al.* (1984) have recently found that a *Drosophila* mutant, rutabaga, that is unable to learn conditioning has a subtle defect in its adenylate cyclase system: the cyclase is insensitive to Ca^{+}/calamodulin. So this notion of phylogenetic generality may be borne out.

ACKNOWLEDGMENTS I am grateful to Eric Kandel and Bob Hawkins for commenting on the manuscript. I thank Louise Katz and Kathrin Hilten for preparing the figures.

References

Abrams, T. W., Carew, T. J., Hawkins, R. D., and Kandel, E. R., 1983, Aspects of the cellular mechanism of temporal specificity in conditioning in *Aplysia:* Preliminary evidence for Ca^{2+} influx as a signal of activity, *Soc. Neurosci. Abstr.* **9:**168.

Abrams, T. W., Bernier, L., Hawkins, R. D., and Kandel, E. R., 1984a, Possible roles of Ca^{2+} and cAMP in activity-dependent facilitation, a mechanism for associative learning in *Aplysia, Soc. Neurosci. Abstr.* **10:**269.

Abrams, T. W., Castellucci, V. F., Camardo, J. S., Kandel, E. R., and Lloyd, P. E., 1984b, Two endogenous neuropeptides modulate the gill and siphon withdrawal reflex in *Aplysia* by presynaptic facilitation involving cAMP-dependent closure of a serotonin-sensitive potassium channel. *Proc. Natl. Acad. Sci. USA* **81:**7956–7960.

Alkon, D. L., Lederhendler, I., and Shoukimas, J. L., 1982, Primary changes of membrane currents during retention of associative learning, *Science* **215:**693–695.

Bailey, C. H., and Chen, M., 1983, Morphological basis of long-term habituation and sensitization in *Aplysia, Science* **220:**91–93.

Bernier, L., Castellucci, V. F., Kandel, E. R., and Schwartz, J. H., 1982, Facilitatory transmitter causes a selective and prolonged increase in adenosine 3':5'-monophosphate in sensory neurons mediating the gill and siphon withdrawal reflex in *Aplysia, J. Neurosci.* **2:**1682–1691.

Brons, J. F., and Woody, C. D., 1980, Long-term changes in excitability of cortical neurons after Pavlovian conditioning and extinction, *J. Neurophysiol.* **44:**605–615.

Brostrom, M. A., Brostrom, C. O., Breckenridge, B. McL., and Wolff, D. J., 1978, Calcium-dependent regulation of brain adenylate cyclase, *Adv. Cyclic Nucleotide Res.* **9:**85–99.

Brunelli, M., Castellucci, V., and Kandel, E. R., 1976, Synaptic facilitation and behavioral sensitization in *Aplysia:* Possible role of serotonin and cyclic AMP, *Science* **194:**1178–1181.

Byrne, J. Castellucci, V. F., and Kandel, E. R., 1974, Receptive fields and response properties of mechanoreceptor neurons innervating skin and mantle shelf of *Aplysia, J. Neurophysiol.* **37:**1041–1064.

Byrne, J., Castellucci, V. F., and Kandel, E. R., 1978, Contribution of individual mechanoreceptor sensory neurons to defensive gill-withdrawal reflex in *Aplysia, J. Neurophysiol.* **41:**418–431.

Carew, T. J., Walters, E. T., and Kandel, E. R., 1981, Classical conditioning in a simple withdrawal reflex in *Aplysia californica, J. Neurosci.* **1:**1426–1437.

Carew, T. J., Hawkins, R. D., and Kandel, E. R., 1983, Differential classical conditioning of a defensive withdrawal reflex in *Aplysia californica, Science* **219:**397–400.

Carew, T. J., Abrams, T. W., Hawkins, R. D., and Kandel, E. R., 1984, A test of Hebb's postulate at identified synapses which mediate classical conditioning in *Aplysia, J. Neurosci.* **4:**1217–1224.

Castellucci, V. F., and Kandel, E. R., 1976, Presynaptic facilitation as a mechanism for behavioral sensitization in *Aplysia, Science* **194:**1176–1178.

Castellucci, V. F., Kandel, E. R., Schwartz, J. H., Wilson, F. D., Nairn, A. C., and Greengard, P., 1980, Intracellular injection of the catalytic subunit of cyclic AMP-dependent protein kinase simulates facilitation of transmitter release underlying behavioral sensitization in *Aplysia, Proc. Natl. Acad. Sci. U.S.A.* **77:**7492–7496.

Castellucci, V. F., Bernier, L., Schwartz, J. H., and Kandel, E. R., 1983, Persistent activation of adenylate cyclase underlies the time course of short-term sensitization in *Aplysia, Soc. Neurosci. Abstr.* **9:**169.

Clark, G. A., McCormick, D. A., Lavond, D. G., Thompson, R. F., 1984, Effects of lesions of cerebellar nuclei on conditioned behavioral and hippocampal neuronal responses, *Brain Res.* **291:**125–136.

Crow, T. J., and Alkon, D. L., 1978, Retention of an associative behavioral change in *Hermissenda, Science* **201:**1239–1241.

Davis, W. J., Villet, J., Lee, D., Rigler, M., Gillette, R., and Prince, E., 1980, Selective and differential avoidance learning in the feeding and withdrawal behavior of *Pleurobranchaea californica, J. Comp. Physiol.* **138:**157–165.

Dudai, Y., and Zvi, S., 1984, Adenylate cyclase in the *Drosophila* memory mutant rutabaga displays an altered Ca^{2+} sensitivity, *Neurosci. Letters* **47:**119–124.

Gelperin, A., 1975, Rapid food-aversion learning by a terrestrial mollusk, *Science* **189:**567–570.

Glanzman, D. L., Abrams, T. W., Hawkins, R. D., and Kandel, E. R., 1984, Extracts of L_{29} interneurons produce spike-broadening in sensory neurons of *Aplysia, Soc. Neurosci. Abstr.* **10:**510.

Gold, M. R., and Cohen, D. H., 1981, Modification of the discharge of vagal cardiac neurons during learned heart rate change, *Science* **214:**345–347.

Hawkins, R. D., and Abrams, T. W., 1984, Evidence that activity-dependent facilitation underlying classical conditioning in *Aplysia* involves modulation of the same ionic current as normal presynaptic facilitation, *Soc Neurosci. Abstr.* **10:**268.

Hawkins, R. D., Castellucci, V. F., and Kandel, E. R., 1981a, Interneurons involved in mediation and modulation of the gill-withdrawal reflex in *Aplysia.* I. Identification and characterization, *J. Neurophysiol.* **45:**304–314.

Hawkins, R. D., Castellucci, V. F., and Kandel, E. R., 1981b, Interneurons involved in mediation and modulation of the gill-withdrawal reflex in *Aplysia.* II. Identified neurons produce heterosynaptic facilitation contributing to behavioral sensitization, *J. Neurophysiol.* **45:**315–326.

Hawkins, R. D., Abrams, T. W., Carew, T. J., and Kandel, E. R., 1983a, A cellular mechanism of classical conditioning in *Aplysia:* Activity-dependent amplification of presynaptic facilitation, *Science* **219:**400–405.

Hawkins, R. D., Carew, T. J., and Kandel, E. R., 1983b, Effects of interstimulus interval and contingency on classical conditioning in *Aplysia, Soc. Neurosci. Abstr.* **9:**168.

Hebb, D. O., 1949, *The Organization of Behavior,* New York, Wiley & Sons, Inc.

Horridge, G. A., 1962, Learning of leg position by headless insects, *Nature (London)* **193:**697–698.

Hoyle, G., 1980, Learning, usual natural reinforcements, in insect preparations that permit cellular neuronal analysis, *J. Neurophysiol.* **11:**323–354.

Kamin, L. J., 1969, Predictability, surprise, attention, and conditioning, in: *Punishment and Aversive Behavior* (B. A. Campbell and R. M. Church, eds.), Appleton-Century-Crofts, New York, pp. 279–296.

Kandel, E. R., and Schwartz, J. H., 1982, Molecular biology of an elementary form of learning: Modulation of transmitter release by cyclic AMP, *Science* **218:**433–443.

Kandel, E. R., Abrams, T., Bernier, L., Carew, T. J., Hawkins, R. D., and Schwartz, J. H., 1983, Classical conditioning and sensitization share aspects of the same molecular cascade in *Aplysia, Cold Spring Harbor Symp. Quant. Biol.* **48:**821–830.

Kistler, H. B., Jr., Hawkins, R. D., Koester, J., Steinbusch, H. W. M., Kandel., E. R., and Schwartz, J. H., 1985, Distribution of serotonin-immunoreactive cell bodies and processes in the abdominal ganglion of mature *Aplysia, J. Neurosci.* **5:**72–80.

Klein, M., and Kandel, E. R., 1978, Presynaptic modulation of voltage-dependent Ca^{2+} current: Mechanism for behavioral sensitization in *Aplysia californica, Proc. Natl. Acad. Sci. U.S.A.* **75:**3512–3516.

Klein, M., and Kandel, E. R., 1980, Mechanism of calcium current modulation underlying presynaptic facilitation and behavioral sensitization in *Aplysia, Proc. Natl. Acad. Sci. U.S.A.* **77:**6912–6916.

Klein, M., Shapiro, E., and Kandel, E. R., 1981, Synaptic plasticity and the modulation of the Ca^{2+} current, *J. Exp. Biol.* **89:**117–157.

Livingstone, M. S., Sziber, P. P., and Quinn, W. G., 1984, Loss of calcium/calmodulin sensitivity responsiveness in adenylate cyclase of rutabaga, a *Drosophila* learning mutant, *Cell* **37:**205–215.

Lukowiak, K., and Sahley, C., 1981, The *in vitro* classical conditioning of the gill withdrawal reflex of *Aplysia californica, Science* **212:**1516–1518.

Mpitsos, G. J., and Collins, S. D., 1975, Learning: Rapid aversive conditioning in the gastropod mollusk *Pleurobranchaea, Science* **188:**954–957.

Ocorr, K. A., Walters, E. T., and Byrne, J. H., 1983, Associative conditioning analog in *Aplysia* tail sensory neurons selectively increases cAMP content, *Soc. Neurosci. Abstr.* **9:**169.

Pavlov, I. P., 1927, *Conditioned Reflexes,* Oxford University Press.

Perlman, A. J., 1979, Central and peripheral control of siphon withdrawal reflex in *Aplysia californica, J. Neurophysiol.* **42:**510–529.

Pinsker, H., Kupfermann, I., Castellucci, V., and Kandel, E. R., 1970, Habituation and dishabituation of the gill-withdrawal reflex in *Aplysia, Science* **167:**1740–1742.

Pinsker, H. M., Hening, W. A., Carew, T. J., and Kandel, E. R., 1973, Long-term sensitization of a defensive withdrawal reflex in *Aplysia, Science* **182:**1039–1042.

Quinn, W. G., Harris, W. A., and Benzer, S., 1974, Conditioned behavior in *Drosophila melanogaster, Proc. Natl. Acad. Sci. U.S.A.* **71:**708–712.

Rescorla, R. A., and Wagner, A. R., 1972, A theory of Pavlovian conditioning: Variations in the effectiveness of reinforcement and nonreinforcement, in: *Classical Conditioning II: Current research and theory* (A. H. Black and W. F. Prokasy, eds.), Appleton-Century-Crofts, New York, p. 64–99.

Rodbell, M., 1980, The role of hormone receptors and GTP-regulatory proteins in membrane transduction, *Nature (London)* **284:**17–22.

Sahley, C., Rudy, J. W., and Gelperin, A., 1981, An analysis of associative learning in a terrestrial mollusc. I. Higher-order conditioning, blocking, and a transient US pre-exposure effect, *J. Comp. Physiol.* **144:**1–8.

Salter, R. S., Krinks, M. H., Klee, C. B., and Neer, E. J., 1981, Calmodulin activates the isolated catalytic unit of brain adenylate cyclase, *J. Biol. Chem.* **256:**9830–9833.

Siegelbaum, S., Camardo, J. S., and Kandel, E. R., 1982, Serotonin and cAMP close single K^+ channels in *Aplysia* sensory neurones, *Nature (London)* **299:**413–417.

Walters, E. T., and Byrne, J. H., 1983, Associative conditioning of single sensory neurons suggests a cellular mechanism for learning, *Science* **219:**405–408.

Walters, E. T., Carew, T. J., and Kandel., E. R., 1979, Classical conditioning in *Aplysia californica, Proc. Natl. Acad. Sci. USA* **76:**6675–6679.

Walters, E. T., Byrne, J. H., Carew, T. J., and Kandel, E. R., 1983, Mechanoefferent neurons innervating the tail of *Aplysia:* II. Modulation by sensitizing stimulation, *J. Neurophysiol.* **50:**1543–1559.

Woolacott, M. H., and Hoyle, G., 1977, Neural events underlying learning in insects: Changes in pacemaker, *Proc. R. Soc. London Ser. B* **195:**395–415.

13

The Logic of *Limax* Learning

ALAN GELPERIN, J. J. HOPFIELD, AND D. W. TANK

1. Introduction

We wish to understand the neuronal computations performed on sensory inputs that result in the categorization of those inputs, their storage as memory states, and their associative combination. Our experimental and theoretical work is focused on the neuronal computations performed on odor and taste inputs in the CNS of *Limax maximus,* a terrestrial mollusk convenient for behavioral, neurophysiological, and neurochemical experiments (Gelperin, 1983). The questions posed in this specific system are designed to illuminate issues of learning and memory storage with panphyletic generality.

Several aspects of the learning behavior of *Limax* have encouraged us to attempt a unique synthesis of the approaches of behavioral biology, neurobiology, and neural modeling. While *Limax* displays many of the learning phenomena of higher organisms, including primates, it accomplishes these learning tasks using only about 10,000 cells in its CNS. We are using behavioral experiments to define the major types of learning exhibited by *Limax* and for each type of learning to establish the critical interevent timing relations that allow learning to occur. Neurophysiological and neurochemical experiments delimit the areas of CNS necessary for learning and describe the types of cellular elements and synaptic interactions available for modification during learning. The neural modeling asks how a collection of relatively simple neurons might interact synaptically to collectively accomplish the learning tasks. The model avoids assumptions about precise anatomical details, instead asking questions about the computational consequences of a simple set of rules governing synaptic interactions.

ALAN GELPERIN • AT & T Bell Laboratories, Murray Hill, New Jersey 07974 and Department of Biology, Princeton University, Princeton, New Jersey 08544. J. J. HOPFIELD • AT & T Bell Laboratories, Murray Hill, New Jersey 07974 and Divisions of Chemistry and Biology, California Institute of Technology, Pasadena, California 91125. D. W. TANK • AT & T Bell Laboratories, Murray Hill, New Jersey 07974.

We will first describe the behavioral and neurophysiological attributes of *Limax* learning. Then a detailed model of the *Limax* learning that incorporates a memory storage network is described, with clear indication of how the model relates to present data and raises questions amenable to experimental answer.

2. *Limax* Learning: Intact Animals and Isolated Brains

2.1. Associative Conditioning of Intact Animal

Limax is a generalist herbivore. It employs a variety of mechanisms to optimize its food choices when confronted with plant prey containing bitter and/or toxic chemical defenses. One such optimization mechanism involves learning to avoid plant odors associated with toxicosis (Gelperin, 1975) or a bitter taste (Sahley *et al.*, 1981a; see also Whelan, 1982; Gouyon *et al.*, 1983). The training procedure by which an attractive plant odor (e.g., carrot, potato, mushroom) is paired with a bitter taste (e.g., quinidine sulfate) is formally identical to a Pavlovian conditioning paradigm, where the attractive plant odor is the conditioned stimulus (CS) and the bitter taste is the unconditioned stimulus (US). Before training, the CS is attractive, eliciting both approach and ingestion. After training, the CS is repellent, eliciting avoidance and rejection. These changes in response to the CS odor due to conditioning are easily and quantifiably measured in an odor-choice chamber described elsewhere (Sahley *et al.*, 1981a).

The *Limax* CNS *in situ* can perform several associative logic operations on taste and odor inputs as revealed by a series of first-order and higher-order conditioning experiments (Sahley *et al.*, 1981b). The essence of these neural computations can be extracted by positing that before training there are several innately attractive food odors (A+, B+, and C+) and innately bitter and repellent tastes (e.g., Q−) available. First-odor conditioning is obtained by paired presentation of A+ and Q−. The result is that A (but not B) is now avoided (A−). The control procedures that demonstrate that this result depends on the association of A and Q are fully described in Sahley *et al.* (1981a).

Describing the conditioning results at this level of abstraction facilitates presentation of the higher-order conditioning data of Sahley *et al.* (1981b), as shown in Fig. 1. Second-order conditioning involves a two-phase training procedure wherein A− acts as a US during the second phase of conditioning. During compound conditioning, the animal learns to avoid both A and B in one training trial when A and B are presented together as a compound CS. The demonstration of blocking (Kamin, 1969) involves a two-phase training procedure with first-order conditioning in phase one and compound conditioning in phase two. The unexpected outcome is that animals do not acquire an aversion to the second CS introduced in phase two of conditioning.

Another striking parallel between results of *Limax* and vertebrate conditioning experiments has emerged from studies on the effect of extinguishing the aversion to food A after second-order conditioning of food B. Theoretically, the experiment can distinguish whether or not, after second-order conditioning, the aversion reac-

1st ORDER CONDITIONING:	$A^+ + Q^- \rightarrow A^-, B^+$
2nd ORDER CONDITIONING:	$A^+ + Q^-;\ B^+ + A^- \rightarrow A^-, B^-, C^+$
	$A^+ + Q^-;\ (A^-\ B^+) \rightarrow A^-, B^-, C^+$
COMPOUND CONDITIONING:	$(A^+\ B^+) + Q^- \rightarrow A^-, B^-, C^+$
BLOCK OF CONDITIONING:	$A^+ + Q^-;\ (A^-\ B^+) + Q^- \rightarrow A^-, B^+$
EXTINCTION AFTER CONDITIONING:	$A^+ + Q^-;\ (A^-\ B^+)\ ;\ A^-, A^- \ldots \rightarrow A^+, B^+$
	$A^+ + Q^-;\ B^+ + A^-\ ;\ A^-, A^- \ldots \rightarrow A^+, B^-$
APPETITIVE CONDITIONING:	$X^- + F^+ \rightarrow X^+, Y^-$

FIGURE 1. Summary of the logic operations performed by the *Limax* CNS as revealed by conditioning experiments. Stimuli in parenthesis are presented simultaneously.

tion to food B is mediated via the internal representation of food A. If extinguishing the aversive reaction to food A alone also results in the disappearance of the aversive reaction to food B, then the conditioned aversion to food B must depend on the integrity of the internal representation of food A. Conversely, if extinguishing the conditioned aversion to food A after using food A to produce second-order conditioned aversion to food B *does not* diminish the aversion response to food B, then clearly the conditioning to food B does not depend on the internal representation of food A. Rescorla (1984) has pointed out that either outcome can be produced depending on the timing of stimulus presentation during phase two of the second-order conditioning procedure. If A and B are presented simultaneously during phase two of the second-order conditioning procedure, extinction of A also extinguishes B. If A and B are presented sequentially during phase two of the second-order conditioning procedure, then extinction of A does not extinguish B.

The *Limax* data obtained using the postconditioning extinction procedure are just what one would predict from the precedents in the vertebrate conditioning literature (Sahley *et al.,* 1984). If odor A and odor B are given simultaneously during phase two of a second-order conditioning experiment and then the aversive response to odor A is extinguished by repeated presentations of odor A alone, the aversive response to odor B is also eliminated (Fig. 1). Conversely, if odor A and odor B are given sequentially during phase two of the second-order conditioning procedure, then extinction of the aversive response to odor A does not diminish the aversive response to odor B.

Limax also can acquire appetitive approach responses to neutral or weakly aversive odors if a very attractive taste (0.5 M fructose) is paired with exposure to the neutral odor (Sahley *et al.,* 1982). We do not know yet whether the higher-order phenomena documented with aversive conditioning also obtain with appetitive conditioning. Delaney has recently shown that *Limax* can learn to avoid diets deficient in a single essential amino acid (Delaney and Gelperin, 1983) using an as yet unidentified postingestive consequence of eating the deficient diet. This learning occurs after one day's intake of the deficient diet and is retained for at least three weeks. The list of neuronal computations probably is not complete as yet.

2.2. Training of Isolated Brain

The reliability of the associative learning combined with the robustness of molluscan neurons and synapses (Reingold and Gelperin, 1980) led to the first experiments aimed at training the isolated central nervous system of *Limax* (Chang and Gelperin, 1980). Rather than odor–taste pairings, taste–taste pairings were applied to the normal taste input pathway via the lip chemoreceptors. The taste stimuli used were standardized extracts of the foodplants used in the whole animal training. The output used to assess learning was the neural substrate of ingestion, termed feeding motor program (FMP). Feeding motor program is a pattern of cyclic, coordinated motoneuron activity recorded from buccal ganglion nerve roots which can be recognized unambiguously by a naive observer given the buccal nerve recordings and a measuring algorithm (Gelperin, Chang, and Reingold, 1978). Isolated lip–brain preparations were differentially conditioned to supress FMP responses to a food extract that initially triggered FMP by pairing lip application of the food extract with lip application of a bitter-tasting substance such as colchicine, tannic acid, or quinidine. The evidence for associative learning derives from the findings that the FMP response decrement is selective to the food extract paired with the bitter taste and that there is a critical time interval within which the food extract and bitter taste must be applied for response suppression to occur (Culligan and Gelperin, 1983). Changes at the taste receptors themselves as a causative mechanism are made less likely by the finding that the lip–brain preparation can learn after being trained using one lip and tested using the other, naive lip.

Another approach to studying the mechanism of memory storage in the isolated brain is to train the animal, prepare its brain for neurophysiological recording, and assess *in vitro* the retention of the memory that had been entered *in vivo* (Gelperin and Culligan, 1984). The intact animals were given taste–taste training using the same food extracts and quinidine solution later used to stimulate their lip chemoreceptors after reducing the animal to a lip–brain preparation. Cold anesthesia of both the animal and brain was used to protect the synapses from alteration and depletion during dissection. One half of the animals given paired training yielded lip–brain preparations showing a selective FMP response suppression to the food extract paired with quinidine. None of the animals given unpaired training yielded lip–brain preparations showing a selective FMP response suppression (Gelperin and Culligan, 1984). These results make it likely that the whole animal and isolated brain training procedures are accessing the same stimulus representation and associative learning mechanisms. Further tests of this idea are possible using a two-phase training procedure, with phase one presented to the intact animal and phase two applied to the isolated lip–brain preparation.

2.3. Training Cultured Brains

The memory retention times shown by the isolated lip–brain preparations (8–12 hr) are much shorter than the retention times shown by the intact animal (2–3 weeks). To test whether this difference results from the impoverished biochemical environment experienced by the isolated brain living only in saline, we developed techniques for dissecting the animal under semisterile conditions and setting up the lip–brain preparation in a sterile culture medium. Under these conditions, FMP responses can be elicited for several days in response to taste stimuli applied to the lips. Test solutions are conveyed to and removed from the lip chambers by tubing connected to a peristaltic pump. The preparation is shown diagrammatically in Fig. 2.

Before training, the responses of the preparation to two different attractive food extracts (A+, B+) are assessed. Stimulus A+ is applied to the lips for 30 sec and the FMP response recorded. Stimulus B+ is applied 1 hr later for 30 sec and the FMP response recorded. After another hour, A+ is again applied for 30 sec, followed immediately by application of a concentrated solution of quinidine sulphate to the lips for 10 min. The preparation is then rested for several hours before testing.

Testing is started with a 30-sec application of B+, repeated every 30 min until

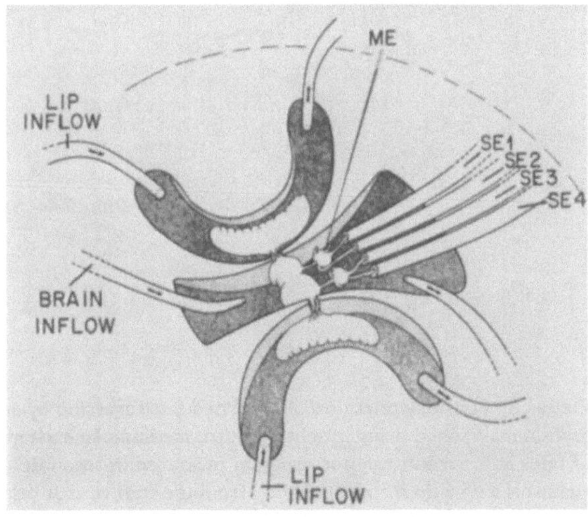

FIGURE 2. Schematic diagram of the isolated nervous system of *Limax maximus*. Cerebral ganglia, buccal ganglia, and the lip chemosensory regions are left attached to one another via nerves. The two lip halves are placed in separate chambers that allow independent perfusion with food-derived stimuli. SE 1, 2, 3, 4 = Suction electrodes 1, 2, 3, 4; ME = microelectrode.

a reproducible level of FMP response is obtained. Then the response to A is determined leaving at least 30 min after the preceding B+ application.

An example of a result obtained with this training and testing procedure is shown in Fig. 3. Initially, at t = 0 and t = 1 hr, the responses of the preparation to 30-sec applications of rat chow extract (A+) and apple juice (B+) were determined. Both stimuli elicited strong FMP responses from the preparation. At t = 2 hr, a 30-sec application of A+ was followed immediately by application of quinidine sulfate solution to the lips for 10 min. For unknown reasons, the FMP evoked in response to A+ was potentiated by the subsequent quinidine exposure.

The preparation was then rested for several hours before its responsiveness to B+ was determined. At t = 10 hr, a 30-sec application of B+ elicited a short but clear bout of FMP. A 30-sec application of A at t = 11 hr did not elicit any FMP. The complete suppression of the FMP response to A was retained for almost two days, even though the preparation received several tests of its response to A which contributed to extinction of the learned response suppression. At t = 45 hr, a weak FMP response consisting of two bites was evoked by A. This response was much weaker than the immediately preceding response to B+ at t = 44 hr. The magnitude of the response to A increased gradually over the next two days. The result indicates that it should be possible to study the transition of the memory trace from short-term to long-term form in the isolated brain preparation, as well as the effect of selective biochemical manipulation of the culture medium to enhance or retard synthetic events thought to be important in memory function.

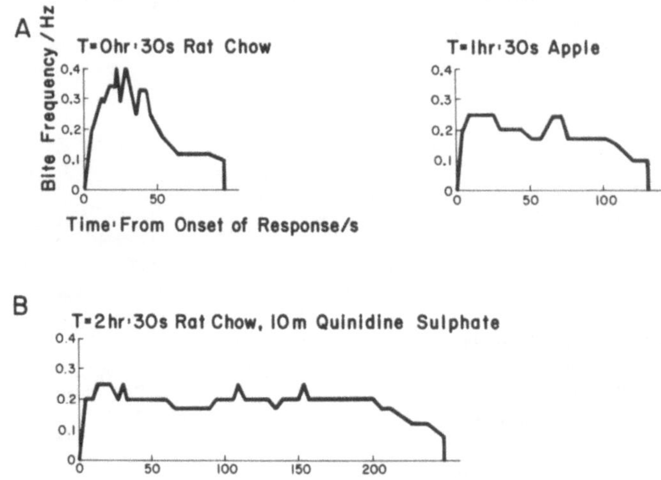

FIGURE 3. Four-day retention of a learned taste aversion by an isolated lip–brain–buccal ganglia preparation maintained in an enriched culture medium. In each graph the instantaneous frequency (in Hz) of bites in the feeding motor program produced in response to a stimulus applied to the lips is plotted against the time (in seconds) elapsed from the start of that response. The time elapsed from the start of the experiment and the type of stimulus applied are shown above each graph. (A) Control responses of the preparation to rat chow (CS1) and apple (CS2) prior to training. (B) Response during the training

C

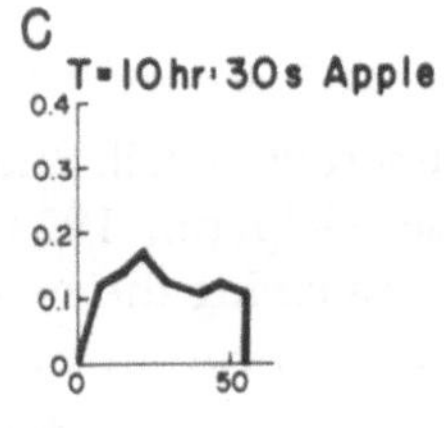

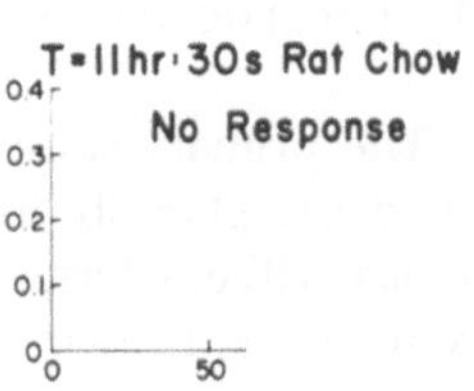

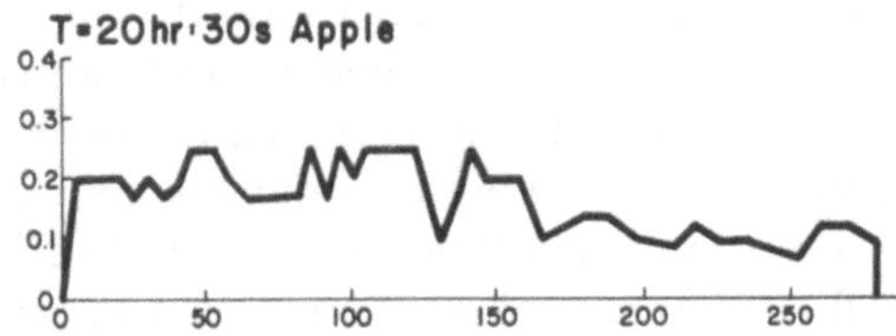

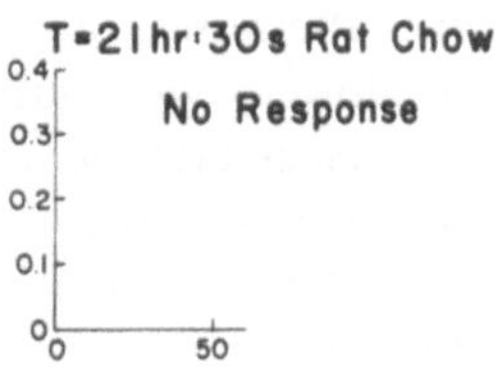

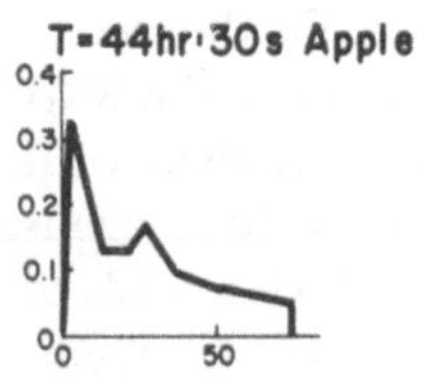

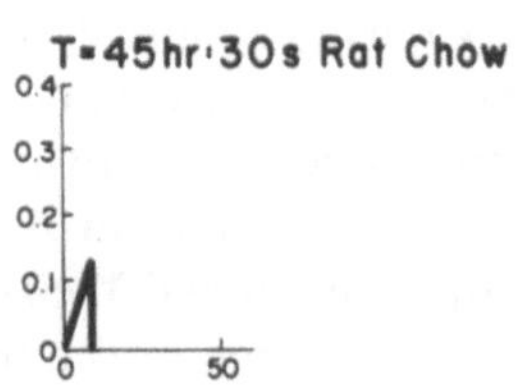

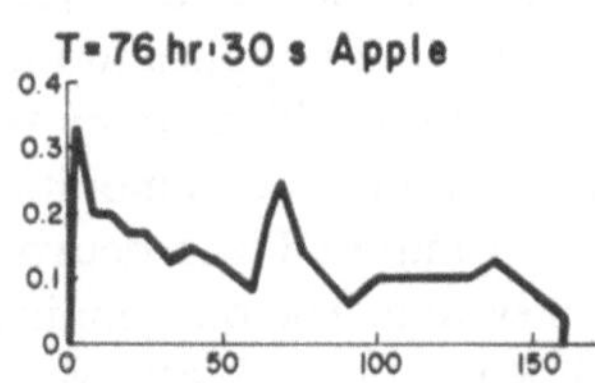

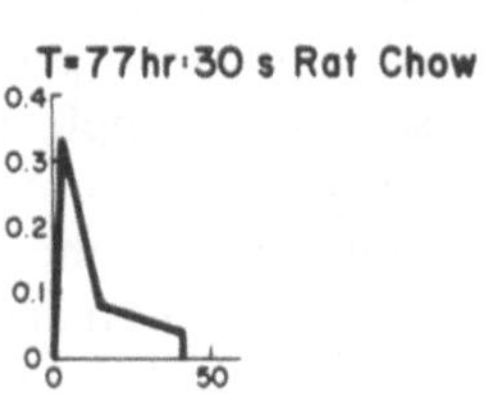

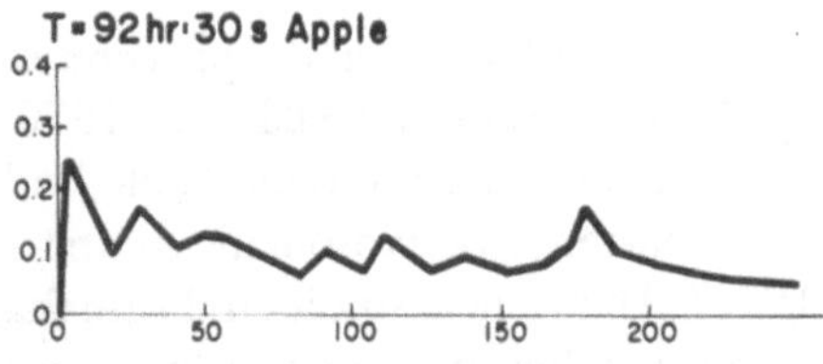

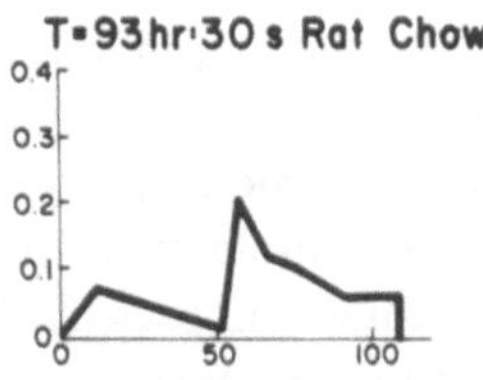

procedure in which a 30-sec application of rat chow (CS1) was followed immediately by a 10-min application of quinidine sulfate (US). (C) Responses to apple and rat chow, applied at intervals following the training procedure. The response of the preparation to rat chow was suppressed completely for at least 19 hr following training. It remained suppressed relative to the response to apple for the duration of the experiment.

2.4. Interneurons

The primary sensory neurons for taste and smell are numerous, small, and located peripherally in the lips (Benedeczky, 1977) and nose (Gelperin, 1974; Kataoka, 1976; Chase and Kamil, 1983). Rather than start by examining the sensory neurons directly (see review by Croll, 1983), we studied interneurons that receive and integrate synaptic input from peripheral chemoreceptors. Our initial study focused on the metacerebral giant cells (MGCs), a pair of serotonergic interneurons known to modulate the expression of FMP (Gelperin, 1981). A preparation of the noses, olfactory nerves, and cerebral and buccal ganglia was developed that allowed odor puffs to be delivered to the olfactory epithelium while an *en passant* electrode recorded from the olfactory nerve and an intracellular record was obtained from an MGC (Egan and Gelperin, 1981). These experiments demonstrated that olfactory afference can provide synaptic excitation to the MGC sufficient to elicit spikes from a quiescent cell or increase spike rate in an active cell. We have not yet determined whether this input from the olfactory nerve monosynaptically excites the MGC, although fiber tracts from the olfactory nerve were shown to arborize near the MGC (Egan and Gelperin, 1981). Input from one olfactory nerve can affect both the ipsilateral and contralateral MGCs. Further work is necessary to determine if the functional effect of the synaptic input to the MGCs differs when attractive or repellent odors are applied to the nose. Activity of the MGC homologue in *Aplysia* is correlated directly with the level of food arousal (Kupfermann and Weiss, 1982) and bilateral lesion of the *Aplysia* MGCs leads to a specific change in the biting response with no change in other aspects of feeding or other behaviors (Rosen *et al.*, 1983).

Kemenes *et al.* (1982) studied the responses of an identified cerebral neuron (C2) to input from lip chemoreceptors in *Helix*. The C2 neuron increased its rate of action potential production by 2.5 to 3 times in response to lip application of 3% sucrose or apple juice, but was only weakly activated by distilled water or mechanical stimulation. It will be important in further work of this type to monitor FMP output from the buccal ganglion and test the effect of imposed interneuron activity (e.g., driven activity in C2) on the feeding control system. The burgeoning collection of such feeding interneurons (Benjamin, 1983; Rosen *et al.*, 1982; Weiss *et al.*, manuscript submitted) provides ample material for studying changes in chemosensory integration due to learning (Davis *et al.*, 1983; compare with Chapter 14, this volume).

A large population of chemosensory interneurons is located in the procerebral lobes of the cerebral ganglia in *Limax* (Veratti, 1900; Zs-Nagy and Sakharov, 1970). These interneurons are intrinsic to the procerebral lobe and are remarkable both for the complexity of their arborizations and the existence of axosomatic as well as axoaxonic synapses (Zs-Nagy and Sakharov, 1970). Anatomical studies indicate that the olfactory nerves provide input to the procerebral lobes, but the three lip nerves do not (Chetail, 1963). Although there are no monoamine-containing somata in the procerebral lobe, monoaminergic fibers containing both serotonin and dopamine innervate the intrinsic interneurons. This suggests that monoaminergic neurons extrinsic to the procerebral lobe might modulate the processing of olfactory information by the intrinsic interneurons. Consistent with this anatomical picture is the finding that the taste input–FMP output reflex pathway does not require the procerebral lobes for its operation and that a brain conditioned *in vitro* retains its taste–taste association after procerebral lobe amputation (I. Cooke and K. Delaney, unpublished observations). It will be very interesting to test the effect of procerebral lobe amputation on odor-elicited synaptic events as recorded from the MGC using the nose–brain preparation.

Another approach to the identification and mapping of interneurons in the feeding control system is the use of immunohistochemical staining, most easily implemented using the whole-mount procedure of Beltz and Kravitz (1982). The entire population of neurons using a particular transmitter can be revealed if an antibody to that transmitter or its synthetic enzyme is available. Work in a variety of molluscan systems has implicated serotonin, dopamine, acetylcholine, gamma-aminobutyric acid, histamine, small cardioactive peptide B, FMRFamide, cholecystokinin, and metenkephalin as involved in some aspect of the neural control system for feeding, An example of *Limax* buccal neurons showing immunoreactivity for a FMRFamidelike substance is shown in Fig. 4. The large neuron is cell B1, which sends a very large axon into the cerebrobuccal connective, through the cerebral ganglion to the abdominal ganglion, and out a peripheral nerve to the heart and kidney region (D. J. Prior and A. Gelperin, unpublished observations). The cerebral ganglia also contained scattered, small immunoreactive neurons. Nerve fibers with FMRFamidelike immunoreactivity were observed in buccal ganglion nerve roots, the cerebrobuccal connectives, and some cerebral ganglion nerve roots (Cooke and Gelperin, 1984). This result immediately suggests experiments testing the family of FMRFamidelike peptides (Greenberg *et al.,* 1983) for a controlling or modulatory function on the FMP reflex and its associatively conditioned modification.

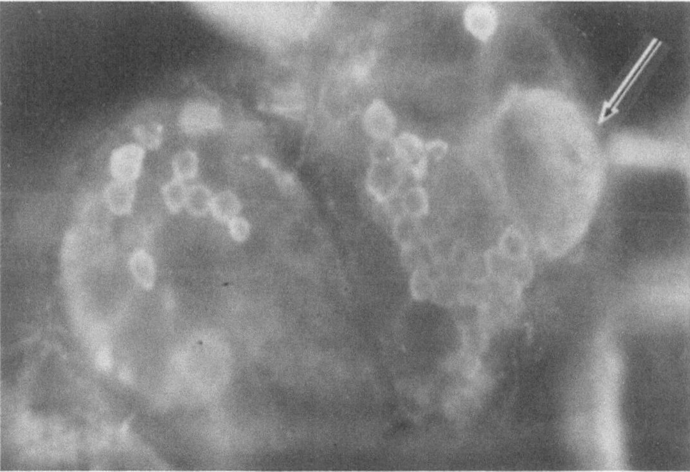

FIGURE 4. A *Limax* buccal ganglion stained with fluorescent-labeled FMRFamide antibody as a whole mount. The arrow indicates cell B1 in the lateral lobe of the ganglion.

2.5. Neurochemical Approaches

A neurochemical analysis of the cerebral and buccal ganglia of *Limax* using the technique of high-pressure liquid chromatography indicated the presence of substantial amounts of dopamine and serotonin (Wieland and Gelperin, 1983). Addition of exogenous dopamine to the isolated cerebral and buccal ganglia resulted in the generation of FMP, while addition of the dopamine blocker ergonovine blocked the ability of lip chemostimuli to trigger FMP. There appears to be a set of dopaminergic synapses whose function is obligatory for operation of the taste input–FMP output reflex, and whose modulation may play a role in learned alterations of the reflex. To examine the effects of altering dopaminergic synapses on the FMP reflex and learning ability of the isolated lip–brain preparation, dopamine synthesis is blocked by applying alpha-methyl-*p*-tyrosine (Wieland *et al.*, 1983). Since measurement of total dopamine in the ganglia is a very insensitive measure of depletion caused by alpha-methyl-*p*-tyrosine, we are now examining dopamine released into the medium bathing the ganglia in response to a short pulse of high (50 mM) potassium medium (Wieland *et al.*, 1984). After removal of the outer connective tissue sheath, a pair of cerebral and buccal ganglia can liberate up to 5 pmole of dopamine in 30 min of high-potassium stimulation, an effect blocked in low-calcium saline containing 5 mM cobalt. Now the effect of alpha-methyl-*p*-tyrosine on the synaptically releasable pool of dopamine can be assessed, first neurochemically and then physiologically. Lip–brain preparations are also being prepared with altered dopaminergic transmission by prior injection of the neurotoxin 6-hydroxydopamine.

In addition to studies aimed at directly manipulating the dopamine pool available for synaptic release, other transmitters known or suspected of modulating dopaminergic transmission are also of direct relevance. Met-enkephalin is known to cause dopamine release in several molluscan ganglia (Stefano, 1982), hence its effect on the feeding system in *Limax* is of particular interest.

The effect of serotonin on the lip–brain preparation parallels the effect of intracellular stimulation of the MGCs, namely, modulation of the intensity of the output. Exogenous serotonin application does not trigger FMP, but will speed up and intensify motor output elicited by lip input (Wieland and Gelperin, 1983). Potassium-stimulated release of serotonin from the isolated lip–brain preparation has also been measured (2.5 pmole per 30 min) so as to have an internal control for the specificity of treatments aimed at the selective alteration of dopamine.

A quite different approach to altering chemistry before studying the learning ability of the isolated brain is to use dietary supplements of neurotransmitter precursors given to the animal for several weeks before testing its brain physiologically. This strategy works well for those transmitters whose synthesis is limited by precursor availability, such as acetylcholine (Blusztajn and Wurtman, 1983). Slugs raised on a high choline diet (1.14% wt/wt) have a significantly higher level of choline in their blood than slugs raised on a low choline diet (0.03% wt/wt) (Barry and Gelperin, 1982a). Transmission at an identified cholinergic synapse is augmented by an increase of only 1.5 μm choline in the medium bathing the synapse, an increase in choline concentration well within the range caused by ingestion of the high-choline diet (Barry and Gelperin, 1983b). Several lines of evidence indicate that the synaptic augmentation caused by increased choline availability in this concentration range is due to effects on acetylcholine synthesis in the presynaptic neuron (Barry and Gelperin, 1984). These observations relate directly to the search for learning and memory storage mechanisms because the slugs fed the high-choline diet showed much better retention of the conditioned odor–taste association than did the slugs fed the low-choline diet (Sahley *et al.,* 1985). Since both high-choline and low-choline slugs showed the learning after one day, the effect appears to be on retention.

3. LIMAX Learning: The Neural Model

3.1. Components of the Model

The neural model of associative learning attempts to simulate the experimental results of associative learning experiments done with the real mollusk. The model uses neural elements whose properties and interactions are realistic abstractions of real neurons (Hopfield, 1982, 1983). The model is currently embodied in a computer simulation, the LIMAX simulation program. The relationships between the components of the model (LIMAX), both neural hardware and processing algorithms, and the components of the real animal *(Limax)* will be discussed after presenting the model and its operation. LIMAX has three major parts (Fig. 5): (1) an array of lip chemoreceptors, (2) a taste categorizer/taste memory network, and (3) a learning control and motor output network. These three components operate as follows.

The lip chemoreceptors are taken to be 100 sensory neurons, each responding to several components of the food-plant extracts used as chemostimuli in the behavioral experiments. The result of presenting a food extract to the array of sensory neurons is to drive some of the receptors strongly and others less so. The sensory code for a particular food is thus described as a set of 100 random numbers, each lying between zero and unity, with an average of 20 values greater than 0.5. This sensory signal was then "clipped": sensory neurons with large outputs produced maximum outputs (value of 1), while the neurons with small outputs generated no output (value of 0). This procedure transforms the sensory input from a particular food from continuous numbers into a list of 100 1s and 0s, with about 20 1s. A taste input from a mixture of foods is produced in an analogous way as the input from a single food, hence the sensory signal will represent an admixture of the two foods in which the total number of maximally active sensory cells was maintained near 20.

The aversive taste input pathway for quinine is represented in a special way in LIMAX in that inputs on this pathway are innately aversive and do not have to be learned to be recognized. The aversive input can vary in strength, but does not become less aversive due to paired application with an attractive taste input. This aversive input pathway is represented in Fig. 5 as a single quinine-recognition sensory cell. This is convenient and rational because the ultimate logic of LIMAX will

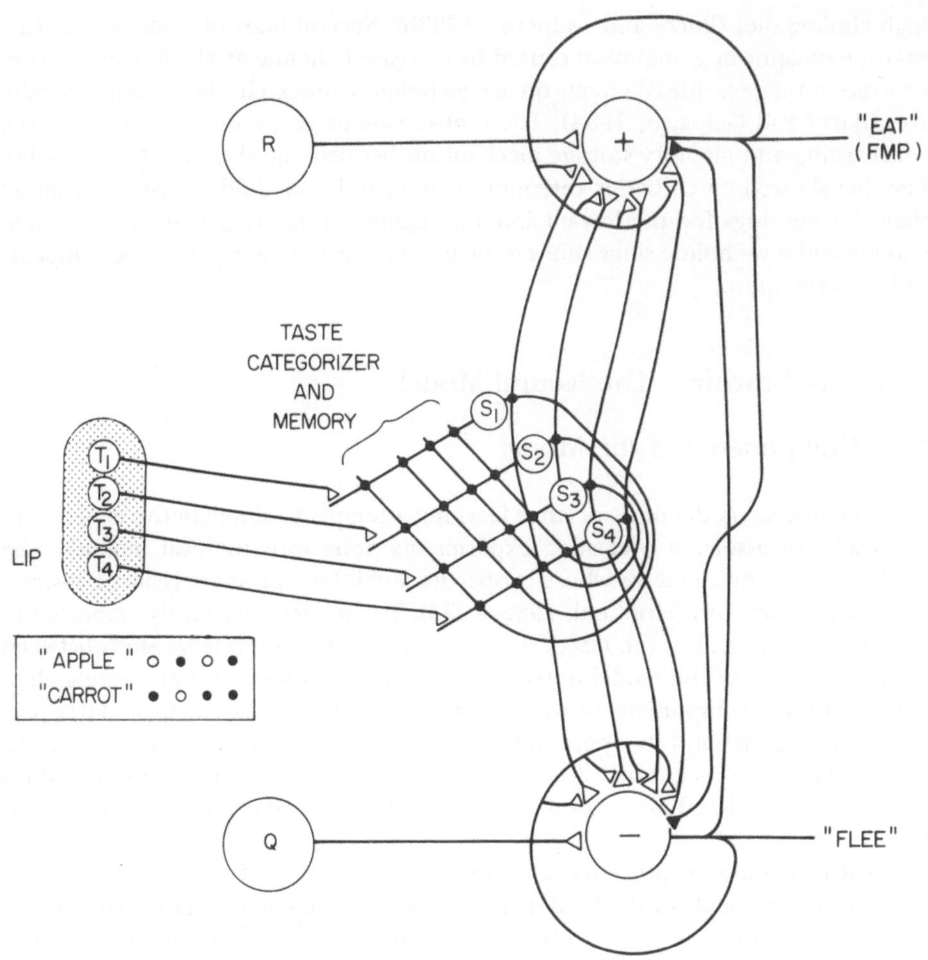

FIGURE 5. A diagram of the components of the LIMAX simulation program.

not depend on any particular attributes of quinine. Any innately aversive stimulus (quinine, sodium chloride, electric shock) could be used to activate this pathway. An innately rewarding taste input pathway is also available.

The taste categorizer/taste memory network receives its input from the chemoreceptor array and has the ability to modify the strengths of the synaptic connections between its elements to produce a set of stable activity patterns. A particular stable activity pattern in the taste categorizer network is an internal representation of the taste input from a food class. A detailed description of how this system works is given in Appendix 1.

The taste categorizer network performs several functions that are crucial to the capability of the LIMAX simulation to learn appropriately. First, internal representations of foods that LIMAX has been exposed to in the past are contained in the organization of the synapses in this network. When presented with a particular food that elicits a pattern of activity in the taste receptors that is close to, but not necessarily exactly the same as, on of the familiar food classes, the categorizer network "recognizes" the similarity. It does so by assuming a firing pattern of its constituent neurons that represents the known food class. The ability of the categorizer network to converge to a unique stable activity state from similar but different activity states is an inherent feature of its organization (see Appendix 1) and is the reason for its description as a "categorizer." Second, when presented with a novel food that is not rapidly or easily categorized as a known familiar food, the network will, if presented with the new food a sufficient number of times, build an internal representation of this new food class and include it as a new category of its known foods.

The learning control/motor output network determines the response of the animal when presented with a food: will it "eat" or "flee." We consider these possibilities mutually exclusive and have modeled their neural hardware with cross-inhibition between a command neuron for "eating" and a command neuron for "fleeing." These command neurons, called the "plus" (+) and "minus" (−) cells, are meant to represent the observed motor responses of the animal, but not the central pattern generators responsible for the control of particular muscles. Note that only one of the two command neurons can be active at any time.

The learning control network is functionally located in the synapses between the taste categorizer and the motor center and controls the learned association of taste categories with one of the two possible motor responses. Every neuron in the categorizer network synapses on the "plus" and "minus" cells. How the firing pattern of the neurons in the categorizer influences the activity of the "plus" and "minus" cells determines the ultimate activity of the organism because it is the pair of "plus" and "minus" cells that coordinate in a mutually exclusive way whether LIMAX is in the "eat" or "flee" state. Hence when food A is presented to LIMAX, it will elicit a "flee" response if the synapses from the neurons in the categorizer that are active in the internal representation of food A are strongly excitatory on the "minus" cell and weakly excitatory on the "plus" cell. It is actually the summed averages of these synaptic influences from the active categorizer neurons onto the "plus" and "minus" cells that determine the actual motor output.

Now one can see what synaptic modification is required when an initially attractive food A is presented to LIMAX paired with an aversive input from quinine. If food A is attractive, this means that the firing pattern of the categorizer that is the internal representation of food A provides strong excitatory drive to the "plus" cell and weak (or at least less strong) synaptic drive to the "minus" cell. In order to make food A aversive, this pattern of synaptic drive must be reversed: The active categorizer neurons for food A must provide stronger synaptic drive to the "minus" cell than the "plus" cell. This change in synaptic strengths between categorizer neurons and the "plus" and "minus" neurons is the essential step in the conditioning.

Special care must be taken to insure that the conditioning shows the temporal specificity observed in real experiments. We only want LIMAX to learn that food A is aversive if food A is presented before or concurrent with the presentation of an aversive stimulus. An aversive stimulus, whether it is quinine or a previously aversively conditioned food, is one which, when presented alone, leads the organism into the "flee" state. Hence if we want LIMAX to learn that attractive food A is to be aversive, in a way consistent with the observed temporal specificity of the presentation of CS and US, we want to change the synaptic strengths between the active categorizer neurons representing food A and the "plus" and "minus" cells only if food A occurs before or during a transition of the state of the organism from the "eat" to the "flee" state. In LIMAX, this temporal specificity is embodied in the logic of the synapse modification rule. Synapse modification occurs only when there is a well-defined temporal relationship between activity in the presynaptic terminal of a "categorizer" cell and the postsynaptic cell (+ or − cell). This rule is illustrated in Fig. 6. Although synapse modification requires the presence of presynaptic activity, the sensitivity to synaptic change is not uniform with time after a presynaptic action potential. The time course of synapse change sensitivity has the form shown in Fig. 6B as the consequence of a typical timecourse of excitation of the "minus" cell illustrated in Fig. 6A. This sensitivity defines a time window: if the presentation of the CS precedes this window as in forward conditioning (Fig. 6C), a maximal increase occurs in the synaptic strength. If the CS is presented after this window as in backward conditioning (Fig. 6E), the synapse modification is greatly reduced, or if there is sufficient delay, completely eliminated.

Although this modeling appears to involve the use of Hebbian synapses, this is strictly true only in the sense of an operational definition. The operational definition of a Hebbian snyapse is that if the presynaptic and postsynaptic elements are active at the same time, then the strength of the synapse is increased. From the point of view of cellular and biochemical hardware, this learning element is usually considered as a two-terminal device consisting solely of the presynaptic and postsynaptic terminals. In the gill withdrawal conditioning system of *Aplysia,* a three-terminal synapse has been suggested as the basic learning element (Kandel and

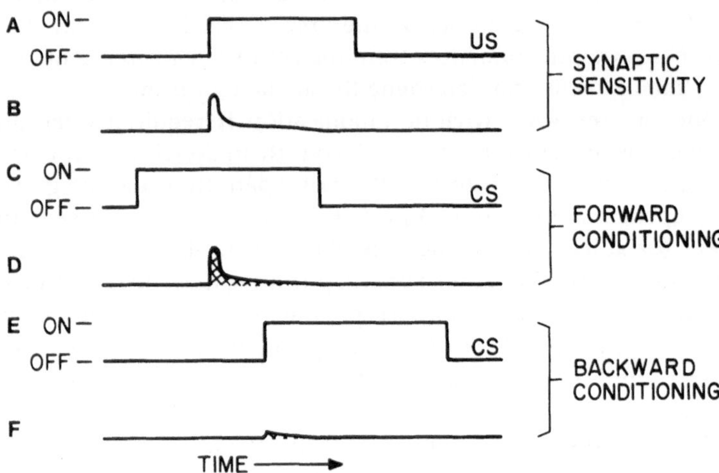

FIGURE 6. Diagram of the temporal specificity of the synapse modification algorithm operating on the synapses between the taste categorizer network and the plus and minus neurons.

Schwartz, 1982; Chapter 12, this volume). In addition to the presynaptic and postsynaptic cells, the third terminal is provided by the facilitator synapse on the presynaptic terminal. The efficacy of transmission between the presynaptic and postsynaptic cell is increased when activation of the presynaptic terminal briefly precedes activation of the facilitator. From our operational point of view, it is possible to think of the Hebbian synapse as a special case of a three-terminal learning element. If the postsynaptic cell synapses on the presynaptic terminal as a facilitator synapse, then the strength of the synapse between the presynaptic and postsynaptic cell will increase with simultaneous activation of the presynaptic cell. In this special case, the postsynaptic cell also plays the role of the facilitator (Hawkins and and Kandel, 1984). The time-order dependence we have described and use in our model, not present in the sample Hebb description, can likewise be described with either two- or three-terminal hardware.

3.2. Learning with LIMAX

A typical learning experiment with the LIMAX simulation involves the following steps: LIMAX is presented with a new food A, which may or may not resemble one of the known foods already in its memory. The taste neurons responding to A drive the categorizer network, which converges either to a previously established category or to a new one for food A. Let food A initially be either neutral or attractive. This means that the synaptic drive from the active neurons in the categorizer representing food A to the "plus" and "minus" cells is either unbiased or stronger to the "plus" cell. The state of LIMAX is now either neutral (i.e., not in either the eat or flee state) or eat. Now LIMAX is presented with quinine, which drives it into the "flee" state. As illustrated in Fig. 6, this is the correct time sequence for learning and LIMAX modifies the synapses from the categorizer network to the "minus" cell so that the active inputs to the "minus" cell are strengthened. The synapses from the categorizer network to the "plus" cell are also modified so that active synapses are weakened. After a few such pairings of food A and quinine, presentation of food A alone causes LIMAX to "flee," i.e., food A has become aversive.

For the system to work well, two additional features are necessary. First, when an association has been learned well, further learning of the same association should cease. Without this control, one association presented often and strongly would obliterate all other memories. This control can be achieved by limiting the synapse-strengthening algorithm so that the synaptic strength between any two neural elements cannot exceed a fixed value. Second, we implemented an algorithm whose effect is to hold constant the total synaptic strength of all the inputs to each neural element. When one synapse to an element was set to unity, another synapse chosen at random was set to zero. As a result, older memories will become weakened and slowly destabilized as new ones are learned. A food learned a long time ago and not tasted again will not have its memory degraded with time per se, but only as a result of learning many new tastes afterward.

The simulation of first-order conditioning just described effectively models the learning experiments with the real mollusk. Second-order conditioning is accomplished by the learning algorithm in a way which emphasizes one of the important computational properties of the categorizer network. Take the case in which LIMAX is presented with food B. As before, after the categorization process, there is an internal representation of the category into which food B is placed that is manifest in the particular firing pattern of the neurons of the categorizer network. If food B is unbiased, the state of LIMAX is neutral, whereas if food B is attractive, the state of LIMAX will be "eat." Now LIMAX is presented with food A, which has been aversively conditioned previously by pairing A and quinine. The effect of driving the categorizer with the sensory neuron pattern that codes for food A while the state of the categorizer is still food B is to change the firing pattern of the categorizer neurons so that it partially resembles both A and B. Since food A has been strongly aversively conditioned, the categorizer inputs to the "minus" cell will be strongly driven, causing the state of LIMAX to change to "flee" while the categorizer firing pattern still partially resembles the representation of food B. Hence, the conditions for the learning algorithm are satisfied, and the neurons that are firing in the categorizer network will have their synapses with the "minus" cell strengthened and their synapses with the "plus" cell weakened. The internal representation of the category that includes food B will be changed so that it is more aversive. After a few trials, food B will achieve an aversive status equal to food A, that of a food aversively conditioned in a first-order learning paradigm.

Note that any system that uses the same neural hardware for input of both the CS1 and the aversively conditioned CS2 in a second-order learning experiment will face the question as to how the information in CS1 is stored while the new information representing CS2 is input. In LIMAX, partial information about CS1 is retained in the firing pattern of the categorizer which is, however, progressively being influenced by CS2. This approach is in contrast to registerlike memories where the representation of CS1 is stored physically separate from the representation of CS2, or the simpler case (Hawkins and Kandel, 1984) where CS1 and CS2 are represented by physically distinct input pathways.

What is the effect of extinguishing the aversiveness of food A after using food A to produce second-order aversive conditioning of food B? Our system does not show extinction, but we can achieve the same effect in a different way by presenting food A followed by a rewarding taste (0.5 M fructose) a few times, after which A will be attractive. Recall that foood B is aversive because of the strong synaptic connections between the categorizer neurons that are on in the representation of B and the "flee" neuron. Since the categorizer neurons that are active in the B representation are generally not the same neurons as are on in the food A representation (although a few of them may be in common), extinguishing food A has little effect on the aversive behavior to food B.

Second-order conditioning can be done in a different way. Once again, we begin by establishing a first-order conditioned aversive response to food A, and then present a 50–50 mixture of foods A and B. This mixture is neither A nor B, but similar to each. Within the present modeling, this food would drive about five neurons that are on in both A and B, five that are on in A but not B, five that are on in B and not A, and five that are on in neither A nor B. While this food mixture AB is present, the synapse modifications appropriate to it are made in the categorizer network and AB itself becomes a new food category, which is neither A nor B. At the same time "flee" will come on during this exposure because the food AB contains a very significant half of its active neurons that are common with A and which therefore are excitatorily connected to the minus cell. When the state changes to "flee," some connections to "flee" are strengthened from neurons that are active in state AB but not in state A. If this conditioning is only done once and the period of time of the AB stimulus presentation substantial, AB will become a known category in which the dominant connections responsible for the behavior "flee" are from neurons active in state A to the "minus" cell. Notice that the system does not know category B; it knows only categories A and AB. Now when LIMAX is presented with food B, its initial response will be through using the known food category that is closest to B, namely the category AB, since B is not familiar. Its initial behavior will then be "flee," and therefore classified as second-order conditioning.

The effect of extinction of A (again carried out by learning a positive association with A) will now be quite different. The extinction of A will succeed in eliminating the connections between the neurons that are active in state A and "flee," and will replace them by excitatory connections between these same neurons and "eat." When next food B (unfamiliar) is presented and invokes the similar category AB, there will be little tendency to flee because the excitatory connections between neurons active in A, and which were responsible for the aversive behavior to category AB, have been chiefly eliminated and replaced by excitation of "eat."

3.3. Relation of LIMAX and *Limax*

The way we have modeled the primary chemosensory input represents an explicit choice between two alternatives: the labeled-line theory and the across-fiber pattern hypothesis. Although these theories are not strictly mutually exclusive, many primary sense cells in the gustatory and olfactory pathways each respond to several components of a biologically relevant, complex food stimulus (Croll, 1983; Derby and Ache, 1984; Dethier and Crnjar, 1982). The much smaller body of data on response properties of chemosensory interneurons likewise shows cells with broad response properties such as those used in the modeling rather than food feature detectors. Given our choice of modeling the sensory input using the across-fiber pattern to represent a particular food taste or odor, the choice of 100 cells in the categorizer provides about 15 distinguishable food categories. Although systematic studies to determine how many different distinguishable food categories *Limax* possesses have not been done, more than 30 different food plants are acceptable to *Limax* (Frömming, 1952; Gain, 1891).

The input pathway for quinine has special properties deriving from its use to channel several types of aversive stimuli that cause food aversion learning into the LIMAX neural network. Toxicosis (Gelperin, 1975), bitter taste (Sahley *et al.*, 1981a), and shock (Delaney and Gelperin, 1984) can all promote rapid learning by real *Limax* so a multimodal aversive input pathway is required. One implication of directing the aversive input directly to the learning control circuit is that the aversive properties of quinine should not decrease no matter how many times quinine is paired with an attractive food. This is a testable notion.

Our use of the taste categorizer network to process the CS inputs (food A and B) provides a natural solution to the need for an internal representation of the first CS (food A) which can interact with a second CS arriving later during second-order conditioning. Additional evidence for an enduring "taste memory" is that first-order conditioning can occur with times of 5 to 30 min intervening between the CS (food A) and US (quinidine) (Chang and Gelperin, 1980; Culligan and Gelperin, 1983).

The model has three possible behavioral outputs: feed, flee, and neutral. The neutral state is more accurately described as the absence of activation of either feed of flee. Recent recordings of anterior pedal nerve output before and after *in vitro* conditioning of the isolated *Limax* lip–brain preparation indicate that a neural correlate of foot withdrawal appears due to associative conditioning as FMP disappears (Delaney and Gelperin, 1984). This will allow more explicit measurement of the appearance of the flee state as the feed state is suppressed during learning.

3.4. Questions Raised by the Model

Within this simple model, there are different ways in which second-order conditioning can be represented. Different training protocols can result in different physical representations of the synaptic changes responsible for second-order conditioning. Such differences can then show up in behavioral differences, which make sense only when described in terms of what the hardware is doing.

This model of the sensory processing of *Limax* captures the essence of many of the learning experiments that have been carried out. It emphasizes what simple collective systems can do with a minimum of detailed prewiring and without feature detectors (grandmother cells). The model raises many questions to which answers are not yet known. To begin, are the categories in *Limax* food learning plastic, or are the categories fixed (as in immunology)? Does a learning experience involve merely assigning a food to one of these categories and labeling that category? Are the categories A, B, and AB indeed all different? Is the essential memory of foods electrical reverberation (as we have assumed) or chemical? Is this memory the same in first-order conditioning (where time delays can be very long) and in second-order conditioning (where time delays must be short)? How close together in "taste" can two foods be and still be distinguished? Does food memory gradually fade with time, or is it an "all or none" effect? The attempt to model the system generates a set of behavioral and electrophysiological questions whose answers are essential to carrying such modeling further.

4. Future Directions

4.1. *Limax* Learning

There are a variety of questions best answered by behavioral experiments on intact animals. The existence of higher-order appetitive conditioning needs to be determined by applying the paradigms used to produce higher-order aversive conditioning, but with neutral or aversive odors as the CS and 0.5 M fructose as the US. Behavioral experiments can explore the nature of the stimulus categories formed during exposure to food odors and tastes, and are the most effective way to accurately determine the time course of the synaptic sensitivity to associative modification depicted graphically in Fig. 6. Finally, the transition of the memory trace from a short-term form that is labile to a long-term form that is resistant to disruption can be studied by cooling animals to 2°C for a brief period at different times after the conditioning is established.

Neurophysiological experiments can capitalize on the ability to produce robust conditioning of isolated brains while recording from interneurons in the feeding control circuit. The metacerebral giant cell and the cerebral dopamine neurons are logical loci from which to record changes in synaptic drive due to conditioning. Electrical recording can also be used to localize the neurons that store the taste or odor memory (CS trace). To determine whether some part of the *Limax* CNS stores information in the way that LIMAX does, multichannel electrical or optical recording methods will be used to record action potentials from tens of neurons simultaneously before and after memory storage.

Neurochemical experiments have been started to determine the number and prevalence of phosphoproteins recovered from extracts of *Limax* cerebral ganglia (B. Oestreicher, T. Yamane, and A. Gelperin, in preparation). As these studies attain cellular resolution and focus on phosphoproteins related to ion channel function, they can contribute in an important way to the elucidation of memory storage machinery.

4.2. LIMAX Learning

The LIMAX program is being developed toward a more nearly complete representation of the *Limax* associative learning behavior and to better approximation of both the electrophysiology and biochemistry of real neurons.

The first major aspect of such a development is to model the cellular electrophysiology by "neurons" having a graded response (not off–on) and communicating by means of action potentials (Hopfield, 1984). This will enable us to view the spiking behavior of any LIMAX "neuron" during processing, and thus to see the electrophysiological correlates of information processing in LIMAX. Our understanding of the general principles of these networks suggests that the relevant behaviors will persist and become easier to work with in the system with graded neuronal responses.

Within the present model, it appears that relatively simple changes in the synapse modification algorithm and the cellular anatomy will be able to produce both extinction and blocking. This issue is also being pursued with neurons that spike and have graded responses.

The model so far considered has only a short-term memory (through a combination of synapse change and continued electrical activity) and long-term memory (through synapse change). The addition of an intermediate form of memory, as in a temporary modification of synapses that can later either be made permanent or allowed to decay, seems likely to be of importance both for the long-term associations made in food learning and for a description of memory consolidation.

Such models will always be incorrect in many details. Nevertheless, the importance of having a complete model is to see how cellular rules of neuron response and synapse modification result in detailed behaviors. Any particular complete model predicts the outcome of scores of behavioral experiments on the basis of cellular suppositions. A few of these will have been used to construct the model. Others are obvious "correct" behaviors, not yet studied experimentally, and for which such a model makes detailed predictions. Finally and perhaps most interestingly, any given model will also have contrived situations for which the model predicts peculiar behavior (the associative learning analogue of perceptual illusions). The comparison of LIMAX with *Limax* in such situations should be particularly informative.

4.3. Conclusion

It is an exciting time in the study of memory storage mechanisms. Three specific and detailed models are available (Chapter 12, this volume; Acosta-Urquidi *et al.*, 1984; Lynch and Baudry, 1984) with different biochemical mechanisms mediating the transduction of critically timed patterns of electrical activity into enduring changes of transmitter action. The biophysical and biochemical measurements of *Limax* memory mechanisms can both test the generality of the existing models and perhaps provide important extensions of them needed to attain the more complex conditioning displayed by *Limax*.

The logic of *Limax* learning is apparent at several levels. At the behavioral level, the adaptive value of the learning mechanism is easy to visualize (Whelan, 1982; Gouyon *et al.*, 1983). In some sense *Limax* may be genetically programed to learn reliably about odors and tastes as predictors of positively or negatively rewarding consequences. The operational logic of the learning is summarized in Fig. 1, although further work may well significantly extend this list. The actual synaptic logic of *Limax* is yet to be revealed, but the power of the simple synaptic alteration rules used by LIMAX to accomplish the same tasks with fewer neurons encourages us in our search.

ACKNOWLEDGMENTS. We thank I. Cooke and K. Delaney for critical comments on the manuscript. Work done at Princeton University was supported by NIH Grant MH39160. Work done at the California Institute of Technology was supported by NSF Grant DMR-8107494.

Appendix 1. A Neural Network for Categorization and Memory Storage

This model results from a search for emergent properties of neural networks that depend on the collective interactions of the neurons rather than localized details of their interactions. It explores what neurons having relatively simple properties and interacting according to relatively simple rules can accomplish collectively. In particular, a neural network and the algorithm for storing memories within it are described, first using neurons with only two activity states, on and off. This network also categorizes inputs to it based on its past experiences. Finally, it is shown that a network of neurons having graded activity states retains the same collective ability to form categories and store memories.

Consider a network of two-state neurons, each of which is either firing as rapidly as possible ($V_i = 1$) or is inactive ($V_i = 0$). At any time the state of the network of N such neurons can be described by a list on N 1s and 0s, representing the instantaneous activity state of each of the N neurons.

Each neuron is connected to every other neuron in the network. Each neuron i receives inputs from other neurons j through synapses of efficacy T_{ij}. The input to neuron i at time t is thus

$$I_i(t) = \sum_j T_{ij} V_j(t) + E_i(t)$$

where $E_i(t)$ is the input to neuron i from sources external to the network, such as direct inputs from sensory neurons. The connection T_{ij} can be either excitatory ($T_{ij} > 0$) or inhibitory ($T_{ij} < 0$).

Each neuron may change its activity state repeatedly as time goes on. The model ascribes to each neuron a behavior that tends to turn neuron i on if its input has been above threshold (H) for a while and to turn it off if it has been below threshold. The detailed description of this process lets each neuron inspect its inputs at random times with a specified mean inspection rate. At each such inspection time the *ith* neuron sets

$$\begin{aligned} V_i &\rightarrow 1 \quad \text{if } I_i > H \\ V_i &\rightarrow 0 \quad \text{if } I_i \leq H \end{aligned}$$

Since each neuron integrates itself at random times, there is no synchrony in the times at which different neurons change state. Unlike computer processing and much neural modeling, the processing described here is asynchronous. The lag between changes in the inputs to neuron i and the change of state of neuron i incorporates the effects of membrane charging times, propagation delays, and synaptic delays.

Stable activity patterns (states) of the neural network can be specified by the initial values of synaptic strengths between all of the elements of the network. The appropriate values of the synaptic strengths needed to make a set of states (V^s; $s = 1,2,3, \ldots$) the stable states of the network are calculated using the following storage prescription

$$T_{ij} = \sum_s (2V_i^s - 1)(2V_j^s - 1)$$

The storage prescription is used to calculate the synaptic strengths to each neuron (i) from every other neuron (j) in the network necessary to made the network activity patterns embodied in states V^1, V^2, V^3, . . ., the stable states of the network. Having entered the appropriate matrix of synaptic strengths into the network to specify stable states V^1, V^2, V^3, etc., the network, started with its neurons in a random state, will converge to one of the specified stable states, specifically the stable state that most closely resembles the starting state (Hopfield, 1982). If the network is started with partial information about one of the stable states, i.e., an activity pattern that resembles but is not identical to one of the prespecified stable states, the network will converge to the stable state most like the starting state. After setting the initial synaptic strengths, the only process operating during the convergence to a stable state is the simple asynchronous input-evaluation algorithm just described. It is an intrinsic, emergent property of a network operating in this way that the network in fact converges to a stable state.

The neural network can come prewired with a matrix of synaptic strengths specifying a set of stable activity patterns, analogous to genetically specified memory states. Alternatively, the network can be equipped with an interface to the real world in the form of sensory receptors and a synapse modification algorithm that causes activity states of the network imposed by activity in the sensory receptors to become inherent stable states of the (categorizer) network.

Computer simulations of the operation of such neural networks have revealed that optimum stability requires a certain amount of global inhibition distributed throughout the network without regard to the specified stable states. Also, as more and more stable states (memories) are specified, the performance of the network during recall begins to deteriorate. About 0.15 N memories can be simultaneously stored in the network before the error in recall is severe. Finally, it was found that 30 neurons was the lower size limit for a network able to store stable activity patterns in this way.

Note that the memories in this network are represented in the collected set of synapse strengths and activity patterns of the constituent neurons. Each neuron is active in several memory states, and memory states will have some active neurons in common. Thus particular neurons and synapses cannot be identified with particular memories.

The neural network simulation has recently been extended to make the neurons more realistic and to show that the basic computational abilities of the two-state networks are not lost. Rather than two-state neurons, we used neurons with a graded (sigmoidal) input–output curve. Spike output was taken to be virtually zero with membrane potential (Vm) below a threshold and output was maximum at highly depolarized values of Vm. At intermediate values of Vm, the output was a steeply graded function of the input. Also, the neurons were endowed with membrane capacitance with which they temporally summed their synaptic inputs. When neuron j in the network fires an action potential, its effect on other cells in the network is to dump charge on their input capacitance. The amount of charge (and its sign) that is placed on the input capacitance of neuron i is determined by the value of T_{ij}. Hence the magnitude of a given T_{ij} is still to be considered as a measure of synaptic strength, while the sign of T_{ij} indicates if the synapse is excitatory or inhibitory. The effect of the change in the cell's charge is to change the input voltage (membrane potential). However, this change is transitory: the membrane potential decays exponentially back to its resting state with a certain time constant.

Our simulations of networks containing spiking neurons have shown that they retain the ability of the two-state networks previously described to categorize input patterns.

References

Acosta-Urquidi, J., Alkon, D. L., and Neary, J. T., 1984, Ca^{2+}-dependent protein kinase injection in a photoreceptor mimics biophysical effects of associative learning, *Science* **224:**1254–1257.

Barry, S. R., and Gelperin, A., 1982a, Dietary choline augments blood choline and cholinergic transmission in the terrestrial mollusc, *Limax maximus, J. Neurophysiol.* **48:**541–547.

Barry, S. R., and Gelperin, A., 1982b, Exogenous choline augments transmission at an identified cholinergic synapse in the terrestrial mollusc *Limax maximus, J. Neurophysiol.* **48:**431–450.

Barry, S. R., and Gelperin, A., 1984, Acetylcholine turnover in an autoactive molluscan neuron, *Cell. Molec. Neurobiol.* **4:**15–29.

Beltz, B., and Kravitz, E. A., 1982, Mapping of serotonin-like immunoreactivity in the lobster nervous system, *J. Neurosci.* **3:**585–602.

Benedeczky, I., 1977, Ultrastructure of the epithelial sensory region of the lip in the snail *Helix pomatia* L, *Neuroscience* **2:**781–789.

Benjamin, P. R., 1983, Gastropod feeding: Behavioral and neural analysis of a complex multicomponent system, in: *Neural Control of Rhythmic Movements* (A. Roberts and B. L. Roberts eds.), Cambridge University Press, New York, pp. 159–193.

Blusztajn, J. K., and Wurtman, R. J., 1983, Choline and cholinergic neurons, *Science* **221:**614–620.

Chang, J. J., and Gelperin, A., 1980, Rapid taste aversion learning by an isolated molluscan central nervous system, *Proc. Natl. Acad. Sci. U.S.A.* **77:**6204–6206.

Chase, R., and Kamil, R., 1983, Neuronal elements in snail tentacles as revealed by horseradish peroxidase backfilling, *J. Neurobiol.* **14:**29–47.

Chetail, M., 1963, Etude de la regeneration du tentacule oculaire chez un arionidae *(Arion rufus)* et un Limacidae *(Agriolimax agrestis), Arch. Anat. Microsc. Morphol. Exp.* **52:**129–203.

Cooke, I., and Gelperin, A., 1984, Immunocytochemical mapping of the neural control system for feeding in *Limax maximus, Soc. Neurosci. Abstr.* **10:**691.

Croll, R., 1983, Gastropod chemoreception, Biol. Rev. **58:**293–319.

Culligan N., and Gelperin, A., 1983, One-trial associative learning by an isolated molluscan CNS: Use of different chemoreceptors for training and testing, *Brain Res.* **266:**319–327.

Davis, W. J., Gillette, R., Kovac, M. P., Croll, R. P., and Matera, E. M., 1983, Organization of synaptic inputs to paracerebral feeding command interneurons of *Pleurobranchaea californica* III. Modifications induced by experience, *J. Neurophysiol.* **49:**1557–1572.

Delaney, K., and Gelperin, A., 1983, The slug *Limax maximus* shows postingestive food aversion learning to amino acid deficient diets, *Soc. Neurosci. Abstr.* **9:**914.

Delaney, K., and Gelperin, A., 1984, Rapid food-aversion learning with shock as UCS in *Limax maximus, Soc. Neurosci. Abstr.* **10:**509.

Derby, C. D., and Ache, B. W., 1984, Quality coding of a complex odorant in an invertebrate, *J. Neurophysiol.* **51:**906–924.

Dethier, V. G., and Crnjar, R. M., 1982, Candidate codes in the gustatory system of caterpillars, *J. Gen. Physiol.* **79:**549–570.

Egan, M., and Gelperin, A., 1981, Olfactory inputs to a bursting serotonergic interneuron in a terrestrial mollusc, *J. Molluscan Stud.* **47:**80–88.

Frömming, E., 1952, Uber die Nahrung von *Limax maximus, Anz Schadlingskde* **25:**41–43.

Gain, W. A., 1891, Notes on the food of some of the British mollusks, *J. Conchol.* **6:**349–361.

Gelperin, A., 1974, Olfactory basis of homing behavior in the giant garden slug, *Limax maximus, Proc. Natl. Acad. Sci. U.S.A.* **71:**966–970.

Gelperin, A., 1975, Rapid food-aversion learning by a terrestrial mollusk, *Science* **189:**567–570.

Gelperin, A, 1981, Synaptic modulation by identified serotonin neurons, in: *Serotonin Neurotransmission and Behavior* (B. Jacobs and A. Gelperin, eds.), MIT Press, Cambridge, pp. 288–304.

Gelperin, A., 1983, Neuroethological studies of associative learning in feeding control systems, in: *Neuroethology and Behavioral Physiology* (F. Huber and H. Markl, eds.), Springer-Verlag, Berlin, pp. 189–205.

Gelperin, A., and Culligan, N., 1984, *In vitro* expression of *in vivo* learning by an isolated molluscan CNS, *Brain Res.* **304:**207–213.

Gelperin, A., Chang, J. J., and Reingold, S. C., 1978, Feeding motor program in *Limax.* I. Neuromuscular correlates and control by chemosensory input, *J. Neurobiol.* **9:**295–300.

Gouyon, P. H., Fort, Ph., and Caraux, G., 1983, Selection of seedlings of *Thymus vulgaris* by grazing slugs, *J. Ecol.* **71:**299–306.

Greenberg, M. J., Painter, S. D., Doble, K. E., Nagle, G. T., Price, D. A., and Lehman, H. K., 1983, The molluscan neurosecretory peptide FMRFamide: Comparative pharmacology and relationship to the enkephalins, *Fed. Proc.* **42:**82–86.

Hawkins, R. D., and Kandel, E. R., 1984, Is there a cell biological alphabet for learning? *Psycho. Rev.* **91:**375–391.

Hopfield, J. J., 1982, Neural networks and physical systems with emergent collective computational abilities, *Proc. Natl. Acad. Sci. USA* **79:**2554–2558.

Hopfield, J. J., 1984, Neurons with graded response have collective computational properties like those of two state neurons, *Proc. Natl. Acad. Sci. USA* **81:**3088–3092.

Hopfield, J. J., Feinstein, D. I., and Palmer, R. G., 1983, "Unlearning" has a stabilizing effect in collective memories, *Nature (London)* **304:**158–159.

Kamin, L. J., 1969, Predictability, surprise, attention, and conditioning, in: *Punishment and Aversive Behavior* (R. Church and B. A. Campbell, eds.), Appleton-Century-Crofts, New York pp. 279–296.

Kandel, E. R., and Schwartz, J. H., 1982, Molecular biology of learning: Modulation of transmitter release, *Science* **218:**433–443.

Kataoka, S., 1976, Fine structure of the epidermis of the optic tentacle in a slug, *Limax flavus* L, *Tissue Cell* **8:**47–60.

Kemenes, G., Hernádi, L., and Salánki, J., 1982, Identification of cerebral moto-neuron responding to lip stimulation in *Helix pomatia, Acta. Biol. Acad. Sci. Hung.* **33:**215–229.

Kupfermann, I., and Weiss, K. R., 1982, Activity of an identified serotonergic neuron in free moving *Aplysia* correlates with behavioral arousal, *Brain Res.* **241:**334–337.

Lynch, G., and Baudry, M., 1984, The biochemistry of memory: A new and specific hypothesis, *Science* **224:**1057–1063.

Reingold, S. C., and Gelperin, A., 1980, Feeding motor program in *Limax.* II. Modulation by sensory inputs in intact animals and isolated central nervous system, *J. Exp. Biol.* **85:**1–19.

Rescorla, R. A., 1984, Comments on three Pavlovian paradigms, in: *Primary Neural Substrates of Learning and Behavioral Change* (D. Alkon and J. Farley, eds.) Cambridge University Press, New York, pp. 25–45.

Rosen, S. C., Weiss, K. R., Cohen, J. L., and Kupfermann, I., 1982, Interganglionic cerebral-buccal mechanoafferents of *Aplysia:* Receptive fields and synaptic connections to different classes of neurons involved in feeding behavior, *J. Neurophysiol* **48:**271–288.

Rosen, S. C., Kupfermann, I., Goldstein, R. S., and Weiss, K. R., 1983, Lesion of a serotonergic modulatory neuron in *Aplysia* produces a specific defect in feeding behavior, *Brain Res.* **260:**151–155.

Sahley, C. L., Gelperin, A., and Rudy, J. W., 1981a, One-trial associative learning modifies food odor preferences of a terrestrial mollusc, *Proc. Natl. Acad. Sci. U.S.A.* **78:**640–642.

Sahley, C. L., Rudy, J. W., and Gelperin, A., 1981b, An analysis of associative learning in a terrestrial mollusc: Higher-order conditioning, blocking and a transient US pre-exposure effect, *J. Comp. Physiol.* **144:**1–8.

Sahley, C. L., Hardison, P., Hsuan, A., and Gelperin, A., 1982, Appetitively reinforced odor-conditioning modulates feeding in *Limax maximus, Soc. Neurosci. Abstr.* **8:**823.

Sahley, C. L., Rudy, J. W., and Gelperin, A., 1984, Associative learning in a mollusc: A comparative analysis, in: *Primary Neural Substrates of Learning and Behavioral Change* (D. Alkon and J. Farley, eds.) Cambridge University Press, New York, pp. 243–258.

Sahley, C. L., Barry, S. R., and Gelperin, A., 1985, Dietary choline augments associative memory function in *Limax maximus, J. Neurobiol.,* in press.

Stefano, G. B., 1982, Comparative aspects of opioid-dopamine interaction, *Cell. Mol. Neurobiol.* **2:**167–178.

Veratti, E., 1900, Ricerche sul sistema nervoso dei *Limax, Memorie Reale Instituto Lombardo Scienze Lettere* **18:**9.

Whelan, R. J., 1982, Response of slugs to unacceptable food items, *J. Appl. Ecol.* **19:**79–87.

Wieland, S. J., and Gelperin, A., 1983, Dopamine elicits feeding motor program in *Limax maximus, J. Neurosci.* **3:**1735–1745.

Wieland, S. J., Zaininger, H., Jahn, E. G., and Gelperin, A., 1983, Dopamine and serotonin in *Limax* feeding: Distribution and metabolism, *Soc. Neurosci. Abstr.* **9:**75.

Wieland, S. J., Jahn, E. G., and Gelperin, A., 1984, Measurement and control of dopamine and serotonin release from *Limax* ganglia *in vitro, Soc. Neurosci. Abstr.* **10:**690.

Zs-Nagy, I., and Sakharov, D. A., 1970, The fine structure of the procerebrum of pulmonate molluscs, *Helix* and *Limax, Tissue Cell* **2:**399–411.

14

Neural Mechanisms of Behavioral Plasticity in an Invertebrate Model System

W. Jackson Davis

1. Introduction

The relatively recent capacity of neuroscientists to unravel the central neural networks mediating simple behaviors, aptly illustrated in the preceding chapters of this book, has enabled a revolutionary paradigmatic advance in neurobiology. Namely, complex, higher-order behavioral phenomena, many of which previously defied even attempts at precise definition, can now be approached mechanistically in relatively simple and defined neural networks. Such behavioral phenomena include, for example, motivation, drive, mood, choice, attention, and learning, both nonassociative (i.e., habituation, sensitization) and associative (i.e., classical and operant conditioning). Many of these phenomena may be considered forms of behavioral plasticity, i.e., changes in behavior that are acquired because of experience.

For the past 15 years, my colleagues and I have employed this general approach to seek neural explanations for three such behavioral phenomena, namely, motivation, associative learning, and choice. Our investigations of the neural mechanisms of these behavioral phenomena have involved the feeding behavior of the carnivorous marine gastropod mollusk *Pleurobranchaea*. We chose this "model system" to study behavioral plasticity because it exhibits the requisite higher-order behavioral phenomena and because it has a tractable nervous system that permits cellular approaches to behavioral questions. In this chapter, I will

W. Jackson Davis • The Thimann Laboratories and The Long Marine Laboratories, University of California at Santa Cruz, Santa Cruz, California 95064.

summarize these studies, beginning with a brief description of *Pleurobranchaea's* feeding behavior and its underlying neural circuitry by way of essential background. I will then show how this information has enabled an understanding of the neural basis of motivation, learning, and choice, at the level of the single, identified central neuron. Finally, I will discuss how these forms of plasticity are unified by convergent neurophysiological mechanisms at the command neuron level.

2. Neural Circuitry Underlying Feeding Behavior

Pleurobranchaea's feeding behavior consists of several distinct components, including orientation to the food stimulus (Fig. 1A), extension of the feeding proboscis (Fig. 1B), and the explosive bite-strike (Fig. 1C), by which the animal captures and ingests food items (Davis and Mpitsos, 1971). If the food is not captured by the bite-strike, the proboscis eversion movement is repeated rhythmically at a period of one to several seconds. If the bite-strike succeeds in capturing the food, it is ingested and rhythmic swallowing movements continue while the proboscis is inside the body.

FIGURE 1. Components of the feeding behavior of the carnivorous marine mollusk *Pleurobranchaea californica.* (A) Orientation to a food stimulus presented to anterior chemosensory structures. (B) Extension of the feeding proboscis. (C) The bite-strike. (From Mpitsos and Davis, 1973.)

FIGURE 2. Scanning electron micrographs (SEMs) showing exteroreceptors in the tentacle of *Pleurobranchaea*. (A) Low-magnification SEM overview of the luminal surface of the sensory epithelium, showing the dense packing of receptor cells. Bar, 50 μm. (B) Higher-power SEM showing details of an individual receptor, illustrating the cilia and their expanded, discoid terminals. Bar, 5 μm. (From Matera and Davis, 1982.)

Electromyograms made from buccal muscles during behavior in intact (Croll and Davis, 1981) and reduced (Croll and Davis, 1982; Croll *et al.*, 1985a) specimens have revealed the pattern of motor activity that underlies feeding behavior. The motor pattern consists of cyclical bursting in individual muscles and alternation of bursts in functionally antagonistic muscles—the same basic motor pattern that underlies many rhythmic behaviors, including, for example, walking in humans. Subtle but distinct shifts in the motor pattern cause an opposite behavioral result, namely egestion (Croll *et al.*, 1985a). Both motor patterns are endogenous to the CNS, and both may employ common neural oscillators (Croll *et al.*, 1985a). The difference between the two motor programs entails their activation by independent command systems (Croll *et al.*, 1985b,c).

The neural circuitry that mediates the feeding behavior begins with primary sensory receptors (chemoreceptors and mechanoreceptors) or anterior sensory structures, the oral veil, tentacles and rhinophores (Davis and Matera, 1982; Matera and Davis, 1982). The major exteroreceptor is a ciliated neuron (Fig. 2A), whose cilia are each distally expanded into a disklike structure (Fig. 2B) that may serve to increase the surface area available for interactions with chemical stimuli. These discociliated primary receptors are distributed only in those body regions that cause feeding behavior when contacted by appropriate chemical stimuli (Davis and Matera, 1982), suggesting that they are the chemoreceptors that mediate reception of food stimuli.

The discociliated exteroreceptors located on the oral veil project directly to the cerebropleural ganglion (brain). Comparable receptors on the tentacles and rhinophores project axons without interruption to peripheral ganglia located at the base of each sensory appendage, namely, the tentacle and rhinophore ganglia. Here the receptor axons terminate in complex neuropilar regions that serve as the first integrating station for chemical and mechanical stimuli. Dye-injection studies

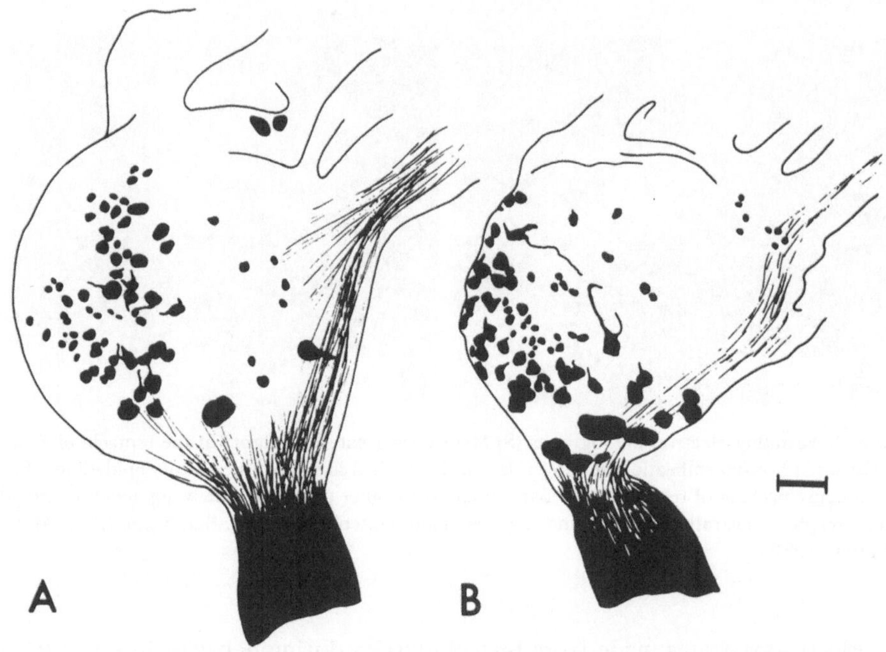

FIGURE 3. Tracings of photographs of whole mounts of the rhinophore (A) and tentacle (B) ganglia following centrifugal cobaltous chloride backfills of the rhinophore and tentacle nerves, respectively. This procedure stains the somata of interneurons that collect primary afferent inputs and relay them to the brain. Bar, 100 μm. (From Bicker *et al.*, 1982b.)

have shown that the neurons responsible for this initial integration consist of a population of about 100 sensory interneurons located in each peripheral ganglion (Fig. 3). The same interneurons that collect and integrate information about food stimuli then relay it directly to the brain (Bicker *et al.*, 1982a,b) where it is processed by the central neural network that controls feeding behavior.

The interface between sensory inputs associated with food and the central neural circuitry that produces the feeding movements has not been explored systematically. It is known, however,that the food and mechanical stimuli excite a population of central neurons—the paracerebral neurons or PCNs (Fig. 4A)—that independently have been shown capable of initiating feeding behavior. Intracellular stimulation of a single paracerebral neuron in a whole animal preparation induces cyclic feeding movements (Fig. 4B), and these same neurons discharge in cyclic bursts of action potentials during feeding behavior that is induced by natural food stimulation (Fig. 4C). In the reduced preparation and in the isolated CNS,

FIGURE 4. Location and functional properties of paracerebral feeding command interneurons (PCNs) in the brain of *Pleurobranchaea.* (A) Tracing of a photograph of a whole mount of the dorsal brain following cobaltous chloride backfilling of the cerebrobuccal connective (cbc). Anterior is toward the top of the picture. Abbreviations (clockwise from bottom): cbc, cerebrobuccal connective; mn, mouth

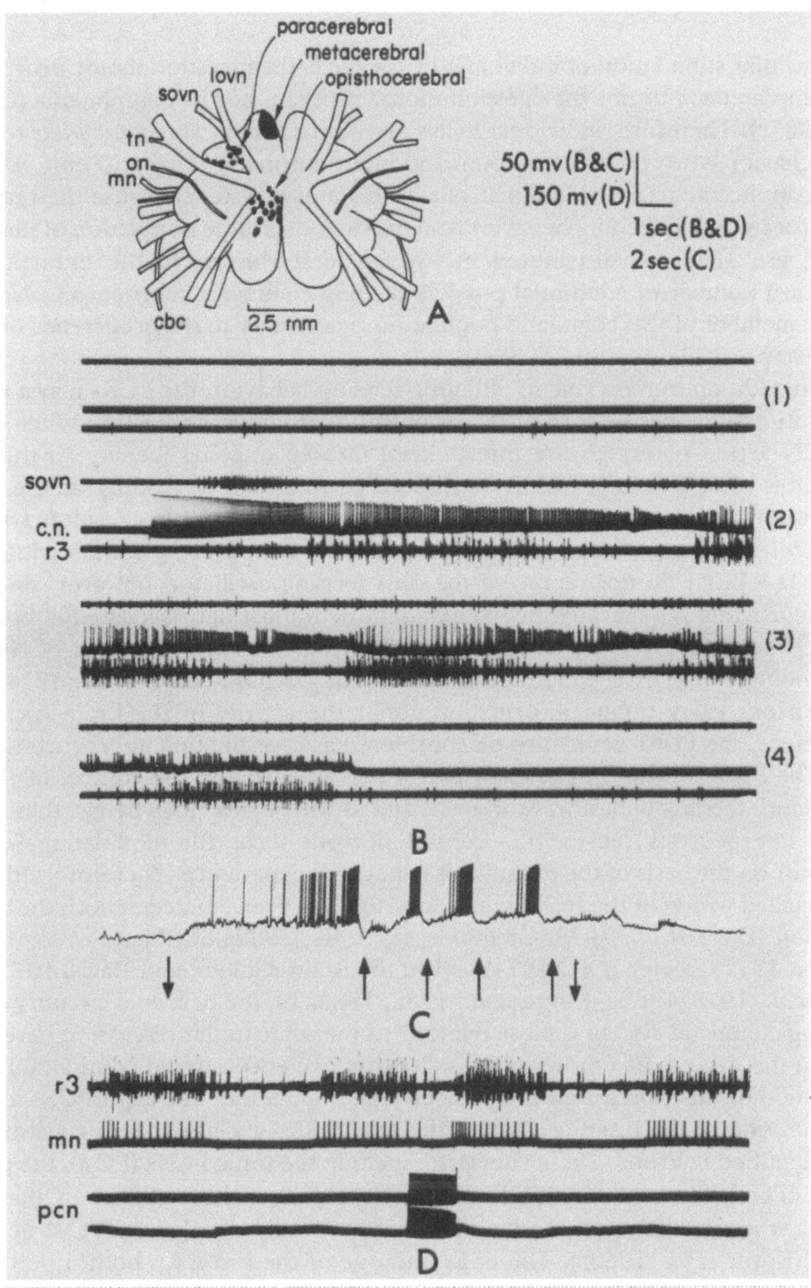

nerve; on, optic nerve; tn, tentacle nerve; sovn, small oral veil nerve; lovn, large oral veil nerve. (B) Four continuous traces showing command effect of the PCNs in the isolated CNS preparation. The PCN (CN) is depolarized tonically starting in the second trace, and ending in the fourth trace. This stimulus elicits the cyclic feeding motor output, which is recorded extracellularly from two feeding nerves (SOVN and r 3). (C) Discharge pattern of a PCN in a whole animal preparation during food-induced feeding. Upward arrows indicate time of visible bites, whereas the two downward arrows demarcate the beginning and end of food stimulation. (D) Reset of the cyclic efferent rhythm induced in feeding nerves by interpolated PCN bursts. The rhythm was induced by stimulating the stomatogastric nerves; mn, intracellular recording from a motor neuron. (From Gillette *et al.,* 1978, 1982c.)

intracellular stimulation of individual PCNs elicits the ingestion motor program or its components, but not the egestion motor program nor its components (Croll *et al.*, 1985c). Therefore, individual PCNs are both *sufficient* and *appropriate* to feeding behavior—two criteria for command interneurons. A third criterion, *necessity*, is not applicable to individual neurons in this motor system because the responsibility for initiating feeding behavior is shared among a large population of neurons. There are 16 PCNs distributed in two bilateral clusters in the brain and an unknown number of additional possible feeding command interneurons. No individual member of this command population is necessary to the occurrence of feeding behavior (Gillette *et al.*, 1982c).

In addition to their role in initiating feeding behavior, the PCNs play a role in generating the cyclic pattern of motor output that underlies feeding. When bursts of PCN action potentials are interpolated into an ongoing feeding rhythm, the rhythm is reset (Fig. 4D)—either advanced or retarded, depending on the phase position of the interpolated burst (Gillette *et al.*, 1982c). Such a result is typically interpreted to imply that the stimulated neuron(s) are part of the underlying oscillator. The PCNs do not represent the only feeding oscillator, however, since the feeding rhythm is generated by independent neural oscillators located in both the brain (Davis *et al.*, 1984) and buccal ganglion (Davis *et al.*, 1973). The PCNs send descending axons from the brain to the buccal ganglion (Davis *et al.*, 1974), and these axons carry timing information about the activity of the brain oscillator. Therefore, the PCNs could also be considered as coordinating interneurons.

The PCNs in the brain clearly play a pervasive role in the control of *Pleurobranchaea's* feeding behavior. Moreover, and as will be described below, these command interneurons represent a central nervous locus for modulating feeding behavior on the basis of the organism's behavioral experience. Therefore, although the detailed wiring of the feeding motor system has been studied in both the buccal ganglion (Davis *et al.*, 1973; Gillette *et al.*, 1980, 1982a,b; Gillette *et al.*, 1982c; Siegler, 1977; Siegler *et al.*, 1974) and in the brain (Gillette and Davis, 1977; Gillette *et al.*, 1978, 1982c, Kovac *et al.*, 1982, 1983a,b), the neuronal circuitry of the feeding command system is most relevant to the topic under discussion here.

In the process of unraveling this circuitry, we encountered considerable specialization of function within the paracerebral population. The eight PCNs located on each side of the brain are subdivided into four subclasses, each consisting of two identified neurons. These subclasses include the tonic PCNs (PC_Ts), the phasic PCNs (PC_ps)(Kovac *et al.*, 1982), the polysynaptic excitors (PSEs) of the PC_ps (Kovac *et al.*, 1983a), and the type II electrotonically coupled (to PC_ps) neurons (ET_{II}s)(Kovac *et al.*, 1983a). The PSEs and one of the two ET_{II}s both supply polysynaptic chemical excitatory inputs to the PC_ps, in addition to their independent feeding command role. In addition, a sizable population of brain neurons has been discovered that supplies monosynaptic and polysynaptic inhibition to the PCNs (Kovac *et al.*, 1983b).

For each of the identified neurons composing the command circuitry of the feeding system, a battery of tests has been performed, as illustrated for a single ET_{II} in Fig. 5. These tests include intracellular injection of the neuron with the marker dye Lucifer Yellow in order to determine the architecture of each cell (Fig. 5A), intracellular stimulation in various ionic environments to determine its recip-

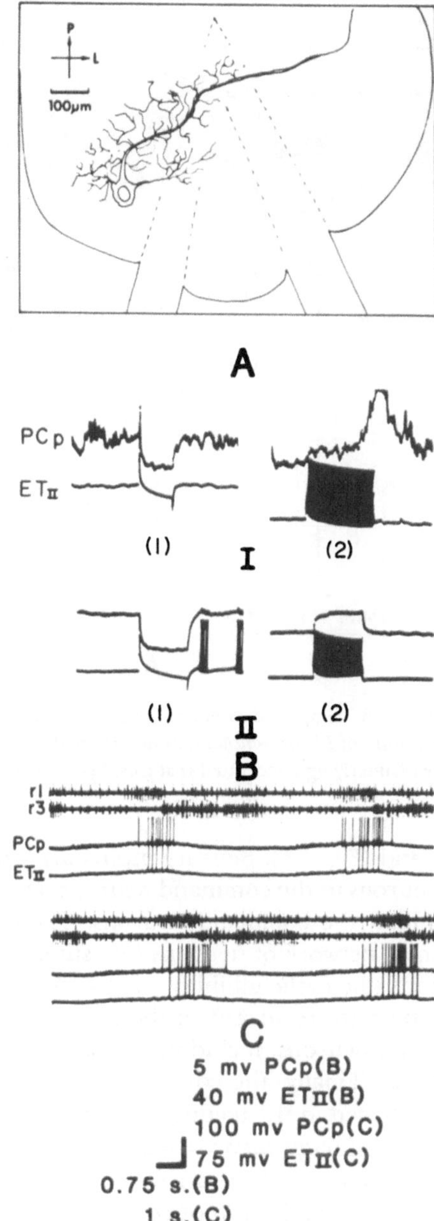

FIGURE 5. Tests performed (in this case on an ET_{II}) to identify each neuron in the feeding command system. (A) Lucifer Yellow injection of the cell seen in dorsal aspect. P, posterior. L, lateral. (B) Physiological effects on a PC_p. (AI) Hyperpolarization (1) and depolarization (2) of the ET_{II} in normal sea water. Note short-latency electrical and long-latency chemical polysynaptic effect on PC_p. (AII) Same experiment in zero Ca^{2+} sea water. Note loss of chemical polysynaptic effect of PC_p. (C) Discharge pattern of ET_{II} of a PC_p during fictive feeding in the isolated CNS preparation. r 1 and r 3, extracellular recordings from buccal root 1 and 3. (From Kovac *et al.*, 1983a.)

rocal synaptic relationships with PC_ps (Fig. 5B), and determination of the normal firing pattern of the neuron during "fictive feeding," i.e., during the generation by the isolated nervous system of the motor output pattern that normally underlies feeding behavior (Fig. 5C). Based on such data, it has been possible to construct the neural "wiring diagram" of the feeding command system. Even in this "simple" invertebrate motor system, the command circuitry is astonishingly complex (Fig. 6A). Reduced to fundamentals, however, the wiring diagram is easier to under-

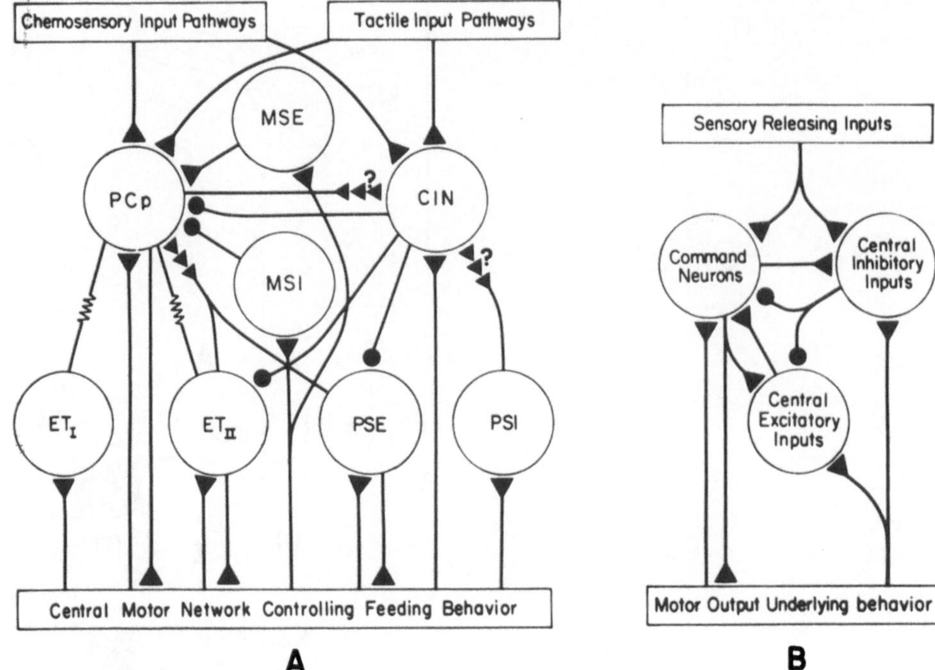

FIGURE 6. Simplified summary wiring diagrams of known synaptic relationships in the feeding command circuitry of *Pleurobranchaea*, as determined from experiments such as those shown in Fig. 5. (A) Details. (B) Underlying conceptual features. See text for further commentary. (From Kovac *et al.*, 1983b.)

stand (Fig. 6B). Sensory inputs are carried by the aforementioned sensory interneurons to the command neurons, and the same sensory inputs excite neurons that inhibit the command neurons. These inhibitory neurons include in particular a small network of neurons that supplies cyclic inhibition to the command interneurons (the cyclic inhibitory network or CIN) (Gillette *et al.*, 1982). These inhibitory inputs are recurrent to the PC_ps and also innervate neurons that excite the command neurons, including the subclasses of the PCNs mentioned above (PSEs and ET_{II}s). Finally, the command neurons show reciprocal connections with neurons contained in the feeding motor network (Gillette *et al.*, 1978; Davis *et al.*, 1984).

Although many details of this neural circuitry remain to be established, the studies to date have uncovered a number of potentially general principles regarding the organization of command and motor systems (e.g., Davis, 1976; Davis and Kovac, 1981). In addition, the available data provide an adequate framework for addressing the neural mechanisms of higher-order behavioral phenomena in the feeding motor network, to which we now turn.

3. Neural Mechanisms of Motivation

On an intuitive level, motivation expresses an organism's drive or willingness to perform a particular behavior. Progressing beyond intuition to a precise and generalizable definition, however, has proved challenging. For present purposes, motivation is defined as the probability of expression of a recognizable behavior in the presence of the stimulus for that behavior. This probability can be defined quantitatively by the strength of stimulus that is required to elicit the behavior. The lower the motivation, the greater the required strength of the effective stimulus. In other words, the probability of behavioral expression is inversely proportional to the stimulus threshold of the behavior, which affords a convenient quantitative measure of motivation.

This definition of motivation can be illustrated with *Pleurobranchaea's* feeding behavior. The various components of the feeding behavior (Fig. 1) can be induced by squirting liquefied squid homogenate onto the animal's anterior chemosensory structures. The minimal concentration of squid homogenate that is capable of eliciting a given component of the feeding behavior is defined as the *behavioral threshold* of the feeding behavior. The threshold thus describes the stimulus strength at which the probability of expression of the behavior is 1. This "feeding threshold" is inversely proportional to feeding motivation as defined above; the lower the feeding threshold, the higher the animal's feeding motivation. The feeding threshold therefore furnishes a simple, objective, and operational measure of feeding motivation.

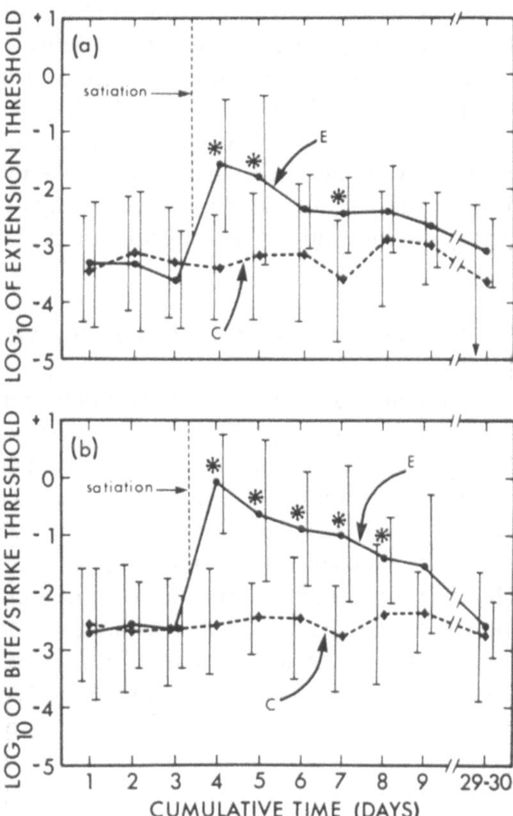

FIGURE 7. Mean feeding response thresholds ($\log_{10}$ of the minimal concentration of squid homogenate necessary to induce the indicated response) in satiated (E) and control (C) animals (n = 12–14). Satiation of Es occurred on day 3. Asterisks indicate significant differences between satiated and control animals ($P \leq 0.05$). (From Davis *et al.*, 1977.)

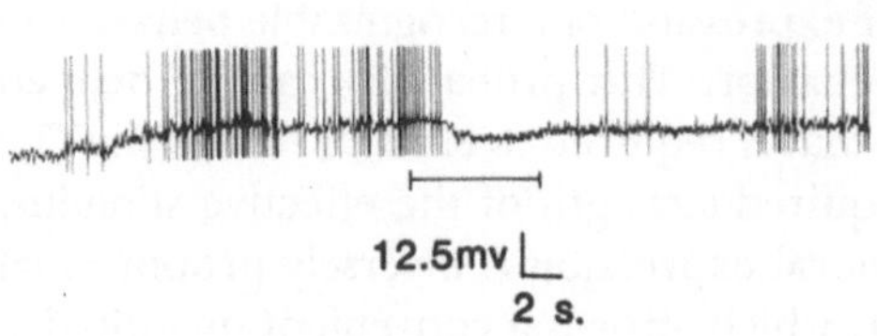

FIGURE 8. Intracellular recording from a paracerebral feeding command interneuron of a whole animal preparation following satiation of the specimen. Bar beneath record indicates time of application of food stimuli to the anterior chemosensory structures. Note inhibition of the command neuron in comparison with hungry specimens (compare with Fig. 10D).

This definition enables precise measurement of behavioral motivation in the feeding system. When an animal is satiated with food, the drive to eat declines to zero. This change in motivation is expressed by the increase in the threshold of components of feeding behavior (Fig. 7). Immediately following a large meal, the threshold increases to infinity, i.e., no food stimulus can induce feeding, but the threshold gradually returns to normal over a time period of about one week (Davis *et al.*, 1977).

How is this altered motivational state represented in the CNS? As described earlier, stimulation of hungry whole animal specimens with food excites the feeding command interneurons (Fig. 4C), leading to feeding behavior. Following satiation, however, the same stimulus inhibits, rather than excites, the command interneurons (Fig. 8), thereby blocking the feeding behavior. The inhibitory postsynaptic potentials (IPSPs) that compose this inhibition are indistinguishable in amplitude and waveform from IPSPs produced by the CIN (Davis *et al.*, 1983). Therefore, it appears that the same neurons that normally cyclically inhibit the command neurons, imparting the temporal structure of their activity during feeding, become tonic suppressors of feeding when the motivation to feed declines. One may speculate that stretch receptors located in the wall of the gut impinge upon the CIN neurons, changing their activity pattern to reflect the amount of food in the gut, and leading to alterations in the behavior that we recognize as changes in motivation. This speculation remains to be tested directly.

4. Neural Mechanisms of Associative Learning

An animal's drive to perform a behavior, i.e., motivation, can be changed in other ways besides satiation, such as associative conditioning. When *Pleurobranchaea* is trained using a conventional avoidance paradigm, for example, it can be rapidly taught not to feed in response to a particular food substance (Mpitsos and Collins, 1975; Mpitsos *et al.*, 1978; Davis *et al.*, 1980). Conditioning is accomplished by exposing hungry animals to food and conditional aversive electric shock. That is, if on presentation of food stimuli the specimens show any component of the feeding behavior (or fail to withdraw from the food stimulus), they receive the punishing shock. After a few such trials each separated by 1 hr, the animal learns to suppress feeding behavior when the food stimulus is presented, i.e., its measured feeding motivation decreases. As in the case of satiated animals, this decrease in motivation is expressed quantitatively as an increase in the feeding response threshold. Immediately after training, the feeding response threshold is elevated, and it remains elevated for one or more weeks while gradually returning to normal (Fig. 9A,B).

This learned modification in feeding behavior is specific to the food substance with which shock was paired. Thus, when animals are trained against squid homogenate, their feeding response thresholds to another food substance, namely homogenate of sea anemone, is unchanged (Fig. 9C,D). To control for nonassociative contributions to this change in feeding motivation, a variety of control procedures have been performed. Figure 9, for example, shows that when control animals are given the same amounts of food and aversive electric shock as experimental animals, but unpaired in time by 0.5 hr, feeding motivation is not elevated as seen in conditioned animals. Therefore, the increase in feeding threshold seen in conditioned animals is indicative of associative learning, whereas the return to normal feeding thresholds indicates extinction of the learned modification.

How is this form of associative learning represented in the CNS? Intracellular recordings made from the feeding command interneurons of whole animal preparations following associative training show the same massive inhibition in response to food stimulation as seen in satiated animals (Fig. 10A,B). Feeding motor output can still be elicited by depolarizing the command interneuron (Fig. 10C), however. Therefore, associative conditioning does not incapacitate the entire feeding circuitry, but is instead manifest near the input end of the circuitry, at the command level. Control specimens are indistinguishable from untrained, hungry animals, i.e., their feeding command interneurons are excited by food stimuli (Fig. 10D). As in the case of satiation of animals with food, the IPSPs following associative training are the same in amplitude and waveform as those generated by the CIN (Davis *et al.*, 1983). Therefore, we have proposed that the CIN is converted by behavioral associative training to a suppressor of the feeding rhythm (Davis *et al.*, 1983), as occurs also following satiation.

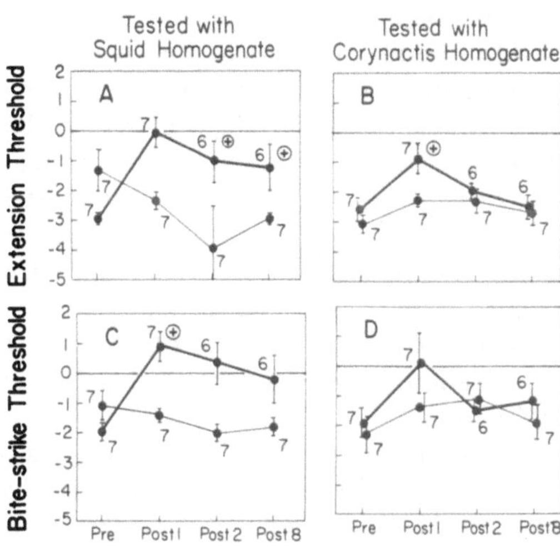

FIGURE 9. Mean feeding response thresholds of food avoidance conditioned (heavy curves) and control (light curves) specimens, before and after conditioning. The sample size (number of specimens) is shown next to each point. Animals were conditioned against squid and then tested either with squid or a sea anemone *(Corynactis)* homogenate as indicated. +, indicates a significant difference between conditioned and control values ($P \leq 0.05$). Note stronger aversion to squid than sea anemone following training, indicating selectivity in the learning. (From Davis *et al.*, 1980.)

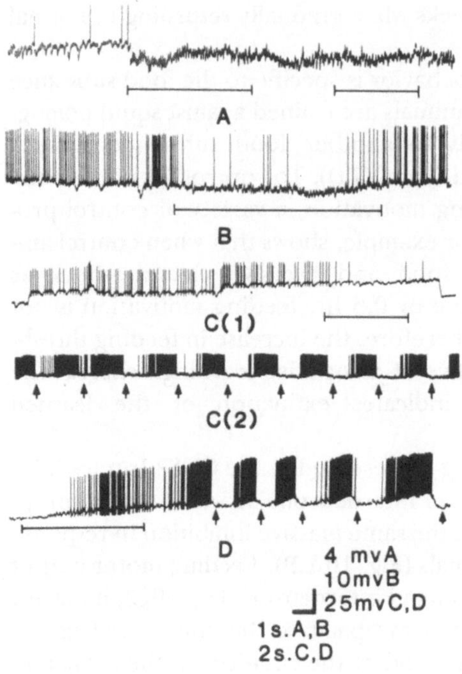

FIGURE 10. Intracellular responses of PC_ps in conditioned (A–C) and control (D) whole animal preparations. (A) Synaptic inhibition of a "resting" PC_p caused by application of food stimuli (bars beneath record). (B) Synaptic inhibition of a PC_p that was tonically depolarized by injected current. (C) Synaptic inhibition of a PC_p by food (1), followed by intracellular depolarization of the same PC_p (2), which induced cyclic bursting and visible bites (upward arrows). (D) Intracellular recording from a PC_p of a control (unpaired food and shock) whole animal preparation. Upward arrows indicate occurrence of visible bites. Bar shows time of delivery of food stimulus. (From Davis and Gillette, 1978.)

Since both food satiation and associative learning entail the same changes at the command interneuron level, the underlying neurophysiological mechanisms may be described as convergent at this level in the nervous system. Satiation and learning are vastly different behavioral experiences, however; one is nonassociative (satiation), whereas the other requires that the nervous system form a new association between a stimulus and a response (learning). Therefore, the underlying mechanisms must be correspondingly different at some level in the nervous system, i.e., presynaptic to the command interneurons.

In an effort to separate nonassociative from associative contributions to motivational state, we have recently developed an isolated brain preparation for the study of associative learning (Kovac *et al.*, 1985). Toward this end, an animal is first trained in the foregoing avoidance learning task. Its brain is then removed, but left attached to the anterior chemosensory structures that normally induce feeding behavior (Fig. 11A). When food stimuli are then applied to the chemosensory structures, the feeding command interneurons show the same inhibition as seen in satiated animals (Fig. 11B). In other words, the surgically isolated brain "remembers" the task learned by the whole animal. The "engram," in this case conditioned modifications in central nervous circuitry, are retained despite the removal of the brain from the trained host animal. In contrast, food stimuli cause excitatory responses in brains taken from hungry, naive animals (Fig. 11C), control animals, i.e., specimens that received unpaired food and shock (Fig. 11D), and animals that were previously satiated with food (Fig. 11E). This latter result is especially significant to our effort to understand learning in cellular terms, because it means that associative and nonassociative causes of changed feeding motivation can be separated from one another for purposes of neural analysis by reducing the preparation to the isolated brain.

Using the isolated brain of previously trained animals, we have made a beginning toward understanding learning in cellular terms. Any such understanding must account both for the increase in synaptic inhibition in feeding command interneurons following training, and also the decrease in excitation in the same neurons (Davis and Gillette, 1978; Davis *et al.*, 1983). With respect to the increase in inhibition, recordings made from pairs of command interneurons show 1:1 IPSPs (Fig. 11F), indicating inhibition from common presynaptic neurons. This finding is consistent with, and required by, the hypothesis that these IPSPs originate in the CIN.

With respect to the decline in command neuron excitation following food avoidance training, results on the isolated brain preparation have revealed two distinct mechanisms. First, since some command interneurons (i.e., the PSEs and ET_{II}s) supply excitatory inputs to others, and since all the command interneurons are innervated in common by the CIN, one cause of the decreased excitation fol-

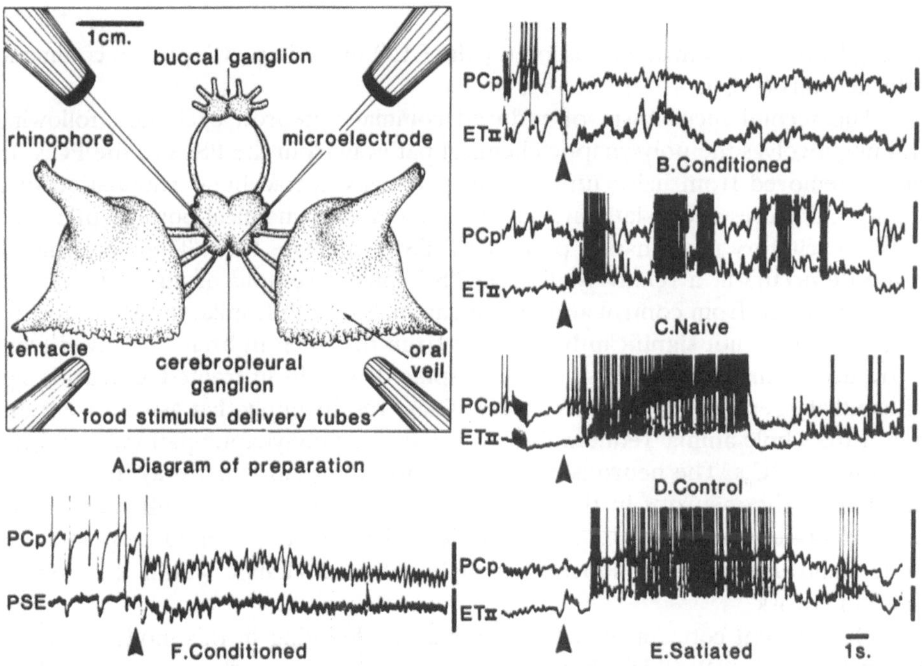

FIGURE 11. Experiments on feeding command interneurons of the nearly isolated brain preparation of *Pleurobranchaea.* (A) Diagram of the preparation as seen from above and anterior. (B) Intracellular responses of a PC_p and ET_{II} following food avoidance training of the whole animal. Arrows indicate time of application of conditioned food stimuli. (C–E) Comparable recordings in brains isolated from naive, control and satiated specimens. (F) Comparable recordings from a different trained specimen showing a PC_p and PSE. Bars to the right of each trace represent 10 mv. (Modified from Kovac *et al.*, 1985.)

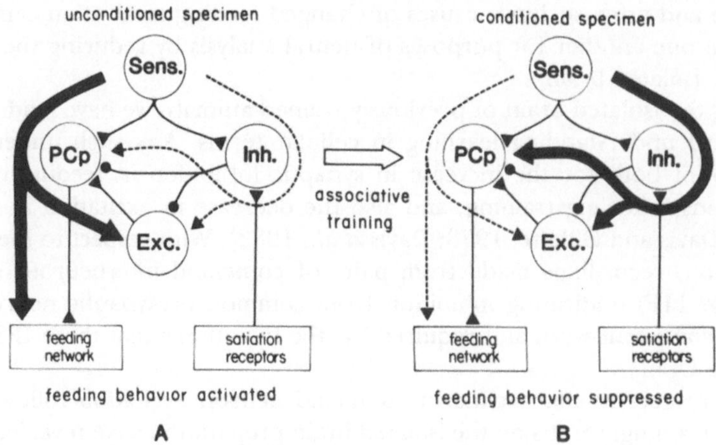

FIGURE 12. Cellular model of associative training in *Pleurobranchaea.* (A, B) Schematically illustrate the activity of the feeding command circuitry before and after associative training. See text for further commentary. Sens, sensory inputs associated with food; Exc, excitatory inputs to command interneurons; Inh, inhibitory inputs to command interneurons. (From Davis *et al.,* 1983.)

lowing learning is central inhibition (by the CIN) of excitatory inputs to command interneurons (PC_ps).

The second mechanism of reduced command neuron excitation following training involves the polysynaptic chemical pathway from the PSEs to the PC_ps. In brains removed from naive (untrained) specimens, intracellular stimulation of a single PSE causes a long-latency, relatively nongraded chemical polysynaptic compound excitatory postsynaptic potential (EPSP) in single PC_ps. This polysynaptic response occurs at a relatively discrete PSE spike threshold of about 18 Hz. In brains removed from control and satiated animals, the PSE spike threshold for the PC_p response is not significantly different from that seen in brains removed from untrained animals. In contrast, the PSE spike threshold in trained animals is significantly larger—approximately 29 Hz. These results imply that associative training of the whole animal reduces the efficacy of the polysynaptic pathway from the PSEs to the PC_ps. The neurons composing this polysynaptic pathway include corollary discharge neurons in the buccal ganglion (Davis *et al.,* 1984), which have been identified (Kovac *et al.,* in preparation). Therefore, it should be possible to understand this learning-induced neurophysiological change within this defined neural network.

Our present concept of the neural basis of learning in this model system is illustrated in Fig. 12, which shows how the feeding command circuitry (Fig. 6B) is altered by avoidance conditioning. Prior to conditioning (Fig. 12A), food stimuli activate feeding behavior because sensory information is channeled through predominantly excitatory pathways. As a consequence of associative conditioning (Fig. 12B), food stimuli suppress feeding, because sensory information is now channeled through predominantly inhibitory pathways. The question of how learning is accomplished in cellular terms is reduced to finding out how this rechanneling of information through an existing central neural network takes place, and in particular to determine how contingent sensory stimuli (i.e., food and shock) lead to this rechanneling.

5. Neural Mechanisms of Choice

There is yet another situation in which an animal's motivation to perform a particular behavior is changed, and that is when it is doing something else. When a hungry organism finds food, for example, its interest in sex or other behaviors might be reasonably expected to wane. Under such circumstances the organism may be said to have made a choice. In this context, choice is defined operationally as the expression of one behavior when the stimuli for more than one behavior are available.

In *Pleurobranchaea,* choices are dictated by the animal's "behavioral hierarchy." A behavioral hierarchy is a "priority sequence" that is established experimentally by repeated application of a choice paradigm and that enables predictions about behavioral responses in conflict situations. For example, when *Pleurobranchaea* is feeding, righting behavior is suppressed. Hungry animals can be induced to lie on their backs for up to 1 hr by stimulating them with homogenized food stimuli (Davis *et al.,* 1974a).

Food stimuli also interrupt mating behavior. Similarly, during feeding the same moderate tactile stimulus that formerly caused withdrawal has no behavioral effect (Davis *et al.,* 1974b). Conversely, when withdrawal behavior is elicited by a strong tactile stimulus, feeding can be interrupted. Thus in this case the two independent behaviors are mutually inhibitory (Kovac and Davis, 1980a). Feeding behavior is suppressed, however, during egg laying (Davis *et al.,* 1974b), through the action of egg-laying hormone (Ram *et al.,* 1977). And all behaviors are subordinate to escape swimming. The behavioral hierarchy of *Pleurobranchaea,* as established by these experiments utilizing a choice paradigm, is summarized in Fig. 13.

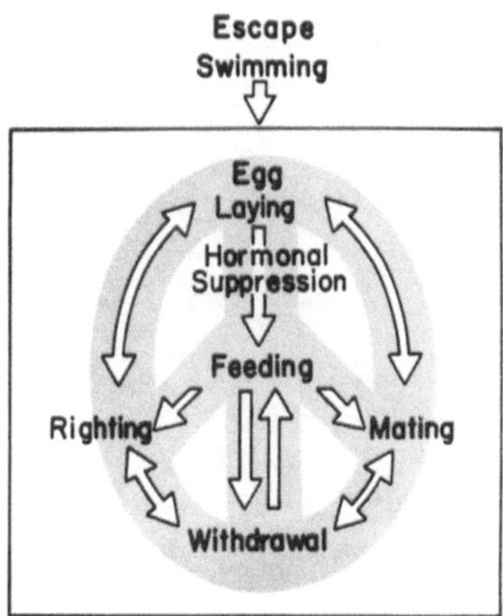

FIGURE 13. The behavioral hierarchy of *Pleurobranchaea.* Unidirectional arrows indicate the predominance of one behavioral act over another, whereas bidirectional arrows signify mutual compatibility. Double arrow indicates reciprocal inhibition. (From Kovac and Davis, 1980a.)

How are such behavioral choices mediated by the nervous system? The first clues were obtained by behavioral experiments in which the level of feeding motivation was varied by satiating animals. Following satiation, food stimuli still suppress righting behavior, but withdrawal behavior is unaffected by the presence of the same food stimuli. From this observation we concluded that chemosensory stimuli directly inhibit the righting behavior, but the suppression of withdrawal during feeding must result from the action of neurons that are part of, or coactive with, the feeding motor network (Davis *et al.*, 1977).

To test this hypothesis, individual, identified neurons within the feeding motor system were stimulated intracellularly while eliciting the withdrawal motor output (Kovac and Davis, 1977, 1980b). A systematic intracellular survey of the feeding network eventually disclosed a pair of reidentifiable corollary discharge neurons in the buccal ganglion (Davis *et al.*, 1973, 1974c) that mediate this form of behavioral choice. Stimulation of either of these neurons eliminates most of the withdrawal motor output (Fig. 14). Therefore, a single one of these neurons is *sufficient* to most of the suppression of the withdrawal motor program (Kovac and Davis, 1980b). These same neurons are also *necessary* to the suppression of withdrawal. That is, they normally discharge tonically during fictive feeding, but when they are silenced by hyperpolarization, withdrawal motor ouput is no longer suppressed (Kovac and Davis, 1980b).

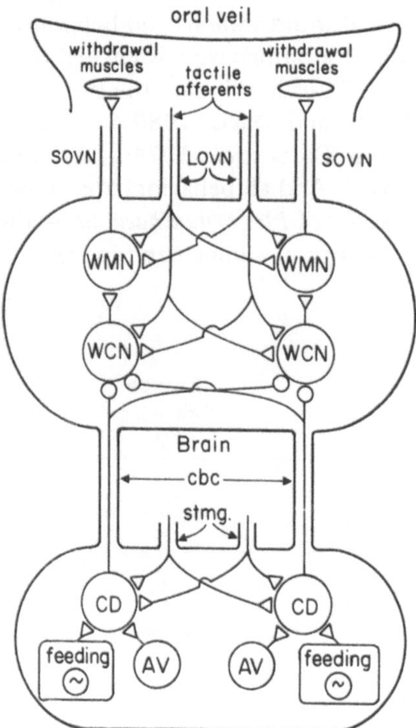

FIGURE 14. Neural network underlying one form of behavioral choice in *Pleurobranchaea*, namely, the suppression of withdrawal behavior during feeding. Withdrawal command neurons (WCN) in the brain are inhibited by identified corollary discharge (CD) feeding neurons in the buccal ganglion. Additional abbreviations (from top to bottom): SOVN, small oral veil nerve; LOVN, large oral veil nerve; WMN, withdrawal motor neurons; AV, anterior ventral neurons. (From Kovac and Davis, 1980b.)

How does this pair of feeding neurons exercise its effect on the withdrawal motor system? Intracellular recordings from the withdrawal motoneurons showed that a particular class of EPSPs, produced presumably by a withdrawal command interneuron, were abolished by stimulating the corollary discharge neuron(s) (Kovac and Davis, 1980b). Therefore, this particular form of choice is mediated by inhibitory interactions between the competing motor networks; this inhibition is mediated by higher-order interneurons that are a part of the feeding motor system; and this inhibition operates on the withdrawal command interneurons (Kovac and Davis, 1977, 1980). The neuronal circuitry and interactions that are known to underlie this form of behavioral choice are summarized in the wiring diagram of Fig. 14.

6. Convergent Neural Mechanisms for Diverse Behavioral Phenomena

The studies reviewed briefly above suggest a number of general conclusions regarding the neural mechanisms of complex behavioral phenomena. First, diverse types of behavioral plasticity all appear to operate by modulating motivational systems. Therefore, the concept of motivation furnishes a key unifying principle for diverse forms of behavioral plasticity.

Second, these motivational systems converge on a final common pathway in the CNS, namely, the command neurons. At this level, various forms of behavioral plasticity are neurally indistinguishable. This discovery is perhaps not surprising, since several forms of plasticity (e.g., satiation and food avoidance learning) manifest the same behaviorally. This finding does imply that the neurophysiological distinctions between various forms of behavioral plasticity must be sought presynaptic to the command level, and it suggests the need for carefully controlled experiments to separate general nonassociative motivational changes from specific associative processes in the study of learning.

Third, our studies of the neuronal bases of behavioral plasticity emphasize the key role of central "behavioral initiator" or "command" neurons. Neither is this conclusion especially surprising, since the locus of behavioral initiation would seem *a priori* a likely juncture for the modulation of behavior by diverse physiological processes and behavioral experience. This conclusion does suggest, however, that the concept of the command (initiator, driver, etc.) neuron is a key one for understanding the neurophysiological mechanisms of higher-order behavioral phenomena.

Fourth, the studies reviewed here emphasize the essential role of neural circuit analysis to the understanding of complex behavioral phenomena. The unraveling of neural circuits underlying behavior can be a tedious and time-consuming task, and in recent years has fallen somewhat out of favor as a research approach. Many of the most basic questions that can be asked about the nervous system, however, dealing for example with hormonal modulation, development, and plasticity, absolutely depend on accurate and detailed circuit analysis for their answer.

Finally, the present chapter, along with others in this volume, underscores the utility of the model system approach. Many fundamental neurobiological questions simply cannot be addressed in higher animals, owing to the tremendous complexity of the neural circuits underlying even simple behaviors. Indeed, the neural circuitry subserving feeding behavior in the relatively "simple" mollusk *Pleurobranchaea* may approach the limit of what is tractable to a reasonably complete cellular neurophysiological analysis. Underlying the model systems approach is the hope—indeed, expectation—that the revealed principles will have more general applicability. Based on the studies reviewed in the present chapter, for example, one might suggest that students of behavioral plasticity in vertebrate animals look first to the central neural sites of behavioral initiation for the underlying neuronal substrates. Only when such comparative studies have been completed will it be possible to accurately assess the assumptions underlying the model systems approach.

ACKNOWLEDGMENTS. Original work from our laboratory has been supported by NIH Research Grants NS 09050 and MH 23254 and NSF Grant BNS-8110235. This paper was written during the tenure of an Alexander von Humboldt Senior Scientist Award. I thank Prof. H. Markl (Universität Konstanz) and Prof. F. Huber (Max-Planck Institüt für Verhaltensphysiologie, Seewiesen) for hospitality and excellent facilities.

References

Bicker, G., Davis, W. J., Matera, E. M., Kovac, M. P., and Stormo-Gipson, J., 1982a, Mechano- and chemo-reception in *Pleurobranchaea californica.* I. Extracellular analysis of afferent responses, *J. Comp. Physiol.* **149:**221–234.

Bicker, G., Davis, W. J., and Matera, E. M., 1982b, Mechano- and chemoreception in *Pleurobranchaea californica.* II. Neuroanatomical and intracellular analysis of centripetal pathways, *J. Comp. Physiol.* **149:**235–250.

Croll, R. P., and Davis, W. J., 1981, Motor program switching in *Pleurobranchaea:* I. Behavioural and electromyographic study of ingestion and egestion in intact specimens, *J. Comp. Physiol.* **145:**277–287.

Croll, R. P., and Davis, W. J., 1982, Motor program switching in *Pleurobranchaea:* II. Ingestion and egestion in the reduced preparation, *J. Comp. Physiol.* **147:**143–154.

Croll, R. P., Davis, W. J., and Kovac, M. P., 1985a, Neural mechanisms of motor program switching in the mollusc *Pleurobranchaea.* I. Central motor programs underlying ingestion, egestion and the "neutral" rhythm(s), *J. Neurosci.* **5**(1)**:**48–55.

Croll, R. P., Kovac, M. P., and Davis, W. J., 1985b, Neural mechanisms of motor program switching in the mollusc *Pleurobranchaea.* II. Role of the ventral white cell and other identified buccal neurons, *J. Neurosci.* **5**(1)**:**56–63.

Croll, R. P., Kovac, M. P., Davis, W. J., and Matera, E. M., 1985c, Neural mechanisms of motor program switching in the mollusc *Pleurobranchaea.* III. Role of the paracerebral neurons and other identified brain neurons, *J. Neurosci.* **5**(1)**:**64–71.

Davis, W. J., 1976, Organizational concepts in the central motor networks of invertebrates, in: *Neural Control of Locomotion* (R. Herman, S. Grillner, P. S. G. Stein, and D. Stuart, eds.), Plenum, New York, pp. 265–292.

Davis, W. J., and Gillette, R., 1978, Neural correlate of behavioral plasticity in command neurons of *Pleurobranchaea, Science* **199:**801–804.

Davis, W. J., and Kovac, M. P., 1981, The command neuron and the organization of movement, *Trends Neurosci.* **4**:73–76.

Davis, W. J., and Matera, E. M., 1982, Chemoreception in gastropod mollusks: Electron microscopy of putative receptor cells, *J. Neurobiol.* **13**:79–84.

Davis, W. J., and Mpitsos, G. J., 1971, Behavioral choice and habituation in the marine mollusk *Pleurobranchaea californica* MacFarland (Gastropoda, Opisthobranchia), *Z. vergl. Physiol.* **75**:207–232.

Davis, W. J., Siegler, M. V. S., and Mpitsos, G. J., 1973, Distributed neuronal oscillators and efference copy in the feeding system of *Pleurobranchaea*, *J. Neurophysiol.* **36**:258–274.

Davis, W. J., Mpitsos, G. J., and Pinneo, J. M., 1974a, The behavioral hierarchy of the mollusk *Pleurobranchaea*. I. The dominant position of the feeding behavior, *J. Comp. Physiol.* **90**:207–224.

Davis, W. J., Mpitsos, G. J., and Pinneo, J. M., 1974b, The behavioral hierarchy of the mollusk *Pleurobranchaea*. II. Hormonal suppression of feeding associated with egg-laying, *J. Comp. Physiol.* **90**:225–243.

Davis, W. J., Mpitsos, G. J., Siegler, M. V. S., Pinneo, J. M., and Davis, K. B., 1974c, Neuronal substrates of behavioral hierarchies and associative learning in the mollusk *Pleurobranchaea*, *Am. Zool.* **14**:1037–1050.

Davis, W. J., Mpitsos, G. J. Pinneo, J. M., and Ram, J. L., 1977, Modification of the behavioral hierarchy of *Pleurobranchaea*. I. Satiation and feeding motivation, *J. Comp. Physiol. A* **117**:99–125.

Davis, W. J. Villet, J., Lee, D., Rigler, M., Gillette, R., and Prince, E., 1980, Selective and differential avoidance learning in the feeding and withdrawal behaviors of *Pleurobranchaea*, *J. Comp. Physiol. A* **138**:157–165.

Davis, W. J., Gillette, R., Kovac, M. P., Croll, R. P., and Matera, E. M., 1983, Organization of synaptic inputs to paracerebral feeding command interneurons of *Pleurobranchaea californica*. III. Modifications induced by experience, *J. Neurophysiol.* **49**:1557–1572.

Davis, W. J., Croll, R. P., Kovac, M. P., and Matera, E. M., 1984, Brain oscillator(s) underlying rhythmic cerebral and buccal motor output in the mollusc, *Pleurobranchaea californica*, *J. Exp. Biol.* **110**:1–15.

Gillette, R., and Davis, W. J., 1977, The role of metacerebral giant neuron in the feeding behavior of *Pleurobranchaea*, *J. Comp. Physiol. A* **116**:129–159.

Gillette, R., Kovac, M. P., and Davis, W. J., 1978, Command neurons in *Pleurobranchaea* receive synaptic feedback from the motor network they excite, *Science* **199**:798–801.

Gillette, R., Gillette, M. U., and Davis, W. J., 1980, Action potential broadening and endogenously sustained bursting are substrates of command ability in a feeding neuron of *Pleurobranchaea*, *J. Neurophysiol.* **43**:669–685.

Gillette, R., Gillette, M. U., and Davis, W. J., 1982a, Substrates of command ability in a buccal neuron of *Pleurobranchaea*. I. Mechanisms of action potential broadening, *J. Comp. Physiol.* **146**:449–459.

Gillette, R., Gillette, M. U., and Davis, W. J., 1982b, Substrates of command ability in a buccal neuron of *Pleurobranchaea*. II. Potential role of cyclic AMP, *J. Comp. Physiol.* **146**:461–470.

Gillette, R., Kovac, M. P., and Davis, W. J., 1982c, Control of feeding motor output by paracerebral neurons in the brain of *Pleurobranchaea californica*, *J. Neurophysiol.* **47**:885–908.

Kovac, M. P., and Davis, W. J., 1977, Behavioral choice: Neurophysiological mechanism in *Pleurobranchaea*, *Science* **198**:632–634.

Kovac, M. P., and Davis, W. J., 1980a, Reciprocal inhibition between feeding and withdrawal behaviors in *Pleurobranchaea*, *J. Comp. Physiol. A* **139**:77–86.

Kovac, M. P., and Davis, W. J., 1980b, Neural mechanism underlying behavioral choice in *Pleurobranchaea*, *J. Neurophysiol.* **43**:469–487.

Kovac, M. P., Davis, W. J., Matera, E. M., and Gillette, R., 1982, Functional and structural correlates of cell size in paracerebral neurons of *Pleurobranchaea californica*, *J. Neurophysiol.* **47**:909–927.

Kovac, M. P., Davis, W. J., Matera, E. M., and Croll, R. P., 1983a, Organization of synaptic inputs to paracerebral feeding command interneurons of *Pleurobranchaea californica*. I. Excitatory inputs, *J. Neurophysiol.* **49**:1517–1538.

Kovac, M. P., Davis, W. J., Matera, E. M., and Croll, R. P., 1983b, Organization of synaptic inputs to paracerebral feeding command interneurons of *Pleurobranchaea californica*. II. Inhibitory inputs, *J. Neurophysiol.* **49**:1539–1556.

Kovac, M. P., Davis, W. J., Matera, E. M., Morielli, A., and Croll, R. P., 1985, Learning: Neural analysis in the isolated brain of a previously trained mollusc, *Pleurobranchaea californica, Brain Res.* (in press).

Matera, E. M., and Davis, W. J., 1982, Paddle cilia (discocilia) in chemosensitive structures of the gastropod mollusk *Pleurobranchaea californica, Cell Tissue Res.* **222:**25–40.

Mpitsos, G. J., and Collins, S. D., 1975, Learning: Rapid aversive conditioning in the gastropod mollusc *Pleurobranchaea, Science* **188:**954–957.

Mpitsos, G. J., and Davis, W. J., 1973, Learning: Classical and avoidance conditioning in the mollusk *Pleurobranchaea, Science* **180:**317–320.

Mpitsos, G. J., Collins, S. D., and McClellan, A. D., 1978, Learning: A model system for physiological studies, *Science* **199:**497–506.

Ram, J. L., Salpeter, S., and Davis, W. J., 1977, *Pleurobranchaea* egg-laying hormone: Localization and partial purification, *J. Comp. Physiol. B* **119:**171–194.

Siegler, M. V. S., 1977, Motor neurone coordination and sensory modulation in the feeding system of the mollusc *Pleurobranchaea californica, J. Exp. Biol.* **71:**27–48.

Siegler, M. V. S., Mpitsos, G. J., and Davis, W. J., 1974, Motor organization and generation of rhythmic feeding output in the buccal ganglion of *Pleurobranchaea, J. Neurophysiol.* **37:**1173–1196.

Part IV
Neurotransmitters and Neuromodulators

15

Neuropeptides and the Control of Egg-laying Behavior in *Aplysia*

EARL MAYERI AND BARRY S. ROTHMAN

1. Introduction

The marine mollusk *Aplysia* is proving to be a useful experimental system for studying the roles of neuropeptides in neuronal function and their roles in the neural regulation of behavior. The main emphasis of investigations in our laboratory has been on the bag cell system, a group of peptide-secreting neuroendocrine cells that induce egg laying in *Aplysia* and also have profound effects on other central neurons.

The bag cells are particularly well-suited for detailed cellular and molecular studies. Both the peptide-secreting cells and the neurons that are targets for peptide action are morphologically defined and can be studied with cellular electrophysiological techniques in great detail. Because of the large mass of the bag cells, large amounts of peptides can be purified for chemical characterization. This feature has also recently permitted application of recombinant DNA techniques to isolate a family of genes that encode the precursor proteins for neuroactive peptides. One of the precursors is for neuropeptides used by the bag cells as neurohormones and/or neurotransmitters. The details of these molecular genetic studies are described in Chapter 26.

In the present chapter, we summarize data indicating that the bag cells are a putative multitransmitter neural system that utilizes two or more neuropeptides that are enzymatically derived from a common precursor protein. These neuropeptides mediate various long-lasting effects of the bag cells on the central nervous system and may serve to regulate the occurrence of the stereotyped pattern of

EARL MAYERI AND BARRY S. ROTHMAN • Department of Physiology, University of California, San Francisco, California 94143.

behavior associated with egg laying. We begin by describing the bag cells as a multitransmitter neural system and then discuss the role of the peptides in the control of behavior during egg laying.

2. The Bag Cells as a Multitransmitter Neural System

2.1. The Bag Cells Are Neuroendocrine Cells that Induce Egg Laying

As shown in Fig. 1, the bag cells consist of two clusters of 400 cells each, located in the abdominal ganglion. The bag cells are morphologically and biochemically similar to other neuroendocrine cells in invertebrates and vertebrates. They have extensively branched axonal processes that are specialized for the release of much more chemical messenger than is the case for most other nerve cells. The distinctive anatomical features of neuroendocrine cells were recognized in the 1930s by E. and B. Scharrer, who originated the concept of neurosecretion. They suggested that neurosecretory cells are neurons that release their chemical messengers into the general circulation to act as hormones on peripheral tissues. Although the physiological and biochemical properties of neurosecretory cells were for a long time poorly understood, it is now certain that they are like other neurons in the way they generate action potentials and synthesize, transport, and release chemical messengers (see Gainer *et al.*, 1977). Neurosecretory cells differ, however, from other neurons in that their cell bodies and axons are packed with electron-dense granules that are stained selectively by certain dyes. These granules are now thought to contain peptides that are released as chemical messengers.

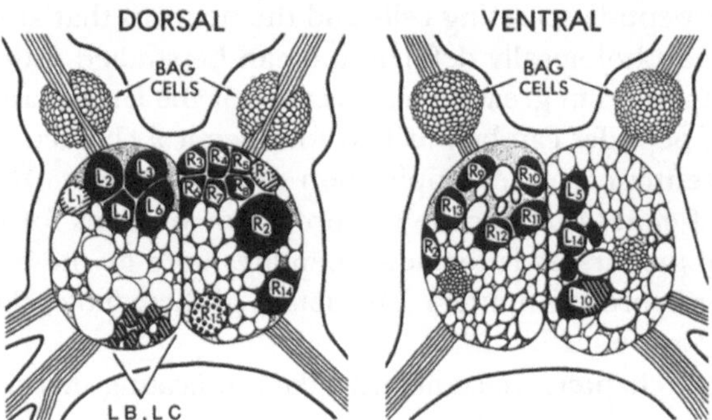

FIGURE 1. Schematic diagram of the dorsal and ventral surfaces of the abdominal ganglion of the marine mollusk *Aplysia californica* showing the locations of the two bag cell clusters (400 cells each) and identified target cells and cell clusters that respond to bag cell activity. The target cells are textured according to the type of response induced by bag cell activity. Not shown are the extensively branched axonal processes of the bag cells, which are located in the sheath surrounding each bag cell cluster and extend to sheath overlying the cell bodies of other ganglion neurons. Cells of the LB and LC identified clusters are collectively referred to in the text as left lower quadrant (LLQ) cells; cells L2, L3, L4, L6 are referred to as left upper quadrant (LUQ) cells. ▦, Burst augmentation, ▨, transient excitation; ▧, prolonged excitation; ■, slow inhibition. (From Mayeri *et al.*, 1979b.)

The function of the bag cells was discovered by Irving Kupfermann (1967). He isolated the abdominal ganglion in a small chamber and electrically stimulated impulse activity in the bag cells. When the medium bathing the ganglion was injected into an intact animal, it induced egg laying, suggesting that the bag cells had released an egg-laying hormone. Kupfermann and Kandel (1970) and others have shown that the bag cells are ordinarily silent, but when electrically stimulated with a brief (1–2 sec) train of pulses, they discharge impulses repetitively for about 20 min. During this burst discharge, the impulses occur in near unison in all the bag cells within a cluster. In intact animals, an all-or-none burst discharge of the bag cells occurs at the initiation of each episode of egg laying (Pinsker and Dudek, 1977). In subsequent studies, the egg-laying hormone (ELH) was partially characterized by Arch (1976), and its amino acid sequence determined by Strumwasser and co-workers (Chiu *et al.*, 1979). It is a 36 amino-acid peptide.

2.2. The Bag Cells Also Produce Several Types of Long-lasting Effects on Abdominal Ganglion Neurons

By making simultaneous intracellular recordings from a bag cell and various other identified neurons in the isolated abdominal ganglion, Mayeri and co-workers found that an electrically stimulated burst discharge of the bag cells has profound effects on central neurons (Mayeri and Simon, 1975; Branton *et al.*, 1978a; Mayeri *et al*, 1979a,b). These effects are widespread, occurring in aproximately one half the neurons of the ganglion, and they last for up to several hours. The effects thus last much longer than those produced by conventional neurotransmitters, such as acetylcholine acting at the vertebrate neuromuscular junction, but they are similar in duration to those produced by peptide hormones acting on peripheral tissue. The effects also last about as long as components of the behavior pattern that accompanies egg laying. Moreover, since the burst discharge in the isolated ganglion is apparently identical to the burst discharge in intact animals, one can reasonably expect that the effects studied in the isolated ganglion are similar to the ones that occur in intact animals during egg laying.

The bag-cell-induced responses are shown in Fig. 2. They have been classified into five types: (1) Burst augmentation occurs in the endogenously active, bursting pacemaker cell, R15. During this response there is an increase in the amplitude of the bursting pacemaker potential and the number of spikes and spike rate during each burst. (2) Prolonged excitation occurs in cells located in the left lower quadrant (LLQ) of the dorsal surface of the ganglion. These normally silent cells depolarize by a few millivolts and discharge at a steady rate. (3) Prolonged inhibition occurs in cells located in the left upper quadrant (LUQ) and in other cells. These cells are hyperpolarized and their bursting pacemaker activity is consequently reduced. (4) Transient excitation occurs in cells L1 and R1. They are depolarized by several millivolts and sometimes fire a short train of spikes. (5) Lastly, there is a prolonged depolarization of the bag cells themselves. The depolarization begins at the initiation of the bag cell burst discharge and lasts for several tens of minutes after bag cell spike activity has ceased. The first three responses last up to several

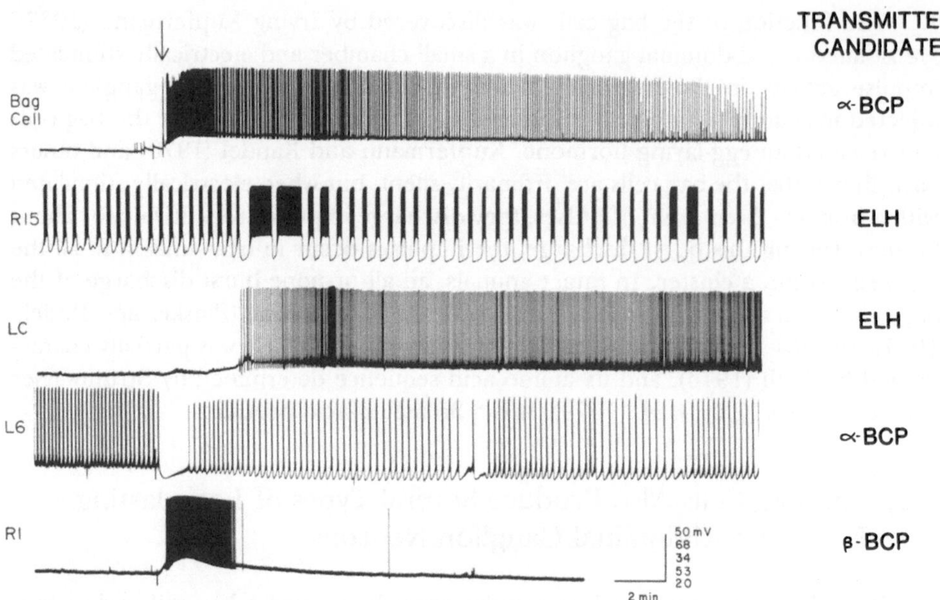

FIGURE 2. Intracellular recordings of a bag cell burst discharge and the types of responses it produces in identified neurons of the abdominal ganglion. The traces are aligned as if they were recorded simultaneously, although they were taken from several preparations. Each neuron shows one of five types of responses to a burst discharge of the bag cells, which was triggered by brief electrical stimulation (arrow) and lasted 20 min. There is burst augmentation (cell R15), prolonged excitation (LC cell), prolonged inhibition (cell L6), and transient excitation (cell R1). Additionally, there is depolarization of each of the bag cells. The peptide transmitter candidate for each response is shown at the right. Impulses in the individual neurons are often seen fused together. (Modified from Mayeri and Rothman, 1982.)

hours, whereas the fourth lasts about 5–10 min. Figure 1 shows the locations of the various target neurons and the type of response produced in each of them.

2.3. ELH and α-Bag Cell Peptide Are Putative Neurotransmitters

To find candidate neurotransmitters for the various responses, bag cell clusters were surgically isolated from animals and extracted in acetic acid. The soluble material was subjected to gel filtration chromatography, followed by ion exchange chromatography or high-performance liquid chromatography. These techniques have allowed us to purify several bag cell peptides to homogeneity (see Rothman *et al.*, 1984). Purified material was assayed for activity by pressure application from a micropipette placed near one of the target neurons or by arterial perfusion of substances into the ganglion through a small artery located at the caudal end of the ganglion (Mayeri and Rothman, 1982b). The arterial perfusion technique has proved especially valuable because small quantities of substances can be applied at known concentration to the entire extracellular and vascular spaces of the ganglion while recording intracellularly from several neurons at a time.

Purification and assay of ELH showed that it mimics two types of bag-cell-induced responses: burst augmentation and prolonged excitation (Branton *et al.*, 1978b; Mayeri and Rothman, 1982a; Mayeri *et al.*, 1985). Alpha bag cell peptide (α-BCP), a 9 amino acid peptide, mimics slow inhibition of LUQ cells and other cells inhibited by the bag cell discharge. Alpha bag cell peptide also depolarizes the bag cells and thus may act as an autoexcitatory transmitter within each of the bag cells clusters (Rothman *et al.*, 1983a; Sigvardt *et al.*, 1983). Lastly, a purified bag cell factor having the amino acid composition of beta bag cell peptide (β-BCP), a 5 amino acid peptide, depolarizes L1 and R1. Although the data are less complete for the role of this peptide, it is a candidate for mediating transient excitation.

Egg-laying hormone and α-BCP fulfill most of the criteria for identification of substances as neurotransmitters. Data from several laboratories have shown that the ELH is synthesized, released, and localized to the bag cells (see Gainer *et al.*, 1977; Berry, 1981; Chiu and Strumwasser, 1981; Scheller *et al.*, 1983b). It is also present in large quantities in bag cell extracts and is the only factor in the bag cells or other parts of the ganglion that mimics the very unusual bag-cell-induced responses in LLQ cells and R15. Stuart and Strumwasser (1980) have found that ELH produces excitation of neurons in the buccal ganglion and other parts of the CNS. Egg-laying hormone also acts directly on the ovotestis to induce ovulation (Rothman *et al.*, 1983b).

Alpha bag cell peptide has been localized to the bag cells; its ionic mechanism of action is identical to that of the bag-cell-induced inhibition of LUQ cells (Sigvardt *et al.*, 1983, 1985; Brownell and Mayeri, 1979), and there is cross-desensitization of the responses produced by α-BCP and the endogenous transmitter; i.e., when a target cell response is desensitized by prolonged application of α-BCP, the inhibition normally produced by discharge of the bag cells is abolished. There is also release of α-BCP (Sigvardt *et al.*, 1985).

2.4. ELH, α-BCP, and Other Bag Cell Peptides Are Derived from a Common Precursor Protein

Recombinant DNA techniques have been utilized by Richard Scheller, Richard Axel, and co-workers (Scheller *et al.*, 1982, 1983a) to isolate a family of genes that encode the precursors for neuroactive peptides used by the bag cells and the atrial gland, a nonneural tissue located in the reproductive tract. As described in detail in Chapter 26, they determined the nucleotide sequences of mRNA coding for three precursor proteins and deduced from them the complete amino acid sequences of the precursors. The three precursor proteins are shown schematically in Fig. 3. Because they arise from members of the same gene family, they contain structurally related sets of peptides. The ELH precursor, also termed here the ELH/BCP precursor, is expressed specifically in the bag cells, and the peptide A and peptide B precursors specifically in the atrial gland.

Encoded on ELH/BCP precursor is not only ELH, but also α-BCP, β-BCP, and several other potential cleavage products. This suggests that the bag cells mediate their effects by two or more peptide transmitters that are enzymatically

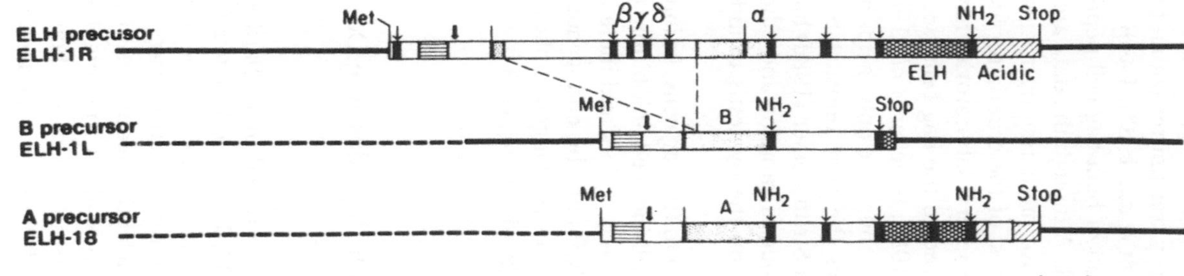

FIGURE 3. Schematic diagram of the precursor protein for egg-laying hormone (ELH) and other bag cell peptides, and for the precursors of peptide A and peptide B which occur in the atrial gland. ELH, α-, and β-bag cell peptide (BCP), three candidate transmitters for the various bag-cell-induced responses, are represented on the ELH/BCP precursor protein as are several other known or potential cleavage products such as γ- and δ-bag cell peptide and acidic peptide. The detailed organization of the three precursors, which are structurally related to one another, is described in the chapter by Richard Scheller. Regions of sequence homology between the three precursors are represented by stippling, cross-hatching, or diagonal lines. Thus α-BCP is homologous to peptides B and A, ELH is homologous to sequences on the B and A precursors, and acidic peptide is homologous to a sequence on the A precursor. The signal sequence for each precursor begins with a methionine residue followed by a hydrophobic region (horizontal lines). Vertical lines above the precursors represent potential or known cleavage sites at single arginine residues, while arrows represent cleavages at dibasic, tribasic, or tetrabasic residues. NH_2 appearing above an arrow represents a site of COOH-terminal amidation. (From Scheller *et al.*, 1983b.)

cleaved from a common precursor protein and released during a bag cell burst discharge. It is now possible to tie together several interesting aspects of the biology and these neuropeptides, from their genetic representation to their physiological roles in the central nervous system and their roles in the control of behavior. Several of the physiological characteristics of the bag cell system are considered next.

2.5. Compared with Conventional Synaptic Transmitters, ELH May Diffuse Long Distances to Target Neurons in the Abdominal Ganglion

The anatomical and physiological characteristics of the bag cell system suggest that ELH (and probably the other bag cell transmitters) diffuse much longer distances to their targets in the abdominal ganglion than conventional transmitters acting at synapses. Available light and electron microscopic data indicate that axons of the bag cells do not make synapses with other neurons. Instead, bag cells have profusely branched axons (Frazier *et al.,* 1967) containing varicosities spaced at regular intervals (Haskins *et al.,* 1981). Conventional synaptic transmitters, such as acetylcholine (ACh) at the vertebrate neuromuscular junction, diffuse a relatively short distance of 200–400 A to their postsynaptic targets, after which they are rapidly inactivated. The anatomical arrangement of the bag cell axons, however, is similar to the "nondirected synapses" of autonomic nerve endings on smooth and cardiac muscle; transmitter is thought to be released from the varicosities of autonomic endings and diffuse much longer distances (up to several microns) to their targets than for transmitter released at conventional synapses. One would therefore expect the physiological characteristics of the ELH-mediated responses to be more similar to those of ACh acting on cardiac muscle of the vertebrate heart than to ACh acting at neuromuscular junction. Examining the properties of the responses, we found that this is indeed the case (Branton *et al.,* 1978a,b; Mayeri and Rothman, 1982a; Mayeri *et al.,* 1985).

The physiological characteristics of prolonged excitation of LLQ cells mediated by ELH are as follows: (1) The response has a slow onset and prolonged duration of action; the latency of onset of response from the first impulse in the bag cell burst discharge to the onset of depolarization in target cells is 15–30 sec, and the response duration is more than 1 hr. (2) There are no rapid deflections in target cells corresponding to individual bag cell impulses, as might occur if there were a conventional synaptic interaction; rather, a repetitive bag cell discharge is required for response to occur. (3) There is overflow. If a bag cell discharge is triggered in one ganglion and the bathing medium is collected and applied to a second ganglion, a response will occur in LLQ cells. (4) The amplitude and time course of the neurally evoked response is mimicked by a low concentration of ELH. In order to mimic the manner in which transmitter is normally thought to be released and to estimate the local concentration of transmitter near the target neurons, ELH was arterially perfused into the ganglion at known concentrations. The concentration that duplicates the neurally evoked response is in the range of 0.2–1 μM. (5) There is a lack of rapid (seconds-to-minutes) desensitization of ELH in target neurons. Rapid desensitization would indeed seem to be incompatible with a response mechanism that lasts more than an hour. (6) Egg-laying hormone acts selectively. When arterially perfused at (or above) the estimated physiological concentration, ELH produces the appropriate response in the subgroup of cells normally affected by bag cell activity, but has no effect on other neurons. This result is to be expected if ELH comes into contact with most, and perhaps all, ganglion neurons, but affects only those cells having receptors for the neuropeptide.

These properties differ from those of ACh at frog neuromuscular junction, but are similar to those of ACh acting on cardiac muscle. Most notably, there is no overflow of ACh at the neuromuscular junction: unhydrolyzed ACh cannot be detected in bathing medium collected from neuromuscular junction after stimulation (Dale *et al.*, 1936), but it can be detected in bathing medium collected from cardiac muscle (Loewi, 1921). Overflow is to be expected if transmitter is released in sufficient quantities and inactivated slowly enough to normally diffuse long distances. In addition, the effective concentration of ACh at postsynaptic receptors of neuromuscular junction (300 mM for one quantum of ACh) (Hartzell *et al.*, 1975) is several orders of magnitude lower than that of ACh on cardiac muscle (1 μM) (Glitsch and Pott, 1978) or ELH on LLQ cells (0.2–1 μM). Moreover, at neuromuscular junction, there is rapid desensitization of locally applied ACh, but desensitization is lacking for ACh applied to cardiac muscle (Glitsch and Pott, 1978).

What are the functional consequences for neurotransmitters such as ELH that diffuse long distances within the central nervous system? One is that diffusion provides a means by which a few neurons can affect large arrays of other neurons without the need for synaptic contacts with every one of them. Because fewer axons and dendrites are needed, more channels of communication can in theory be packed into a given space in the brain. In a sense, the speed of communication provided by axons and dendrites is sacrificed for an increase in the capacity of nerve cells to communicate with one another (Mayeri and Rothman, 1982a). The effects of the diffusing, or "nonsynaptic" transmitter (see Dismukes, 1979) can be global, affecting every cell it contacts, or selective as in the case of ELH. The characteristics of nonsynaptic communication seem well-suited for mediating complex behavioral processes that are long lasting and likely to involve alterations in electrical signaling in many networks in the central nervous system.

2.6. In Contrast to ELH, α-BCP Has Rapid Effects and Is Rapidly Degraded

Available evidence suggests that α-BCP is released nonsynaptically along with ELH, but that it is more rapidly degraded in the vascular and interstitial spaces of the ganglion. Thus, it may diffuse a shorter distance than ELH before being inactivated, but a longer distance than conventional synaptic transmitters. Although the latency of onset of the bag-cell-induced inhibition mediated by α-BCP is much longer (several seconds) than that of conventional synaptic transmission, it is much shorter than the latencies of the responses mediated by ELH. The pharmacological effects of α-BCP are also more rapid than ELH; when applied to the target cell by focal pressure application through a micropipette, the time course of the response is quite similar to what occurs when a rapidly acting transmitter such as ACh is applied in the same manner. (By contrast, application of ELH often requires 15–30 sec before any change is observed.) Unlike ELH, where the response persists for an hour or more even after the peptide has been washed away from the target cell, the effects of α-BCP end a few seconds after the end of application.

Two kinds of experiments indicate that after release α-BCP is more rapidly degraded than ELH. First, if a bag cell burst discharge is triggered in one ganglion and the bathing medium is collected and applied to a second ganglion, the response that is characteristic of ELH excitation occurs in LLQs, but the response that is characteristic of α-BCP inhibition does *not* occur in LUQs and other cells. If the experiment is repeated with several inhibitors of proteolytic enzymes added to the bathing medium, the inhibitory activity characteristic of α-BCP is readily demonstrated (Sigvardt *et al.*, 1983, 1985). This suggests that normally α-BCP is rapidly inactivated by proteolytic enzymes after release from the bag cells and, unlike ELH, does not enter the general circulation to act on more distant targets. Addition of the protease inhibitors prevents the inactivation of α-BCP and its activity is therefore detected in the bathing medium. A second indication of degradation is that α-BCP is apparently rapidly inactivated when arterially perfused into the ganglion. When perfused in the absence of protease inhibitors in the perfusing medium, the peptide is threefold to tenfold less potent in producing inhibition than when perfused in the presence of protease inhibitors.

2.7. α-BCP May Regulate the Release of ELH and Other Bag Cell Peptides

In addition to inhibiting various target neurons, α-BCP also causes depolarization of the bag cells themselves. Arterial perfusion of a α-BCP at a concentration sufficient to produce a large hyperpolarization of LUQ cells (1 μM) also produces 5–10 mV depolarization of bag cells. If the amplitude of depolarization is sufficient to reach threshold, the all-or-none burst discharge of the bag cells in initiated (Rothman *et al.*, 1983a). These and other data suggest that α-BCP is an autoexcitatory transmitter that plays a critical role in producing the sustained depolarization that results in the burst discharge of the bag cells.

The role of a α-BCP in generating the burst discharge is as follows: When a subgroup of bag cells within one of the clusters is electrically depolarized to near threshold by an extracellular stimulating electrode, the cells fire a few spikes followed by a sustained depolarization of several millivolts lasting several tens of seconds. It is postulated that α-BCP is released from the stimulated cells to cause the sustained depolarization of the stimulated cells and unstimulated cells near them. In instances when the depolarization is sufficiently large, an impulse will be generated in one or a few members of the subgroup, resulting in the release of more α-BCP, and further depolarization. By this regenerative (or positive feedback) mechanism, larger and larger numbers of cells within the cluster discharge repetitively and bag cell burst discharge spreads to the entire cluster and from one cluster to the other. Electronic connections among the bag cells (Kupfermann and Kandel, 1970; Blankenship and Haskins, 1979) aid in the passive spread of the depolarization and also synchronize the impulse activity of the cells within a cluster. This regenerative mechanism involving α-BCP thus accounts for the all-or-none nature of the burst discharge; it is similar to the role of the influx of sodium ions in producing an all-or-none action potential when a nerve cell is depolarized above threshold.

2.8. There Are Multiple Active Forms of α-BCP

Three neuroactive forms of α-BCP have been purified from bag cell extracts: α-BCP(1-9), the nine-residue peptide encoded on the ELH/BCP precursor protein, and two NH_2-terminal fragments, α-BCP(1-8) and α-BCP(1-7) (Rothman *et al.*, 1983a). The amino acid sequence of α-BCP(1-9) is shown in Fig. 4. Interestingly, the nine-residue peptide is only 1/30 as potent as the eight-residue peptide and 1/10 as potent as the seven-residue peptide in inhibiting LUQ cells. Although more biochemical study is needed, it is thought that the precursor protein is processed to yield initially the 9 amino-acid peptide, and that there is cleavage of COOH-terminal residues either before or after release to yield the 8 and 7 amino acid forms. The functional significance of multiple neuroactive forms of peptides is unclear. At present, there are few examples for neuropeptides, but for peripheral hormones, most notably gastrin and cholecystokinin, it is a well-known phenomenon.

2.9. α-BCP, β-BCP, and γ-BCP Are Structurally Similar, but Have Different Activities

Beta and gamma BCP are each five amino acid peptides that are also encoded on the ELH/α-BCP precursor. As shown in Fig. 4, α-, β-, and γ-BCP share a sequence of four amino acids. Beta and gamma BCP differ from one another only

	1	2	3	4	5	6	7	8	9
α-BCP	ALA	PRO	ARG	LEU	ARG	PHE	TYR	SER	LEU
β-BCP			ARG	LEU	ARG	PHE	HIS		
γ-BCP			ARG	LEU	ARG	PHE	ASP		

FIGURE 4. Amino acid sequences of α-, β-, and γ-bag cell peptides. There are three neuroactive forms of α-BCP, the sequences (1-9), (1-8), and (1-7). Although β- and γ-BCP are sequentially homologous to α-BCP (boxed region), their activities differ (see Fig. 5).

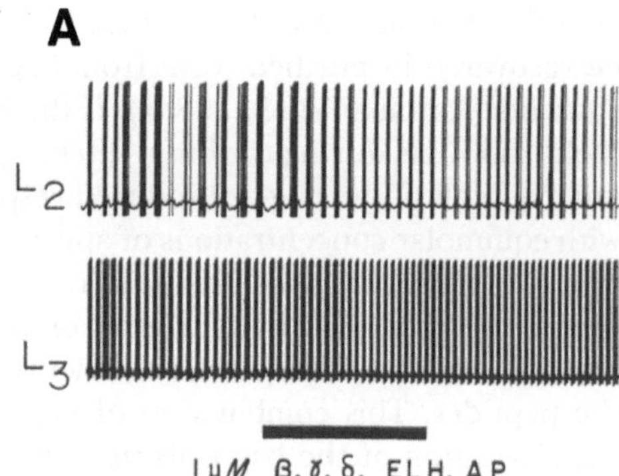

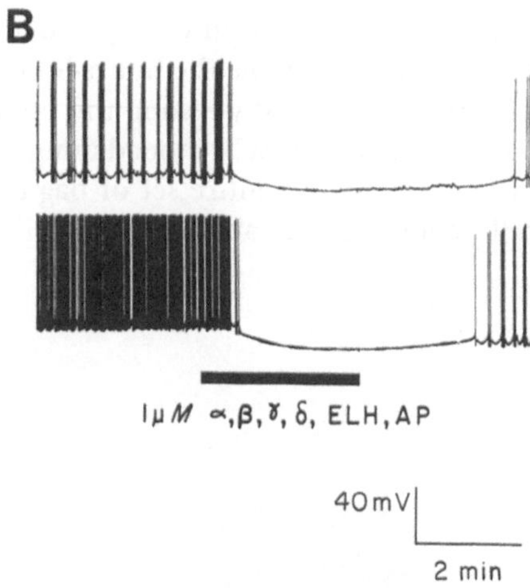

FIGURE 5. Of several bag cell peptides, only α-BCP duplicates bag-cell-induced inhibition of LUQ cells, L2 and L3. The effects of various bag cell peptides is assessed by recording simultaneously from L2 and L3 and arterially perfusing the ganglion with a solution of the indicated peptides at 1 μM concentration each (bar). (A) β-, γ-, and δ-BCP, ELH, and AP have no effect on ongoing endogenous bursting pacemaker activity. (B) Arterial perfusion of the same solution plus α-BCP (1-7), the putative bag cell transmitter for these cells, produces inhibition. (K. Sigvardt *et al.*, unpublished observations.)

in their COOH-terminal residue. Their effects, however, do not simply mimic those of α-BCP. As shown in Fig. 5A, cells that are inhibited by α-BCP are unaffected by five peptides encoded on the ELH/BCP precursor: β-, γ- and δ-BCP, ELH, and acidic peptide (AP). (AP and δ-BCP are 24 and 8 amino acid peptides, respectively. They are structurally unrelated to the other bag cell peptides.) When α-BCP is added to the group of peptides, inhibition is produced, as in Fig. 5B. Although the pharmacology of the peptides is incomplete, it appears that the functional significance of the structurally related α-, β-, and γ-BCPs is that they have different, selective actions on various central neurons.

2.10. Bag Cell Transmitters Have Coordinate Actions

Available biochemical, biophysical, and immunocytochemical data suggest that the bag cells are a homogenous group of neurons and that the precursor is processed uniformly in each cell to yield equimolar amounts of each of the released peptides. Consistent with this idea, ELH, AP, and α-BCP in its various forms can be recovered in purified form from bag cell clusters in approximately equimolar amounts (Rothman *et al.*, 1983a). If the bag cell peptides are also released in equimolar amounts during the burst discharge, then one might be able to mimic simultaneously all of the bag-cell-induced responses by arterially perfusing the ganglion with equimolar concentrations of appropriate bag cell transmitters. Figure 6 shows that two types of bag-cell-induced responses, prolonged excitation and slow inhibition, are mimicked by perfusion of equimolar concentrations of ELH, α-BCP, and AP, but that L1, which is transiently excited by the bag cells, is unaffected by the peptides. This combination of peptides also mimics burst augmentation and depolarization of the bag cells (not shown). Acidic peptide perfused by itself has no effect on ganglion neurons even at high concentration, although it is released by the bag cells and was therefore included in the experiment. Although α-BCP duplicates the duration of bag-cell-induced inhibition in two LUQ cells (L2 and L4), it does not mimic the hours-long duration of inhibition in two others (L3 and L6). We are presently testing the possibility that another bag cell transmitter may participate with α-BCP in mediating the latter response. Ultimately it may be possible to mimic the entire set of bag cell actions by applying a combination of bag cell peptides to the abdominal ganglion.

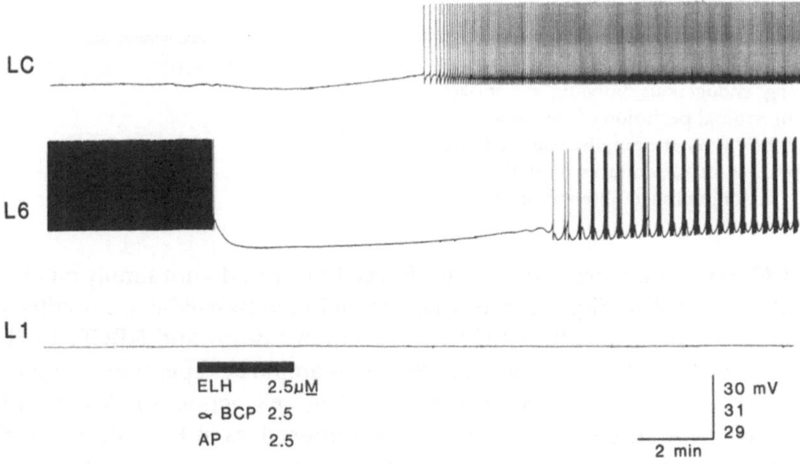

FIGURE 6. Several types of bag-cell-induced responses can be duplicated by arterial perfusion of ELH and α-BCP at equimolar concentration. Recording simultaneously from three target neurons, a solution of peptides is arterially perfused into the ganglion (bar) in order to simulate the manner in which transmitters released from the bag cells are thought to diffuse into the vascular and interstitial spaces of the ganglion. The solution contains ELH, α-BCP (1-7), and acidic peptide at 2.5 μM concentration each and six protease inhibitors. All three peptides are encoded on the ELH/BCP precursor protein. The peptides mimic bag-cell-induced prolonged excitation of an LC cell (due to ELH) and part of the slow inhibition of L6 (due to α-BCP), but not transient excitation of cell L1. Although not shown here, burst augmentation of cell R15 is duplicated by ELH and depolarization of the bag cells is duplicated by α-BCP. Acidic peptide (AP) is synthesized and released by the bag cells, but is without effect on any of the ganglion neurons so far tested. (K. Sigvardt *et al.*, unpublished observations.)

3. Genes, Neuropeptides and Egg-laying Behavior

3.1. Egg-laying Behavior Is Made Up of Several Components

The behavior associated with egg laying is one example of innate behavior patterns associated with reproduction in both invertebrates and vertebrates. These behavior patterns are genetically predetermined, and species specific. In addition, they are often complex, involving several motor and sensory systems, and last for minutes or hours (Tinbergen, 1951). During the egg-laying season, animals lay eggs approximately once every 7 days. At the start of egg laying, the animal locomotes to an appropriate location, often vertical surface, and orients itself vertically, after which it remains stationary. An egg string, emerging from the genital pore of the animal, is conveyed via the external genital groove to the mouth and is deposited on the surface with head-weaving and head-tamping movements (Arch and Smock, 1977; Stuart and Strumwasser, 1980; Cobbs and Pinsker, 1982a,b). The egg string is often several tens of centimeters long and requires several hours to be deposited. During egg laying, there is also inhibition of feeding, and based on studies on the isolated abdominal ganglion (Mayeri *et al.*, 1979a,b) and semi-intact preparations (Brownell and Schaefer, 1982), there are likely to be increases in cardiac output and respiratory movements.

3.2. The Actions of Peptides Derived from the ELH/BCP Precursor on CNS Neurons May Serve to Regulate Egg-laying Behavior

From studies on intact animals, three lines of evidence indicate that the behavior pattern is initiated by the bag cell burst discharge: (1) A burst discharge invariably precedes each episode of egg laying; (2) stimulation of a burst discharge produces egg laying (Pinsker and Dudek, 1977); and (3) injection of bag cell extract induces egg laying and the associated behavior pattern (Arch and Smock, 1977). During the burst discharge, ELH, α-BCP, and other peptides derived from the ELH/BCP precursor are released into the general circulation and/or the abdominal ganglion to act on various targets. The targets for many components of the behavior pattern, such as head movements and locomotion, are likely to be neural circuits located in the head ganglia. Others, mainly visceromotor components, are neural circuits located in the abdominal ganglion.

The actions of the bag cells on abdominal ganglion neurons is an appropriate model for understanding how the entire behavior is regulated. The actions are well-defined and often on neural circuits that have previously been studied in considerable detail (see Mayeri *et al.*, 1979a,b; Mayeri, 1979). For example, prolonged excitation occurs in members of the LB and LC identified clusters (LLQ cells), which include motorneurons to the gill, siphon, and vasculature. Prolonged inhibition occurs in neurosecretory cells (cells R3-R14) and in motoneurons (L_{14} cells) that cause the release of ink from the purple gland. A command interneuron that controls increases in cardiac output (L10) is first inhibited and then excited. Prolonged excitation occurs in mechanoreceptor neurons that innervate the foot (L1, R1). Burst augmentation occurs in a cell implicated in the control of water balance (R15). Although these bag cell actions have been studied in the isolated abdominal ganglion, in a semi-intact preparation, Brownell and Schaefer (1982) have found that the bag cell burst discharge causes increases in spontaneous respiratory movements which are controlled by increases in spontaneous activity of the identified command interneuron, interneuron II. Thus the overall function of the bag cell actions may be to facilitate certain reflex pathways and inhibit others. In other instances, sustained increases or decreases in spike rate of target neurons produced by bag cell activity may result in alterations in muscle tone or homeostatic function.

4. The Atrial Gland and Peptide A and Peptide B Precursor Proteins

Two other members of the ELH gene family are expressed in the atrial gland. These are the peptide A and peptide B precursor proteins depicted in Fig. 3. Although the function of the atrial gland is unknown, it is likely to play an important role in reproduction because it contains large amounts of neuroactive peptides. The neuroactive peptides were discovered and isolated by Arch and colleagues (1978) and the amino acid sequences for peptides A and B were determined by Heller *et al.* (1980). They are 34 amino acid peptides that are structurally related to each other and to α-, β-, and γ-BCP. Like α-BCP, they excite the bag cells. An ELH-like peptide encoded on the peptide A precursor has been isolated and sequenced by Rothman *et al.* (1982). It has the same activity on identified neurons as bag cell ELH. Thus the peptide A precursor, like the ELH/BCP precursor, has at least two neuroactive proteins encoded on it (see Rothman *et al.*, 1984).

What might be the function of the atrial gland? An early hypothesis was that one or more atrial gland peptides enter the general circulation from the gland and initiate egg laying by activating a bag cell discharge. However, the release of atrial gland peptides has not as yet been detected in the general circulation. Moreover, the anatomy of the atrial gland suggests that it is an exocrine organ that releases peptides into the reproductive tract (Arch *et al.*, 1978, 1980). *Aplysia* is a hermaphrodite and the reproductive tract has separate grooves for allosperm, autosperm, and the egg string. An alternative possibility is that one or more of the peptides are pheromones that are used for a particular aspect of reproductive physiology and behavior. For example, the peptide might be released onto the egg string just before it emerges from the reproductive tract or released into the sea water surrounding the animal to serve as a chemical signal for induction of egg laying or attraction of other animals.

5. Concluding Remarks

5.1. What Is the Functional Significance of Multiple Transmitters Derived from a Common Precursor Protein?

In addition to the ELH/BCP precursor, recent studies indicate that the precursors for many other biologically active peptides contain more than one active molecule (see Herbert *et al.*, 1981). It is therefore worth considering the possible functional significance of a neural system such as the bag cells that utilize multiple transmitters derived from a common precursor. First, the coordinate actions of multiple bag cell transmitters permit a broad range of effects to occur in target neurons. In theory, the various responses could be produced by only a single transmitter; indeed, there are many examples of neurons that release a single transmitter such as ACh (e.g., from cell L10) to produce multiple actions on postsynaptic neurons (Kandel, 1976). What is particularly distinctive of the bag cell system, however, is that ELH and α-BCP have very different physiological roles: ELH has a longer duration of action, diffuses longer distances to its central and peripheral targets, and is inactivated more slowly than α-BCP. These differences may provide an additional degree of freedom for coordinating long- and short-lasting events in the egg-laying program than would be the case if the bag cells used only one transmitter.

Second, one peptide (α-BCP) may regulate the release of ELH and other peptides derived from the precursor protein. This autoexcitatory feature of α-BCP provides a means for discharging large amounts of chemical messenger in one episode. The all-or-none burst discharge of the bag cells is similar to the episodic release of biologically active peptides from the hypothalamus and pituitary of vertebrates. It is therefore conceivable that this feature of the bag cell system may occur in other peptide-secreting cells as well.

Third, a common precursor ensures that all the chemical messengers are synthesized and released together; this results in a fixed set of changes in neural signaling and presumably a certain degree of stereotypy in the egg-laying behavior pattern. The use of a common precursor might still allow a degree of plasticity in the system. Perhaps during development, or because of some physiological event, there is an alteration in the processing of the precursor or in the posttranslational modification of the resultant peptides that results in changes in the activities of the released peptides and corresponding changes in electrical signaling and behavior.

Lastly, the key molecular determinants of the egg-laying behavior pattern can be regarded as being encoded on the precursor protein. There may be a direct relationship between the evolution of the gene coding for the precursor protein and the evolution of the behavior pattern. Perhaps the present behavior pattern evolved from a more simple one that was regulated by a single peptide transmitter. The appearance of additional transmitters on the precursor protein provided an evolutionarily adaptive means for increasing the ways the bag cells could regulate electrical signalling in the central nervous system and thereby increase the complexity of the behavior pattern. Thus as a result of gene duplication, mutation, and translocation, new peptide sequences were added to the precursor and new neuronal activities and behavioral capabilities were brought into play.

References

Arch, S., 1976, Biochemical isolation and physiological identification of the egg laying hormone in *Aplysia californica, J. Gen. Physiol.* **68:**197–210.

Arch, S., and Smock, T., 1977, Egg laying behavior in *Aplysia californica, Behav. Biol.* **19:**45–54.

Arch, S., Smock, T., Gurvis, R., and McCarthy, C., 1978, Atrial gland induction of the egg-laying response in *Aplysia californica, J. Comp. Physiol.* **A128:**67–70.

Arch, S., Lupatkin, J., Smock, T., and Beard, M., 1980, Evidence for an exocrine function of the *Aplysia* atrial gland, *J. Comp. Physiol.* **A141:**131–137.

Berry, R. W., 1981, Proteolytic processing in the biogenesis of the neurosecretory egg-laying hormone in *Aplysia.* 1. Precursors, intermediates and products, *Biochemistry* **20:**6200–6205.

Blankenship, J. E., and Haskins, J. T., 1979, Electrotonic coupling among neuroendocrine cells in *Aplysia, J. Neurophysiol.* **42:**347–355.

Branton, W. D., Mayeri, E., Brownell, P., and Simon, S. B., 1978a, Evidence for local hormonal communication between neurones in *Aplysia, Nature (London)* **274:**70–72.

Branton, W. D., Arch, S., Smock, T., and Mayeri, E., 1978b, Evidence for mediation of a neuronal interaction by a behaviorally active peptide, *Proc. Natl. Acad. Sci. U.S.A.* **75:**5732–5736.

Brownell, P., and Mayeri, E., 1979, Prolonged inhibition of neurons by neuroendocrine cells in *Aplysia, Science* **204:**417–420.

Brownell, P. H., and Schaefer, M. E., 1982, Activation of a long-lasting motor program by bag cell neurons in *Aplysia, Neurosci. Abstr.* **8:**736.

Chiu, A. Y., and Strumwasser, F., 1981, An immunohistochemical study of the neuropeptide bag cells of *Aplysia, J. Neurosci.* **1:**812–826.

Chiu, A. Y., Hunkapiller, M. W., Heller, E., Stuart, D. K., Hood, L. E., and Strumwasser, F., 1979, Purification and primary structure of the neuropeptide egg-laying hormone of *Aplysia californica, Proc. Natl. Acad. Sci. U.S.A.* **76:**6656–6660.

Cobbs, J. S., and Pinsker, H. M., 1982a, Role of bag cells in egg deposition of *Aplysia brazilliana.* I. Comparison of normal and elicited behaviors, *J. Comp. Physiol.* **147:**523–535.

Cobbs, J. S., and Pinsker, H. M., 1982b, Role of bag cells in egg deposition of *Aplysia braziliana.* II. Contribution of egg movement to elicited behaviors, *J. Comp. Physiol.* **147:**537–546.

Dale, H. H., Feldberg, W., and Vogt, M., 1936, Release of acetylcholine at voluntary motor nerve endings, *J. Physiol.* **86:**353–380.

Dismukes, R. K., 1979, New concepts of molecular communication among neurons, *Behav. Brain Sci.* **2:**409–448.

Frazier, W. T., Kandel, E. R., Kupfermann, I., Waziri, R., and Coggeshall, R. E. 1967, Morphological and functional properties of identified neurons in the abdominal ganglion of *Aplysia californica, J. Neurophysiol.* **30:**1288–1351.

Gainer, H., Loh, Y. P., and Sarne, J. Y., 1977, Biosynthesis of neuronal peptides, in: *Peptides in Neurobiology* (H. Gainer, ed.), Plenum Press, New York, pp. 183–219.

Glitsch, H. G., and Pott, L., 1978, Effects of acetylcholine and parasympathetic nerve stimulation on membrane potential in quiescent guinea-pig atria, *J. Physiol.* **279:**655–668.

Hartzell, H. C., Kuffler, S. W., and Yoshikami, D., 1975, Postsynaptic potentiation: Interaction between quanta of acetylcholine at the skeletal neuromuscular synapse, *J. Physiol. (London)* **251:**427.

Haskins, J. T., Price, C. H., and Blankenship, J. E., 1981, A light and electron microscope investigation of the neurosecretory bag cells of *Aplysia, J. Neurocytol.* **10:**729–747.

Heller, E., Kaczmarek, L. K., Hunkapiller, M. W., Hood, L. E., and Strumwasser, F., 1980, Purification and primary structure of two neuroactive peptides that cause bag cell afterdischarge and egg-laying in *Aplysia, Proc. Natl. Acad. Sci. U.S.A.* **77:**2328–2332.

Herbert, E., Birnberg, N., Lissitsky, J-C., Ciuelli, O., and Uhler, M., 1981, Proopiomelanocortin: A model for the regulation of expression of neuropeptides in pituitary and brain, *Neurosci. Comment.* **1:**16–27.

Kandel, E. R., 1976, *Cellular Basis of Behavior,* Freeman, San Francisco, California.

Kupfermann, I., 1967, Stimulation of egg laying: Possible neuroendocrine function of bag cells of abdominal ganglion of *Aplysia californica, Nature (London)* **216:**814–815.

Kupfermann, I., and Kandel, E. R., 1970, Electrophysiological properties and functional interconnections of two symmetrical neurosecretory clusters (bag cells) in abdominal ganglion of *Aplysia, J. Neurophysiol.* **33:**865–876.

Loewi, O., 1921, Uber humorale Ubertragbarkeit der Herznerven wirkung. I. Mitteilung *Pflueger's Arch. Ges. Physiol.* **189:**239–242.

Mayeri, E., 1979, Local hormonal modulation of neural activity in *Aplysia,* (FASEB Symposium on "Modulation of Synaptic Excitability"), *Fed. Proc.* **38:**2103–2108.

Mayeri, E., and Rothman, B. S., 1982a, Nonsynaptic peptidergic neurotransmission in the abdominal ganglion of *Aplysia,* in: *Neurosecretion—Molecules, Cells and Systems* (D. S. Farner and K. Lederis, eds.), Plenum Press, New York, pp. 307–318.

Mayeri, E., and Rothman, B. S., 1982b, Peptide effects on mollusckan neurons, in: *Strategies for Studying the Roles of Peptides in Neuronal Function* (J. L. Barker, ed.), Short Course Syllabus, Society for Neuroscience, Bethesda, Maryland, pp. 121–130.

Mayeri, E., and Simon, S., 1975, Modulation of synaptic transmission and burster neuron activity after release of a neurohormone in *Aplysia, Neurosci. Abstr.* **1:**584.

Mayeri, E., Brownell, P. H., Branton, W. D., and Simon, S. B., 1979a, Multiple, prolonged actions of neuroendocrine bag cells on neurons in *Aplysia.* I. Effects on bursting pacemaker neurons, *J. Neurophysiol.* **42:**1165–1184.

Mayeri, E., Brownell, P. H., and Branton, W. D., 1979b, Multiple, prolonged actions of neuroendocrine bag cells on neurons in *Aplysia.* II. Effects on beating pacemaker and silent neurons, *J. Neurophysiol.* **42:**1185–1197.

Mayeri, E., Rothman, B. S., Brownell, P., Branton, W. D., and Padgett, L., 1985, Nonsynaptic characteristics of neurotransmission mediated by ELH in the abdominal ganglion of *Aplysia, J. Neurosci.* (in press).

Pinsker, H. M., and Dudek, F. E., 1977, Bag cell control of egg-laying in freely-behaving *Aplysia, Science* **197:**490–493.

Rothman, B. S., Brown, R. O., Mayeri, E., and Shively, J. E., 1982, Isolation of novel, neuroactive ELH-like peptides from the atrial gland of *Aplysia, Soc. Neurosci. Abstr.* **8:**14.

Rothman, B. S., Mayeri, E., and Scheller, R. H., 1984, The bag cell neurons of *Aplysia* as a possible peptidergic multitransmitter system: From genes to behavior, in: *Gene Expression in Brain* (C. Zomzely-Neurath and W. A. Walker, eds.), Wiley and Sons, New York.

Rothman, B. S., Weir, G., and Dudek, F. E., 1983b, Egg-laying hormone: Direct action on the ovotesis of *Aplysia, Gen. Comp. Endocrinol.* **52:**134–141.

Rothman, B. S., Mayeri, E., and Scheller, R. H., 1984, The bag cell neurons of *Aplysia* as a possible peptidergic multitransmitter system: From genes to behavior, in: *Gene Expression in Brain* (C. Zomzely-Neurath and W. A. Walker, eds.), Wiley and Sons, New York.

Scheller, R. H., Jackson, J. F., McAllister, L. B., Schwartz, J. H., Kandel, E. R., and Axel, R., 1982, A family of genes that codes for ELH, a neuropeptide eliciting a stereotyped pattern of behavior in *Aplysia, Cell* **28:**707–719.

Scheller, R. H., Jackson, J. F., McAllister, L. B., Rothman, B. S., Mayeri, E., and Axel, R., 1983a, A single gene encodes multiple neuropeptides mediating a stereotyped behavior, *Cell* **32:**7–22.

Scheller, R. H., Rothman, B. S., and Mayeri, E., 1983b, A single gene encodes multiple peptide neurotransmitter candidates involved in a stereotyped behavior, *Trends Neurosci.* **6:**340–345.

Sigvardt, K., Rothman, B. S., and Mayeri, E., 1983, Analysis of inhibition produced by the candidate neurotransmitter, α-bag cell peptide, in identified neurons of *Aplysia, Soc. Neurosci. Abstr.* **9:**311.

Sigvardt, K. A., Rothman, B. S., Brown, R. O., and Mayeri, E., 1984, The bag cells of *Aplysia* as a multitransmitter system: identification of alpha bag cell peptide as a second neurotransmitter (submitted).

Stuart, D. K., and Strumwasser, F., 1980, Neuronal sites of action of a neurosecretory peptide, egg-laying hormone, in *Aplysia californica, J. Neurophysiol.* **43:**499–519.

Tinbergen, N., 1951, *The Study of Instinct,* Oxford University Press, New York.

16

The Central Nervous System of *Aplysia californica*

A Model System for Cellular Studies of Central Neurotransmission

JOYCE K. ONO AND RICHARD E. MCCAMAN

1. Introduction

Recent analyses of neuronal networks have focused on some of the quantitative features of a network, such as the temporal organization of generated patterns and the plasticity of the outputs. It is becoming clear that knowledge of the conductance mechanisms and temporal characteristics of synaptic responses mediated by specific neurotransmitters are important for understanding how neuronal networks generate their particular outputs and manifest plasticity. Studies in our laboratory are directed toward determining the role of chemically mediated responses in neuronal networks. We are interested in ascertaining what chemicals are used as neurotransmitters, identifying the neurons that utilize specific chemicals as neurotransmitters, characterizing the ionic conductance mechanisms that underlie the responses mediated by a transmitter, and evaluating the effects of various pharmacological agents on the specific receptors activating these conductances.

In order to characterize chemical transmission between neurons, we must have a convenient way to conduct in-depth studies of both presynaptic and postsynaptic neurons, i.e., respectively, neurons that utilize a specific transmitter and neurons that respond to the released transmitter. We have therefore focused on

JOYCE K. ONO AND RICHARD E. MCCAMAN • Section of Neuropharmacology, Division of Neurosciences, Beckman Research Institute of the City of Hope, Duarte, California 91010.

two types of neuronal aggregates that have been defined previously by Kandel and Wachtel (1968) as divergent and convergent neuronal aggregates. A divergent neuronal aggregate, consisting of one presynaptic neuron making direct synaptic connections with a number of follower cells, would be useful for identifying chemicals utilized as neurotransmitters, particularly if the followers exhibit a variety of responses. A transmitter candidate should mimick appropriately each synaptic response when applied onto each of the postsynpatic neurons, thus providing a rigorous test of the mimicry criterion (Werman, 1966) for demonstrating that a compound is a neurotransmitter. On the other hand, convergent neuronal aggregates, consisting of a single follower neuron that receives direct inputs from several presynaptic neurons, present an opportunity for understanding how a follower neuron differentiates and/or integrates various synaptic inputs. Convergent neuronal aggregates also offer an excellent opportunity to compare the pharmacology, the biophysical properties, and the interactions of the different transmitter receptors present on a single cell.

The nervous system of the marine mollusk, *Aplysia californica* offers many useful divergent and convergent neuronal aggregates that may be utilized to characterize neurotransmitter systems. An ideal preparation should contain many reidentifiable neurons that are sufficiently large in size to permit chemical studies at the single-cell level. Moreover, the biophysical properties of both the presynaptic and postsynaptic neurons should be conducive for ascertaining connectivity. That is, current injected into the presynaptic neuron soma should be capable of generating a conducted action potential and intracellular recordings through microelectrodes in the postsynaptic neuron soma should permit observations of synaptic events as well as alterations of membrane potential at the synaptic site. The preparation should also offer attractive behavioral and physiological properties at the organismic level to provide a functional framework for the cellular studies. The *Aplysia* preparation manifests many of these desirable characteristics, hence, its popularity as an experimental animal for neurobiological questions that require analyses at the cellular level.

Much of our past efforts have been devoted to developing sensitive microchemical methods appropriate for chemical analyses at the single cell level. The use of these specific quantitative methods has been successful in identifying neurons that contain a specific transmitter candidate and in obtaining in-depth chemical information for an identifiable neuron. Recently, we have incorporated immunohistochemical methods to aid in more efficiently identifying *Aplysia* neurons that may contain high concentrations of neuroactive compounds. Another area of investigation that we have pursued in conjunction with identifying neurotransmitters involves characterizing conductance mechanisms underlying the responses induced by various transmitter candidates. These studies include a detailed examination of the effects of various antagonists as well as various agonists. Studies of divergent and convergent neuronal aggregates in *Aplysia* have aided in establishing the neurotransmitter role of histamine and have revealed possible interactions between receptors to different transmitters that may underlie certain pharmacological effects.

2. Chemical Studies of Individual Neurons

We have utilized sensitive radioenzymatic assays for the quantitative measurement of several biogenic amines and have identified individual *Aplysia* neurons that contain high concentrations of either acetylcholine (ACh) (see McCaman and McCaman, 1976; McCaman and Ono, 1982), histamine (Hm) (Weinreich *et al.*, 1976; Ono and McCaman, 1980), or serotonin (5-HT) (Weinreich *et al.*, 1973; Ono and McCaman, 1984; McCaman *et al.*, 1984). Considerable dopamine (Da) and some GABA and octopamine (Oa) are present in the different ganglia, but of the many hundreds of individual neurons assayed, none contained significant amounts of these substances.

Data accumulated over several years on the amount of 5-HT, ACh, or Hm in the "same" reidentifiable neurons in *Aplysia* demonstrate a large variance that is not attributable to methodology, animal size, season, or the length of time the animal had been kept in the laboratory. Similar large variations in the concentration of 5-HT were observed in the paired, 5-HT-containing giant cerebral neurons found in other gastropod mollusks such as *Lymnaea stagnalis* and *Helisoma trivolis* (McCaman *et al.*, 1984). In all the species surveyed, the variation in 5-HT concentration between the paired cells from the same animal was minimal in comparison to about a tenfold increase in the variance when data from the "same" cells obtained from different animals was compared (see Table I). However, the identified serotonergic neurons obtained from a highly inbred strain of *Lymnaea* exhibit a level of variance as small as for paired cells within an animal. Thus, it appears that the transmitter level in certain neurons may be determined in part by genetic factors (Audersirk *et al.*, 1985). Any attempts to study alterations in transmitter levels of identified neurons from different animals would require an enormous number of animals to compensate for this variance in wild stocks. Alternatively, we have exploited the low variance between homologous, paired neurons in a given animal to compare the effects of incubating 5-HT and its precursors on the levels of endogenous 5-HT in individual neurons (McCaman *et al.*, 1984). Similar paired cell experiments have proved effective for evaluating techniques that permit *in tandem* physiological and chemical experiments on the same neuron (Ono and McCaman, 1979, 1980).

We have begun to use immunohistochemical techniques on whole mounts of *Aplysia* ganglia as a more efficient approach to identify neurons that may use specific compounds as transmitters. In the past most histochemical procedures have utilized sections of the various ganglia, but this approach severely limits identifying the stained neuron for parallel physiological or chemical studies. One of the problems with immunohistochemical methods is the possibility that the antiserum may cross-react with tissue constituents other than the intended. We selected an antiserum to 5-HT for our initial study since (1) it was commercially available, (2) reliable chemical information was already available for specific cells which contained 5-HT and cells which did not, and (3) a specific microchemical method was already in hand for direct chemical verification of any new immunopositive cells (Ono and McCaman, 1984). Figure 1a is an example of an abdominal ganglion containing neurons stained specifically with an immunohistochemical procedure utilizing an antiserum made to 5-HT conjugated to bovine serum albumin. We were able to visualize "new" 5-HT-containing neurons in the cerebral and pedal ganglia and the 5-HT content of these neurons was confirmed in parallel chemical experiments.

The immunohistochemical localization of 5-HT also provided some insights into the source of 5-HT associated with certain nonserotoninergic neurons by revealing the presence of 5-HT-containing fibers in close association with the somata of these neurons. R2, the most readily identifiable ACh-containing neuron in the abdominal ganglion, was observed to be surrounded by a network of 5-HT-immunoreactive fibers (Fig. 1b), as were many other neurons. These observations were surprising since early histological studies had suggested that invertebrate neuronal cell bodies were devoid of innervation. In our previous studies, microchemical studies on the extracts of certain individual neurons revealed the presence of more than one biogenic amine (McCaman and McCaman, 1976). Such observations led many (Brownstein *et al.*, 1974; Cottrell, 1977) to speculate that molluscan neurons may use multiple transmitters. However, in light of recent immunohistochemical studies in *Aplysia* (Goldstein *et al.*, 1984; Ono and McCaman, 1984), it is becoming apparent that *Aplysia* neuronal somata may be "contaminated" by fibers

TABLE I
Comparisons of Variability in the 5-HT Concentrations[a] of Bilaterally Paired 5-HT-Containing Neurons from Several Gastropod Mollusks.[b]

	Aplysia		*Helisoma*		*Lymnaea*		*Inbred Lymnaea*	
Pair	LC1	RC1	LC1	RC1	LC1	RC1	LC1	RC1
1	680	890	1100	770	4180	3130	4960	3130
2	80	120	410	450	3660	2270	7310	3920
3	520	350	440	230	1380	1330	3660	4180
4	180	140	270	180	5740	7310	2380	3130
5	210	270	350	350	1120	2480	4440	5740
6	240	410	—	—	890	710	3660	4180
7	330	330	—	—	2610	1880	4770	6010
8	360	440	—	—	3390	2870	5480	3920
9	270	260	—	—	5480	5740	2870	3660
10	440	260	—	—	—	—	4180	2870
11	290	360	—	—	—	—	—	—
Mean	327	348	516	396	3160	3080	4360	4070
S.E.M.	51	62	151	105	603	708	442	333
X^2 within	6932		16,940		458,606		1,206,210	
X^2	63,373		157,085		7,277,530		1,771,876	
F ratio (between/within)	9.1		9.3		16		1.5	

[a]Neuron cell volume was calculated from cell diameter measurements and assumes that the neuron is spherical. Cell nucleus was not subtracted from the calculated volume.
[b]Data are in micromolar concentrations of 5-HT.

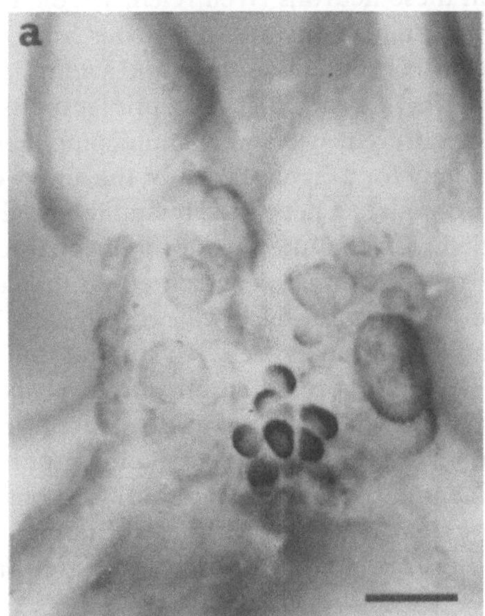

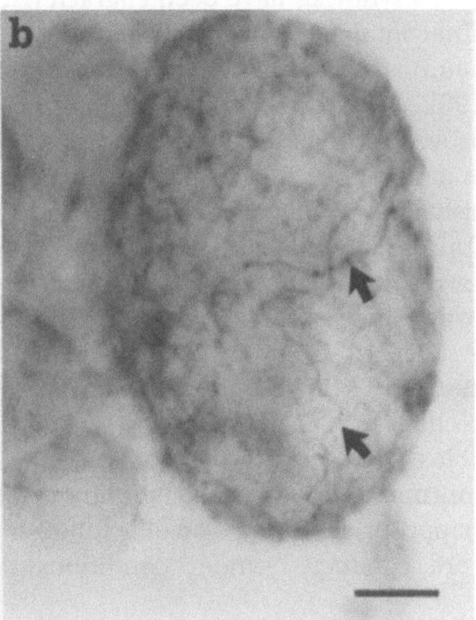

FIGURE 1. (a) Dorsal surface of the *Aplysia* abdominal ganglion depicting 5-HT-immunoreactive neurons of the RB cluster. Calibration bar, 500 μm. (b) Magnification of the ACh-containing R2 neuron in a. Arrows point to 5-HT-immunoreactive varicose fibers on the cell body of R2. Calibration bar, 100 μm.

from other neurons. The immunohistochemical observation of 5-HT-immunoreactive fibers on R2 and the finding that the amount of 5-HT associated with R2 increases when the abdominal ganglion is incubated in 5-HT (a characteristic of serotoninergic fibers and not of serontoninergic somata) suggest that the source of 5-HT is extrinsic. The presence of 5-HT-containing fibers on particular neurons appears to be a characteristic of that neuron and can be detected histochemically or chemically from animal to animal. Further experiments are necessary to determine whether the fibers visualized on neuronal somata by immunohistochemistry have some functional role.

3. The Neurotransmitter Function of Histamine

Identification of neurons with uniquely high concentrations of Hm (Weinreich *et al.,* 1976) was the germinal observation leading to the present compilation of evidence that Hm is a neurotransmitter. The discovery of several follower neurons receiving direct inputs from the Hm-containing neurons, the C2s, identified a divergent neuronal aggregate that has been extremely useful for demonstrating the transmitter role of Hm as well as for characterizing the receptor types and conductance mechanisms underlying the various responses to Hm (Weinreich, 1977; McCaman and McKenna, 1978; McCaman and Weinreich, 1982). Histamine is present in certain neurons at millimolar concentrations and metabolic pathways for its synthesis have been characterized in these neurons (Weinreich, 1976). The Hm-containing neurons make monosynaptic connections with numerous follower neurons as determined by several physiological criteria (McCaman and Weinreich, 1982, 1985; Ono and McCaman, 1980). The C2-mediated synaptic responses proved to be quite diverse and most of the different types were multicomponent in nature (Fig. 2). This provided an opportunity for a rigorous test of the ability of histamine to mimic each of the synaptic responses when directly applied to the appropriate follower. When Hm was applied focally to the soma of individual followers, it was apparent that not all of the receptors were represented on the cell body. These observations are similar to those made previously for receptors to Da (Ascher, 1972) and 5-HT (Gerschenfeld and Paupardin-Tritsch, 1974b), but unlike those for ACh receptors which appear to be well represented on various somata (Kehoe, 1972c). However, by careful searching in the neuropil, it has proved possible to find sites at which the focal application of Hm mimics the synaptic response appropriate to a particular follower (McCaman and Weinreich, 1985). Responses inappropriate for a particular follower were not observed when these experiments were conducted under conditions that insured that transmitter release was

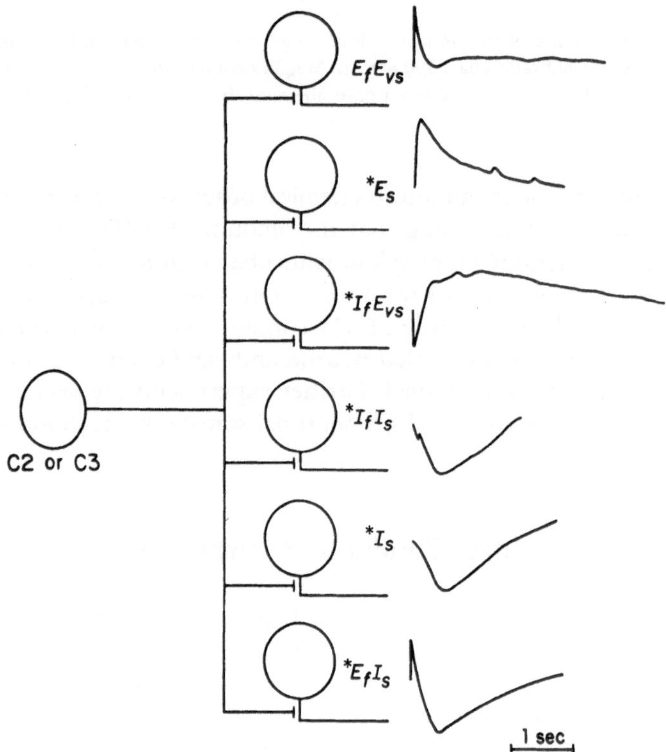

FIGURE 2. Divergent neuronal aggregate represented by the Hm-containing neuron (C2 or C3) and its various followers. Asterisks refer to the synaptic responses which can be mimicked by the application of exogenous Hm.

TABLE II
Summary of Conductance Mechanisms and Pharmacology of the Components of the Histaminergic Synaptic Responses

Response components	Ionic conductance mechanisms	Antagonists
E_f Fast excitatory	↑ gNa	
E_s Slow excitatory	↑ gNa	Pyrilamine, 2-(2-aminoethyl)-thiazole[a]
E_{vs} Very slow excitatory	↓ gK	
I_s Slow inhibitory	↑ gK	Cimetidine[a]
I_f Fast inhibitory	↑ gCl	

[a]Drug blocks synaptic and iontophoretic responses to Hm (McCaman and Weinreich, 1985).

blocked. Several drugs were found that blocked both the synaptically evoked and Hm-evoked responses (see Table II). Histamine transmission may be inactivated by a novel mechanism. Stein and Weinreich (1982) demonstrated that exogenous Hm is quickly taken up by the ganglia and converted to a glutamyl peptide, gamma-glutamylhistamine. More studies are underway to determine if this metabolic pathway plays a role in terminating Hm transmission. The physiological, biochemical, and pharmacological studies completed thus far provide compelling evidence that Hm is indeed a neurotransmitter.

Comparison of the results from studying neurotransmission mediated by Hm in the *Aplysia* CNS with those obtained from studying cholinergic and serotonergic neuronal aggregates suggests several generalizations about aminergic transmission. All three transmitters produce a variety of responses, each response may be composed of one or more components. Each component reflects a change in membrane permeability to either Na^+, Cl^-, or K^+. In most cases the responses are due to increased permeability; however, in the case of 5-HT and Hm, responses due to decreased conductances of a tonically open Na^+ or K^+ channel were also observed. It thus appears that the recognition site for the transmitter molecule is associated with a particular ion channel or ionophore. The recognition site and a particular ion channel constitute a functional unit often referred to as a receptor–ionophore complex. The three types of conductance mechanisms underlying the ACh responses described by Kehoe (1972a,b), for example (see Fig. 4), are mediated by an ACh recognition site associated with either a Na-ionophore (AChR-Na), a Cl-ionophore (AChR-Cl), or a K-ionophore (AChR-K). The multicomponent responses to ACh exhibited in certain neurons arise from combinations of these three types of receptor–ionophore complexes. The six types of synaptic responses mediated by the C2 histaminergic neurons are among the most varied and are due to five types of conductance mechanisms (Table II): two kinds of increased conductances to Na^+, an increased conductance to Cl^-, an increased conductance to K^+, and a decreased conductance mechanism to K^+.

Improvements in our methods for isolating individual neurons for chemical assay were necessary for the identification of additional histaminergic neurons, the C3s. The C3 neurons share some of the same postsynaptic neurons with the orig-

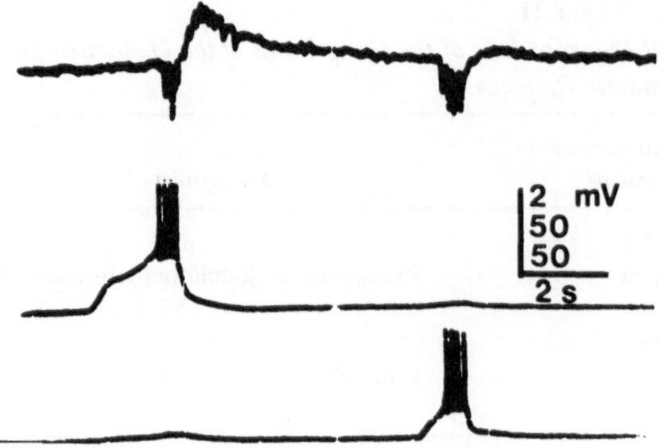

FIGURE 3. Quantitative differences in synaptic responses mediated by two different presynaptic histaminergic neurons on a common follower. The membrane potential of the common follower exhibiting an $I_fE_{vs}PSP$ is preset to −58 mV. The middle trace is the record from the histaminergic neuron C2 and the lower trace is from C3.

inal histaminergic C2 neurons, and at first could only be identified physiologically. The original protocol for isolating individual neurons called for infusing the ganglion with a solution of ethylene glycol and artificial seawater and equilibrating the preparation at −20°C (Giller and Schwartz, 1971). This freeze-substitution procedure appeared to stabilize and retain small molecules and facilitated handling and isolation of individual neurons. However, we discovered that Hm was particularly labile and there was a significant loss of Hm if the ganglion was desheathed prior to infusion with the ethylene glycol solution (Ono and McCaman, 1979). By simply replacing the ethylene glycol with propylene glycol, we were able to retain normal levels of Hm in the C2 neurons isolated from desheathed ganglia. This change in procedure permitted chemical studies on the same neuron after it had been characterized physiologically, eliminating the ambiguities of parallel, correlative studies. The improved protocol was utilized to isolate the C3 neurons after they were identified physiologically and their Hm content subsequently confirmed (Ono and McCaman, 1980).

The C2 and C3 histaminergic neurons and their common postsynaptic neurons constitute a network of convergent and divergent associations. Comparisons of the responses evoked by two presynaptic neurons, each utilizing the same neurotransmitter, demonstrated that qualitatively the convergent inputs produced the same synaptic responses (see also Segal and Koester, 1980). However, in the case of multicomponent responses, it was evident that the relative amplitude of each component response could be quite different for the C2 versus the C3 input, i.e., there was significant quantitative variation in responses to the two inputs (Fig. 3). These variations could result from different electrotonic properties of the synaptic site for each input or from different relative densities of receptor types at each site. Similar variations are also observed in the synaptic responses between the same neurons in different animals. That is, qualitatively the responses are highly reproducible, but quantitatively each component of multicomponent responses may show marked and seemingly independent variations in amplitude. The causal factors and the functional consequences of variations in the strength of the different components of multicomponent responses are largely unknown since we know very little about the distributions of the different receptor–ionophore complexes that mediate multicomponent responses to a single transmitter. For example, are all the different receptor–ionophore complexes mediating a multicomponent response present at the site of synaptic input or are complexes mediating each component spatially segregated? These and related questions represent an area of current investigation.

4. Pharmacology of Receptors to Neurotransmitters

It is clear from studying the identified cholinergic, serotonergic, and now histaminergic connections in *Aplysia,* as well as agonist-induced responses in this preparation and others, that many pharmacological agents (agonists and antagonists) affect a specific type of receptor–ionophore complex. In a now classical study, Kehoe (1972a–c) demonstrated that different agonists and antagonists could differentiate among each of the three types of receptor–ionophore complexes mediating the responses to ACh (Fig. 4). For example, hexamethonium specifically antagonized the AChR-Na without affecting the AChR-K or the AChR-Cl. Importantly, the specificity of hexamethonium for antagonizing the AChR-Na is manifested not only in cells that have the AChR-Na alone, but also in cells in which the AChR-Na is present along with either or both of the other types of complexes. Similarly, Kehoe has observed that there are selective antagonists of the AChR-K (Kehoe, 1972b) and for the AChR-Cl (Kehoe *et al.,* 1976). Different agonists also show different relative affinities for each of these types of ACh receptor–ionophore complexes (Kehoe, 1972b). Independent studies of the agonist and antagonist profiles of different 5-HT- or Hm-activated responses support the general-

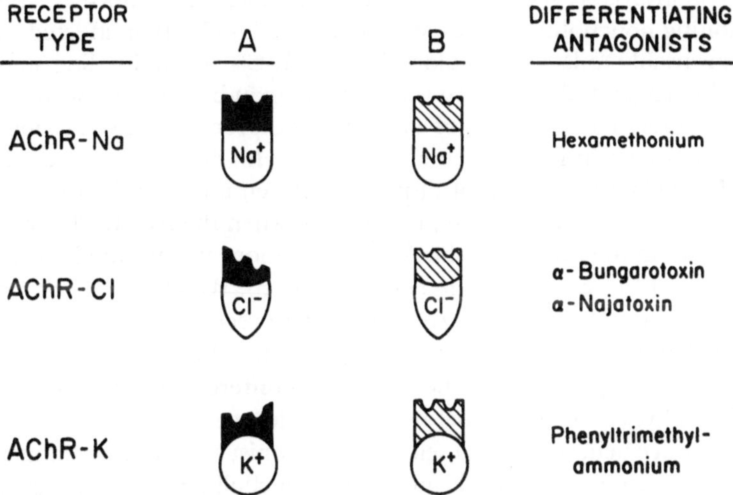

FIGURE 4. Possible models of the three types of receptor–ionophore complexes that mediate the responses to ACh in *Aplysia.* Models in column A depict each ACh recognition site as different molecules associated with Na^+, Cl^-, and K^+ ionophores. Models in column B depict the ACh recognition site as the same molecule and presumes that the association with the different ionophores results in a receptor–ionophore complex that is pharmacologically distinct.

ization that each of the various receptor–ionophore complexes activated by a single transmitter appears to be pharmacologically distinct from other receptors activated by that same transmitter. There are some pharmacological agents such as penicillin and high levels of tetraethylammonium that appear to block all agonist-induced increased permeabilities to Cl^- or K^+, respectively; presumably, these agents act directly on the ionophore component of the complex. There are at least two ways to conceive how each of the receptor–ionophore complexes for a single transmitter can be pharmacologically distinct (Fig. 4): (1) for each transmitter, the recognition site associated with each type of ionophore is a chemically unique entity; or (2) for a given transmitter, the recognition site is the same chemical entity for all the different complexes, but its association with a particular ionophore imparts its pharmacological distinctiveness.

Some pharmacological agents appear to be relatively specific antagonists of a single type of receptor–ionophore for a given transmitter (e.g., hexamethonium for AChR-Na, cimetidine for the HmR-K, etc.). However, some appear to have a "specificity" that is, at first, difficult to appreciate (e.g., α-bungarotoxin, curare, etc.). Although Shain *et al.* (1974) originally reported that α-bungarotoxin (Btx) antagonized all cholinergic responses in *Aplysia,* later studies by Kehoe *et al.* (1976) clearly demonstrated that authentic Btx affects only one type of ACh receptor, the AChR-Cl. We have confirmed Kehoe's observations, but have shown that in some neurons Btx (as well as α-Naja toxin) also antagonizes several Cl-mediated noncholinergic synaptic responses (Ono and Salvaterra, 1981). The toxins are clearly not channel blockers as they do not affect all agonist-induced Cl conductances (e.g., Cl-mediated responses to GABA are not affected). A major insight was gained when we discovered that (1) neurons with toxin-blockable noncholinergic respones (e.g., the HmR-Cl) invariably seem to also have toxin blockable AChR-Cls; and (2) neurons exhibiting noncholinergic respones (e.g., the HmR-Cl) unaffected by the toxins seem invariably to lack toxin-blockable AChR-Cls. Toxin-binding studies using iodinated Btx suggested that there was a single toxin-binding site with agonist and antagonist profiles characteristic of a nicotinic ACh receptor. Even though Btx was shown to block certain Hm-mediated responses, Hm does not compete with toxin binding. To explain the apparent disparity between the toxin-binding results and the physiological observations, we propose that when the AChR-Cl and the HmR-Cl are present on the same cell, the ACh recognition site and the Hm recognition site may control the same chloride channel (Fig. 5). We envision that toxin binds specifically to the AChR-Cl and prevents the Hm recognition site from activating the shared channel. These observations suggest that the pharmacology of central receptors may be complicated by the presence of different receptors or recognition sites in close physical or functional relationship to one another.

Curare, once believed to be a selective antagonist of nicotinic cholineric receptors, has been reported to block numerous noncholinergic responses in *Aplysia* (Ascher, 1972; Gerschenfeld and Paupardin-Tritsch, 1974a; Carpenter *et al.,* 1977). We believe that these apparent "nonspecific" actions of curare may be, in fact, quite specific for cholinergic receptors, but may affect other types of responses due to interactions between different receptor complexes. We have found numerous neurons that display noncholinergic responses (involving

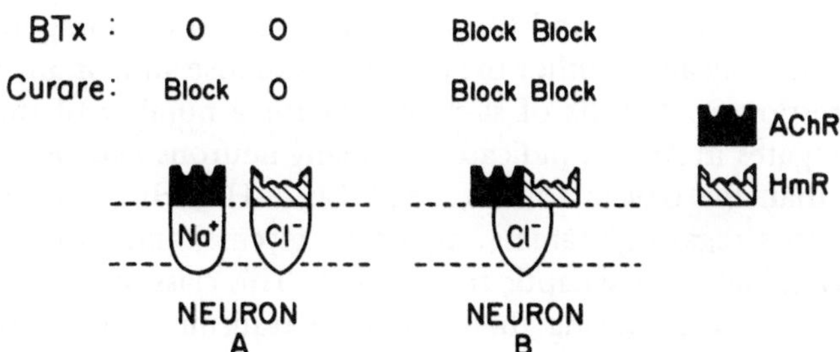

FIGURE 5. Schematic illustration of how possible associations between receptors to ACh and to Hm may produce the pharmacological effects of Btx and curare observed in different *Aplysia* neurons. In neuron A, Btx and curare do not affect the HmR-Cl since a Btx- and curare-blockable AChR-Cl are not present. In neuron B, the HmR-Cl is antagonized by both Btx and curare because the ACh and Hm recognition sites share the same Cl ionophore. The antagonists bind to the ACh recognition site and inactivate the shared Cl ionophore so that Hm can no longer activate the channel.

increased conductances to Na and/or Cl), both synaptic and agonist induced, that are unaffected by curare. Thus, curare is not an indiscriminate antagonist of all Na- and/or Cl-mediated responses. Further, our studies of Hm-mediated responses have revealed that curare antagonism of a Cl-mediated response on one neuron but not on another is correlated with the type of ACh receptor present on these different cells. These observations suggest that curare, like Btx and possibly other drugs, may produce "anomalous" or "nonspecific" effects as a consequence of previously unappreciated interrelationships between different agonist receptors (and ionophores) on neuronal membranes (see Fig. 5).

5. Summary and Conclusions

We have demonstrated how identified neuronal aggregates in the *Aplysia* preparation facilitate studies aimed at identifying neurotransmitter compounds and for characterizing the receptors and conductance mechanisms transmitters activate. The combination of biochemical studies of the identified presynaptic neurons and physiological and pharmacological characterization of the responses of the postsynaptic neurons has provided convincing data for the transmitter roles of ACh, 5-HT, and Hm in *Aplysia*. Histochemical studies utilizing glyoxylic-acid-induced histofluorescence indicate that there may be Da-containing neurons in the various ganglia (Tritt *et al.*, 1983), but such neurons have not been precisely characterized. Our chemical surveys of neuronal clusters as well as individual neurons in *Aplysia* have resulted in identifying the transmitters in less than 5% of the neurons in this relatively simple CNS.

Studies of many neuronal aggregates in *Aplysia* suggest that there are many more neurotransmitter substances that have not yet been identified. When chemical analyses of the presynaptic neurons have failed to yield a transmitter candidate, other established approaches have been employed. Thus we have (1) applied various substances to the postsynaptic neurons to determine which agonists may mimic the synaptic response, and (2) perfused with various antagonists to determine whether any may affect either the synaptic response and/or any relevant agonist-induced responses. Results of such studies for a number of the identifiable neuronal aggregates in *Aplysia* indicate that many neurons may be utilizing transmitters other than the typical amines, i.e., ACh, 5-HT, Hm, Da, Oa, or GABA. Many investigators regard glutamate, aspartate, glycine, and taurine, among others, as viable candidates for synaptic transmitters. This class of substances presents several problems in considering their potential transmitter role. Most of these amino acids are present in all neurons. Frequently, there is little difference in their concentrations in different neurons and their absolute concentrations (0.5–50 mM) are quite high in comparision to the levels of most known transmitters (Borys *et al.*, 1973; Zeman and Carpenter, 1975; Iliffe *et al.*, 1977, McCaman and Stetzler, 1977). Although it may be that neurons that use amino acids as transmitters retain them in special compartments, we do not presently have assays for detecting these differences. However, several immunohistochemical markers for putative glutamergic neurons have been proposed (Altschuler *et al.*, 1981; Storm-Mathiesen *et al.*, 1983). These antisera appear to stain certain vertebrate neurons in brain areas where glutamate is a candidate transmitter. It would be interesting to test whether these immunohistochemical reagents stain identifiable neurons in invertebrates in which direct and rigorous pharmacological and physiological studies are possible.

Immunohistochemical localization of transmitter compounds in whole mounts of ganglia shows promise as the initial basis for focusing on specific neurons to ascertain whether they utilize a specific compound as a neurotransmitter. Whole mounts preserve the anatomical characteristics of the ganglia and thereby faciliate reidentification of specific neurons for parallel chemical, physiological, and pharmacological studies. We have recently begun to look at the distribution of neurons immunoreactive to antibodies against several peptides. Preliminary results indicate that in general, a large number of neurons appear to contain substances that cross-react with antibodies made against several peptides, including those found in the vertebrate central nervous system. Many of these immunoreactive neurons are known to contain more traditional transmitter amines. Since aminergic transmitters can activate many different conductances in a single follower cell, it will be a challenge to determine the function of peptides coexisting with traditional transmitters. Studies at the level of identifiable, single cells may offer unique advantages for defining the function of coexistent peptides.

Some of the generalities derived from studying transmitter receptors in simpler systems may be applicable to central transmission in vertebrates. Transmitters in mollusks cannot be classified as simply excitatory or inhibitory since the receptors that mediate the responses can activate several types of ionic channels. Pharmacological studies of the various ionic conductances activated by a single transmitter indicate that each of these receptor–ionophore complexes may be pharmacologically unique. There are some indications from intracellular studies on mammalian neurons that certain amines such as GABA (Andersen *et al.*, 1980) and ACh (Dodd and Dingledine, 1979) can produce different conductances on various neurons or on different parts of a neuron. Whether the multiple types of vertebrate receptors recently defined by binding assays may be correlated with different functional types may be determined by future studies.

Rigorous tests of the specificity of numerous pharmacological agents in *Aplysia* and other invertebrates indicate that many drugs do not display the expected specificity when tested on central receptors. Thus, it may be misleading to conclude from pharmacological data alone that a synaptic response is mediated by a certain transmitter. There are several antagonists that so far appear to be specific for certain cholinergic, serotonergic, and histaminergic responses in *Aplysia*. However, the observation that the snake toxins antagonize noncholinergic responses in *Aplysia* suggests that there may be interactions and relationships among receptors for different transmitters on a single cell. These studies suggest that a pharmacological agent may bind to a specific recognition site for one transmitter receptor, yet affect the function of receptors to other transmitters. Further studies of individual neurons with known receptor profiles are necessary to determine the nature of the functional coupling between certain receptors.

Studies of the basic processes underlying central neurotransmission in simpler invertebrate nervous systems have revealed the wealth of signals and the mechanisms that underlie them which neurons might use to communicate with each other. Our hope in this reductionist approach is that principles may emerge from such studies that can be used as guides for determining how more complex systems may be organized and function.

ACKNOWLEDGMENTS. We thank Dr. Marilyn McCaman for collaboration and Judith Stetzler for technical assistance on portions of this work. The authors' research was supported by grants from the National Institutes of Health, NS 18862 (J.K.O.) and NS 18857 (R.E.M.).

References

Altschuler, R. A., Neises, G. R., Harmison, G. C., Wenthold, R. J., and Fex, J., 1981, Immunocytochemical localization of aspartate aminotransferase immunoreactivity in cochlear nucleus of the guinea pig, *Proc. Natl. Acad. Sci. U.S.A.* **78:**6553–6557.

Andersen, P., Dingledine, R., Gjerstad, L., Langmoen, I. A., and Mosfeldt Laursen, A., 1980, Two different responses of hippocampal pyramidal cells to application of gamma-amino butyric acid, *J. Physiol. (London)* **305:**279–296.

Ascher, P., 1972, Inhibitory and excitatory effects of dopamine on *Aplysia* neurones, *J. Physiol. (London)* **225:**173–209.

Audersirk, G., Audersirk, T., McCaman, R., and Ono, J., 1985, Evidence for genetic influences on neurotransmitter content of identified neurons of *Lymnaea stagnalis, Comp. Biochem. Physiol.* (in press).

Borys, H. K., Weinreich, D., and McCaman, R. E., 1973, Determination of glutamate and glutamine in individual neurons of *Aplysia californica, J. Neurochem.* **21:**1349–1351.

Brownstein, M. J., Saavedra, J. M., Axelrod, J., Zeman, G. H., and Carpenter, D. O., 1974, Coexistence of several putative neurotransmitters in single identified neurons of *Aplysia, Proc. Natl. Acad. Sci. U.S.A.* **71:**4662–4665.

Carpenter, D. O., Swann, J. W., and Yarowsky, P. J., 1977, Effect of curare on responses to different putative neurotransmitters in *Aplysia* neurons, *J. Neurobiol.* **8:**119–132.

Cottrell, G. A., 1977, Identified amine-containing neurons and their synaptic connexions, *Neurosciences* **2:**1–18.

Dodd, J. S., and Dingledine, R., 1979, Acetylcholine as an excitatory and inhibitory transmitter in the mammalian central nervous system, in: *Progress in Brain Research, The Cholinergic Synapse,* Volume 49 (S. Tucek, ed.), Elsevier, Amsterdam, Netherlands, pp. 254–266.

Gerschenfeld, H. M., and Paupardin-Tritsch, D., 1974a, Ionic mechanisms and receptor properties underlying the response of molluscan neurons to 5-hydroxytryptamine, *J. Physiol. (London)* **243**:427–456.

Gerschenfeld, H. M., and Paupardin-Tritsch, D., 1974b, On the transmitter function of 5-hydroxytryptamine at excitatory and inhibitory monosynaptic junctions, *J. Physiol. (London)* **243**:457–481.

Giller, E., and Schwartz, J. H., 1971, Choline acetyltransferase in identified neurons of abdominal ganglion of *Aplysia californica, J. Neurophysiol.* **34**:93–107.

Goldstein, R., Kistler, H. B., Steinbusch, H. W. M., and Schwartz, J. H., 1984, Distribution of serotonin-immunoreactivity in juvenile *Aplysia, Neurosciences* **11**:535–547.

Iliffe, T. M., McAdoo, D. J., Beyer, C. B., and Haber, B., 1977, Amino acid concentrations in the *Aplysia* nervous system: Neurons with high glycine concentrations, *J. Neurochem.* **28**:1037–1042.

Kandel, E. R., and Wachtel, H., 1968, The functional organization of neural aggregates in *Aplysia,* in: *Physiological and Biochemical Aspects of Nervous Integration* (F. D. Carlson, ed.), Prentice Hall, New Jersey, pp. 17–65.

Kehoe, J., 1972a, Ionic mechanisms of a two-component cholinergic inhibition in *Aplysia* neurones, *J. Physiol. (London)* **225**:85–114.

Kehoe, J., 1972b, Three acetylcholine receptors in *Aplysia* neurones, *J. Physiol. (London)* **225**:115–146.

Kehoe, J., 1972c, The physiological role of three acetylcholine receptors in synaptic transmission in *Aplysia, J. Physiol. (London)* **225**:147–172.

Kehoe, J., Sealock, R., and Bon, C., 1976, Effects of α-toxins from *Bungarus multicinctus* and *Bungarus caeruleus* on cholinergic responses in *Aplysia* neurones, *Brain Res.* **107**:527–540.

McCaman, M. W., Ono, J. K., and McCaman, R. E., 1984, 5-Hydroxytryptamine measurements in molluscan ganglia and neurons using a modified radioenzymatic assay, *J. Neurochem.* **43**:91–99.

McCaman, R. E., and McCaman, M. W., 1976, Biology of individual cholinergic neurons in the invertebrate CNS, in: *Biology of Cholinergic Function* (A. M. Goldberg and I. Hanin, eds.), Raven Press, New York, pp. 485–513.

McCaman, R. E., and McKenna, D., 1978, Monosynaptic connections between histamine-containing neurons and their various follower cells, *Brain Res.* **141**:165–171.

McCaman, R. E., and Ono, J. K., 1982, *Aplysia* cholinergic synapses: A model for central cholinergic function, in: *Progress in Cholinergic Biology: Model Cholinergic Synapses* (I. Hanin and A. M. Goldberg, eds.), Raven Press, New York, pp. 23–43.

McCaman, R. E., and Stetzler, J., 1977, Determination of taurine in individual neurones of *Aplysia californica, J. Neurochem.* **29**:739–741.

McCaman, R. E., and Weinreich, D., 1982, On the nature of histamine mediated slow hyperpolarizing synaptic potentials in identified neurons from the cerebral ganglion of *Aplysia californica, J. Physiol. (London)* **328**:485–506.

McCaman, R. E., and Weinreich, D., 1985, Histaminergic synaptic transmission in the cerebral ganglion of *Aplysia*, *J. Neurophysiol.* (in press).

Ono, J. K., and McCaman, R. E., 1979, Measurement of endogenous transmitter levels after intracellular recording, *Brain Res.* **165**:156–160.

Ono, J. K., and McCaman, R. E., 1980, Identification of additional histaminergic neurons in *Aplysia:* Improvements of single cell isolation techniques for *in tandem* physiological and chemical studies, *Neurosciences* **5**:835–840.

Ono, J. K., and McCaman, R. E., 1984, Immunocytochemical localization and direct assays of serotonin-containing neurons in *Aplysia, Neuroscience* **11**:549–560.

Ono, J. K., and Salvaterra, P. M., 1981, Snake toxin effects on cholinergic and noncholinergic responses of *Aplysia californica, J. Neurosci.* **1**:259–270.

Segal, M., and Koester, J., 1980, Different cholinergic synapses converging onto neurons in *Aplysia* produce the same synaptic action, *Brain Res.* **199**:459–465.

Shain, W., Greene, L. A., Carpenter, D. O., Sytkowski, A. J., and Vogel, Z., 1974, *Aplysia* acetylcholine receptors: Blockade by and binding of α-bungarotoxin, *Brain Res.* **72**:225–240.

Stein, C., and Weinreich, D., 1982, An *in vitro* characterization of gamma-glutamylhistamine synthetase: A novel enzyme catalyzing histamine metabolism in the central nervous system of the marine mollusk, *Aplysia californica, J. Neurochem.* **38**:204–214.

Storm-Mathiesen, J., Leknes, A. K., Bore, A. T., Vaaland, J. L., Edminson, P., Haug., F. S., and Ottersen, O. P., 1983, First visualization of glutamate and GABA in neurons by immunocytochemistry, *Nature (London)* **301:**517–520.

Tritt, S. H., Posner Lowe, I., and Byrne, J. H., 1983, A modification of the glyoxylic acid induced histofluorescence technique for demonstration of catecholamines and serotonin in tissues of *Aplysia californica, Brain Res.* **259:**159–162.

Weinreich, D., 1976, The distribution of histamine, histidine and histidine decarboxylase in ganglia, nerves and single identified neuronal cell bodies of *Aplysia californica,* in: *Neurobiology of Invertebrates. Gastropoda Brain* (J. Salanki, ed.), Akademiai Kiado, Budapest, pp. 191–206.

Weinreich, D., 1977, Synaptic responses mediated by identified histamine-containing neurones, *Nature (London)* **267:**854–856.

Weinreich, D., McCaman, M. W., McCaman, R. E., and Vaughn, J. E., 1973, Chemical, enzymatic and ultrastructural characterization of 5-hydroxytryptamine-containing neurons from the ganglia of *Aplysia californica* and *Tritonia diomedia, J. Neurochem.* **20:**969–976.

Weinreich, D., Weiner, C., and McCaman, R., 1976, Endogenous levels of histamine in single neurons isolated from CNS of *Aplysia californica, Brain Res.* **84:**341–345.

Werman, R., 1966, Critertia for identification of a central nervous system transmitter, *Comp. Biochem. Physiol.* **18:**745–766.

Zeman, G. H., and Carpenter, D. O., 1975, Asymmetric distribution of aspartate in ganglia and single neurons of *Aplysia, Comp. Biochem. Physiol.* **52:**23–26.

17

Neurotransmitter Modulation of the Stomatogastric Ganglion of Decapod Crustaceans

EVE MARDER AND SCOTT L. HOOPER

1. Introduction

A large number of substances including amines, amino acids, and peptides have been assigned neurotransmitter roles in both vertebrate and invertebrate nervous systems. Although explanations for this large number of neurotransmitters have been proposed, it is still far from understood how all these neurotransmitter systems function and interact within the nervous system. Some clues to the answer to this question should come from current work on the neurotransmitter systems involved in the modulation of the output of a "relatively simple system," the stomatogastric ganglion (STG) of decapod crustaceans.

The STG contains approximately 30 neurons and is the central pattern generator (CPG) responsible for timing the movements of the stomach of lobsters and crabs (Maynard, 1972; Selverston *et al.*, 1976). This means that the STG, even when entirely isolated from sensory input, continues to produce rhythmic motor discharges in which the motor neurons that innervate antagonistic muscles fire in alternation. There are two subsets of neurons in the STG, those that innervate the large gastric mill muscles (termed the gastric system) and those that innervate the muscles of the pyloric region (the pyloric system). The pyloric system has been an extremely attractive preparation to investigators interested in the underlying basis of pattern generation because it is a robust system, the neurons are easy to record from, and the pyloric motor output produced by the isolated stomatogastric system

EVE MARDER AND SCOTT L. HOOPER • Biology Department, Brandeis University, Waltham, Massachusetts 02254.

appears stereotyped and therefore relatively easy to analyze (Selverston *et al.*, 1976). Indeed, due to the work of a number of investigators, we have today a good first approximation of how the "ideal" pyloric pattern is produced (Selverston *et al.*, 1976; Hartline, 1979; Hartline and Gassie, 1979; Selverston and Miller, 1980; Miller and Selverston, 1982a,b).

The STG in the animal is not an isolated ganglion. Instead, it receives information from other ganglia, and its output is modified according to sensory information or other environmental factors. It is striking that although there are only 30 neurons in the STG, there are between 100 and 250 fibers in the stomatogastric nerve (STN) that connects the STG to the rest of the animal (Fig. 1) (Maynard, 1971; King, 1976). In this chapter we will present the accumulated evidence that shows that a large number of different neurotransmitters are present in fibers in the STN. We will then show that each of these substances produces different and specific changes in the pyloric rhythm. Although a number of the neurotransmitters that are found in fibers in the STN increase the frequency of the pyloric rhythm, the phase relations of the firing patterns of the pyloric neurons are completely different for each.

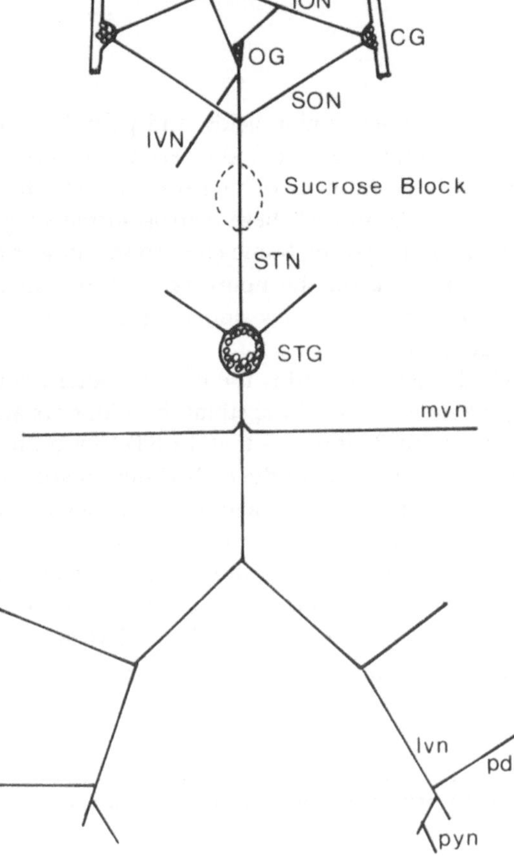

FIGURE 1. The stomatogastric nervous system. The four ganglia of the stomatogastric nervous system are diagrammatically represented. The top of the page is anterior. Abbreviations: CG, commissural ganglion; OG, esophageal ganglion; ION, inferior esophageal nerve; SON, superior esophageal nerve; IVN, inferior ventricular nerve; STN, stomatogastric nerve; STG, stomatogastric ganglion; mvn, medial ventricular nerve; lvn, lateral ventricular nerve; pdn, pyloric dilator nerve; pyn, pyloric nerve. The dashed line labeled "Sucrose Block" indicates the position of a vaseline well placed around the STN that allows the investigator to reversibly isolate the STG by blocking impulse traffic in the STN with isotonic sucrose in the well. (Russell, 1976, 1979.)

Thus, the central thesis of this chapter is that there is not one "ideal" pyloric pattern, but rather many stable firing patterns that are variations on the pyloric theme. We show that different neurotransmitters, present in inputs to the STG, elicit different variations of the theme. These data suggest as a working model that many of the large number of neurotransmitters found in all nervous systems may serve the function of tuning the output of neural circuits. Thus, a neural circuit, although defined by the anatomical relationships of its component neurons, may nonetheless act functionally as if it were many different circuits. Neurotransmitters that change the excitability of individual neurons or modulate the efficacy of synapses provide mechanisms to modulate the output of the circuit.

2. Studying the Pyloric System

The STG is routinely studied by dissecting the stomatogastric nervous system out of the animal and then pinning it out in a Sylgard-lined Petri dish (Mulloney and Selverston, 1974; Selverston *et al.,* 1976). Extracellular recording electrodes are placed on the motor nerves and intracellular recordings are made from the somata of the neurons of the STG.

Three different procedures have been used over the years to isolate the STG from all inputs. The first of these is to cut the STN. In a number of species, including *Panulirus interruptus* and *Cancer irroratus,* the STG isolated by cutting the STN continues to produce rhythmic pyloric motor patterns for many hours if kept in chilled physiological saline. In contrast, in *Homarus americanus,* isolation of the STG by cutting the STN frequently results in complete cessation of pyloric activity (Beltz *et al.,* 1984). If impulse activity in the STN is blocked by placing a vaseline well filled with isotonic sucrose around the desheathed STN then most frequently pyloric activity will cease in all species studied (Moulins and Cournils, 1982; F. Nagy and J. Miller, personal communication). If the STN is not desheathed, but action potentials are blocked with sucrose, in *Panulirus interruptus* and *Cancer irroratus* the STG will usually continue to produce pyloric activity, although the frequency and intensity of this activity is decreased. Thus these three methods of isolating the STG appear to differ in their severity. This is an interesting observation, since, as will be seen in Section 4, a number of neurotransmitters can initiate pyloric cycling in quiescent preparations, and may participate in the maintenance of normal pyloric activity if these neurotransmitters are released in a slow, continuous fashion from input fibers.

An example of the pyloric motor pattern produced by an isolated STG from the crab, *C. irroratus,* is shown in Fig. 2A. Activity of the pyloric dilator (PD) motor neurons alternates with activity of the lateral pyloric (LP) neuron and the pyloric (PY) neurons (Fig. 2A). For the purposes of simplicity, in this chapter, we will restrict ourselves to a consideration of the time of firing of the PD, LP, and PY neurons. The major features of the pyloric motor pattern produced by the isolated STG are similar in all of the decapod species studied, with neurons firing in the LP, PY, PD sequence that then repeats (Fig. 2A).

The output of the STG is modulated by other neuronal inputs in all species

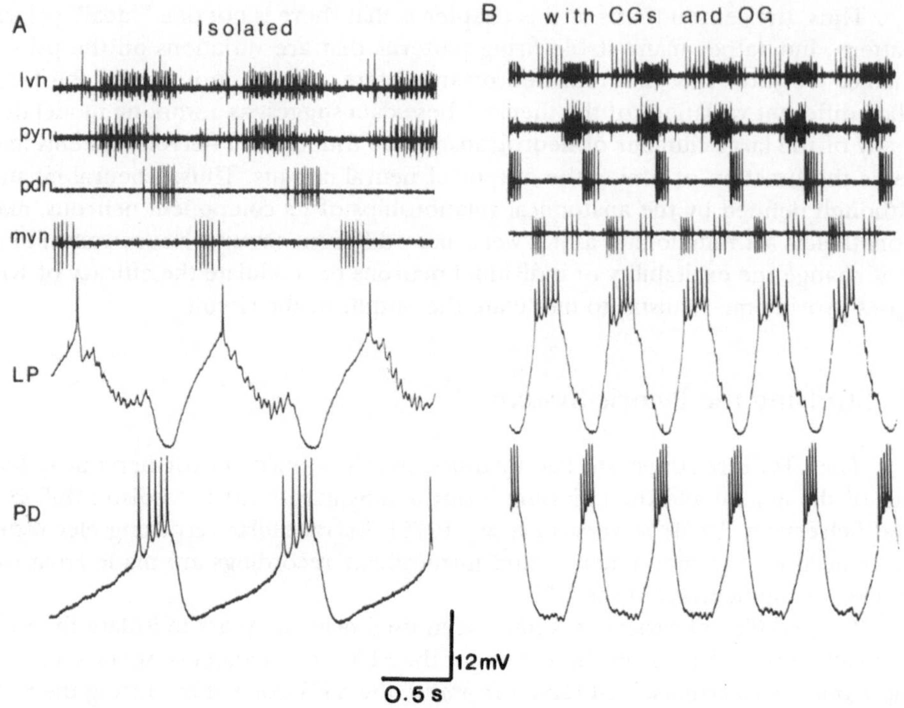

FIGURE 2. The pyloric motor pattern. (A) Pyloric motor pattern produced by the STG of the crab *C. irroratus* isolated from the anterior ganglia by sucrose block. (B) Activity of the same ganglion after removal of sucrose block. The STG was dissected and prepared for intracellular and extracellular recording as described in Selverston *et al.* (1976). In both panels the top four traces are extracellular recordings made with bipolar stainless steel pin electrodes placed on the nerves and insulated from the bath with vaseline. In both panels the bottom two traces are intracellular recordings made with glass microelectrodes (20–40 MΩ, filled with 2.5 M KCl). Lateral pyloric (LP) neuron activity is seen as the largest unit on the lvn and in the first intracellular recording. Pyloric dilator (PD) neuron activity is seen on the pdn, on the lvn, and as the second intracellular recording. Pyloric neuron (PY) activity is seen on the pyn. The mvn carries the axon of the ventricular dilator (VD) neuron. The preparation was continuously superfused with chilled (12°C) *C. irroratus* saline (mM/1): NaCl, 440; KCl, 11; $MgCl_2$, 26; $CaCl_2$, 13; Tris base, 11; maleic acid, 5.2; pH 7.6. In this and all following figures, the most hyperpolarized point of the membrane potential excursion shown in the intracellular recordings will be referred to as E_{max}. In A, E_{max}: LP, −70 mV; PD, −78 mV. In B, E_{max}: LP, −65 mV; PD, −81 mV.

studied. If the STG is left attached to the paired commissural ganglia (CGs) and the single esophageal ganglion (OG) (Fig. 1), the resultant pyloric motor pattern is increased in frequency and in intensity (Russell, 1976, 1979; Russell and Hartline, 1982; Miller and Selverston, 1982a,b; Eisen and Marder, 1982), as is illustrated for *C. irroratus* in Fig. 2B.

What does the pyloric rhythm look like in intact animals, and how closely do the recordings produced by preparations in the dish mimic normal activity in the animal? Recordings from intact and behaving animals show a great deal more cycle by cycle variation in frequency and phase relationships than is seen in recordings from isolated ganglia (Rezer and Moulins, 1983). Additionally, these recordings show clearly that the pyloric rhythm in the intact animal operates in several different long-lasting modes of quite different intensity and frequency (Rezer and Moulins, 1983). Thus, presumably in the animal, activity of inputs to the STG that travel in the STN, at least some of which are likely to originate in the CG, are continuously modulating the output of the STG.

3. Explaining the "Ideal" Pyloric Pattern

Historically the first attempts to understand how the approximately 14 neurons of the pyloric system generate the pyloric motor pattern consisted of identifying the neurons, specifying the synaptic connections among these neurons (Maynard, 1972; Maynard and Selverston, 1975), trying to establish which of these neurons could be considered intrinsically bursting neurons (Selverston and Miller, 1980; Miller and Selverston, 1982a), and determining the neurotransmitters released by these neurons (Marder, 1974, 1976; Lingle, 1980; Eisen and Marder, 1982; Marder and Eisen, 1984a). Briefly, the pyloric pacemaker consists of the single glutamatergic anterior burster (AB) interneuron that is rhythmically active and the two cholinergic PD motor neurons to which the AB neuron is electrically coupled. These neurons synchronously depolarize and together inhibit the glutamatergic LP and PY constrictor neurons, which are then constrained to fire out of phase with the pacemaker group (Fig. 2). The mutually inhibitory synaptic connections between the LP neuron and the PY neurons force them to fire out of phase with each other as well. The pattern of synaptic connections for the pyloric system of *P. interruptus* is illustrated in Fig. 3. [Many of these synaptic connections are likely to be the same in the other species, but it should be stressed that the connectivity among the STG neurons in the other species we use has not been studied as extensively nor have the synaptic connections been verified by using the Lucifer Yellow photoinactivation technique (Miller and Selverston, 1979; Eisen and Marder, 1982), and that not all features of this circuit can be assumed to be correct in all species.]

Much, however, remains to be explained. (1) The pyloric rhythm operates in a large frequency range (from less than 0.5 Hz to more than 2.5 Hz). How is the frequency controlled? A number of neurotransmitters increase the frequency of bursting of isolated AB neurons (Marder and Eisen, 1984b), but much further work is needed to elucidate completely how pyloric cycle frequency is controlled. (2) Although the LP, PY, and PD neurons usually fire in the same sequence, the exact time in the pyloric cycle that the LP and PY neurons start and finish firing can vary (Eisen and Marder, 1984). One mechanism that accounts for shifts in the firing phase of the PY neurons has been proposed (Eisen and Marder, 1984), but this is only a start in understanding how the firing phase of all the pyloric neurons is controlled.

To date, there is evidence suggesting that dopamine, octopamine, histamine, acetylcholine, serotonin, a proctolinlike peptide, and a FMRFamidelike peptide are present in somata in the CGs or OG, in fibers in the STN, and in the neuropil of

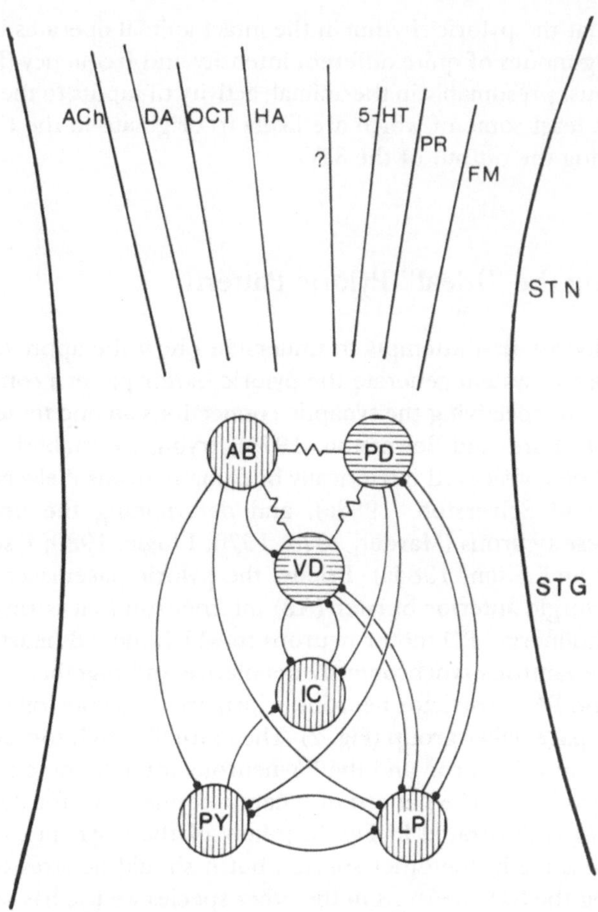

FIGURE 3. Neurotransmitter organization of the stomatogastric system. The synaptic connectivity of the pyloric system of the lobster *P. interruptus* is drawn in the bottom part of the figure. Resistor symbols indicate electrical junctions. Filled circles indicate chemical inhibitory synapses. Neurons shown with vertical lines are glutamatergic, while neurons shown with horizontal lines are cholinergic (Marder and Eisen, 1984a). Transmitters found in inputs traveling in the STN are shown on the top of the figure. ACh, acetylcholine; DA, dopamine; OCT, octopamine; HA, histamine; 5-HT, serotonin; PR, proctolin; FM, FMRFamide.

the STG (Fig. 3). In this chapter we will demonstrate that all of these neurotransmitters influence the overall frequency of the pyloric rhythm. Some of them appear to affect directly the time of firing of the LP and PY neurons in characteristic and different ways. A comparison of three agents that increase the overall frequency of the pyloric rhythm shows that they do so while eliciting different pyloric motor patterns. This demonstrates unambiguously that modulation of frequency and the phase of the pyloric rhythm can occur independently (see also Eisen and Marder, 1984).

4. Neurotransmitters Present in Inputs to the STG and Physiological Action of Input Neurotransmitters on STG Neurons and Pyloric Output

4.1. Dopamine and Octopamine

The amines, dopamine and octopamine, were the first of the neurotransmitters present in STN fibers to be intensively studied. Using the Falck-Hillarp technique, Kushner and Maynard (1977) visualized dopamine-containing neurons, fibers, and neuropilar processes in *P. interruptus* and *H. americanus.* In addition to the large dopamine-containing neuron in the CGs (Kushner and Barker, 1983), previously described by Goldstone and Cooke (1971), that projects away from the STG, Kushner and Maynard (1977) described fluorescence, attributed to dopamine, in several small somata in the CG, in fibers in the superior esophageal nerves (SONs) and STN, and in neuropilar processes in the STG. Biochemical experiments (Barker *et al.,* 1979; Kushner and Barker, 1983) demonstrated that the CGs, STG, and STN synthesized and accumulated [^{3}H]dopamine and [^{3}H]octopamine after incubation with [^{3}H]tyrosine in ganglia dissected from *P. interruptus.* For some reason, possibly because these data were the first to provide any information on the characterization of the input transmitters, these data left many researchers with the notion that the only modulatory control of the STG was the dopaminergic system, which we now know is certainly not the case.

The physiological action of dopamine on the STG of *P. interruptus,* although studied extensively (Anderson and Barker, 1981; Raper, 1979; Marder and Eisen, 1984b; Eisen and Marder, 1984, Flamm and Harris-Warrick, 1984; Harris-Warrick and Flamm, 1984) is far from completely understood. The action of dopamine on the STG is complex and varies during the time of application (Fig. 4), which has caused some confusion in the literature. Immediately after the application of 10^{-4} M dopamine to an STG showing pyloric activity, the PD neurons hyperpolarize and the frequency of the pyloric cycle decreases (Fig. 4B). In some cases, pyloric cycling ceases completely immediately following dopamine application. A second effect of dopamine is seen after a delay of 5–10 min or after washing with normal saline (Fig. 4C). This second effect is characterized by an increased frequency and intensity of pyloric bursting and lasts for about 10–20 min following wash with normal saline. Marder and Eisen (1984b) showed that dopamine hyperpolarized and inhibited isolated PD neurons, but enhanced the intrinsic burstiness of isolated AB neurons. Flamm and Harris-Warrick (1984) extended this study and found that 10 min dopamine applications also excited isolated LP and PY neurons. Therefore, it is likely that early after dopamine applications the action on the PD neurons is most prominent, and only several minutes later do the excitatory effects on other neurons become pronounced enough to overcome the inhibition of the PD neurons. When dopamine is applied to a slowly cycling or quiescent preparation of *P. interruptus,* it will initiate robust pyloric activity after a delay of 5–10 min, presumably due to its strong excitatory action on the AB neuron (Marder and Eisen, 1984b).

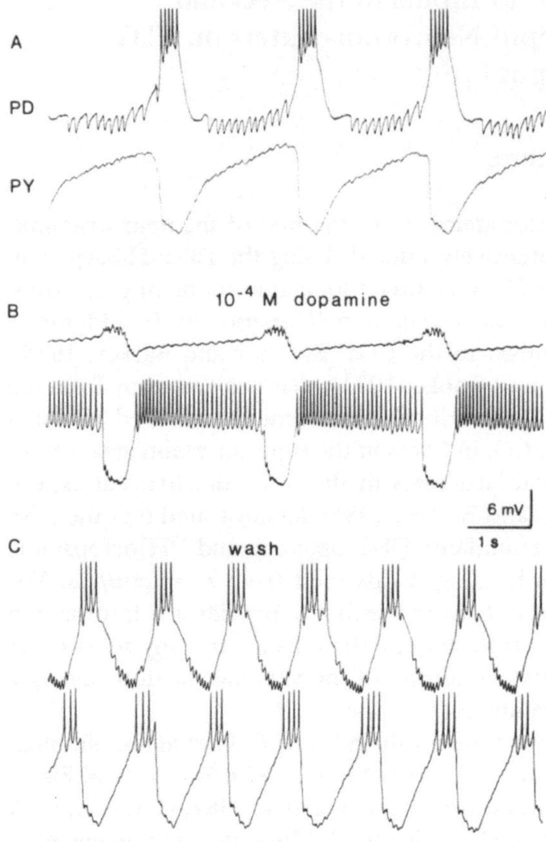

FIGURE 4. The effects of dopamine on the pyloric motor pattern of *P. interruptus*. (A) In control saline. (B) 2 min after application of 10^{-4} M dopamine (3-hydroxy-tyramine, Sigma Chemical Co.). (C) 5 min after washing with control saline. In this and in all subsequent experiments, substances were bath applied by dissolving them in the appropriate saline immediately before use and then introducing them into the continuous superfusion system. *P. interruptus* saline (mM/1): NaCl, 479; KCl, 12.8; $CaCl_2$, 13.7; Na_2SO_4, 3.9; $MgSO_4$, 10; Tris base, 11; maleic acid, 4.8; pH 7.5. E_{max} in A: PD, −61 mV; PY, −54 mV. E_{max} in B: PD, −66 mV; PY, −44 mV. E_{max} in C: PD, −63 mV; PY, −50 mV.

Dopamine has very different effects when applied to the STG of the crab *C. irroratus*. In the presence of 10^{-4} M dopamine, the neurons of the isolated crab STG fire tonically and all normal pyloric rhythmicity is lost. At lower concentrations (10^{-8} M), dopamine has acceleratory and excitatory effects similar to the delayed actions of dopamine in the lobster.

The application of 10^{-4} M octopamine to the STG of *P. interruptus* also produces effects that change during the time of application (Fig. 5). Soon after octopamine application to an isolated STG, the frequency of the pyloric rhythm increases, and the PY neurons start firing (Fig. 5; 2 min). After 5–10 min of octopamine application (Fig. 5; 7 min), the LP neuron fires in long bursts that interrupt normal pyloric activity due to the strong inhibition of the PD neurons by the LP neuron. More detailed studies of the actions of octopamine are being carried out by Harris-Warrick and his colleagues (Harris-Warrick, and Flamm, 1984; Flamm and Harris, Warrick, 1984).

In the crab, the effects of 10^{-4} M octopamine are much less pronounced, but also result in an increase in pyloric frequency and in increased LP neuron firing. In both species these effects are reversible following extensive washing in control saline.

Dopamine and octopamine are very closely related in structure, but a comparison of Figs. 4 and 5 demonstrates clearly that they have very different physiological actions and most likely act via different receptor systems.

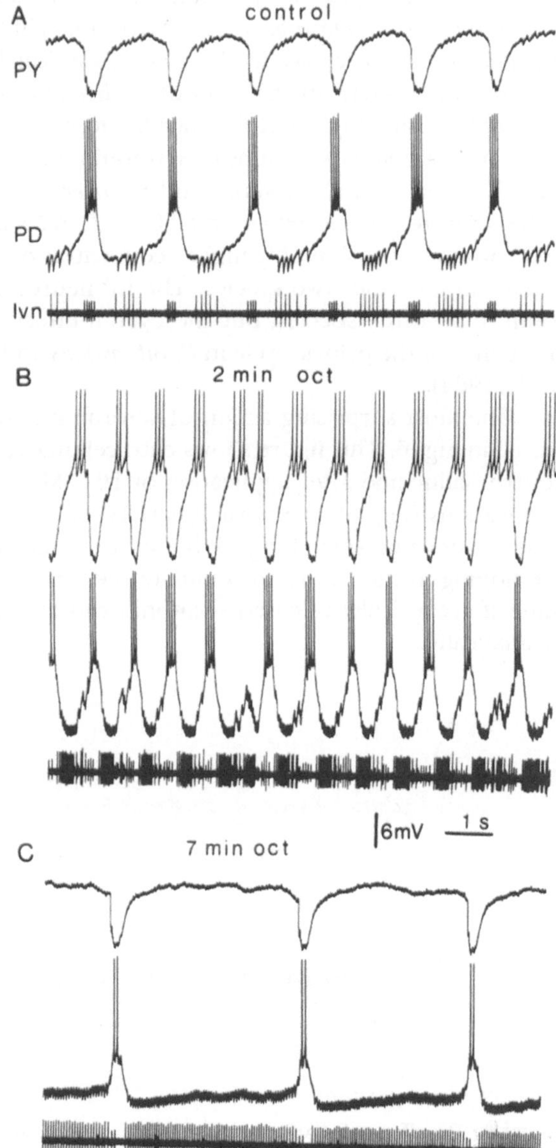

FIGURE 5. The effects of octopamine on the pyloric motor pattern of *P. interruptus*. (A) In control saline. (B) 2 min in 10^{-4} M octopamine (Sigma). (C) 7 min in octopamine. E_{max} was the same in all panels: PD, −61 mV; PY, −54 mV.

4.2. Serotonin

The recent advent in immunohistochemical techniques for the localization of neurotransmitters made possible our description of the distribution of serotonin-containing structures in the stomatogastric system of three decapod crustaceans (Beltz *et al.*, 1984). This study demonstrates clearly that all species of decapods are not identical with regard to their serotonergic pathways. Using an antiserum directed against serotonin, we stained somata in the CGs of three species of crustaceans, *P. interruptus, C. irroratus,* and *H. americanus.* In *C. irroratus* and *H. americanus,* but not *P. interruptus,* we stained fibers in the STN and found densely staining neuropilar processes in the STG. Additionally, we verified that endogenous serotonin is present in the STG of *C. irroratus* and *H. americanus,* but not in *P. interruptus,* using high-performance liquid chromatography (HPLC).

The physiological actions of serotonin on the pyloric activity of the three species, *P. interruptus, C. irroratus,* and *H. americanus* were also different (Beltz *et al.*, 1984). Serotonin was active on STG of *P. interruptus* at concentrations as low as 10^{-9}, whereas considerably higher concentrations were required for physiological activity in the other two species. The LP neuron appeared to be sensitive to serotonin in all three species, but the overall effect of serotonin applications on the frequency of the pyloric cycle in *P. interruptus* and *H. americanus* was variable (Beltz *et al.*, 1984).

The most surprising action of serotonin is seen in the crab, and this is illustrated in Fig. 6. This figure shows extracellular recordings of the pyloric activity in control saline and after application of 10^{-4} M serotonin. Subsequent to serotonin applications (Fig. 6), the normal pattern of LP, PD alternation was replaced by a stable pattern in which long LP bursts are interspersed with several PD bursts. We are hoping in the future to stimulate the serotonergic pathways directly to determine if synaptically released serotonin can also transform the pyloric pattern as dramatically.

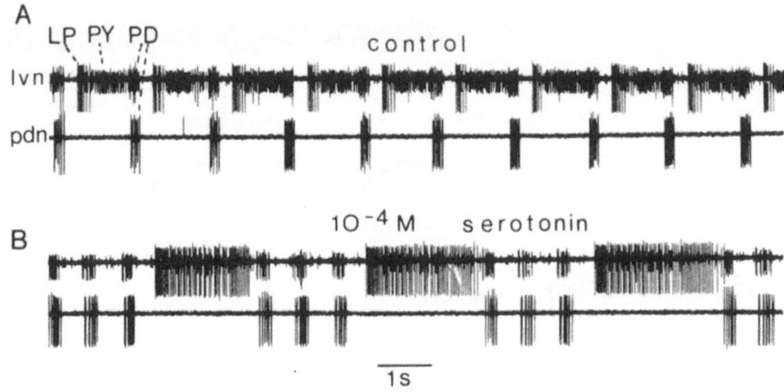

FIGURE 6. The effects of serotonin on the pyloric motor pattern in *C. irroratus.* (A) Control. (B) 10 min in 10^{-4} M serotonin.

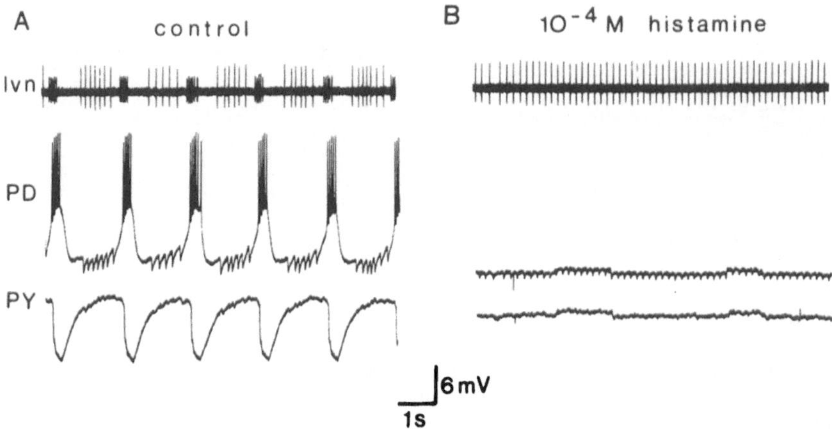

FIGURE 7. The effects of histamine on the pyloric motor pattern of *P. interruptus.* (A) Control, E_{max}: PD, −58 mV; PY, −54 mV. (B) 5 min in 10^{-4} M histamine dihydrochloride (Sigma). The large unit firing on the lvn is the LP neuron. LP neuron-evoked inhibitory postsynaptic potentials (IPSPs) are seen in the PD neuron as well. E_{max}: PD, −60 mV; PY, −47 mV.

4.3. Histamine

The inferior ventricular nerve (IVN) connects the brain to the OG (Fig. 1). It has long been known that stimulation of the IVN disrupts and then enhances activity of pyloric motor neurons (Dando and Selverston, 1972; Sigvardt and Mulloney, 1982a,b). Recently Claiborne and Selverston (1984a,b) have found two somata in the brain of *P. interruptus* that contain histamine. These neurons have axons in the IVN and in the STN. Direct electrical stimulation of these somata elicits the same action on neurons of the STG as does stimulation of the IVN.

Claiborne and Selverston (1984a,b) demonstrated that iontophoretically or bath-applied histamine inhibits many of the neurons of the STG of *P. interruptus,* including the PD neurons. The recordings in Fig. 7 show that bath application of histamine completely inhibited rhythmic pyloric activity, and only the LP neuron continued firing. The inhibitory action of histamine mimics only one component of the synaptic potential evoked by stimulation of the IV neurons, and there is a possibility that the IV neurons contain an additional neurotransmitter.

4.4. Acetylcholine

The pyloric rhythm in all species is excited by muscarinic cholinergic agonists such as oxotremorine and pilocarpine. When applied to slowly cycling preparations, pilocarpine increases pyloric cycle frequency. The most exciting action of the muscarinic agonists is seen, however, when they are applied to quiescent preparations (Marder and Paupardin-Tritsch, 1978; Anderson, 1980; Marder and Eisen, 1984b; Nagy *et al.,* 1984). As seen in Fig. 8, pilocarpine initiates pyloric

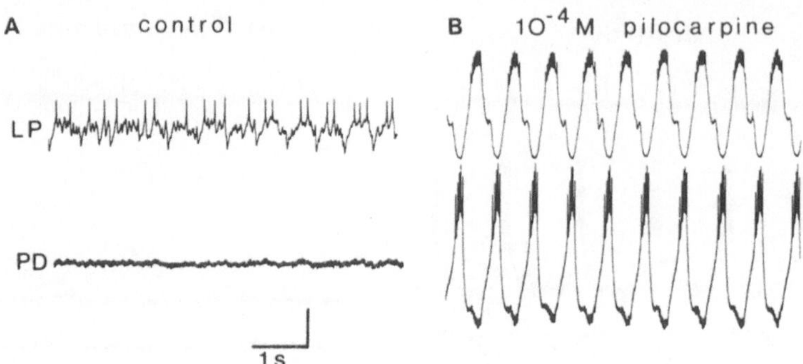

FIGURE 8. Pilocarpine initiation of pyloric activity in a quiescent ganglion from *C. irroratus*. (A) Control. E_{max}: PD, −53 mV; LP, −51 mV. (B) 5 min in 10^{-4} M pilocarpine (Sigma). E_{max}: PD, −53 mV; LP, −54 mV.

cycling and elicits slow membrane potential oscillations that can be recorded in PD neurons (Marder and Paupardin-Tritsch, 1978) and AB neurons (Marder and Eisen, 1984b).

The physiological significance of this response to muscarinic agonists is demonstrated by the discovery by Nagy and Dickinson (1983) of an identified neuron, the anterior pyloric modulator (APM), whose action is similar to that of the muscarinic agonists. The APM has been found in the OG of several species of crustaceans, *Jasus lalandii, H. gammarus*, and *Palinurus vulgaris*. The APM modulates the activity of many pyloric neurons by enhancing the ability of these neurons to produce "plateau" potentials. The action of the APM is mimicked by pilocarpine and oxotremorine and blocked by atropine (a muscarinic cholinergic antagonist) (Dickinson and Nagy, 1983). Thus these data are consistent with the interpretation that the APM is a cholinergic neuron, although direct biochemical evidence is still lacking.

4.5. FMRFamidelike Peptide

FMRFamide (molluscan cardioexcitatory peptide) also has excitatory actions on the pyloric rhythm in both *P. interruptus* and *C. irroratus*. As with pilocarpine, FMRFamide applications can initiate pyloric cycling in quiescent preparations; as with pilocarpine, it increases the frequency of slowly cycling preparations. Unlike pilocarpine, FMRFamide increases the firing of PY neurons (Fig. 9). In the crab, this increased PY neuron firing is often sufficient to inhibit the firing of the LP neuron (see also Fig. 12).

The physiological relevance of these FMRFamide actions is suggested by our discovery of FMRFamidelike immunoreactivity in the stomatogastric nervous systems of *P. interruptus* and *C. irroratus*. Using antibodies raised against FMRFamide, we have stained neurons and a neuropil in the CG and fibers in the commissures in both *P. interruptus* and *C. irroratus* (Marder *et al.*, 1984). In *C. irroratus* there are several somata in the OG that show FMRFamidelike staining and fibers in the STN that terminate in an intensely staining neuropil in the STG (Fig. 10). The staining in *P. interruptus* is much less intense than that found in the crab, but staining neuropils are found in the OG and the STG as well. We do not yet know whether the immunoreactivity is due to the presence of FMRFamide or a FMRFamidelike peptide, but if it is due to the presence of a FMRFamidelike peptide, it is likely that exogenously applied FMRFamide can activate the receptors for that peptide on neurons of the STG.

4.6. Proctolinlike Peptide

Another peptide, proctolin, appears to be present in fibers entering the STG. We have recently carried out a series of immunohistochemical studies with an antibody raised against the peptide proctolin by the Kravitz laboratory. We find somata that show proctolinlike staining in the CGs of both *P. interruptus* and *C. irroratus*, as well as in fibers in the SONs, inferior esophageal nerves (IONs), and STN. In *C. irroratus* several neurons in the OG also stain. In both species, the neuropil of the

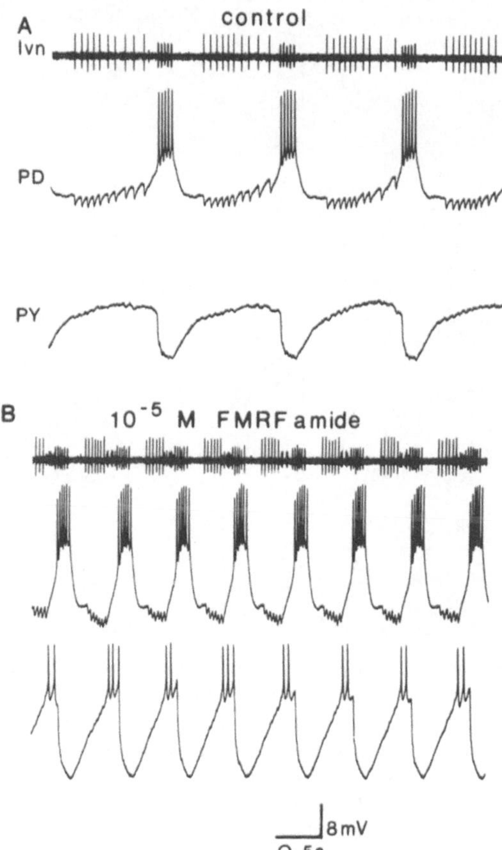

FIGURE 9. The effects of FMRFamide on pyloric activity in *P. interruptus*. (A) Control. (B) In the presence of FMRFamide (Peninsula Laboratories). E_{max} for both panels: PD, −57 mV; PY, −49 mV.

STG shows widespread staining (Marder, *et al.*, 1984). Experiments with HPLC suggest that the proctolinlike immunoreactivity we have visualized is due to native proctolin (Marder *et al.*, 1984).

Like pilocarpine and FMRFamide, proctolin can initiate pyloric cycling in quiescent preparations (Hooper and Marder, 1984a,b) or increase the frequency of pyloric cycling when applied to slowly cycling preparations from either *C. irroratus* or *P. interruptus.* Figure 11A shows the results of applying 10^{-7} M proctolin to a STG from *P. interruptus.* Before the proctolin application, the frequency of

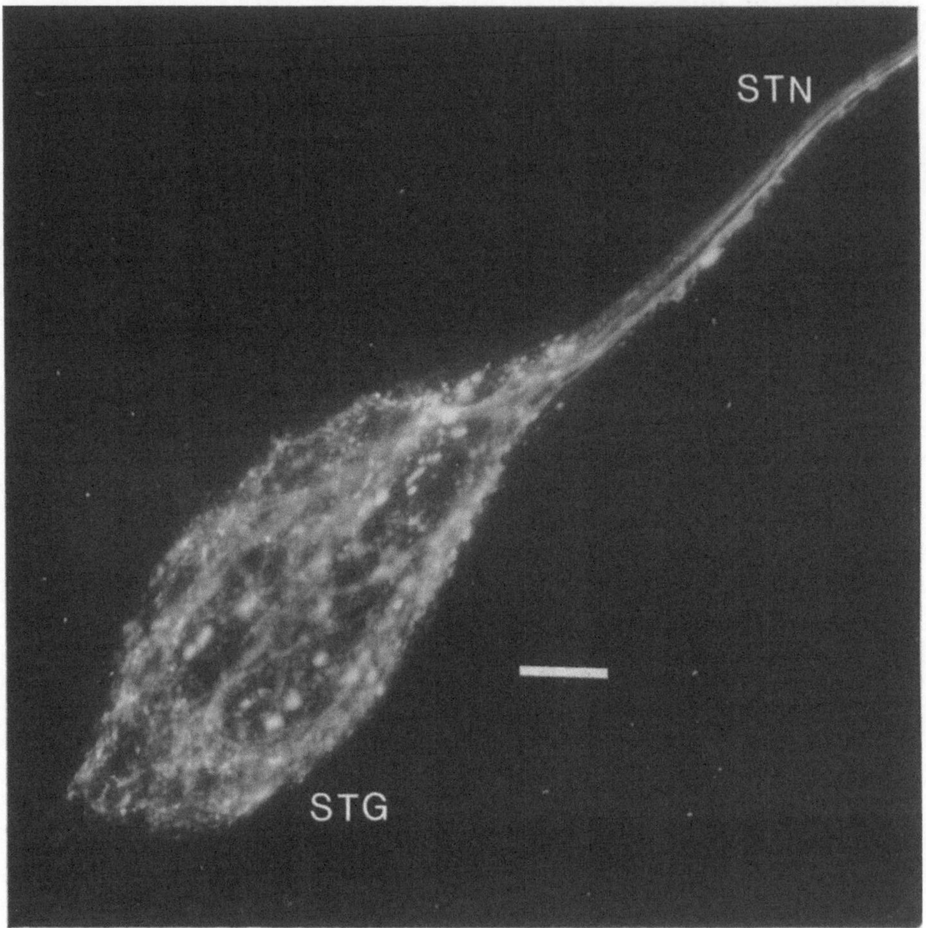

FIGURE 10. FMRFamidelike immunoreactivity in the STG of *C. irroratus.* Immunoreactivity is seen in several fibers in the STN and in processes that ramify throughout the neuropil. The somata of the STG neurons are completely unstained and are therefore entirely invisible in this picture (they surround the neuropil). Ganglion was processed as a whole mount as described in Beltz *et al.* (1984) and Hooper and Marder (1984b). In brief, the ganglion was fixed in 4% paraformaldehyde, rinsed, reacted with the primary antibody (rabbit anti-FMRFamide, 1:300, Cambridge Research Biochemicals), rinsed, reacted with a fluorescein-labeled secondary antibody (Boehringer-Manneheim goat anti-rabbit), and then mounted in 80% glycerin, 20% carbonate. Scale bar: 50 μm.

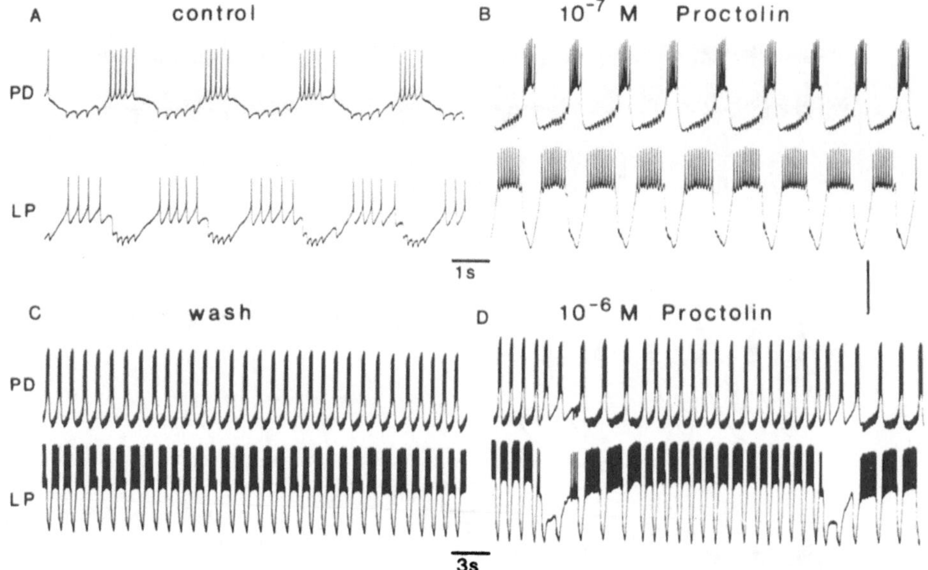

FIGURE 11. The effects of proctolin on the motor patterns in *P. interruptus.* (A) Control saline. E_{max}: PD, −48 mV; LP, −52 mV. (B) In 10^{-7} M proctolin (Bachem, Inc.). E_{max}: PD, −50 mV, LP, −55 mV. (C) 30 min wash in control saline, frequency remained elevated. E_{max}: PD, −50 mV; LP, −56 mV. (D) In 10^{-6} M proctolin. Gastric system neurons are activated, as can be seen by periodic interruption of normal pyloric activity. E_{max}: PD, −47 mV; LP, −53 mV.

the pyloric rhythm was 0.41 Hz. After the application of the 10^{-7} M proctolin, the LP neuron was more active and the frequency of the pyloric rhythm was 0.87 Hz. When 10^{-6} M proctolin was applied subsequently, it stimulated activity of the gastric system of the ganglion, which periodically interrupted the normal pattern of pyloric activity (Fig. 11B). The effects of proctolin are incompletely reversible in that formerly quiescent ganglia often continue to cycle after washing for several hours. The action of proctolin on STG of the crab (Fig. 12) is similar to that found in *P. interruptus,* except that the gastric system is seldom sufficiently activated to interfere with the pyloric pattern.

5. Conclusions

Although at first glance the physiological actions of a number of the substances studied in this paper appear similar, in fact, each of these substances produces a different and characteristic effect on the pyloric motor pattern. This is illustrated in the recordings shown in Fig. 12. In previous sections it was shown that pilocarpine, proctolin, and FMRFamide all increase the frequency of the pyloric rhythm. However, a direct comparison of the action of these three substances on the same preparation demonstrates that they evoke pyloric motor patterns with different phase relations. Pilocarpine applications increase the fre-

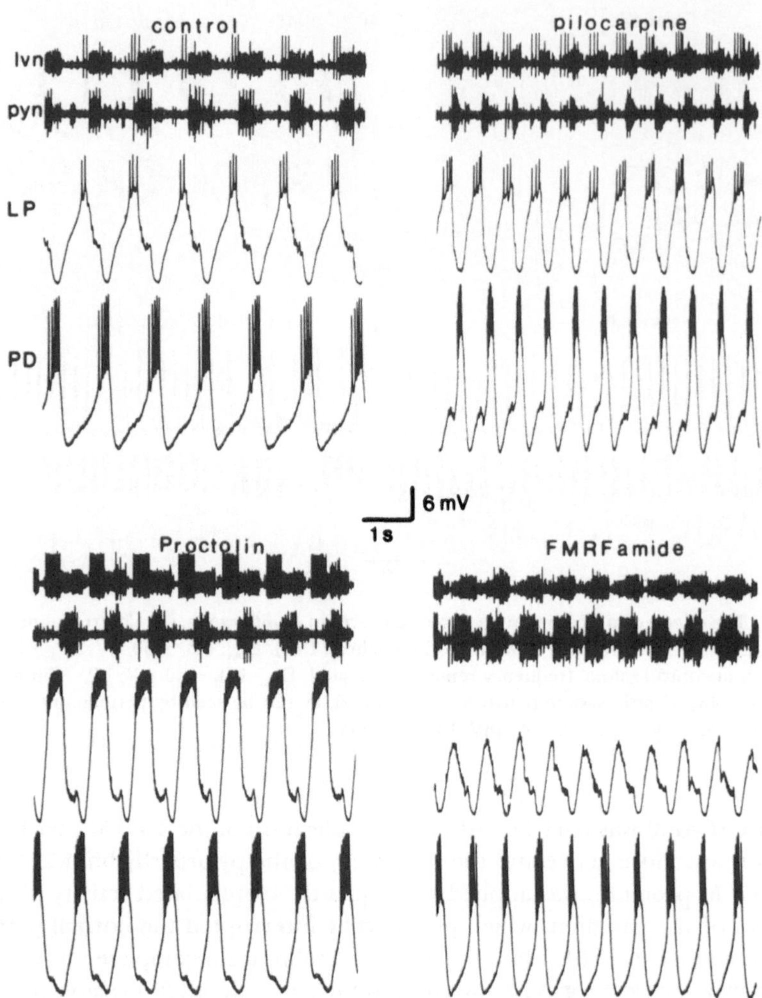

FIGURE 12. Comparison of the effects of pilocarpine, proctolin, and FMRFamide on pyloric activity of *C. irroratus*. Pilocarpine (10^{-4} M) induced an increase in frequency. Proctolin (10^{-6} M) results in increased LP neuron firing. FMRFamide (10^{-6} M) results in increased PY neuron firing and decreased LP neuron firing. E_{max}: (control and FMRFamide) PD, −65 mV; LP, −74 mV; (pilocarpine): PD, −57 mV; LP, −69 mV; (proctolin): PD, −63 mV; LP, −70 mV.

quency of the pyloric rhythm, but the firing pattern of the LP, PY, and PD neurons is quite similar to that seen in the control. FMRFamide also increases the frequency of the pyloric rhythm, but in this case the PY neurons are excited, the LP neuron is therefore inhibited, and no longer fires action potentials. Proctolin again increases the frequency of the pyloric rhythm, but elicits plateaulike activity in the LP neuron, resulting in a pyloric motor pattern with a dominant LP component. These data indicate very clearly that it is possible to vary the relative phasing of neurons in the pyloric pattern independent of changes in frequency, as has also been suggested before (Eisen and Marder, 1984).

Thus, it now appears that different neurotransmitters may be responsible for producing variations in the pyloric motor pattern. It is important to bear in mind that the neurotransmitters discussed in this manuscript most certainly represent only a partial listing of what will be eventually found in fibers traveling in the STN. It is to be expected that as more antisera are developed against neurotransmitters and peptides a more extensive catalog of the neurotransmitters present in all of the inputs to the STG will become possible. It is also important to bear in mind that not all of the substances indicated in Fig. 3 are necessarily found in the STN of any one species of decapod. For example, although serotonin is physiologically active on STG of *P. interruptus,* we were unable to demonstrate its presence in the neuropil of the STG in *P. interruptus,* and it is likely to serve a hormonal role in this animal (Beltz *et al.,* 1984). Pilocarpine, FMRFamide, proctolin, and histamine have similar actions on the STGs of *C. irroratus* and *P. interruptus.* There, however, appear to be differences in the actions of dopamine, serotonin, and octopamine in various species that may also correlate with the differences in amine synthesis in neurosecretory structures in these animals (Cooke and Sullivan, 1982). These differences may point to shifts in the use of these amines as neuronally released or hormonally released modulators of the STG in the different animals.

Although we have demonstrated the capability of the neurons of the pyloric system to produce a number of different stable motor patterns when different neurotransmitters are bath applied, is this physiologically relevant, or do these data demonstrate only the presence of receptors on neurons of the STG that respond to these substances? The ultimate answer to this question must await the time when we can stimulate the neurons in the CGs, OG, and elsewhere that project into the STG and contain these neurotransmitters. It will then be possible to determine directly whether bath application of these substances mimics the action of the neurons that contain these substances. If some or many of the inputs to the STG release their neurotransmitter at any distance from the postsynaptic targets, as is likely, then bath application of exogenous neurotransmitter may, in fact, mimic the physiological effects of the neurons themselves.

In summary, it appears that the pyloric system of decapod crustaceans can generate a range of motor patterns. Although the pattern generator is composed of very few neurons, different input neurotransmitters may be used in order to produce motor patterns of different frequencies and different firing patterns. In other preparations as well, different neurotransmitters may act on a single neuronal circuit to call from the circuit a variety of outputs. Many synapses may act by releasing neurotransmitter at some distance from their final sites of action; it is possible that a large number of neurotransmitters makes possible considerable but precise modulation of function.

ACKNOWLEDGMENTS. We thank Dr. Judith S. Eisen for collaboration during the years that led up to this work and for considerable intellectual contributions to this study. We thank E. A. Kravitz and K. K. Siwicki for the gift of the anti-proctolin antibody and Michael O'Neil for photographic assistance. This research was supported by NIH NS-17813, Biomedical Support Grant RR07044 to Brandeis University, and a Sloan fellowship to E.M.

References

Anderson, W. W., 1980, Synaptic mechanisms generating nonspiking network oscillations in the stomatogastric ganglion of the lobster, *Panulirus interruptus*, PhD. dissertation, University of Oregon.

Anderson, W. W., and Barker, D. L., 1981, Synaptic mechanisms that generate network oscillations in the absence of discrete postsynaptic potentials, *J. Exp. Zool.* **216:**187–191.

Barker, D. L., Kushner, P. D., and Hooper, N. K., 1979, Synthesis of dopamine and octopamine in the crustacean stomatogastric nervous system, *Brain Res.* **161:**99–113.

Beltz, B. S., Eisen, J. S., Flamm, R., Harris-Warrick, R. M., Hooper, S. L., and Marder, E., 1984, Serotonergic innervation and modulation of the stomatogastric ganglion of three decapod crustaceans (*Homarus americanus, Cancer irroratus* and *Panulirus interruptus*), *J. Exp. Biol.* **109:**35–54.

Claiborne, B. J., and Selverston, A. I., 1984a, Histamine as a neurotransmitter in the stomatogastric nervous system of the spiny lobster, *J. Neurosci.* **4:**708–721.

Claiborne, B. J., and Selverston, A. I., 1984b, Localization of stomatogastric IV neuron cell bodies in lobster brain, *J. Comp. Physiol.* **154:**27–32.

Cooke, I. M., and Sullivan, R. E., 1982, Hormones and neurosecretion, in: *The Biology of Crustacea,* Volume 3 (H. L. Atwood and D. C. Sandeman, eds.), Academic Press, New York, pp. 205–290.

Dando, M. R., and Selverston, A. I., 1972, Command fibers from the supraesophageal to the stomatogastric ganglion in *Panulirus argus*, *J. Comp. Physiol.* **78:**138–175.

Dickinson, P. S., and Nagy, F., 1983, Control of a central pattern generator by an identified modulatory interneurone in *Crustacea*. II. Induction and modification of plateau properties in pyloric neurones, *J. Exp. Biol.* **105:**59–82.

Eisen, J. S., and Marder, E., 1982, Mechanisms underlying pattern generation in lobster stomatogastric ganglion as determined by selective inactivation of identified neurons. III. Synaptic connections of electrically coupled pyloric neurons, *J. Neurophysiol.* **48:**1392–1415.

Eisen, J. S., and Marder, E., 1984, A mechanism for the production of phase shifts in a pattern generator, *J. Neurophysiol.* **51:**1374–1394.

Flamm, R. E. and Harris-Warrick, R. M., 1984, Neuronal targets of dopamine, octopamine, and serotonin in the pyloric central pattern generator of the stomatogastric ganglion of the lobster, *Panulirus interruptus, Soc. Neurosci. Abst.* **10:**149.

Goldstone, M., and Cooke, I., 1971, Histochemical localization of monoamines in the crab central nervous system, *Z. Zellforsch. Mikrosk. Anat.* **116:**7–19.

Harris-Warrick, R. M. and Flamm, R. E., 1984, Aminergic modulation of the pyloric rhythm in the stomatogastric ganglion of *Panulirus interruptus, Soc. Neurosci. Abst.* **10:**149.

Hartline, D. K., 1979, Pattern generation in the lobster *Panulirus* stomatogastric ganglion. II. Pyloric network simulation, *Biol. Cybern.* **33:**223–236.

Hartline, D. K., and Gassie, D. V., 1979, Pattern generation in the lobster *Panulirus* stomatogastric ganglion. I. Pyloric neuron kinetics and synaptic interactions, *Biol. Cybern.* **33:**209–222.

Hooper, S. L. and Marder, E., 1984a, The physiological effects of proctolin and FMRFamide on the stomatogastric ganglion of *Panulirus interruptus* and *Cancer irroratus, Soc. Neurosci. Abst.* **10:**148.

Hooper, S. L., and Marder, E., 1984b, Modulation of a central pattern generator by two neuropeptides, proctolin and FMRFamide *Brain Res.* **305:**186–191.

King, D. G., 1976, Organization of crustacean neuropil. I. Patterns of synaptic connections in lobster stomatogastric ganglion *J. Neurocytol.* **5:**207–237.

Kushner, P. D., and Barker D. L., 1983, A neurochemical description of the dopaminergic innervation of the stomatogastric ganglion of the spiny lobster, *J. Neurobiol.* **14:**17–28.

Kushner, P. D., and Maynard, E. A., 1977, Localization of monoamine fluorescence in the stomatogastric nervous system of lobsters, *Brain Res.* **129:**13–28.

Lingle, C. J., 1980, Sensitivity of decapod foregut muscles to acetylcholine and glutamate, *J. Comp. Physiol.* **138:**187–199.

Marder, E., 1974, Acetylcholine as an excitatory neuromuscular transmitter in the stomatogastric system of the lobster, *Nature (London)* **251:**730–731.

Marder, E., 1976, Cholinergic motor neurones in the stomatogastric system of the lobster, *J. Physiol.* **257:**63–86.

Marder, E., and Eisen, J. S., 1984a, Transmitter identification of pyloric neurons: Electrically coupled neurons use different transmitters, *J. Neurophysiol.* **51:**1345–1361.

Marder, E., and Eisen, J. S., 1984b, Electrically coupled pacemaker neurons respond differently to the same physiological inputs and neurotransmitters, *J. Neurophysiol.* **51:**1362–1373.

Marder, E., and Paupardin-Tritsch, D., 1978, The pharmacological properties of some crustacean neuronal acetylcholine, γ-aminobutyric acid, and l-glutamate responses, *J. Physiol.* **280:**213–236.

Marder, E., Hooper, S. L., and Siwicki, K. K., 1984, Distribution of Proctolin-like and FMRF-like immunoreactivity in the stomatogastric system of decapod crustacea, *Soc. Neurosci. Abstr.* **10:**688.

Maynard, D. M., 1972, Simpler networks, *Ann. N.Y. Acad. Sci.* **193:**59–72.

Maynard, D. M., and Selverston, A. I., 1975, Organization of the stomatogastric ganglion of the spiny lobster. IV. The pyloric system, *J. Comp. Physiol.* **100:**161–182.

Maynard, E. A., 1971, Electron microscopy of stomatogastric ganglion in the lobster, *Homarus americanus, Tissue Cell* **3:**137–160.

Miller, J. P., and Selverston, A. I., 1979, Rapid killing of single neurons by irradiation of intracellularly injected dye, *Science* **206:**702–704.

Miller, J. P., and Selverston, A. I., 1982a, Mechanisms underlying pattern generation in lobster stomatogastric ganglion as determined by selective inactivation of identified neurons. II. Oscillatory properties of pyloric neurons, *J. Neurophysiol.* **48:**1378–1391.

Miller, J. P., and Selverston, A. I., 1982b, Mechanisms underlying pattern generation in lobster stomatogastric ganglion as determined by selective inactivation of identified neurons. IV. Network properties of the pyloric system, *J. Neurophysiol.* **48:**1416–1432.

Moulins, M., and Cournil, I., 1982, All-or-none control of the bursting properties of the pacemaker neurons of the lobster pyloric pattern generator, *J. Neurobiol.* **13:**447–458.

Mulloney, B., and Selverston, A. I., 1974, Organization of the stomatogastric ganglion in the spiny lobster. I. Neurons driving the lateral teeth, *J. Comp. Physiol.* **91:**1–32.

Nagy, F., and Dickinson, P. S., 1983, Control of a central pattern generator by an identified interneurone in crustacea. I. Modulation of the pyloric motor output, *J. Exp. Biol.* **105:**33–58.

Nagy, F., Benson, J. A., and Moulins, M., 1984, Cholinergic activation of burst generating oscillations mediated by opening of Ca^{2+} channels in lobster pyloric neurons, *Soc. Neurosci. Abst.* **10:**148.

Raper, J. A., 1979, Nonimpulse mediated synaptic transmission during the generation of a cyclic motor program, *Science* **205:**304–306.

Rezer, E., and Moulins, M., 1983, Expression of the crustacean pyloric pattern generator in the intact animal, *J. Comp. Physiol.* **153:**17–28.

Russell, D. F., 1976, Rhythmic excitatory inputs to the lobster stomatogastric ganglion, *Brain Res.* **101:**582–588.

Russell, D. F., 1979, CNS Control of pattern generation in the lobster stomatogastric ganglion, *Brain Res.* **177:**598–602.

Russell, D. R., and Hartline, D. K., 1982, Slow active potentials and bursting motor patterns in pyloric network of the lobster, *Panulirus interruptus, J. Neurophysiol.* **48:**914–937.

Selverston, A. I., and Miller, J. P., 1980, Mechanism underlying pattern generation in lobster stomatogastric ganglion as determined by selective inactivation of identified neurons. I. Pyloric system, *J. Neurophysiol.* **44:**1102–1121.

Selverston, A. I., King, D. G., Russell, D. F., and Miller, J. P., 1976, The stomatogastric nervous system: Structure and function of a small neural network, *Prog. Neurobiol.* **7:**215–290.

Sigvardt, K. A., and Mulloney, B., 1982a, Properties of synapses made by IVN command-interneurons in the stomatogastric ganglion of the spiny lobster, *Panulirus interruptus, J. Exp. Biol.* **97:**153–168.

Sigvardt, K. A., and Mulloney, B., 1982b, Properties of synapses made by IVN command-interneurons in the stomatogastric ganglion of the spiny lobster, *Panulirus interruptus, J. Exp. Biol.* **97:**137–152.

18

The Well-Modulated Lobster

The Roles of Serotonin, Octopamine, and Proctolin in the Lobster Nervous System*

EDWARD A. KRAVITZ, BARBARA BELTZ, SILVIO GLUSMAN, MICHAEL GOY, RONALD HARRIS-WARRICK, MICHAEL JOHNSTON, MARGARET LIVINGSTONE, THOMAS SCHWARZ, AND KATHLEEN KING SIWICKI

1. Introduction

When serotonin or octopamine are injected into freely moving lobsters, animals assume static poses that last for prolonged periods of time (up to several hours). With serotonin injection, animals stand in a flexed posture high on the tips of their walking legs with their claws spread apart and slightly open in front of them and their abdomens loosely tucked underneath them. With octopamine injection, animals lie close to the substrate in an extended posture with their walking legs and claws pointed forward and lifted off the substrate and their abdomens gently arching upward (Livingstone *et al*, 1980). Such poses are normally seen in lobster behavior. For example, lobsters assume serotoninlike poses when startled, during

*This is a partially revised and updated version of an article that appeared in the October 1984 (Volume **22**:133–147) issue of the *Journal Pesticide Biochemistry and Physiology*. It is reprinted with permission of the publisher, Academic Press, Inc.

EDWARD A. KRAVITZ, BARBARA BELTZ, SILVIO GLUSMAN, MICHAEL GOY, RONALD HARRIS-WARRICK, MICHAEL JOHNSTON, MARGARET LIVINGSTONE, THOMAS SCHWARZ, AND KATHLEEN KING SIWICKI • Harvard Medical School, Department of Neurobiology, Boston, Massachusetts 02115.

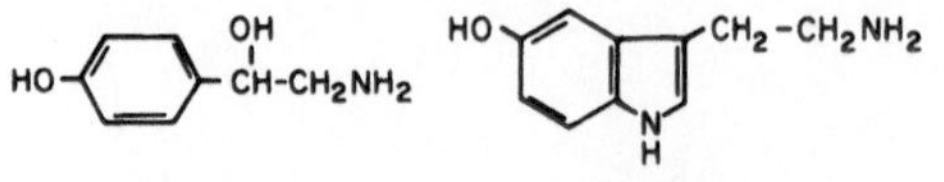

OCTOPAMINE

SEROTONIN

ARG-TYR-LEU-PRO-THR

PROCTOLIN

FIGURE 1. Lobster neurohormones.

agonistic encounters (at the beginning of a "fight" and when a "winner" emerges), and during part of the mating cycle (male only). Animals assume octopaminelike poses during agonistic encounters (the "loser"), during mating (female only), and, in young animals, while "playing-dead" in threatening situations (Scrivener, 1971; Atema and Cobb, 1980).

Since posturing is an essential part of the behavioral repertoire of lobsters, our recent studies have concentrated on trying to learn how and where amines fit into the neuronal circuitry underlying this behavior. In animals injected with serotonin, most or all of the postural flexor muscles in the lobster body are contracted, whereas with octopamine, most or all of the postural extensors are contracted. The question then is whether amines produce opposite postures by direct actions on flexor and extensor muscles, by direct or indirect actions on the excitatory and inhibitory motoneurons that innervate the muscles, or by some combined action at the two sites. The results, to be described in Section 4, suggest that the last possibility is the correct one. Moreover, the actions at the two sites are complex, and we have been able to identify individual amine-containing neurons within the lobster central nervous system that may play important roles in the actions at the two sites.

In this chapter we present a review of our studies of serotonin, octopamine, and proctolin (Fig. 1), a pentapeptide closely associated with the amines, and the roles they serve in the lobster nervous system. The review will cover the following areas: (1) biosynthesis and further metabolism of serotonin and octopamine in the lobster nervous system; (2) distribution and cellular localization of serotonin, octopamine, and proctolin; (3) actions of amines and proctolin on exoskeletal muscles and explorations of the molecular basis of these actions; and (4) actions of amines on patterned activation of motoneurons of the ventral nerve cord.

2. Biosynthesis and Further Metabolism of Serotonin and Octopamine in the Lobster Nervous System

Our first studies were aimed at learning which amines were used in the lobster nervous system. Accordingly, radioactive precursor compounds (tyrosine, tryptophan) were incubated with lobster nervous tissues and the products formed were isolated and identified. With tyrosine incubations, dopamine and octopamine were the major products formed. No norepinephrine synthesis was detected. With tryptophan, serotonin was the major amine product detected. The pathways of amine metabolism have been elaborated, mainly from experiments with radioactive tracers, but some of the enzymes have been isolated and characterized (Evans *et al.*, 1976a,b; Wallace, 1976; Livingstone *et al.*, 1981; Kennedy, 1977, 1978).

In general, amines are formed in lobsters by enzymatic pathways similar to those found in the vertebrate nervous sytem (Fig. 2). Serotonin is formed from tryptophan in two steps: (1) hydroxylation to 5-hydroxytryptophan; and (2) decarboxylation to serotonin. Octopamine synthesis also involves two enzymes: (1) decarboxylation of tyrosine to tyramine; and (2) β-hydroxylation of tyramine to octopamine. The only enzyme from the synthetic pathways that has been isolated, characterized, and partially purified from lobster tissues is the tyramine-β-hydroxylase (Wallace, 1976).

In contrast to the biosynthesis, the pathway of further metabolism of amines in lobsters is very different from the vertebrate pathway: neither of the enzymes monoamine oxidase or catechol-*O*-methyl transferase are present in lobster tissues. Instead, amines are taken up by many different tissues (including tissues where no amines are synthesized) and converted to singly or doubly conjugated products (Fig. 3). One of the conjugated products is a sulfate ester. A soluble enzyme has been isolated and partially purified from lobster nervous tissues that catalyzes the transfer of a sulfate group from the activated intermediate 3′-phosphoadenosine-5′-phosphosulfate to the ring hydroxyl of serotonin or octopamine (Kennedy, 1978). A second enzyme adds β-alanine to the amino group of the amines. This reaction, also catalyzed by a soluble enzyme (β-alanylamine synthetase), requires ATP along with β-alanine for full activity and is likely to operate by the intermediate formation of β-alanyl adenylate (Kennedy, 1977, 1978). Beta-alanyl, sulfate double conjugates of the amines are formed upon incubation of intact tissues with amines. Whether amine conjugates are formed solely for inactivation and ultimate excretion of amines, or whether they serve some other purpose in lobsters is not known. These novel compounds are expensive cellular metabolites: the formation of each molecule of amine double conjugate requires the consumption of three molecules of ATP.

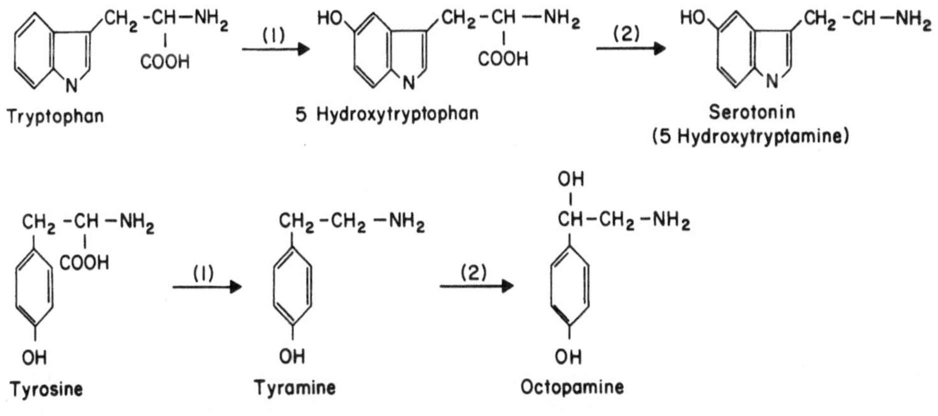

$CH_2-CH-NH_2$ (COOH) Tryptophan → (1) → HO, $CH_2-CH-NH_2$ (COOH) 5 Hydroxytryptophan → (2) → HO, $CH_2-CH-NH_2$ Serotonin (5 Hydroxytryptamine)

$CH_2-CH-NH_2$ (COOH), OH Tyrosine → (1) → $CH_2-CH_2-NH_2$, OH Tyramine → (2) → OH, $CH-CH_2-NH_2$, OH Octopamine

FIGURE 2. The pathways of serotonin and octopamine biosynthesis in lobsters.

3. Distribution and Cellular Localization of Serotonin, Octopamine, and Proctolin

3.1. Amines

Sensitive assays for serotonin and octopamine were adapted for use with lobster tissues and the lobster nervous system was screened for the sites of highest endogenous content of amines. The central and peripheral nervous systems of lobsters were divided into small segments for analyses. The lobster central nervous system consists of a chain of ventrally located ganglia (Fig. 4). The ganglia, clusters of unipolar neuronal cells and their associated neuropil regions, are linked together by axon-filled connectives and to the periphery by nerve roots containing sensory and motor axons. We found that while amines were widely distributed in the lobster central nervous system, the highest concentrations were along pairs of nerve roots emerging from ganglia in the thoracic region (thoracic second roots) of the nervous system (Fig. 4). Furthermore, amines were not uniformly distributed along these nerve roots, but were concentrated at two locations: close to where the roots emerge from the central ganglia, near a bifurcation where clusters of neurosecretory neurons are seen (Fig. 4, inset), and at the distal ends of the roots within a well-known neurosecretory structure called the pericardial organ (Alexandrowicz, 1953; Alexandrowicz and Carlisle, 1953; Maynard and Welsh, 1959). Amine concentrations are low in the lobster nervous system (probably because so few cells are found; see Section 3.). Some examples are (in pmole/μg protein): first thoracic ganglion, serotonin = 0.11, octopamine = 0.23; region of neurosecretory cells of nerve root from first thoracic ganglion, serotonin = 0.4, octopamine = 6.0. A more complete description of the distribution is found in Evans *et al.* (1976a) and Livingstone *et al.* (1981). Finally, amine synthesis from radioactive precursor compounds is detected wherever amines are found in the lobster nervous

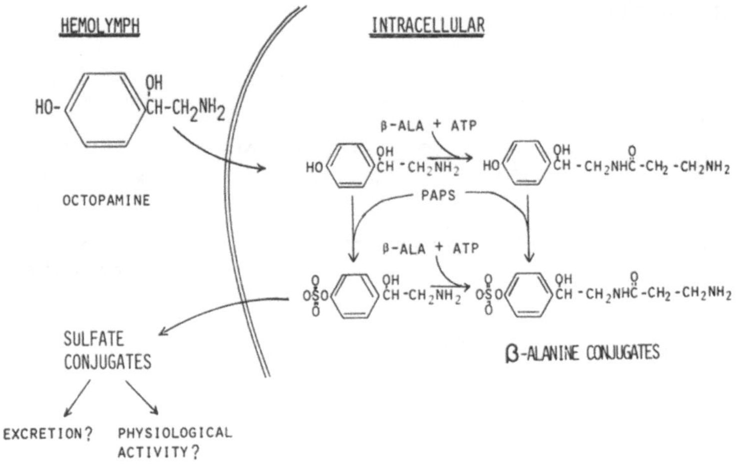

FIGURE 3. Further metabolism of amines in lobster tissues. The illustration shows the structural formulae for octopamine and its metabolites. Similar metabolites are formed from serotonin and dopamine.

system, and the highest rates of synthesis correlate well with the highest concentrations of amine.

Both of the sites along thoracic nerve roots where amines are found in high concentrations appear to be neurosecretory regions. Electron microscopic examinations reveal the presence of structures having the typical appearance of neurosecretory terminals (Evans *et al.*, 1976b; Livingstone *et al.*, 1981; Sullivan *et al.*, 1977). Nerve endings, often full of large dense-cored granules along with smaller vesicle-sized particles, lie within microns of the external surface of nerve roots sur-

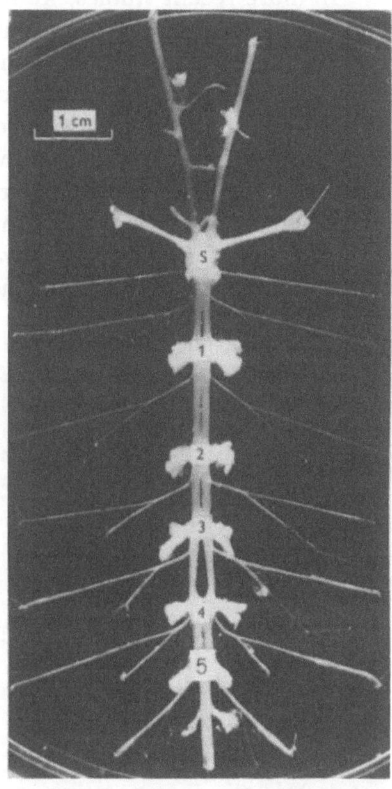

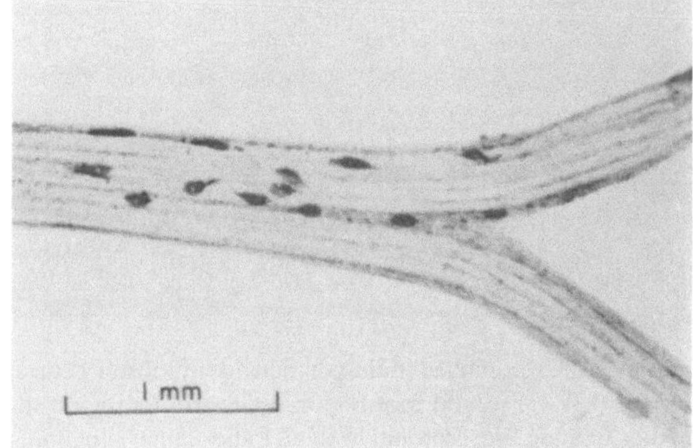

FIGURE 4. Thoracic region of the ventral nerve cord—location of the amine-associated neurosecretory neurons (top). The ganglia are labeled S = subesophageal, 1-5 = thoracic ganglia 1-5. Each ganglion has a pair of thin second thoracic roots associated with it. The arrows indicate the regions of high concentration of amine. In the boxed-in area cell bodies are concentrated. On the bottom are shown a cluster of neurosecretory neurons stained with the dye neutral red from a third thoracic segment. (Both figures are reprinted from Evans *et al.*, 1976a, with permission.)

rounded only by connective tissue or other terminals. Amines can be released from nerve roots by a Ca^{2+} dependent depolarization with K^+ (Evans *et al.*, 1976b; Livingstone *et al.*, 1981; Sullivan *et al.*, 1977) and are found circulating, at low concentrations (10^{-10} to 10^{-8} M range), in the hemolymph of lobsters (Livingstone *et al.*, 1980). By electron microscopic autoradiography, serotonin and octopamine have been localized to separate morphologically distinguishable categories of nerve terminals within these neurosecretory regions (Livingstone *et al.*, 1981). The autoradiographic studies, however, did not label any of the neurosecretory cell bodies found along the thoracic nerve roots where the amines are concentrated.

In more recent studies, we have turned to immunocytochemical methods to find cell bodies showing serotoninlike and octopaminelike immunoreactivity (Beltz and Kravitz, 1983; B. Trimmer and E. A. Kravitz, unpublished observations). For serotonin, we have used a rabbit antibody prepared against serotonin conjugated to bovine serum albumin and a secondary goat anti-rabbit antibody labeled with fluorescein for visualization, and we have mapped the lobster nervous system for cells, axonal processes, and nerve endings showing serotoninlike immunoreactivity.

When the thoracic root regions that are rich in serotonin are examined with this technique, a dense plexus of varicosities is observed (Fig. 5A). In confirmation of the autoradiographic studies, however, none of the neurosecretory cells in the

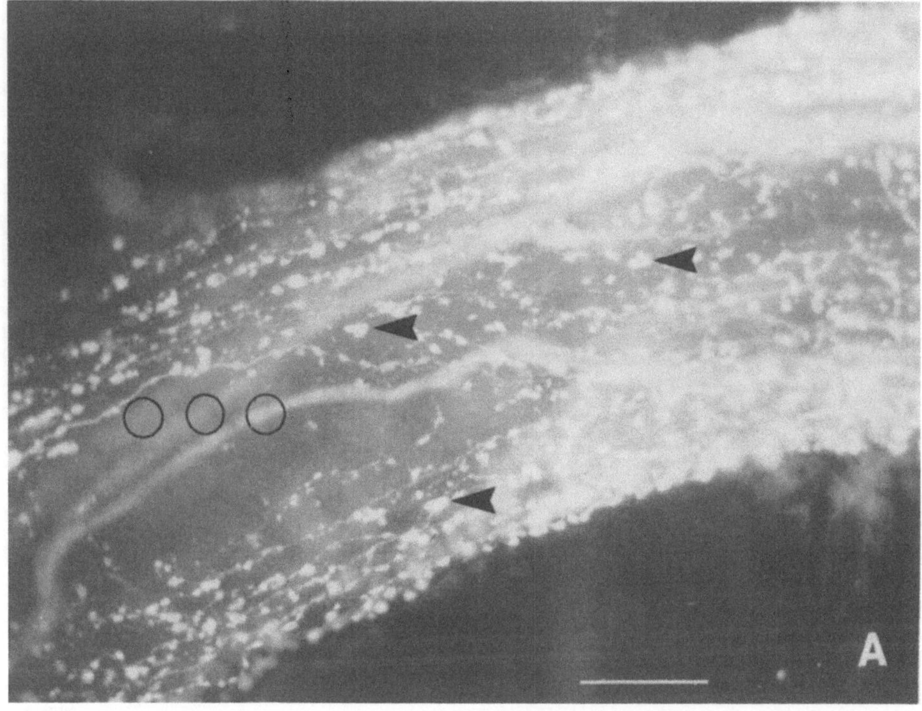

FIGURE 5. Serotoninlike immunoreactivity in lobster central and peripheral tissues. (A) A photograph from a thoracic second root region similar to the region shown in the inset diagram of Fig. 2. (B) Cell bodies staining for serotonin in the A1 abdominal ganglion. A large unpaired weakly staining cell is also observed in this ganglion. (C) A view of a second thoracic ganglion showing a single fluorescent process

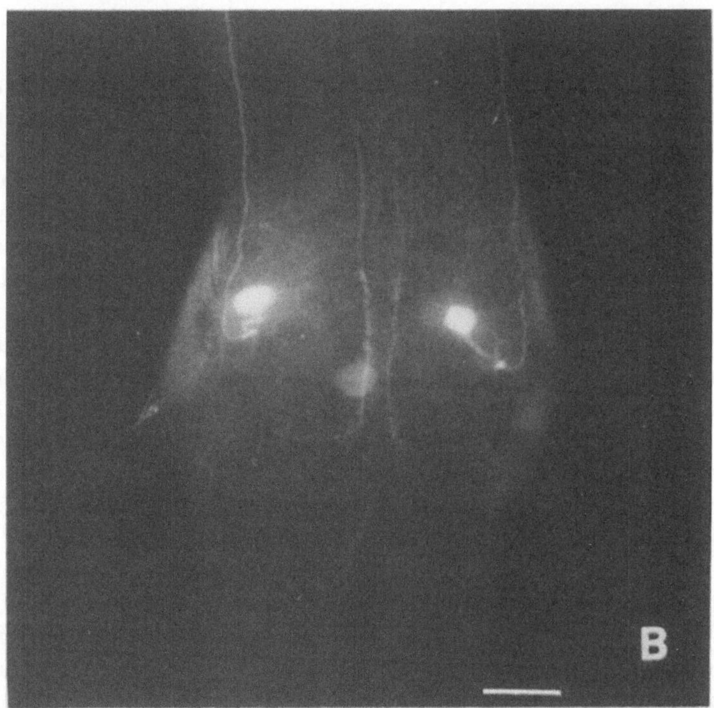

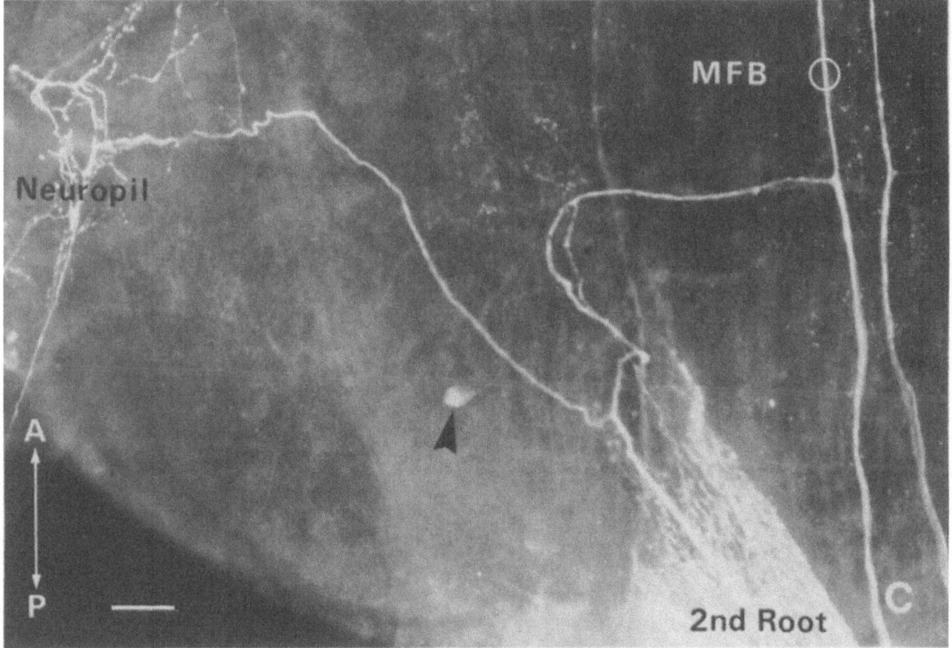

that gives rise to arbors of central and peripheral varicosities. This process originates in one of the paired cells in the A1 ganglion (see Fig. 7). Experimental details are outlined in the text and fully described in Beltz and Kravitz, 1983. Scale bar, 50 μm. (Reprinted with permission.)

root are stained. In examining central ganglia with this method, we find the cell bodies showing staining for serotonin (eg., Fig. 5B). Over 100 cells are found, widely distributed in the nervous system, with each ganglion containing at least one immunoreactive cell. The staining method reveals, in rich detail, the locations of presumed serotonergic cells, the arbors of dendritic and axonal processes from these cells, and their terminal fields of innervation. In Fig. 6, we show a portion of the map made of these cells and their processes. Some important features shown by the immunocytochemical methods are the following: (1) central cells send processes out thoracic second roots that contribute endings to the peripheral nerve plexuses—some of the endings in the thoracic roots of the fifth thoracic ganglion (T5) have been traced to cells in the first abdominal ganglion (A1); endings in the more anterior thoracic roots could not be traced to individual cells by immunocytochemistry, but were shown to originate in part from small nerve branches coming off axon bundles derived from the large paired immunoreactive cells seen in the T5 and A1 ganglia; (2) single processes coming off these same axon bundles give rise to central neuropil regions and thoracic root peripheral nerve plexuses (Fig. 5C)—this has been clearly seen in the T5 and T4 (on Fig. 6) and the T3, T2, and T1 ganglia (not shown), but the cellular origins of these processes were not traceable by immunocytochemical methods; (3) presumed dendritic arbors (identified by the absence of varicosities) also show serotoninlike immunoreactivity—such processes have been seen most clearly in the T5 ganglion; and (4) the major axonal projections of most serotonin immunoreactive cells are forward to more anterior ganglia—these cells, therefore, appear to be particularly concerned with interganglionic communication.

The immunocytochemical results suggested that of the 100 or so serotonin-staining cells we found in the lobster nervous system, the large paired neurons in the T5 and A1 ganglia might be particularly interesting and important. Accordingly, in recent studies we began detailed morphological and physiological analyses

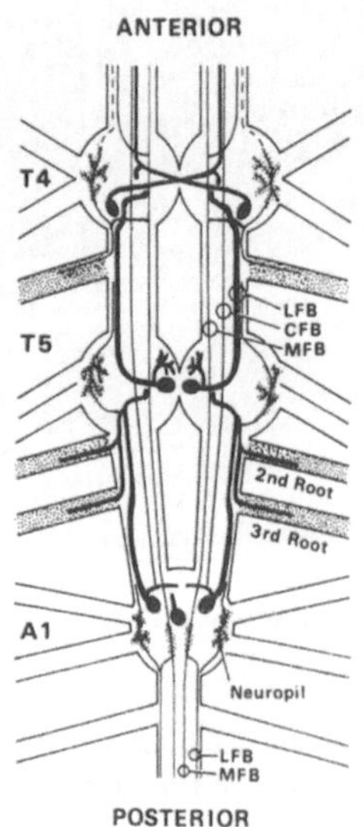

FIGURE 6. Schematic diagram of the immunoreactive cell bodies, fibers, and neuropil of a part of the ventral nerve cord. This is a composite drawing of whole mount preparations of ten ventral nerve cords. Cell bodies are drawn as large, filled, round or elongate circles. Heavy black lines represent immunoreactive fibers that have been traced to their cell bodies of origin. Fine line indicate immunoreactive fibers that have not been connected with cell bodies. Each of the fine lines of the lateral fiber bundles (LFBs), central fiber bundles (CFBs), and midline fiber bundles (MFBs) represents several fibers. Dashed lines indicate fibers that have not been directly visualized in these immunohistochemical preparations, but which we believe exist because the patterns of staining are similar from ganglion to ganglion. Stippled regions represent fine processes and varicosities of neuropil and plexus regions. (Further details are available in Beltz and Kravitz, 1983, from which this figure has been reprinted in a slightly modified form with permission.)

of these four cells. First, on the basis of the immunocytochemical location of the cell bodies and their major axonal projections, a method was devised to physiologically identify the cells. The identification procedure was verified by combining the injection of Lucifer Yellow into cells with the processing of tissues for serotonin immunocytochemistry using the peroxidase–antiperoxidase procedure. Then, to obtain more complete morphological pictures of the cells and their axonal arbors, we pressure injected either horseradish peroxidase or cobalt hexamine chloride into identified cells. The markers were allowed to diffuse for a 1- to 3-day period and tissues were processed to reveal the extent of filling of the neurons. Markers diffused 2–3 cm distances and the extensive branching patterns of the neurons were traced using a projection microscope (Fig. 7). The following features were confirmed: (1) processess of the T5 and A1 cells accounted for all the serotonergic axons emerging via thoracic second roots to form the peripheral nerve root plexuses (see Fig. 5C); (2) the single process on either side of each thoracic ganglion that gave rise to a central neuropil region and a peripheral root nerve plexus came from the A1 cells (compare Fig. 5C, Fig. 7); and (3) the A1 cells projected at least as far as the T2 ganglion and the T5 cells to at least T1, giving off processes and nerve terminals to central neuropil regions and peripheral second root nerve plexuses in every ganglion. The picture that emerges of these four cells, therefore, is of large neurons with fields of innervation that cover the entire thoracic region of the nervous system and that also supply the serotonin believed to be released from peripheral neurosecretory regions into the general circulation as a neurohormone (see next paragraph). Later in this article we shall return to how such cells might play a role in the regulation or control of posture in lobsters. More recently we have generated antibodies to octopamine conjugated to bovine serum albumin (B. Trimmer and K. K. Siwicki, unpublished data). The immunocytochemical mapping of neurons staining with these antibodies has just begun.

As mentioned above, from the two amine-rich regions along the thoracic second roots (the sites of the peripheral nerve plexuses), amines can be released into incubation media by depolarization. In intact lobsters, these two regions are found in hemolymph sinuses: the proximal region is in the ventral sinus through which hemolymph passes on its way to the gills; the distal region is in the pericardial sinus surrounding the heart. The pericardial sinus receives hemolymph from the gills just before it is drawn into the heart for distribution throughout the body (see Fig. 8). It is not clear why there are two separate sites of amine release along single nerve roots, but released material can be very effectively circulated throughout the lobster in this way. A similar organization is seen in other Crustacea and many amine-sensitive peripheral tissues have been found that are likely targets for circulating amines (for a detailed review of crustacean neurohormonal systems, including amines, see Cooke and Sullivan, 1982). Octopamine enhances clotting in the hemolymph (lobster: Battelle and Kravitz, 1978), both amines affect the rate and amplitude of the heartbeat (crab, lobster: Battelle and Kravitz, 1978; Florey and Rathmayer, 1978), scaphognathite beating is increased by serotonin treatment (crab: Berlind, 1977), the rate of firing of particular neurons in the stomatogastric system is changed by amines (lobster: Anderson and Barker, 1977; Barker *et al.*, 1979), and there are a complex set of actions on exoskeletal muscles (crayfish, lobster: Battelle and Kravitz, 1978; Dudel, 1965; Fischer and Florey, 1983; Glusman

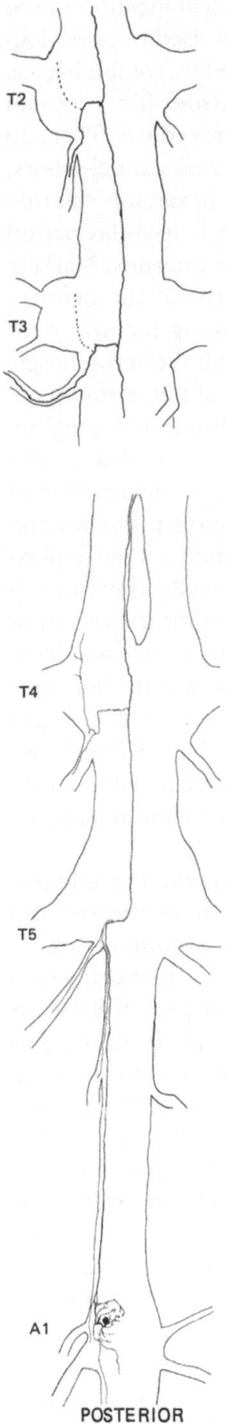

FIGURE 7. Reconstruction of an A1 serotonin-immunoreactive cell (one of the paired cells). Horseradish peroxidase was injected into the cell body and allowed to diffuse for 2–3 days. The upper and lower parts of the figure (A1, T5, T4 ganglia; T3, T2 ganglia) are from two different injections, but the main features have been seen repeatedly in injections of A1 cells. Details are outlined in text. (From B. S. Beltz, unpublished observations).

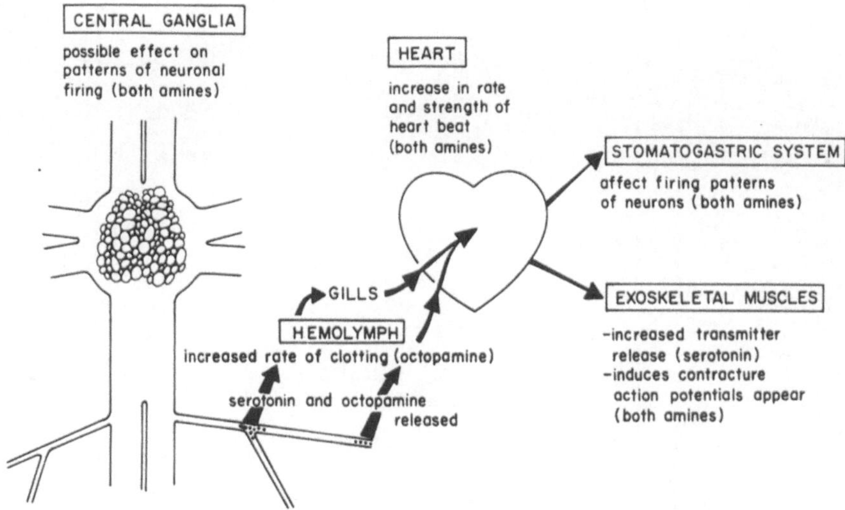

FIGURE 8. Peripheral and central targets of amines. (Reprinted from Kravitz *et al.*, 1980, with permission.)

and Kravitz, 1982; Florey and Florey, 1954; Grundfest and Reuben, 1961; Breen and Atwood, 1983). The last set of actions (on exoskeletal muscles) are of particular relevance to the postures we have described and have served as a primary focus of our efforts.

3.2. The Pentapeptide Proctolin

Since its original isolation from the hindgut of the cockroach, *Periplaneta americana* in 1975 (Brown and Starratt, 1975; Starratt and Brown, 1975), proctolin has been found widely distributed in invertebrate nervous systems. By bioassay, radioimmunoassay, and several chromatographic techniques, proctolinlike material has been detected in Crustacea (Sullivan, 1979; Kingan and Titmus, 1983; Schwarz *et al.*, 1980) and authentic proctolin has been isolated from lobster nervous tissues by use of immunological, high-performance liquid-chromatographic and mass-spectrometric methods (Schwarz *et al.*, 1984). Using immunocytochemical methods, some 1500 neurons containing proctolinlike material have been found in the lobster nervous system (K. K. Siwicki, unpublished observations). Of particular interest in the context of this article, however, is the observation that the pairs of large serotonin-immunoreactive cells in the T5 and A1 ganglia also stain for proctolin. This is shown in Fig. 9 where whole mounts of two T5 ganglia have been processed for proctolinlike and serotoninlike immunoreactivity. This observation has been confirmed by cutting serial sections of the T5 and A1 ganglia and staining alternate sections for proctolin and serotonin, and by dissecting single T5 and A1 cell bodies and measuring their proctolin contents by high-performance liquid chromatography followed by radioimmunoassay for proctolin. Each cell body contains approximately 10 fmole of proctolin by this procedure. The major axonal projections of these cells can be recognized in tissues processed by immu-

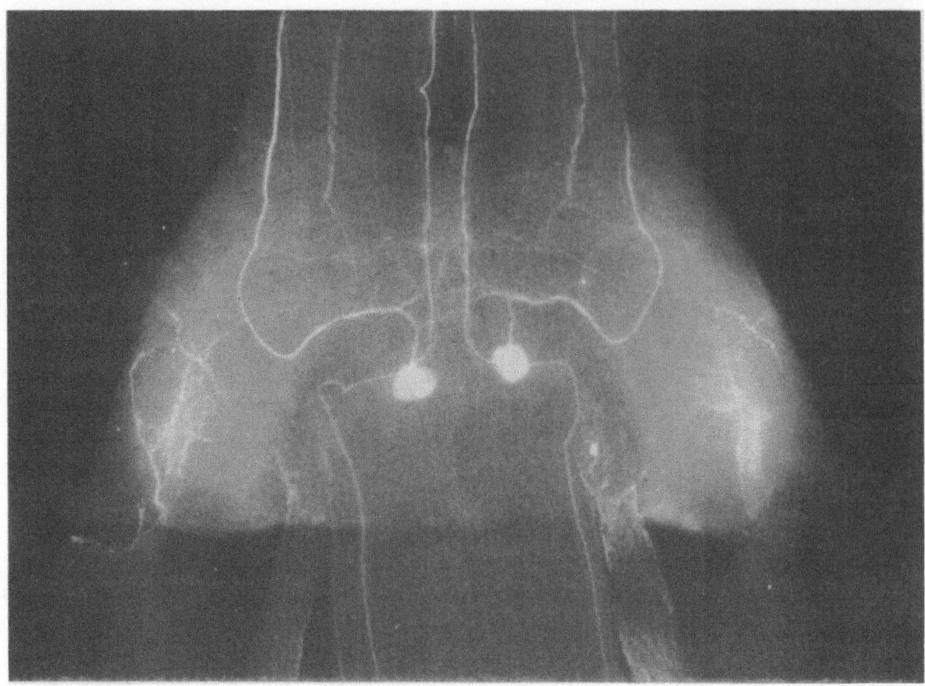

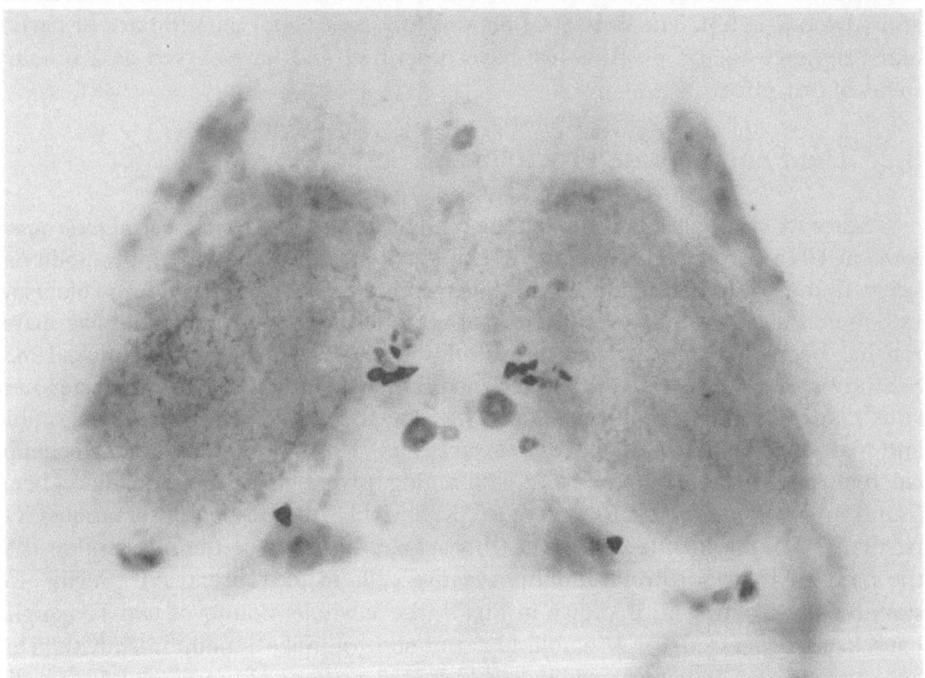

FIGURE 9. Immunocytochemical staining of the T5 thoracic ganglion for serotonin (upper-immunofluorescence) and proctolin (lower-immunoperoxidase technique). Note the large paired cells staining in the posterior-medial quadrant of the ganglion in both preparations. These cells have been dissected out and shown to contain authentic proctolin. (From K. K. Siwicki, unpublished data.)

nocytochemical methods for proctolin (K. K. Siwicki, unpublished data), and thoracic second roots show dense plexuses of proctolin-staining terminals in the same root regions from which octopamine and serotonin are released. Proctolinlike immunoreactive material also can be released from these root regions by depolarization with K^+ and this release is Ca^{2+} dependent (Schwarz *et al.*, 1984). Thus the likelihood exists that the T5 and A1 cells liberate both proctolin and serotonin when stimulated. Proctolin, like serotonin (and octopamine), therefore probably serves at least partially as a neurohormonal substance released into the general circulation for actions on remote target tissues. The half life of proctolin in lobster hemolymph is on the order of 6 min.

4. The Actions of Amines and Proctolin on Exoskeletal Muscles and Explorations of the Molecular Basis of These Actions

Lobster exoskeletal muscles have a dual innervation. They are innervated by excitatory axons likely to use glutamate and inhibitory axons using GABA as transmitter compounds. The preparation that we routinely use, the opener muscle of the dactyl of the walking leg, is innervated by a single excitatory and a single inhibitory axon, each of which (1) branches many times to innervate different muscle fibers, and (2) has multiple endings on individual fibers. The muscle fibers lack sodium channels and usually do not have action potentials. Occasionally an abortive calcium action potential can be seen. As is shown in Section 4, this is one aspect of neuromuscular physiology that is dramatically altered by amines.

It has been long known that serotonin can enhance contractions of crustacean neuromuscular preparations (Florey and Florey, 1954). Grundfest and Reuben (1961) reported that serotonin increased the size of excitatory junctional (synaptic) potentials (EJPs) and produced action potentials in crustacean muscle fibers, and Dudel (1965) showed that the increase in EJP size caused by serotonin in crayfish preparations was presynaptic in origin. We, and other investigators, have carried out further studies of these actions and have shown that a complex set of hormonal controls exists at lobster neuromuscular junctions. At least three substances are involved: octopamine, serotonin, and the peptide proctolin. All of the actions of these substances fall into the general category of synaptic modulation: the three are external agents, not found in significant quantities within opener muscle preparations, that have dramatic long-term effects on the synaptic machinery that regulates the activity of these preparations. Serotonin acts on excitatory and inhibitory nerve terminals and on muscle fibers, octopamine has small effects on excitatory nerve terminals and on muscle fibers (see Florey and Rathmayer, 1978; Fischer and Florey, 1983; Breen and Atwood, 1983), and proctolin only affects the muscle fibers (Fig. 10). Serotonin increases transmitter release from excitatory nerve terminals by as much as fivefold. The effect is slow in onset (minutes) and persists long after washing serotonin out of the bath. The increased EJP size declines to the control size in two steps: one with a fast $T_{1/2}$ of 1–2 min; one lasting much longer ($T_{1/2}$ = 30 min). Some preliminary pharmacological evidence suggests that the two effects result from actions at distinct presynaptic sites, but further studies will be needed to substantiate these results. In exploring the mechanism of the slower of

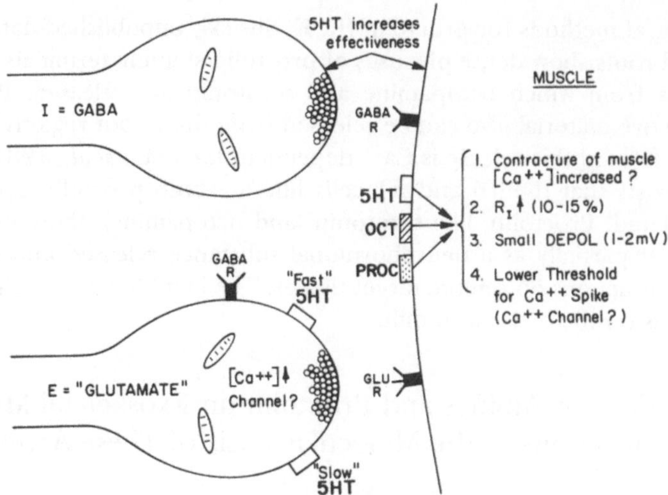

FIGURE 10. Summary diagram of sites of action and effects of serotonin (5-HT), octopamine (Oct) and proctolin (PROC) on the lobster opener muscle neuromuscular preparation. (Reprinted in a slightly modified form from Kravitz *et al.*, 1980, with permission.)

the two components, we have accumulated evidence to suggest that at least a part of the increase in quantal release results from a metabolic change within excitatory nerve terminals. The change is likely to be an alteration in either the buffering or storage of Ca^{2+}, or the sensitivity of the transmitter release machinery to existing intraterminal Ca^{2+} levels (for details, see Glusman and Kravitz, 1982). Serotonin also increases transmitter release from inhibitory nerve terminals. This is also a long-lasting action, but its mechanism has not been studied in any detail.

On muscle fibers, all three substances (serotonin, octopamine, and proctolin) are effective, and all have similar actions: they induce long-lasting contractures of muscle fibers, which are accompanied by little or no change in membrane potential or muscle fiber input resistance; and they produce action potentials in muscle fibers where none were seen before (Fig. 11). The serotonin-induced generation of action potentials has been studied in preliminary voltage clamp experiments (Kravitz *et al.*, 1980). Although no change was seen in an outward voltage-sensitive K^+ conductance, a substantial increase (up to threefold) was seen in an inward Ca^{2+} current. We cannot tell yet whether this inward current increase results from a direct action of serotonin on Ca^{2+} channels, or from an indirect action on Ca^{2+}-activated K^+ channels (for example). We suspect that the action is directly on Ca^{2+} channels since we also observe an increased inward current when Ba^{2+} or Sr^{2+} replace Ca^{2+} in the bathing medium, but further studies will be needed before any firm conclusions can be drawn.

Recently we have begun to search for possible molecular correlates of the neurohormonally induced physiological changes. Toward this end, we have carried out two types of experiments. In one type we have incubated intact nerve muscle preparations with $[^{32}P]PO_4$ in the presence or absence of serotonin, octopamine, or proctolin, and searched for changes in the labeling patterns of phosphorylated proteins subsequently extracted from tissues. In the second type of experiment we have measured tissue levels of cyclic nucleotides in response to the two amines and the peptide (Goy *et al.*, 1984). A particular virtue of using muscle preparations for this type of study is that our detailed knowledge of lobster muscle physiology allows us to isolate hormone-induced changes from changes related to contraction (which can be induced by K^+, glutamate, or caffeine), or depolarization (K^+, glutamate), or intracellular Ca^{2+} levels (caffeine, calcium ionophores).

These studies have shown that incubation of muscles with serotonin leads to a selective increase in phosphorylation of a 29,000 dalton mol. wt. protein. Incubation with octopamine in the presence of IBMX (a phosphodiesterase inhibitor) also can cause phosphorylation of this protein, whereas incubation with proctolin never has been observed to cause phosphorylation of this protein (see Fig. 12). The effectiveness of the neurohormonal agents in changing cyclic AMP levels parallels their effectiveness in causing phosphorylation of the 29,000 dalton protein (Goy *et al.*, 1984). Serotonin, in the presence of IBMX, causes up to 15-fold increases of cyclic AMP levels (Goy *et al.*, 1984); octopamine produces much smaller effects (twofold to fourfold); while proctolin causes no detectable increases (Fig. 12). None of the substances alter cyclic GMP levels. The parallel between changes in cyclic AMP levels and phosphorylation of the 29,000 dalton protein suggests that the two are

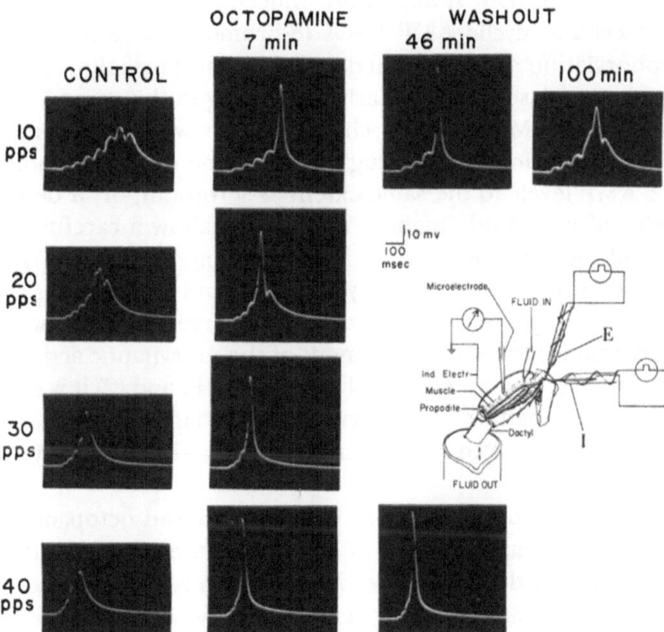

FIGURE 11. Action potentials generated in muscle fibers after treatment with octopamine. Stimuli at different frequencies were delivered to the excitatory nerve innervating an opener muscle preparation (left side of figure). Several minutes after octopamine treatment (5×10^{-6} M), large action potentials are seen. After 100 min of washout in control saline, action potentials begin to disappear in this experiment. (Reprinted with permission from Kravitz *et al.*, 1981.)

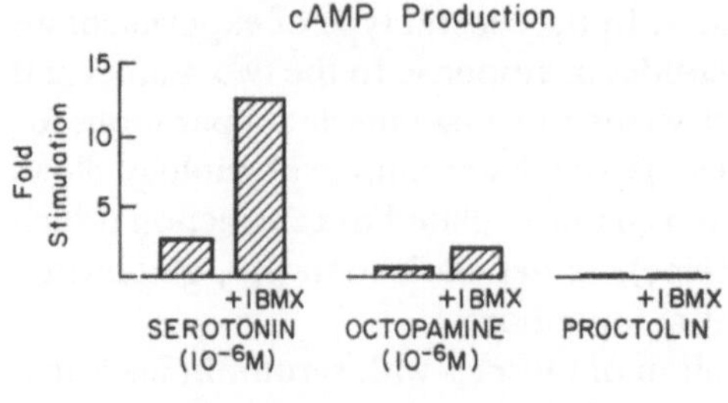

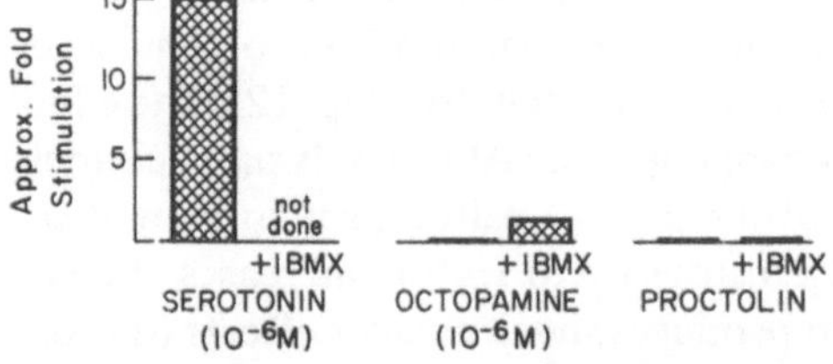

FIGURE 12. Effects of serotonin, octopamine, and proctolin on cyclic AMP synthesis and phosphorylation of the 29,000 dalton protein in lobster opener muscle preparations. See text for description. (Further details are available in Battelle and Kravitz, 1978, and Goy *et al.*, 1984.)

related. Incubations of muscle homogenates with [^{32}P]ATP in the presence of cyclic AMP lead to phosphorylation of an identical protein, thereby lending further support to this suggestion (Goy *et al.*, 1984). Indeed, in intact tissues, any agent that increases intracellular cyclic AMP levels (8-bromocyclic AMP, IBMX, forskolin) causes phosphorylation of the 29,000 dalton protein. Forskolin, an activator of the enzyme adenylate cyclase, is particularly interesting in this regard. With this agent we can titrate cyclic AMP levels to whatever point we desire in intact tissues by varying the concentration of the drug and the time of incubation. Thus we can elevate cyclic AMP levels to the same extent as serotonin, or, if desired, bring the levels an order of magnitude higher. This should allow a careful examination of which aspects of the neurohormonally induced changes in physiological properties of these preparations are related to cyclic AMP, and possibly to phosphorylation of the 29,000 dalton protein. Such studies are currently under way. Preliminary results suggest that (1) the slow component of the presynaptic actions of serotonin on excitatory nerve terminals may involve cyclic AMP, and (2) it is unlikely that any of the physiological changes we observe at postsynaptic sites (contracture, Ca^{2+} action potentials) involved cyclic nucleotides (M. F. Goy, unpublished observations).

To return to the opposite actions of serotonin and octopamine on contractions of flexor and extensor muscles in lobsters, as the above results show, we see no opposing actions of the two amines in studies on isolated opener muscle preparations. Nor do we see any opposing actions when we examine postural flexor and extensor muscle pairs from the abdomen or from the walking legs (Harris-Warrick and Kravitz, 1984). Instead, both amines seem to prime peripheral muscles to react more vigorously when stimulated. This action is seen most dramatically with serotonin, which produces a long-term increase of transmitter release from excitatory and inhibitory nerve terminals as well as enhancement of the contractility of muscle fibers. Opposite actions of the amines, therefore, must reside in the their actions on motoneurons innervating the muscles. Accordingly, we turned next to studies of amine effects on ganglia of the ventral nerve cord.

5. Actions of Amines on Motoneurons of the Ventral Nerve Cord

Preparations were dissected consisting of a chain of central ganglia linked through one or more nerve roots to postural flexor or extensor muscles. The ganglia were pinned in one chamber, isolated by a vaseline seal (through which the intact nerve root passed) from the muscles. In our first experiments we recorded (1) with intracellular electrodes from single muscle fibers and (2) with extracellular electrodes from the intact nerve trunks. At the same time we could separately superfuse serotonin or octopamine onto the ventral nerve cord. The inset diagram in Fig. 13 shows the experimental arrangement and some typical recordings from a muscle fiber (right side of figure).

The firing of individual axons in nerve trunks can be correlated with the appearance of EJP's and inhibitory junctional potentials IJP's in muscle fibers. In

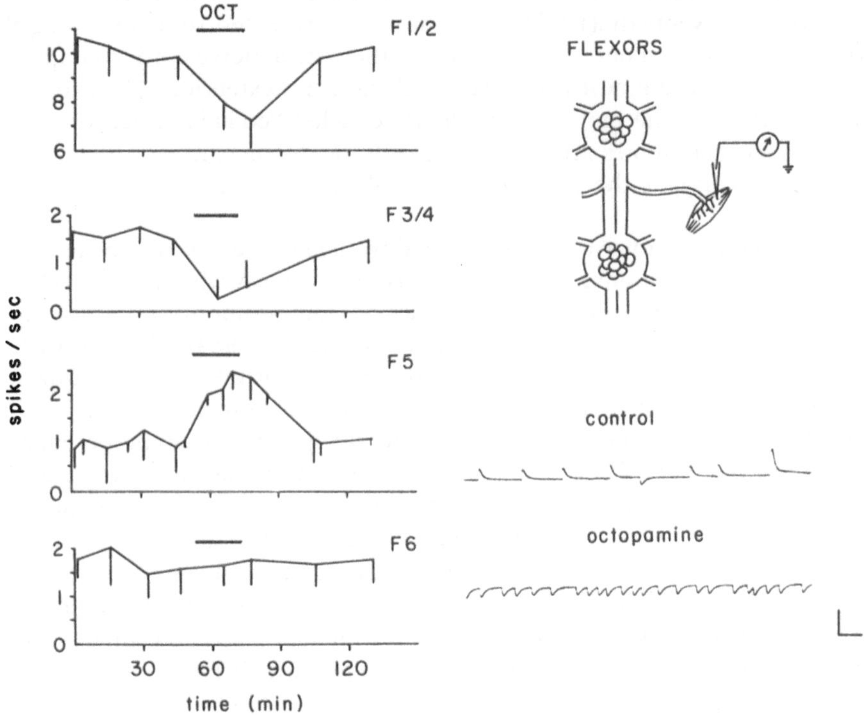

FIGURE 13. Effect of octopamine (3×10^{-5} M) on motoneurons innervating the superficial flexor muscles. The inset diagram illustrates the experimental arrangement. On the lower right a sample intracellular recording is shown. The left side of the figure shows the average action potential frequency (one S.D. shown) in the five excitatory (F1, 2, 3, 4, 6) and one inhibitory (F5) motoneurons innervating the muscle. (Reprinted from Kravitz *et al.*, 1983, with permission.)

this way the five excitatory axons (F1, F2, F3, F4, and F6) and the inhibitory axon (F5) innervating the muscle can be positively identified. In the example illustrated (Fig. 13), a flexor muscle preparation was used and, with octopamine treatment, we see a dramatic reduction of the firing of excitatory neurons innervating the muscle (with the exception of F6, which does not change) and an increase in the firing of the inhibitory neuron (see sample record, F5 graph on Fig. 12). With serotonin treatment we see essentially the opposite result (not shown). The inhibitory neuron innervating the muscle decreases in firing, while excitatory neurons, although showing more variability in their responsiveness, increase in firing. With extensor muscle preparations (also not shown) we see a reversed pattern: octopamine increases the firing of excitatory neurons and decreases the firing of the inhibitor; serotonin increases the firing of the inhibitor and decreases the firing of the excitors. The general patterns of changes in firing that we see are illustrated in Fig. 14. In other words, serotonin directs the readout of a central motor pattern causing flexion, octopamine of a central motor pattern causing extension. The activation of these patterns leads to increases and decreases in the firing of large numbers of neurons in each ganglion. We see identical results with individual ganglia or when a chain of ganglia are used. A good candidate for the circuitry activated by the amines already has been well described in the crustacean literature. In the 1960s, Evoy and Kennedy (1967) and co-workers presented results showing that individual axons, teased out of connectives of the ventral nerve cord, could trigger the readout of central motor programs for flexion and extension. The axons producing activation of the motor programs were called "command neurons." Our results suggest that amines are interacting with the "command neuron" circuitry in some way to trigger the readout of central motor programs for flexion and extension.

In more recent studies we have recorded from identified excitatory and inhibitory cell bodies in abdominal ganglia to try to establish whether the cells are directly activated by amines or whether particular amine-responsive synaptic inputs lead to activation of the cells (Harris-Warrick and Kravitz, 1984). Thus far, we have recorded from two neurons, an excitatory neuron innervating slow extensor muscles (M-15) and the principal inhibitory neuron to the slow flexors (I-1), and we find similar results with both cells. We see no change in membrane potential or input resistance of the cells with amine treatment, but the threshold for activation of the cells is reduced by octopamine and increased by serotonin. These changes in threshold for activation are prevented if the concentration of Ca^{2+} is lowered or Co^{2+} is added to the bathing medium. In addition, octopamine produces an increase and serotonin a decrease in spontaneous excitatory synaptic input to the cells. Not all synaptic input is enhanced by octopamine, however. When histograms are generated of the size of synaptic potentials against their frequency, a new size

	SLOW	FLEXORS	SLOW	EXTENSORS
	E	I	E	I
SEROTONIN	↑	↓	↓	↑
OCTOPAMINE	↓	↑	↑	↓

FIGURE 14. Actions of serotonin and octopamine on the firing of excitatory and inhibitory motoneurons innervating postural flexor and extensor muscles. (Reprinted from Kravitz *et al.*, 1983, with permission.)

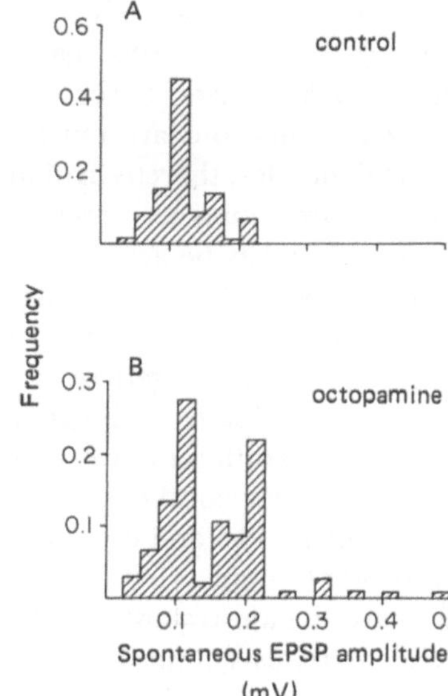

FIGURE 15. Histogram of spontaneous excitatory postsynaptic potential (EPSP) amplitudes in the cell I-1 before (A) and during (B) superfusion with 3×10^{-5} M octopamine. Abcissa: peak amplitude of EPSP, in 25 μV bins. Ordinate: fraction of total EPSPs with that amplitude. (Reprinted from Harris-Warrick and Kravitz, 1984, with permission.)

class of synaptic potential is observed after octopamine treatment. This result is illustrated for the inhibitory cell to the flexors in Fig. 15. We also have activated known synaptic inputs to the cells (from stretch receptor neurons) and find that amines cause no change in the size of the resulting synaptic potentials. The size of action potentials generated by backfiring the axons of these cells and recorded in the cell bodies is also unaffected. Taken together, these results suggest that at least for two postural motoneuron cell bodies (M-15, I-1), the amines may act through cells that are presynaptic to the motoneurons to produce the alterations in firing observed in intact ganglia or in intact animals. Clearly, further studies will be needed to verify this suggestion.

6. Summary

If we now return to the type of cell shown in Fig. 7, we can see how cells with peripheral and central endings might function in the generation of a flexed posture. When the cell is activated, serotonin would be released from peripheral endings to serve as a neurohormone and from central endings to have specific synaptic actions on target neurons. In the periphery, serotonin primes excitatory and inhibitory nerve terminals on flexor and extensor muscles, and the muscles themselves, to respond more vigorously. In the central nervous system, serotonin activates the motor program for flexion: excitatory neurons to flexors and the inhibitory neuron to the extensors increase in firing; excitatory neurons to extensors and the inhibitory neuron to the flexors decrease in firing. We have observed considerable concentration differences in the responsiveness at the two sites (central and peripheral). The levels of serotonin found in the circulation are low (10^{-10}–10^{-8} M), in the range of the threshold responses of exoskeletal muscles to serotonin. On the other hand, activation of central motor programs require 10^{-6}–10^{-5} M concentrations of serotonin. This may represent the concentration needed to overcome permeability barriers to serotonin in ventral ganglia. It also could be that receptors on central neurons are much less sensitive to serotonin than receptors on exoskeletal muscles, thereby avoiding possible activation of central serotonergic target sites by circulating serotonin. The high concentrations needed to activate central sites could easily be achieved in synaptic areas in the central nervous system. The central arbors of nerve endings showing serotoninlike immunoreactivity could be such synaptic areas. We are at present examining whether activation of the T5 or A1 cells actually triggers or modulates the central motor patterns for flexion and thereby participates in the generation of posturing in lobsters.

In many different invertebrate species, amines and other neurohormones have important behavioral effects. This can be illustrated with just a few examples: the studies of Kandel and co-workers have implicated serotonin in models of learning in *Aplysia* (Kandel, 1979; Klein *et al.*, 1980); O'Shea and Evans and their co-workers have demonstrated roles for octopamine and proctolin in the cockroach and locust nervous systems and elegant mapping studies have been carried out in the cockroach system (Evans, 1980; Bishop and O'Shea, 1982; O'Shea and Bishop, 1982; Evans and O'Shea, 1978; O'Shea, 1982); Kristan and co-workers have demonstrated prominent effects of serotonin in swimming behavior in leeches (Willard, 1981; Nusbaum and Kristan, 1982; Kristan and Nusbaum, 1982–83); and Truman and co-workers have elaborated on the roles of various neurohormones in development and metamorphosis in *Manduca* (Truman, 1978; Truman and Schwartz, 1980). The fundamental importance of the behaviors governed by the neurohormonal substances in these animals and the ability to understand how neurohormones function at a cellular level make these animal models invaluable to our understanding of the functioning of nervous systems.

References

Alexandrowicz, J. S., 1953, Nervous organs in the pericardial cavity of the decapod *Crustacea, J. Mar. Biol. Assoc. U.K.* **31:**563–580.

Alexandrowicz, J. S., and Carlisle, D. B., 1953, Some experiments on the function of the pericardial organs in *Crustacea, J. Mar. Biol. Assoc. U.K.* **32:**175–192.

Anderson, W. W., and Barker, D. L., 1977, Activation of a stomatogastric motor pattern generator by dopamine and L-dopa, *Soc. Neurosci. Abstr.* **3:**522.

Atema, J., and Cobb, J. S., 1980, Social behavior of lobsters, in: *The Biology and Management of Lobsters,* Volume I, (J. S. Cobb and B. F. Phillips, eds), Academic Press, New York, pp. 409–450.

Barker, D. L., Kushner, P. D., and Hooper, N., 1979, Synthesis of dopamine and octopamine in the crustacean stomatogastric nervous system, *Brain Res.* **161:**99–113.

Battelle, B. A., and Kravitz, E. A., 1978, Targets of octopamine action in the lobster: Cyclic nucleotide changes and physiological effects in haemolymph, heart and exoskeletal muscle, *J. Pharmacol. Exp. Ther.* **205:**438–448.

Beltz, B. S., and Kravitz, E. A., 1983, Mapping of serotonin-like immunoreactivity in the lobster nervous system, *J. Neurosci.* **3:**585–602.

Berlind, A., 1977, Neurohumoral and reflex control of scaphognathite beating in the crab *Carcinus maenus, J. Comp. Physiol.* **116:**77–90.

Bishop, C. A., and O'Shea, M., 1982, Neuropeptide proctolin (H-Arg-Tyr-Leu-Pro-Thr-OH): Immunocytochemical mapping of neurons in the central nervous system of the cockroach, *J. Comp. Neurol.* **207:**223–238.

Breen, C. A., and Atwood, H. L., 1983, Octopamine—A neurohormone with presynaptic activity dependent effects at crayfish neuromuscular junctions, *Nature (London)* **303:**716–718.

Brown, B. E., and Starratt, A. N., 1975, Isolation of proctolin, a myotropic peptide, from *Periplaneta american*, *J. Insect. Physiol.* **21:**1879–1881.

Cooke, I. M., and Sullivan, R. E., 1982, Hormones and neurosecretion, in: *The Biology of Crustacea,* Volume 3 (H. L. Atwood, and D. C. Sandeman, eds.), Academic Press, New York, pp. 205–290.

Dudel, J., 1965, Facilitatory effects of 5-hydroxytryptamine on the crayfish neuromuscular junction, *Naunyn-Schmied. Arch. Exp. Pathol. Pharmacol.* **249:**515–528.

Evans, P. D., 1980, Biogenic amines in the insect nervous system, *Adv. Insect. Physiol.* **15:**317–437.

Evans, P. D., and O'Shea, M. (1978) The identification of an octopaminergic neuron and the modulation of a myogenic rhythm in the locust. *J. Exp. Biol.* **73:**235–260.

Evans, P. D., Kravitz, E. A., Talamo, B. R., and Wallace, B. G., 1976a, The association of octopamine with specific neurons along lobster nerve trunks, *J. Physiol.* **262:**51–70.

Evans, P. D., Kravitz, E. A., and Talamo, B. R., 1976b, Octopamine release at two points along lobster nerve trunks, *J. Physiol.* **262:**71–89.

Evoy, W. H., and Kennedy, D., 1967, The central nervous organization underlying control of antagonistic muscles in the crayfish. I. Types of command fibers, *J. Exp. Zool.* **165:**223–238.

Fischer, L., and Florey, E., 1983, Modulation of synaptic transmission and excitation–contraction coupling in the opener muscle of the crayfish, *Astacus leptodactylus,* by 5-hydroxytryptamine and octopamine, *J. Exp. Biol.* **102:**187–198.

Florey, E., and Florey, E., 1954, Uber die mogliche Bedeutung von Enteramin (5-oxytryptamin) als nervoser Aktimssubstanz bei cephalopoden und dekapoden crustacean, *Z. Naturforsch.* **96:**58–69.

Florey, E., and Rathmayer, M., 1978, The effects of octopamine and other amines on the heart and on neuromuscular transmission in decapod crustaceans: Further evidence for a role as a neurohormone, *Comp. Biochem. Physiol.* **61C:**229–237.

Glusman, S., and Kravitz, E. A., 1982, The action of serotonin on excitatory nerve terminals in lobster nerve-muscle preparations, *J. Physiol.* **325:**223–241.

Goy, M. F., Schwarz, T. L., and Kravitz, E. A., 1984, Serotonin-induced protein phosphorylation in a lobster neuromuscular preparation, *J. Neurosci.* **4:**611–626.

Grundfest, H., and Reuben, J. P., 1961, Neuromuscular synaptic activity in lobster, in: *Nervous Inhibition* (E. Florey, ed.), Pergamon Press, Oxford, pp. 92–104.

Harris-Warrick, R. M., and Kravitz, E. A., 1984, Cellular mechanisms for modulation of posture by octopamine and serotonin in the lobster, *J. Neurosci.* **4:**1976–1993.

Kandel, E. R., 1979, Cellular insights into behavior and learning, *Harvey Lect.* **73:**19–92.

Kennedy, M. B., 1977, Amine metabolism: A different pathway in lobsters, *Soc. Neurosci. Abstr.* **3:**252.

Kennedy, M. B., 1978, Products of biogenic amine metabolism in the lobster: Sulfate conjugates, *J. Neurochem.* **30:**315–320.

Kingan, T., and Titmus, M., 1983, Radioimmunologic detection of proctolin in Arthropods, *Comp. Biochem. Physiol. C* **74:**75–78.

Klein, M. E., Shapiro, E., and Kandel, E. R., 1980, Synaptic plasticity and the modulation of the Ca^{2+} current, *J. Exp. Biol.* **89:**117–157.

Kravitz, E. A., Glusman, S., Harris-Warrick, R. M., Livingstone, M. S., Schwarz, T., and Goy, M. F., 1980, Amines and a peptide as neurohormones in lobsters: Actions on neuromuscular preparations and preliminary behavioral studies, *J. Exp. Biol.* **89:**159–175.

Kravitz, E. A., Glusman, Livingstone, M. S., and Harris-Warrick, R. M., 1981, Serotonin and octopamine in the lobster nervous system: Mechanism of action at neuromuscular junctions and preliminary behavioral studies, in: *Serotonin Neurotransmission and Behavior* (B. Jacobs and A. Gelperin, eds.), MIT Press, Cambridge, p. 189.

Kravitz, E. A., Beltz, B. S., Glusman, S., Goy, M. F., Harris-Warrick, R. M., Johnston, M. F., Livingstone, M. S., Schwarz, T. L., and Siwicki, K. K., 1983, Neurohormones and lobsters: Biochemistry to behavior, *Trends Neurosci.* **6:**346–349.

Kristan, W. B., and Nusbaum, M. P., 1982–1983, The dual role of serotonin in leech swimming, *J. Physiol. (Paris)* **78:**743–747.

Livingstone, M. S., Harris-Warrick, R. M., and Kravitz, E. A., 1980, Serotonin and octopamine produce opposite postures in lobsters, *Science* **208:**76–79.

Livingstone, M. S., Schaeffer, S. F., and Kravitz, E. A., 1981, Biochemistry and ultrastructure of serotonergic nerve endings in the lobster: Serotonin and octopamine are contained in different nerve endings, *J. Neurobiol.* **12:**27–54.

Maynard, D., and Welsh, J. H., 1959, Neurohormones of the pericardial organs of brachyuran *Crustacea, J. Physiol.* **149:**215–227.

Nusbaum, M. P., and Kristan, W. B., Jr., 1982, The swim initiating ability of intersegmental serotonin-containing leech interneurons, *Soc. Neurosci. Abstr.* **8:**161.

O'Shea, M., 1982, An identified neuron approach with special reference to proctolin, *Trends Neurosci.* **5:**69–73.

O'Shea, M., and Bishop, C. A., 1982, Neuropeptide proctolin associated with an identified skeletal motoneuron, *J. Neurosci.* **2:**1242–1251.

Schwarz, T. L., Harris-Warrick, R. M., Glusman, S., and Kravitz, E. A., 1980, A peptide action in a lobster neuromuscular preparation, *J. Neurobiol.* **11:**623–628.

Schwarz, T. L., Lee, G. M.-H., Siwicki, K .K., Standaert, D. G., and Kravitz, E. A., 1984, Proctolin in the lobster: The distribution, release, and chemical characterization of a likely neurohormone, *J. Neurosci.* **4:**1300–1311.

Scrivener, J. C. A., 1971, Agonistic behavior of the American lobster *Homarus americanus* (Milne Edwards), *Fisheries Research Board of Canada,* Technical Report, #235.

Starratt, A. N., and Brown, B. E., 1975, Structure of the pentapeptide proctolin, a proposed neurotransmitter in insects, *Life Sci.* **17:**1253–1256.

Sullivan, R. E., 1979, A proctolin-like peptide in crab pericardial organs, *J. Exp. Zool.* **210:**543–552.

Sullivan, R. E., Friend, B. J., and Barker, D. L., 1977, Structure and function of spiny lobster ligamental nerve plexuses: Evidence for synthesis, storage and secretion of biogenic amines, *J. Neurobiol.* **8:**581–605.

Truman, J. W., 1978, Hormonal control of invertebrate behavior, *Hormones Behav.* **10:**214–234.

Truman, J. W., and Schwartz, L. M., 1980, Peptide hormone regulation of programmed death of neurons and muscle in an insect, in: *Peptides: Integrators of Cell and Tissue Function* (F. E. Bloom, ed.), Raven Press, New York, pp. 55–67.

Wallace, B. G., 1976, The biosynthesis of octopamine—Characterization of lobster tyramine β-hydroxylase, *J. Neurochem.* **26:**761–770.

Willard, A. L., 1981, Effects of serotonin on the generation of the motor program for swimming by the medicinal leech, *J. Neurosci.* **1:**936–944.

19

Neurosecretory Role of Crustacean Eyestalk in the Control of Neuronal Activity

HUGO ARÉCHIGA, UBALDO GARCÍA, AND LEONARDO RODRÍGUEZ-SOSA

1. Functional Organization of the Eyestalk Neurosecretory System

From early histological work with methylene blue staining, the existence of neurosecretory cells was postulated in different regions of the eyestalk. Histochemical work rendered similar results (see Gabe, 1966). The most conspicuous system is that composed by the sinus gland, a neurohemal organ located in the distal part of the eyestalk, between the medulla externa and the medulla interna in many species (Fig. 1A). Its basic structure, as seen in Fig. 1B, is that of a bunch of neurosecretory endings, which are the dilated terminals of axons coming from other regions of the eyestalk, to end in apposition to a blood sinus. From morphological and physiological work, the notion was evolved of the sinus gland as the common end of secretory neurons all over the eyestalk and even of incoming fibers from other central ganglia. However, more recently, from experiments with cobalt backfills, a more restricted origin has been advocated, limiting the source of neurosecretory fibers to the sinus gland, to a group of 100–150 cell bodies clustered in the medulla terminalis and known since long ago as the X organ, or Hanstrom's organ (Andrew *et al.*, 1978; Jaros, 1978). Only a small number of cells outside this cluster were backfilled from the sinus gland.

A variety of cell shapes and sizes is present in the X organ, as seen in Fig. 1C.

HUGO ARÉCHIGA, UBALDO GARCÍA, AND LEONARDO RODRÍGUEZ-SOSA • Department of Physiology and Biophysics, Center of Investigation and of Advanced Studies of the IPN, Mexico, D.F.

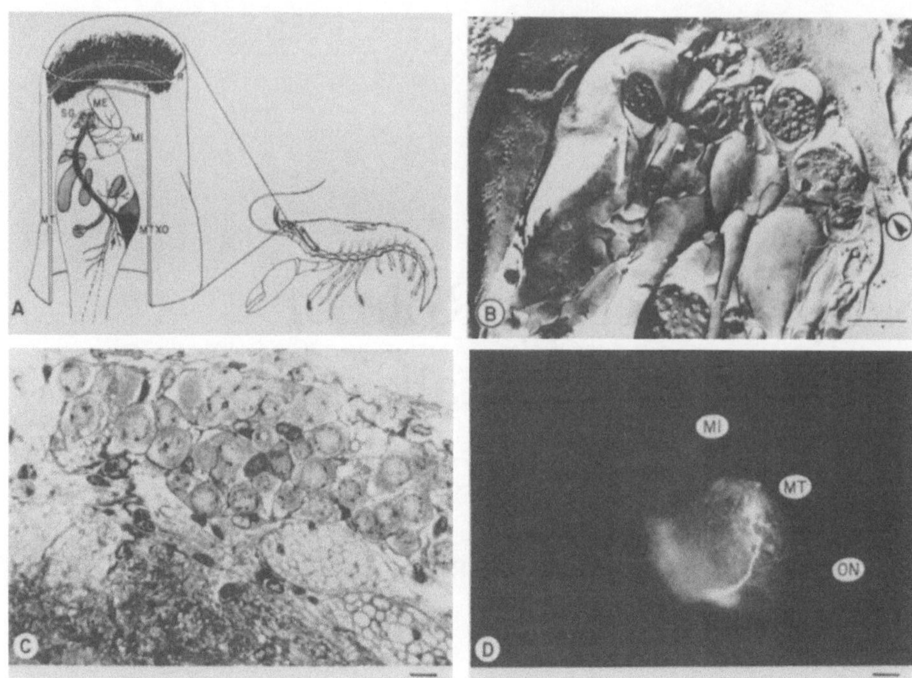

FIGURE 1. Main components of the neurosecretory system of the eyestalk. (A) Schematic representation of the crayfish eyestalk. Dashed areas indicate the position of main clusters of neurosecretory cells and their axonal projections to the sinus gland. (B) Freeze-fracture of the sinus gland of *Procambarus*, showing the great enlargement of the neurosecretory axons, to form the terminals, which are loaded with granules. Arrow indicates blood sinus. Exocytotic figures are occasionally seen. Scale, 5 μm (From Aréchiga and Chávez, unpublished). (C) Cell bodies in the X organ. Notice the variety in shapes of somata and the emergence of the axon bundle that projects to the sinus gland. Scale, 17.5 μm. (D) Profile of an X organ cell, intracellularly stained with Lucifer Yellow. Scale, 60 μm. MT, medulla terminalis; MTXO, medulla terminalis X organ; MI, medulla interna; ME, medulla externa; SG, sinus gland; R, retina; ON, optic nerve. (From Glantz *et al.*, 1983.)

They are characteristically monopolar with a profuse branching in the medulla terminalis. Their axons run initially along the rim of the medulla and then distally to end in the sinus gland (Fig. 1D).

The X organ–sinus gland system is regarded as the main neurosecretory structure of the eyestalk, which in turn appears to be the source of most of the known neurohormones in crustaceans.

The basic electrophysiology of these neurons is similar to that described for other neurosecretory cells. As seen in Fig. 2, cell bodies and axons generate overshooting spikes. Those in somata are longer than 5 msec, a duration greater than that of spikes recorded from other crustacean cell bodies and axons. This property, plus the fact that somatic spikes may persist in sodium-free medium, led Iwasaki and Satow (1971) to propose that calcium is the ion providing the electromotive force for the spike. A similar situation has been described by Cook (1977) for the spikes generated at the neurosecretory terminals in the sinus gland. The spikes in the axons are of shorter duration and abolished in sodium-free solutions, hence the notion that sodium is the active ion in their generation. Some cell bodies in the X organ have been found to be nonspiking; the antidromic stimulation of the axon gives rise only to a decrementally conducted signal, presumably originated in a spiking locus out of the cell body (Glantz *et al.*, 1983).

Some X organ neurons, both in intact animals and in the isolated eyestalk, exhibit spontaneous activity. The relationship between the pattern of electrical activity and the secretory function has not been yet defined. However, by correlating rate of electrical stimulation with secretory granule depletion (Bunt and Ashby, 1968) or with hormonal release, within a range of 0.1 to 10 Hz, a good correlation exists suggesting a one to one following (Aréchiga *et al.*, 1977; Cooke *et al.*, 1977). In fact, firing rates of 1–10/sec are not seldom seen in eyestalk neurosecretory cells.

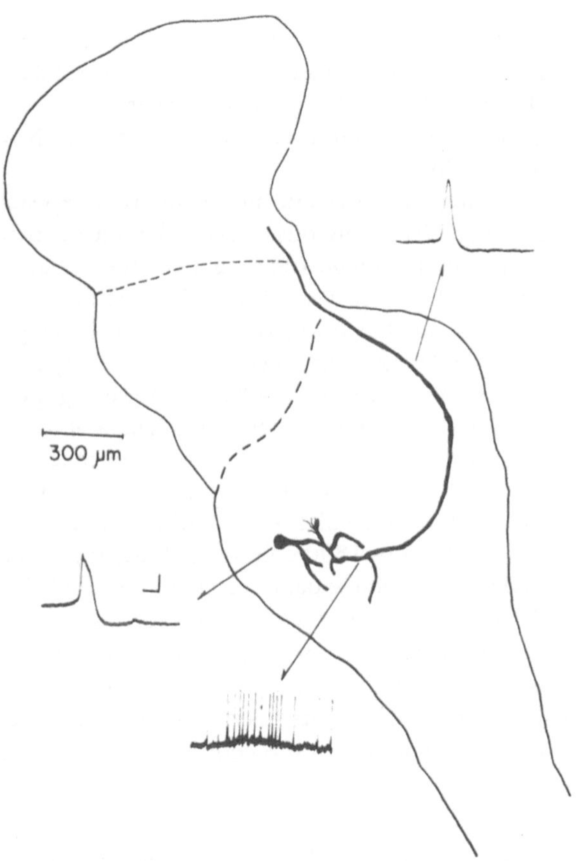

FIGURE 2. Types of electrical activity recorded at different levels of the X organ neurons. Notice duration of spike at cell soma. Burst of spikes in the branching region triggered by a light pulse. Calibrations: for single spikes, 30 mV, 10 msec at cell body; 30 mV, 2 msec at axon; for burst, 100 msec. (Modified from Glantz *et al.*, 1983).

From electrophysiological evidences, Iwasaki and Satow (1971) suggested the possibility of an electrotonic coupling between neurons in the X organ. Recently, Aréchiga *et al.* (1985a) have documented the existence of dye coupling with intracellular injection of Lucifer Yellow, and demonstrated the presence of gap junctions by freeze fracture, both in the X organ and the sinus gland. The coupling is also apparent between glial cells. Although the functional implications of this anatomical arrangement are not wholly clear yet, it is conceivable that neurons producing the same secretion can function as a unit. Since several neurohormones are known to be produced in the X organ, its functional organization as segregated neuroendocrine units appears likely.

2. Initial Characterization of a Neurodepressing Hormone in the Eyestalk

A hormonal control of neuronal activity in crustaceans was postulated on the basis of ablation and extract injection experiments. The eyestalk was supposed to secrete a substance lowering locomotor activity in several species of decapod crustaceans (for reviews of the early literature, see Aréchiga and Naylor, 1976; Aréchiga and Huberman, 1980b).

By recording from single sensory and motor neurons, eyestalk extracts were found to induce a lowering of spontaneous or evoked unitary activity in the crayfish *Procambarus bouvieri,* the crab *Carcinus maenas* (Aréchiga *et al.,* 1974), and the prawn *Nephrops norvegicus* (Aréchiga *et al.,* 1980). The region of the eyestalk where the highest content of this substance was reported is the sinus gland. The stimulation of isolated sinus glands either by electrical pulses or by incubation in high-potassium solutions was shown to release a substance lowering spontaneous activity in motoneurons of isolated abdominal ganglia (Aréchiga *et al.,* 1977). The substance responsible for this effect turned out to be dializable and inactivated by incubation in proteolytic enzymes (Aréchiga *et al.,* 1974). On this basis it was considered to be a peptide of low-molecular weight, which in Sephadex columns coeluted with markers between 1000–1300 daltons mol. wt. It showed no net electric charge when subjected to high-voltage paper electrophoresis (Aréchiga *et al.,* 1977; Huberman *et al.,* 1979). The peptide was also identified in the hemolymph of crayfish in levels high enough to induce a lowering of neuronal activity (Aréchiga *et al.,* 1979a). On this basis it was proposed that a neurodepressing hormone (NDH), acting as a modulator of neuronal activity in crustaceans, is released from the eyestalk. This chapter will present some more recent information about its properties.

3. Biological Assessment of the Neurodepressing Hormone (NDH)

So far, NDH activity has only been estimated by bioassay. Three levels of assay have been used.

3.1. Intact Animal Assays

As mentioned above, the initial assays involved whole animals and the effect on locomotor activity was determined. The equivalent to 0.5 eyestalk was found to result in a significant lowering of spontaneous locomotor activity when injected to *Carcinus,* shortly before the projected onset of the nocturnal phase of high locomotor activity (Williams *et al.,* 1979b).

Neurodepressing hormone has also been assessed by single-unit recording in intact animals. The same equivalent amounts of tissue are necessary to reduce the spontaneous or induced firing rate in identified neurons in either tethered or unrestricted animals.

3.2. Isolated Ganglia

Neurodepressing hormone has been tested also on isolated abdominal ganglia by recording the spontaneous activity of a great number of motoneurons. In all the six ganglia, the assayed units have been found to be depressed by NDH, although the sensitivity varies from one unit to another. By separately recording the activity of the six motoneurons in the pool of superficial flexors to abdominal muscles, the response to a pulse of NDH is quantitatively different in the various units. The differences may be related to the depth of the cell bodies in the ganglion (Wine *et al.,* 1974), since they are embedded in a thick mesh of connective tissue and surrounded by glial elements. This reduces the penetration of the peptide. It is known that the perineurium of crayfish central nervous system constitutes an important barrier to the diffusion of ions or large molecules (Abbot *et al.,* 1975; Kristensson *et al.,* 1972). Since the effect of NDH takes a few seconds to establish, it is unlikely that an equilibrium has been attained between the bath and the tissue.

The duration of the response is as long as the permanence of the substance. There is no apparent desensitization. As seen in Fig. 3, the effect may persist for minutes or even longer, at doses within the physiological range. Repeated application of NDH pulses result in depressions of similar magnitude. The isolated abdominal ganglion offers the advantage of the great number of bioassays that can be performed on a single preparation in a short time. The responsiveness of the preparation is constant for many hours. Also, the rate of activity in several units is steady for long periods of time. However, given the arrangement of the ganglia, the dose necessary to obtain effects by applying solutions to ensheathed preparations is not very different from that of the intact preparation. By desheathing the ganglia, the sensitivity increases, but the viability of the preparation is reduced.

3.3. Isolated Neurons

The most sensitive bioassay for NDH activity so far used is the isolated stretch receptor of the crayfish *Procambarus.* Since the pioneer work by Wiersma *et al.*

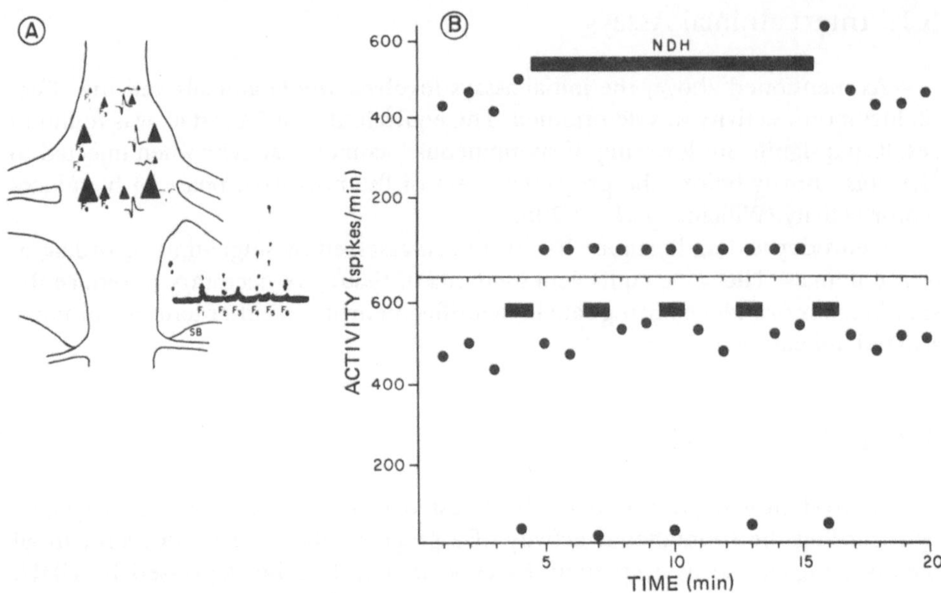

FIGURE 3. NDH effects on the abdominal superficial flexor motoneurons. (A) Schematic representation of the pool of six motoneurons sending their axons in the superficial third root (SB). Frequency histogram of the spontaneous firing of spikes in each neuron (From Aréchiga and Cerbón, 1981). (B) Lack of desensitization of the response to NDH with pulses of different duration and the same concentration (0.5 equivalent eyestalk). Upper graph, depression of neuronal firing by applying an 11-min pulse. Lower graph, depressions by applying a succession of 1-min pulses. Notice the slight rebound after long pulse. Ordinate, spontaneous firing rate in motoneurons. F 1–6 (spikes/min). Abscissae, time, in minutes.

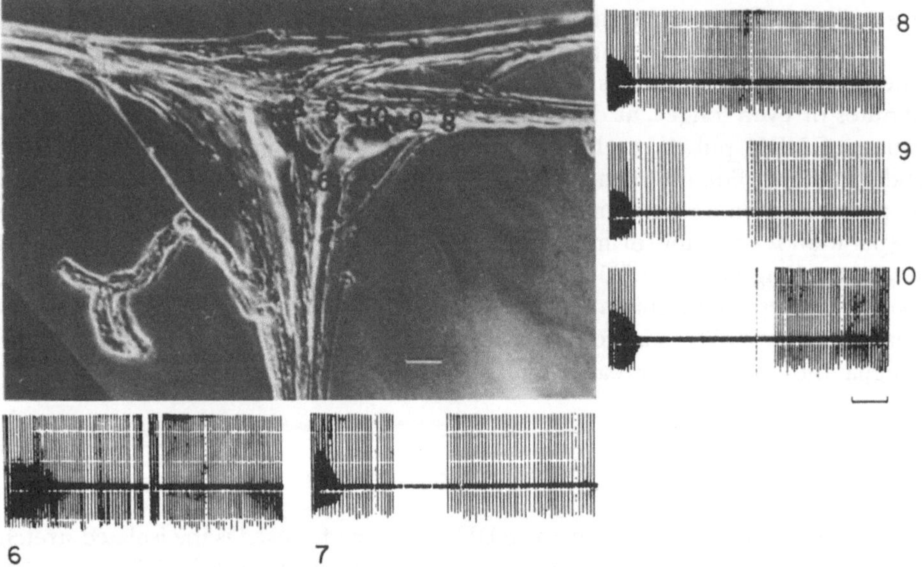

FIGURE 4. Effect of NDH application by pressure injection on different parts of isolated abdominal stretch receptor. Indicated by numbers. Points 1–5 were ineffective pulses along the axon (not shown). Scale, 60 μm. Time scale, 2 sec.

(1953), this preparation has turned out to be highly suitable for physiological and pharmacological purposes. As seen in Fig. 4, its sensitivity to pressure-injected NDH topically applied onto different parts of the neuron is greater in the somatodendritic region, and is 1–2 orders of magnitude higher than that of the superficial flexor motoneurons. By application to the bathing fluid, it is active at doses 0.01 to 0.001 equivalent eyestalks. In the following sections, unless otherwise stated, NDH was bioassayed on isolated stretch receptors.

4. Biochemical Identity of NDH

A great number of putative neurohormones have been postulated to be released from the eyestalk of crustaceans. However, only two have so far been chemically characterized and another is partly known in terms of its amino acid composition. This latter is the hyperglycemic hormone, a peptide of 57 residues (see Keller and Wunderer, 1978; Kleinholz, 1975), although small differences can be anticipated in its structure in different crustacean species given the fact that it does not cross-react between crustacean groups. One common feature in all species so far studied is that it does not dialize. This property clearly distinguishes it from NDH, which in all studied species is readily dializable.

The other two known peptides are the erythrophore-concentrating hormone (ECH) and the distal pigment light-adapting hormone (DPLH). The former is an octapeptide (Fernlund, 1974) that, similar to NDH, lacks a net electric charge and in high-voltage paper electrophoresis remains in the origin. However, both elute with different retention times in high-pressure liquid chromatography. As can be seen in Fig. 5, the upper trace is the elution of 2 μg of synthetic ECH in a gradient of acetonitrile. The lower trace is the elution in the same gradient of NDH purified from 100 eyestalks of the shrimp *Penaeus vannamei*. On the other hand, there is no physiological cross-reaction between the two peptides. While NDH has no effect on tegumentary chromatophores, ECH does not affect the discharge of neurons known to be affected by NDH. Moreover, as will be discussed later, ECH is released predominantly at night while the opposite is true for NDH (Aréchiga *et al.*, 1985b).

The third identified peptide from the eyestalk is the DPLH. It is an octadecapeptide (Fernlund, 1976) with different elution time to NDH in Sephadex G-25 and G-15 columns. Again, as illustrated in Fig. 5, DPLH does not affect neuronal activity. Neurodepressing hormone in turn has no effect on the distal pigment cells.

As to other neuroactive principles contained in the eyestalk, Mancillas *et al.* (1981), by immunocytochemistry, identified the presence of enkephalin-like somatostatin-like, and substance P-like immunoreactivity in various eyestalk regions of the lobster. More recently, Fingerman *et al.* (1983) have reported enkephalin-like and substance P-like immunoreactivity in the eyestalk of *Uca*. Neither of these peptides matches the effect of NDH on its various neuronal targets. In an extensive survey (H. Aréchiga and V. Anaya, unpublished data) of the effects of neuroactive substances on crayfish neurons, only the delta factor (Schoenenberger and Monnier, 1977) shows some effect, although at concentrations much higher than those calculated for NDH. Abdominal motoneurons of crayfish have been found sensi-

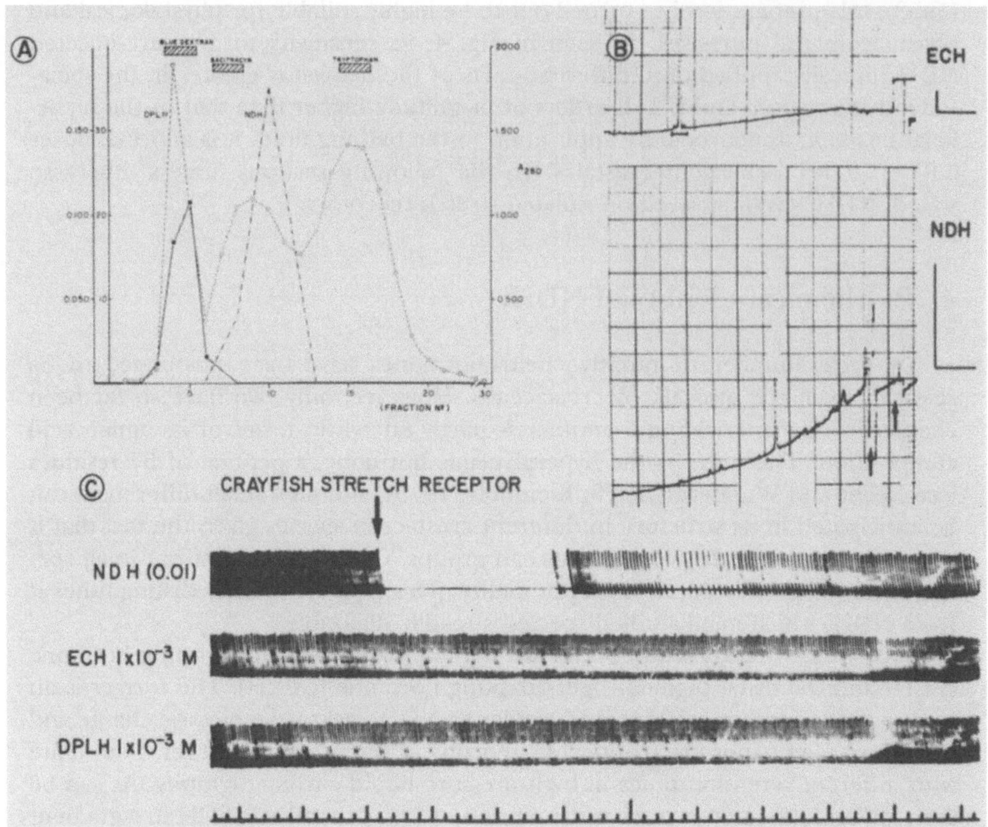

FIGURE 5. Separation of NDH from other eyestalk peptides. (A) Sephadex G-25 and G-15 suffice to separate NDH from DPLH. Notice markers. Ordinates from left to right, distal pigment index (●-●), NDH units (·--·), absorbence at 280 nm (○—○). Abscissae, fraction number. (From H. Aréchiga and T. Villanueva, unpublished data). (B) Difference in retention times in HPLC between ECH and NDH. (From H. Aréchiga *et al.*, unpublished data). Peaks corresponding to fractions containing ECH and NDH activities are indicated. Notice that gain for NDH record is five times larger than that for ECH. (C) Lack of effect of ECH and DPLH on the stretch receptor. Arrow indicates time of application. Time scale, 3 min. ECH for this experiment was obtained commercially. DPLH was donated by Prof. K. Ranga Rao. Time scale, 58 sec.

tive to another sleep-inducing peptide (Nagasaki *et al.*, 1976) obtained from rat brain. Preliminary tests of the factor S (Pappenheimer *et al.*, 1975) did not show a depressing effect on the crayfish stretch receptor (J. García-Arrarás, unpublished data). Other inhibitory substances present in the eyestalk are GABA and taurine (Arámburo, 1983).

4.1. Cross-Reactivity

Neurodepressing hormone cross-reacts between crustacean groups. Indeed, the sensitivity of *Nephrops* and *Carcinus* to NDH obtained from either species is almost equal (Aréchiga *et al.*, 1979b). Neurodepressing hormone from the shrimp *P. vannamei* is as effective on the freshwater crayfish *Procambarus* as the hormone obtained from animals of the same species. There is no evidence for NDH sensitivity or the presence of the hormone in other zoological groups.

5. Biosynthesis and Transport of NDH

As stated earlier, most of NDH in the crayfish is located in the X organ–sinus gland system. One strategy followed to identify the elements involved in its biosynthesis has been to deplete the eyestalk, either by focal electrical stimulation of the sinus gland or by incubation in high-K^+ solutions. Once NDH content is depleted, the eyestalk or segments of it are incubated in a proper amino acid mixture to allow for NDH content to be restored. Biosynthesis of NDH in the eyestalk is virtually confined to the medulla terminalis (Fig. 6). The localization of NDH-containing cell bodies, as identified by dissociating neurons from different regions of the medulla terminalis and determining their NDH content by bioassay, has allowed us to locate the main source of NDH in the medial region of the X organ (Fig. 7) within the cluster composed by the few tens of cells that, as mentioned above, have

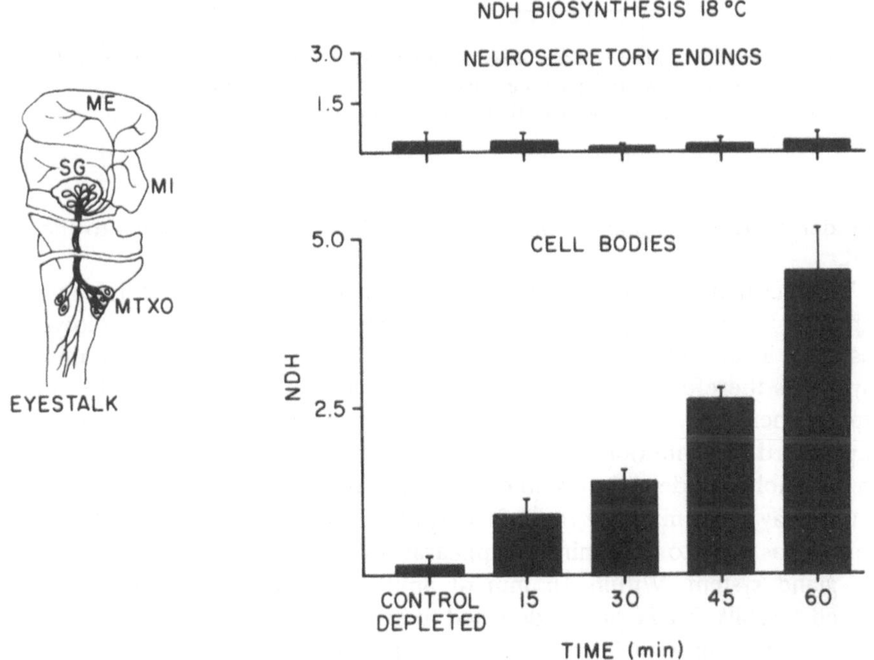

Figure 6. NDH biosynthesis in crayfish X organ somata. After transecting the axons in the X organ–sinus gland tract, NDH accumulates in the cell bodies and is absent in the sinus gland. Ordinate, NDH content in arbitrary units. Abscissae, time after incubation in amino acid mixture (From Aréchiga *et al.*, 1985c) (see text). ME, medulla externa; MI, medulla interna; MTXO, medulla terminalis X organ; SG, sinus gland.

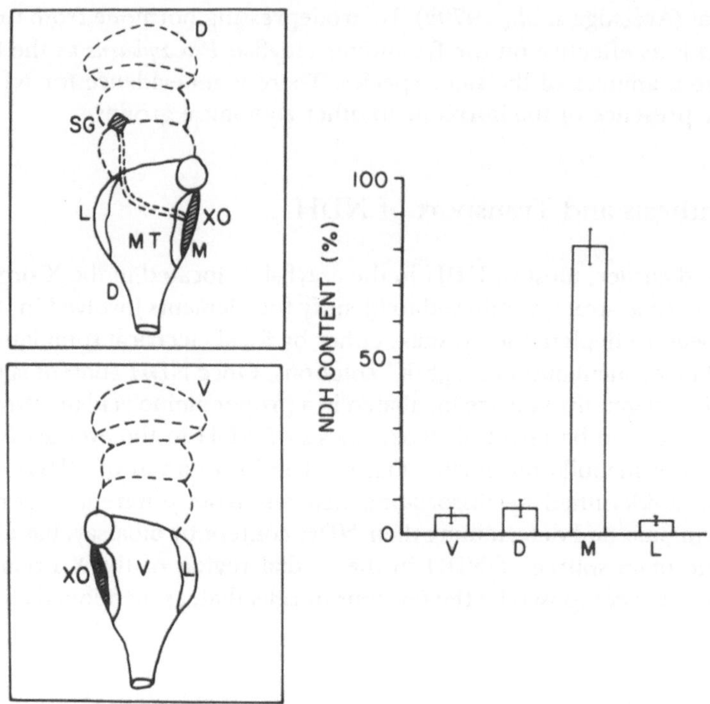

FIGURE 7. NDH content in cell bodies isolated from different regions of the crayfish eyestalk, as indicated by shaded areas. XO, X organ; MT, medulla terminalis; SG, sinus gland. Upper diagram illustrates dorsal view. Lower diagram, ventral view. D, dorsal; V, ventral; M, medial; L, lateral. Ordinate, percent of NDH content.

been identified as the main conglomerate of neurons sending axons to the sinus gland.

The fact that the sinus gland is incapable of synthesizing NDH indicates that, as appears to be the common feature for peptidergic neurons, the biosynthesis takes place in the cell soma and the product is transported to the terminals. The biosynthesis therefore appears to depend on a ribosomal protein synthesis mechanism (Gainer *et al.*, 1982). Actually, NDH biosynthesis in the medulla terminalis is suppressed by inhibition of ribosomal-dependent protein synthesis with agents such as cycloheximide, puromycin or emetine (Aréchiga *et al.*, 1985c).

One way to estimate the rate of transport of NDH from the cell somata to the terminals has been to determine its appearance in various regions of the X organ–sinus gland system. Within 15 min of amino acid incubation of a previously depleted eyestalk, NDH can be detected in small amounts in the medulla terminalis, but none is present in the terminals. After 1 hr, most of the NDH is in the sinus gland. Knowing the length of the X organ neuron axons as measured from the intracellular dye injection experiments, an estimation of the approximate rate of transport has been made in ~100 mm/24 hr, which corresponds to a rapid axoplasmic flow that is dependent on cytoskeletal translocation mechanisms (see

Ochs, 1982). This property has also been indicated by the fact that agents such as colchicine and D_2O, known to inhibit tubulin polymerization and therefore blocking axoplasmic flow, are capable of preventing the transfer of NDH from the X organ to the sinus gland (Aréchiga *et al.*, 1985c)

6. NDH Release

Since most NDH is stored in the sinus gland terminals, its release has been mainly studied at that level. It can be induced by depolarizing the nerve terminals, either by electric pulses of by incubation in high-K^+ solutions. The process is calcium dependent, since it does not occur in Ca^{2+}-free solutions, and it appears to involve an exocytotic process. It corresponds with a decrement in the number of neurosecretory granules present in the nerve terminals (see Aréchiga *et al.*, 1977).

Another possible site of NDH release can be the proximal part of the neuron at the soma or branching in the medulla terminalis. This is suggested by the fact that concentrations of K^+ similar to those evoking NDH release from the sinus gland do release the peptide from isolated medulla terminalis and even from solitary cell bodies (Fig. 8) with only a short stump of the axon. Actually, extraterminal release of neurosecretory substances has been demonstrated in crustaceans for octopamine (Evans *et al.*, 1976), but so far not for a peptide. The possible physiological significance of this phenomenon remains to be elucidated.

7. Synaptic Input to the X Organ

There are various morphological and physiological evidences indicating that the secretion of peptides from the X organ–sinus gland system is under synaptic control. Electron microscopy studies have disclosed the presence, in the neuropil, of synaptic contacts (Andrew and Saleuddin, 1978). The nature of the neurons making these connections has not been determined. Sensory stimulation is known to affect the release of eyestalk peptides. Light is the physiological stimulus triggering the secretion of the DPLH (Aréchiga and Mena, 1975; Aréchiga, 1977) and quite likely of the chromatophoric hormone(s) participating in the dispersion of the various tegumentary pigments. Conversely, the release of peptides like the ECH and others related to the nocturnal adaptation is expected to be triggered by darkness.

In a recent search for the possible synaptic control of the light responses by intracellularly recording from X organ neurons in the crayfish, excitatory postsynaptic potentials (EPSPs) were detected in many neurons in response to light, which also enhanced the ongoing activity in some cells. The opposite effect, i.e., inhibition by light, was also detected in some neurons. The detailed circuitry involved in the pathway from the photoreceptors to the X organ cells is still to be determined, but from preliminary estimations of latencies, it does not appear to be much more complex than that of the visual interneurons themselves (Glantz *et al.*, 1983).

It is likely that the activity of the neurosecretory cells in the eyestalk can be

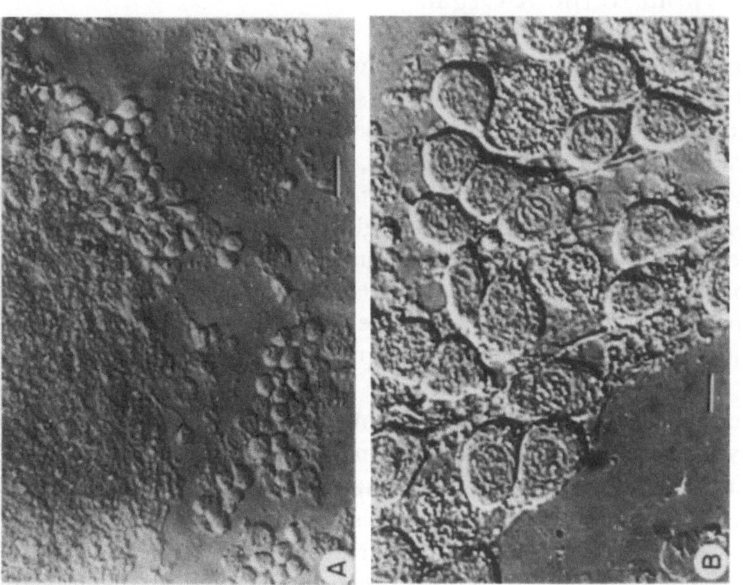

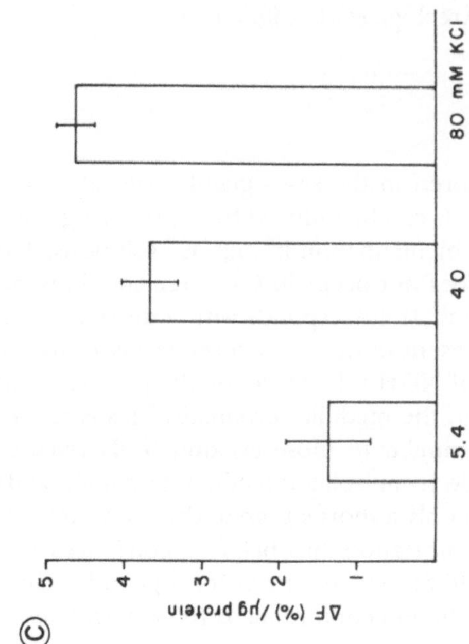

FIGURE 8. NDH release from isolated X organ neurons of crayfish. Photographs at left illustrate the population of cell somata at the beginning of the isolation (A), and once they are detached from the medulla terminalis (B). The right part of the figure represents NDH release from isolated cells, incubated in solutions with different potassium concentrations as indicated. Ordinate, reduction in firing rate of crayfish stretch receptor %/μg of protein in the cell population. Scales, (A) 44 μm; (B) 12 μm.

modulated transsynaptically from at least two sources: (1) the visual system intrinsic to the eyestalk and (2) other afferent influences conveyed by the multisensory channels represented in the vast number of efferent fibers in the optic nerve. It has also been postulated that extraretinal influences can modulate neurohormone release from the eyestalk (Arěchiga *et al.*, 1985b).

Electrophysiologically, the stimulation of the optic nerve has been shown to evoke inhibitory postsynaptic potentials (IPSPs), which can be blocked with picrotoxin. This suggests a GABAergic nature; however, their equilibrium potential is far from the chloride equilibrium potential, the value to be expected of a GABA-induced conductance change (Iwasaki and Satow, 1973). This indicates the need for further analysis.

Another agent known to affect the eyestalk neurohormones is serotonin (5-HT). It has been shown to enhance the release of the hyperglycemic hormone (Bauchau and Mengeot, 1966; Keller and Beyer, 1968). The release of chromatophoric hormones is also known to be affected by 5-HT, norepinephrine, dopamine, GABA, and met-enkephalin (see Quackenbush and Fingerman, 1984, Fingerman *et al.*, 1985).

By multiunit recording from the sinus gland terminals, 5-HT has been found to enhance the firing rate of certain elements, whereas GABA exerts an inhibitory action. Serotonin, in turn, increases the basal rate of NDH release, whereas GABA lowers it (Arěchiga *et al.*, 1985c). The role of GABA as an inhibitory transmitter in this system can also be inferred from the fact that in a recent survey of GABA content in the eyestalk by reverse-phase liquid chromatography, U. García *et al.* (unpublished data) have found that 18% of all GABA in the crayfish eyestalk (6.92 nmole) is in the medulla terminalis, and it can be released by incubation in high-potassium solutions in a calcium-dependent manner (Fig. 9). This can be a common

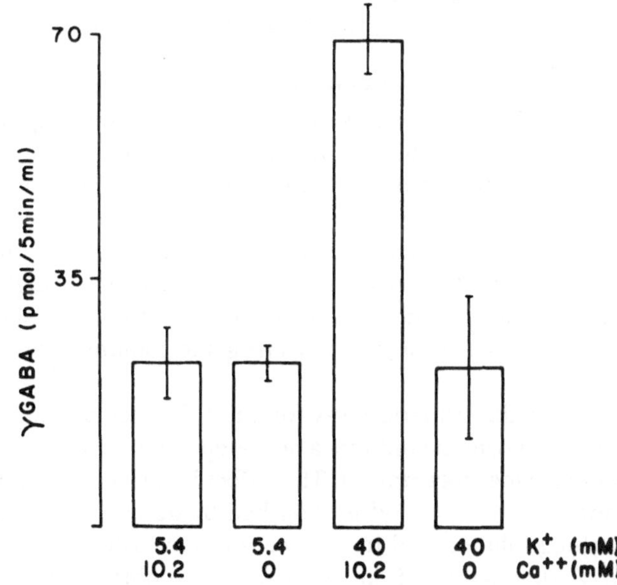

FIGURE 9. Gamma-GABA release from the crayfish eyestalk. The figures at bottom correspond to the combination of K^+ and CA^{2+} in the solution in which the eyestalks were incubated. Ordinate, GABA released, as indicated. Each bar corresponds to five experiments, with S.E.

feature in crustaceans, since Arámburo (1983) has recently identified a similar amount of GABA in the eyestalk of the shrimp *P. vannamei.*

The role of GABA as an inhibitory transmitter in the neurosecretory cells of the eyestalk does not appear restricted to the medulla terminalis X organ, since a group of putative neurosecretory cells in the medulla externa of the crayfish, recently studied by Kirk *et al.* (1983), are also inhibited by GABA, which may be the transmitter mediating the inhibitory effect of light on these cells.

8. Sites and Modes of Action of Neuroactive Peptides

As mentioned above, one peculiar feature of NDH is the wide range of its inhibitory action. From sensory receptors to interneurons and motoneurons, virtually all classes of nervous elements tested so far appear to be responsive to the peptide. It is effective at different cellular sites. As mentioned earlier, on the isolated abdominal stretch receptor, the most sensitive spots are on the soma and initial dendrite area. The axon hillock and the length of axon are much less sensitive. There is a clear regional difference of sensitivity, as compared with the profile obtained for other agents such as KCl or local anesthetics, which are more active on the axonal part of the neuron, or glutamate, which selectively affects the muscle fiber attached to the receptor (H. Aréchiga and V. Anaya, unpublished data). However, NDH has been found to lower the excitability of the isolated medial giant axon. In central neurons, PSPs evoked by sensory stimulation are reduced by NDH without affecting the presynaptic spikes (J. J. Wine *et al.*, unpublished data).

So far, most of NDH effects have been studied on sensory systems and those subserving locomotion, but the vegetative system may also be a target for its hormonal action. Mancillas *et al.* (1980) have described a depressing effect of eyestalk extracts on the spontaneous activity of motoneurons in the stomatogastric ganglion; Fleischer (1981) found that eyestalk ablation led to enhanced pyloric activity; and eyestalk extracts were shown to inhibit the spontaneous rhythm. Whether NDH is responsible for these effects is a matter for further study. On the other hand, NDH injection at doses capable of reducing locomotor activity does not affect the heart rate of crayfish (H. Aréchiga and V. Anaya, unpublished data). It is clearly necessary to further define the establishment of NDH activity on the crustacean nervous system.

As to its mechanism(s) of action, NDH's effect on the spontaneous electrical activity of motoneurons in isolated abdominal ganglia in the crayfish has been found to be abolished by ouabain or potassium-free solutions, thus suggesting that NDH may act through an electrogenic sodium system (Aréchiga and Huberman, 1980a).

A similar situation was found in the medial giant axon of crayfish. Neurodepressing hormone induces a dose-dependent hyperpolarization and enhances posttetanic hyperpolarization. This effect is exerted without changes in the input resistance of the axon and is abolished by oubain, by substituting Li^+ for Na^+ in the bathing solution and in K^+-free solutions. All these evidences strengthen the view of a stimulation of an electrogenic sodium pump (Aréchiga and Huberman, 1980a).

Another concurrent piece of evidence is the biochemical characterization of a Na^+-K^+ ATPase in the crayfish abdominal ganglia (Aréchiga and Cerbón, 1981). This enzyme appears to be activated in a temperature-dependent manner by NDH (J. Cerbón and H. Aréchiga, unpublished data). Whether this mechanism of action is the only one for NDH or whether other mechanisms operate as well is a matter yet unsolved. The intervention of cyclic nucleotides in mediating NDH action has also been suggested (Aréchiga and Huberman, 1980a).

9. Interaction between Peptides and Other Blood-borne Agents in Crustaceans

The modulation of neuronal activity is also achieved by pathways different from NDH, and even the same neurons are targets for more than one blood-borne agent. One case in point is the set of motoneurons supplying the superficial flexor abdominal muscles. As mentioned above, this pool of six motoneurones is composed of five excitatory axons and one inhibitory. They are also sensitive to 5-HT, with a different pattern of NDH responsiveness (see Chapter 18, this volume). It is conceivable that at least these two agents are simultaneously acting on the excitability of the same neurons.

Another example is provided by the effect of amino acids on the tonic discharge of the slow-adapting abdominal stretch receptor. It is well known that it receives efferent inhibitory innervation and GABA was proposed as the neurotransmitter of this synapse (see Craelius and Fricke, 1981). As in other systems, the stretch receptor is sensitive to other amino acids such as taurine and glycine, and at least GABA and taurine have been identified in the hemolymph at concentrations at least threshold to reduce the tonic discharge of the receptor (U. García *et al.*, unpublished). The possible integration of influences exerted by the same agent acting either transynaptically or by a humoral route is still to be defined. From pharmacological evidence, it is clear that unlike the effect of the inhibitory amino acids, which are blocked by picrotoxin, the effect of NDH persists, but is blocked by ouabain (as mentioned earlier), suggesting different receptors and mechanisms. As in other neurons, GABA has been proposed to act on the stretch receptor by opening chloride channels, whereas NDH, as stated above, stimulates an ATPase system. The possible physiological interactions between these systems are still unclear.

Other possible interactions are with those peptides whose function has not been characterized as yet, but which have been identified by immunocytochemistry (Mancillas *et al.*, 1981; Fingerman *et al.*, 1983), such as enkephalin, substance P, and somatostatin. Enkephalin appears to reduce the release of ECH from the eyestalk of *Uca* (Quackenbush and Fingerman, 1984). This suggests the possibility in crustaceans of systems of peptides controlling peptide secretion, as observed in vertebrate systems. Leu-enkephalin is also active in inhibiting the spontaneous activity of abdominal motoneurons (H. Aréchiga and V. Anaya, unpublished data). Another possibility is peptides acting at nonneural structures in a concerted manner to those active on the nervous system. Such a model of integration has been pointed out for the modulation of visual input in crustaceans. The concurrence of effects of NDH reducing neuronal excitability in visual neurons and DPLH promoting the shielding of photoreceptors and thus reducing the amount of light reaching the rhabdomes (Aréchiga, 1977; Aréchiga and Huberman, 1980b) contribute to adjust the output of retinal photoreceptors, depending on the intensity of illumination and time of day. More recently, an effect on muscle has been described for proctolin (Chapter 18, this volume). In all likelihood, more neuroendocrine substances modulate neuronal excitability in crustaceans. This is the case of a peptide identified in the eyestalk and other regions of the central nervous system, which enhances the spontaneous activity both in abdominal superficial flexor motoneurons and in the tonic stretch receptor (Aréchiga and Huberman, 1981).

10. Physiological Role of Neuroactive Peptides in Crustaceans

Since the first studies of NDH action, given its prolonged activity and the fact that it acts on those neurons that in long-term recordings were found to display a circadian rhythm of activity (see Aréchiga, 1979; Aréchiga *et al.*, 1980), it was postulated that it could act as a modulator conveying an inhibitory influence and therefore partaking of the integration of the diurnal phase of the activity rhythms. This point was explored more recently by determining NDH activity at different times of day, and, as might be expected for a nocturnal animal such as crayfish, its content in the eyestalk was found to be higher at daytime than at night (Aréchiga and Huberman, 1980a). There is at present sufficient evidence to postulate that NDH secretion is under the influence of a circadian rhythm, and it is under study whether the rhythmicity is a property inherent to the neuroendocrine cells or whether they are under the influence of circadian clocks located elsewhere. It is interesting to note in this regard that Williams *et al.* (1979a) described in isolated eyestalks evidences of rhythmical changes with circatidal periodicity in a particular class of neurons in the X organ (Hanstrom's organ, in their nomenclature) of the crab *Carcinus maenas.* It is likely, therefore, that NDH participates in the control of the low-activity phase of the circadian cycle, together with other neuropeptides acting both on neural and nonneural targets. That the rhythms of activity may be also under the influence of nonpeptide channels appears quite likely as well. For instance, the level of GABA in the crayfish hemolymph is higher at daytime than at night (U. García, unpublished data).

The role of NDH in controlling the low-activity phase leaves unsettled the point of the possible mechanisms underlying the nocturnal phase of high activity; it is significant in this regard that the excitatory peptide is found in the crayfish eyestalk in higher levels at night than at day time (H. Aréchiga *et al.,* unpublished data). The interactions between humoral and axonal channels in the control of circadian rhythmicity of crustaceans are still to be defined (see Larimer and Smith, 1980).

The role of NDH in diurnal rhythmicity may well also be valid for other rhythms, such as the tidal one. In fact, early experiments by Naylor *et al.* (1973) in

Carcinus showed different potency of the locomotion reduction by eyestalk extracts depending on the phase of tidal cycle at which the samples were taken. It is also known that locomotor activity in crustaceans undergoes profound changes during the molt cycle (Bliss, 1962). No information is available as to the correlative changes of NDH secrection.

From the aforementioned data, the picture that is emerging is that of a complex, finely tuned interplay of modulatory influences on the crustacean nervous system in which blood-borne agents interact with synaptic mediators adjusting neuronal activity to levels compatible with the physiological integrations at a given time.

ACKNOWLEDGMENTS. The authors are greatly indebted to Víctor Anaya, María Teresa de la Vega, and Paula Vergara for their assistance during the experiments. This work was partly funded by grant No. PCCBNAL 790004 of CONACyT to H. A. and predoctoral fellowships of CONACyT to U. G. and L. R. S.

References

Abbot, N. J., Moreton, R. B., and Pichon, Y., 1975, Electrophysiological analysis of potassium and sodium movements in crustacean nervous system, *J. Exp. Biol.* **63:**85–115.

Andrew, R. D., and Saleuddin, A. S. M., 1978, Structure and innervation of a crustacean neurosecretory cell, *Can. J. Zool.* **56:**423–430.

Andrew, R. D., Orchard, I., and Saleuddin, A. S. M., 1978, Structural reevaluation of the neurosecretory system in the crayfish eyestalk, *Cell Tissue Res.* **190:**235–246.

Arámburo, C., 1983, Correlación entre actividad neurodepresora y diferentes entidades químicas presentes en el tallo ocular de *Penaeus vannamei* (Boone), Ph. D. Thesis, Universidad Nacional Autónoma de México.

Aréchiga, H., 1977, Modulation of visual input in the crayfish, in: *Identified Neurones and Behavior of Arthropods* (G. Hoyle, ed.), Plenum Press, New York, pp, 387–403.

Aréchiga, H., 1979, Circadian modulation of behavior in crustaceans, *Neurosci. Res. Prog. Bull.* **17:**672–679.

Aréchiga, H., and Cerbón, J., 1981, The influence of temperature and deuterium oxide on the spontaneous activity of crayfish motoneurons, *Comp. Biochem. Physiol.* **69A:**631–636.

Aréchiga, H., and Huberman, A., 1980a, Peptide modulation of neuronal activity in crustaceans, in: *The Role of Peptides in Neuronal Function,* Marcel Dekker, Inc. New York, pp. 317–349.

Aréchiga, H., and Huberman, A., 1980b, Hormonal modulation of circadian rhythmicity in crustaceans, in: *Comparative Aspects of Neuroendocrine Control of Behavior* (C. Valverde and H. Aréchiga, eds.), S. Karger Basel, pp. 16–34.

Aréchiga, H., and Huberman, A., 1981, A neuropeptide inducing enhancement of neuronal activity in crayfish. *Soc. Neurosci. Abstr.* **8:**103.

Aréchiga, H., and Mena, F., 1975, Circadian variations of hormonal content in the nervous system of the crayfish, *Comp. Biochem. Physiol.* **52A:**581–584.

Aréchiga, H., and Naylor, E., 1976, Endogenous factors in the control of rhythmicity in decapod crustaceans, in: *Biological Rhythms in the Marine Environment.,* (P. J. De Coursey, ed.), University of South Carolina Press, pp. 1–16.

Aréchiga, H., Huberman, A., and Naylor, E., 1974, Hormonal modulation of circadian neural activity in *Carcinus maenas* (L.), *Proc. R. Soc. London B* **187:**299–313.

Aréchiga, H., Huberman, A., and Martínez-Palomo, A., 1977, Release of a neuro-depressing hormone from the crustacean sinus gland, *Brain Res.* **128:**93–108.

Aréchiga, H., Cabrera-Peralta, C., and Huberman, A., 1979a, Functional characterization of the neurodepressing hormone in the crayfish, *J. Neurobiol.* **10:**409–422.

Aréchiga, H., Williams, J. A., Pullin, R. S. V., and Naylor, E., 1979b, Cross sensitivity to neurodepressing hormone in two different groups of crustaceans, *Gen. Comp. Endocrinol.* **37:**350–357.

Aréchiga, H., Atkinson, R. J. A., and Williams, J. A., 1980, Neurohumoral basis of circadian rhythmicity in *Nephrops norvegicus* (L), *Mar. Behav. Physiol.* **7:**185–197.

Aréchiga, H., Chávez, B., and Glantz, R. M., 1985a, Dye coupling and gap junctions between crustacean neurosecretory cells, *Brain Res.* **326:**183–187.

Aréchiga, H., Cortés, J. L., García, U., and Rodríguez-Sosa, L., 1985b, Neuroendocrine correlates of circadian rhythmicity in crustaceans, *Am. Zool.* **25:** (in press).

Aréchiga, H., Flores-López, J., and García, U., 1985c, Control of biosynthesis and release of the crustacean neurodepressing hormone, in: *Comparative Endocrinology Symposium* (B. Lofts and D. Chan, eds.) (in press).

Bauchau, A. G., and Mengeot, J. C., 1966, Sérotonine et glycemie chez les crustacés, *Experientia* **22:**238.

Bliss, D. E., 1962, Neuroendocrine control of locomotor activity in the land crab *Gecarcinus lateralis,* in: *Memoirs of the Society of Endocrinologists,* Vol. 12, *Neurosecretion* H. Heller and R. B. Clark (eds.), New York Academic Press, New York, pp. 391–408.

Bunt, A. H., and Ashby, E. A., 1968, Ultrastructural changes in the crayfish sinus gland following electrical stimulation, *Gen. Comp. Endocrinol.* **10:**376–382.

Cooke, I. M., 1977, Electrical activity of neurosecretory terminals and control of peptide hormone release, in: *Peptides in Neurobiology* (H. Gainer, Ed.), Plenum Press, New York, pp. 345–374.

Cooke, I. M., Haylett, B. A., and Weatherby, T. M., 1977, Electrically elicited neurosecretory and electrical responses of the isolated crab sinus gland in normal and reduced calcium salines, *J. Exp. Biol.* **70:**125–149.

Craelius, W., and Fricke, R. A., 1981, Release of ^{3}H-gamma-aminobutyric acid (GABA) by inhibitory neurons of the crayfish, *J. Neurobiol.* **12:**249–258.

Evans, P. D., Kravitz, E. A., and Talamo, B. R. 1976, Octopamine release at two points along lobster nerve trunks, *J. Physiol. (London)* **262:**71–89.

Fernlund, P., 1974, Structure of the red-pigment concentrating hormone of the shrimp *Pandalus borealis, Biochem. Biophys. Acta* **371:**304–316.

Fernlund, P., 1976, Structure of a light-adapting hormone from the shrimp *Pandalus borealis, Biochim. Biophys. Acta* **439:**17–25.

Fingerman, M., Hanumante, M., and Vacca, I., 1983, Enkephalin-like and substance P-like immunoreactivity in the eyestalk neuroendocrine complex of the fiddler crab, *Uca pugilator, Soc. Neurosci. Abstr.* **9:**439.

Fingerman, M., Hanumante, M. M., and Fingerman, S. W., 1985, The role of neurotransmitter substances in the release of chromatophorotropic hormones in crustaceans, in: *Comparative Endocrinology Symposium* (B. Lofts, ed.) (in press).

Fleischer, A. G., 1981, The effect of eyestalk hormones on the gastric mill in the intact lobster *Panulirus interruptus, J. Comp. Physiol.* **141:**363–368.

Gabe, M., 1966, *Neurosecretion,* Pergamon Press, New York.

Gainer, H., Loh, Y. P., and Neale, E. A., 1982, The organization of posttranslational precursor processing in peptidergic neurosecretory cells, in: *Proteins of the Nervous System: Structure and Function* B. Haber, J. R. Pérez-Polo, and J. D. Coulter, eds.), Alan Liss, Inc., New York, pp. 131–145.

Glantz, R. M., Kirk, M. D., and Aréchiga, H., 1983, Light input to crustacean neurosecretory cells, *Brain Res.* **265:**307–311.

Huberman, A., Aréchiga, H., Cimet, A., De La Rosa, J., and Arámburo, C., 1979, Isolation and purification of a neurodepressing hormone from the eyestalk of *Procambarus bouvieri* (Ortmann), *Eur. J. Biochem.* **99:**203–208.

Iwasaki, S., and Satow, Y., 1971, Sodium- and calcium-dependent spike potentials in the secretory neuron soma of the X-organ of the crayfish, *J. Gen Physiol.* **57:**216–238.

Iwasaki, S., and Satow, Y., 1973, Electrical characteristics of the membrane in neurosecretory neurons, in: *Neuroendocrine Control* (K. Yagi and S. Yoshida, eds.), Wiley, New York, pp. 85–109.

Jaros, P. P., 1978, Tracing of neurosecretory neurons in crayfish optic ganglia by cobalt iontophoresis, *Cell Tissue Res.* **194:**297–302.

Keller, R., and Beyer, J., 1968, Zur hyperglykämischen wirkung von Serotonin und Augenstiel extract beim Flusskrebs *Orconectes limosus, Z. Vergl. Physiol.* **59**:78–85.

Keller, R., and Wunderer, G., 1978, Purification and aminoacid composition of the neurosecretory hyperglycaemic hormone from the sinus gland of the shore crab *Carcinus maenas*, *Gen. Comp. Endocrinol.* **34**:328–335.

Kirk, M. D., Prugh, J. I., and Glantz, R. M., 1983, A visually induced GABA mediated IPSP in a crustacean neurosecretory cell, *J. Neurobiol.* **14**:473–480.

Kleinholz, L. H., 1975, Purified hormones from the crustacean eyestalk and their physiological specificity, *Nature* (London) **258**:256–257.

Kristensson, K., Stromberg, E., Eloffson, R., and Olsson, Y., 1972, Distribution of protein traces in the nervous system of the crayfish *(Astacus astacus* Linné) following systemic and local application, *J. Neurocytol.* **1**:35–47.

Larimer, J., and Smith, J., 1980, Circadian rhythm of retinal sensitivity in crayfish: Modulation by the cerebral and optic ganglia, *J. Comp. Physiol.* **136**:313–326.

Mancillas, J. R., Leff, S., and Selverston, A., 1980, A neuroactive factor from the lobster sinus gland modulates the spontaneous activity of identified neural networks, *Neurosci. Abstr.* **6**:703.

Mancillas, J. R., McGinty, J. F., Selverston, A. I., Karten, H., and Bloom, F. E., 1981, Immunocytochemical localization of enkephalin and substance P in the retina and eyestalk neurones of lobster, *Nature (London)* **293**:576–578.

Nagasaki, H., Iriki, M., and Uchizono, K., 1976, Inhibitory effect of the brain extract from sleep-deprived rats (BE-5DR) on the spontaneous discharges of crayfish abdominal ganglion, *Brain Res.* **109**:202–205.

Naylor, E., Smith, G., and Williams, B., 1973, The effect of eyestalk extracts on the circadian locomotor rhythm of *Carcinus*, in: *Neurobiology of Invertebrates* (J. Salanki, ed.), Publishing House of the Hungarian Academy of Sciences, Budapest, pp. 423–429.

Ochs, S., 1982, *Axoplasmic Transport and Its Relation to Other Nerve Functions*, J. Wiley & Sons, New York.

Pappenheimer, J. R., Koski, G., Fencl, V., Karnovsky, M. L., and Krueger, J., 1975, Extraction of sleep-promoting factor S from cerebrospinal fluid and from brains of sleep-deprived animals, *J. Neurophysiol.* **38**:1299–1311.

Quackenbush, L. S., and Fingerman, M., 1984, Regulation of neurohormone release in the fiddler crab *Uca pugilator*: Effects of Gamma-aminobutyric acid, Octopamine, met-enkephalin and beta-endorphin. *Comp. Biochem. Physiol.* **79**:77–84.

Schoenberger, G. A., and Monnier, M., 1977, Characterization of a delta-encephalogram (-sleep) inducing peptide, *Proc. Natl. Acad. Sci. U.S.A.* **74**:1282–1286.

Wiersma, C. A. G., Furschpan, E., and Florey, E., 1953, Physiological and pharmacological observations on muscle receptor organs of the crayfish *Cambarus clarkii*, Girard, *J. Exp. Biol.* **30**:136–150.

Williams, J. A., Pullin, R. S. V., Naylor, E., Smith, G., and Williams, B. G., 1979a, The role of Hanstrom's organ in clock control in *Carcinus maenas*, in: *Cyclic Phenomena in Marine Plants and Animals* (E. Naylor, and R. G. Hartnoll, eds.), Pergamon Press, Oxford and New York, pp. 459–466.

Williams, J. A., Pullin, R. S. V., Williams, B. G., Aréchiga, *H.*, and Naylor, E., 1979b, Evaluation of the effects of injected eyestalk extract on rhythmic locomotor activity in *Carcinus*, *Comp. Biochem. Physiol.* **62A**:903–907.

Wine, J. J., Mittenthal, J. E., and Kennedy, D., 1974, The structure of tonic flexor motoneurons in crayfish abdominal ganglia, *J. Comp. Physiol.* **93**:315–336.

20

Activation of Neuronal Circuits by Circulating Hormones in Insects

JAMES W. TRUMAN AND JANIS C. WEEKS

1. Introduction

A central question in neuroethology is why do animals behave in the way they do? When confronted with a particular set of environmental stimuli, why does an animal act in one way at one time and then in a completely different fashion the next? Two factors that play a major role in this behavioral plasticity are the prior experience of the animal and the animal's "behavioral state." The latter feature is set by a number of internal conditions, perhaps the most important of which is the hormonal composition of the fluids bathing the nervous system. It has been clear for well over a hundred years since the first studies by Berthold (1849) that chemicals (that later turned out to be hormones) can unlock specific behaviors that are not accessible to the animal under other conditions. These earliest studies involved the expression of sexual behaviors, but it has since been shown that other coordinated behaviors such as those involved in hunger, thirst, aggression, and so forth are all called forth to a greater or lesser extent by their own specific circulating chemicals.

Two types of hormones involved in behavioral modulation are the peptide and the steroid hormones. A major contemporary question in neurobiology relates to how these chemicals interact with neural circuits to bring about coordinated behavioral responses. Over the past few years a number of invertebrates have provided useful preparations in which to study hormone action on the CNS at a cellular level

JAMES W. TRUMAN • Department of Zoology, University of Washington, Seattle, Washington 98195.
JANIS C. WEEKS • Department of Entomology, University of California, Berkeley, Berkeley, California 94720.

(e.g., Mayeri *et al.*, 1979a,b; Levine and Truman, 1982). This chapter will examine the work that has been carried out on the hormonal modification of behavior during the different life stages of the tobacco hawkmoth, *Manduca sexta,* and the silkmoth, *Hyalophora cecropia.* In both species, peptide and steroid hormones act on the CNS to bring about profound behavioral changes throughout the life of the animal.

2. Eclosion Hormone as a Modulator of Behavioral Activity

Most insects have complex life histories during which the individual progresses through a series of distinct stages, each with its own particular morphology and behavior. The transition between stages occurs during a molt, when the cuticular exoskeleton for the next stage is formed. At the culmination of the molt, the cuticle of the preceding stage is shed by a behavior called ecdysis. (In insects that have complete metamorphosis, the ecdysis to the adult stage is often termed *eclosion.*) Ecdysis involves a series of stereotyped motor patterns (e.g., Carlson 1977a,b; Weeks and Truman, 1984a), the principal one being rhythmic peristaltic waves that progress anteriorly along the abdomen. In some cases these abdominal waves are accompanied by other motor patterns that serve to extricate appendages from their surrounding sheaths of old cuticle (e.g., Hughes, 1980; Carlson, 1977a; Weeks and Truman, 1984a,b).

At metamorphic molts (i.e., from the larval to the pupal stage or from the pupal to the adult), ecdysis also signals the abrupt appearance of behaviors appropriate to the new stage. For example, immediately prior to the adult ecdysis of moths, the newly formed adult does not show normal adult behavior even when the surrounding pupal cuticle is manually peeled from around it. Adult behavior patterns then abruptly appear when the insect attempts ecdysis (Blest, 1960; Truman, 1976). In *Manduca,* these abrupt behavioral changes at the pupal and adult stages include alterations in specific sensory-to-motor reflexes that have been amenable to neural analysis (see Section 3.2).

Studies on adult eclosion of giant silkmoths first suggested that eclosion and associated behavioral changes were triggered by a factor released into the circulation from the brain (Truman and Riddiford, 1970). This factor, termed eclosion hormone (EH), was subsequently isolated and partially purified from pre-emergent adult *Manduca* (Reynolds and Truman, 1980). Eclosion hormone is an acidic peptide having an approximate molecular weight of 4000 daltons. The peptide appears to trigger every ecdysis (larval, pupal, and adult) during the life history of insects such as *Manduca* (Truman *et al.*, 1981), as well as mediating other specific neural and nonneural changes associated with the entry into a new life stage (Reynolds and Truman, 1983). Insects become transiently responsive to the peptide only during a brief period near the end of each molt. Within a few hours after the onset of sensitivity, EH is abruptly released into the blood. The peptide has not been found in the circulation at any other times, suggesting that the exclusive role of the peptide is to trigger ecdysis behaviors and, at the metamorphic molts, to activate the behavioral repertoire of the next stage. That EH is directly responsible for these events can be shown by injecting the peptide into insects prior to their normal time of EH release, which causes both precocious ecdysis and early activation of the new behaviors. The primary target of EH is the CNS (see Section 3), although other nonneural targets (e.g., muscles) are known (Reynolds, 1977; Schwartz and Truman, 1982).

An example of the behavioral actions of EH is seen in the giant silkmoth, *H. cecropia* (Truman, 1976). At the end of metamorphosis, the "pharate" adult (a fully developed adult still covered by the cuticle of the former stage) shows few of the behaviors that are typical of the adult stage. Pharate adults that have had the surrounding pupal cuticle removed about 4 to 6 hr before their normal time of ecdysis show a very reduced behavioral repertoire. Spontaneous or induced walking occurs rarely, if at all, reflexes such as the righting reflex are absent, and movements of the legs for extension or grasping at the substrate are poorly developed and occur infrequently. Injection of EH into these peeled pharate adults brings about a marked change in their behavior that is identical to that shown by normal insects in response to their endogenously released hormone. Ten to 15 min after injection, the insects begin a series of abdominal movements called the pre-eclosion behavior (Fig. 1). The behavior begins with an initial 30-min period of frequent rotary move-

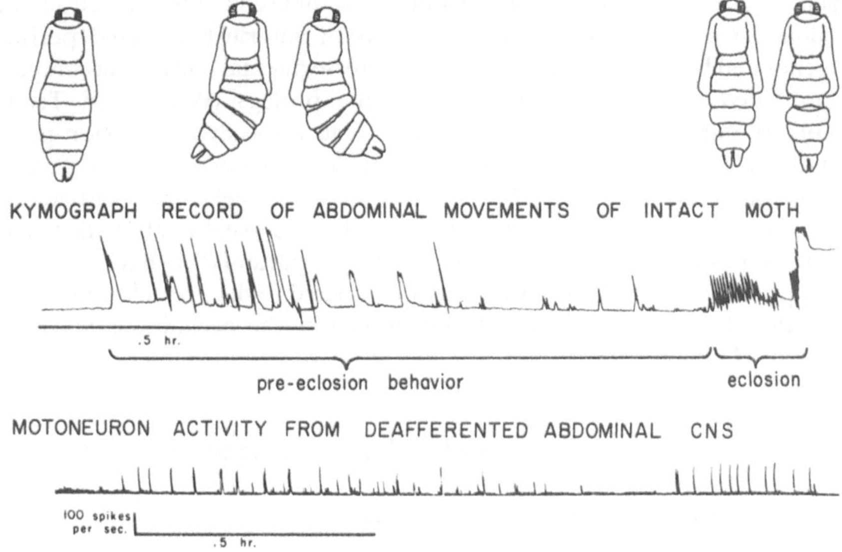

FIGURE 1. Behavioral response of *Hyalophora cecropia* to treatment with eclosion hormone. Top: schematic representations of the abdominal movements shown after peptide treatment; the insect is quiescent prior to treatment (left), shows frequent bouts of rotary movements of the abdomen early in the hormone response (middle), and shows peristaltic abdominal movements characteristic of eclosion late in the response (right). The overlying pupal cuticle has been removed so it will not obscure the movements of the moth. Middle: A record of the abdominal movements of the moth starting about 5 min after EH injection. Vertical excursions represent abdominal movements. During the pre-eclosion behavior the movements had a rotary patterning, whereas during eclosion they were exclusively of the peristaltic type. Bottom: the spontaneous motor activity generated by a completely deafferented abdominal CNS in response to EH treatment. The head and thorax of the moth were removed prior to the start of the recording. The records represent integrated motor activity. During the first period of activity the bursts had a rotary pattern, whereas during the second phase they showed exclusively a peristaltic pattern.

ments of the abdomen that, in animals which have not been peeled, causes connections between the old and new cuticles to be loosened. This active period is followed by a quiescent period of similar length. During the pre-eclosion behavior, the expression of occasional adult behaviors becomes even less frequent. The end of the pre-eclosion behavior is signaled by the onset of eclosion waves that are used to shed the old pupal cuticle. During the few minutes surrounding the onset of eclosion movements, adult behaviors such as spontaneous locomotion and the righting reflex become switched on for the remainder of the insect's life. The behavioral effects of this peptide are profound, transforming a helpless, poorly coordinated animal into a normally behaving adult within the span of 1 to 2 hr. Interestingly, following the expression of pre-eclosion and eclosion behavior, these motor patterns can no longer be elicited in the adult, even by another injection of EH (Truman, 1980a).

Eclosion hormone also triggers the stage-specific ecdysis behaviors in *Manduca.* At larval molts, EH elicits first a pre-ecdysis behavior that loosens the old cuticle, followed by ecdysis (Copenhaver and Truman, 1982). The pre-ecdysis behavior consists of rhythmic, synchronous contractions of tergopleural abdominal muscles, which produce conspicuous furrows along the dorsal body wall. These contractions alternate with contractions of ventral musculature and retraction of the abdominal prolegs. Larval ecdysis consists of anteriorly directed peristaltic waves generated by sequential contraction of intersegmental muscles, accompanied by coordinated "stepping" movements of the prolegs (Weeks and Truman, 1984a,b). At pupal ecdysis, there is a very weak pre-ecdysis behavior possibly related to the larval pattern, after which the cuticle is shed by abdominal peristalsis (Truman *et al.,* 1980; Weeks and Truman, 1984a; J. C. Weeks, unpublished observations). Finally, at adult eclosion, EH again elicits peristaltic abdominal movements (Reynolds *et al.,* 1979). Interestingly, a "preparatory" behavior shown by *Manduca* prior to adult eclosion, which may be related to the pre-eclosion behavior of silkmoths, is not triggered by EH, but instead by the declining blood titer of ecdysteroids (Truman, 1984).

3. Neurophysiological Aspects of Eclosion Hormone Action

3.1. Activation of Ecdysial Motor Patterns

Over the past 12 years, a number of preparations have been used to study how EH acts on the CNS to cause these behavioral effects. The initial demonstration of a direct action of the peptide on the CNS was made in the silkmoth *H. cecropia* (Fig. 1) (Truman, 1978). By recording extracellularly from ganglionic nerves, it was shown that even a totally isolated abdominal CNS would respond to EH by producing the appropriate pre-eclosion and eclosion motor patterns in their proper sequence. More recent work, involving both intracellular and extracellular recording techniques, has focused on the EH-evoked larval and pupal ecdysis motor patterns produced by semi-intact deafferented *Manduca* preparations (Truman and Weeks, 1983; Weeks and Truman, 1984a,b).

Figure 2 presents representative recordings from deafferented *Manduca* preparations. Larvae respond to EH by producing first the pre-ecdysis motor pattern (Fig. 2A) and then ecdysis (Fig. 2B). The dorsal nerve (DN) motor bursts seen during pre-ecdysis are synchronous throughout the abdomen and are produced by motoneurons innervating tergopleural muscles. In intact animals, these bursts generate the rhythmic furrowing of the body wall. Motor bursts that produce proleg retractions, which alternate with furrowing, are seen in the ventral nerve (data not shown). Following the pre-ecdysis pattern, larval preparations switch to the peristaltic ecdysis pattern (Fig. 2B) (Weeks and Truman, 1984a,b). During this pattern, DN motor bursts are metachronal rather than synchronous, and motoneurons to intersegmental muscles are rhythmically active. Deafferented prepupae also

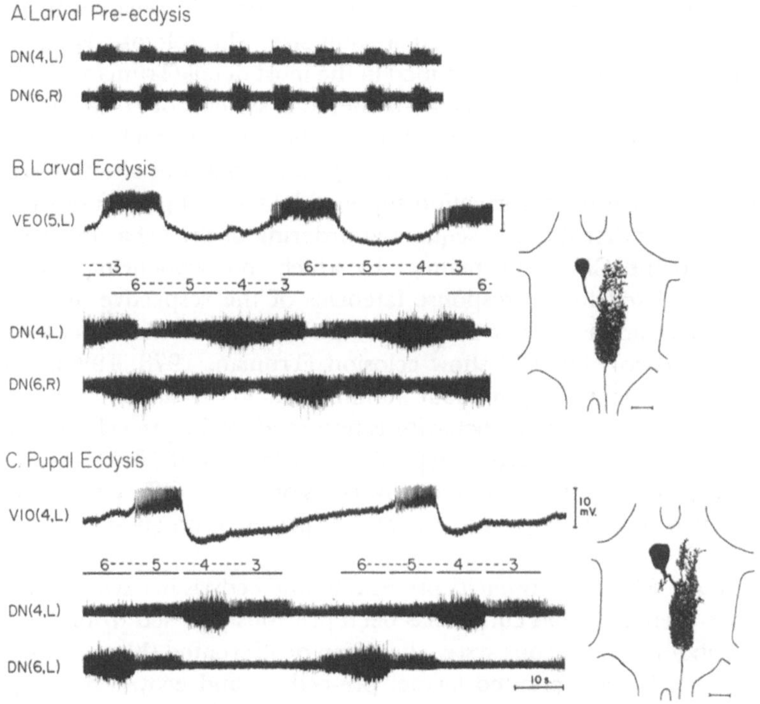

FIGURE 2. EH-evoked motor patterns in *Manduca*. Recordings in A and B were taken from a larva at the time of the molt to the 5th larval instar, following injection of EH extract and deafferentation, while recordings in C were from a prepupa prepared in a similar manner (Weeks and Truman, 1984a). Extracellular recordings are from dorsal nerves (DNs), with the abdominal ganglion and body side indicated in parentheses. Intracellular recordings (top traces in B and C) are from motoneurons innervating intersegmental muscles ventral external oblique (VEO) and ventral internal oblique (VIO). Camera lucida drawings of $CoCl_2$ fills of the recorded motoneurons are shown at right (anterior is up; scale bar = 50 μm). The axons of these motoneurons exit the ganglion and innervate muscles via the DN of the next posterior ganglion. In B and C, numbered bars indicate the progression of the peristaltic waves along the abdomen, based on the DN motor bursts (Weeks and Truman, 1984a). The larval preparation responded to EH by producing first the pre-ecdysis motor pattern (A) and subsequently switched to ecdysis (B). Pupae show a minimal pre-ecdysis pattern (not shown) and then ecdysis (C). (Data from Weeks and Truman, 1984a.)

respond to EH by generating an ecdysis motor pattern (Fig. 2C) that is indistinguishable from that of larvae (Weeks and Truman, 1984a) and that is preceded by a very weak pre-ecdysis pattern (data not shown). Just as in intact animals, deafferented preparations are responsive to EH only for a short period immediately prior to the normal time of ecdysis. The ability of EH to evoke these motor patterns in neurally isolated nervous systems indicates that the hormone mediates its behavioral effects by a direct action upon the insect CNS.

Like many other hormones, EH does not just trigger a single motor pattern, but rather it releases a coordinated sequence of behaviors. This raises the question of how these sequences are brought about. This question was examined in *H. cecropia,* which shows a series of three successive behaviors during the emergence of the adult: the pre-eclosion behavior loosens the old cuticle, eclosion movements shed the old cuticle, and then the wing expansion behavior enables the moth to inflate its soft new wings to their normal adult size. The role of EH in activating the first two behaviors has been examined in the most detail (Truman, 1978, 1980). Both deafferented and isolated abdominal nervous systems of *H. cecropia* respond to EH by generating the motor pattern for first the pre-eclosion behavior and then eclosion (Fig. 1) (Truman, 1978). This finding shows that the onset of eclosion is not dependent on sensory information supplied by the completion of the preceding behavior, but rather that the sequential ordering of the behaviors is somehow programed into the CNS. The manner by which this sequencing comes about appears to be related to the response latencies of the respective motor pattern generators. There are a number of ways in which to induce the insect to omit the pre-eclosion behavior, but still show eclosion (Truman, 1978, 1980b). In these cases eclosion movements begin about 90 min after EH treatment, essentially the same latency as is seen when the behavior is preceded by the pre-eclosion behavior. Similarly, the pre-eclosion behavior typically has a latency of 10–15 min, irrespective of whether or not it is followed by eclosion. This difference in latencies between the two motor programs insures that pre-eclosion always occurs before eclosion (Fig. 3).

In *Manduca,* EH similarly elicits pre-ecdysis and ecdysis behaviors in their normal sequence even if the old cuticle has been previously peeled from the body, or if the entire abdominal nervous system has been deafferented (Weeks and Truman, 1984a); (Fig. 2). In deafferented larvae, pre-ecdysis and ecdysis motor patterns show characteristic latencies after EH exposure (Copenhaver and Truman, 1982; Weeks and Truman, 1984a). For pupal ecdysis, the latency to the start of ecdysis movements is similar to that seen for larvae even though pupae show a minimal pre-ecdysis pattern (Weeks and Truman, 1984a). These results in *Manduca* are consistent with the hypothesis that the sequence of behaviors released by EH is not simply based on the completion of the first behavior (i.e., the loosening of the old cuticle) then serving to trigger the next.

At this time it is not known whether other sequential behaviors in these insects might be based on different response latencies to a hormonal signal. In order to understand how such behavioral series arise, one must not only identify the neural circuitry responsible for the motor patterns, but also the nature of the processes

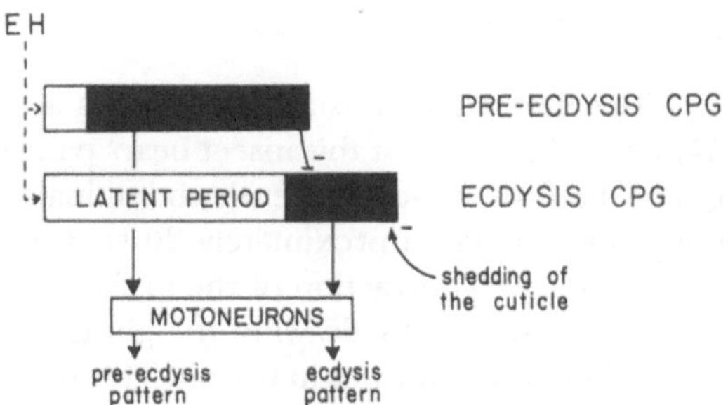

FIGURE 3. A working model for interactions that underlie the temporal organization of the pre-ecdysis and ecdysis behaviors. EH is thought to act directly on the individual CPGs (or higher-order interneurons) to activate the motor patterns (boxed area). In response to EH each CPG shows a characteristic latent period (unshaded portion of the box) followed by a period of active expression of the motor pattern (shaded portion). During these active periods the interneurons of the CPG drive motoneurons to generate the appropriate motor output. The sequential appearance of the two motor patterns is due primarily to the different intrinsic response latencies of the two CGPs and the ability of the first motor program to inhibit the second. The successful shedding of the old cuticle provides information important for turning off the ecdysis CPG. Time runs from left to right.

that generate delays of many minutes to hours. One possible site for such delays is the biochemical sequence of events that occur in target cells in response to the hormone. In the ecdysis system, an examination of this possibility awaits the identification of the EH target neurons.

Unlike the activation of ecdysis behaviors, which appears to be rather rigidly tied to the appearance of EH in the blood and relatively insensitive to sensory cues, the termination of the motor patterns is influenced by sensory information regarding the successful execution of the behavior (Fig. 3). For instance, premature peeling of the cuticle does not interfere with the initiation of ecdysis behaviors, but in contrast, damage to the old cuticle that prevents its being shed can prolong the ecdysis attempt for hours (Truman *et al.*, 1980). Similarly, deafferented preparations that lack sensory feedback show a prolongation of ecdysis or eclosion motor patterns (Truman, 1978; Weeks and Truman, 1984a). Thus, during normal execution of the behaviors, motor output from the CNS is controlled to greater or lesser degrees by hormonal or afferent influences.

3.2. Activation of Stage-specific Behaviors

Besides its ability to trigger stereotyped sequences of ecdysis motor patterns, EH also activates new stage-specific behaviors. The ability to show these behaviors is turned on rather abruptly within a span of 1 to 2 hr and the behaviors then stay functional throughout the duration of the stage, which can be weeks or months. At this time we do not know how EH provides behavioral access to relatively complex motor patterns such as those involved in locomotion or for copulation, but we have some understanding of how relatively simple reflex pathways may be irreversibly altered after peptide exposure. We will deal with two reflexes that are changed at different ecdyses; in one case a nonfunctional pathway is turned on, whereas in the other case the sign of a reflex is changed.

3.2.1. The Gin-trap Reflex

The first example is a reflex that is activated at the time of pupal ecdysis in *Manduca.* The pupa of this insect bears pairs of large pits on the anterolateral margin of the fifth through seventh abdominal segments (Fig. 4A). These pits, termed gin-traps, contain approximately 20 sensory hairs. Deflection of the hairs results in the reflexive contraction of the ipsilateral longitudinal muscles in the next anterior segment, thereby drawing the gin-trap under a shelf of cuticle and trapping or crushing whatever was in the pit. This reflex is thought to act as a defense against small insects that might try to burrow into the pupa in its underground pupation chamber. The neural pathway underlying the gin-trap reflex was first worked out by Bate (1973a,b) and appears to involve the gin-trap sensory neurons, an interganglionic interneuron and the longitudinal muscle motoneurons (Fig. 4B).

An important feature of the gin-trap reflex is that it is restricted to the pupal stage. The cuticular pit that forms the trap is constructed during the larval-pupal transformation, although the sensillae inside the pit apparently come from a pre-

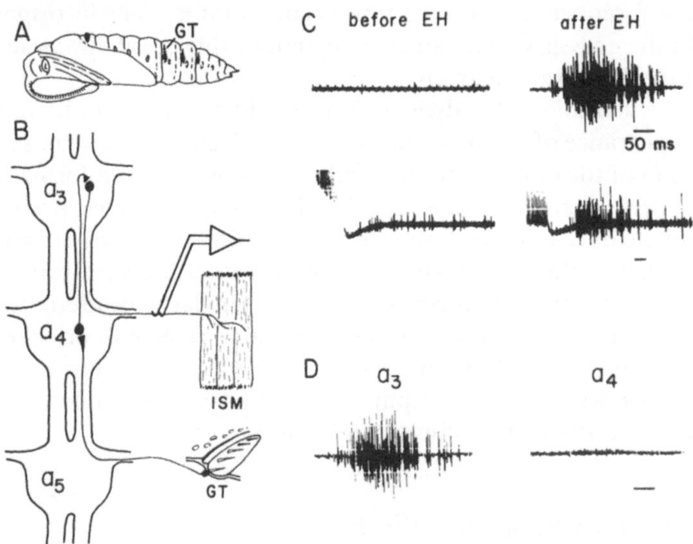

FIGURE 4. Activation of the gin-trap reflex by EH. (A) The pupa of *Manduca sexta* showing the location of the abdominal gin-traps (GT). (B) Schematic representation of the proposed circuitry underlying the gin-trap reflex and location of the extracellular electrodes recording motoneuron activity. a_3, a_4, a_5, ganglia for abdominal segments 3, 4, and 5; ISM, intersegmental muscles. (C) Response of the intersegmental muscle motoneurons to tactile stimulation of the gin-trap receptors (top) and electrical stimulation of the gin-trap nerve (bottom). Before exposure to EH (left) neither stimulus evoked a response. One hour after exposure to EH (right) both stimuli evoked the characteristic firing response of the motoneurons. The tactile stimulation was supplied at the start of the record. (D) The result of the selective exposure of individual ganglia to EH. Recordings from the lateral nerve of a_4 as shown in B. Left, exposure of a_3 to EH activates the entire reflex pathway as shown by the motoneuron response to tactile stimulation of the gin-trap receptors approximately 1 hr after EH exposure. Right, exposure of a_4 to EH in another preparation does not activate the pathway. Bar equals 50 msec. (Data from Levine and Truman, 1983.)

existing pool of larval sensory neurons (Bate, 1973a). Just prior to pupal ecdysis, removal of the old larval cuticle from over the pit area and stimulation of the sensillae is ineffective in triggering the reflex closure. The reflex becomes operational only when the insect undertakes pupal ecdysis behavior. Early injection of EH into intact animals or semi-intact preparations results in the precocious activation of the gin-trap pathway (Fig. 4C) (Levine and Truman, 1983). In semi-intact preparations this activation can occur even if the hormone exposure is insufficient to elicit the ecdysis motor pattern (Levine and Truman, 1983). Thus, the activation of this reflex is not tied to the performance of ecdysis behavior, and is attributable to a direct action of EH on the CNS at target sites that are apparently different from those that turn on ecdysis.

Examination of neurons involved in the gin-trap reflex indicated that prior to EH exposure, the sensory and motoneurons functioned normally. Consequently, the activation of the pathway by EH presumably involved changes in the interganglionic interneuron. It has been possible to record an extracellular unit that is likely to be this cell (Bate, 1973b; Levine and Truman, 1983), but because its ganglionic location is not yet known, it has not been possible to examine the effect of EH on the cell while recording intracellularly. However, its anatomical arrangement with input sites in one ganglion and output sites in the next has provided some insight into the site of action for the peptide (Fig. 4D). Local application of EH to the ganglion containing the sensory-to-interneuron synapses was ineffective in activating the pathway. However, peptide exposure of only the next anterior ganglion, containing the interneuron–motoneuron synapses, caused the entire reflex pathway to be activated (Levine and Truman, 1983). Thus, it appears that prior to pupal ecdysis, the reflex does not work because of a block at the level of the synapse between the interneuron and the motoneurons. The nature of this block and how EH releases it is not yet known, but in the future this small circuit should provide an ideal system in which to examine the effects of EH at a cellular level.

3.2.2. Alteration of the Stretch Receptor Reflex

Adult eclosion in *Manduca* is accompanied by the modification of a different sensory-to-motor reflex. In this case the reflex is not turned on, but rather its sign is changed from inhibitory to excitatory (Levine and Truman, 1982). During larval and pupal life, and throughout adult development, the abdomen of the insect is cylindrical in shape and capable of lateral bending. After eclosion the adult abdomen becomes dorsoventrally flattened, with very limited ability to show lateral flexions. Thus, most of the movements of the adult abdomen are confined to the dorsal-ventral plane. This change in shape of the abdomen is reflected in alterations in the response of the motor system to proprioceptive input. Each abdominal segment contains three bilateral pairs of stretch receptors (SRs). In the larva, pupa, and developing adult, activation of a dorsolateral SR (SR-3) causes a characteristic motor reflex in which the motoneurons that supply ipsilateral and contralateral longitudinal muscles near the SR are excited or inhibited, respectively. This circuitry is appropriate to allow lateral flexions of the abdomen. However, at adult eclosion this circuit changes.

Of particular interest has been the SR's effects on an identified motoneuron, MN-1, which innervates a longitudinally oriented muscle located near the dorsal midline of the segment (Levine and Truman, 1982) (Fig. 5). Prior to adult eclosion, the SR-3 causes excitation of the MN-1 that supplies ipsilateral muscles and causes inhibition of the cell controlling contralateral muscles. At the time of eclosion, however, the effect of the SR-3 on the contralateral MN-1 abruptly switches from inhibitory to excitatory. In this arrangement, activation of an SR-3 results in excitation of both ipsilateral and contralateral MN-1s, an appropriate arrangement for the dorsoventral movements of the adult abdomen. Intracellular recordings from MN-1 immediately before and after adult emergence resulted in the model shown in Fig. 5. In the pharate adult, stimulation of the contralateral SR-3 resulted in a biphasic synaptic response in MN-1. This response is thought to be a composite of an excitatory monosynaptic contact onto MN-1 by SR-3 and a polysynaptic pathway mediated by an inhibitory interneuron. The inhibitory pathway is presumably retained from the larval stage, whereas the excitatory synapse arises anew during adult development (Levine and Truman, 1982). Stimulation of the SR in the pharate adult results in the coactivation of both pathways which then converge on MN-1. With repetitive stimulation, the inhibitory component predominates resulting in the net inhibition seen in the pharate adult. The removal of the inhibitory component at eclosion brings about an abrupt change in the sign of the reflex. It is postulated, but not yet demonstrated, that EH acts on this circuit (presumably on the interneuron) to turn off the inhibitory pathway (Fig. 5) and thus to unmask the excitatory circuit that will be used for the remainder of adult life.

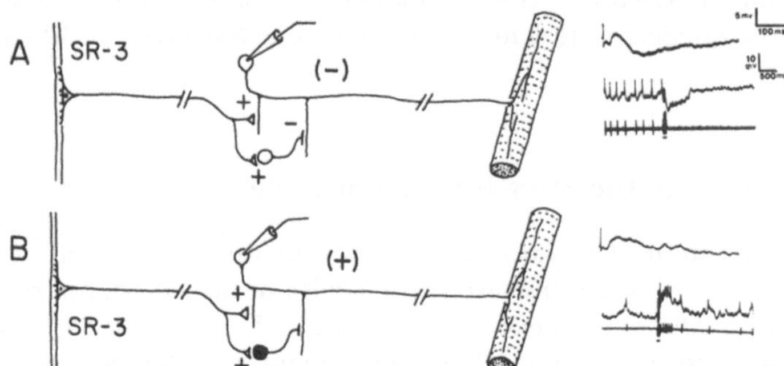

FIGURE 5. The effect of stretch receptor (SR-3) stimulation on the firing behavior of motoneuron MN-1 either (A) at the end of metamorphosis but before EH exposure, or (B) immediately after the moth has emerged in response to EH. Left, a schematic representation of the arrangement of the neurons thought to be involved in the reflex. + and − refer to excitation or inhibition respectively; the value in parenthesis indicates the net synaptic effect on the activity on MN-1. The darkened inhibitory interneuron in B indicates that this pathway has somehow been rendered nonfunctional. Right, (top) intracellular records from MN-1 showing its response to single SR-3 spikes; prior to eclosion MN-1 receives biphasic excitation/inhibition (A), whereas after eclosion it receives only excitation (B). Intracellular record from (middle) the soma of MN-1 and (bottom) an extracellular record from its axon in response to a train of stimuli to SR-3 (bar). Prior to eclosion, SR-3 trains result in net inhibition, whereas after eclosion the cell is excited. (Data from Levine and Truman, 1982.)

In many animals, a particular stimulus will elicit a behavioral response only under certain circumstances, e.g., female rats respond to males by showing lordosis behavior only when they have been primed by the proper hormonal state (reviewed in McEwen *et al.*, 1979). The gin-trap and SR reflexes in *Manduca* provide very simple models for how hormones might change the gating between sensory stimuli and a motor output. In these two insect examples, the behavioral modifications are permanent, and in the case of the inhibitory SR interneuron, the cell is suspected to die (Levine and Truman, 1982). Obviously, in other hormonally modulated systems, key interneurons could be reversibly taken in and out of a circuit depending on the presence or absence of a hormone, thereby allowing episodic gating of particular behaviors throughout an animal's life.

4. Hormonal Requirements for the Ecdysis Response

In any system in which hormones influence behavioral expression, several questions are relevant. First, what are the temporal requirements for the presence of the hormone relative to the duration of the behavioral response? Second, is there an absolute requirement for the hormone, or in special circumstances can the behavior be elicited in the hormone's absence? Third, which neurons are targets for the hormone and where in the behavioral circuits are they located? Finally, what is the physiological basis for critical periods and for circadian or longer-term periodicities in hormonal responsiveness? In the following sections, we will review what is known about these topics in the ecdysis systems.

Studies on isolated and deafferented CNS preparations clearly show that in order for EH to have prolonged behavioral effects it need be present for only a short period of time. For example, in the deafferented larval and pupal preparations, the insects were first injected with EH and then 15 to 30 min later opened up, throughly rinsed, and prepared for electrophysiological recording. Despite the fact that the nerve cord was repeatedly flushed with saline throughout the experiment, these preparations nevertheless initiated ecdysis motor patterns after the characteristic 1–2 hr latency (Weeks and Truman, 1984a). In some experiments, deafferented pupae were not injected with EH, but were instead allowed to release their own hormone. Even when the body cavity was perfused continuously with saline at a rate that exchanged the bath volume every minute, the preparations still initiated their motor patterns after a latency similar to other preparations that were not rinsed (J. C. Weeks, unpublished observations).

A systematic study of the relationship between the duration of hormone exposure and the subsequent motor response was carried out on the isolated abdominal CNS of *H. cecropia* (Truman, 1978). Eclosion hormone was added to the bath and after various times was washed out by repeated changes of the bathing medium. Peptide exposures of from 5 to 10 min were consistently sufficient to trigger the subsequent pre-eclosion and eclosion motor patterns which then lasted for a number of hours. These results taken with those in *Manduca* indicate that after EH exerts its effects on the CNS, the hormone need not be present in the blood during the time that the motor patterns are played out.

In many systems the behavioral changes caused by a particular hormone fade after the hormone is removed. Eclosion hormone differs in this respect in that in many cases its effects are essentially irreversible. For instance, as discussed above, the various stage-specific behaviors that are turned on by EH remain functional for the duration of the stage be it days, weeks, or in the case of diapausing pupae, months. Likewise for the performance of ecdysis itself, once the nervous system has responded to the hormone, it is refractory to subsequent exposures (Truman, 1980a). Thus the cellular actions of EH include not only the activation of the ecdysis motor patterns, but also some modification of the nervous system that then renders the ecdysis circuits insensitive to further hormone exposure. The only environment that makes the system responsive again is the endocrine state leading to the next molt (see Section 6).

A related question concerns whether EH-mediated behaviors can ever be released in the absence of the hormone. All of our evidence to date suggests that circulating EH is the normal pathway by which ecdysis behaviors and stage-specific behavioral changes are activated. Interestingly, however, the removal of the brain during adult development to remove the source of EH at adult emergence does not prevent the insects from subsequently eclosing (Truman, 1971). However, when this motor pattern is seen in brainless animals, it typically appears later than expected and usually in isolation rather than as part of the normal behavioral sequence including the pre-eclosion and wing expansion behaviors (Truman, 1971). Thus, although EH is not absolutely necessary for the display of the behavior, it is essential if this motor pattern is to be part of a coordinated sequence of behaviors.

5. Target Sites for Eclosion Hormone Action

A mapping of target sites for EH would require the use of a labeled ligand whose binding would indicate which neurons have cell surface receptors specific for the hormone. Such a ligand is not yet available, but it has been possible to obtain a general feeling for the distribution of potential target sites within the CNS based on the experiments described above, and on the ability of reduced portions of the nervous system to generate ecdysis motor patterns in response to the hormone.

By analogy to other behavioral circuits that have been investigated, one can rank the neurons involved in the different EH-mediated behaviors into a functional hierarchy. One simple scheme would include at the bottom level the motoneurons, whose activity would be governed by various premotor interneurons, including those belonging to different central pattern generator (CPG) circuits (e.g., for pre-ecdysis and ecdysis behaviors) (Fig. 3). The separation of motoneurons and CPG neurons into different hierarchical levels is supported, at least in the case of larval and pupal ecdysis in *Manduca,* by the finding that resetting experiments to test whether the motoneurons contribute to pattern generation were uniformly negative (Weeks and Truman, 1984b; J. C. Weeks, unpublished observations). The activity of the CPGs could be turned on or off by higher-order interneurons, perhaps similar to descending interneurons known to control the expression of other rhythmic insect behaviors (e.g., Bentley, 1977). Assuming such a scheme, what is the segmental distribution of these different neuronal classes, and which might be EH targets?

The abdominal motoneurons are segmentally iterated. The only careful examination of whether motoneurons might be EH targets was in the gin-trap system (see Section 3.2.1; Levine and Truman, 1982), and they were judged not to be targets based on electrophysiological data. One level up from the motoneurons are the interneurons. In the case of ecdysis, the segmental distribution of the CPG can be tested by severing connectives to isolate abdominal ganglia during ecdysis. In the isolated CNS of *H. cecropia,* pairs of ganglia can respond to EH by generating the motor bursts characteristic of first the pre-eclosion and then the eclosion patterns (Fig. 6), but single ganglia were not adequately tested for this ability (Truman, 1979). During pupal ecdysis in *Manduca,* however, even single abdominal ganglia that had already been exposed to EH continued rhythmic bursting after isolation (Fig. 7) (J. C. Weeks, unpublished observations). In this case, it is clear that each ganglion has an autonomous CPG.

An initial test of the distribution of EH target sites involved the application of EH to single ganglia in the intact nerve cord or to individual ganglia that had been isolated by connective sectioning. Using the first technique, Truman (1979) found that EH applied to just one ganglion could drive normal eclosion bursts throughout the whole chain. This indicates that each ganglion contains EH target sites that can activate eclosion, but that ganglia need not have been exposed to the hormone in order to participate in the behavior. In contrast, in *Manduca* prepupae, the addition of EH to one ganglion in the intact chain (Levine and Truman, 1983) or to isolated single ganglia (J. C. Weeks, unpublished observations) does not turn on

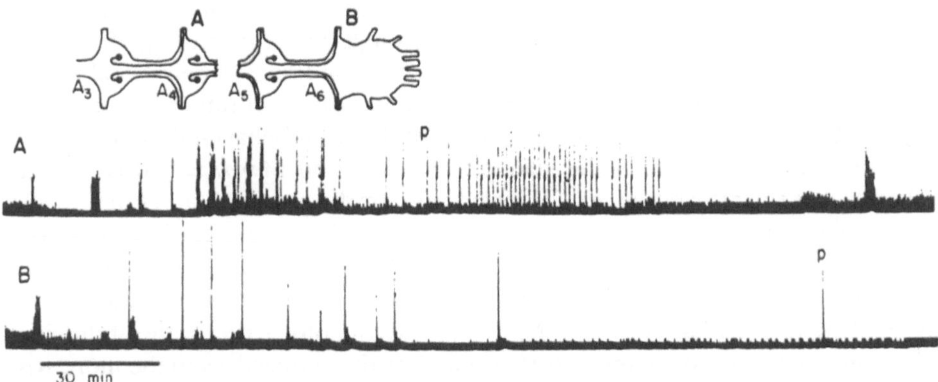

FIGURE 6. Demonstration of the ability of various parts of the chain of abdominal ganglia of *H. cecropia* to generate the pre-eclosion and eclosion motor patterns in response to eclosion hormone. Traces A and B are simultaneous integrated records of the spontaneous motor activity recorded from the dorsal nerves at the indicated positions on the nerve cord. The connectives were transected behind A_4 prior to hormone addition. Both pairs showed pre-eclosion bursts followed by eclosion bursts; p indicates the first eclosion burst recorded in each root. The record starts 25 min after the addition of the peptide. (From Truman, 1979.)

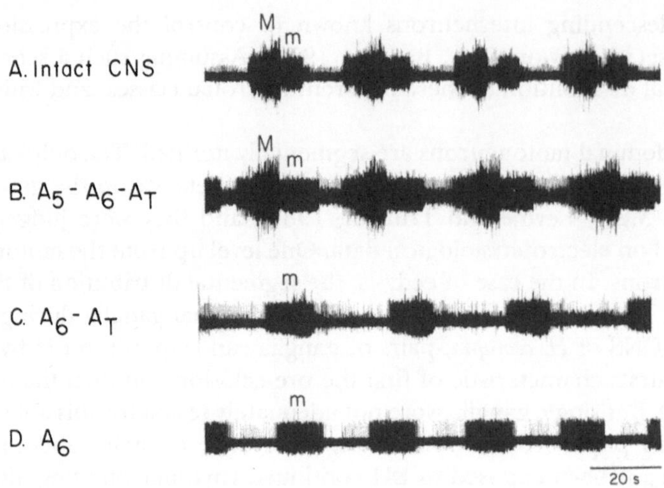

FIGURE 7. Ability of single *Manduca* ganglia to generate the ecdysis motor pattern. Each trace is an extracellular recording from the left dorsal nerve (DN) of the 6th abdominal ganglion (A_6) of a deafferented prepupal preparation which had been previously injected with EH extract. (A) Ecdysis motor pattern with all connectives intact. As reported previously (Weeks and Truman, 1984a), the DN pattern in each segment consists of paired bursts: the major (M) burst followed by the minor (m) burst. The major burst is produced by the axons of motoneurons whose cell bodies are located one ganglion anterior (see Fig. 2), while the minor burst is produced by homoganglionic motoneurons. (B) Pattern after cutting the connectives between ganglia in A_4 and A_5, leaving just A_5, A_6, and the terminal segment (AT). (C) Cutting anterior to A_6 eliminates the major burst, which originates in A_5. However, the homoganglionic minor burst continues. (D) When totally isolated from other ganglia, A_6 continues to produce rhythmic minor bursts.

the ecdysis pattern. However, application of EH to single ganglia in a chain can activate the gin-trap circuit that is mediated through that segment (Fig. 4) (Levine and Truman, 1983). This supports the idea that the widespread behavioral effects of EH are not mediated via a single set of target neurons, but rather that different target sites are associated with the activation of different behavioral pathways.

The difference between the silkmoth eclosion and *Manduca* ecdysis results indicates that the distribution of EH target sites, or the ability of target neurons in single ganglia to activate the motor patterns, differs in the two species. In *H. cecropia* one can conclude that, minimally, each pair of ganglia has both a competent CPG and adequate EH target sites to turn it on. In *Manduca,* the situation is not so simple. Once EH has activated the pupal ecdysis motor pattern, the abdominal CNS continues to produce the pattern even after the connectives to the more rostral CNS are severed (Weeks and Truman, 1984a). However, if the connectives are first cut and then EH is added, the abdominal CNS will not initiate the pattern (J. C. Weeks, unpublished observations). This observation suggests that higher-order interneurons, perhaps in the thoracic or cephalic nervous system, may be necessary sites of EH action in order to activate the segmentally repeated CPGs located in the abdomen.

Thus, the class or classes of neurons, i.e., motoneurons, CPG interneurons, or higher-order interneurons, that are EH targets at ecdysis is not yet certain, but we support the possibility that interneurons are the primary targets (Fig. 3). Certainly, electrophysiological evidence from the gin-trap and SR circuits suggests that the major action of EH in these cases is exerted on interneurons rather than sensory or motor neurons. However, in some other invertebrate (e.q., Mayeri *et al.*, 1979a,b) and vertebrate (Arnold, 1981; Kelley, 1978) systems, hormones or modulators have been found to act on neurons located at a variety of different hierarchical levels.

6. The Basis of Critical Periods

In many hormonally modified systems there are critical periods during which hormones must act in order to have an effect upon the nervous system. This has been especially well documented for steroid-induced changes in the CNS during vertebrate development (reviewed in Arnold and Gorski, 1984). In *Manduca* there are also critical periods during which EH must act in order to release the ecdysis motor patterns (Copenhaver and Truman, 1982; Reynolds *et al.*, 1979; Truman *et al.*, 1980). The nature of these critical periods is not yet understood, but certain factors can probably be excluded. Since insects only become sensitive to EH near the end of a molt when they are encased within the old cuticle, one might expect that stimuli from the old cuticle prime the insect to display ecdysis behaviors. However, results from peeled intact insects as well as deafferented and isolated CNS preparations show that sensory information unique to the molt is not needed for the display of the ecdysis behaviors. Similarly, one would expect that such inputs are not likely to be involved in the appearance of the critical periods.

Another possibility is that the ecdysis circuitry is dismantled after the behavior is finished and is reconstructed only in response to the endocrine environment of the next molt. This possibility may actually exist during the pupal-adult transition of moths (Truman and Weeks, 1983), but based on electrophysiological data this seems unlikely during the larval and pupal stages. At each larval ecdysis, the cuticle is shed by abdominal peristaltic movements upon which are superimposed rhythmic proleg retractions (Weeks and Truman, 1984a,b). Each performance of the behavior during the various larval stages appears identical, as one would expect if each were generated by the same CPG circuit. At pupal ecdysis the behavior appears different because the prolegs have been lost and the cuticle is shed by peristalsis alone. However, quantitative cinematographic and electrophysiological analyses indicate that the motor patterns produced at larval and pupal ecdysis are indistinguishable; this similarity extends even to the point of driving the proleg retractor motoneurons to fire rhythmically during pupal ecdysis despite the prior degeneration of the prolegs and their muscles (Weeks and Truman, 1984a,b). If the ecdysis CPG were built up anew prior to pupal ecdysis, one would not expect it to contain these useless larval attributes.

In developing moths a new ecdysis CPG does appear to be constructed; one indication of this is that the intersegmental coordination of the peristaltic motor pattern at eclosion is very different than that seen at earlier ecdyses (J. W. Truman and J. C. Weeks, unpublished observations). Furthermore, when developing moths are examined at various times during adult differentiation, it is seen that even several days before an animal is sensitive to EH it is capable of showing many of the thoracic and abdominal movements characteristic of eclosion (Fig. 8) (Truman, 1976). Likewise, in crickets, decapitation and the application of the appropriate stimuli can evoke ecdysislike movements even during the intermolt periods (Carlson, 1981). These observations suggest that the ecdysis circuitry is in place well before the appearance of EH responsiveness. Therefore, the answer to critical periods probably does not reside at the gross level of making new circuits, but probably with some aspect of the cellular physiology of the appropriate target cells. This might involve the periodic appearance and disappearance of EH receptors in the cell membranes, or some other periodic change in the biochemical response of the target cells.

Although we do not know the cellular changes that underlie the critical periods of CNS responsiveness to EH, we do know the factors that control the time of onset of sensitivity. There is a second circulating factor that is involved in the ecdysis response, the steroid hormone 20-hydroxyecdysone (20-HE). The decline in this steroid late in every molt is essential for the eventual ecdysis of the insect. If the 20-HE titers remain high, EH will not be released and the CNS remains unresponsive even to injected peptide (Truman *et al.*, 1983). At the end of adult development the steroid decline directs the final maturation of the adult motor programs, culminating in the ability of these circuits to be activated by EH at eclosion. In addition, the falling ecdysteroid titer may enable some of the adult circuits to activate spontaneously even if the peptide signal is not later provided.

Although steroids set the onset of responsiveness, the loss of responsiveness

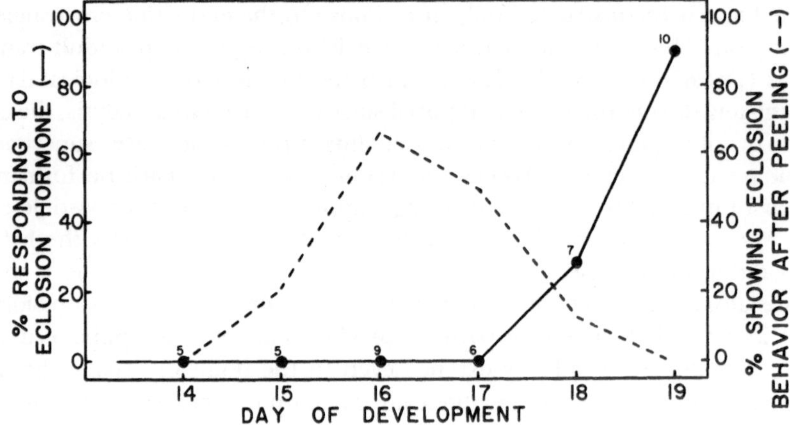

FIGURE 8. The ability of developing *Antheraea pernyi* moths to show eclosion movements at various times during the 19 days of adult development. The dashed line represents the percentage of animals that show spontaneous eclosion movements after the surrounding pupal cuticle is manually removed. The solid lines show the percent of animals that respond to eclosion hormone by showing the behavior. The ability to perform the behavior occurs about 3 days before that particular motor pattern becomes hormonally sensitive. (From Truman, 1976.)

that occurs after eclosion appears to result from the action of EH. The mechanism for the loss of sensitivity is not clear, but preliminary experiments suggest that it may be due to changes in membrane receptors or in other steps in the biochemical response of the cells to EH (Truman, 1980a).

One final aspect of the periodic nature of EH action throughout the insect's life relates to the controls over release of the peptide. Adult eclosion in *Manduca* and the silkmoths is controlled by a circadian clock located in the moth's brain that can be transplanted from individual to individual by transplantation of the brain (Truman and Riddiford, 1970). In these animals eclosion occurs at a particular time of day (the eclosion gate) that is tied to the photoperiod and temperature cycle. At the end of adult development when the circulating ecdysteroid titers drop to a level permissive for eclosion to occur, EH release then occurs only when the next eclosion gate opens (Truman *et al.*, 1983). Corresponding to this cephalic control of eclosion, at this molt EH is released from neurosecretory cells located in the brain (Copenhaver and Truman, 1983). By contrast, the ecdyses of larvae and pupae do not show circadian gating and occur anytime during the day or night once the ecdysteroid titers have dropped sufficiently. A different set of neurosecretory cells, located in the ventral chain of ganglia, is responsible for releasing hormone under these conditions (Truman *et al.*, 1981). Thus critical periods and circadian gates act to restrict the expression of these behaviors to very precise times in the insect's life.

7. Conclusions and Future Directions

The ecdysis system of insects has been attractive for study because it allows the effects of eclosion hormone on deafferented and isolated pieces of the CNS to be directly related to the hormone's behavioral effects on intact animals. The most dramatic effects of the peptide are the release of the ecdysial motor patterns. The circuitry underlying these motor patterns is still largely unknown and a major challenge deals with the identity and role of the peptide-sensitive neurons in relation to the network that composes the central pattern generator.

Besides activating ecdysis motor patterns, EH also renders numerous other motor circuits competent for use by the insect after ecdysis has occurred. These hormonally modified circuits, which include the gin-trap and the SR-3 reflexes, provide the opportunity to examine the action of the hormone at the cellular level. The key neurons that are responsible for modifying these reflexes have been inferred indirectly from physiological data, but the job remains to identify and record from them so that the effects of EH on their properties can be studied directly. Besides the physiological demonstration that these cells are altered by EH, it is important to develop a labeled hormone that can be used to confirm that these key cells bind the hormone as well.

Another insight provided by this system is the realization that the identification of EH target neurons that mediate changes in different behavioral circuits is not the endpoint of this work. Many questions will remain, including the origin of differential latencies, which are important for organizing behavioral sequences, and of critical periods, which underlie the waxing and waning of responsiveness to the peptide. These phenomena are likely not to be explained simply in terms of synaptic connections or circuit properties. An increased emphasis on the biochemical response of individual target cells to EH will be necessary in order to generate a full understanding of how this peptide regulates CNS function.

ACKNOWLEDGMENTS. Unpublished results reported here were supported by grants from NIH, NSF, and the McKnight Foundation.

References

Arnold, A. P., 1981, Logical levels of steroid hormone action in the control of vertebrate behavior, *Am. Zool.* **21**:233–242.

Arnold, A. P., and Gorski, R. A., 1984, Gonadal steroid induction of structural sex differences in the central nervous system, *Annu. Rev. Neurosci.* **7**:413–442.

Bate, C. M., 1973a, The mechanism of the pupal gin trap. I. Segmental gradients and the connections of the triggering sensilla, *J. Exp. Biol.* **59**:95–107.

Bate, C. M., 1973b, The mechanism of the pupal gin trap. III. Interneurones and the origin of the closure mechanism, *J. Exp. Biol.* **59**:121–135.

Bentley, D., 1977, Control of cricket song patterns by descending interneurons, *J. Comp. Physiol.* **116**:19–38.

Berthold, A. A., 1849, Transplantation der Hoden, *Arch. Anat. Physiol.* **16**:42–46.

Blest, A. D., 1960, The evolution, ontogeny, and quantitative control of settling movements of some new world saturniid moths, with some comments on distance communication by honey bees, *Behaviour* **16**:188–253.

Carlson, J. R., 1977a, The imaginal ecdysis of the cricket *(Teleogryllus oceanicus)*. I. Temporal structure and organization into motor programmes, *J. Comp. Physiol.* **115**:299–317.

Carlson, J. R., 1977b, The imaginal ecdysis of the cricket *(Teleogryllus oceanicus)*. II. The role of identified motor units, and control by sensory and central factors, *J. Comp. Physiol.* **115**:319–336.

Carlson, J. R., 1981, Temporal variation in the availability of an ecdysial motor programme during the last instar and early adult stages of the cricket *Teleogryllus Oceanicus*, *J. Comp. Physiol.* **27**:159–193.

Copenhaver, P. F., and Truman, J. W., 1982, The role of eclosion hormone in the larval ecdysis of *Manduca sexta*, *J. Insect Physiol.* **28**:695–701.

Copenhaver, P. F., and Truman, J. W., 1983, Identification and electrophysiology of the peptidergic neurosecretory cells that contain eclosion hormone in *Manduca sexta*, *Soc. Neurosci. Abstr.* **9**:453.

Hughes, T. D., 1980, The imaginal ecdysis of the desert locust, *Schistocerca gregaria*. I. A description of the behaviour, *Physiol. Entomol.* **5**:47–54.

Kelley, D. B., 1978, Neuroanatomical correlates of hormone sensitive behaviors in frogs and birds, *Am. Zool.* **18**:477–488.

Levine, R. B., and Truman, J. W., 1982, Metamorphosis of the insect nervous system: Changes in the morphology and synaptic interactions of identified cells, *Nature* (London) **299**:250–252.

Levine, R. B., and Truman, J. W., 1983, Peptide activation of a simple neural circuit, *Brain. Res.* **279**:335–338.

Mayeri, E., Brownell, P., Branton, W. D., and Simon, S. B., 1979a, Multiple, prolonged actions of neuroendocrine bag cells on neurons in *Aplysia*. I. Effects on bursting pacemaker neurons, *J. Neurophysiol.* **42**:1165–1184.

Mayeri, E., Brownell, P., and Branton, W. D., 1979b, Multiple, prolonged actions of neuroendocrine bag cells on neurons in *Aplysia*. II. Effects on beating pacemaker and silent neurons, *J. Neurophysiol.* **42**:1186–1197.

McEwen, B. S., Davis, P. G., Parsons, B., and Pfaff, D. W., 1979, The brain as a target for steroid hormone action, *Annu. Rev. Neurosci.* **2**:65–112.

Reynolds, S.E., 1977, Control of cuticle extensibility in the wings of adult *Manduca* at the time of eclosion: Effects of eclosion hormone and bursicon, *J. Exp. Biol.* **70:**27–39.

Reynolds, S. E., and Truman, J. W., 1980, Eclosion hormones, in: *Neurohormonal Techniques in Insects* (T. A. Miller, ed.), Springer-Verlag, New York, pp. 196–215.

Reynolds, S. E., and Truman, J. W., 1983, Eclosion hormone, in: *Insect Endocrinology* (R. G. H. Downer and H. Laufer, eds.), A. R. Liss, Inc., New York, pp. 217–233.

Reynolds, S. E., Taghert, P. H., and Truman, J. W., 1979, Eclosion hormone and bursicon titres and the onset of hormonal responsiveness during the last day of adult development in *Manduca sexta* (L.), *J. Exp. Biol.* **78:**77–86.

Schwartz, L. M., and Truman, J. W., 1982, Peptide and steroid regulation of muscle degeneration in an insect, *Science* **215:**1420–1421.

Truman, J. W., 1971, Physiology of insect ecdysis. I. The eclosion behaviour of saturniid moths and its hormonal release, *J. Exp. Biol.* **54:**805–814.

Truman, J. W., 1976, Development and hormonal release of adult behavior patterns in silkmoths, *J. Comp. Physiol.* **107:**39–48.

Truman, J. W., 1978, Hormonal release of stereotyped motor programmes from the isolated nervous system of the Cecropia silkmoth, *J. Exp. Biol.* **74:**151–174.

Truman, J. W., 1979, Interaction between abdominal ganglia during the performance of hormonally triggered behavioural programmes in moths, *J. Exp. Biol.* **83:**239–253.

Truman, J. W., 1980a, Cellular aspects of eclosion hormone action on the CNS of insects, in: *Receptors for Neurotransmitters, Hormones, and Pheromones in Insects* (D. B. Satelle, L. M. Hall, and J. G. Hildebrand, eds.), Elsevier/North Holland Biomedical Press, Amsterdam, pp. 223.-232.

Truman, J. W., 1980b, Organization and hormonal release of stereotyped motor programs from the CNS of an insect, *Front. Horm. Res.* **6:**1–15.

Truman, J. W., 1984, The preparatory behavior rhythm of the moth *Manduca sexta:* An ecdysteroid-triggered circadian rhythm that is independent of the brain, *J. Comp. Physiol.* **155:**521–528.

Truman, J. W., and Riddiford, L. M., 1970, Neuroendocrine control of ecdysis in silkmoths, *Science* **167:**1624–1626.

Truman, J. W., and Weeks, J. C., 1983, Hormonal control of the development and release of rhythmic ecdysis behaviors in insects, in: *Neural Origin of Rhythmic Movements* (A. Roberts and B. Roberts, eds.), Cambridge University Press, Cambridge, *Soc. Exp. Biol. Symp.* **37:**223–241.

Truman, J. W., Taghert, P. H., and Reynolds, S. E., 1980, Physiology of pupal ecdysis in the tobacco hornworm, *Manduca sexta.* I. Evidence for control by eclosion hormone, *J. Exp. Biol.* **88:**327–337.

Truman, J. W., Taghert, P. H., Copenhaver, P. F., Tublitz, N. J., and Schwartz, L. M., 1981, Eclosion hormone may control all ecdyses in insects, *Nature (London)* **291:**70–71.

Truman, J. W., Rountree, D. B., Reiss, S. E., and Schwartz, L. M., 1983, Ecdysteroids regulate the release and action of eclosion hormone in the tobacco hornworm, *Manduca sexta* (L), *J. Insect Physiol.* **29:**895–900.

Weeks, J. C., and Truman, J. W., 1984a, Neural organization of peptide-activated ecdysis behaviors during the metamorphosis of *Manduca sexta.* I. Conservation of the peristalsis motor pattern at the larval-pupal transformation, *J. Comp. Physiol.* **155:**407–422.

Weeks, J. C., and Truman, J. W., 1984b, Neural organization of peptide-activated ecdysis behaviors during the metamorphosis of *Manduca sexta.* II. Retention of the proleg motor pattern despite loss of the prolegs at pupation, *J. Comp. Physiol.* **155:**423–433.

21

Are Skeletal Motoneurons in Arthropods Peptidergic?

MICHAEL O'SHEA

1. Introduction

There is growing evidence for widespread involvement of peptides in the control of muscle performance in the invertebrates (O'Shea and Schaffer, 1984). In this chapter, I will consider the role of neuropeptides as possible muscle effectors in arthropods.

The first arthropod peptide that was completely chemically characterized is the myotropic pentapeptide proctolin (Arg-Tyr-Leu-Pro-Thr). Proctolin was initially identified as the biologically active substance capable of producing contracture of the hindgut or proctodeum of the cockroach *Periplaneta americana* (Brown, 1967). This factor was subsequently isolated, purified, sequenced, and proposed as the neuromuscular transmitter of motoneurons that innervate hindgut muscles (Brown and Starratt, 1975; Starratt and Brown, 1975). The first direct evidence that a peptide may act as a neuromuscular transmitter in insects was provided by the observation that proctolin applied to proctodeum muscle produces contractures similar to those produced by stimulating the gut motor nerve (Brown, 1975). Soon after proctolin was synthesized and was made generally available, it became clear that its bioactivity was not confined to gut musculature. For example, at low concentration it was shown to cause contracture of certain insect skeletal muscle (Piek and Mantel, 1977) and it was also shown to be a cardioactive peptide (Miller, 1979).

In spite of the growing evidence suggesting roles for proctolin on skeletal muscle, there was reluctance in accepting the fact that it might be a transmitter for skeletal motoneurons. By 1980, however, the circumstantial evidence that procto-

MICHAEL O'SHEA • University of Geneva, Laboratory of Neurobiology, CH-1211 Geneva 4, Switzerland.

lin was a skeletal muscle transmitter was strong. Also, there seemed to be no *a priori* reason to suppose that peptides could be involved in gut neuromuscular transmission, but not skeletal neuromuscular transmission. On the contrary, gut muscle in insects is striated and quite similar to slow skeletal muscle. Moreover, for the cockroach hindgut, the visceral motoneurons are located in the CNS and appear to be structurally similar to skeletal motoneurons (Keshishian and O'Shea, 1984). This coupled to the nanomolar sensitivity of some insect skeletal muscle to proctolin ought to have suggested the hypothesis that proctolin, and perhaps other peptides, participates in the process of skeletal neuromuscular transmission. In fact, May *et al.* (1979) showed that tonic skeletal muscle fibers in the locust would generate depolarizing potentials to local iontophoretic application of proctolin and that areas of high proctolin sensitivity corresponded to regions where nerve–muscle contact is made.

Questions concerning precisely how proctolin is delivered to target structures, such as skeletal muscle and heart, are only now being resolved. The mere fact of sensitivity is not sufficient to establish that a tissue is ever normally exposed to the active compound. Demonstration that a tissue is a target of a bioactive compound depends largely on localizing the source of the compound. Cellular localization of proctolin therefore became an important prerequisite for resolving the question of whether proctolin acts on skeletal muscle as a circulating hormone or as a local transmitter.

Similar questions abut the role of proctolin as a transmitter or hormone are not confined to insect preparations. The presence of proctolin in crustaceans is particularly well established in lobster, crayfish, and crab (Schwarz *et al.*, 1984; Bishop *et al.*, 1984a; Kingan and Titmus, 1983; Sullivan, 1979). Consistent with its actions in insects, proctolin has a variety of effects on several types of crustacean muscles. It increases cardiac muscle contraction in the crab *Portunus sanguinolentus* (Sullivan, 1979) and the lobster *Humarus americanus* (Miller and Sullivan, 1981), it enhances contractions of stomatogastric muscle in *Panulirus interruptus* (Lingle, 1979), and causes long-lasting contracture of the opener muscle of the finger or dactyl of the lobster walking leg (Schwarz *et al.*, 1980; Kravitz *et al.*, 1980). In general, however, it remains undetermined in most of these examples precisely how these muscles are normally exposed to the peptide.

My reluctance to accept the idea that peptides could be transmitters of skeletal motoneurons can be attributed to the belief that the actions of arthropod motoneurons are adequately accounted for by three nonpeptide transmitters: glutamate, GABA, and octopamine. It has become generally accepted that L-glutamic acid is the transmitter of fast and slow excitatory motoneurons, and that GABA is the transmitter of inhibitory motoneurons (Pitman, 1984). In insects, in addition to excitatory and inhibitory motoneurons, there is a class of octopaminergic modulatory motoneurons (O'Shea and Evans, 1979; Evans and O'Shea, 1978). Thus it seemed that questions of which transmitters were involved in arthropod skeletal neuromuscular transmission were largely resolved. Any additional transmitters, peptides for example, would presumably have to be colocalized with glutamate, GABA, or octopamine. Also, if the actions of the peptide were different from the actions of the recognized transmitters, some aspect of muscle response to nerve stimulation would not be accounted for by the recognized transmitters. Evoking a peptide as a transmitter of a skeletal motoneuron therefore would both violate the one neuron–one transmitter idea and challenge the accepted view that neurally evoked muscle responses were due to the three recognized transmitters. The undeniable high and specific sensitivity of some arthropod skeletal muscle to proctolin is therefore more easily explained by assuming that proctolin is not a transmitter, but a circulating hormone. Evidence accumulated in recent years, however, suggests strongly that it can also function as a skeletal motoneuronal transmitter (Bishop and O'Shea, 1982; O'Shea and Bishop 1982; Adams and O'Shea, 1983; Witten *et al.*, 1984; Bishop *et al.*, 1984; Schwartz *et al.*, 1984). In some cases, the evidence suggests, it also acts on muscle as a neurohormone (Schwarz *et al.*, 1980, 1984).

The impetus for our improved understanding of proctolin's physiological role was the development of proctolin antibodies (Bishop *et al.*, 1981; Eckert *et al.*, 1981; Kingan and Titmus, 1983; Schwarz *et al.*, 1984). This allowed the localization and identification of proctolin immunoreactive neurons in the insect and crustacean nervous systems (Bishop and O'Shea, 1982; Bishop *et al.*, 1984; Schwarz *et al.*, 1984). It is now clear that proctolin is widely distributed in neurons of the arthropod CNS. It is probably primarily associated with efferent neurons, some of which have been identified as skeletal motoneurons (O'Shea and Bishop, 1982; Witten *et al.*, 1984a; Bishop *et al.*, 1984a,b) and others release proctolin into the circulation for hormonal action (Schwarz *et al.*, 1984). Proctolin immunoreactivity has also been associated with interneurons in the locust (Keshishian and O'Shea, 1984) and has recently been implicated in the neural regulation of a central pattern generator in crustaceans (Chapter 17, this volume). In summary, the current evidence suggests multiple roles for proctolin as a gut muscle transmitter, a skeletal muscle cotransmitter, a circulating hormone, and a transmitter or modulator within the CNS.

In this chapter, our evidence that proctolin, and perhaps other peptides, acts as a skeletal neuromuscular cotransmitter will be reviewed. As a result of our recent findings, we have started to think about arthropod skeletal motoneurons in an entirely new way. We believe that several peptidergic subpopulations of motoneurons may exist. In each hypothesized subpopulation, a rapidly-acting transmitter, like glutamate or GABA, would be associated with different peptide cotransmitters. The existence of peptide neuromuscular cotransmitters other than proctolin has not yet been established. We are, however, now using skeletal muscle bioassays to detect and purify new putative neuromuscular peptide cotransmitters. Some success in this approach was recently achieved when the two new myoactive peptides MI and MII were found (O'Shea *et al.*, 1984) and sequenced (Witten *et al.*, 1984b). It remains to be seen whether these peptides, like proctolin, actually participate in neuromuscular transmission.

2. Proctolinergic Skeletal Motoneurons

We now have evidence in three different arthropod systems for proctolin as a skeletal neuromuscular transmitter. The first to be characterized involves the identified slow coxal or Ds motoneuron of the cockroach *Periplaneta americana* (O'Shea and Bishop, 1982; Adams and O'Shea, 1983). Recently we have also established the presence of proctolin in another well-characterized insect motoneuron, the slow extensor of the locust extensor tibialis (SETi) muscle (Witten *et al.*, 1984). The third example is in the crayfish *Procambarus clarkii* where we have obtained evidence for proctolin as a transmitter of three of the five excitatory motoneurons innervating the abdominal tonic flexor muscles (Bishop *et al.*, 1984a).

The first direct evidence that proctolin is associated with efferent neurons other than gut motoneurons was its localization to the lateral white neurons of the cockroach abdominal ganglion (O'Shea and Adams, 1981). This neuron projects to the cardiac nerve and probably functions in regulating cardiac muscles or it releases proctolin into the circulation. The proctolin content of the lateral white (LW) neuron was established by isolating the cell body and performing high-pressure liquid chromatography (HPLC) on the cell extract. Fractions from the HPLC purification were applied to the proctolin-sensitive skeletal muscle of the locust—the extensor muscle of the hindleg tibia (the extensor tibialis or ETi muscle).

It is important to understand a little more about the proctolin-sensitive ETi muscle because it has played a central role in the identification of proctolinergic neurons and because it is innervated by the proctolinergic SETi neuron. The ETi muscle is the main jumping muscle and is the largest and strongest muscle in the locust (Hoyle, 1983). It was first shown to be sensitive to proctolin by Piek and Mantel (1977). They showed that at nanomolar concentration, proctolin induces a myogenic rhythm of contraction and relaxation. In fact, this muscle normally generates a spontaneous myogenic rhythm without application of proctolin (Hoyle and O'Shea, 1974). The spontaneous rhythm is accelerated by proctolin and this response has been used as a sensitive bioassay, in conjunction with HPLC, to identify proctolin in extracts made from single uniquely identified neurons (O'Shea and Adams, 1981; O'Shea and Bishop, 1982). At low concentration (about 10^{-10} M), the primary effect of proctolin on the locust ETi muscle is to accelerate the weak myogenic rhythm. At higher concentration (about 10^{-9} M), however, a strong and maintained tonic contracture of the muscle is produced. The myogenic contractile rhythm is generated by a small part of the large ETi muscle consisting of a bundle of about 20 fibers at the proximal end of the main muscle (Fig. 1). The strong tonic contraction produced by proctolin is not generated only by the myogenic bundle, but appears to be due to proctolin sensitivity that extends to other slow fibers of the ETi muscle. The cardiaclike myogenic bundle probably functions to aid in the movement of hemolymph and air into the highly specialized and elongated locust hindleg.

In addition to being proctolin sensitive, the myogenic rhythm is also octopamine sensitive (Evans and O'Shea, 1977). Octopamine, in contrast to proctolin, slows and inhibits the myogenic oscillations. In 1977, Peter Evans and I considered how octopamine might normally be delivered to the extensor muscle. We found that stimulation of one of the muscle's motoneurons, the dorsal unpaired median (DUM) neuron, caused an inhibition of the rhythm just like the application of octopamine. This fact, the presence of octopamine in DUM cells, and other evidence supported the view that octopamine was the modulatory transmitter of the DUM neuron innervating the ETi muscle (Evans and O'Shea, 1977, 1978; O'Shea and

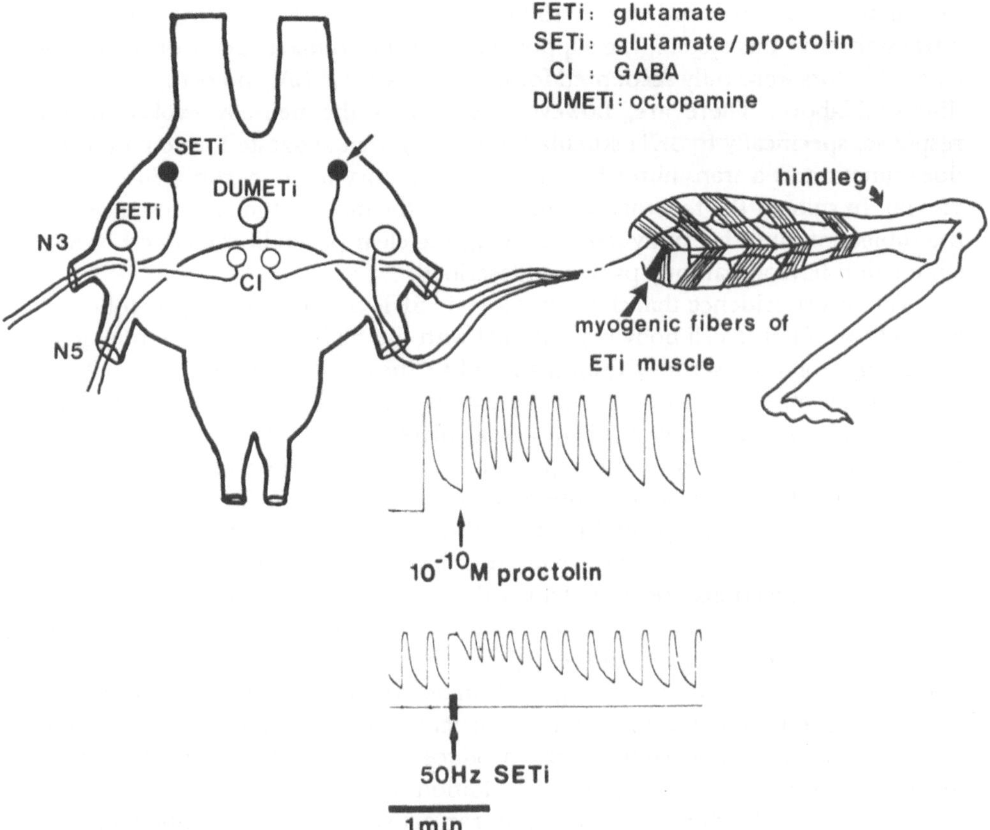

FIGURE 1. Proctolinlike response of the extensor tibialis (ETi) muscle to stimulation of the slow extensor (SETi) motoneuron. The diagram shows the outline of the third thoracic ganglion and the positions of the identified motoneurons that innervate the hindleg extensor muscle (see drawing on right) through nerve 5 (N5) and nerve 3 (N3). The motoneurons are FETi, the fast extensor; SETi, the slow extensor; CI, the common inhibitor; and the DUMETi, the dorsal unpaired median motoneuron. Their transmitters are indicated on the table. The response of the ETi muscle to proctolin (upper trace) and to a 10-sec burst of SETi spikes at 50 Hz (lower trace) are similar. Following SETi stimulation, there is a long-lasting increase in the frequency of the myogenic rhythm. See text for further details.

Evans, 1979). To summarize, the myogenic rhythm of the ETi muscle is stimulated by proctolin and inhibited by octopamine. One of the neurons innervating the muscle (the DUM neuron) uses octopamine as a transmitter and one of its actions is to down-regulate the myogenic rhythm (Evans and O'Shea, 1977). In view of this, it would seem reasonable to propose proctolin as a transmitter of another neuron innervating the extensor muscle. This neuron might therefore be expected to accelerate or up-regulate the myogenic rhythm.

All of the neurons involved in the innervation of the hindleg extensor ETi muscle have been uniquely identified and extensively studied. These are the octopaminergic DUMETi neuron (Evans and O'Shea, 1977) discussed above, the slow (SETi) and fast (FETi) excitatory motoneurons, and the common inhibitory (CI) motoneuron (Burrows and Hoyle, 1973). A reason for not proposing proctolin as a transmitter in this system (the supposition that the complete array of neuromuscular effectors were fully accounted for by glutamate, GABA, and octopamine) was discussed above. There are, however, aspects of the neurally evoked muscle response, specifically to SETi stimulation (Fig. 1), that suggested proctolin in fact does function as a transmitter for the slow excitatory motoneuron (Witten *et al.*, 1984a). In spite of the fact that this neuron and the muscle it innervates have been the subject of intensive study (Hoyle, 1983), the clear proctolinlike effects of SETi stimulation have apparently passed undescribed.

The direct evidence that the metathoracic SETi motoneuron is proctolinergic is as follows. First, a cell body copositional with the SETi soma is immunoreactive to our proctolin antiserum (Keshishian and O'Shea, 1984; Witten *et al.*, 1984a). None of the other motoneurons of the extensor muscle (FETi, CI, or DUMETi) are proctolin immunoreactive. The second piece of evidence is the presence of proctolin immunoreactivity in nerve terminals on the slow fibers of the ETi muscle, including the rhythmic myogenic fibers. Third, proctolin is released from the SETi terminals when depolarized by high-potassium saline in the presence of calcium, but not when calcium is removed from the saline (Witten *et al.*, 1984a).

These observations are consistent with a role for neurally released proctolin in the direct control of the locust extensor muscle. If the SETi motoneuron is proctolinergic, we should expect to see a proctolinlike acceleration of the myogenic rhythm when SETi is stimulated (Fig. 1). Stimulation of the SETi motoneuron produces effects on the extensor muscle consistent with a dual-transmitter function. The glutamate-mediated contractions produced by SETi stimulation have been thoroughly described (Hoyle, 1983). In addition to these effects, which do not significantly outlast the duration of motoneuron activity, we have identified a persistent effect of SETi stimulation that is qualitatively and quantitatively similar to the action of proctolin (Witten *et al.*, 1984a). After a brief delay following a burst of SETi action potentials and the related transient contracture, the myogenic rhythm is accelerated (Fig. 1). The duration and decay of this increased frequency and its dependence on the amount of neural stimulation are exactly predicted by proctolin's actions. Thus based on proctolin's presence, release, and postsynaptic action, we now believe that the SETi cell is a dual-transmitter motoneuron. It probably uses glutamate for transient contractions and proctolin for longer-lasting effects.

A difficulty with this interpretation, as it applies to the SETi-induced activation of the myogenic rhythm, is that the extensor muscle is highly specialized. Indeed, the homologous muscles, innervated by homologous motoneurons, in the front and middle legs do not generate myogenic rhythms. It is likely therefore that proctolin serves a more general neuromuscular transmitter function than the regulation of specialized rhythmic myogenic contractures. One more general motor function for proctolin may be the generation of persistent "catch" contractures in skeletal muscle (Adams and O'Shea, 1983).

The ability of an insect motoneuron to induce muscle "catch" has been established for the slow coxal depressor (Ds) motoneuron of the cockroach (Chesler and Fourtner, 1981). Recently this neuron (Fig. 2) was shown to contain proctolin in its cell body and motor terminals and its catch-inducing action was ascribed to the

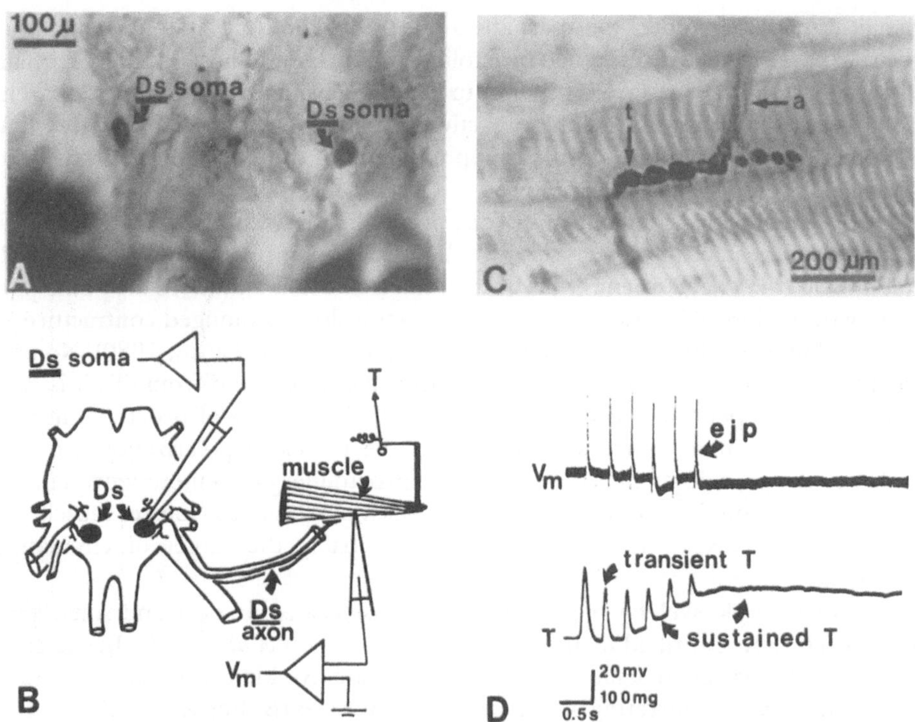

FIGURE 2. An identified peptidergic motoneuron in the cockroach. (A) The proctolin-immunoreactive somata of the Ds motoneurons are indicated in a dorsal view of the metathoracic ganglion. (B) The electrophysiological preparation is illustrated showing microelectrodes placed in the presynaptic peptidergic neuron Ds and its postsynaptic target muscle. (C) Immunoreactive Ds axon (a) and terminals (t) on the postsynaptic target. (D) The postsynaptic result of stimulating Ds in a brief burst of six action potentials. The upper trace is an intracellular muscle recording (Vm) and the lower trace is a measure of muscle tension (T). Note the biphasic response in the lower trace. The sustained tension is due to peptide release and is not associated with muscle depolarization. (Staining in C provided by Jane Witten.)

release of proctolin (O'Shea and Bishop, 1982; Adams and O'Shea, 1983). The muscle response to Ds stimulation is biphasic. Rapid transient contractions coupled 1:1 with motor action potentials and with excitatory junctional potentials (EJPs) are followed by a delayed slow and sustained contracture. The biphasic response has been attributed to the action of two transmitters, L-glutamic acid and proctolin. Proctolin produces a slow and sustained "catch" contracture of the coxal depressor muscles without depolarizing the muscle cells, whereas the transient repsonse is produced by the presumed glutamate-mediated transient EJP. The muscles innervated by the Ds neuron are also innervated by at least four other excitatory motoneurons, none of which contain proctolin. Muscle responses to these nonproctolinergic motoneurons are not biphasic but only transient.

The Ds and SETi neurons are currently the only uniquely identified peptidergic skeletal motoneurons in insects. We know, however, from immunohistochemical mapping studies that perhaps 5 to 10% of the total motoneuronal pool in cockroach and locust contained proctolin (Bishop and O'Shea, 1982; Keshishian and O'Shea, 1984). These cells have yet to be uniquely identified, but it is apparent from analysis of proctolin-immunoreactive terminals on identified muscles that proctolin is largely associated with a subpopulation of the slow motoneurons (Witten *et al.*, 1984a).

Consistent with its actions in insects, proctolin has a variety of effects on several types of crustacean muscle (see above). The action of proctolin on lobster skeletal muscle (the dactyl opener) has been studied in detail and there are some parallels with the insect Ds system. Proctolin causes a slow, prolonged contracture of the dactyl muscle without significant depolarization (Schwarz *et al.*, 1980). No presynaptic action of proctolin on motoneurons is measured, and from this it is concluded that the peptide acts on the muscle itself. The effects of proctolin, in particular the long time-course and absence of the muscle depolarization are very similar to the peptide-mediated effects of the proctolinergic Ds motoneuron in the insect preparation. Proctolinergic motoneurons innervating the lobster muscle are not thought to exist; the dactyl muscle is probably the target of circulating proctolin.

Proctolinergic skeletal motoneurons in crustaceans do exist and some have been specifically identified (Schwarz *et al.*, 1984; Bishop *et al.*, 1984). In the crayfish, proctolin has been associated with three of the five identified excitatory skeletal motoneurons innervating abdominal flexor muscles (Bishop *et al.*, 1984a). This sytem will provide an exceptionally accessible neuromuscular preparation for studying peptidergic transmission. The role of proctolin in the abdominal flexor system is currently under study, but preliminary findings suggest a modulatory cotransmitter function (Bishop *et al.*, 1984b). Proctolin released by the motoneurons does not directly cause contraction of the flexor muscle, but enhances the contracture caused by the rapidly acting transmitter (glutamate). This contrasts with the function of proctolin in the insect examples in which released proctolin has direct contractile effects.

The three examples of proctolinergic skeletal motoneurons examined so far have revealed quite different functions for the peptide cotransmitter. The SETi motoneuron regulates a myogenic rhythm, the Ds and the SETi motoneuron cause "catchlike" contractures, and the crayfish proctolinergic flexor motoneurons modulate the action of the conventional neuromuscular transmitter. Clearly, it is too early to generalize about the functional or behavioral principles that govern the use of peptide cotransmitters at the neuromuscular junction. We need to know much more about the functions of proctolinergic motoneurons and we need to investigate the possible neuromuscular functions of other peptides. It is easy to imagine that several peptidergic subpopulations of motoneurons exist. We know that only a small proportion of the insect skeletal motoneurons express proctolin. Although it tends to be associated with slow motoneurons, most slow motoneurons do not contain this peptide. Do other peptidergic subpopulations exist? If they do exist, how many are there likely to be and what are the functional consequences of multiple peptide cotransmitters in arthropod neuromuscular systems?

3. Do Peptides Other Than Proctolin Participate in Neuromuscular Transmission?

As a prelude to determining whether peptides other than proctolin function as neuromuscular transmitters, we are attempting to isolate and characterize the chemical nature of new myoative neuropeptides in insects. We examined the corpus cardiacum, a major neurosecretory structure of the cockroach. Using a skeletal muscle bioassay and HPLC, two myoactive factors, MI and MII, were isolated and sequenced (MI = pGlu-Val-Asn-Phe-Ser-Pro-Asn-Trp-NH_2; MII = pGlu-Leu-Thr-Phe-Thr-Pro-Asn-Trp-NH_2) (O'Shea *et al.*, 1984; Witten *et al.*, 1984b). The bioassay used to detect the presence of these peptides (which are now available from Peninsula Laboratories) is the same assay used to detect proctolin. The myogenic rhythm of the locust ETi muscle is accelerated by peptides MI and MII (Fig. 3). We do not yet know the significance of this since peptides MI and MII probably are not present in locust and are not likely therefore to be involved in the innervation of this locust muscle. The sequences of MI and MII peptides are, however, closely homologous to a known locust peptide called adipokinetic hormone (AKH) (Stone *et al.*, 1976) and a crustacean peptide called red-pigment-concentrating hormone (RPCH) (Fernlund and Josefsson, 1972). The activity of MI and MII on the locust muscle probably therefore reflects the close structure–function relationships among these peptides. Indeed, AKH also causes an activation of the locust myogenic rhythm. These peptides also produce "catchlike" contracture of the

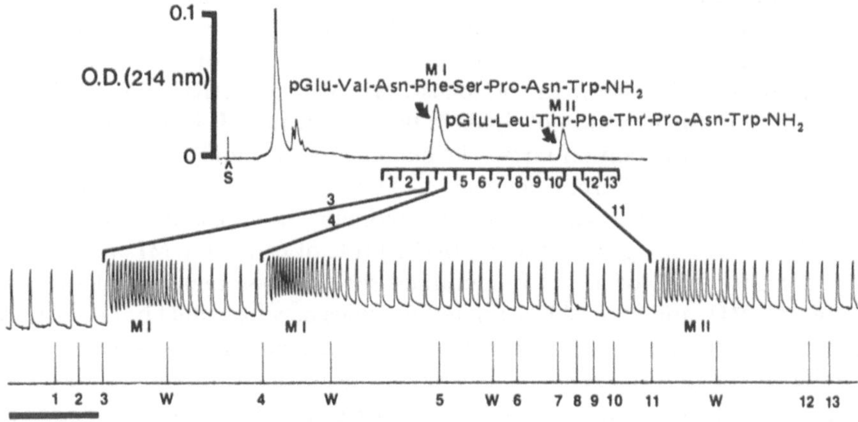

FIGURE 3. Identification of two myoactive factors (MI and MII, sequences shown) in an HPLC fractionation of an extract of 10 corpora cardiaca. The upper part shows the chromatographic record; S indicates the start. Eluting buffer was 1 mM ammonium acetate (pH 4.5) and a linear gradient of between 25% and 50% acetonitrile. Fractions numbered 1 to 13 (1-ml fractions) were collected and bioassayed on the locust ETi muscle (lower part of this figure and Fig. 1). The numbers under the lowest trace indicate when that fraction was applied. The W indicates saline wash. The monitor of muscle movement (contraction is an upward deflection) shows that fractions 3, 4, and 11 are most active and cause an increase in beating frequency and a tonic contraction of the locust ETi muscle. Time scale (horizontal bar) is 5 min.

muscle, another activity reminiscent of the action of proctolin. The activity of MI, MI, and AKH on the ETi muscle can be distinguished from the similar action of proctolin. The major difference is in the persistence of the effects. The elevated myogenic frequency and catch produced by the AKH-related peptides outlast the effects produced by proctolin. A structural basis for this is suggested by the fact that the longer-acting peptides, unlike proctolin, are terminally modified with an N-terminal pyro-Glu and a carboxy-terminal amide. Peptides MI, MII, and AKH may be protected from the inactivating of aminopeptidases and carboxypeptidases.

The ability of AKH-related peptides to contract skeletal muscle suggest that, like proctolin, they too may be found in insect motoneurons. Recently an AKH antibody was developed (Schooneveld *et al.*, 1983) that may also recognize MI and MII peptides. This has provided the first direct evidence for the presence of AKH-related peptides in neurons of the CNS as well as in the neurosecretory corpus cardiacum (Schooneveld *et al.*, 1983; Witten *et al.*, 1984a). It is possible therefore that as well as serving hormonal functions, the AKH-related peptides are transmitters. In the cockroach we have recently obtained immunohistochemical evidence that some of the cells containing AKH-related peptides in the CNS are efferent neurons (Witten *et al.*, 1984a). Our best evidence that these peptides participate in neuromuscular transmission is provided by the presence of the immunoreactive motor terminals on gut musculature. We have not yet identified skeletal muscles innervated by AKH-immunoreactive neurons. But by analogy with the gut and skeletal motor transmitter proctolin, and because of the pronounced ability of MI and MII to contract skeletal muscle, we believe that these peptides may indeed have skeletal motor functions.

The cockroach AKH-related peptides MI and MII are known to be released from the corpora cardiaca into the hemolymph and function as circulating hormones (O'Shea *et al.*, 1984). In the locust, MI, MII, and AKH injected into the circulation cause the elevation of lipids in the hemolymph. It is possible that these peptides exert their action on target skeletal muscle hormonally. The presence of AKH-immunoreactive efferent neurons in the CNS, however, argues that a non-proctolin peptidergic subpopulation of motoneurons exists. We may speculate therefore that, by analogy with the established role of proctolin, MI and MII function as cotransmitters of as yet unidentified skeletal motoneurons. If this is the case, the MI, MII, and other AKH-related peptides may have dual but coordinated hormonal and transmitter actions. Functional coordination may be achieved during prolonged or vigorous motor behavior such as flight (Mayer and Candy, 1969). The same peptide, for example, would be involved as a metabolic hormone to provide energy for the increased demand and may also be a transmitter that alters the physiology of specific muscles used in the behavior.

4. Concluding Remarks

A role for neuropeptides in skeletal neuromuscular transmission has now been established. There is also evidence that neuropeptides can influence skeletal muscle by hormonal action. These results are significant for a number of different reasons.

First, questions related to the precise physiological actions of neuropeptides may be quite conveniently addressed at the neuromuscular junction. The nerve–muscle synapse in both vertebrates and invertebrates has long been important in analyzing the fine details of the synaptic physiology of the classical transmitters acetylcholine and GABA. There is now good reason to hope that our understanding of peptidergic systems will be similarly facilitated by model neuromuscular preparations. Second, these findings greatly increase the complexity of the arthropod neuromuscular system. They force us to accept and account for a complex new world of multiple transmitters. Proctolin accounts for only about 5% of the motoneurons in the insects we have studied, and an even smaller proportion of efferent neurons contain AKH-like peptides. These figures suggest the number of peptide neuromuscular transmitters may be very large indeed. We now must explain a great deal, including, why so many transmitters? Why do neurons require more than one transmitter? It seems likely that detailed physiological studies in model neuromuscular and neuronal systems, such as in arthropods, can answer these questions. It seems unlikely that they will be answered in any other way.

ACKNOWLEDGMENTS. I am indebted to many associates and colleagues for sharing their thoughts, insights, and experimental observations with me. I especially thank Mike Adams, Jane Witten, Mary-Kate Worden, Haig Keshishian, and Marty Schaffer upon whose work many of my speculations are based. Cindy Bishop deserves special thanks also. She provided the vital key to the subsequent discoveries when she successfully produced an exceptional staining antibody against proctolin.

References

Adams, M. E., and O'Shea, M., 1983, Peptide cotransmitter at a neuromuscular junction, *Science* **221**:286–289.

Bishop, C. A., and O'Shea, M., 1982, Neuropeptide proctolin (H-Arg-Tyr-Leu-Pro-Thr-OH): Immunological mapping of neurons in the central nervous system of the cockroach, *J. Comp. Neurol.* **207**:223–238.

Bishop, C. A., O'Shea, M., and Miller, R. J., 1981, Neuropeptide proctolin (H-Arg-Tyr-Leu-Pro-Thr-OH): Immunological detection and neuronal localization in the insect central nervous system, *Proc. Natl. Acad. Sci. U.S.A.* **78**:5899–5902.

Bishop, C. A., Wine, J. J., and O'Shea, M., 1984a, Neuropeptide proctolin in postural motoneurons of the crayfish, *J. Neurosci.* **4**(8):2001–2009.

Bishop, C. A., Nagy, F., Wine, J. J., and O'Shea, M., 1984b, Neural release and physiological action of an identified peptide contained in crayfish motor neurons, *Soc. Neurosci. Abstr.* **10**:151.

Brown, B. E., 1967, Neuromuscular transmitter substance in insect visceral muscle, *Science* **155**:595–597.

Brown, B. E., 1975, Proctolin: A peptide transmitter candidate in insects, *Life Sci.* **17**:1241–1252.

Brown, B. E., and Starratt, A. N., 1975, Isolation of proctolin, a myotropic peptide from *Periplaneta americana*, *J. Insect Physiol.* **21**:1879–1881.

Burrows, M., and Hoyle, G., 1973, Neural mechanisms underlying behavior in the locust *Schistocerca gregaria*. III. Topography of limb motoneurons in the metathoracic ganglion, *J. Neurobiol.* **4**:167–186.

Chesler, M., and Fourtner, C. R., 1981, Mechanical properties of a slow muscle in the cockroach, *J. Neurobiol.* **12**:391–402.

Eckert, M., Agricola, H., and Penzlin, H., 1981, Immunocytochemical identification of proctolinlike immunoreactivity in the terminal ganglion and hindgut of the cockroach *Periplaneta americana* (L.), *Cell Tissue Res.* **217**:633–645.

Evans, P. D., and O'Shea, M., 1977, An octopaminergic neurone modulates neuromuscular transmission in the locust, *Nature (London)* **270**:257–259.

Evans, P. D., and O'Shea, M., 1978, Identification of an octopaminergic neurone and the modulation of a myogenic rhythm in the locust, *J. Exp. Biol.* **73**:235–260.

Fernlund, P., and Josefsson, L., 1972, Crustacean color-change hormone: Amino acid sequence and chemical synthesis, *Science* **177**:173–175.

Hoyle, G., 1983, *Muscles and Their Neural Control,* Wiley-Interscience, New York.

Hoyle, G., and O'Shea, M., 1974, Intrinsic rhythmic contractions in insect skeletal muscle, *J. Exp. Zool.* **189**:407–412.

Keshishian, H., and O'Shea, M., 1984, The distribution of a peptide neurotransmitter in the postembryonic grasshopper CNS, *J. Neurosci.* (in press).

Kingan, T. G., and Titmus, M., 1983, Radioimmunologic detection of proctolin in Arthropods, *Comp. Biochem. Physiol. C.* **74**:75–78.

Kravitz, E. A., Glusman, S., Harris-Warrick, R. M., Livingstone, M. S., Schwarz, T. L., and Goy, M. F., 1980, Amines and a peptide as neurohormones in lobster: Action on neuromuscular preparations and preliminary behavioral studies, *J. Exp. Biol.* **89**:159–175.

Lingle, C. J., 1979, The effects of acetylocholine, glutamate, and biogenic amines on muscle and neuromuscular transmission in the stomatogastric system of the spiny lobster, *Panulirus interruptus,* Ph.D. thesis, University of Oregon, Eugene, Oregon.

May, T. E., Brown, B. E., and Clements, A. N., 1979, Experimental studies upon a bundle of tonic fibers in the locust extensor tibialis muscle, *J. Insect Physiol.* **25**:169–181.

Mayer, R. J., and Candy, D. J., 1969, Control of hemolymph lipid concentration during locust flight: An adipokinetic hormone from the corpora cardiaca, *J. Insect. Physiol.* **15**:611–620.

Miller, T., 1979, Nervous vs. neurohormonal control of insect heart beat, *Am. Zool.* **19**:77–86.

Miller, M. W., and Sullivan, R. E., 1981, Some effects of proctolin on the cardiac ganglion of the Maine lobster, *Hommarus americanus* (Milne Edwards), *J. Neurobiol.* **12**:629–639.

O'Shea, M., and Adams, M. E., 1981, Pentapeptide (proctolin) associated with an identified neuron, *Science* **213**:567–569.

O'Shea, M., and Bishop, C. A., 1982, Neuropeptide proctolin associated with an identified skeletal motoneuron, *J. Neurosci.* **2**:1242–1251.

O'Shea, M., and Evans, P. E., 1979, Potentiation of neuromuscular transmission by an octopaminergic neuron in the locust, *J. Exp. Biol.* **79**:169–190.

O'Shea, M., and Schaffer, M., 1985, Neuropeptide function: The invertebrate contribution, *Annu. Rev. Neurosci.* **8**:171–198.

O'Shea, M., Witten, J., and Schaffer, M., 1984, Isolation and characterization of two myoactive neuropeptides: Further evidence of an invertebrate peptide family, *J. Neurosci.* **4**:521–529.

Piek, T., and Mantel, P., 1977, Myogenic contractions in locust muscle induced by proctolin and by wasp, *Philanthus triangulum,* venum, *J. Insect Physiol.* **23**:321–325.

Pitman, R. M., 1984, Nervous system, in: *Comprehensive Insect Physiology, Biochemistry and Pharmacology,* Volume II (G. A. Kerkut, ed.), Pergamon Press, New York, pp. 000–000.

Schooneveld, H., Tesser, G. I., Veenstra, J. A., and Romberg-Privee, H. M., 1983, Adipokinetic hormone and AKH-like peptide demonstrated in the corpora cardiaca and nervous system of *Locusta migratoria* by immunohistochemistry, *Cell Tissue Res.* **230**:67–76.

Schwarz, T. L., Harris-Warrick, R. M., Glusman, S., and Kravitz, E. A., 1980, A peptide action on a lobster neuromuscular preparation, *J. Neurobiol.* **11**:623–628.

Schwarz, T. L., Lee, G. M.-H, Siwicki, K. K., Standaert, D. G., and Kravitz, E. A., 1984, Proctolin in the lobster: The distribution, release and chemical characterization of a likely neurohormone, *J. Neurosci.* **4**:1300–1311.

Starratt, A. N., and Brown, B. E., 1975, Structure of the pentapeptide proctolin, a proposed neurotransmitter in insects, *Life Sci.* **17**:1253–1256.

Stone, J. V., Mordue, W., Batley, K. E., and Morris, H. R., 1976, Structure of locust adipokinetic hormone, a neurohormone regulates lipid utilization during flight, *Nature (London)* **263**:207–211.

Sullivan, R. E., 1979, A proctolin-like peptide in crab pericardial organs, *J. Exp. Zool.* **210**:543–552.

Witten, J., Worden, M. K., Schaffer, M. H., and O'Shea, M., 1984a, New classification of insect motoneurons: Expression of different peptide transmitters, *Soc. Neurosci. Abstr.* **10**:151.

Witten, J., Schaffer, M. H., O'Shea, M., Carter Cook, J., Hemling, M. E., and Rinehart, K. L., 1984b, Structures of two cockroach neuropeptides assigned by fast atom bombardment mass spectometry, *Biochem. Biophys. Res. Comm.* (in press).

PART V
CELLULAR AND MEMBRANE BIOPHYSICS

22

Electrical Behavioral Correlates of Calcium and Potassium Currents in Molluscan Nerve Cells

MAURICE GOLA

1. Introduction

When the original report by Hodgkin and Huxley was published in 1952, the two-channel model of nerve excitability derived from the squid giant axon was expected to explain the electrical properties of many nerve cells as well as those of their respective axons, cell bodies, dendrites, and terminals. During the past 15 years, this model has been superseded by the discovery of several voltage-gated ionic conductances, particularly for potassium and calcium.

In invertebrate nerve cells, in which most of the data available have been collected, most somata possess ionic conductances in addition to the sodium and potassium conductances found in the squid axon. Voltage(V)-dependent calcium conductance changes large enough to be regenerative have been observed in a number of preparations from protozoa to mammals (see reviews by Kostyuk, 1981; Tsien, 1983). At least four different K channels have been identified in several invertebrates: the V-dependent H-H–type K current g_{K-V}; the fast transient outward current g_A (Connor and Stevens, 1971); the slow K current g_s (Partridge and Stevens, 1976); and the calcium-activated K current g_{K-Ca} (Meech, 1978; Lux and Heyer, 1979). Various evidence, direct and indirect, indicated that such channels are fairly common in a variety of cell types.

MAURICE GOLA • Institute of Neurophysiology and Psychophysiology, C.N.R.S., Marseilles, France.

Each ionic conductance plays a role in the integrative and encoding functions of individual neurons. In cells such as nerve axons and fast-twitch muscles, which specialize in the rapid conduction of information, the inward current is mainly carried by sodium ions through the fast Na channels. By contrast, Ca-dependent action potentials are found in cells or cell parts that are more specially concerned with integration.

The current across the membrane can be separated into a variety of components flowing through specific channels. However, with several conductances active in the same membrane, the task of sorting out their respective roles is difficult, and it is not yet possible to attempt a complete description of the controls exerted by the various ionic conductances on the electric behavior of a given cell. Moreover, the conductances are not homogeneously distributed over the cell surface; sodium channels predominate in axons, whereas calcium channels are generally restricted to soma membranes and nerve terminals (Junge and Miller, 1974). In order to elucidate the role of each individual channel, one must therefore take into consideration (1) the interactions between the individual ion fluxes and their local electrical correlates and (2) the distribution of the specific conductances among the various morphologically defined compartments of the cell and even within a single compartment.

In solving these questions, the main approach used so far relies on theoretical studies of either isopotential membrane models or whole cell models derived from experimental data mainly collected from invertebrate nerve cells. Another way of estimating the physiological role of given channels is the pharmacological dissection of the various channels with specific inhibitors (Thompson, 1977). For instance, poststimulus hyperpolarizations are generally eliminated by blockers of the Ca channels, which points to the involvement of a Ca-activated K current (Hotson and Price, 1980). However, this approach suffers from limitations because (1) specific inhibitors of the various K channels are still lacking and (2) one cannot directly establish how the remaining currents in current-clamped cells may have been affected by specific blockers since these currents tend to be coupled, at least with regard to their voltage dependence.

Identifiable neurons that can be studied repeatedly can help to answer the above questions (Connor, 1975). In mollusks such as *Aplysia* and *Helix,* the set of nonsynaptic ionic conductances are believed to produce a wide repertoire of physiological behavior. These conductances are for the present restricted to the channels listed above, specific for sodium, calcium, and potassium ions. Therefore, channels selected from this limited palette and strategically distributed over the cell surface should be able to generate various behaviors, such as frequency adaptation, intrinsic burstiness, and long-lasting spikes.

Another conclusion that has emerged from numerous investigations of identifiable neurons is that neurons are likely to be all individually different with regard to their intrinsic properties, their anatomical connections, and their responsiveness to transmitters and hormones. Restricting our focus to the intrinsic properties, it appears that the quantitative adjustment of these channels is crucial for expressed behavior as is the type of channels present in the electrogenic membrane. One type of channel may be abundant in one nerve cell and completely absent from another.

Heyer and Lux (1976) have noted that the relative amounts of V- and Ca-dependent K currents vary from cell to cell. Therefore, the strategy for evaluating the role of given channels in a given electrical pattern implies both qualitative and quantitative evaluations of the membrane properties of the cell under investigation.

The aim of the present study is to assess how some of the already-known nonsynaptic ionic conductances contribute to the shape of the signals in response to direct soma stimulation. As the ratio between the various channels seems to play a major role in the signal patterning, identifiable cells were subjected to a preliminary evaluation of their channel content and cells were mainly selected on the basis of their g_{K-V}/g_{K-Ca} and g_{Na}/g_{Ca} ratio.

2. Calcium-Related Electrical Phenomena

Five neurons were selected from *Helix* ganglia (see Fig. 1). Voltage-clamp experiments were conducted to determine the relative Na, Ca, and K channel content of soma membranes. Neither the fast-transient K current I_A nor the slow K current I_s, neither of which play a significant role in the spike configuration (except in interspike trajectory), were assessed. Separation of Na and Ca currents was made in TEA-4AP-treated cells in order to block most of the K currents; they were evaluated as the current part suppressed either by O-Na saline or by adding Ca channel blockers. Calcium-gated K currents were evaluated as the outward current part

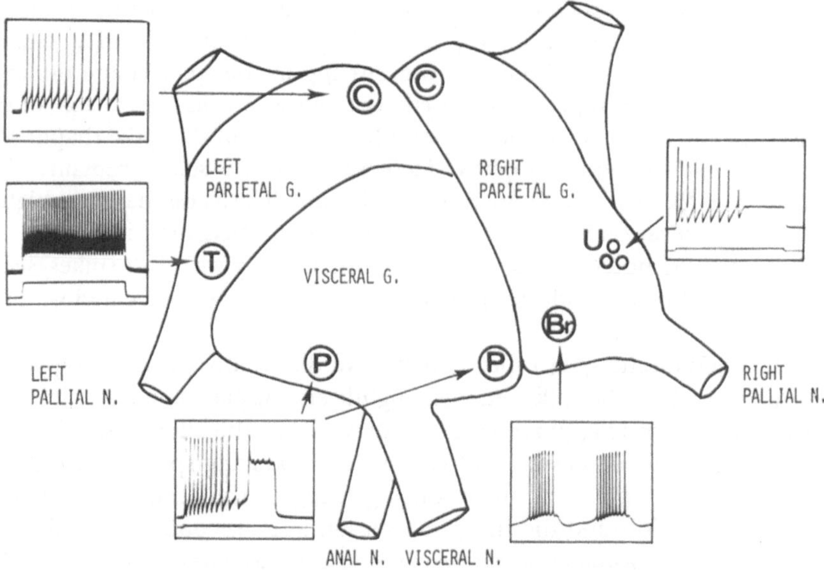

FIGURE 1. Diagram of the subesophageal ganglion complex of *Helix pomatia* with cell body positions and electrical activities of the selected cells.

TABLE I

Ionic Components in Selected Cell Somata[a]

Cell Type	I_{Na}	I_{Ca}[b]	I_{K-V}	I_{K-Ca}[c]
C: low-frequency tonic	+	+	+	+
T: high-frequency tonic[d]	+		+	
Br: endogeneous burster	+	+	+	+
P: plateauing cell		+	+	+
U: fast-adaptating cell		+		+

[a]+ indicates presence of channel; no + indicates that the corresponding channel is either lacking or constitutes less than 10% of the inward or outward currents.

[b]The relative I_{Ca} contribution to the total inward current ranges from 0 in T cells to 1 in P and U cells.

[c]The relative I_{K-Ca} contribution to the total outward current is 0 in T cells, $\frac{1}{3}$ in C cells, $\frac{1}{2}$ in P and Br cells, and almost 1 in U cells.

[d]T cells are under the control of g_{Na} and g_{K-V} H-H–type channels, whereas U cells lack such channels and possess only g_{Ca} and g_{K-Ca} channels (Lux and Hofmeier, 1982a).

suppressed by Ca blockers. Voltage-gated K currents were further extracted by adding TEA and 4-AP to the Ca blocker containing saline.

Results are expressed in qualitative terms in Table I. The electrical behavior of the various cells is briefly indicated which refers either to spontaneous activity (Br) or to the firing mode in response to maintained depolarization (see also Fig. 1).

Under resting conditions, the Na-dependent spikes of T cells are of short duration (3.5 msec at half amplitude) and they have fast rising and falling phases (about 40 and 20 V/sec, respectively). By contrast, U cells, which are Ca-dependent, exhibit large action potentials (up to 115 mV) of long duration (20 to 24 msec at half amplitude), (Lux and Hofmeier, 1982a). The rising and falling phases are slow: 8 to 6V/sec, respectively. The other cells (C, P, and Br) have durations ranging from 6 to 8.5 msec, and rising and falling phases of 10 to 14 V/sec.

Of greater interest are the changes occurring in the spike shape under sustained stimulation. With repeated short-current pulses, T cells are capable of firing at relatively high frequency without noticeable alteration of the spike shape. Above 10 Hz, the spike amplitude slightly decreases, but the shape remains almost unchanged (Fig. 2). With maintained depolarization, T cells discharge at high frequency with moderate adaptation: they were thus termed high-frequency tonic neurons. It must be noted that T cells fire spontaneously with 1–2 spikes/sec and that spikes start abruptly, indicating that the pacemaker area is located within the proximal axon.

U cells when discharging repeatedly behave differently. At relatively low frequencies (0.5 to 1 Hz), the spike enlarges slightly and displays a more or less prominent two-step repolarizing phase (Fig. 2). Above 1 Hz, the amplitude of the spike decreases sharply and U cells are not capable of producing spikes above 2.5 to 3 Hz. In maintained depolarization producing firing in this frequency range, the spike attenuates rapidly and the firing vanishes in a few seconds, hence the classification of U cells as fast-adaptating neurons. U cells are either silent with a large resting potential or regularly firing at a slow rate (0.5 to 1 Hz). In the latter case they display a regular slow-pacemaker potential that indicates that the spikes are generated within the cell body.

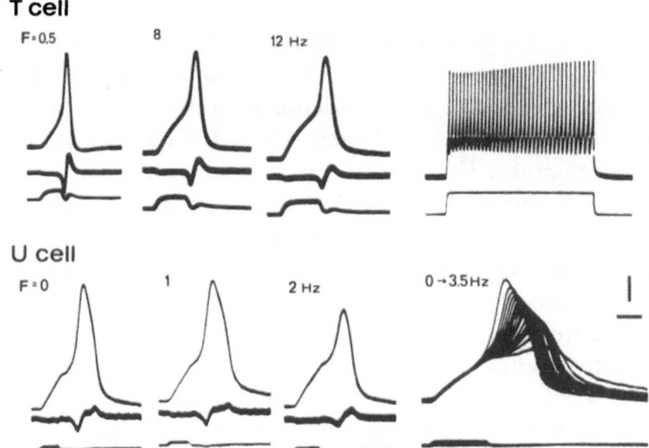

FIGURE 2. Behavior of current-stimulated T and U cells. Spikes in T cells are produced by Na- and V-dependent K currents and those in U cells are dependent on Ca and Ca-activated K currents. T cells are able to fire short-lasting spikes at large frequency while U cells display fast adaptation due to spike attenuation occurring in low-frequency range. Vertical bar: T cell, 20 mV, 40 V/sec (inverted spike derivation in middle traces); U cell, 20 mV, 20 V/sec. Horizontal bar: T cell, 10 msec and 0.5 sec; U cell, 20 msec and 10 msec in the superposed recordings.

P and C cells are always silent. In both cell types, soma stimulation produces overshooting spikes with a more or less pronounced shoulder on the falling phase. The shoulder is consistently more prominent than in T and U cells. Repeated stimulation of P and C cells produces spike enlargement that makes the two-step repolarizing phase more obvious. In C cells the spike duration increases with stimulation frequency and results in a twofold increase at frequencies above 5 Hz. In P cells the spike enlargement is much more pronounced; the spike duration is already

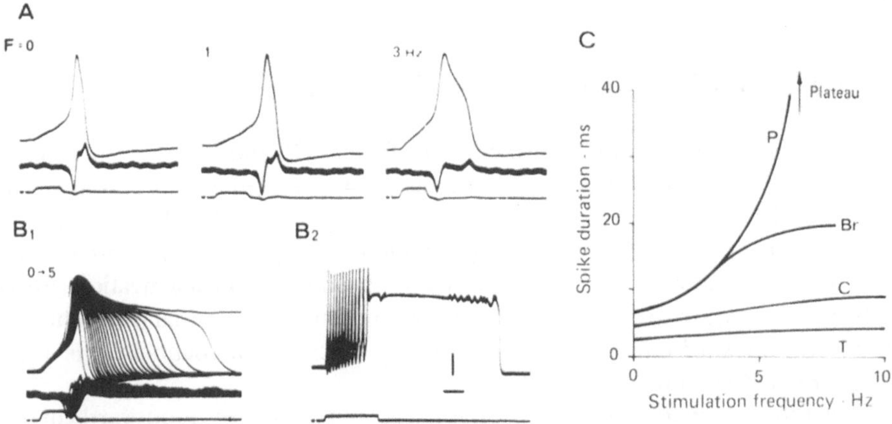

FIGURE 3. Plateau production in stimulated P cells. (A) Shows spike enlargement in stimulated P cell. (B_1, B_2) Plateaus occur for stimulation frequencies ranging from 5 to 6 Hz (B_1) or in response to a sustained stimulus (B_2). Vertical bar: 20 mV and 20 V/sec. Horizontal bar: A and B_1, 10 msec; B_2, 0.5 sec. (C) Spike duration (measured at half amplitude) versus stimulation frequency in the selected cells. U cells not shown.

twice its rest value for stimulation frequencies of 2.5 to 3 Hz (Fig. 3A). At higher frequencies there is a sharp increase of the spike duration that reaches 20 to 25 msec at 3–4 Hz. Finally, between 5 and 6 Hz, the second phase of repolarization aborts and the spike converts to a steady depolarization or plateau (Fig. 3B,C). The plateau occurs at a positive value (between +5 and +10 mV) and it may last tens of seconds or even minutes. The plateau duration depends on previous production of spikes or plateaus and on the stimulation frequency. The plateaus are easily aborted by short inward current pulses. However, in a given P cell, the plateau duration is almost unpredictable which points strongly to the existence of an unstable state of equilibrium. The underlying inward current, mainly carried by Ca ions, is almost fully counterbalanced by K and leakage currents. Transition from the plateau to the resting potential would be triggered by a slight change in the ionic current balance.

The plateaus as well as the elongated spikes in P cells persist in zero sodium saline or in presence of TTX. They are blocked by adding Ca-channel blockers and the plateau level depends on the Ca content of the saline. Therefore, P cells fulfill all the criteria (Hagiwara, 1973) for the involvement of Ca current in the plateau production.

A point to be underlined is the rather low-firing frequency (5–6 Hz) needed for the transition from the firing mode to the plateau. In bursting cells, which resemble P cells in their ionic conductance content (see Table I), plateaus never occur. Upon repeated stimulation, the spike enlarges as with stimulated P cells (Heyer and Lux, 1978). However, the spike duration stabilizes at 4–5 Hz (20 to 25 msec duration): the two-step repolarizing phase is then very prominent, but the transition to the plateau does not occur, i.e., the repolarizing outward current still dominates the long-lasting depolarizing Ca current. In view of the analogy between P and bursting cells with regard to inward and outward currents, this result is rather surprising. It will be shown to reflect a qualitative difference between the Ca-gated K currents in those cells.

2.1. Ca- and V-Dependent Outward Currents

Figure 4 shows sets of outward currents from T, P, and U cells. (Corresponding currents from C and bursting cells, not illustrated, resemble those from P cells.) There are clear-cut differences between the three illustrated cell types. The outward current from T cells increases regularly with voltage; its rising phase becomes faster at large depolarizations and it displays subsequent slow inactivation. Recovery from inactivation needs several tens of seconds at rest potential (Aldricht *et al.*, 1979). This outward current is not affected by blocking Ca channels: it is a pure V-dependent K current.

The pure Ca-dependent K current is illustrated by the set of outward currents from U cells. Its main features are: slow rising phase, particularly at large depolarization (Kostyuk *et al.*, 1980; Woolun and Gorman, 1981; Lux and Hofmeier, 1982a,b), N-shaped relationship to the membrane potential (Meech and Standen, 1975), and sensitivity to Ca-channel blockers. In contrast to the V-dependent K

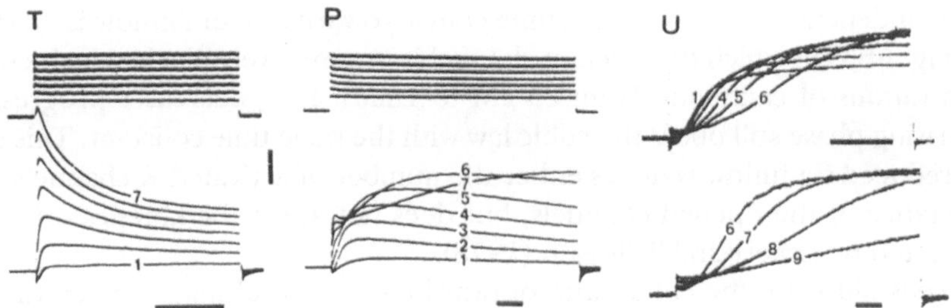

FIGURE 4. V- and Ca-dependent K currents. T cell: pure V-dependent K currents; P cell: compound V- and Ca-dependent K currents; U cell: pure Ca-dependent K currents. T cell: holding potential, −50 mV; test potentials (1 to 7) from −5 to +125 mV, with 22 mV steps. P cell: holding potential, −50 mV; test potentials (1 to 7) from −10 to +110 mV, with 20 mV steps. U cell: I_{K-Ca} obtained by subtracting the total outward current from the leakage current remaining in 1 mM Cd^{2+}. Holding potential, −50 mV; test potentials (1 to 9) from +50 to +130 mV, with 10 mV steps. Vertical bar: T, 0.5 μA; P, 1 μA; U, 0.2 μA. Horizontal bars: 40 msec.

current in T cells, I_{K-Ca} in U cells does not display inactivation. The sigmoid rising phase of U cells I_{K-Ca} may be approximated by a cubic law, i.e., by two exponential curves with τ- and $\tau/2$-associated time constants (Lux and Hofmeier, 1982b) (Fig. 5A). Like other V-dependent channels, τ depends on the membrane potential, but unexpectedly it increases sharply at large depolarizations instead of displaying the usual bell-shaped relationship. From this observation, Lux and Hofmeier have concluded that I_{K-Ca} is actually V-dependent, although calcium must play a role in the channel opening (see also Kostyuk *et al.*, 1980). One interesting conclusion is that the initial inflow of Ca is necessary for g_{K-Ca} opening, but the channel may remain opened for minutes in depolarized cells in spite of the fact that I_{Ca} vanishes during long depolarizations (Brehm and Eckert, 1978).

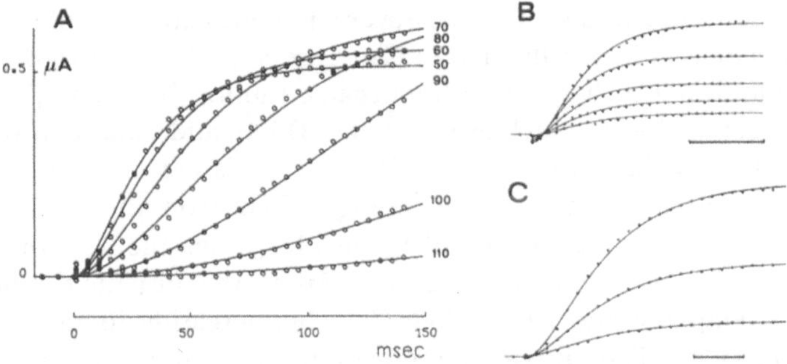

FIGURE 5. Ca-dependent K current in U cells. (A) Open symbols, I_{K-Ca} at different membrane potentials obtained by subtracting the total outward current from the Cd-resistent leakage current. Continuous curves, best fit of experimental data by squared exponential curves. The time constant decreased from 22 msec at +50 mV to 800 msec at + 110 mV. (B and C) Progressive blockade of I_{K-Ca} by increasing concentrations of cadmium (from 25 μM to 0.5 mM). Experimental points, current in control and Cd-containing salines minus leakage current in 0.5 mM Cd. Continuous curves as in A with τ = 23 msec in B (test potential +55 mV) and τ = 56 msec in C (test potential, +80 mV); the progressive blockade of I_{Ca} by cadmium ions does not change the time course of I_{K-Ca} activation. Horizontal bars: 50 msec.

The independence of the I_{K-Ca} time course toward calcium inflow is illustrated in the Fig. 5B,C in which the calcium channel is progressively blocked by increasing concentrations of cadmium (from 25 μM to 1mM). I_{K-Ca} decreases progressively, but its rising phase still obeys the cubic law with the same time constant. This means that a reduced Ca inflow reduces either the number of activated K channels or the conductance of the opened channels, but does not affect the kinetics of opening as suggested by Eckert and Tillotson (1978).

P cells illustrate the compound outward current resulting from activation of both V- and Ca-dependent K currents (see Fig. 4). At membrane potentials ranging from −30 to +20 mV, both currents are indistinguishable because they have almost similar time courses. At positive potentials, the compound outward current can be separated into a fast phase corresponding to the I_{K-V} of T cells and a slower phase corresponding to the I_{K-Ca} of U cells. Above +100 mV, I_{K-Ca} vanishes because the Ca driving force is then reduced, leaving the I_{K-V} component.

Current kinetics in T and U cells already give some insights to their contribution to the spike repolarizing phase: an abrupt repolarization will be caused by I_{K-V}, a smooth and slow repolarization will be characteristic of I_{K-Ca}. The two-step repolarizing phase seen in the other selected cells is due to their compound outward current as well as to large Ca currents.

2.2. Separation of I_{K-V} and I_{K-Ca} in P Cells

From the work of Lux and Hofmeier (1982a,b), much is known about the properties of I_{K-Ca} in U cells. In these cells, I_{K-Ca} is characterized by the voltage dependence of its time course which becomes very slow at large depolarizations. The second property of I_{K-Ca} is the lack of inactivation during long-lasting depolarizations. It was of interest to see whether I_{K-Ca} in other identifiable cells is identical to that in U cells. For this purpose, it was necessary to separate the I_{K-Ca} component from I_{K-V} and leakage currents. Several methods are available, all based on the fact that increased intracellular free Ca triggers the I_{K-Ca}. Blocking of I_{K-Ca} is obtained by blocking Ca channels with inorganic cations Cd^{2+}, La^{3+}, Co^{2+}, Mn^{2+}, or selective organic blockers such as verapamil, D600, and diltiazem. I_{K-Ca} from a P cell obtained by blocking the Ca channels with Cd^{2+} is shown in Fig. 6C. Because I_{K-Ca} is obtained by subtracting the remaining current from the total current, use of Ca blockers actually gives the sum of I_{Ca} plus I_{K-Ca}, i.e., the I_{K-Ca} time course is obscured by I_{Ca}, particularly below +20 mV. Moreover, organic Ca blockers are not selective: verapamil and D600 affect I_{K-V} by increasing its inactivation rate (Kostyuk *et al.*, 1975). They also act directly on the I_{K-Ca} at concentrations that do not affect I_{Ca} and this effect is irreversible (Gola and Ducreux, unpublished results).

Impeding free intracellular Ca changes blocks I_{K-Ca} activation and can be used for separation purposes without requiring I_{Ca} block. Substituting Ba for Ca still leads to an inward current through Ca channels, but displacement of intracellular Ca by the more electronegative Ba suppresses I_{K-Ca}. However, Ba blocks part of I_{K-V} (Connor, 1977; Gorman and Hermann, 1979) and the inward Ba current has a slower inactivation rate than I_{Ca}. Intracellular free Ca can be clamped at low levels

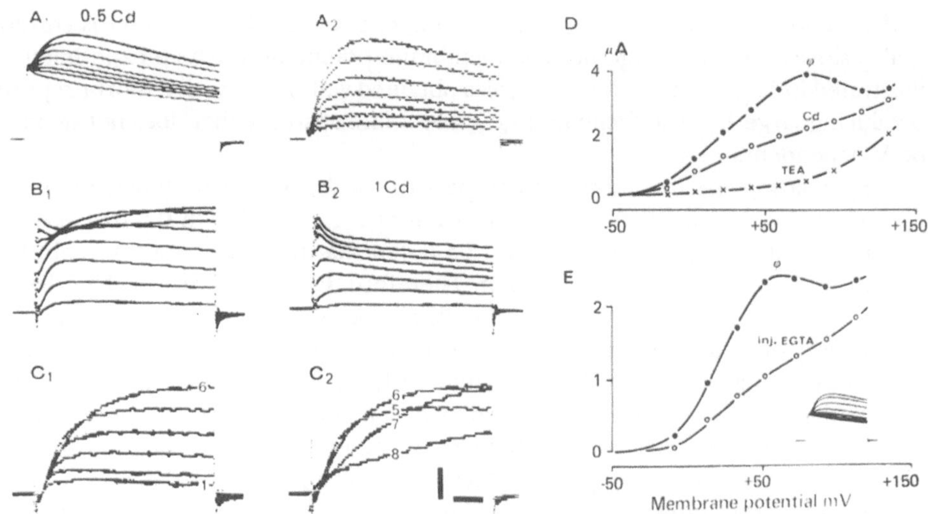

FIGURE 6. Separation of V- and Ca-dependent K currents in P cells. (A) Progressive blockade of I_{K-Ca} by 0.5 mM Cd. Holding potential, −60 mV; test potential, +60 mV. Current suppressed by Cd shown in A_2. (B) Sets of outward currents in control saline (B_1) and in presence of 1 mM Cd (B_2). Holding potential, −50 mV; test potentials from −10 mV to +130 mV, with 20 mV steps. (C) Currents suppressed by Cd (series B_1 minus series B_2); C_1 from −10 (1) to +90 mV (6); C_2 from +70 (5) to +130 mV (8). Horizontal bar: A, 50 msec; B and C, 20 msec. Vertical bar: A_1 and B, 1.2 μA; A_2 and C, 0.6μA. (D) Current-voltage relationships of P cells. (D_1) Cell bathed in control saline (φ) in presence of Cd (I_{K-Ca} blocked) and after adding TEA and 4-AP to the Cd saline; (E) before and after 3-min intracellular injection of EGTA (inset, progressive block of I_{K-Ca} during iontophoresis).

by injecting Ca chelators such as EDTA or EGTA (Meech, 1974). Apart from the fact that the resulting I_{Ca} has a slower inactivation rate, the eliminated current gives a good estimate of I_{K-Ca}. Corresponding I-V curves are shown in Fig. 6E for a P cell.

Figure 6C shows that I_{K-Ca} from P cells (the same holds true for C and Br cells) shares the same properties as I_{K-Ca} from U cells. It has a rather slow onset particularly prominent at positive potentials and vanishes as the potential approaches the Ca equilibrium potential, near +150 mV. In both cell types, the onset of I_{K-Ca} can be approximated by second-order equations.

The difference between U cells and the other I_{K-Ca}-containing cells becomes apparent in long-lasting depolarizations. In contrast to noninactivating I_{K-Ca} in U cells, that in P cells decreases during prolonged depolarizations. The rate of decrease is rather low (time constants in hundreds of milliseconds), but is clearly apparent at all membrane potentials (see Fig. 9C).

2.3. Facilitated Ca-dependent K Current

Lux and Hofmeier (1982a,b) showed that a conditioning pulse producing a large Ca entry facilitates a subsequent activation of I_{K-Ca}. The I-V relationship no longer demonstrates the N-shape observed without conditioning pulses. These authors concluded that I_{K-Ca} is actually V-dependent, but that its onset is considerably slowed down at large positive membrane potentials. Such an experiment is illustrated in Fig. 7A,B in which I_{K-Ca} was obtained without or with a short prepulse producing large I_{Ca}. The facilitated I_{K-Ca} has a fast upstroke that does not seem to be V-dependent.

By delaying the test pulse with respect to the I_{Ca}-producing pulse, the extra I_{K-Ca} decreases exponentially, i.e., facilitation of I_{K-Ca} vanishes in 0.5 to 1 sec following Ca entry (Fig. 7C,D). It should be noted that the steady-state I_{K-Ca} in the test pulse is not changed by facilitation. Therefore, facilitation of I_{K-Ca} by calcium does not modify its V-dependence and must not be considered as an alternative or parallel way of channel opening. For the present, only tentative hypotheses can be put forward to account for (1) the change in I_{K-Ca} activation kinetics by calcium and (2) the persistence of the V-dependence of its final level (for instance, reduction of the energy barrier between the closed and open states of the macromolecular structure of the K ionophore, which would result in increased forward and backward reaction rates). Very little is known about the ways by which calcium (as well as other divalents ions) (Gorman and Hermann, 1979) interacts with I_{K-Ca}. However, the unusual time constant V-relationship of I_{K-Ca} suggests the existence

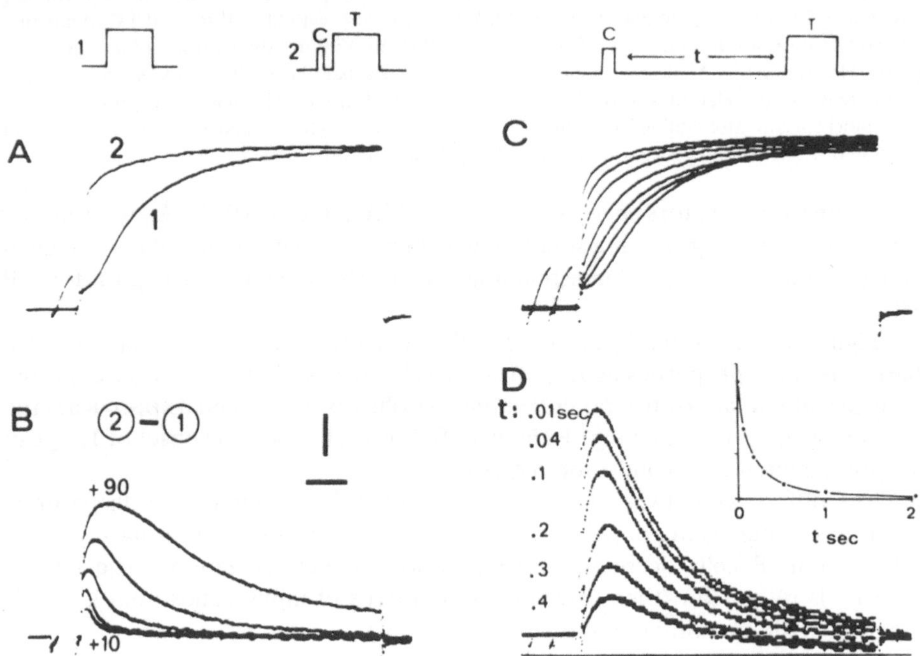

FIGURE 7. Facilitated I_{K-Ca} in U cells. (A) Superposed outward currents during +70 mV test pulses (T), without (trace 1) or with (trace 2) a short (25 msec, +20 mV) prepulse (C). (B) Extra currents produced by the prepulse (trace 2 minus trace 1) for test pulses ranging from +10 to +90 mV, with 20 mV steps. (C and D) Decay of facilitation following the conditioning prepulse. Superposed current traces in C obtained by delaying the test pulse. (D) Decay of extra currents in C with increasing interpulse delay (indicated on the recordings). Facilitation of I_{K-Ca} decreases exponentially (inset in D) after the conditioning pulse. Vertical bar: A–C, 0.2 μA; D, 0.1 μA. Horizontal bar: 50 msec.

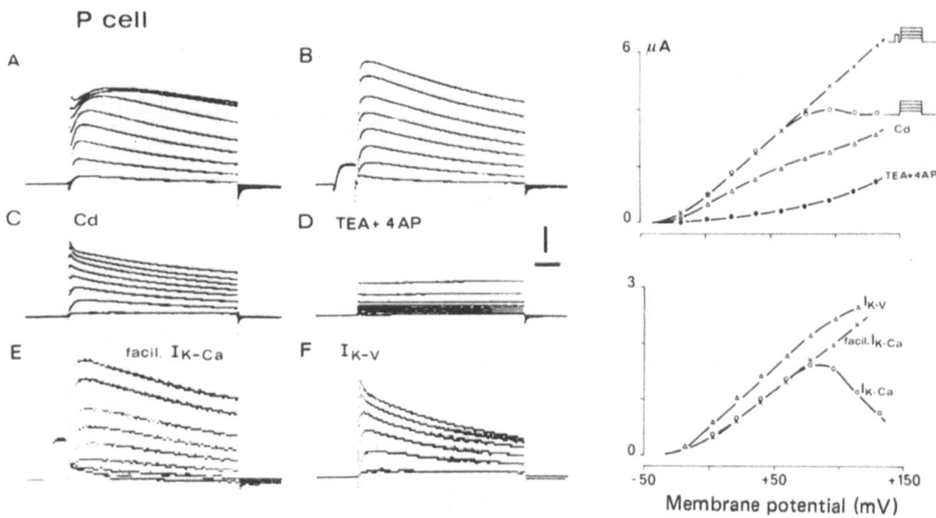

FIGURE 8. Facilitated I_{K-Ca} in P cells (A–D) superposed outward currents during test pulses ranging from -20 to $+120$ mV (with 20 mV steps). (A) Test pulses in control saline; (B) test pulses preceded by a conditioning pulse (30 msec, $+30$ mV); (C) currents in presence of 0.5 mM Cd^{2+}; (D) leakage currents in 60 mM TEA and 10 mM 4-AP; (E) facilitated I_{K-Ca} obtained by subtracting records B and C; and (F) I_{K-V} obtained by subtracting C and D. Vertical bar: A-D, 1.2 μA; E and F, 0.6 μA. Horizontal bar: 50 msec. Diagrams: I-V display of the experimental data.

of several steps (Kostyuk *et al.*, 1980), some of them being rate limiting and others controlling the number of open channels.

Due to Ca facilitation, successive spikes in U cells produce increasing Ca influx which gradually speeds up the activation of I_{K-Ca}. The outward current tends to reduce the spike amplitude even at low frequencies (2 to 3 Hz) and finally blocks the firing.

In other I_{K-Ca} containing cells (Br, P, and C cells) facilitation of I_{K-Ca} by previous Ca entry also exists. Families of outward currents with or without prepulses are shown in Fig. 8A,B for a P cell. It is clear that the I_{K-Ca} component is consistently speeded up by the conditioning prepulse. As in U cells, facilitation is related to the voltage of the prepulse, being larger at voltages corresponding to large inward Ca currents (between $+10$ and $+30$ mV) and vanishing at large depolarizations. However, in sustained stimulation, these cells do not display spike attenuation as expected from facilitated K currents, but spike enlargement or even long-lasting plateaus which are usually ascribed to I_K inactivation (Aldricht *et al.*, 1979).

2.4. Late Depression of I_{K-Ca}

When the test pulse is delayed with respect to the prepulse, I_{K-Ca} facilitation vanishes in P cells as in U cells, with a 200–250 msec time constant. However, for long interpulse delays (0.5 to 1 sec), facilitation converts to depression and most of the I_{K-Ca} is suppressed (Eckert and Lux, 1977). Records in Fig. 9 show com-

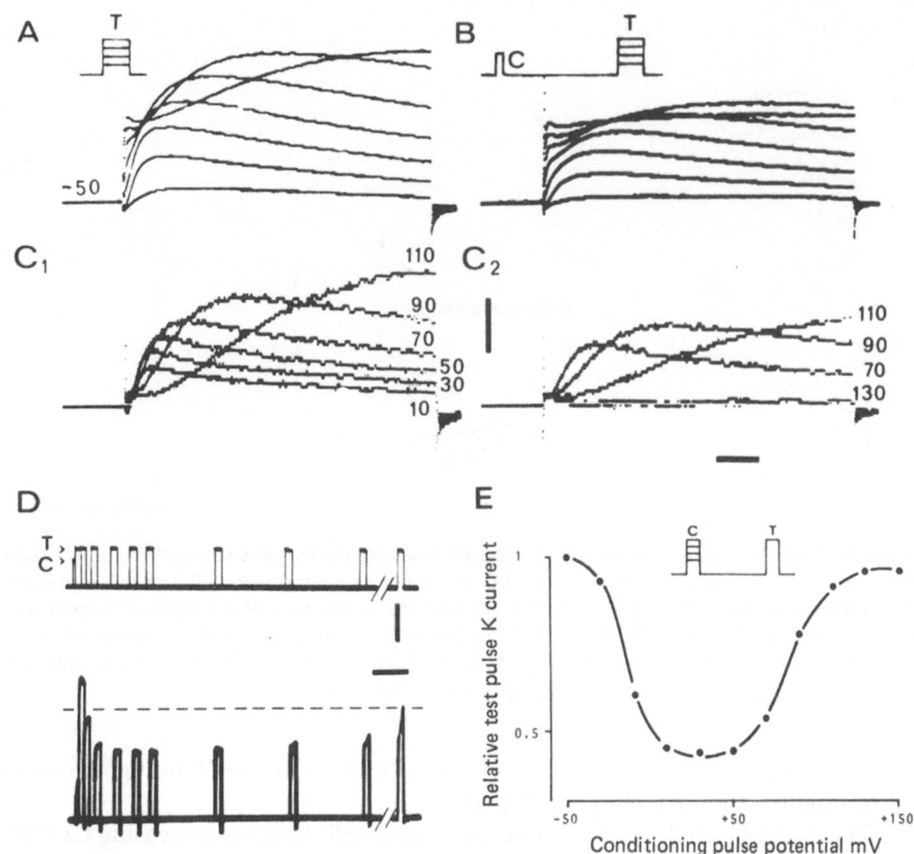

FIGURE 9. I_{K-Ca} depression in P cells. (A and B) Sets of compound V T- and Ca-dependent K currents during test pulses (T) from −10 to +110 mV (voltage steps, 20 mV); in B the test pulses are preceded by a conditioning pulse (C, +20 mV, 25 msec) applied 1 sec before. (C) Current suppressed by the conditioning pulse (series A minus series B). Vertical bar: A and B, 1 μA; C_1, 0.5 μA; C_2, 0.75 μA. Horizontal bar: 50 msec. (D) Time course of early facilitation, late depression and slow recovery of I_{K-Ca} during test pulses (+90 mV, 100 msec) following a short conditioning pulse (+20 mV, 30 msec); control test current outlined by the dashed line. Vertical bar: 100 mV and 0.5 μA. Horizontal bar: 0.5 sec. (E) Inactivation of test pulse outward current as a function of conditioning pulse potential; interpulse delay, 0.8 sec.

pound outward currents in P cells without (Fig. 9A) and with (Fig. 9B) a conditioning prepulse (+20 mV, 25 msec) applied 1 sec before the test pulses. It can be noted that the initial outward current peak related to I_{K-V} is only slightly depressed (less than 5%), whereas the slowly rising phase of I_{K-Ca} is consistently decreased. The currents eliminated by the early prepulse (records A minus B) are shown in Fig. 9C. They fulfill all the properties of I_{K-Ca} described in U cells: slow sigmoid onset and decreased amplitude at large depolarizations (near +130 mV).

Late depression of outward currents has already been described in mollusks; experiments such as those of Fig. 9 performed in P, C, and Br cells indicate that Ca-related depression affects only the Ca-activated component of the current, but not its V-dependent component. Full recovery from inactivation takes 20 to 30 sec (Fig. 9D).

The late depression is Ca-related. Depression of I_{K-Ca} shown in Fig. 9B depends on the prepulse amplitude. Maximum depression occurs with voltage prepulses at +20 to +30 mV that produce large Ca inflow (Fig. 9E). I_{K-Ca} recovers its control level as the prepulse approaches +120 to +140 mV, i.e., the equilibrium potential for Ca ions (Meech and Standen, 1975). Under these conditions, I_{K-V} still displays a small inactivation that points to the V-dependence of its inactivating process.

It is worth noting that once I_{K-Ca} has been inactivated by an early Ca entry, facilitation of I_{K-Ca} by a conditioning prepulse is no longer effective. This indicates that the channels have entered a state of inactivation that cannot be reversed by an additional Ca inflow. Since the opening of K-Ca channels requires a large calcium influx, the late depression of I_{K-Ca} might reflect a persistent inactivation of Ca channels. However, the late depression occurs 1 sec after a brief calcium influx, a delay large enough for Ca channels to recover more than 90% of their activation possibility. Using inorganic calcium blockers, we found that an almost linear relationship exists between the amplitudes of the Ca and K-Ca currents. Therefore, the residual small inactivation of Ca channels, which persists 1 sec after calcium entry, cannot account for the large depression of K-Ca current. More probably, I_{K-Ca} inactivation develops as soon as the channels open, like usual inactivation of V-dependent channels and full recovery from inactivation is required for subsequent channel opening. This suggestion is backed up by the observation that I_{K-Ca} in P, C, and Br cells declines slowly in long-depolarizing steps, as is already obvious in the extracted currents shown in Fig. 9C for P cells (however, see Kostyuk *et al.*, 1980).

Using this method, we note that delayed depression of I_{K-Ca} exists in all g_{K-Ca}-containing cells except U cells. Therefore two questions must be formulated: (1) Why does I_{K-Ca} in U cells not inactivate? (2) Inactivation of I_{K-Ca} in P cells likely accounts for spike enlargement and plateaus. Why do plateaus not occur in bursting Br cells or even in C cells that have similar ionic conductance contents?

For the time being there are no clear-cut answers to these questions. We observe that in contrast to facilitation which needs short prepulses (2- to 5-msec duration), late depression increases with prepulse durations over several hundreds of milliseconds. A brief Ca inflow is effective in facilitating I_{K-Ca} activation, whereas a large Ca inflow is needed for its depression. Therefore it was tempting to postulate that increasing Ca inflow in U cells (with a long prepulse or repeated short prepulses) would lead to a consistent I_{K-Ca} inactivation. This was never observed.

Another striking difference exists between P and Br cells. In P cells, I_{K-Ca} inactivation is half maximal with prepulse duration of 15 msec. In Br cells, such prepulses lead to 10% reduction of I_{K-Ca}. Half inactivation is effective with prepulses of 40 to 80 msec, or several (8 to 10) 15-msec prepulses. One might suggest that the lower inactivation in bursting cells is related to a reduced inward calcium current or to a lower ratio of I_{Ca}/cell volume so that intracellular free calcium might not accumulate sufficiently. P and Br cells have comparable soma diameter (160 to 230 μm) as well as maximum inward Ca current. Therefore, if inactivation is directly linked to the free intracellular calcium, differences must exist in the cellular mechanisms involved in the maintenance of a low level of free calcium in these cells.

3. Conclusion

From the available data, an overall portrait of electrical behavioral correlates of some of the somatic ionic conductances can be proposed. Neurons such as T cells, resembling axons in their channel content, do not require much discussion; they are able to fire long series of spikes at rather high frequency, without any sign of fatigue. On the contrary, cells governed by the I_{Ca}–I_{K-CA} system are characterized by low-firing frequencies, although it must be noted that they possess very clear soma pacemaker properties, whereas the pacemaker area of Na-dependent cells is located deep in the neuropil. The large facilitation exhibited by I_{K-Ca} in U cells accounts for the prominent spike attenuation that occurs above 2.5–3 Hz firing frequency. In this frequency range, the first spike acts as prepulse in double-pulse experiments producing Ca inward current that facilitates I_{K-Ca} activation in subsequent spikes. The temporal overlap between the inward and outward currents increases progressively and the amplitude of the regenerative response will therefore be determined by their relative magnitude. This process will go on during a spike train until the facilitated outward current overtakes the inward Ca current and stops the firing. Such a process may be the basis of graded activity found in several excitable structures, particularly in nerve terminals and secretory tissues.

In the cat petrosal ganglion, two types of primary sensory neurons have been identified: chemoreceptors with long-lasting Ca spikes and baroreceptors with fast Na spikes; the latter fire spikes at high frequency, whereas chemoreceptors fire only one or a few spikes in response to maintained depolarization (Belmonte and Gallego, 1983). The strong adaptation exhibited by chemoreceptors is most likely attributable to properties identical to those of U cells rather than to slow restorative processes such as slow-K conductances. Interestingly, a third cell type, slow baroreceptors, have compound Na-Ca spikes and fire series of spikes at slow rate like C cells.

The difference in I_{K-Ca} in the selected cells is schematically summarized in Fig. 10. Figure 10A shows the changes in I_{K-Ca} in P and U cells, respectively, when short prepulses of increasing amplitude are applied 1 sec or just before test pulses in order to detect late depression and early facilitation of I_{K-Ca}. Kinetics of I_{K-Ca} changes occurring, with increasing latency, after a short prepulse producing large Ca entry are displayed in Fig. 10B and the dependence of I_{K-Ca} amplitude on the early prepulse duration is shown in Fig. 10C.

To sum up, P, C, Br, and U cells have a large I_{K-Ca} that displays similar facilitation upon Ca entry. Except for U cells, these neurons have a late I_{K-Ca} depression which is easily triggered in P and C cells by short Ca entry and which is less pronounced in Br cells. Depression is maximal about 1 sec after Ca entry; therefore at low-frequency firing, the successive spikes in P, C, and Br cells become progressively lengthened because inactivation enhances the expression of the inward Ca

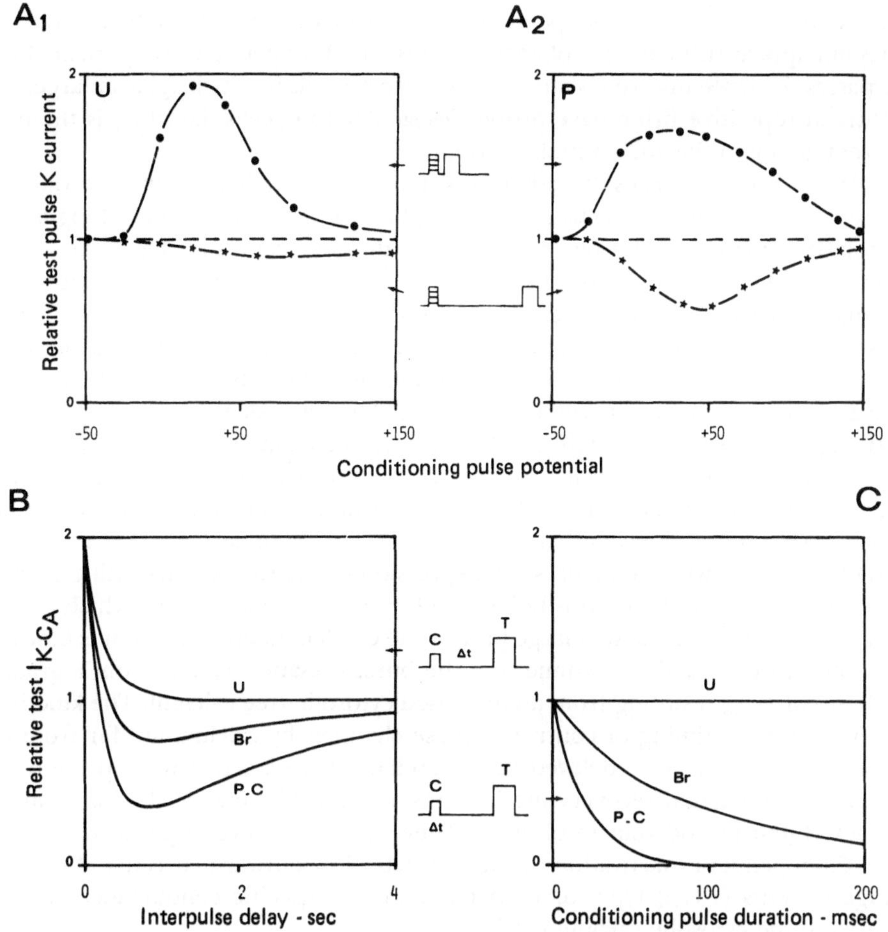

FIGURE 10. Summary of I_{K-Ca} properties. (A) Test pulse K current as a function of conditioning pulse potential in U and P cells; interpulse delay 10 msec (early facilitation in both cells) and 0.8 sec (late depression in P cell). (B) Kinetics of test pulse I_{K-Ca} changes following a short conditioning pulse. (C) Relative test pulse I_{K-Ca} as a function of conditioning pulse duration with interpulse delay of 0.8 sec.

current. In hormone-producing cells in *Lymnaea,* spikes may have a distinct Ca component when cells are in active state; they then display prolonged spiking activity with prominent spike enlargement, although the firing frequency is rather low (0.7 Hz), (Kits and Bos, 1982). Since g_{K-Ca} has been inferred in these cells, its inactivation may take part in spike shape and long-lasting discharge.

In P cells, transition to long-lasting depolarized plateaus results both from the existence of a large Ca current and from the depressed outward K current. Calcium-dependent long-lasting plateaus have been reported in crustacean (Dickinson and Nagy, 1983) as well as in mammalian neuroendocrine cells (Theodosis *et al.,* 1983). C cells have similar properties to P cells except that the spike enlargement in repetitive firing is of limited amplitude and that I_{K-V} is almost twice that in P cells; depression of the Ca-dependent component of the C cell outward current does not appear sufficient for plateaus because the V-dependent component always surpasses the large inward Ca current. Most nerve cells in which spike enlargement occurs in repetitive firing have properties similar to C cells, i.e., I_{K-V} is the main current acting in the spike repolarization.

In contrast to C cells, bursting cells have almost similar inward and outward current components to P cells and one might expect them to exhibit long-lasting depolarized plateaus. The fact that plateaus do not occur and that spike lengthening is of limited amplitude must be ascribed to the reduced I_{K-Ca} depression produced by Ca inflow (see Fig. 10C). This difference may result from either the existence of two different g_{K-Ca} channels exhibiting or not late depression, or in the Ca-sequestering limits of the cell. Another hypothesis is suggested by the existence of inactivating and noninactivating Ca channels (Akaike *et al.*, 1978; Llinas and Sugimori, 1980) that might produce contrasted changes in the cytosol-free Ca concentration in P and Br cells. However, Ca-current inactivation and subsequent recovery from inactivation in both cells have similar time courses. Whatever the exact mechanism, it appears that during P-cell firing, late depression prevails over early facilitation, which enhances the expression of the slowly inactivating calcium current. To a lesser extent, this holds true for endogenous bursters, which implies that at the end of the burst, a large part of the Ca-dependent K current is blocked. It is therefore difficult to assume that the burst cessation results from a gradual build-up of I_{K-Ca} resulting from an increased cytosolic free calcium. The kinetic of opening and the timing of control of these channels by calcium are far from the expected properties of a delayed Ca-dependent process acting to stop the burst. Therefore, the actual recovery mechanisms acting at the end of the burst and in the interburst period remain obscure. They do involve several processes such as slow Ca-dependent inactivation of Ca channels, slow outward currents of the Partridge–Stevens (1976) type, or even more or less specific conductance changes related to the cytosolic calcium level.

From data obtained in a number of preparations, we note that virtually all cells that exhibit Ca spikes also have a Ca-dependent component that may in various ways control repetitive discharge or pacemaker activity. The simultaneous presence of Ca channels and Ca-dependent K channels has been described extensively in several molluscan neurons and identified further in a variety of preparations: crustacean neurons (Tazaki and Cooke, 1979), striated and smooth muscle fibers (Stefani and Uchitel, 1976; Mironneau and Savineau, 1980), cat motoneurons (Barret *et al.*, 1980), sympathetic neurons (McAfee and Yarowsky, 1979), and cerebellar Purkinje cells (Llinas and Sugimori, 1980).

An interesting observation is that in the selected *Helix* cells, there is an almost linear relationship between the relative I_{Ca} and I_{K-Ca} components (Fig. 11). In an extensive study of many *Helix* nerve cells we failed to find cells having large I_{Ca} and deprived of I_{K-Ca}. These observations suggest that Ca channels and Ca-dependent K channels are intimately associated within the membrane in the same macromolecular structure. An alternative hypothesis has been suggested by LUX and his coworkers (Hofmeier and Lux, 1978; Lux and Heyer, 1979); Ca-dependent K con-

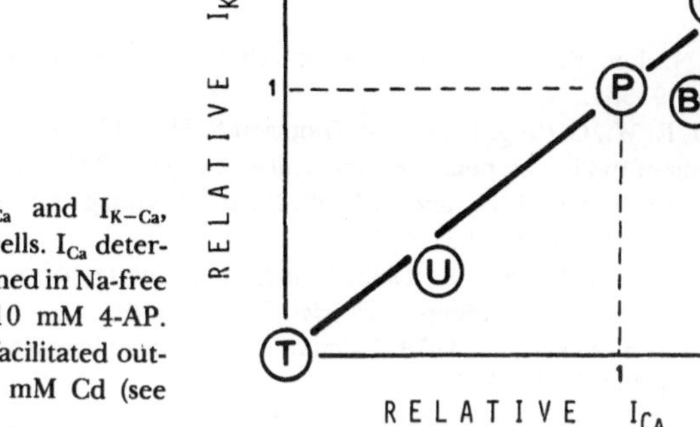

FIGURE 11. Relationship between I_{Ca} and I_{K-Ca}, expressed relative to their values in P cells. I_{Ca} determined as the peak inward current obtained in Na-free saline containing 60 mM TEA and 10 mM 4-AP. I_{K-Ca} measured near +80 mV as the facilitated outward current part suppressed by 0.5 mM Cd (see Fig. 8).

ductances might be based on the same channels which initially pass calcium and become transformed into K permeable ones. This hypothesis has been challenged by Kostyuk *et al.* (1980).

Primary sensory neurons in dorsal root ganglia seem to lack Ca-activated K currents, although they have Ca spikes (Matsuda *et al.*, 1976; Gorke and Pierau, 1980). These observations do not definitively rule out the concept of associated or combined Ca-K channels because evidence is based on membrane potential recordings in which the presence of I_{K-Ca} is generally inferred from the existence of Ca-dependent poststimulus hyperpolarizations that actually involve a basic increase in resting K conductance by calcium rather than specific activation of I_{K-Ca} (Lux and Heyer, 1979; Hofmeier and Lux, 1981). However, recent data indicate that horizontal cells from goldfish retina, studied under voltage-clamp conditions, possess a typical Ca current with Ca-dependent inactivation, but are fully deprived of Ca-dependent K channels (Tachibana, 1983).

More information concerning (1) the distribution of Ca and Ca-dependent K channels in restricted cell compartments and (2) the relationship between Ca entry and K channel control (facilitation or depression) are still needed to reconcile these apparently incompatible observations. An alternative hypothesis on these lines could be that Ca-dependent K channels are basic components of excitable and nonexcitable membranes, but they are rate limited by the numbers of available Ca channels. The advent of isolated patch membranes allowing the possibility of controlling the ionic environment on both sides and of observing single-channel events and gating currents should lead to an improved understanding of the relations between calcium and potassium movements which, as we hope to have illustrated, play such a substantial role in the encoding properties of nerve cells.

ACKNOWLEDGMENTS. I am indebted to Dr. C. Ducreux and E. Ehile for permission to use results from unpublished works, and to Mrs. M. Andre, H. Chagneux, and Mr. G. Jacquet for helpful technical assistance. I wish to thank Dr. E. Labos (Semmelweis University, Budapest), who pointed out the existence and location of plateauing cells to our attention.

This work was supported by CNRS and a DGRST grant No. 82.E.0678.

References

Akaike, N., Lee, K. S., and Brown, A. M., 1978. The calcium current in *Helix* neurons, *J. Gen. Physiol.* **71**:509–531.

Aldricht, R. W., Getting, P. A., and Thomson S. H., 1979, Mechanism of frequency dependent broadening of molluscan neurone soma spike, *J. Physiol.* **291**:531–544.

Barret, E. F., Barret, J. N., and Crill, W. E., 1980, Voltage sensitive outward-currents in cat motoneurones, *J. Physiol.* **304**:251–276.

Belmonte, C., and Gallego, R., 1983, Membrane properties of cat sensory neurones with chemoreceptor and baroreceptor endings, *J. Physiol.* **342**:603–614.

Brehm, P., and Eckert, R., 1978, Calcium entry leads to inactivation of calcium channel in *Paramecium, Science* **202**:1203–1206.

Connor, J. A., 1975, Neural repetitive firing. A comparative study of membrane properties of crustacean walking leg axons, *J. Neurophysiol.* **38**:922–932.

Connor, J. A., 1977, Time course separation of two inward currents in molluscan neurons, *Brain Res.* **119**:487–492.

Connor, J. A., and Stevens, C. F., 1971, Voltage clamp studies of a transient outward membrane current in gastropod neural somata, *J. Physiol.* **213**:21–30.

Dickinson, P. S., and Nagy, F., 1983, Control of a central pattern generator by an identified modulatory interneurone in crustacea. II. Induction and modification of plateau properties in pyloric neurones, *J. Exp. Biol.* **105**:59–82.

Eckert, R., and Lux, H. D., 1977, Calcium-dependent depression of a late outward current in snail neurones, *Science* **197**:472–475.

Eckert, R., and Tillotson, D., 1978, Potassium activation associated with intraneutronal free calcium, *Science* **200**:437–439.

Gorke, K., and Pierau, F. K., 1980, Spike potentials and membrane properties of dorsal root ganglion cells in pigeons, *Pflug. Arch.* **386**:21–28.

Gorman, A. L. F., and Hermann, A., 1979, Internal effects of divalent cations on potassium permeability in molluscan neurones, *J. Physiol.* **296**:393–410.

Hagiwara, S., 1973, Calcium spikes, *Adv. Biophys.* **4**:71–102.

Heyer, C. B., and Lux, H. D., 1976, Control of the delayed outward potassium currents in bursting pace-maker neurons of *Helix pomatia, J. Physiol.* **262**:349–382.

Heyer, C. B., and Lux, H. D., 1978, Unusual properties of the Ca-K system responsible for prolonged action potentials in neurons from the snail *Helix pomatia*, in: *Abnormal Neuronal Discharges* (N. Chalazonitis and M. Boisson, eds.), Raven Press, New York, pp. 311–327.

Hodgkin, A. L., and Huxley, A. F., 1952, A quantitative description of membrane current and its application to conduction and excitation in nerve, *J. Physiol.* **117**:500–544.

Hofmeier, G., and Lux, H. D., 1978, Inversely related behaviour of potassium and calcium permeability during activation of calcium-dependent outward currents in voltage-clamped snail neurons, *J. Physiol.* **287**:28–29P.

Hofmeier, G., and Lux, H. D., 1981, The time course of intracellular free calcium and related electrical effects after injection of $CaCl_2$ into neurons of the snail, *Helix pomatia, Pflug. Arch.* **391**:242–251.

Hotson, J. R., and Prince, D. A., 1980, A calcium-activated hyperpolarization follows repetitive firing in hippocampal neurons, *J. Neurophysiol.* **43**:409–419.

Junge, D., and Miller, J., 1974, Different spike mechanisms in axon and soma of molluscan neurons, *Nature (London)* **252**:155–156.

Kits, K. S., and Bos, N. P. A., 1982, Na^+- and Ca^{++}-dependent components in action potentials of the ovulation hormone producing caudo-dorsal cells in *Lymnaea stagnalis* (Gastropoda), *J. Neurobiol.* **13**:201–216.

Kostyuk, P. G., 1981, Calcium channels in the neuronal membrane, *Biochim. Biophys. Acta* **650**:128–150.

Kostyuk, P. G., Krishtal, O. A., and Doroshenko, P. A., 1975, Outward current in isolated snail neurones. III. Effect of verapamil, *Comp. Biochem. Physiol.* **51C**:269–274.

Kostyuk, P. G., Doroshenko, P. A., and Tsydrenko, A. Y., 1980, Calcium-dependent potassium conductance studied on internally dialyzed nerve cells, *Neuroscience* **5**:2187–2192.

Llinas, R., and Sugimori, M., 1980, Electrophysiological properties of *in vitro* Purkinje cell dendrites in mammalian cerebellar slices, *J. Physiol.* **305:**197–213.

Lux, H. D., and Heyer, C. B., 1979, A new electrogenic calcium-potassium system, in: *The Neurosciences Fourth Study Program* (F. O. Schmitt and F. G. Worden, eds.), MIT Press, Cambridge, pp. 601–615.

Lux, H. D., and Hofmeier, G., 1982a, Properties of a calcium- and voltage-activated potassium current in *Helix pomatia* neurons, *Pflug. Arch.* **394:**61–69.

Lux, H. D., and Hofmeier, G., 1982b, Activation characteristics of the calcium-dependent outward potassium current in *Helix, Pflug. Arch.* **394:**70–77.

Matsuda, Y., Yoshida, S., and Yonezawa, T., 1976, A Ca-dependent regenerative response in rodent dorsal root ganglion cells cultured *in vitro, Brain Res.* **115:**334–338.

McAfee, D. A., and Yarowsky, P. J., 1979, Calcium-dependent potentials in the mammalian sympathetic neurons, *J. Physiol.* **290:**507–523.

Meech, R. W., 1974, Prolonged action potentials in *Aplysia* neurones injected with EGTA, *Comp. Biochem. Physiol.* **48A:**397–402.

Meech, R. W., 1978, Calcium-dependent potassium activation in nervous tissue, *Annu. Rev. Biophys. Bioeng.* **7:**1–18.

Meech, R. W., and Standen, N. B., 1975, Potassium activation in *Helix aspersa* neurons under voltage clamp: A component mediated by calcium influx, *J. Physiol.* **249:**211–239.

Mironneau, J., and Savineau, J. P., 1980, Effects of calcium ions on outward membrane currents in rat uterine smooth muscle, *J. Physiol.* **302:**411–425.

Partridge, L. D., and Stevens, C. F., 1976, A mechanism for spike frequency adaptation, *J. Physiol.* **256:**315–332.

Stefani, E., and Uchitel, O. D., 1976, Potassium and calcium conductance in slow muscle fibres of the toad, *J. Physiol.* **255:**435–448.

Tachibana, M., 1983, Ionic currents of solitary horizontal cells isolated from goldfish retina, *J. Physiol.* **345:**329–351.

Tazaki, K., and Cooke, I. M., 1979, Ionic bases of slow, depolarizing responses of cardiac ganglion neurons in the crab, *Portunus sanguinolentus, J. Neurophysiol.* **42:**1022–1047.

Theodosis, D. T., Legendre, P., Vincent, J. D., and Cooke, I., 1983, Immunocytochemically identified vasopressin neurons in culture show slow calcium-dependent electrical responses, *Science* **221:**1052–1054.

Thomson, S. H., 1977, Three pharmacologically distinct potassium channels in molluscan neurones, *J. Physiol.* **255:**465–488.

Tsien, R. W., 1983, Calcium channels in excitable cell membranes, *Annu. Rev. Physiol.* **45:**341–358.

Woolun, J. C., and Gorman, A. L. F., 1981, Time dependence of the calcium-activated potassium current, *Biophys. J.* **36:**297–302.

23

Calcium and cAMP: Second Messengers in Gastropod Neurons

JOHN A. CONNOR AND PHILIP HOCKBERGER

1. Introduction

Basic research in cell biology over the past 20 years has demonstrated that intracellular free calcium ions (Ca_i^{2+}) and cyclic nucleotides (cyclic AMP and cyclic GMP) are critical elements in the internal signaling system of virtually all types of animal and plant cells. There are four common features that distinguish these "second messengers" from other molecules and ions found inside cells: (1) each exists in very low concentrations under basal conditions; (2) the level increases in response to an extracellular stimulus, i.e., first messenger; (3) rapid regulation occurs following elevation; and (4) its presence is a rate-limiting factor in the cell's response to the stimulus. Thus the internal concentration of a second messenger is the result of a dynamic balance between processes that are continuously elevating the level matched against those which are reducing it.

An important consequence of a rapid control system like this one is that it permits an almost instantaneous coupling between stimulus and response. It is therefore not surprising that this coupling would be under the maintenance of a sophisticated set of feedback and feedforward control systems often involving *both* Ca^{2+} and cyclic nucleotides (cf. Berridge, 1975, 1979; Rasmussen and Goodman, 1977; Rasmussen, 1981). For example, in muscle and secretory cells where elevation of Ca_i^{2+} triggers excitation-contraction and excitation-secretion coupling, respectively, cAMP levels enhance the coupling mechanism. In heart muscle, cAMP

JOHN A. CONNOR AND PHILIP HOCKBERGER • Department of Molecular Biophysics, AT&T Bell Laboratories, Murray Hill, New Jersey 07974 and Department of Physiology and Biophysics, University of Illinois, Urbana, Illinois 61801.

acts directly on the membrane Ca channels to increase Ca influx during excitation (Reuter, 1979; Bean *et al.,* 1984). However, in most other cells, the interrelationship between Ca_i^{2+} and cyclic nucleotides is not well understood and has only been inferred from ^{45}Ca flux measurements where the temporal relationship cannot be resolved faster than a few seconds.

Recently several new techniques have been developed for measuring Ca_i^{2+} changes with greater precision and resolution (Caswell, 1979; Ashley and Campbell, 1979). Three methods in particular, i.e., Ca-sensitive electrodes, photoproteins, and metallochromic dyes, have received the greatest attention, each method having unique advantages and disadvantages for measuring calcium *in situ.* We have primarily used the metallochromic indicator arsenazo III for Ca_i^{2+} measurements in gastropod neurons for several reasons which have been thoroughly discussed elsewhere (Scarpa, 1979; Brinley, 1979). In short, this dye is readily available, negatively charged (enabling it to be injected easily into cells), nontoxic, membrane impermeable, selective for Ca^{2+} in the submicromolar range ($K_D = 40 \times 10^{-6}$ M), and binds Ca^{2+} rapidly and reversibly which allows measurements of dynamic as well as steady-state changes in Ca_i^{2+}. We have used this indicator to examine changes in the resting level of Ca_i^{2+}, cellular calcium-buffering compo-

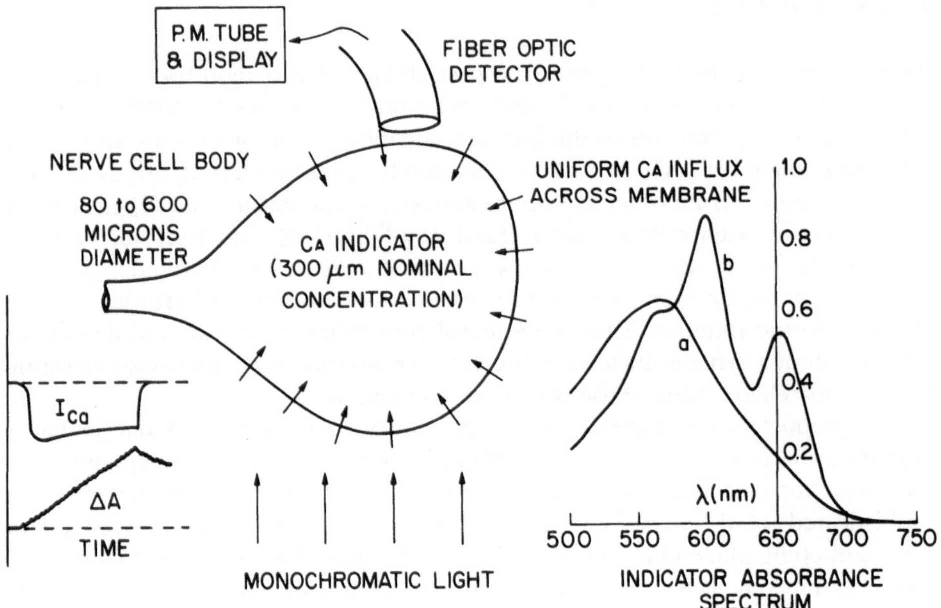

FIGURE 1. Schematic representation of dye absorbance measurement system. Monochromatic light of four different wavelengths (time shared on a single light guide) is brought to the preparation and picked up after passing through the neuron by a second light guide which supplies a photomultiplier tube. The tube output is demultiplexed giving the time-resolved transmittance at each wavelength. Calcium influx across the cell membrane gives rise to the absorbance spectra change (a to b) in arsenazo III shown in the right-hand inset. The approximate relation between I_{Ca} under voltage clamp and the time course of the absorbance change for a differential wavelength pair such as 660–690 nm is given in the left inset.

nents and capacity, as well as the elevation and distribution of Ca_i^{2+} in giant molluscan neurons under various experimental conditions. Figure 1 summarizes the important features of the measurement system. In addition, we have used the arsenazo III technique as a probe for Ca_i^{2+} changes following cyclic nucleotide elevation. This chapter will outline the results of our investigations and discuss the role of second messengers in gastropod neurons.

2. Intracellular Free Ca^{2+} Changes in Gastropod Neurons

2.1. Ca Influx Detected with Arsenazo III

We have been looking at internal free Ca and pH (pH_i) changes in neurons for a number of years using indicator dyes, such as arsenazo III (for Ca) and phenol red (for pH), which provide a very rapid measure of ion changes. In practice the time resolution is set by electronic filtering requirements due to light source and detector noise and the small size of biological signals. Time resolution greater than around 20 msec generally requires signal averaging. Dyes such as arsenazo III and phenol red are relatively nontoxic and in the case of arsenazo III stable within the cytoplasm. Figure 2A illustrates this point by showing absorbance and voltage-clamp data recorded on consecutive days from an arsenazo-filled *Limax maximus* neuron. The cell was stored overnight at 9°C in normal physiological saline. There was a partial loss or degradation of dye over this time span, but the quantity remaining was adequate to detect changes in Ca_i^{2+}, as evidenced by the increase in dye absorbance at 660 nm.

Membrane currents were sufficiently stable over the experimental period to also enable comparison of the voltage-clamp records. In this cell as in all gastropod neurons the net membrane current was outward after several milliseconds even though Ca influx (absorbance increase) occurred throughout the step. Research in several labs has shown that Ca ions enter the cell through channels that are activated by depolarization. This inward Ca current, as well as inward Na current, was simply masked in the voltage-clamp record of Fig. 2A by a larger outward K current activated simultaneously. These voltage-clamp fluxes have been described in considerable detail elsewhere (Connor, 1977, 1979; Ahmed and Connor, 1979; Adams *et al.*, 1980; Gorman and Thomas, 1980; see also Chapter 22, this volume).

Identity of the absorbing species as calcium–arsenazo III complex (Ca-Arz) in Fig. 2A was established by the spectrum of the absorbance change. The characteristic spectrum for Ca binding is very different from the spectra for complexes with Mg^{2+} or the common monovalent cations. The absorbance signal *in situ* is actually a difference measurement, resting (low) Ca-Arz, against (high) Ca-Arz due to influx of Ca^{2+}. Figure 2B shows that the system is sensitive enough to resolve the Ca influx due to single-action potentials.

Although records such as those of Fig. 2 show that there is a measurable increase in cytoplasmic calcium even during action potentials, they say very little about the absolute size of the increase. For events on the millisecond time scale, such as action potentials and synaptic activity, standard calibration techniques are

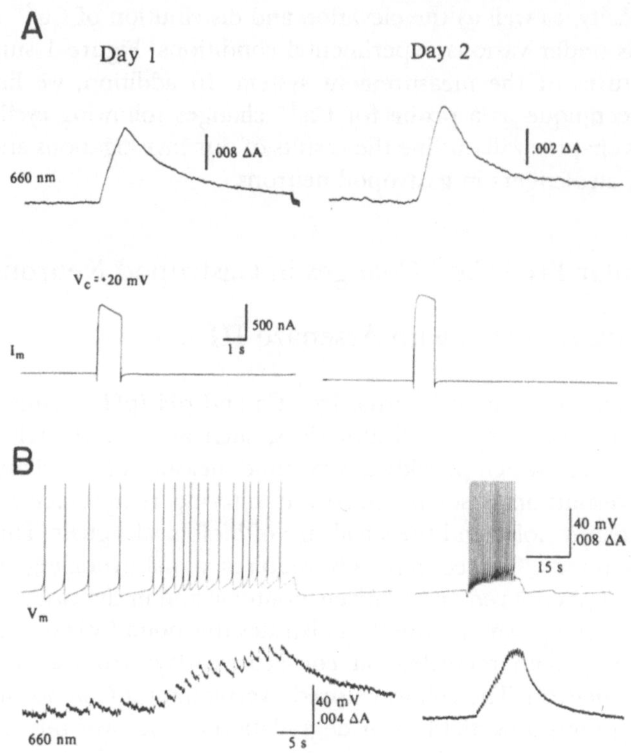

FIGURE 2. (A) Arsenazo absorbance and voltage-clamp records taken from the same neuron on two consecutive days. Membrane current is largely unchanged. The absorbance change was smaller on the second day representing a loss or deterioration of dye. A longer time-constant filter was used for the second day absorbance record which accounts for the greater roundedness of the peak ΔA. (B) Action potentials and the associated dye absorbance changes (day 1 recordings).

of little use because Ca^{2+} will be far from a uniform internal distribution in a nerve cell body of diameter greater than 20 μm or so. That is, the concentration near the site of influx, the plasma membrane, will be much higher than that deeper within the cell both during influx and for some time after influx has ceased. Therefore it remains unclear what levels of Ca^{2+} are reached during action potentials or voltage-clamp pulses, how large the gradient is, and how fast the gradients collapse. At present there is no single technology that will provide straightforward answers to these questions. Ion-sensitive electrodes small enough to use in any but the largest neurons (e.g., *Aplysia* R-2) have response times of seconds and would therefore miss much of the action. Additionally, the tip placement within the cell is generally not known (radial distances on the order of 1 μm may show large differences in Ca^{2+} under non-steady-state conditions as shown below). We have therefore worked at combining mathematical modeling with our indicator dye methods as one promising approach to gaining quantitative information on transient Ca_i^{2+} concentration changes.

2.2. Analysis of Ca_i^{2+} Concentrations Changes

2.2.1. Measurements of Ca^{2+}-Buffering Capacity of Cytoplasm

Figure 3 shows, diagrammatically, the one-dimensional diffusion model we are currently using to analyze Ca movement during electrical activity. This system is a first-order approximation to the experimental situation in which a small light guide is placed immediately over a large cell body sampling a small portion of the cell surface area. Calcium influx is assumed to be uniform across the membrane and measurable under voltage-clamp experiments (cf. Ahmed and Connor, 1979; Ahmed *et al.*, 1980). Once inside the cell, we assume that there is a rapid equilibrium reached in each compartment (ΔX) between free calcium (Ca^{2+}) and calcium bound to cytoplasmic buffers (CaB) and arsenazo III (Ca-Arz). The minimum com-

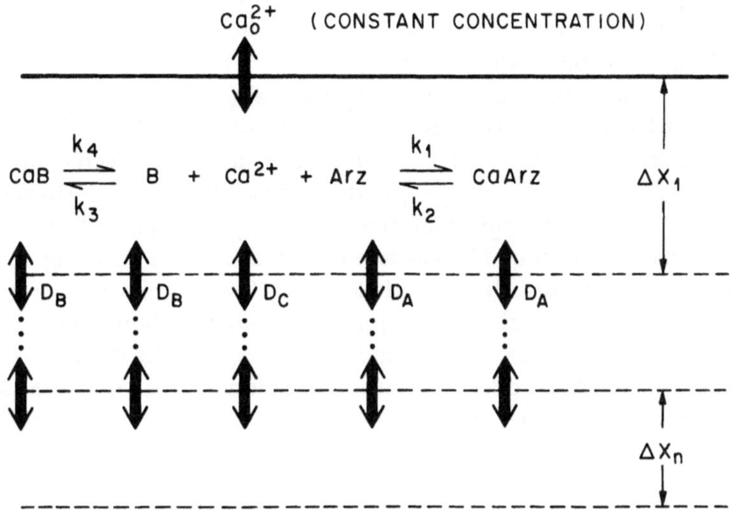

PARAMETERS

BUFFER & INDICATOR CONCENTRATIONS
BUFFER & INDICATOR DISSOCIATION CONSTANTS (RATE CONSTS.)
DIFFUSION COEFFICIENTS
INITIAL FREE CALCIUM CONCENTRATION

SAMPLE EQUATION

$$[CaArz] = k_1 [Arz] \cdot [Ca^{2+}] - k_2 [CaArz] + D_A \frac{\delta^2 [CaArz]}{\delta x^2}$$

FIGURE 3. One-dimensional model for analyzing Ca^{2+} transients in large neurons. Ca enters the cell from an infinite reservoir (Ca_o^{2+}) and, once inside, diffuses as either free species, (Ca^{2+}), or a dye ($CaAr_3$) or buffer (CaB) complex. All species are at local equilibrium.

partment width increases with distance from the membrane and is 0.01 μm near the membrane. We are assuming that the cytoplasmic buffers are initially uniformly distributed, free to diffuse, and that binding is reversible. We are not at this time considering calcium extrusion across the plasma membrane or internal sequestration into organelles. These phenomena must exist in order for the neuron to maintain a steady state with respect to internal free Ca^{2+}, but many of our observations suggest that they may not be of critical importance for brief transient events.

We have measured the *apparent* dissociation constant for the Ca-Arz reaction, K_A and found it to be approximately 40 μM for micromolar levels of Ca^{2+} (Ahmed *et al.*, 1980). There is, however, disagreement on the overall stoichiometry of the reaction (Thomas, 1979; Bauer, 1981; Palade and Vergara, 1983). For computational purposes, we use a 1:1 stoichiometry as indicated in Fig. 3.

We have made estimates of resting free $[Ca_i^{2+}]$ and the cellular buffer capacity of Ca, [B], by injecting known amounts of EGTA (a Ca-chelating agent) into cells and measuring the effect on the arsenazo absorbance change. The procedure is summarized in Fig. 4 and 5. Figure 4A shows the effects of an EGTA injection on cytoplasmic Ca^{2+} and pH under resting conditions. The pH was monitored with a glass microelectrode of the type invented by Thomas (1978). Two wavelength pairs (660–700 and 610–700 nm) were used to monitor dye absorbance since arsenazo III also responds to changes in $[H^+]$, but with a different spectral dependence (Ogan and Simons, 1979). If a simple pH decrease occurs, the absorbance *decrease* of the 610–700 nm pair is larger than the 660–700 pair. This is illustrated in Fig. 4B where a small injection of H^+ was administered and the resulting absorbance change monitored (see also Ahmed and Connor, 1980; Hockberger and Connor, 1983). Where a $[Ca^{2+}]$ change is the major occurrence, the 660–700 nm response is larger. When both Ca_i^{2+} and pH_i change simultaneously, the dye absorbance will be a complicated function reflecting both interactions. Nevertheless, as shown in Fig. 4A, a decrease in the Ca_i^{2+} level was detected by the dye following EGTA injec-

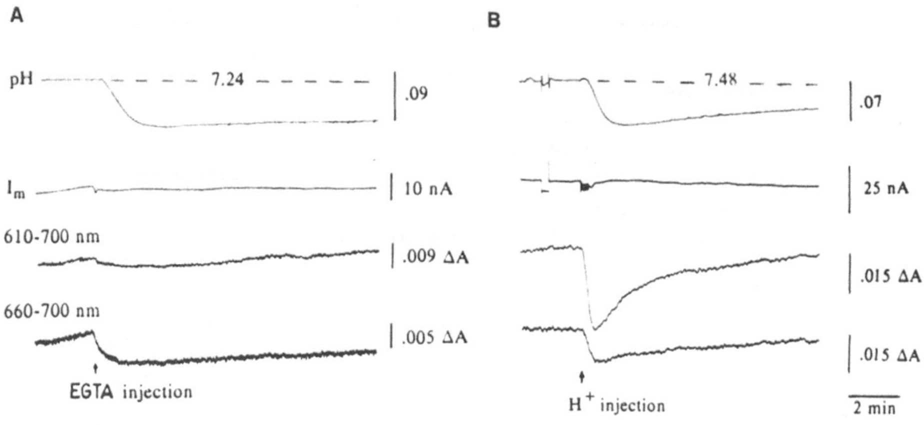

FIGURE 4. (A) Effects of EGTA injection on intracellular Ca^{2+} and pH tracked by arsenazo III and intracellular pH electrode. (B) Records showing the same parameters when H^+ injection is made. EGTA acidifies the cytoplasm presumably because two protons are released for every Ca^{2+} bound.

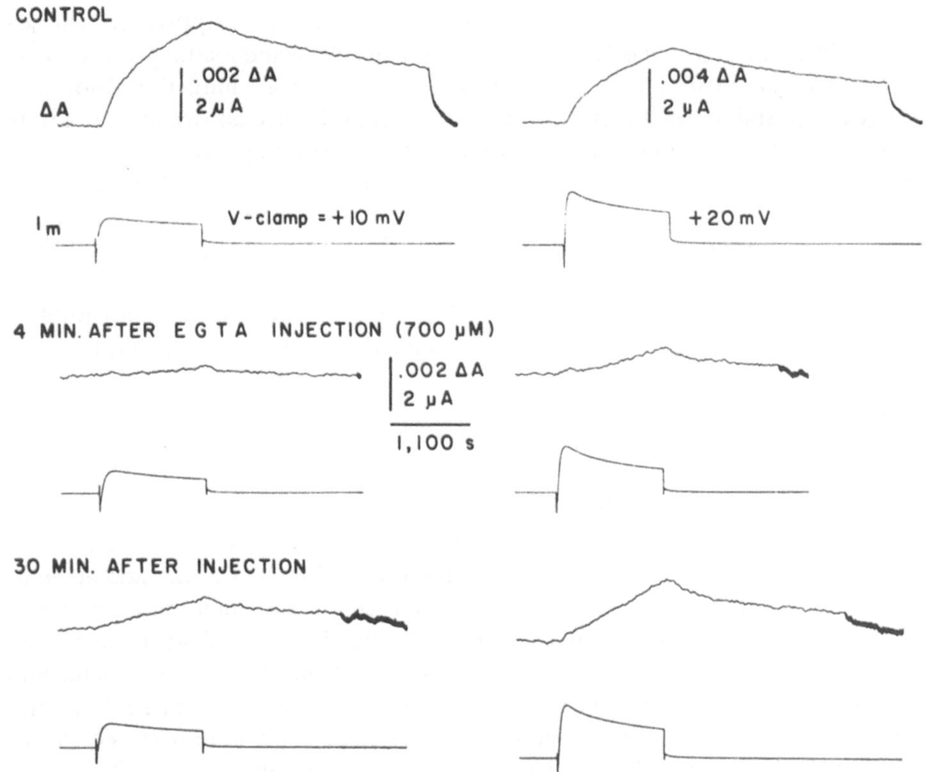

FIGURE 5. Effects on the dye-absorbance change of injecting a measured quantity of EGTA. Top records illustrate voltage-clamp pulses to +10, +20 mV and associated absorbance changes under control conditions. Middle records show the reduction in ΔA (as well as I_m) shortly after injection of 700 μM EGTA. Bottom records show data obtained after equilibrium Ca^{2+} levels should have been restored. (From Z. Ahmed and J. A. Connor, unpublished data.)

tion. We have estimated resting [Ca^{2+}] by null-point injection techniques using EGTA titrated to different Ca^{2+} levels (cf. Dipolo *et al.*, 1976) and have found levels of approximately 100 nM. This is within the range measured by Ca-ion-sensitive electrodes in other molluscan neurons (Alvarez-Leefmans *et al.*, 1981; Levy *et al.*, 1982).

Figure 5 shows the effect of a measured injection of EGTA on arsenazo absorbance and membrane current during a voltage-clamp pulse. The injected amount of EGTA was measured by coinjection of $^{35}SO_4$ (see Ahmed and Connor, 1980). The dye absorbance change and outward membrane current (much of it Ca-activated K current) were reduced markedly and much more severely in the period before re-equilibration of Ca^{2+} had occurred (middle records, Fig. 5). Calcium influx during the voltage-clamp pulses was approximately the same before and after the EGTA injection; therefore, the small dye absorbance changes reflect the increased buffer capacity of the cytoplasm.

When the voltage steps were applied before equilibration was complete (middle trace, Fig. 5), the dye absorbance was almost completely suppressed, whereas at later times the signal was larger. Presumably this reflected loading of the EGTA as [Ca^{2+}] regained its initial resting level. After 30 to 40 min, the absorbance changes were stable. By measuring the reduction in the dye absorbance in a number of cells due to a known injection of EGTA and assuming that the EGTA buffering is additive with endogeneous cytoplasmic buffers for Ca (we have also done this for pH buffering, see Ahmed and Connor, 1980), we have estimated that the endogeneous buffer capacity of *Archidoris* neurons is in the range 100 to 200 μM/ΔpCa. This number cannot be directly compared with buffer capacity measured on a much slower time scale (compare Brinley, 1978) because we are purposefully attempting to neglect slower, high-capacity mechanisms such as sequestration and pumping.

2.2.2. Measurements of Ca^{2+} Diffusion in Cytoplasm

We have used a second line of experiments to investigate the relative mobility of ions and dye molecules in the cytoplasm and to characterize further, Ca_i^{2+} buffering. In these experiments we have used point injections of Ca, Ba, and arsenazo III into neurons and into saline droplets of similar size and compared the rates of dispersion. The experimental layout is shown in Fig. 6. Two small optic fibers were placed above the neuron or droplet to collect light passing through them. Substances were injected beneath one of the fibers (L_0) by barrel-to-barrel iontophoresis. When Ca or Ba was injected, the neurons or droplets were preinjected with arsenazo III that binds either ion with similar affinity, but shows a slightly different absorbance spectrum shift with each (Kendrick *et al.*, 1977).

The saline droplet experiments (Fig. 7, right side) illustrate simple diffusion of the injected molecules in salt water. The concentration of the injected species (tracked by watching the increased absorbance of the dye–metal complex) rose at L_0 during the period when iontophoretic current was on and began to fall upon termination, as the injected species (whether free or dye complex) diffused out of the light path of L_0. A time-delayed absorbance was detected by fiber L_1 as the diffusing substances passed into this detection volume. The records for Ca and Ba are nearly identical. Dispersion of Ba and arsenazo III was slowed by a factor of 5 or more in the neuron as compared with saline. Moreover, the records for Ca and

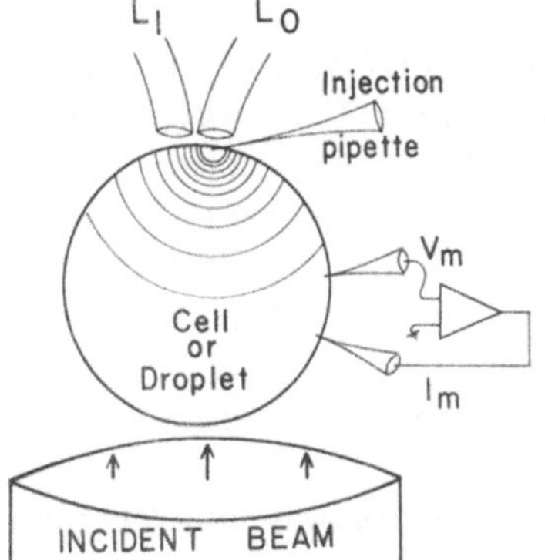

FIGURE 6. Diagram of experimental configuration. Recording light guides (L_0, L_1) are positioned above cell body or saline droplet, ~10 μm from the surface. The incident beam guide was 700 μm diameter; receiving guides approximately 50 μm. (From Connor and Ahmed, 1984.)

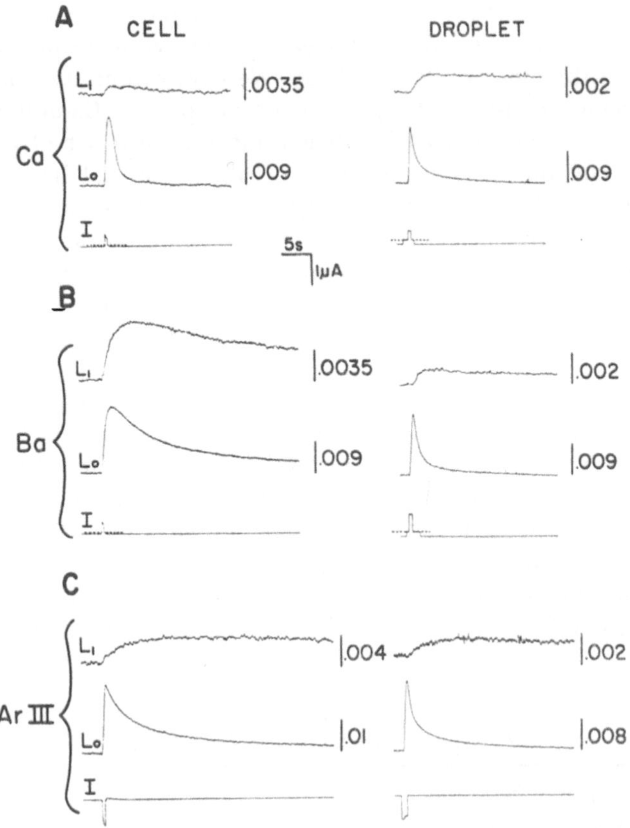

FIGURE 7. Absorbance records indicating concentration changes (local increase and dispersion) of Ca-Arz, Ba-Arz, and Arz in neurons (left column) and in saline droplets (right column). (A and B) The cells and droplets were preloaded with Arz (~300 μM) and then Ca^{2+} or Ba^{2+} was injected from a microelectrode while the measurement was made. (C) The measurements were made during the injection of Arz itself. Iontophoretic current is shown in each of the bottom traces. Generally a braking current was applied and was removed before the injection. (From Connor and Ahmed, 1984).

Ba are very different (left side, Fig. 7). Our interpretation of these and other data from injections (see Connor and Ahmed, 1984; Tillotson and Gorman, 1983) is that Ca is buffered in the cytoplasm by a system with much higher affinity than arsenazo, whereas Ba is poorly buffered. The rapid decay of the absorbance signal when Ca is injected into the cell does not represent rapid dispersion away from L_0, but binding to nearby sites. Ba, in order to disappear from the dye, must diffuse away, a process requiring a much greater time. If Ca were not buffered, the absorbance signal should be as slow as that of Ba.

2.2.3. Simulations of Ca^{2+} Regulation in Cytoplasm

In the following model of Ca regulation (also see Connor and Nikolakopoulou, 1982), we have adopted average estimates of cytoplasmic buffer concentration, dissociation constant, and small ion diffusion coefficients which are in qualitative agreement with the experimental data summarized above. We by no means have been able to make direct quantitative measurements of these parameters *in situ,* nor is it clear at the moment how to do so. We have run most of our simulations using a dissociation constant of 1 μM for the cytoplasmic Ca-buffer system (K_B). This is a small number compared with K_A, and on the order of Ca–calmodulin interactions. Total buffer concentrations of 50 to 200 μM have been used to bracket the estimated buffer capacity range. For the cytoplasmic diffusion coefficient of Ca, D_{Ca}, we have used 2×10^{-6} cm^2/sec, approximately one fifth of the value in water. Model parameters are summarized in Table I. It should be pointed out that we are not using the simplifying assumption of a fixed free-to-bound Ca ratio (compare Smith and Zucker, 1980). This ratio varies widely in time and space as shown below.

In Fig. 8, we compare absorbance records from a neuron given two different loads of Ca (0.2- and 0.5-sec loading pulses) with computations of the changes in Ca-Arz for similar loads in the model system. These results are compared because the absorbance change is just a measure of the change in the number of absorbing molecules in the light path. It can be seen that the simulations, based on an entirely independent set of data, predict the initial time course of the dye-absorbance decay well. Decay half times are indicated on the records. The decay half times measured experimentally in different neurons varied by a factor of up to 3 or 4.

TABLE I
Parameters Used in Diffusion Simulation[a]

Cytoplasmic Parameters

Resting free calcium (U_3 (o,x)): 0.1 μM
Calcium diffusion coefficient (D_c): 2×10^{-6} cm^2/sec
Buffer concentration [B_{tot}]: 100 μM
Buffer dissociation constant [K_B]: 1 μM
($k_3 = 10^9$ sec^{-1}m^{-1}, $k_4 = 10^3$ sec^{-1})
Buffer diffusion coefficient: 10^{-7} cm^2/sec

Indicator Parameters

Arz concentration [Arz_{tot}]: 300 μM
Ca-Arz dissociation constant (K_A): 40 μM
($k_1 = 2.5 \times 1^7$ sec^{-1}m^{-1}, $k_2 = 10^3$ sec^{-1})
Arz diffusion coefficient: 6×10^{-7} cm^2/sec
Extinction coefficient change (λ = 660 nm): 2×10^4

Initial Conditions

$U_1 (o,x) = U_3 (o,x) [Arz_{tot}] / (K_A + U_3 (o,x))$
$U_2 (o,x) = K_A [Arz_{tot}] / (K_A + U_3 (o,x))$
$U_4 (o,x) = U_3 (o,x) [B_T] / (K_B + U_3 (o,x))$
$U_5 (o,x) = K_B [B_T] / (K_B + U_3 (o,x))$

[a] $I_{Ca} = g_{ca} (Vm - E_{Ca}) = zF M_{Ca}$
$E_{Ca} = 29 \log \frac{[Ca_o]}{[Ca_i]}$
where $V_m = 10^{-2}$ V
$g_{Ca} = 8.6 \times 10^{-4}$ S/cm^2
$[Ca_o] = 10^{-2}$ M
$[Ca_i] = U_3 (t,o)$

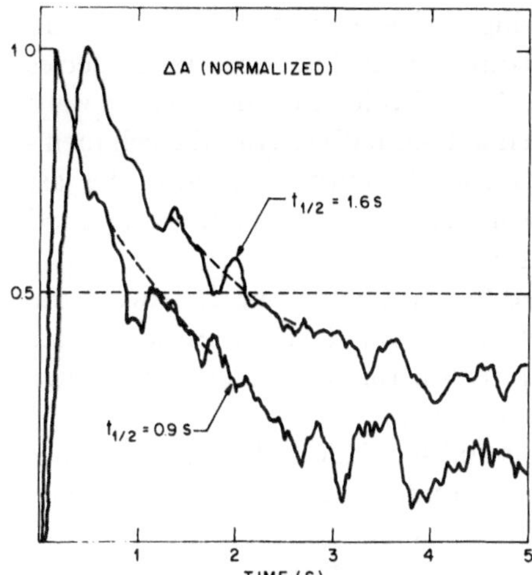

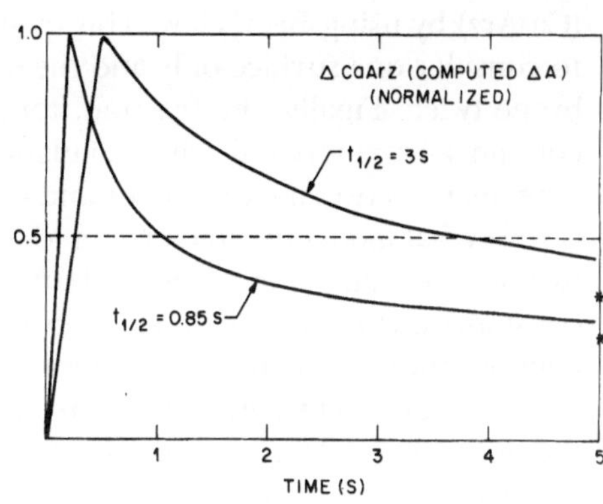

FIGURE 8. (Right) Simulation of indicator response for pulses of Ca^{2+} influx, 0.2 and 0.5 sec duration. Model parameters given in Table I. Asterisks at right side of curves indicate equilibrium levels of signal. (Left) Experimental data for loading pulses of the same duration and approximately the same magnitude. Decay half-times ($t_{1/2}$) are given for each curve.

For the neuron data, the absorbance returned to baseline after each load, although complete recovery required a minute or more. The simulated absorbance change went to a steady-state value (shown by asterisks) greater than the initial condition. This discrepancy represents the fact that simulations are for a finite volume of buffer, whereas the nerve cell has additional processes that actually remove Ca^{2+} from the cytoplasm, e.g., organelle or plasma membrane pumps. Table II compares the amplitudes of measured and computed absorbance changes for several cells. For the cells, the total amount of Ca entering (per cm^2 of membrane) was estimated from voltage-clamp measurements. In dorid neurons, Ca influx does not inactivate seriously in the 250-msec period considered here (Connor, 1977,

TABLE II

Comparison of Experimental and Computed Absorbance Magnitudes[a]

Identification	Diameter (μ)	Total Ca entry at 250 msec (pmole/cm^2)	$(Arz)_T$ (μ)	ΔCa-Arz at 250 msec ÷ 2 (pmole/cm^2)	ΔCa-Arz / ΔCa
RPel-5/7	390	81	330	36	0.44
RPlG-5/14	490	104	220	32	0.31
RPel-5/15	340	90	300	36	0.40
LPel-5/7	400	90	275	52	0.57
RPel-5/14	325	100	263	57	0.57
Simulations		150	250	42.7	0.29
		150	300	45.5	0.31
		150	400	53.4	0.356

[a]Computations are shown for three indicator concentrations.

1979), so total entry was estimated by multiplying peak slow inward current by 250 msec. The dye absorbance change has been converted to total absorbing species (Ca-Arz) by using Beer's law. This quantity has been divided by 2 because we want to consider one surface only and the experimental light path crosses the cell membrane twice. Finally, the fraction, total Ca-Arz/total Ca entry, is given in the last column and compared with simulations (lower data sets). Again there is good agreement between experiment and computation.

Several model parameters could be juggled to give a better match between the two sets of data. For example, buffer concentration could be increased or rate constants changed, but there is no particular reason for picking one parameter over another on the basis of experimental measurement.

The above agreement between experiments and simulations demonstrates that the diffusion model is a good starting point toward understanding Ca^{2+} movements in the cytoplasm. Consequently, it is constructive to look at the Ca^{2+} profiles predicted by the model. Figure 9A shows the buildup of Ca^{2+} during an influx period of 200 msec, whereas figure 9B illustrates the collapse of the Ca^{2+} gradient after the influx. Particularly notable are the large values of Ca^{2+} near the membrane and the slow time course with which the postinflux gradients collapse. The logarithmic scale necessary to display the data clearly, overemphasizes small gradients, but some seconds are required for the Ca^{2+} gradient to flatten to within a factor of 2. In Section 3, where we describe cAMP-induced changes in Ca^{2+}, we deal with time scales that fall in this slow region.

Figure 10 shows the concentration profiles of Ca-cytoplasmic buffer complex (CaB) and Ca-indicator complex (Ca Arz) during the influx period, and illustrates from where the indicator signal arises. Ca^{2+} influx leads to a quick saturation of intrinsic buffer in the initial micron adjacent to the membrane, and as this occurs, the lower-affinity indicator begins binding appreciable Ca producing an absorbance change. As the influx continues, this front of CaB and Ca-Arz moves away from the membrane (or into the cell). The dye signal then arises from an expanding

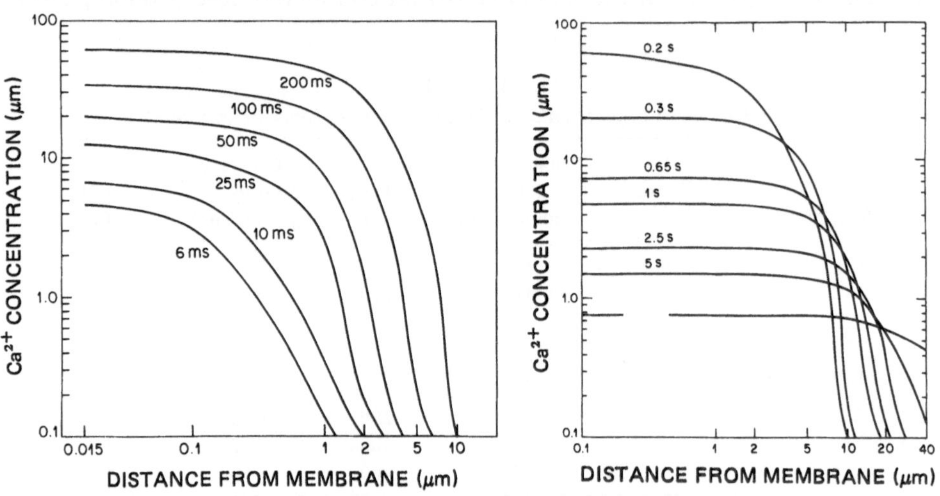

FIGURE 9. Ca^{2+} profiles during (left), and following the influx period.

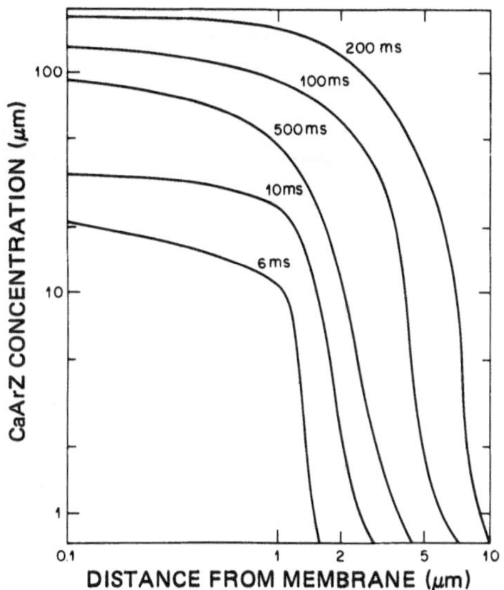

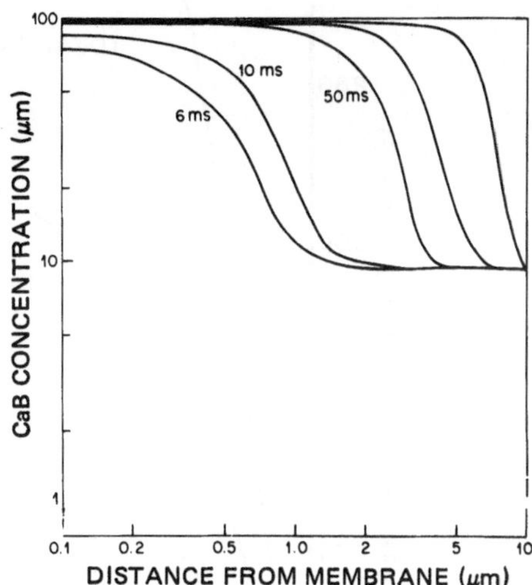

FIGURE 10. (Left) Spatial profiles of Ca-Arz complex (the absorbing species) during a 200 msec influx. (Right) Corresponding profiles of Ca-intrinsic buffer complex during influx. Note that, initially, 10% of B is bound to Ca. Saturation levels are given by horizontal lines at top of each graph.

depth of nearly saturated indicator. Even for loading times as long as 0.5 sec, almost the entire absorbance change arises in a thin (~ 10 μm) layer of indicator. Ca^{2+} and the complexes diffuse further into the cytoplasm and when the influx terminates, there is a major redistribution of Ca between Ca^{2+} and Ca-Arz to CaB, giving an indicator absorbance decrease.

The influx and redistribution are illustrated in Fig. 11 in terms of total quantities. It can be seen that the amount of Ca bound by the intrinsic buffer grows rather slowly after the loading phase. This does not reflect the speed of the Ca buffer on-reaction or the Arz off-reaction (both effectively instantaneous), but arises from the time required for Ca^{2+} and Ca-Arz to diffuse to depths where the intrinsic buffer is not saturated. The equilibrium level of Ca^{2+} and the complexes is a function of the diffusion compartment depth, here 40 μm.

Figure 12 illustrates Ca^{2+} changes near the membrane and indicator dye signal in a situation approximating three sequential action potentials. Here, voltage-clamp pulses of 10-msec duration were applied at 50-msec intervals. Ca^{2+} concentration near the membrane (X = 0.015 μm) increases and decreases very rapidly and shows a moderate summation effect. At very small distances away, the changes are greatly attenuated and slowed. The dye signal, since it reports more or less on the integral of Ca^{2+}, shows far greater summation and slower postinflux recovery. This form of the model then predicts a certain degree of "facilitation" and a rapid onset and offset for Ca-dependent processes that have K_Ms in the micromolar range. We do not know of clearly defined Ca-dependent processes in the cell body that we can monitor at present that would give an independent check of the above model predictions.

Any model of Ca diffusion in cytoplasm formulated at this time will contain

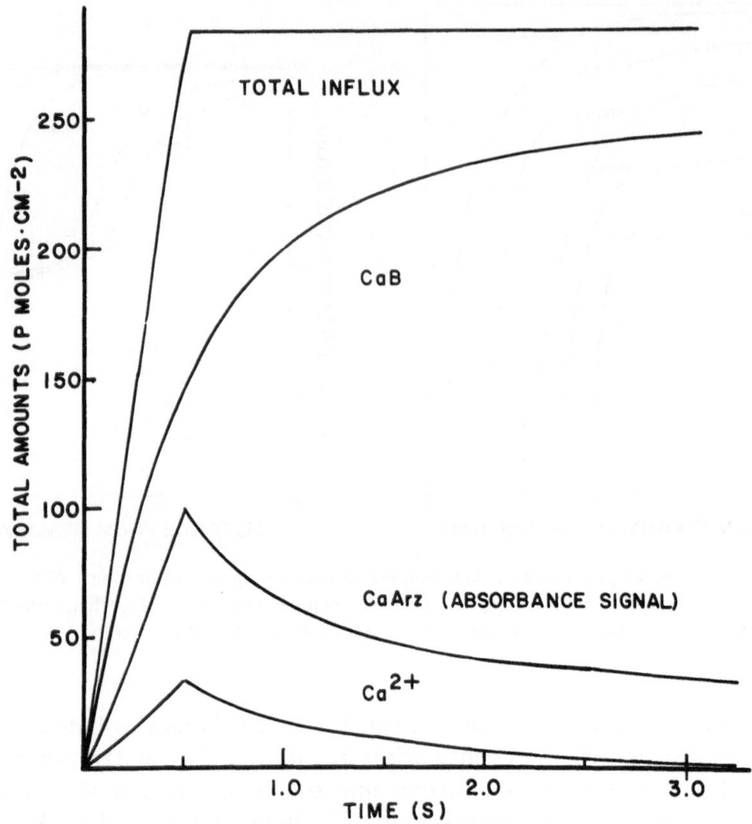

FIGURE 11. Time course of total Ca^{2+} influx, buffer and indicator complexes, and Ca^{2+} for simulation of 0.5-sec loading period. (From Connor and Nikolakopoulou, 1982.)

several parameters estimated with marginal accuracy and one or more downright guesses (such as D_B in this study). This should not detract from the results which are robust even under large parameter variations; for example, given any reasonable buffer capacity and D_{Ca}, the experimentally measured values of Ca current require that Ca^{2+} reach levels of several micromolar near the cell membrane for action potential-type depolarization. "Reasonable" for D_{Ca} would mean any value less than the value for water. Upper bounds for buffer capacity are set by the experiments described earlier and the fact that we measure dye signals of the magnitude we do (see Table I). Values greater than 200–300 μM would predict unreasonably small indicator absorbance changes. Figure 13 illustrates the differences in model $[Ca^{2+}]$ produced by varying D_{Ca} while holding other parameters fixed. A fourfold change in D_{Ca} causes less than a 50% change in $[Ca^{2+},]$ and the maximum values reached for influx periods of 5 to 25 msec (the approximate range of spike durations) always remain above 3 μM. In addition, the time course of intracellular redistribution of Ca^{2+} lies very much in the range of general physiological interest and must be considered carefully as more analytic descriptions of conductance activation, transmitter release, or Ca-dependent phase transitions are formulated.

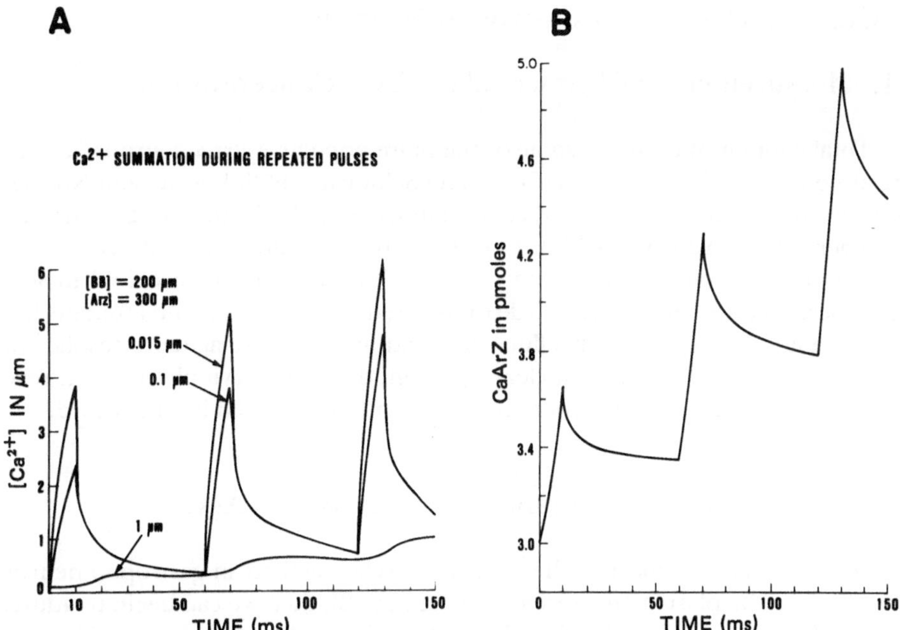

FIGURE 12. Simulations showing Ca^{2+} changes at three distances from the membrane for three 10-msec loads spaced at 50-msec intervals (A). (B) Gives the corresponding dye signal.

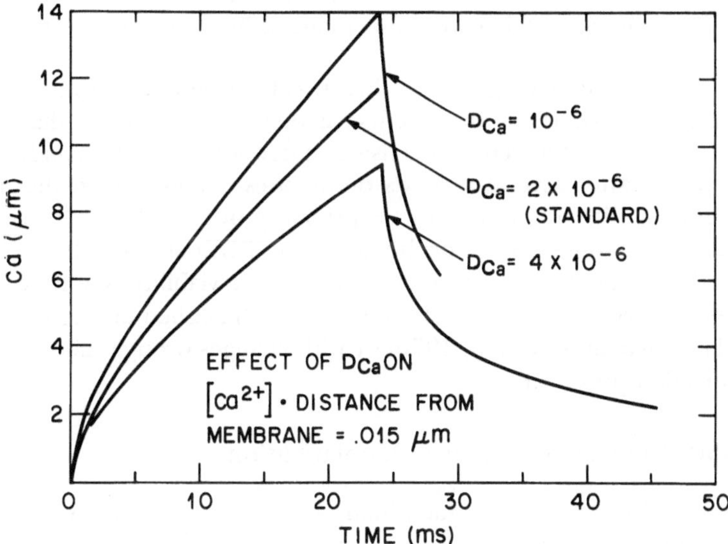

FIGURE 13. Differences in predicted $[Ca^{2+}]$ changes produced by varying the calcium diffusion coefficient.

3. Cyclic Nucleotides in Gastropod Neurons

3.1. Measurements of Intracellular cAMP Concentration

Total content of cAMP in some of the more popular *Aplysia* neurons has been measured in several laboratories (Cedar and Schwartz, 1972; Levitan and Norman, 1980; Bernier *et al.,* 1982) and recently estimated to be in the 10–20 μM range (Hockberger and Yamane, 1984). Certain neurotransmitters as well as electrical stimulation of presynatic fibers can raise this level twofold to threefold in some cells. These measurements were made on isolated cell bodies, and no attempts have yet been made to examine subcellular localization. Thus the actual cytosolic concentration is not known, nor is it clear how high the level in cytoplasm might reach under stimulated conditions. There is still much to be learned in this regard.

3.2. Modulation of Membrane Conductance by cAMP

If we assume that most of the adenylate cyclase activity in gastropod neurons is localized to the plasma membrane (Levitan, 1978), then we can begin to address one possible function of cAMP by asking what effects, if any, this nucleotide might have on membrane conductance parameters. We have been particularly interested in examining any possible nucleotide action on Ca conductance or internal Ca^{2+} regulation since there is ample evidence that these second messengers often affect one another (Berridge, 1975; Rasmussen, 1981). Here again we have used the Ca indicator dye arsenazo III together with voltage-clamp techniques to explore this possibility while we elevated the internal level of cAMP using intracellular injection. We have found that injections induced changes in membrane conductance to Na and Ca ions, changes which were not induced by a variety of other injected nucleotides.

These conductance changes were direct, cell specific, and found in a wide range of gastropod species. A direct effect is defined here as one that occurred concurrently with cAMP injection, persisted under voltage-clamp conditions, and typically reversed after several seconds or minutes depending on the delivered dose. The reversibility was found to be primarily due to the hydrolysis of the injected cAMP by endogenous phosphodiesterase (PDE) and was thus a relatively good indication of nucleotide specificity and cellular viability. We have not concerned ourselves with effects brought about by bath-applied analogues of cAMP (compare Treistman and Levitan, 1976) or with changes brought about only after repeated injections of cAMP.

3.2.1. cAMP-Induced Membrane Depolarization

While injecting cAMP into identified *Archidoris montereyensis* neurons, we noticed that most cells responded with a transient loss of membrane potential typically lasting several seconds (Hockberger and Connor, 1983; also see Liberman *et al.,* 1975). The response was not elicited with comparable injections of 5′AMP,

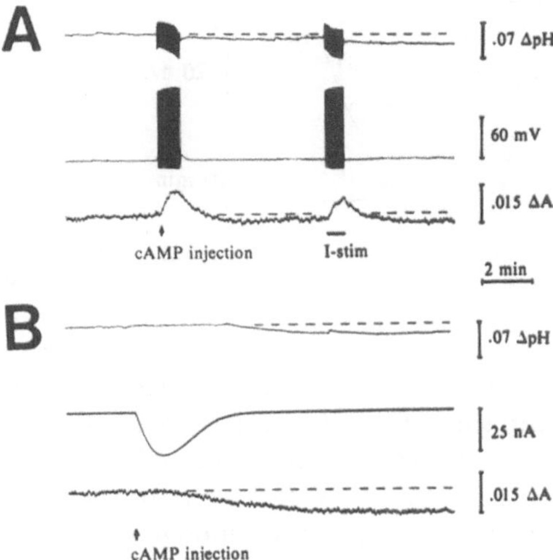

FIGURE 14. Pressure injections of a buffered cAMP solution (pH 7.5) into the same neuron (pH_i 7.46) with and without the membrane voltage clamped. (A) Without voltage clamp, the injection elicited a train of action potentials (middle trace) and a corresponding increase in the 660–700 nm signal (lower trace, τ = 1 sec) indicating a rise in intracellular [Ca^{2+}]. Simultaneous monitoring of pH_i (top trace) showed an induced acidification during and after the train of spikes. Both changes were mimicked by current stimulation (I_{stim}) of a similar train of spikes. (B) With voltage clamped, a much larger injection elicited a large inward current (middle trace), a decreased 660–700 nm signal (lower trace, τ = 1 sec), and pH_i decrease. (From Hockberger and Connor, 1983.)

5′GMP, ATP, GTP, cCMP, or cUMP, but it was mimicked by cGMP. Figure 14A shows records taken during a cAMP injection into an arsenazo III-filled *Archidoris* neuron. The cell also had a Thomas-type pH microelectrode inserted intracellularly to check for potential pH effects on the dye. The injection induced a train of action potentials accompanied by an elevation of cellular Ca (seen as an increase in dye absorbance at 660–700 nm) which reversed when firing ceased. A longer-lasting cellular acidification was also induced and was detected by both the pH electrode and by a delayed absorbance decrease. Action potentials elicited with current stimulation evoked similar changes in cellular Ca and pH.

Repeating the injection with the membrane-voltage clamped near resting potential generated a reversible inward current without a corresponding elevation in Ca level (Fig. 14B). Again, a drop in internal pH was detected by both the pH electrode and the dye since the absorbance decrease at 610–700 nm was larger than the 660–700 nm decrease (see Connor and Hockberger, 1984b). Similar results were found in the 14 largest, identified neurons in *Archidoris montereyensis* indicating that at least in these cells there was no obligatory coupling between the membrane current and cellular Ca levels. Of course under physiological conditions cAMP elevation would raise Ca_i in an indirect fashion when Ca channels open following depolarization.

3.2.2. Estimate of Minimal Effective Dose

Figure 15 shows records where the dose of cAMP injected into an *Archidoris* neuron was monitored photometrically. The injection pipet contained cAMP plus arsenazo III (here used as a simple indicator of dose), and as pulses of increasing pressure were applied, the nucleotide-dye solution was ejected into the cell situated between a fiber optic-photomultiplier detector system (see Hockberger and Connor, 1984a). The cell was under voltage clamp at −40 mV and responded with summating inward currents. Each injection delivered a dose that resulted in nearly

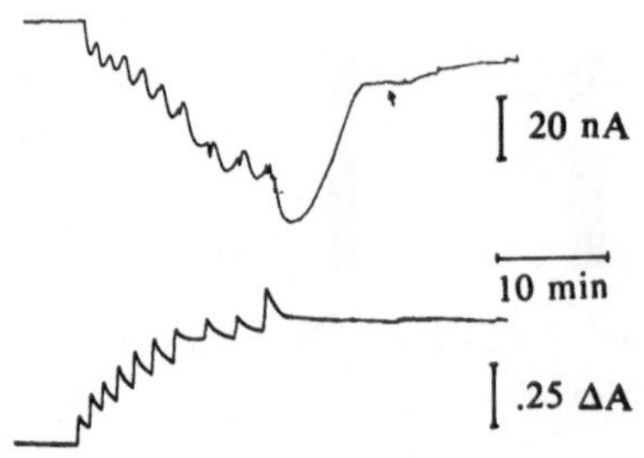

FIGURE 15. Serial pressure injections of cAMP monitored by dye absorbance at the dye isosbestic wavelength.

identical peak absorbance deflections, while absorbance signal recoveries reflected diffusion of the dye from beneath the optical probe. The summation of both the current and absorbance records indicated sustained elevation of cAMP and dye near the injection site. We have used this procedure of overlapping injections to minimize the action of endogenous PDE and achieve an estimate of the minimal effective cAMP dose necessary to elicit a membrane depolarization. Our results indicated that an elevation (averaged over the entire cell volume) of 30–35 μM was sufficient to elicit several action potentials in quiescent *Archidoris* as well as certain *Aplysia californica* neurons (Connor and Hockberger, 1984a).

3.2.3. Analysis of an Induced Inward Current

Inward currents evoked by cAMP injections into neurons clamped at membrane potentials below spike threshold have been found in quite a variety of gastropod species (Kononenko and Mironov, 1981; Hockberger and Connor, 1983, 1984a; Aldenhoff *et al.*, 1983; Green and Gillette, 1983). The amplitude of the current either increased, remained constant, or decreased over the voltage range -30 to -100 mV reflecting a voltage as well as a chemical sensitivity. The duration of the current was greatly prolonged in neurons bathed in PDE inhibitors, e.g., isobutylmethylxanthine, or following injection of nonhydrolyzable cAMP analogues indicating that the reversible nature of the current stems primarily from *in situ* hydrolysis of cAMP and not dilution or current inactivation.

We have analyzed the ionic basis of the cAMP-induced current in *Archidoris* neurons as well as cells R_2, LP1, and L_2–L_6 cluster in *Aplysia* (Connor and Hockberger, 1984a). In those cells, several lines of evidence argued that the current was carried primarily if not entirely by sodium ions. Removing sodium from the bathing medium blocked the response, whereas alteration of extracellular K^+, Cl^-, or Ca^{2+} had no such effect. The reversal potential of the induced current in low Na saline, as shown in Fig. 16, was approximately the value expected from a response mediated entirely by sodium ions. Third, the elevation of internal [Na] measured using Na-sensitive electrodes was nearly the estimated rise based on the integral of the current response (e.g., Fig. 17).

Recently, we have found that Ca ions can contribute to the induced inward current in neurons of the terrestrial slug *Limax maximus* (Hockberger and Connor, 1984b). Unlike the marine gastropod neurons, cAMP injection into arsenazo-filled *Limax* cells produced dye absorbance changes consistent with a corresponding rise in Ca_i^{2+} (Fig. 18). Iontophoretic injection of cAMP induced an inward current

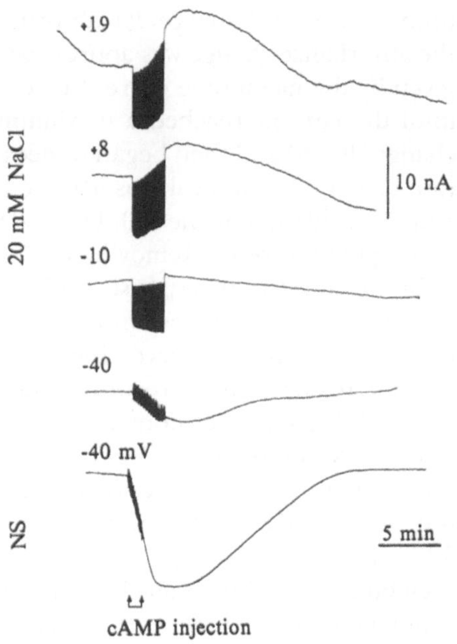

+19
+8
10 nA
20 mM NaCl
-10
-40
-40 mV
NS
5 min
cAMP injection

FIGURE 16. Reversal potential measurement of the nucleotide-induced current responses made in 20 mM NaCl saline. Identical iontophoretic injections were used at several holding potentials, and the current response reversed around −10 mV. (From Connor and Hockberger, 1984a.)

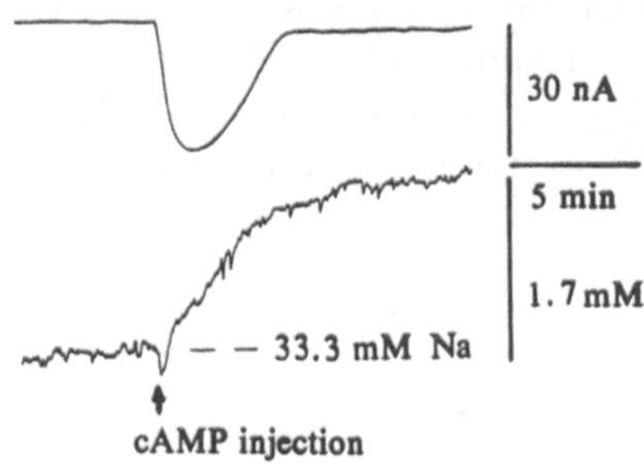

FIGURE 17. Measurements of membrane current and internal [Na] increase following cAMP injection.

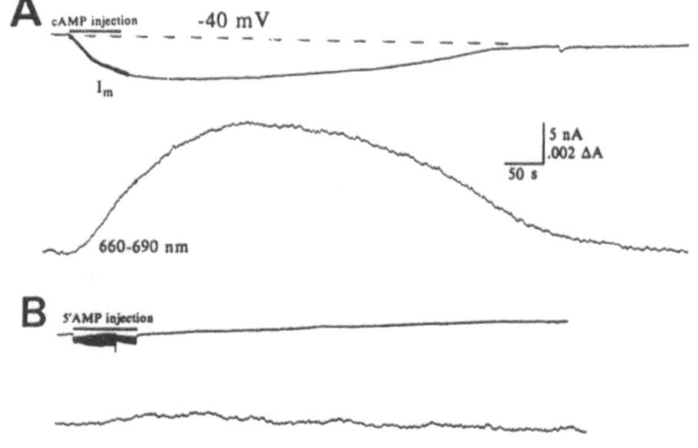

FIGURE 18. (A) Membrane current (upper trace) and arsenazo III—Ca signal resulting from injection of cAMP. (B) Same parameters showing that 5′AMP was ineffective in inducing the response. (From Hockberger and Connor, 1984b.)

(upper trace) and a large dye absorbance increase (Fig. 18A). The time course of the absorbance change was appropriate to reflect calcium influx with a time course given by the membrane current record. That is, the absorbance increased steeply until the current reached a maximum. As the current declined, the absorbance change slowed and then began to decay. As the current became smaller, the decay rate increased. The peak absorbance reflects a balance point between influx rate and Ca regulation by the cell. Figure 18B shows that 5′AMP was ineffective in generating the response. Removing Ca^{2+} from the bath eliminated the dye signal and reduced the current response. Only after also removing Na^{+} from the bath, however, was the current response completely blocked (Fig. 19). This type of Na/Ca response was found in several identified *Limax* neurons, although cellular differences were encountered regarding the percentage of the current carried by Ca ions (0–50%). We have not done a systematic study of neurons in other species to see whether Na or Na/Ca current is more common. Reports using identified *Helix pomatia* neurons have described both types of currents (Kononenko *et al.*, 1983; Aldenhoff *et al.*, 1983), and we have noticed possible seasonal and maturational differences within *Limax* cells. For example, in reproductively immature *Limax* (weighing <5 gm) the same identified neurons described above never showed a Ca component of the current response. This is illustrated in Fig. 20. Figure 20A shows that the cell has normally behaving voltage-activated channels in that a depolarizing pulse to +20 mV gives a large dye signal. Cyclic AMP injection gave an inward current and an acidification (Fig. 20 B), but not a rise in Ca^{2+}. Figure 20C shows that the current was insensitive to V_m below −30 mV. These characteristics make the response identical to the behavior analyzed in *Archidoris* neurons. The results suggest that the Ca component may be regulated by factors that do not affect the Na component of the current response.

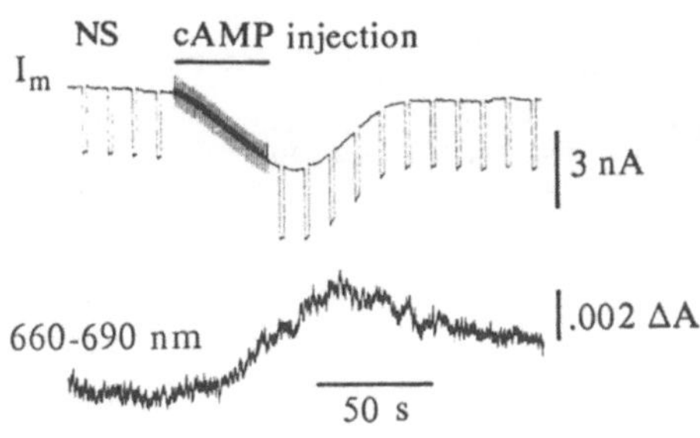

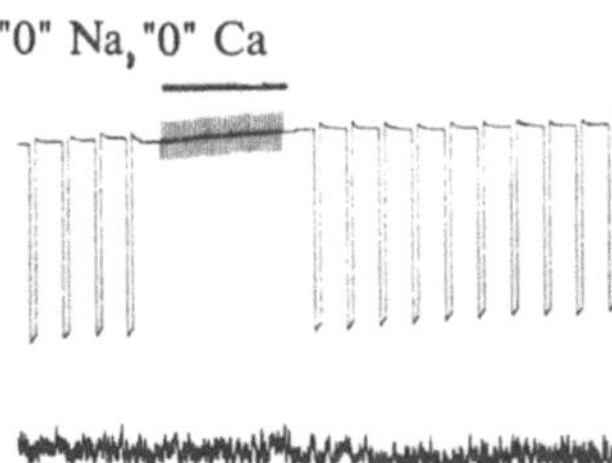

FIGURE 19. Current and dye absorbance records showing that zero Na-Ca saline removes inward current and dye-absorbance change. Conductance measurement pulses (−10 mV) were applied throughout this experiment and illustrate that a small increase occurred during current flow in normal saline. Low-Ca saline generally caused a large increase in conductance.

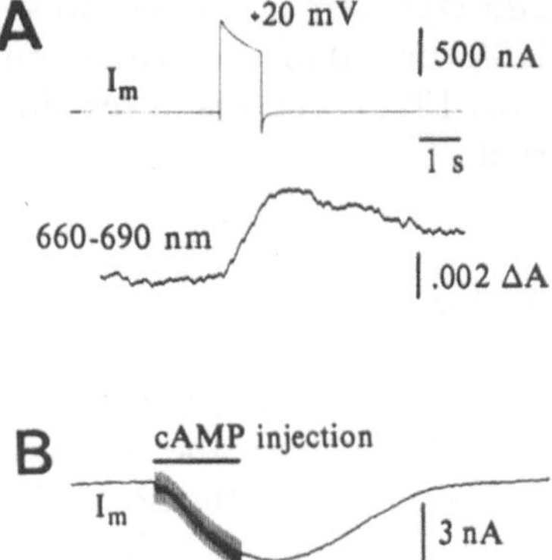

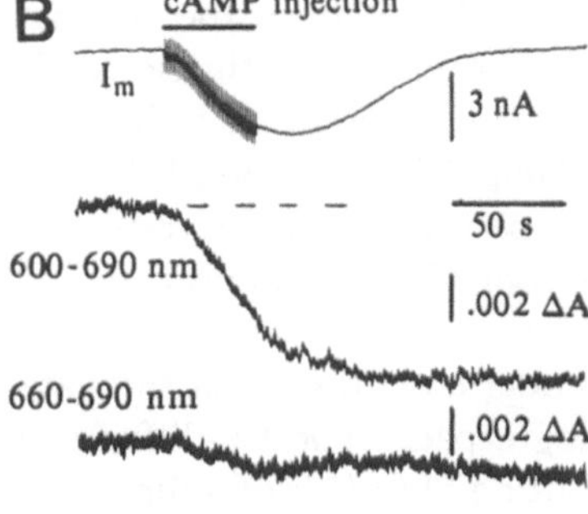

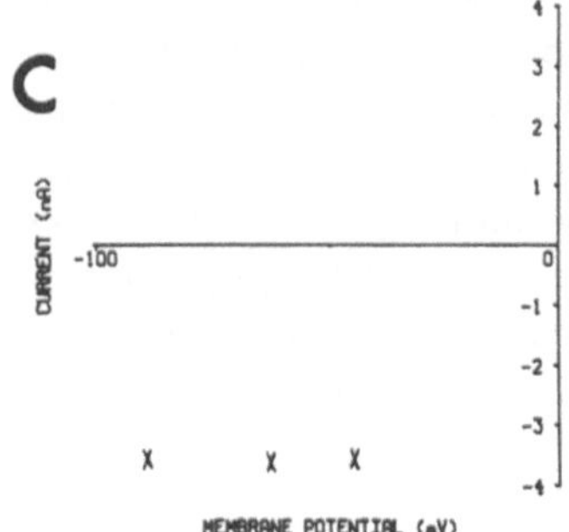

FIGURE 20. Current and dye records taken from immature specimen but same cell type as in Fig. 18. (A) Records show that there is normal Ca influx during voltage pulses. (B) No Ca influx in response to cAMP injection. The absorbance decreases shown indicate an acidification of the cytoplasm. (C) Plot of peak induced current versus membrane-holding potential. (From Hockberger and Connor, 1984b.)

The Na-mediated response exhibited pharmacological in addition to voltage-dependent properties that separated it from previously described sodium currents. The induced current was insensitive to TTX, ouabain, and amiloride at concentrations that either dramatically changed spike threshold (TTX), resulted in loss of resting potential (ouabain), or exceeded levels needed to block Na transport across epithelial tissue (amiloride). However, the current was activated in the membrane potential range where the inward current underlying the negative slope resistance (NSR) region is found (compare Wilson and Wachtel, 1974; Partridge *et al.*, 1979). It is possible that the NSR current becomes activated by cAMP giving rise to the induced inward current. However, the induced current was present below −80 mV (although reduced in the *Aplysia* cells), a membrane potential where it is not clear that NSR current flows. The NSR current is also impermeant to lithium ions (Smith *et al.*, 1975; J. A. Connor and P. Hockberger, unpublished observations), whereas Li can carry the nucleotide-induced current (Kononenko, 1981; Connor and Hockberger, 1984a). In spite of these arguments, it is a curious fact that the NSR current in some *Helix* neurons is carried by both Na and Ca ions (Eckert and Lux, 1976), whereas in certain *Aplysia* neurons it is apparently Na selective (Smith *et al.*, 1975).

4. Summary

Gastropod neurons, like most other cells, employ second messengers to transduce signals from the plasma membrane to the cell interior. In this chapter we describe our investigations of two such messengers, internal free Ca^{2+} and cAMP, probed using the Ca indicator dye arsenazo III, ion-sensing electrodes, and voltage clamp. We have found that both types of messenger exist at low levels inside gastropod neurons, the result of both an efficient buffering system for Ca ions and enzymatic degradation of cAMP. Simulations of Ca^{2+} influx and redistribution have shown that Ca^{2+} likely reaches micromolar levels near the cell membrane under conditions that simulate action potential trains. Estimates of the Ca-buffering capacity and Ca-diffusion coefficient of cytoplasm indicated that the redistribution of Ca^{2+} is of a time scale whereby it could influence conductance activation, transmitter release, or other Ca-dependent processes.

The injection of micromolar amounts of cAMP into neurons induced a dose-dependent depolarization due to the activation of a voltage-dependent Na or Na/Ca current. The unique voltage-sensitivity and pharmacological properties of this current indicated that it was unlike previously described sodium currents. It was also found that where no direct coupling between elevated cAMP and cellular Ca^{2+} occurred, under physiological conditions a rise Ca^{2+} would also follow if action potentials were induced. Thus gastropod neurons, like other cells, exhibit coupling between internal Ca^{2+} and cAMP. This coupling could play a role in transmitter release, spontaneous activity, as well as frequency encoding in these neurons.

References

Adams, D., Smith, S., and Thompson, S., 1980, Ionic currents in molluscan soma, *Annu. Rev. Neurosci.* **3**:141–167.

Ahmed, Z., and Connor, J. A., 1979, Measurement of calcium influx under voltage clamp in molluscan neurons using the metallochromic dye Arsenazo III, *J. Physiol.* **286**:61–82.

Ahmed, Z., and Connor, J. A., 1980, Intracellular pH changes induced by calcium influx during electrical activity in molluscan neurons, *J. Gen. Physiol.* **75**:403–426.

Ahmed, Z., Kragie, L., and Connor, J. A., 1980, Stoichiometry and apparent dissociation constant of the calcium-arsenazo III reaction under physiological conditions, *Biophys. J.* **32**:903–920.

Aldenhoff, J., Hofmeier, G., Lux, H., and Swandulla, D., 1983, Stimulation of a sodium influx by cAMP in *Helix* neurons, *Brain Res.* **276**:289–296.

Alvarez-Leefmans, F. J., Rink, T. J., and Tsien, R. Y., 1981, Free calcium ions in neurons of *Helix aspersa* measured with ion-selective micro-electrodes, *J. Physiol.* **315**:531–548.

Ashley, C., and Campbell, A. (eds.), 1979, *Detection and Measurement of Free* Ca^{2+} *in Cells*, Elsevier/North Holland, New York.

Bauer, P., 1981, Affinity and stoichoimetry of calcium binding by arsenazo III, *Anal. Biochem.* **110**:61–72.

Bean, B., Nowyckg, M., and Tsien, R., 1984, β-Adrenergic modulation of calcium channels in frog ventricular heart cells, *Nature (London)* **307:**371–375.

Bernier, L., Castellucci, V., Kandel, E., and Schwartz, J., 1982, Facilitatory transmitter causes a selective and prolonged increase in adenosine 3′:5′-monophosphate in sensory neurons mediating the gill and siphon withdrawal reflex in *Aplysia, J. Neurosci.* **2:**1682–1691.

Berridge, M. J., 1975, The interaction of cyclic nucleotides and calcium in the control of cellular activity, *Adv. Cyclic Nucleotide Res.* **6:**1–98.

Berridge, M. J., 1979, Modulation of nervous activity by cyclic nucleotides and calcium, in: *The Neurosciences: Fourth Study Program* (F. O. Schmitt and F. G. Worden, eds.), MIT Press, Cambridge, Massachusetts, pp. 873–889.

Brinley, F. J., 1978, Calcium buffering in squid axons, *Annu. Rev. Biophys. Bioeng.* **7:**362–392.

Brinley, F. J., 1979, Techniques for measuring free calcium *in situ* in single isolated cells using aequorin and metallochromic indicators, in: *Detection and Measurement of Free* Ca^{2+} *in Cells* (C. Ashley and A. Campbell, eds.), Elsevier/North Holland, New York, pp. 319–338.

Caswell, A., 1979, Methods of measuring intracellular calcium, *Int. Rev. Cyto.* **56:**145–181.

Cedar, H., Schwartz, J., 1972, Cyclic adenosine monophosphate in the nervous system of *Aplysia calfornica.* II. Effect of serotonim and dopamine, *J. Gen. Physiol.* **60:**570–587.

Connor, J. A., 1977, Time course separation of two inward currents in molluscan neurons, *Brain Res.* **119:**487–492.

Connor, J. A., 1979, Calcium current in molluscan neurons: Measurement under conditions which maximize its visibility, *J. Physiol.* **286:**41–60.

Connor, J. A., and Ahmed, Z., 1984, Diffusion of ions and indicator dyes in neural cytoplasm, *Cell Mol. Neurobiol.* **4:**53–66.

Connor, J. A., and Hockberger, P., 1984a, A novel membrane sodium current induced by injection of cyclic nucleotides into gastropod neurons, *J. Physiol.* **354:**139–162.

Connor, J. A., and Hockberger, P., 1984b, Intracellular pH changes induced by injection of cyclic nucleotides into gastropod neurons, *J. Physiol.* **354:**163–172.

Connor, J., and Nikolakopoulou, G., 1982, Calcium diffusion and buffering in nerve cytoplasm, *Lect. Math. Life Sci.* **15:**79–96.

Dipolo, R., Requena, G., Brinley, F. J., Mullins, L., Scapa, A., and Tiffert, T., 1976, Ionized calcium concentrations in squid axons, *J. Gen. Physiol.* **67:**433–467.

Eckert, R., and Lux, H. D., 1976, A voltage-sensitive persistent calcium conductance in neuronal somata of *Helix, J. Physiol.* **254:**129–151.

Gorman, A. L. F., and Thomas, M. V., 1980, Intracellular calcium accumulation during depolarization in a molluscan neuron, *J. Physiol.* **308:**259–285.

Green, D., and Gillette, R., 1983, Patch- and voltage-clamp analysis of cyclic AMP-stimulated inward current underlying neuron bursting, *Nature (London)* **306:**784–785.

Hockberger, P., and Connor, J. A., 1983, Intracellular calcium measurements with arsenazo III during cyclic AMP injections into molluscan neurons, *Science* **219:**869–871.

Hockberger, P., and Connor, J. A., 1984a, Dose-dependent effects of intracellular cyclic AMP on nerve membrane conductances and internal pH, in: *Primary Neural Substrates of Learning and Behavioral Change* (J. Farley and D. Alkon, eds.), Cambridge University Press, New York, pp. 337–358.

Hockberger, P., and Connor, J. A., 1984b Alteration of calcium conductances and outward current by cyclic AMP in neurons of *Limax maximus, Cell Mol. Neurobio.* **4:**319–338.

Hockberger, P., and Yamane, T., 1984, Measurement of cyclic AMP concentration in *Aplysia* nerve cell bodies, *Soc. Neurosci. Abstr.* **10:**896.

Kendrick, N., Ratzlaff, R., and Blaustein, M., 1977, Arsenazo III as an indicator for ionized calcium in physiological salt solutions: Its use for determination of the CaATP dissociation constant, *Anal. Biochem.* **83:**433–450.

Kononenko, N., 1981, Ionic mechanism of the transmembrane current evoked by injection of cyclic AMP into identified *Helix pomatia* neurons, *Neurophysiology* (USSR), **12:**339–343.

Kononenko, N., and Mironov, S., 1981, Effects of intracellular injection of cyclic AMP on electrical characteristics of identified neurons in *Helix pomatia. Neurophysiology* (USSR), **12:**332–338.

Kononenko, N., Kostyuk, P., and Shcherbatko, A., 1983, The effect of intracellular cAMP injections on stationary membrane conductance and voltage- and time-dependent ionic currents in identified snail neurons, *Brain Res.* **268:**321–338.

Levitan, I., 1978, Adenylate cyclase in isolated *Helix* and *Aplysia* neuronal cell bodies: Stimulation by serotonim and peptide-containing extract, *Brain Res.* **154**:404–408.

Levitan, I., B., and Norman, J., 1980, Differential effects of cAMP and cGMP derivatives on the activity of an identified neuron: Biochemical and electrophysiological analysis, *Brain Res.* **187**:415–429.

Levy, S., Tillotson, D., and Gorman, A. L. F., 1982, Intracellular Ca^{2+} gradient associated with Ca^{2+} channel activation measured in a nerve cell body, *Biophys. J.* **37**:182a.

Liberman, Y. A., Minina, S., and Golubtsov, K., 1975, Study of the metabolic synapse. I. Effect of intracellular microinjection of 3′,5′-AMP, *Biofizika* **20**:451–456.

Ogan, K., and Simons, E., 1979, The influence of pH on arsenazo III. *Anal. Biochem.* **96**:70–76.

Palade, P., and Vergara, J., 1983, Stoichiometries of arsenazo III-Ca complexes, *Biophys. J.* **43**:355–369.

Partridge, L. D., Thompson, S. H., Smith, S. J., and Connor, J. A., 1979, Current-voltage relationships of repetitively firing neurons, *Brain Res.* **164**:69–80.

Rasmussen, H., 1981, *Calcium and cAMP as Synarchic Messengers,* Wiley-Interscience, New York.

Rasmussen, H., and Goodman, D., 1977, Relationships between calcium and cyclic nucleotides in cell activation, *Physiol. Rev.* **57**:421–509.

Reuter, H., 1979, Properties of two inward membrane currents in the heart, *Annu. Rev. Physiol.* **41**:413–424.

Scarpa, A., 1979, Measurement of calcium ion concentrations with metallochromic indictors, in: *Detection and Measurement of Free* Ca^{2+} *in Cells* (C. Ashley and A. Campbell, eds.), Elsevier/North Holland, New York, pp. 85–115.

Smith, S., and Zucker, R., 1980, Aequorin response facilitation and intracellular calcium accumulation in molluscan neurons, *J. Physiol.* **300**:167–196.

Smith, T., Barker, J., and Gainer, H., 1975, Requirements for bursting pacemaker potential activity in molluscan neurons, *Nature (London)* **253**:450–452.

Thomas, M. V., 1979, Arsenazo III forms 2:1 complexes with Ca and 1:1 complexes with Mg under physiological conditions. Estimates of the apparent dissociation constants, *Biophys. J.* **25**:541–548.

Thomas, R. C., 1978, *Ion-sensitive Intracellular Microelectrodes,* Academic Press, New York.

Tillotson, D., and Gorman, A. L. F., 1983, Localization of neuronal Ca^{2+} buffering near plasma membrane studied with different divalent cations, *Cell. Mol. Neurobiol.* **3**:297–310.

Treistman, S., and Levitan, I., 1976, Alteration of electrical activity in molluscan neurons by cyclic nucleotides and peptide factors, *Nature (London)* **261**:62–64.

Wilson, W. A., and Wachtel, H., 1974, Negative resistance characteristic essential for the maintenance of slow oscillations in bursting neurons, *Science* **186**:932–934.

24

Synaptic Facilitation and Residual Calcium

ROBERT S. ZUCKER

1. Introduction

The synapse is the point of contact between a neuron and its target organ—a muscle, a gland, or another neuron. It is the point of intercellular communication, of transfer of information within the nervous system.

Most synapses operate by releasing a chemical substance following electrical activity. Typically, a depolarizing signal, usually an action potential, invades presynaptic terminals, opening voltage-dependent calcium channels. Calcium rushes in, perhaps for only a millisecond if an active spike invades the terminal, and leads directly to the phasic release of neurotransmitter, apparently by exocytosis or the fusion of synaptic vesicles with the presynaptic plasma membrane and the release of their contents. Once liberated, transmitter diffuses rapidly across the narrow cleft to the postsynaptic cell, where it binds to specialized receptor proteins embedded in that cell's membrane. The most common effect of this reaction is the brief opening of synaptic channels, permeable to only one or two of the ions Na^+, K^+, or Cl^-, and resulting in depolarization or hyperpolarization of the membrane. When these channels close, again often within only a millisecond or so, the transmitter is released into the cleft where it is inactivated, recovered presynaptically, or just diffuses away, and the postsynaptic potential (PSP) settles back to the resting level.

2. Short-term Dynamic Behavior of Synapses

This conventional description of synaptic transmission leaves the impression that synapses are static in function—a spike opens calcium channels, transmitter

ROBERT S. ZUCKER • Department of Physiology-Anatomy, University of California, Berkeley, California 94720.

is released, the target cell responds, and that is that. Reality, in fact, is usually quite different. Most synapses respond very dynamically, or plastically, to repeated presynaptic activity. Successive action potentials may each release more and more transmitter, reaching a plateau level of transmission much greater than that shown by an isolated spike. This is called synaptic facilitation. In synapses with a strong facilitation, like crayfish neuromuscular junctions at tonic muscles, this dynamic property is an essential aspect of effective synaptic transmission. Without it, PSPs would remain so small that, even with temporal summation, they would never cause any significant postsynaptic response. Thus, such synapses transmit effectively only in response to high-frequency bursts of presynaptic spikes. This may be a filter for distinguishing meaningful signals from noise.

At other synapses, like the giant synapse in the stellate ganglion of the squid, a second form of synaptic plasticity is dominant. Successive presynaptic action potentials release less and less transmitter, until a steady state attains which is far below threshold for eliciting postsynaptic activity. Such synapses respond effectively only at the beginning of an increase in spike frequency, and thus transmit information about a change in the presynaptic neuron. They also are responsible for some forms of behavioral habituation (Castellucci *et al.*, 1970; Zucker, 1972).

In many (but not all—see Castellucci and Kandel, 1974; Zucker and Bruner, 1977) synapses displaying depression, this form of fatigue is relieved by reducing the amount of transmitter released by spikes, for example, by using a low-calcium medium or reducing the amplitude of the presynaptic depolarization. This has led to the idea that depression is due to the depletion of a store of transmitter immediately available for release (Liley and North, 1953), perhaps a population of synaptic vesicles in contact with release sites on the presynaptic membrane.

When depression is reduced, an underlying facilitation is revealed similar to that seen in synapses showing no depression. Thus many synapses are normally subject to facilitation and depression operating simultaneously. Compared with facilitation, recovery from depression occurs slowly. A pair of closely spaced action potentials will display a net facilitation in a low-calcium medium, but on repetition of the pair at low frequency, the responses to both will decline until a steady state of depression is experienced by the doublet. Nevertheless, the second response in each pair will exceed the first. This is how facilitation is studied in relative isolation at synapses displaying both processes, such as the squid giant synapse.

3. Early Studies on Mechanism

Although the depletion hypothesis has been useful in explaining synaptic depression, facilitation has been less easily understood. Early experiments on the mechanism of synaptic facilitation used the crayfish neuromuscular junction as a preparation. Extracellular recordings of nerve terminal potentials, and depolarization of nerve terminals when spikes were blocked with tetrodotoxin, suggested that facilitation occurs in the absence of changes in the presynaptic spike amplitude (Zucker, 1974a).

Many neurons show spike prolongation during repeated activity, due to potassium channel inactivation. Pharmacologically blocking such potassium channels leaves all action potentials in a train constantly broadened, but synaptic facilitation persists (Zucker and Lara-Estrella, 1979). Afterpotentials following a presynaptic spike were inferred from measurements of excitability changes in terminals, but these, too, could be dissociated from facilitation (Zucker, 1974b). Recent intracellular recordings from motor neuron terminals confirm that facilitation is not caused by changes in presynaptic spikes or afterpotentials (Bittner and Baxter, 1983).

These experiments disproved certain mechanisms for facilitation, but failed to determine what was responsible. All that could be said was that facilitation was a manifestation of an increased probability of release of single quanta by release sites (Zucker, 1973). Experiments on frog neuromuscular junctions had suggested that calcium influx was required for facilitation (Katz and Miledi, 1968), but the exact mechanism remained obscure.

4. Studies on a Model System—Molluscan Cell Bodies

One idea that gained currency was that facilitated PSPs were due simply to an increased presynaptic calcium influx to facilitating action potentials. In molluscan cell bodies, spikes or other repeated depolarizations were accompanied by successively brighter flashes of light from cells filled with the calcium-sensitive photoprotein aequorin (Fig. 1A). However, measurements of calcium current under voltage clamp indicated that the calcium channels were not activated more in successive depolarization (Smith and Zucker, 1980). Likewise, when intracellular calcium transients were measured with the calcium-sensitive dye arsenazo III, no facilitation of calcium influx was detected (Fig. 1B). The facilitation of aequorin signals was attributed to its nonlinear power-law dependence on calcium activity. A residual submembrane calcium causes relatively little maintained aequorin photoemission. But a given influx of calcium, now occurring on a more sensitive portion of the curve relating aequorin response to calcium concentration, leads to a larger incremental light emission (Fig. 1C).

Since calcium is entering at the surface, and then diffusing inward, a cell cannot be regarded simply as a one-compartment system, especially when dealing with nonlinear indicators such as aequorin. The behavior of the two calcium indicators was replicated by a computer simulation of the influx of calcium into a spherical cell, with diffusion into the center, binding to fixed buffer sites, removal by calcium pumps, and reaction with uniformly distributed arsenazo or aequorin according to their known reactions. Evidently, aequorin response facilitation is due to the effects of residual calcium and a nonlinear calcium sensitivity.

5. Extension to a Real Synapse—The Squid Giant Synapse

The analysis of aequorin response facilitation provokes the inference that synaptic facilitation works the same way. However, argument by analogy hasn't been

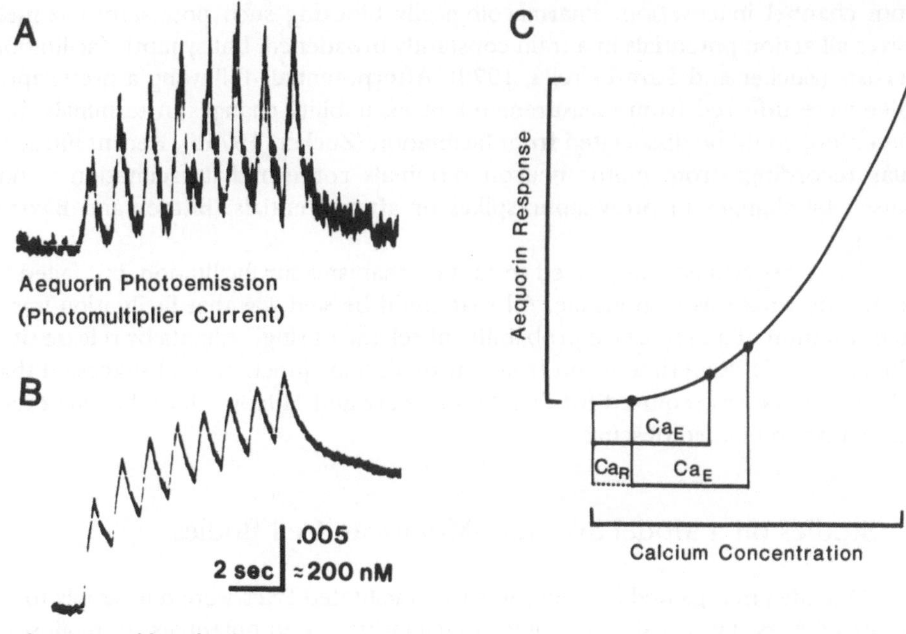

FIGURE 1. Signals recorded from *Aplysia* abdominal ganglion neurons filled with calcium indicators and stimulated under voltage clamp with eight depolarizing pulses, each 300 msec long, repeated once per second. (A) Shows facilitating photoemissions generated by the nonlinear calcium-sensitive photoprotein aequorin. (B) Shows nonfacilitating absorbance changes in a cell filled with the linear calcium-sensitive dye arsenazo III. (C) Shows the relationship between aequorin photoemission and calcium concentration. A constant increment of calcium entering during a depolarization (Ca_E) will elicit a larger aequorin response in the presence of residual calcium (Ca_R) from previous activity. (Indicator responses from Smith and Zucker, 1980.).

scientifically acceptable since the time of Aristotle. Experiments on a real facilitating synapse were required.

The best candidate seemed to be the giant synapse in the squid stellate ganglion. A glutaminergic contact between the presynaptic command neuron coming from the brain and the postsynaptic giant axon in the stellar nerve running to the distal mantle, this synapse is the largest known to man. The postsynaptic element is also the largest axon, the one used so fruitfully in studies of ionic currents underlying spikes. The presynaptic terminal in this synapse is almost 1 mm long and 50 μm in major diameter. It is possible to penetrate this process with several microelectrodes, to voltage clamp the terminal, and to inject substances like calcium ions and arsenazo.

5.1. Nonfacilitating Calcium Currents

It had already been shown that facilitation at the squid giant synapse can occur in the absence of changes in presynaptic spike amplitude or duration and that it is unrelated to afterpotentials (Charlton and Bittner, 1978b). Thus the same was known about this synapse as neuromuscular junctions. Working with Milton Charlton and Stephen Smith at the Marine Biological Laboratory in Woods Hole, Massachusetts, I began by observing the behavior of presynaptic calcium channels to paired spikelike depolarizations that actually led to facilitating excitatory postsynaptic potentials (EPSPs). Blocking sodium currents with tetrodotoxin and potassium currents with 3,4-diaminopyridine and intracellular injection of tetraethylammonium, and using the three-electrode voltage clamp to study calcium currents through local presynaptic channels in the terminal (Charlton *et al.*, 1982), we observed nonfacilitating inward currents during the pulses (Fig. 2).

5.2. Nonlinear Transmitter Release

In this experiment, the facilitated EPSP was 100% larger than the control. A similar increase in transmitter release during a single depolarization was obtained by increasing the amplitude of the depolarizing pulse by a few mV. This opened

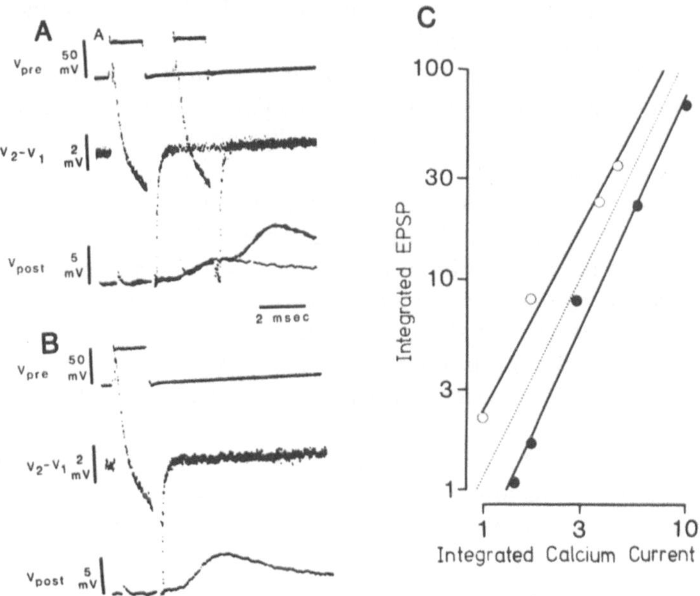

FIGURE 2. Properties of calcium currents recorded from the presynaptic terminal of the squid giant synapse using the three-electrode voltage clamp. (A) Two presynaptic depolarizing pulses (V_{pre}) elicit constant inward calcium currents, recorded as the presynaptic spatial voltage gradient, V_2—V_1. Responses to single and to paired pulses are superimposed. The postsynaptic recording (V_{post}) shows that the second EPSP is almost twice as large as the first. (B) The depolarizing pulse was increased enough so that the postsynaptic response matched the facilitated EPSP. This was caused by a 25% larger presynaptic calcium current, unlike the facilitated EPSP. (C) Shows the relationship between transmitter release (integrated EPSP waveform) versus total calcium influx (integrated presynaptic calcium current) for two synapses. The dotted line indicates a second-power relationship. (A and C are adapted from Zucker, 1982.)

more calcium channels, and led to a calcium influx that was 25% greater. This contrasts sharply with the facilitated EPSP, which occurred in the absence of any such increase in calcium current. This interesting result tells us two things: (1) there is a rather nonlinear relation between calcium influx and transmitter release at this synapse; and (2) facilitation occurs without an increase in calcium current of the magnitude necessary to explain facilitation by a change in calcium influx.

The nonlinear relation between calcium influx and transmitter release was measured in several experiments. The relation could be described as a power-law relationship with an exponent of about 2–3 (Fig. 2C).

5.3. Calcium Concentration Transients

Although calcium influx is unchanged during facilitated transmission, calcium concentration changes inside the terminal might not be constant. For example, there might be a local saturation of submembrane calcium-binding sites known to exist in cytoplasm and to strongly buffer intracellular calcium. Saturation of buffer could lead to a larger free calcium transient near the membrane at the end of the calcium influx during an action potential and to facilitated transmitter release, even though the influx was the same.

To test these ideas, we injected arsenazo into the distal end of the presynaptic terminal and focused a light spot on the slightly pinkish terminal. The light coming through the distal end of the terminal was collected in a fiber optics light guide and used for microspectrophotometric absorbance measurements of calcium concentration changes in the presynaptic terminal. Single action potentials were accompanied by sharply rising calcium concentration changes (Fig. 3), on the order of hundredths of a micromolar in magnitude when averaged throughout the width of the terminal. Of course, the calcium concentration change near calcium channels must be much larger. At the peak of the response, the calcium detected by arsenazo is still near its point of entry and where it is acting to release transmitter. Surprisingly, these transients were quite prolonged, decaying with a time course of seconds.

When two action potentials were elicited at a brief enough interval to evoke facilitating EPSPs, the calcium concentration transients were seen to be quite constant. Moreover, a train of *N* presynaptic spikes was accompanied by a steadily ris-

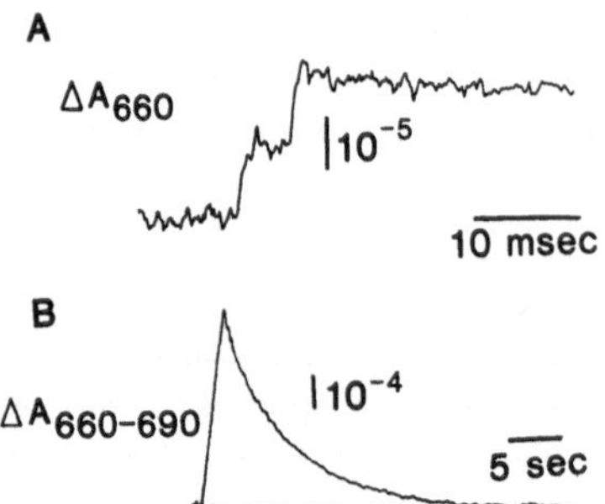

FIGURE 3. Absorbance changes in a squid presynaptic terminal injected with arsenazo III. (A) The presynaptic neuron fired two action potentials in close succession. The presynaptic calcium concentration increments caused by the action potentials were identical. (B) Shows the absorbance change caused by a 2-sec train of presynaptic impulses at 33 Hz (open bar). (The single wavelength signal in A and differential microspectrophotometric recording in B are adapted from Charlton *et al.*, 1982.)

ing absorbance signal, about N times the size of the absorbance peak during one spike. We found no evidence of saturated submembrane calcium buffering, or facilitation of submembrane calcium concentration transients that might underlie synaptic facilitation. Finally, the apparent proportionality between calcium concentration change and calcium influx indicates that the nonlinear dependence of transmitter release on calcium influx really reflects a nonlinear dependence on intracellular calcium activity.

5.4. Injecting Calcium Facilitates Transmission

We next turned to a direct test of the idea that residual submembrane calcium facilitates release of transmitter to subsequent action potentials. We injected calcium iontophoretically into presynaptic terminals, using interbarrel current passage between double-barrel electrodes to avoid any effect of the current injection on presynaptic membrane potential. In our first experiments, we found little or no effect on spike-evoked transmitter release, even though the calcium injection was effective in releasing transmitter tonically, indicating that the calcium had reached presynaptic release sites. We finally realized that this disappointing result was due to the fact that the giant synapse is too big for such an experiment. We can inject calcium at only one point, and so affect only a tiny fraction of the 1-mm long synaptic terminal.

We tackled this problem by restricting spike-evoked synaptic transmission to the small portion of the terminal into which we injected calcium. We did this by perfusing the ganglion with a low-calcium medium containing manganese, a calcium channel antagonist, and restoring transmission focally with an external calcium pipette positioned just adjacent to the intracellular calcium micropipette. Now the experiment worked every time; calcium injection significantly enhanced spike-evoked transmitter release without having any effect on the presynaptic action potential or resting potential (Fig. 4). At last we had positive evidence for a mechanism of synaptic facilitation: residual calcium, which certainly exists in nerve terminals as detected by arsenazo microspectrophotometry, really is capable of facilitating transmitter release.

6. A Model of Calcium Movements

I was quite satisfied with these experimental results until I began to think about the time courses of facilitation, transmitter release, and arsenazo signals. How is it that calcium ions that remain free in the presynaptic terminal for seconds lead to a synaptic facilitation that decays in tens of milliseconds? Even more puzzling, how can transmitter release cease within milliseconds after a presynaptic spike?

Llinás *et al.* (1982) had shown that calcium influx lasts only about 1 msec during the falling phase of the action potential, and that transmitter release terminates shortly thereafter. My earlier work (Smith and Zucker, 1980) on calcium diffusion

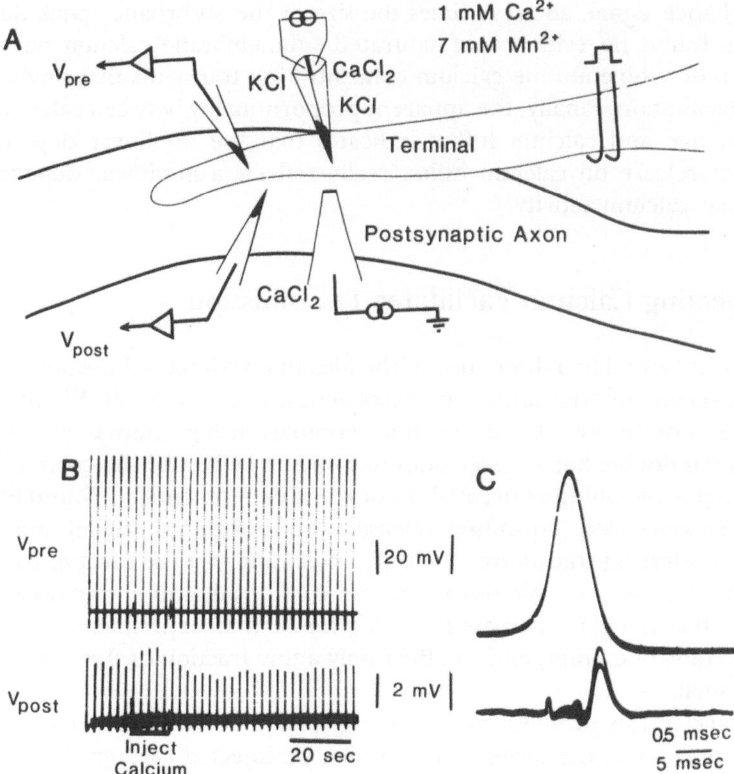

FIGURE 4. Facilitation of synaptic transmission by injected presynaptic calcium. (A) Shows the experimental arrangement for restricting synaptic transmission to the same portion of the synapse into which calcium was injected. (B) Shows the facilitated EPSPs elicited by action potentials during presynaptic calcium injection. Injecting calcium also elicits a tiny increase in transmitter release, indicated by an increased postsynaptic noise. (C) Shows superimposed presynaptic action potentials and EPSPs before and during calcium injection. The spikes are unaffected by injecting calcium. (Adapted from Charlton *et al.*, 1982.)

simulation in *Aplysia* neurons taught me never to think of a cell as a single compartment. So it seemed natural to imagine that calcium diffusion was the key to the drastic differences in decay of transmitter release, synaptic facilitation, and arsenazo absorbance.

I formulated the following scenario: Calcium enters at the surface, and is highly concentrated at the mouths of calcium channels at the end of an action potential. It rapidly diffuses away from transmitter release sites toward the cell interior, terminating phasic release. A tail of residual calcium is present for some tens of milliseconds, and can add to calcium entering in a subsequent action potential to elicit facilitated transmitter release, without itself continuing to release much transmitter. Finally, the calcium that has diffused away from the membrane is still in cytoplasm, and although highly diluted it is still detected by the linear indicator arsenazo which is distributed uniformly throughout the interior of the terminal.

The gradual decline in arsenazo absorbance reflects the removal of calcium from the terminal.

I needed to simulate this scenario quantitatively to test its plausibility. To do so, Norman Stockbridge and I used a set of difference equation approximations to the differential equation for diffusion with binding in cylindrical coordinates. We included processes for removal of free calcium by uptake into organelles and a surface membrane extrusion pump. We chose parameters for extrusion, uptake, binding, and resting free calcium from values reported on experiments on the squid giant axon (see Zucker and Stockbridge, 1983, for discussion of parameters; the uptake rate is that for endoplasmic reticulum in synaptosomes, provided by Dr. Mordecai Blaustein). Calcium influx was assumed to occur over the entire terminal surface with a magnitude and time course similar to that measured under voltage clamp.

In Fig. 5, the results of such a computer simulation are compared with experimental data. The growth of arsenazo absorbance during a tetanus and its subsequent decay are well matched by the simulated change in total intracellular free calcium. This removal of calcium from cytoplasm, by uptake into organelles and active extrusion, gives no inkling of what is happening just underneath the membrane, however. There, calcium concentration is intense at the end of an action potential, and is rapidly diluted by diffusion.

The square of submembrane calcium concentration provides an approximate estimate of the time course of transmitter release, which will be delayed by subsequent presynaptic exocytotic events. The postsynaptic current should be even slower and more delayed, due to the time it takes for synaptic channels to open and close. Sure enough, the postsynaptic current decays somewhat more slowly than the predicted square of submembrane calcium, but both decline within a few milliseconds.

The model may be driven with two calcium influxes separated by a variable interval, and a prediction of synaptic facilitation is generated and compared with measurements at two synapses. The simulation matches pretty well the form and magnitude of synaptic facilitation. The wide variability (Charlton and Bittner, 1978a) of the kinetics of facilitation among individual specimens places the simulation within the reported range of facilitation kinetics. Particularly interesting are what appear to be two components of facilitation when it is plotted on semilogarithmic coordinates. Far from indicating two distinct phenomena, this is simply a consequence of the fact that radial diffusion is a second-order process, and so does not decay exponentially.

I have recently extended this diffusion model to simulations of long trains of action potentials (Zucker, 1984). The model predicts an accumulation of facilitation during a tetanus similar to that observed at neuromuscular junctions (Magleby and Zengel, 1982). A slowly increasing component of facilitation also appears, similar in magnitude and time course to a process called augmentation at neuromuscular junctions. The posttetanic decay of facilitated release proceeds with three apparent components: augmentation and the same two components of facilitation that follow a single spike. Measurements of tetanic and posttetanic facilitation are frustrated at squid giant synapses by an overbearing depression, but the close

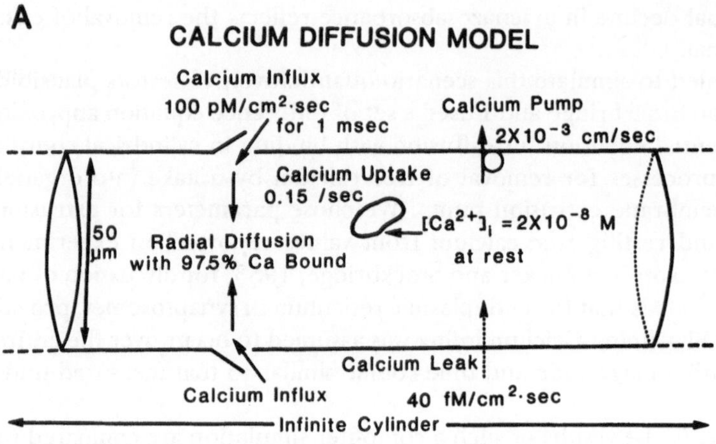

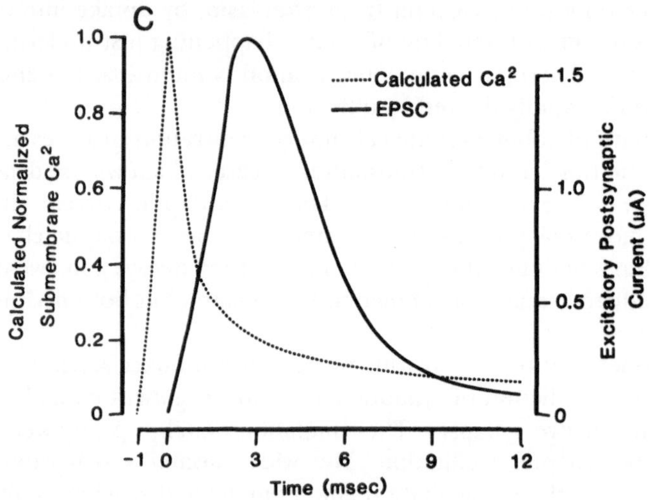

FIGURE 5. (A) Diagram of diffusion model used to simulate the behavior of the squid giant synapse. Simulations used the equations of Zucker and Stockbridge (1983), except that the pump and influx terms of equation 4 are divided by the ratio of bound to free calcium (correcting an error in that paper), and a term is added to equations 2–4 to remove calcium into organelles at a rate proportional to the difference between cytoplasmic calcium and the resting level. (B) The presynaptic neuron is stimulated at 33 Hz for 2 sec ending at time 0, and the change in absorbance in a terminal filled with arsenazo III

resemblance of these model properties to the behavior of frog neuromuscular junctions is certainly encouraging. Perhaps this kind of analysis of calcium movements is applicable to more than just the squid giant synapse.

7. Residual Calcium and MEPSP Frequency

Even in the absence of presynaptic spikes, transmitter is released spontaneously in the form of quantal packages of several thousand molecules, apparently

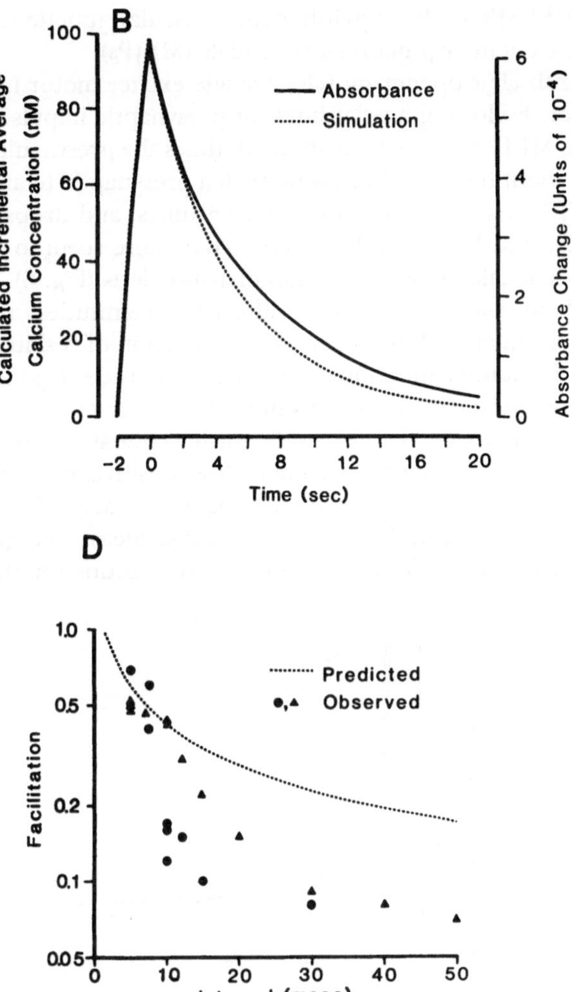

is compared with the computer-simulated change in average free presynaptic calcium concentration. (C) The excitatory postsynaptic current (EPSP) is plotted along with the computed presynaptic submembrane calcium concentration during an action potential ending at 0 msec. (D) The presynaptic neuron is stimulated twice at various intervals, and the facilitation of the second response, defined as the fractional increase of the second EPSP relative to the first, is plotted for two preparations and compared with the computer prediction. (Data from Charlton and Bittner, 1978a, Charlton *et al.*, 1982.)

due to the random exocytosis of synaptic vesicles. These quanta produce miniature EPSPs (MEPSPs) in the postsynaptic cell. If residual submembrane calcium is present to facilitate spike-evoked transmitter release, it ought also to cause some increase in the rate of spontaneous transmitter release or in the frequency of MEPSPs.

Miniature excitatory postsynaptic potentials are practically undetectable at the squid giant synapse, owing to the very large size of the postsynaptic cell and consequently the high postsynaptic input conductance. A more suitable preparation for observing MEPSPs is the crayfish neuromuscular junction, where they are called miniature excitatory junctional potentials (MEJPs).

At the crayfish claw opener muscle, a single exciter motor neuron innervates each muscle fiber. Following a brief burst of presynaptic impulses, there is a dramatic increase in MEJP frequency to about 40 times the prestimulation level, which drops to half in about 50 msec. Following such a presynaptic tetanus, spike-evoked facilitation of EJPs reaches a peak of about 15 times, and drops to half in about 150 msec (Zucker and Lara-Estrella, 1983). Thus there seem to be two effects of presynaptic residual calcium at neuromuscular junctions (Fig. 6).

Why do these two effects have different magnitudes and decay rates? Increased MEJP frequency should be a direct reflection of residual calcium, raised to some power. Excitatory junctional potential amplitude depends on the amplitude on the submembrane calcium concentration change at the end of a test action potential, as augmented by the relatively small residual calcium from previous activity. Therefore, EJP facilitation should be less sensitive to residual calcium, and so be smaller and decline more slowly, than changes in MEJP frequency.

Table I outlines a simple formulation of these ideas. The goal is to see if a unitary residual calcium model can quantitatively account for the differences in

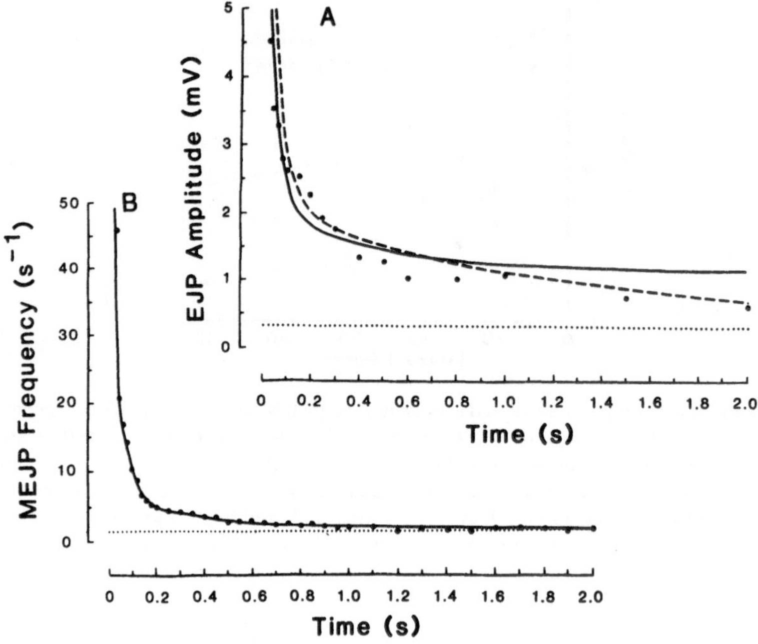

FIGURE 6. The residual calcium theory and posttetanic facilitation of spontaneous and evoked transmitter release at crayfish neuromuscular junctions. The dots show the time courses of increased EJP amplitudes evoked by test impulses (A) and of MEJP frequency (B) following a 5-sec tetanus at 20 Hz ending at time 0. The dotted lines indicate resting (unfacilitated) values. The double exponential function fit to the MEJP frequency (line in B) was used to predict the EJP frequency according to the equations of Table I, assuming resting MEJP frequency is Ca dependent (solid line in A) or Ca independent (dashed line in A).

TABLE I
Posttetanic Effects of Residual Calcium

Equations	*Definitions*
Resting MEJP frequency:	f_I = Ca-independent component of resting MEJP frequency
$f_o = K[Ca_S]^n + f_I$	K = constant relating MEJP frequency to presynaptic Ca
Posttetanic MEJP frequency:	Ca_S = resting presynaptic Ca
$f(t) = K[Ca_S + Ca_R(t)]^n + f_I$	$Ca_R(t)$ = posttetanic residual Ca at time t after tetanus
Single EJP amplitude:	Q = quantum amplitude
$v_o = QTK[Ca_S + Ca_E]^n + QTF_I$	T = period of evoked release
Posttetanic EJP amplitude:	Ca_E = Ca entering during a spike
$v(t) = QTK[Ca_S + Ca_E + Ca_R(t)]^n + QTF_I$	
Q, T, f_o, f(t), v_o, and v(t) are measurable	
Parameter estimation assuming that resting MEJP frequency is Ca dependent	*Parameter estimation assuming that resting MEJP frequency is Ca independent*
Assume $f_I = 0$	Assume $Ca_S = 0$
Let $Ca_S = 1$ (Defines unit of [Ca])	f_o gives f_I
f_o gives K	Let K have same value as at left (Defines unit of [Ca])
Choose n	Choose n
v_o gives Ca_E	v_o gives Ca_E
f(t) gives $Ca_R(t)$ and v(t)	f(t) gives $Ca_R(t)$ and v(t)

posttetanic changes in MEJP frequency and EJP facilitation. Equations are given that express resting MEJP frequency, single EJP amplitude, and posttetanic MEJP frequency and EJP amplitude in terms of resting presynaptic calcium at transmitter release sites, calcium entering during an action potential, and residual posttetanic calcium. Spike-evoked transmitter release is represented as a transient increase in the frequency of MEJPs of a fixed amplitude for a brief period. Both this amplitude and period can be estimated experimentally.

A choice has to be made regarding the resting MEJP frequency; it could be mainly dependent on resting presynaptic calcium activity, or largely calcium-independent. Another choice has to be made for the exponent relating transmitter release to presynaptic calcium. The dependence of release on extracellular calcium suggests a power of 4 or 5 (Dudel, 1981). Once these choices are made, the measured values of resting MEJP frequency, EJP amplitude, and posttetanic MEJP frequency changes can be used to deduce the posttetanic decay of residual calcium and predict the posttetanic decay of EJP facilitation. As Fig. 6 indicates, this formulation provides a satisfactory fit to the data, especially if resting MEJP frequency is largely independent of the resting presynaptic calcium level.

8. Summary and Conclusions

The residual calcium hypothesis was first proposed in 1965 by Katz and Miledi. Recent experimental evidence indicates that presynaptic residual calcium exists following activity, and that increased presynaptic calcium can facilitate synaptic transmission to spikes. Synaptic facilitation occurs without changes in action potentials, is unrelated to afterpotentials, and is not a property of calcium channels nor a consequence of calcium buffer saturation. A simple diffusion model of calcium movements in squid terminals accounts for the highly disparate time courses of transmitter release, facilitation, and arsenazo absorbance changes. The posttetanic increase in MEJP frequency at neuromuscular junctions can also be explained in terms of a residual calcium model and related quantitatively to posttetanic facilitation.

Not all forms of posttetanic enhancement of synaptic transmission are due to residual calcium remaining from what entered during the tetanus. The form of facilitation at neuromuscular junctions called potentiation, lasting for minutes following a very long tetanus, seems to be caused by accumulation of intracellular sodium in nerve terminals (Atwood *et al.*, 1975; Rahamimoff *et al.*, 1980). It can also be elicited at the squid giant synapse by presynaptic injection of sodium ions (Charlton and Atwood, 1977). Exactly how sodium acts to potentiate release, and what the relation is between facilitation and potentiation, remain challenges to our imagination and mysteries open to future exploration.

References

Atwood, H. L., Swenarchuk, L. E., and Gruenwald, C. R., 1975, Long-term synaptic facilitation during sodium accumulation in nerve terminals, *Brain Res.* **100:**198–204.

Bittner, G. D., and Baxter, D. A., 1983, Intracellular recordings from synaptic terminals during facilitation of transmitter release, *Soc. Neurosci. Abstr.* **9:**883.

Castellucci, V. F., and Kandel, E. R., 1974, A quantal analysis of the synaptic depression underlying habituation of the gill-withdrawal reflex in *Aplysia, Proc. Natl. Acad. Sci. U.S.A.* **71:**5004–5008.

Castellucci, V., Pinsker, H., Kupfermann, I., and Kandel, E. R., 1970, Neuronal mechanisms of habituation and dishabituation of the gill-withdrawal reflex in Aplysia, *Science* **167:**1745–1748.

Charlton, M. P., and Atwood, H. L., 1977, Modulation of transmitter release by intracellular sodium in squid giant synapse, *Brain Res.* **134:**367–371.

Charlton, M. P., and Bittner, G. D., 1978a, Facilitation of transmitter release at squid synapses, *J. Gen. Physiol.* **72:**471–486.

Charlton, M. P., and Bittner, G. D., 1978b, Presynaptic potentials and facilitation of transmitter release in the squid giant synapse, *J. Gen. Physiol.* **72:**487–511.

Charlton, M. P., Smith, S. J., and Zucker, R. S., 1982, Role of presynaptic calcium ions and channels in synaptic facilitation and depression at the squid giant synapse, *J. Physiol. (London)* **323:**173–193.

Dudel, J., 1981, The effect of reduced calcium on quantal unit current and release at the crayfish neuromuscular junction, *Pflug. Arch.* **391:**35–40.

Katz, B., and Miledi, R., 1965, The effect of calcium on acetylcholine release from motor nerve terminals, *Proc. R. Soc. London Ser. B* **161:**496–503.

Katz, B., and Miledi, R., 1968, The role of calcium in neuromuscular facilitation, *J. Physiol. (London)* **195:**481–492.

Liley, A. W., and North, K. A. K., 1953, An electrical investigation of effects of repetitive stimulation on mammalian neuromuscular junction, *J. Neurophysiol.* **16:**509–527.

Llinás, R., Sugimori, M., and Simon, S. M., 1982, Transmission by presynaptic spike-like depolarization in the squid giant synapse, *Proc. Natl. Acad. Sci. U.S.A.* **79:**2415–2419.

Magleby, K. L., and Zengel, J. E., 1982, A quantitative description of stimulation-induced changes in transmitter release at the frog neuromuscular junction, *J. Gen. Physiol.* **80:**613–638.

Rahamimoff, R., Lev-Tov, A., and Meiri, H., 1980, Primary and secondary regulation of quantal transmitter release: Calcium and sodium, *J. Exp. Biol.* **89:**5–18.

Smith, S. J., and Zucker, R. S., 1980, Aequorin response facilitation and intracellular calcium accumulation in molluscan neurones, *J. Physiol. (London)* **300:**167–196.

Zucker, R. S., 1972, Crayfish escape behavior and central synapses. II. Physiological mechanisms underlying behavioral habituation, *J. Neurophysiol.* **35:**621–637.

Zucker, R. S., 1973, Changes in the statistics of transmitter release during facilitation, *J. Physiol. (London)* **229:**787–810.

Zucker, R. S., 1974a, Crayfish neuromuscular facilitation activated by constant presynaptic action potentials and depolarizing pulses, *J. Physiol. (London)* **241:**69–89.

Zucker, R. S., 1974b, Excitability changes in crayfish motor neurone terminals, *J. Physiol. (London)* **241:**111–126.

Zucker, R. S., 1982, Processes underlying one form of synaptic plasticity: Facilitation, in: *Conditioning: Representation of Involved Neural Functions* (C. D. Woody, ed.), Plenum Press, New York, pp. 249–264.

Zucker, R. S., 1984, A calcium diffusion model predicts facilitation, but not the time course of transmitter release, during tetanic stimulation, *Biophys. J.* **45:**264a.

Zucker, R. S., and Bruner, J., 1977, Long-lasting depression and the depletion hypothesis at crayfish neuromuscular junctions, *J. Comp. Physio.* **121:**223–240.

Zucker, R. S., and Lara-Estrella, L. O., 1979, Is synaptic facilitation caused by presynaptic spike broadening? *Nature (London)* **278:**57–59.

Zucker, R. S., and Lara-Estrella, L. O., 1983, Post-tetanic decay of evoked and spontaneous transmitter release and a residual-calcium model of synaptic facilitation at crayfish neuromuscular junctions, *J. Gen. Physiol.* **81:**355–372.

Zucker, R. S., and Stockbridge, N., 1983, Presynaptic calcium diffusion and the time courses of transmitter release and synaptic facilitation at the squid giant synapse, *J. Neurosci.* **3:**1263–1269.

PART VI
NEUROGENETICS AND MOLECULAR NEUROBIOLOGY

25

A Monoclonal Antibody to an Internal Molecule Differentiates Three Electrically Coupled Leech Neurons

BIRGIT ZIPSER

1. Introduction

1.1. Antibody Categories

About half a decade ago, neurobiologists began to search for molecules involved in neuron-to-neuron recognition using monoclonal antibodies. The advent of monoclonal antibodies made it possible to immunize mice with the complex mixture of antigens present in whole brain tissue and to make specific antibody probes to single molecules. Generating antibodies has been relatively easy. The challenge has been to identify those antibodies that bind to candidate molecules playing a role in specifying synaptic connectivity. The trend has been to judge the likelihood of an antigen specifying neuronal connectivity by its distribution in organized brain tissue. This strategy is based on the assumption that neuronal connectivity molecules differ in kind or quantity from cell to cell.

Several different laboratories have been making monoclonal antibodies to whole brains or tissue enriched for molecules of interest and have been selecting for antibodies that give rise to promising patterns (Barnstable, 1980; Trisler *et al.*,

BIRGIT ZIPSER • Cold Spring Harbor Laboratory, Cold Spring Harbor, New York 11724.

1981; Vulliamy *et al.*, 1981; Kushner, 1984; Barald, 1982; Wood *et al.*, 1982; Sternberger *et al.*, 1982; Fujita *et al.*, 1982; McKay and Hockfield, 1982). The patterns that have emerged are very interesting. For example, an antibody has been made that stains chick retina according to a topographical gradient (Trisler *et al.*, 1981). Another antibody that was generated stains regions of the rat brain that belong to the limbic system (Levitt, 1982).

My laboratory has been making antibodies against the leech because in the simple leech nervous system, all neurons and their synaptic connections are identifiable, making it a useful system to analyze the molecular basis of neuronal connectivity. So far, my laboratory has screened about 2000 hybridomas that were generated from spleen cells of mice immunized with leech brain tissue and myeloma cells, each secreting a single antibody. We have saved 140 cell lines that produce monoclonal antibodies giving rise to interesting patterns. We are now analyzing and categorizing these patterns.

We recognize three broad categories of staining patterns that indicate the molecular heterogeneity of the leech nervous system. One class of antibody that we are generating binds to 130-Kd glycoproteins on Western blots (Hogg *et al.*, 1983; Flaster *et al.*, 1983). In CNS tissue, some of these antibodies are specific probes for select cell bodies and axons that run in bundles (Hockfield and McKay, 1983); other antibodies bind to 130-Kd antigens that are shared by select axons and glial cells. Possibly, these 130-Kd surface antigens are nCam-like molecules (Edelmann, 1983) subserving specific adhesion between axons or between axons and glial cells, with implications for such phenomena as axonal guidance.

The second kind of antibody we are generating also binds specific sets of neurons. None of these antibodies bind to proteins on Western blots which suggests that they may react with either low-molecular-weight peptides or with other types of molecules such as glycolipids. What has been very exciting is that these antibodies define combinations of markers for several identified neurons. For example, the primary mechanosensory neurons responding to pressure are stained by four different antibodies that bind to different sized neuronal sets (Zipser and McKay, 1981; Zipser, 1982; Loer *et al.*, 1983), thereby specifying an assortment of four specific markers. Does this assortment of markers specify the character of the pressure cells for its synaptic connectivities?

The third kind of antibodies that we are generating bind to all neurons. What is interesting about these antibodies is that they do not necessarily stain all neurons with the same intensity. It is this new type of antibody that will be described here in detail, particularly one that differentiates a trio of synaptically linked neurons from their neighbors which are less strongly stained. Scrutinizing different neurons in the 400-neuron ganglion makes it apparent that different cell bodies are stained according to different intermediate intensities. The differently stained cell bodies are not laid out according to a topographic gradient. The three neurons that are described here as being most strongly stained have their cell bodies at very different ganglionic locations. These three strongly stained neurons are synaptically linked. If other neurons stained according to different intensities fall into appropriate synaptic relationships, this antibody may define a connectivity gradient.

1.2. Lan3-8, a Monoclonal Antibody That Differentiates a Trio of Electrically Coupled Neurons

Lan3-8 is a monoclonal antibody that stains all cell bodies in the 400-neuron leech ganglion, but in the fiber tracts it only stains a subset of the axons, among them the giant S cell axon. The hybridoma line secreting this antibody was generated during the third fusion at a time when we mostly kept only those monoclonal antibodies that were specific for single types or small sets of neurons (Zipser and McKay, 1981). During the initial screening, we decided to save Lan3-8 because it prominently stained the giant axon of the S cells. It seemed useful to have a histological marker for the S cell axon whose synaptic connection and regeneration had been studied in considerable detail (Elliot and Muller, 1982). Later it became apparent, using diluted antibody, that Lan3-8 did not label cell bodies homogeneously either. Different neuronal cell bodies in the ganglion stain with different intensities. It was intriguing that other neurons that are electrically coupled to the S cells were singled out together with the S cells.

A typical 400-neuron midbody ganglion stained with undiluted Lan3-8 supernatant is illustrated in Fig. 1. From this figure the features that make the leech ganglion into an attractive preparation for electrophysiological studies are readily apparent: the ganglion is bilaterally symmetrical and many of the neuronal cell bodies can be identified by their size and ganglionic position. The large antibody-stained neurons in the center of the ganglion are the Retzius cells, and on the right and left of it are the large mechanosensory neurons responding to touch, pressure, and noxious stimulation applied to leech skin. The annulus erector motor neurons at the posterior aspect of the ganglion are also readily apparent. Similarly, the large L motor neurons can be seen on the dorsal aspect of the ganglion close to the posterior roots.

Concentrated Lan3-8 supernatant is a general cell body stain for ganglionic neurons (Fig. 1A and B); however, in the fiber tracts it already discriminates between different axons (Fig. 1B). In the nerve roots, to the right and left of the ganglion, only a subset of axons are stained. Similarly, only a subset of axons are stained in the connective, the fiber tract between adjacent ganglia. Among them is the giant axon of the S cell, which runs medially in the smaller Faivre's nerve of the connective. The axon originates from a small ventral cell body and then can be seen (Fig. 1B) in the center of the ganglion where it bifurcates, projecting one branch into the anterior and another branch into the posterior connective. Other unidentified smaller axons in the thick right and left fiber bundles of the connective are also stained. However, Lan3-8 is not an indiscriminate axonal stain since the connective is not solidly stained.

When the antibody is used in lower dilutions, it also begins to discriminate between different cell bodies. The antibody differentiates the cell bodies of three neurons that are linked to each other through electrical synapses. They are the pair of "coupling interneurons" (Muller and Scott, 1981) and the S cell. The pair of antibody-stained coupling interneurons are readily apparent in the photograph in Fig. 2. The small cell bodies of the coupling interneurons next to the posterior

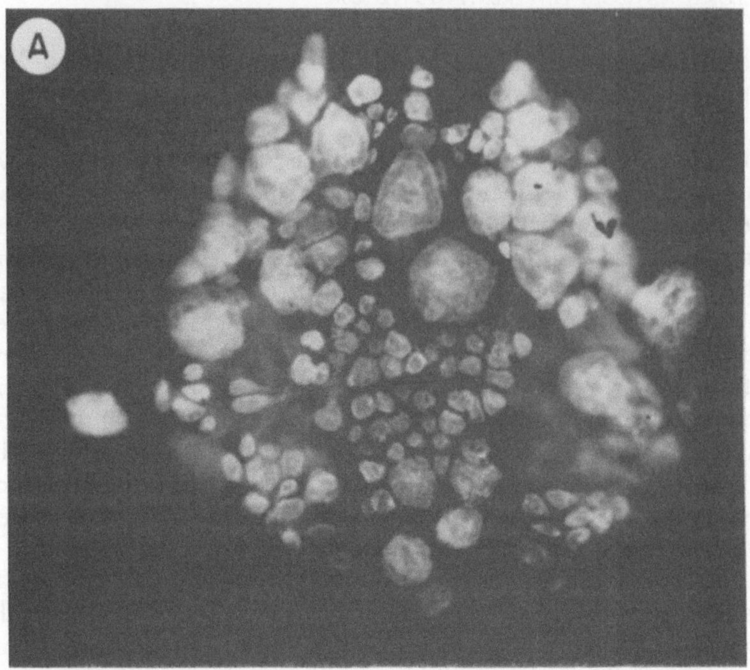

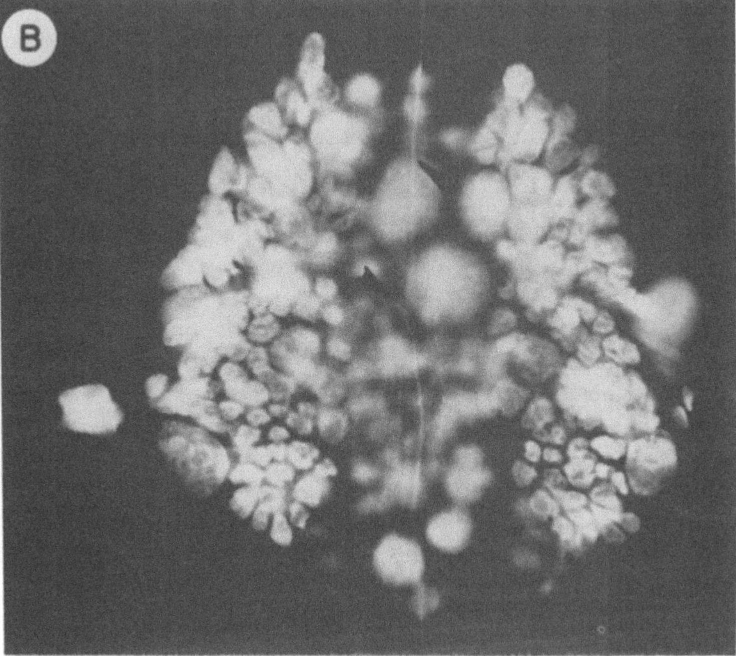

FIGURE 1. Monoclonal (mAb) antibody Lan3–8 labels the 400-neuron leech ganglion. A desheathed leech ganglion, fixed with 4% paraformaldehyde in 0.1 M phosphate buffer, pH 7.4, was incubated first with mAB Lan3–8 and then with a rhodamine-conjugated second antibody. (A, B) Show antibody-stained neurons on the ventral and dorsal aspects of the ganglion. The prominently stained S-cell axon (B, arrows) bifurcates in the center of the ganglion and runs up into the anterior connective and down into the posterior connective.

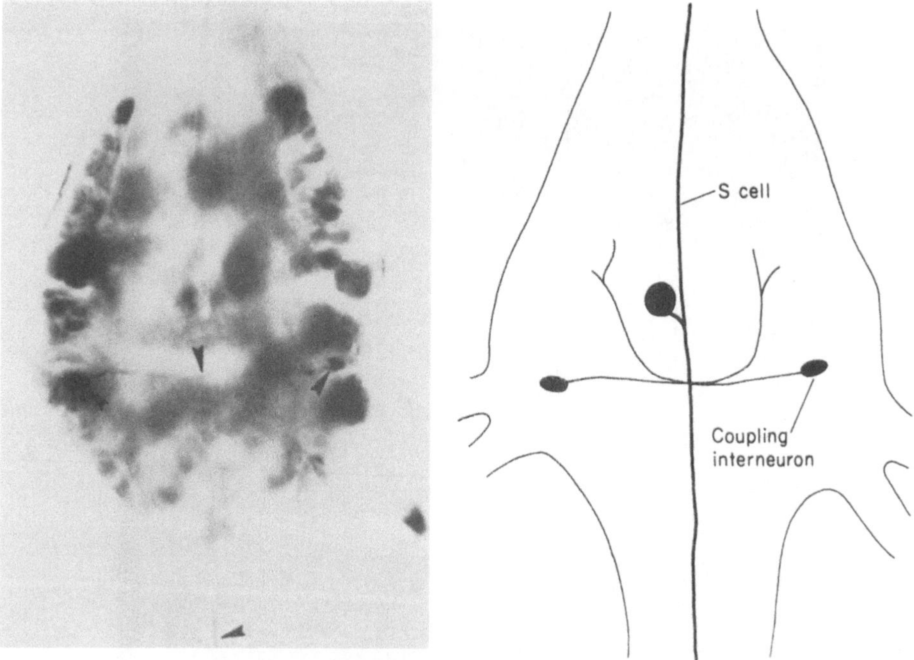

FIGURE 2. Lan3–8 differentiates the coupling interneurons and S-cell axon. (A) Cell bodies and axons of the coupling interneurons (arrows in the middle of the ganglion) are stained with Lan3–8 and a horseradish peroxidase-conjugated second antibody. The stained S-cell axon comes into focus below the ganglion in the posterior connective (arrow). (B) A diagrammatic representation of the S cell and the coupling interneurons. (Adapted from Muller and Scott, 1981.)

roots are differentiated from surrounding, less strongly stained cell bodies. Their axons are also differentiated from other axons and can be seen to project toward the central neuropil where they make connections with the S-cell axon. The prominently stained S-cell axon comes into focus where the anterior and posterior connective fiber tracts leave the ganglion.

The identities of the small pair of coupling interneurons and its synaptically linked S-cell axons were established in a double-labeling experiment. First, the S cell was injected with Lucifer Yellow. The coupling neurons did not have to be injected separately since Lucifer Yellow injected into the S cell readily spreads into them. After the dye injection, the ganglion was fixed in 4% paraformaldehyde, rinsed and permeabilized. It was then incubated first with the monoclonal antibody, followed by rhodamine-conjugated second antibody. The cell body of the coupling interneuron, stained with Lucifer Yellow in Fig. 3A1, is also labeled with Lan3-8 which is visualized by the rhodamine-conjugated second antibody (Fig. 3A2). Other cell bodies surrounding the coupling interneuron are only weakly stained. Nevertheless, the combined glow of the weakly stained axons and cell bodies in the 250-μm-thick whole mount create a substantial background against which the coupling interneuron barely stands out. Similarly, the S-cell axon, identified in the connective through its Lucifer Yellow staining (Fig. 3B1), is differentiated from other axons through antibody staining (Fig. 3B2).

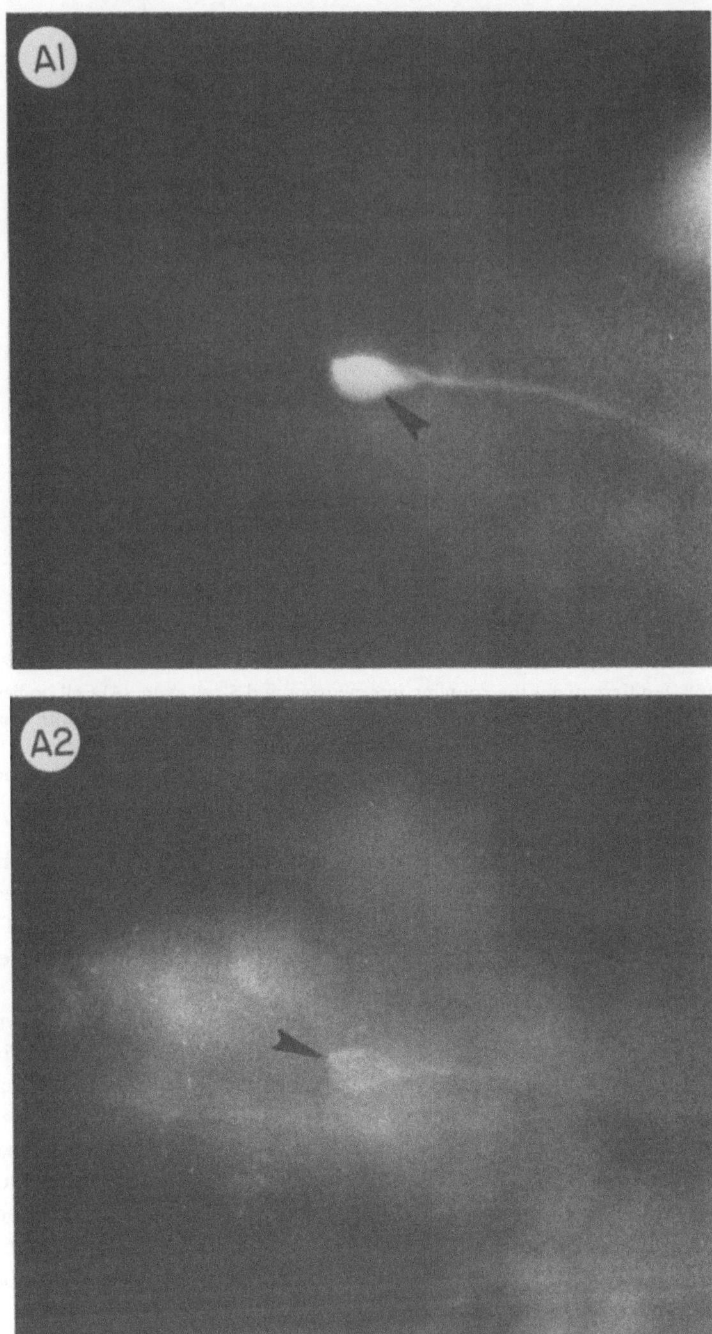

FIGURE 3. Antibody-stained S cell and coupling interneurons are double-labeled with Lucifer Yellow. Lucifer Yellow injected into the S cell spreads into the coupling interneurons (A1) and diffused down the S-cell axon into the connective (B1). The Lucifer-Yellow-injected tissue was fixed and stained with

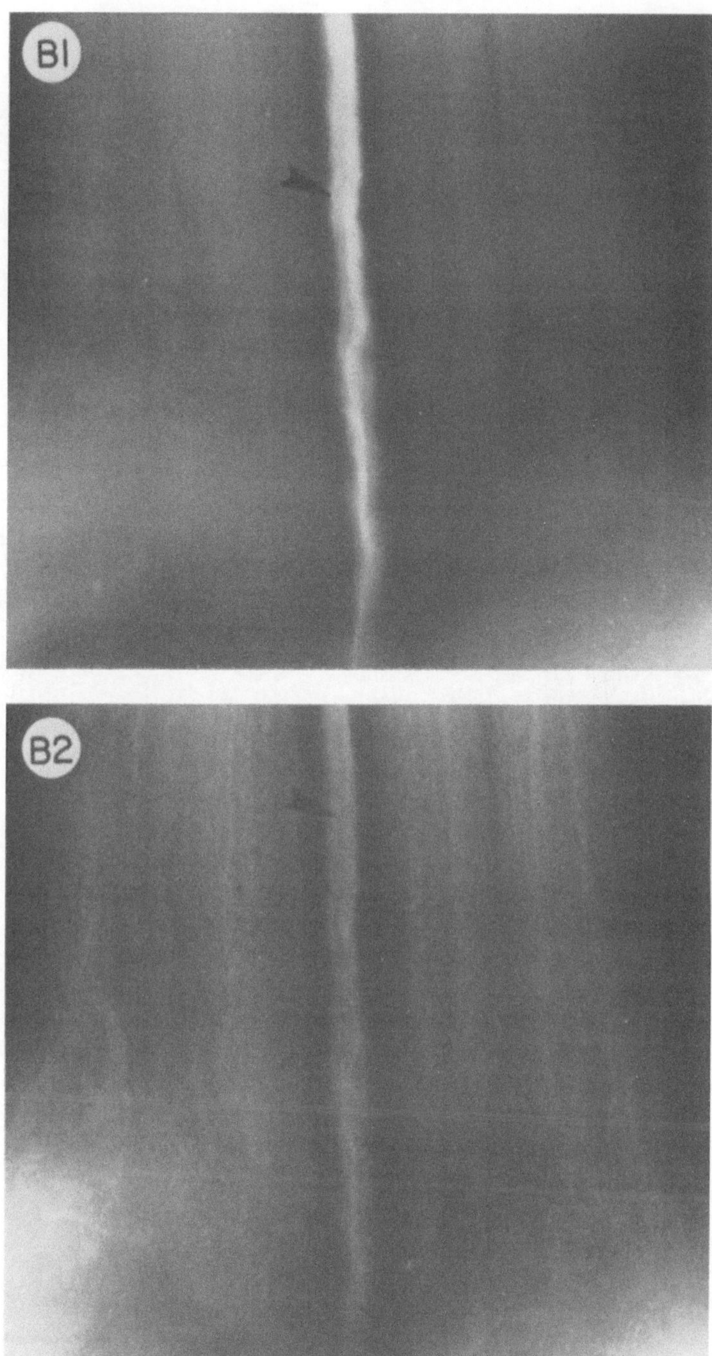

Lan3–8 and a rhodamine-conjugated second antibody. The coupling interneurons (A2, arrow) and S-cell axon (B2, arrow) are differentiated from surrounding cell bodies and axons.

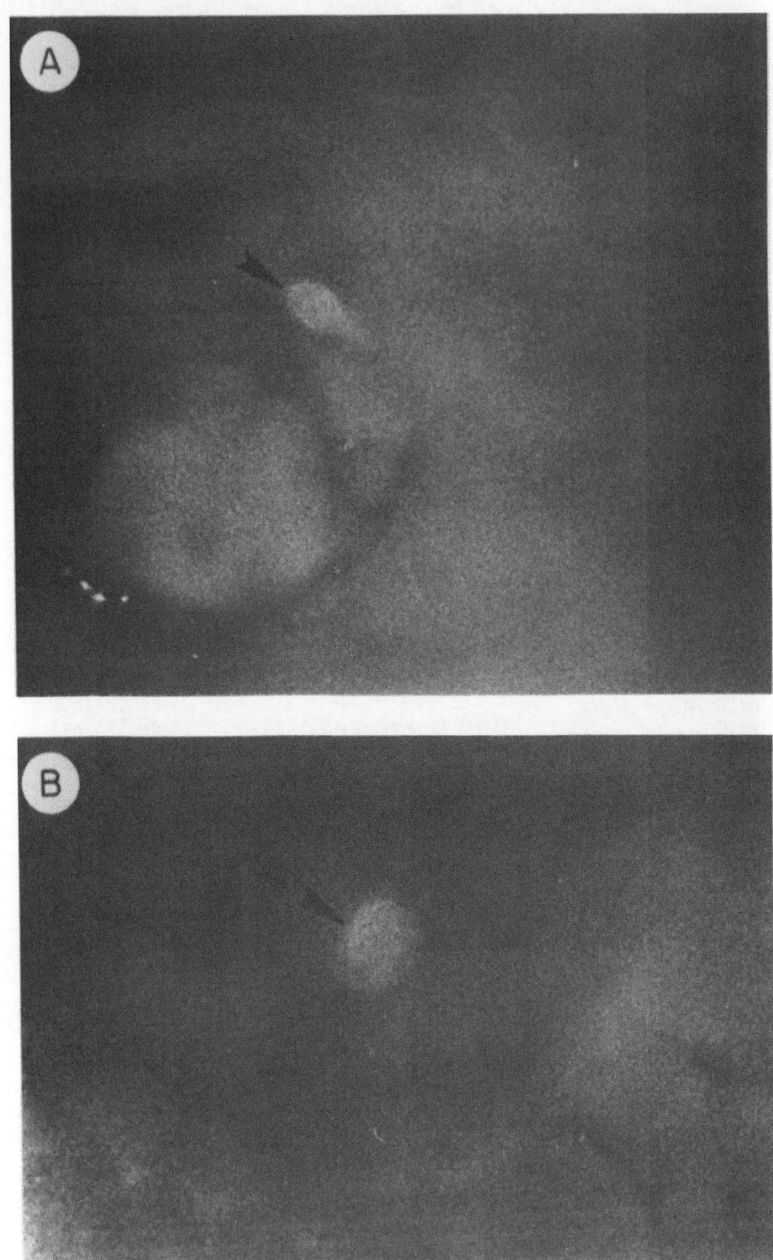

FIGURE 4. Another example of Lan3–8 staining the synaptically linked trio of neurons. Lan3–8 stains

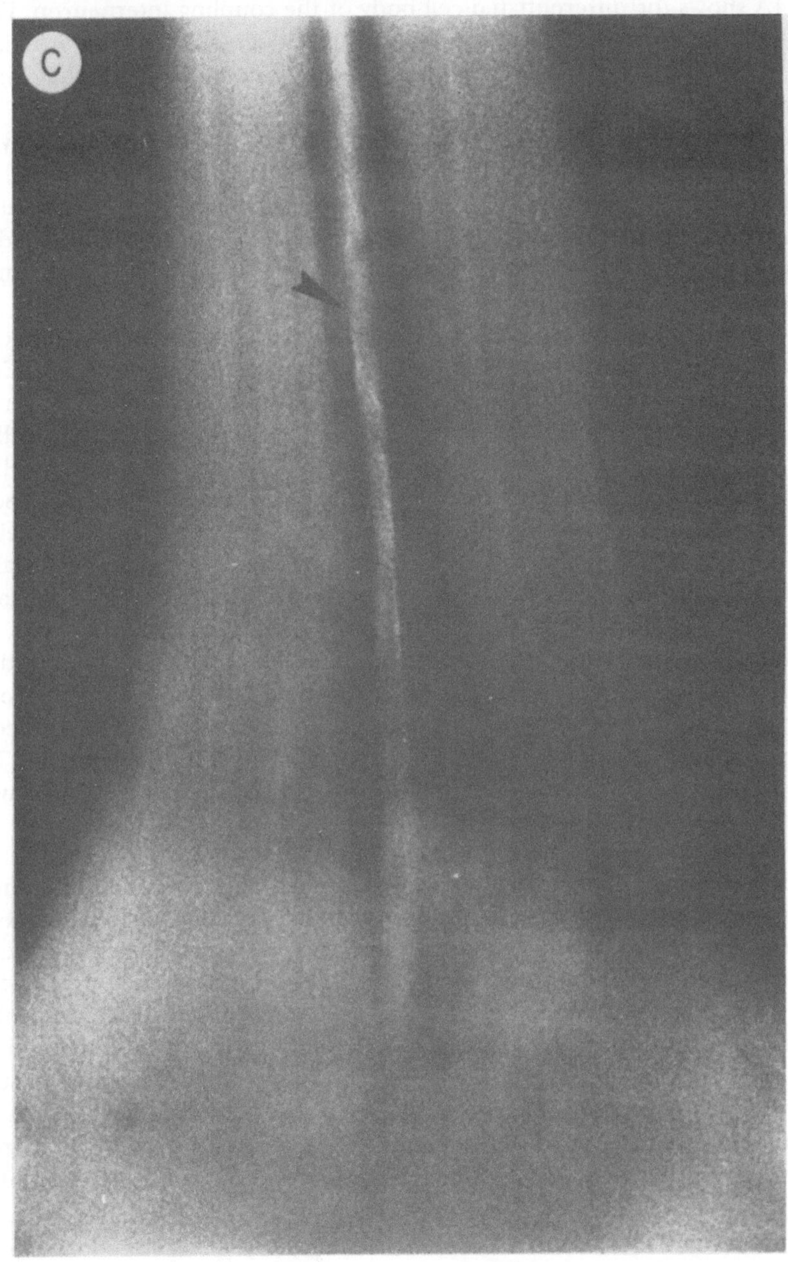

the cell body of the coupling interneuron (A), the cell body of the S cell (B), and the axon of the S cell (C).

Figure 4 shows another example of the extent to which Lan3-8 differentiates the S cell and its coupling interneurons from other neurons that surround them. Figure 4A shows the differentiated cell body of the coupling interneuron, Fig. 4B shows the differentiated cell body of the S cell, and Fig. 4C shows the prominently stained S-cell axon in the connective. Thus, a monoclonal antibody that binds to all neurons can be made to highlight a trio of synaptically connected cells.

Another variable controlling the intensity of staining, besides antibody titer, is method of fixation. Bouin's fixation impairs antibody staining. By contrast, fixing a ganglion with 3.7% formalin in leech ringer potentiates the intensity of antibody staining to the extent that all neuronal cell bodies are now equally stained. The gradient in neuronal cell-body staining shown here exists for 4% paraformaldehyde-fixed ganglia. What is the molecular basis of the gradient of staining?

The biochemical nature of Lan3-8 reactive antigen has been studied with immunoblots prepared from sodium dodecyl sulfate (SDS) acrylamide gels. Central nervous system extracts prepared by boiling nerve cord in SDS sample buffer (Laemmli, 1970) were run on SDS acrylamide gels and electroblotted on nitrocellulose paper (Towbin *et al.*, 1979). Lan3-8 stains a single 68-K band on the nitrocellulose replica of the SDS acrylamide gel (Hogg *et al.*, 1983). It does not bind to several different lectins and so it is unlikely that the Lan3-8 antigen is a glycoprotein (M. Flaster, unpublished observation). It is not detectable immunocytochemically on the surface of neurons. Neuronal cell bodies from live ganglia or fixed ganglia were incubated with the antibody without detergent. In the absence of detergent facilitating antibody penetration across cell membrane, Lan3-8 did not bind to the tissue. Other antibodies such as Lan3-2 known to bind to surface antigens do stain neuronal cell bodies in the absence of detergent. Other experiments showed that the Lan3-8 antigen is not extractable from live neurons with the detergent Triton X-100. The Lan3-8 antigen may be one of the cytoskeletal proteins or may be a component of the cytoskeleton.

There are several possible explanations for the differences observed in Lan3-8 neuronal staining. The antigen could occur in different neurons in different concentrations. Alternatively, the concentration of the Lan3-8 antigen may be the same in different neurons, but it may be modified posttranslationally. An example of a posttranslation modification that affects antibody binding is reported by Sternberger for rat neurofilaments (Sternberger and Sternberger, 1983). Phosphorylation of neurofilaments prevents access to their amino acid antigenic determinants. Possibly, the antigens in synaptically linked neurons and the more weakly stained surrounding neurons may differ in their posttranslational modification. A third explanation is that leech neurons contain a family of 68-K proteins that are distributed along different ratios in different neurons and the Lan3-8 has different binding affinities for individual proteins.

1.3. Significance of Lan3-8 Differential Staining

We have generated a monoclonal antibody that is reactive with an internal, possibly cytoskeletal protein and stains different neurons according to different intensities. Three synaptically linked neurons are more strongly stained than their immediately surrounding neighbors to which they are not connected. The fact that synaptically connected neurons are more strongly stained suggests that the different staining properties are related to function, maybe the neuron's degree of activity or even connectivity. The three neurons that are differentiated through Lan3-8 are present in all standard midbody ganglia. One of them is the S cell, an unpaired interneuron. It sends one axon into the anterior and one axon into the posterior connective. The axons of two S cells from adjacent ganglia meet halfway in the connective between the two ganglia and make an electrical synapse (Muller, 1979). The fused S-cell axons constitute a fast, thorough conducting septate system along the nerve cord. The other two neurons differentiated by Lan3-8 are the pair of coupling interneurons. The coupling interneurons also repeat in all midbody ganglia. One of their functions is to transmit first-order mechanosensory input directly from the touch cells to the S cells (Muller and Scott, 1981). The function of the S cells has remained elusive. Their giant fused axons do not appear to be involved in a simple escape reflex, coordinating electrical activity along the nerve cord. Perhaps the S cells are involved in mediating a higher level survival mechanism such as "arousal."

The S cells and their presynaptic coupling interneurons are differentiated from other neurons by an antibody that binds to an internal marker. It would be unorthodox to postulate that the Lan3-8 antigen was directly involved in specifying connectivity between the S cells and their coupling interneurons since connectivity factors are generally postulated to be either surface molecules or molecules that can be secreted. Nevertheless, it is possible that internal proteins can influence the extent to which two cells are coupled to each other through electrical synapses.

ACKNOWLEDGMENTS. I thank P. Kushner, M. Flaster, and J. Cuddihy for their comments on the manuscript, and R. Schwarz, D. Balke, and D. Greene for preparation of the manuscript. The author's work, summarized in this chapter, was supported by NSF grant BNS 78-24672, a Whitehall grant, and NIH grants NS 17984, NS20057.

References

Barald, K., 1982, Monoclonal antibodies to embryonic neurons, in: *Neuronal Development* (N. C. Spitzer, ed.), Plenum Press, New York, pp. 101–119.

Barnstable, C. J., 1980, Monoclonal antibodies which recognize different cell types in the rat retina, *Nature (London)* **286:**231–235.

Edelman, G. M., 1983, Cell adhesion molecules, *Science* **219:**450–457.

Elliot, E. J., and Muller, K. J., 1982, Synapses between neurons regenerate accurately after destruction of ensheathing glial cells in the leech, *Science* **215:**1260–1262.

Flaster, M. S., and Zipser, B., 1983, The macroglial cells of the leech are molecularly heterogeneous, *Annu. Meet. Soc. Neurosci.* **343.12:**1183.

Flaster, M., Schley, C., and Zipser, B., 1983, Generating monoclonal antibodies against excised gel bands to correlate immunocytochemical and biochemical data, *Brain Res.* **277:**196–199.

Fujita, S. C., Zipurski, S. L., Benzer, S., Ferrus, A., and Shotwell, S. L., 1982, Monoclonal antibodies against the *Drosophila* nervous system, *Proc. Natl. Acad. Sci. U.S.A.* **79:**7929–7933.

Hockfield, S., and McKay, R., 1983, Monoclonal antibodies demonstrate the organization of axons in the leech, *J. Neurosci.* **3**:369–375.

Hogg, N., Flaster, M., and Zipser, B., 1983, Cross-reactivities of monoclonal antibodies between select leech neuronal and epithelial tissues, *J. Neurosci. Res.* **9**:445–457.

Kushner, P., 1984, A library of monoclonal antibodies to torpedo cholinergic synaptosomes, *J. Neurochem.* **43**:775–786.

Laemmli, U. K., 1970, Cleavage of structural protein during the assembly of the head of bacteriophage T4, *Nature (London)* **227**:680–685.

Levitt, P., 1983, A monoclonal antibody to limbic system neurons, *Science* **223**:299–300.

Loer, C., Schley, C., Zipser, B., and Kristan, W., 1983, A monoclonal antibody specific to the leech pressure-sensitive mechanosensory neurons, *Annu. Meet. Soc. Neurosci.* **179.8**:605.

McKay, R. D., and Hockfield, S. J., 1982, Monoclonal antibodies distinguish antigenically discrete neuronal types in the vertebrate central nervous system, *Proc. Natl. Acad. Sci. U.S.A.* **79**:6747–6752.

Muller, K. J., 1979, Synapses between neurones in the central nervous system of the leech, *Biol. Rev.* **54**:99–134.

Muller, K. J., and Scott, S. A., 1981, Transmission at a "direct" electrical connexion mediated by an inteneurone in the leech, *J. Physiol.* **311**:565–583.

Sternberger, L. A., and Sternberger, N. H., 1983, Monoclonal antibodies distinguish phosphorylated and nonphosphorylated forms of neurofilament in situ, *Proc. Natl. Acad. Sci. U.S.A.* **80**:6126–6130.

Sternberger, L. A., Harwell, L. W., and Sternberger, N. H., 1982, Neurotypy: Regional individuality in rat brain detected by immunocytochemistry with monoclonal antibodies, *Proc. Natl. Acad. Sci. U.S.A.* **79**:1326–1330.

Towbin, H., Staehelin, T., and Gordon, J., 1979, Electrophoretic transfer of proteins from polyacrylamide gels to nitrocellulose sheets: Procedure and some applications, *Proc. Natl. Acad. Sci. U.S.A.* **76**:4350–4354.

Trisler, G. D., Schneider, M. D., and Nirenberg, M., 1981, A topographic gradient of molecules in retina can be used to identify neuron position, *Proc. Natl. Acad. Sci. U.S.A.* **78**:2145–2149.

Vulliamy, T., Rattray, S., and Mirsky, R., 1981, Cell-surface antigen distinguishes sensory and autonomic peripheral neurones from central neurones, *Nature (London)* **291**:418–420.

Wood, J. N., Hudson, L., Jessell, T. M., and Yamamoto, M., 1982, A monoclonal antibody defining antigenic determinants on subpopulations of mammalian neurones and trypanosoma cruzi parasites, *Nature (London)* **296**:34–38.

Zipser, B., 1982, Complete distribution patterns of neurons with characteristic antigens in the leech central nervous system, *J. Neurosci.* **2**:1453–1464.

Zipser, B., and McKay, R., 1981, Monoclonal antibodies distinguish identifiable neurons in the leech, *Nature (London)* **289**:549–554.

Zipser, B., Stewart, R., Flanagan, T., Flaster, M., and Macagno, E., 1983, Do monoclonal antibodies stain sets of functionally related leech neurons? *Cold Spring Harbor Symp. Quant. Biol.* **48**:551–556.

26

Neuropeptide Gene Expression and Behavior in *Aplysia*

RICHARD H. SCHELLER AND MARK SCHAEFER

1. Neuropeptides Control Stereotyped Behavior in *Aplysia*

The relatively simple nervous system of the marine mollusk *Aplysia californica* is especially suitable for examining the molecular and cellular basis of fundamental neural processes common to both invertebrates and vertebrates. The *Aplysia* nervous system is amenable to this kind of study because it is composed of only some 20,000 central neurons organized into four easily accessible pairs of symmetrical ganglia and a single asymmetric abdominal ganglion. Extensive physiological studies by previous investigators have helped elucidate how individual neurons interact to achieve the level of sophistication required to coordinate the physiological, behavioral, and learning processes of this organism (reviewed by Kandel, 1979). One common theme emerging from studies of invertebrate nervous systems is the wide diversity of neurons and the precise interconnections they make to provide the neural circuitry necessary to control even relatively simple behavioral components. This diversity is thought to be due to a large extent to differential expression of the structural genes of the nervous system. Indeed, an important question is how only an estimated 10,000 genes, in the case of *Aplysia,* can provide the necessary genetic information to give rise to some 20,000 neurons as well as to the numerous cells composing the other tissues of the organism. In recent years, advances in the field of molecular biology have allowed neurobiologists to approach questions such as this by examining the genetic foundation of neural systems. Recombinant DNA techniques allow one to identify specific genes expressed in nerve cells and to obtain large enough quantities of the proteins they

RICHARD H. SCHELLER AND MARK SCHAEFER • Department of Biological Sciences, Stanford University, Stanford, California 94305.

encode to examine the structure and function of the nervous system in exquisite detail.

Our laboratory is using a variety of molecular, cellular, and physiological techniques to determine how neuropeptides control the physiology and behavior of an organism (Scheller *et al.*, 1984). We are focusing these studies on the *Aplysia* nervous system in order to investigate questions that are difficult or impossible to study in higher organisms. We are particularly interested in the neuroendocrine control of *Aplysia* egg laying, a stereotyped reproductive behavior. Egg laying in this organism consists of a complex behavioral repertoire that includes cessation of walking and inhibition of feeding, followed by head waving, and finally egg expulsion. These behavioral events apparently coincide with a variety of physiological responses including, for example, changes in respiratory function, blood pressure, and heart rate. The individual behavioral and physiological components of egg laying are mediated by several neuropeptides released from a number of neurosecretory cells throughout the *Aplysia* nervous system. The largest group of neurosecretory cells, the bag cells, lie in association with the abdominal ganglion and are organized into two symmetrical and homogenous clusters of about 400 neurons each (Fig. 1). The bag cells are electrically silent except for periodic bursts of activity when they fire in all-or-none fashion for 20–30 min and secrete a number of peptides. These bag-cell peptides are thought to act both as conventional transmitters and hormones to control certain physiological and behavioral components of egg laying. In addition, it is likely that peptides from other cells mediate some of the egg-laying behavioral components. The function and physiological properties of these cells and others expressing the same or similar genes to those expressed in the bag cells are poorly understood. The primary goal of the research in our laboratory is to identify the genes encoding neuropeptides of the *Aplysia* nervous system and to determine how gene expression modulates neural function to control egg laying as well as other physiological and behavioral events.

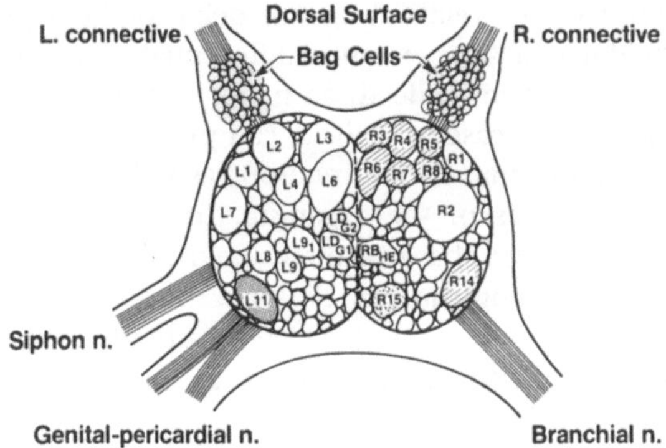

FIGURE 1. Schematic representation of the dorsal surface of the abdominal ganglion. Representative cells are labeled L or R (designating left or right hemiganglion) along with an identifying number. Peptidergic neurons are indicated with various shadings. The bag cells are grouped in two large clusters of about 400 neurons each on the rostral portion of the ganglion. Cells R3-8 and R14 (▨) each send a single axon down the branchial nerve which terminates on the efferent vein of the gill at the base of the heart. R15 (⁘) is a neurosecretory cell thought to be involved in controlling salt and water balance of the animal. L11 (▩) is a cholinergic cell which also uses one or more neuropeptides in intercellular communication.

2. Cloning the Genes Encoding the *Aplysia* Neuropeptides

A large portion of the proteins synthesized in neurosecretory cells are neuropeptides. For example, during the *Aplysia* reproductive period, nearly half the protein synthesized in the bag cells may consist of a single neuropeptide precursor (Arch, 1972). The synthesis of relatively large quantities of one or a few neuronal proteins normally corresponds to the presence of large quantities of a particular mRNA species.

The strategy we are employing to identify genes encoding neuropeptides takes advantage of the abundance of mRNA devoted to their synthesis. Individual cells thought to synthesize a particular peptide are dissected and their RNA is extracted. In the presence of a radiolabeled precursor (dCTP), copy DNA (cDNA) is synthesized from the RNA using the enzyme reverse transcriptase. A cDNA library (contained within a bacteriophage vector) that was previously made from RNA extracted from the entire abdominal ganglion (a cDNA library) or from sperm DNA (a genomic library) is then plated onto Petri dishes. Nitrocellulose filters that bind DNA are briefly placed on the surface of the dishes and removed. Next the DNA is denatured and fixed to the filter by baking. Since filters lifted from the same plate are all exact duplicates, the DNA on individual filters can be hybridized with multiple radiolabeled probes and compared. Typically, the DNA on each of two duplicate filters is allowed to hybridize with one cDNA probe made from RNA extracted from the cell synthesizing the peptide whose mRNA transcript one is attempting to locate; the other probe is made from the RNA of a cell that is known not to synthesize the peptide of interest. After hybridization the filters are washed, dried, and autoradiographed. One then compares the pattern of spots on the two filters and attempts to locate a radiolabeled clone on one filter that does not appear on the other. Due to the abundance of peptide-encoding transcripts in *Aplysia* neurons, this clone will most likely represent the cDNA encoding the peptide of interest. After a repeat of the screening process to verify the initial screen, the clone is grown up in larger quantities for restriction mapping and nucleotide sequencing. The amino acid sequence of the protein encoded by the clone can then be inferred from the nucleotide sequence. This differential screening technique is described in more detail by Nambu and Scheller (1983). Using versions of this basic strategy, a number of genes encoding a variety of putative peptides in several *Aplysia* neurons have been or are currently being identified in our lab.

2.1. Genes Encoding Bag Cell and Atrial Gland Neuropeptides

A family of genes has recently been isolated that encode several *Aplysia* neuropeptides that are thought to be functionally related (Scheller *et al.*, 1982, 1983a; reviewed by Mahon and Scheller, 1983). These peptides are synthesized in both the bag cells and the atrial gland, an exocrine gland associated with the reproductive tract. The family consists of five to nine genes all related by nucleotide sequences encoding peptides homologous to egg-laying hormone (ELH), a 36

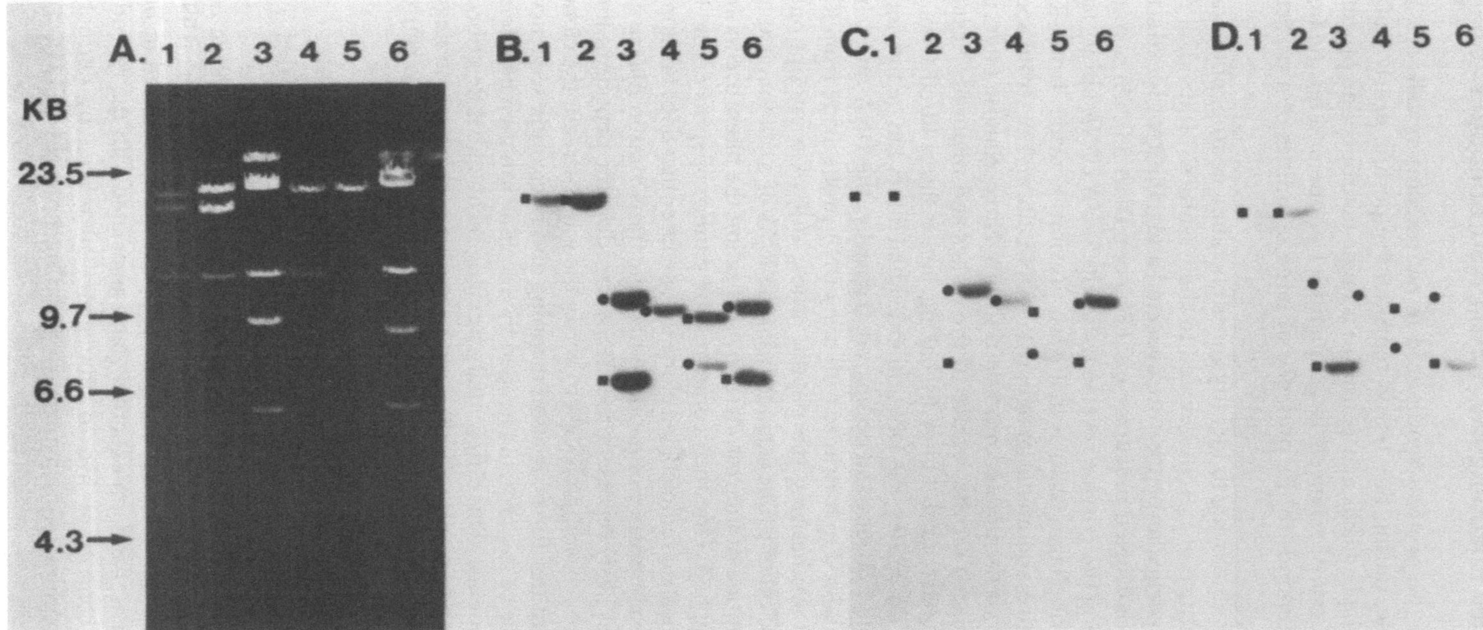

FIGURE 2. Hybridization analysis of ELH lambda clones. (A) Six different ELH-homologous recombinant clones were cleaved with the restriction enzyme Eco RI and subjected to electrophoresis on an 0.7% agarose gel. The gel was stained with ehtidium bromide and photographed. A Hind III digest of wild-type lambda was used as a size marker. (Lane 1) ELH-18; (lane 2) ELH-13, (lane 3) ELH-8; (lane 4) ELH-7; (lane 5) ELH-3; (lane 6) ELH-1. (B) The gel shown in A was transferred to nitrocellulose and hybridized with ^{32}P-labeled cDNA from either the bag cells or the atrial gland. Hybridization was carried out in 30% formamide-TNE at 48°C. (C) A gel identical to that shown in A was hybridized with ^{32}P-labeled cDNA from the bag cells in 70% formamide-TNE at 48°C. (D) A third gel was hybridized to ^{32}P-labeled cDNA from the atrial gland under the conditions described in C. The squares indicate the positions of restriction enzyme fragments that hybridize to atrial gland cDNA, and the dots indicate the positions of fragments that hybridize specifically to bag cell cDNA under stringent hybridization conditions.

amino acid peptide involved in the control of certain aspects of egg laying. DNA hybridization techniques were used to determine in which tissue(s) each of the six ELH genes are expressed. Each clone was digested with the restriction enzyme Eco RI and the cleavage products separated on an agarose gel. The products were then analyzed by blot hybridization to ^{32}P-labeled cDNA synthesized from either bag cell or atrial gland poly(A$^+$) RNA. At least one Eco RI fragment from each clone hybridized to the cDNA, and in three of the six clones two fragments hybridized, demonstrating that at least one pair of genes are linked (Fig. 2A,B). Under less stringent conditions, all of the fragments annealed to both bag cell and atrial gland cDNA. But as the hybridization stringency was increased, they annealed to cDNA from either the bag cells or atrial gland, but not to both (Fig. 2C,D). These data indicate that a family of related genes are expressed in the bag cells and atrial gland and that particular genes are expressed in one tissue but not the other.

The structural organization of six clones encoding the egg-laying peptide genes has been determined using restriction endonuclease mapping and nucleotide sequencing. As shown in Fig. 3, each clone contains either a single gene or a pair of genes transcribed in opposite directions. Three clones (ELH-1, 3, and 8) have a pair of genes, one expressed only in the atrial gland which encodes the B peptide, and one expressed only in the bag cells which encodes the egg laying peptides. Another clone (ELH-7) contains only the ELH gene, and two others (ELH-18 and 13) encode a single peptide also unique to the atrial gland known as A peptide (Scheller *et al.*, 1983a,b). The considerable homology of the clones is readily apparent by comparing the positions of the restriction enzyme sites. Annealing and sequencing experiments indicate that the genes are in fact about 90% homologous.

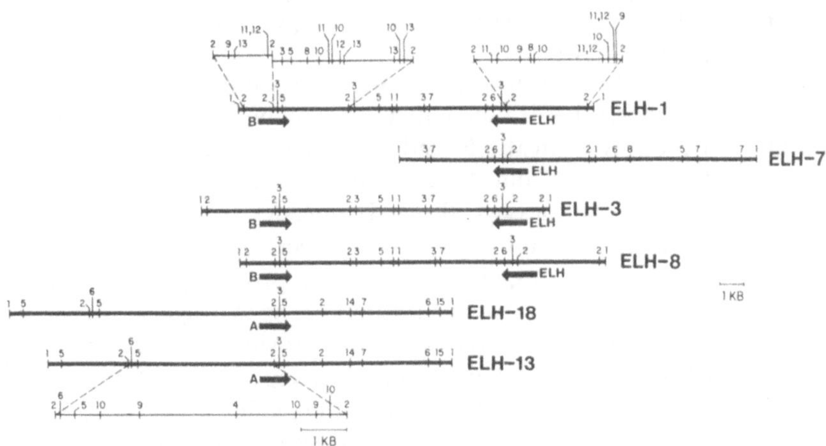

FIGURE 3. Restriction enzyme maps of the ELH-family recombinant clones. The restriction maps were developed by partial and double digests of clones and isolated fragments with a variety of restriction enzymes. The arrows indicate the positions of sequences homologous to mRNA's and point in the direction of transcription. The numbers indicate positions of restriction enzyme sites as follows: 1-Eco RI, 2-Pst I, 3-Xho I, 4-Stu I, 5-Pvu II, 6-Hind III, 7-Bgl I, 8-Xba I, 9-Ava II, 10-Hinc II, 11-Hae II, 12-Hha I, 13-Hpa II, 14-Bam HI, 15-Sal I.

This homology has important functional consequences and implies lineage from a common ancestral gene.

Of the five known RNA transcripts encoded by the ELH family, at least three distinct mRNA species are found in the bag cells and two occur in the atrial gland. The three transcripts expressed in the bag cells encode ELH, but not the A or B peptides. Conversely, the two atrial gland transcripts encode the A and B peptides, but not ELH. Thus, this gene family is a useful model system for examining the basis of tissue-specific expression of structurally related genes.

It has recently been shown that the A, B, and ELH transcripts contain exons in the 5′ untranslated region of the message that are not encoded by the ELH family clones shown in Fig. 3 (A. C. Mahon *et al.*, unpublished observations). This implies that transcription is initiated at least several thousand nucleotides upstream of the precursor coding region. Our laboratory is currently identifying these regions and trying to determine the mechanisms whereby the genes encoding ELH are expressed in the nervous system and the genes encoding the A and B peptides are expressed in the atrial gland.

2.2. A Single Gene Encodes the Peptides of Neuron R3-14

The *Aplysia* abdominal ganglion contains a number of neurosecretory cells that are believed to modulate cardiovascular activity. Neurons R3-14 are believed to use both neuropeptides and the amino acid glycine as neuromodulators. *In vitro* translation of R3-14 poly (A^+) RNA indicates that these cells contain a prevalent mRNA encoding a 14-kd protein. The gene encoding this protein has been isolated using the techniques described earlier (Nambu *et al.*, 1983). Unlike the ELH gene family, the *Aplysia* haploid genome contains a single copy of the sequence. As shown in Fig. 4, the gene is interrupted by two large introns and spans approximately 7 kb of genomic DNA.

The first intron is located just upstream of the coding sequence of the gene beginning two nucleotides before the initiator methionine. The second intron lies about a third of the way from the 5′ end of the coding region. As a result, the mature transcript is derived from three exons. Exon I encodes the complete 5′ untranslated region of the message. Exon II encodes the hydrophobic "signal sequence" found in most membrane-bound and secreted proteins, as well as the majority of the negatively charged portion of the protein precursor. Exon III encodes the remainder of the translated portion of the message.

This arrangement of introns and exons may provide a clue about the evolutionary origin of the R3-14 neuropeptide gene (Taussig *et al.*, 1984a). It is possible, for example, that the present-day gene arose by the joining of two domains encoding portions of proteins that have essentially opposite charge. This would result in a protein with charge characteristics suitable for efficient packaging in secretory vesicles. In addition, the region of exon I corresponding to the 5′ untranslated region of the message along with upstream control sequences in the intron are thought to be important in determining which cells and tissues the gene is expressed in.

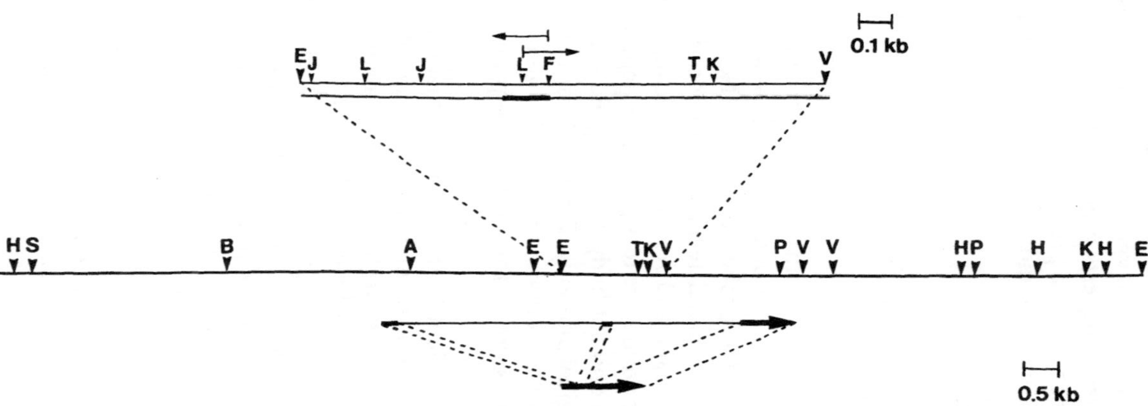

FIGURE 4. The R3-14 peptide gene. Exons and introns composing the R3-14 gene are schematically depicted. The primary transcript and the mRNA are indicated by arrows which point in the direction of transcription. The letters designate restriction enzyme sites as follows: E-Eco RI, B-Bam HI, H-Hind III, K-Kpn I, P-Pst I, A-Sal I, S-Sma I, V-Pvu II, T-Sst I, J-Ava II, L-Hae II, F-Hinf I.

2.3. The Gene-encoding Peptides in Neuron L11

Neuron L11 is a large cholinergic and peptidergic cell located on the dorsal surface of the abdominal ganglion. A cDNA clone specific to L11 has been identified and characterized (Taussig *et al.*, 1984b). The gene encodes a 14.7-kd protein which is thought to be the precursor of one or more neuroactive peptides. The nucleotide sequence of the gene suggests a neuropeptide precursor 125 amino acids in length which is similar in size to a prevalent protein seen *in vivo.* The number of copies of the gene in the genome and the structural organization of the gene is presently being determined.

3. The ELH Gene Family Is Expressed in Neurons throughout the *Aplysia* Nervous System

Two techniques can be used to identify neurons expressing particular genes: indirect immunofluorescence using antibodies to individual peptides and *in situ* hybridization of radiolabeled cDNA probes to cellular RNA. Indirect immunofluorescence is a very sensitive and useful technique for locating axons and nerve terminals serving as sites of storage and release of particular neuropeptides. The technique involves generating antibodies to a particular peptide, reacting these to tissue, and then reacting a second antibody that is tagged with a fluorescent label such as rhodamine to the original antibody. Figure 5 indicates how this technique has been used to locate processes containing molecules immunoreactive to ELH antibodies. The technique is less useful for locating the primary sites of synthesis of these molecules, the cell bodies.

In situ hybridization allows identification of individual cells expressing a specific gene. One labels cloned DNA fragments of the gene of interest with [^{125}I]-dCTP by nick translation and then hybridizes these fragments to thin sections of fixed tissue. The unhybridized radiolabeled DNA fragments are then washed off and the sections coated with liquid emulsion and exposed. Finally, the sections are developed, stained with a histological reagent, and analyzed. Grains indicate the positions of sequences homologous to the probe. Figure 6 illustrates *in situ* hybridizations of radiolabeled ELH gene probes to particular cells.

In situ hybridization data indicate that a number of individual and clusters of cells contain mRNA transcribed from the ELH gene family (McAllister *et al.*, 1983). The approximate positions of identified cells and processes are shown schematically in Fig. 7. Immunoreactive cell bodies were consistently found in stereotyped positions in the cerebral and buccal ganglia. Neurons were also labeled in the pedal ganglion, but both the number of cells and their positions were highly

FIGURE 5. The use of indirect immunofluorescence to locate processes containing ELH. Rhodamine-coupled goat anti-rabbit antibodies reacted with rabbit antibodies to ELH were used to visualize immunoreactive molecules in cells and processes containing the peptide. (A) Shows the cerebral ganglion; (B) the abdominal ganglion; and (C) aborization within the pedal ganglion.

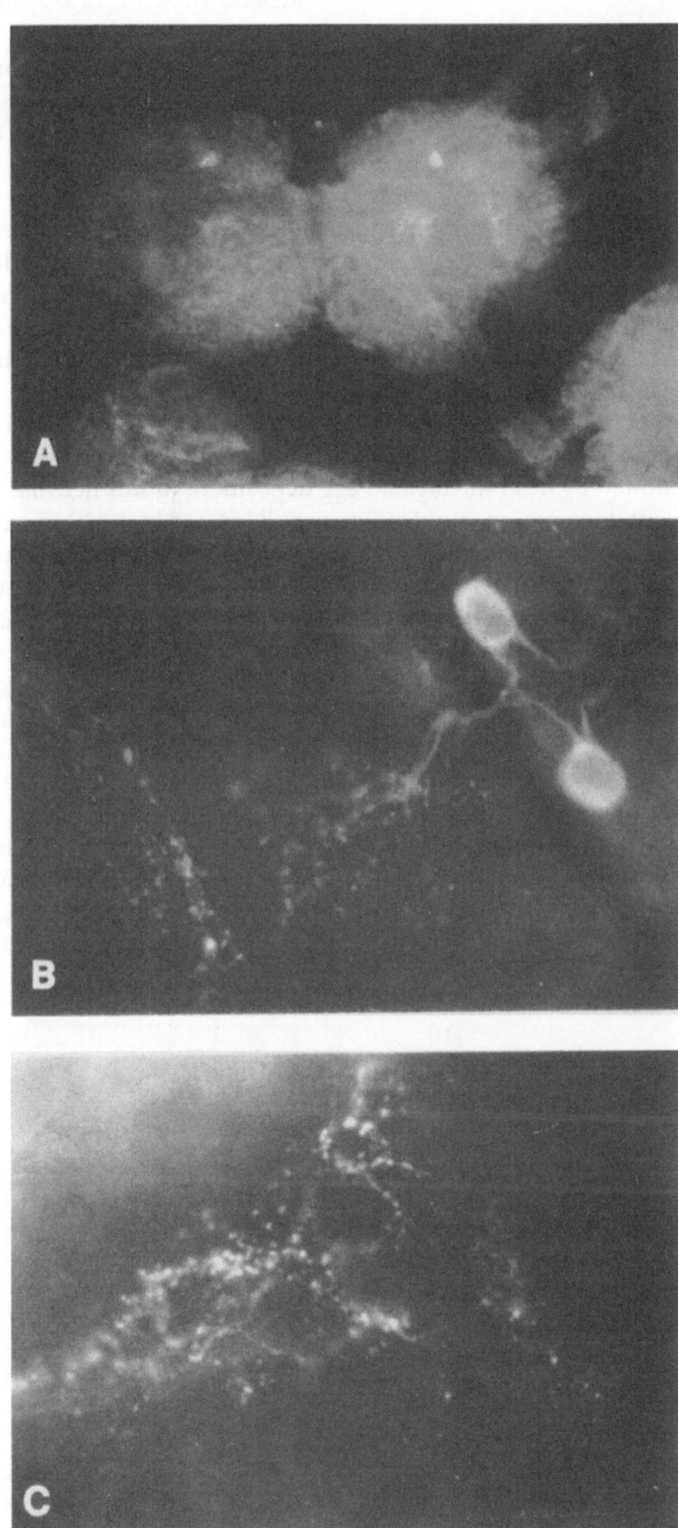
A
B
C

variable. Many of the processes seen throughout the nervous system appear to enter the sheath or neuropil while others seem to surround cell bodies. Thus, it is assumed that the peptides are being released at the neuropil, onto cell bodies, and into the vascularized sheath. Although existing hypotheses regarding the role of the *Aplysia* neuropeptides in egg laying focus solely on the bag cell and atrial gland peptides, it is possible that the interneurons expressing the ELH gene family in other ganglia mediate some or even most of the behavioral and physiological components of egg laying. The functional relationship between the bag cells and the other ELH-expressing neurons is not known. Our laboratory is currently investigating whether the activity of these neurons is related to discharge of the bag cells, and if so, what their function is.

As described earlier, egg laying involves a complex behavioral repertoire including cessation of walking, inhibition of feeding, and increased respiratory pumping, followed by head waving and egg deposition. Motor neurons controlling head movement, locomotion, and feeding are all located in the head ganglia, while respiratory functions are controlled by the abdominal ganglion. It is not necessarily the case that all of the behavioral components of egg laying are mediated by peptides released from the bag cells. There is considerable evidence that ELH acts hormonally to induce the release of eggs from the ovotestis (Dudek and Tobe, 1979; Rothman *et al.*, 1983a) and that discharge of the bag cells leads to enhanced respiratory pumping (Brownell and Schaefer, 1982), but evidence for bag cell mediation of the other behavioral components of egg laying is lacking. Bag-cell extracts do suppress the serotonin-mediated triggering of locomotion (Mackey and Carew, 1983) and other components of egg laying (Kupfermann, 1970), but whether the peptides responsible for the inhibition originate in the bag cells or other neurosecretory cells in the head or abdominal ganglia is not known.

The bag-cell clusters are the source of relatively large quantities of peptides, but their axons do not extend to the head ganglia. This suggests that if the bag

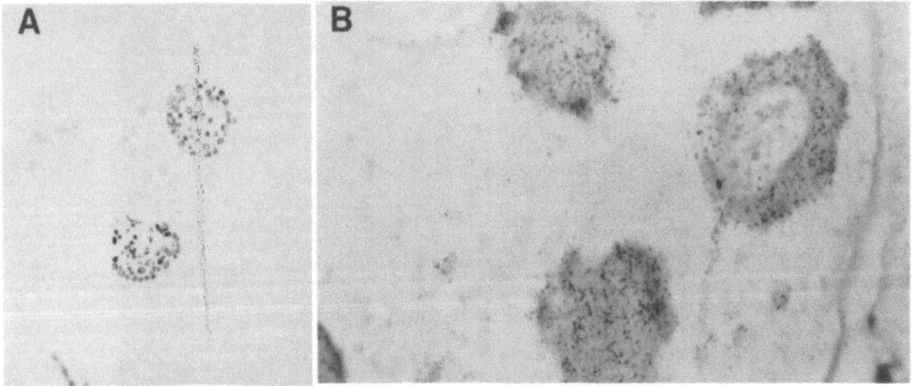

FIGURE 6. The use of *in situ* hybridization to locate cells expressing genes encoding ELH. Radiolabeled DNA encoding the ELH gene was hybridized to fixed abdominal ganglion tissue sections. (A) An autoradiogram of a section through a bag-cell cluster magnified 20 X. (B) A single bag cell magnified 600 X.

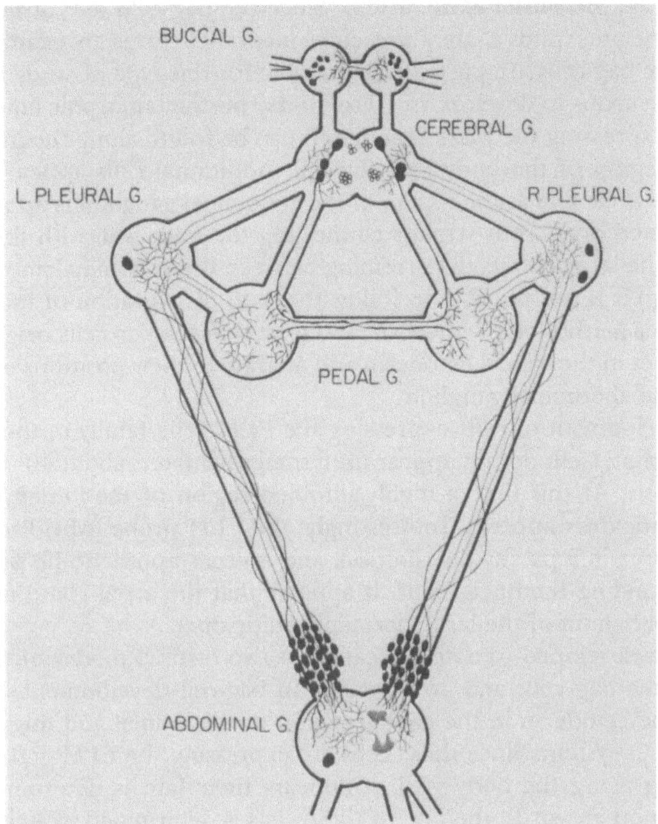

FIGURE 7. A network of interneurons express members of the ELH gene family. The approximate positions of the interneurons expressing members of the ELH gene family are illustrated. The positions of these cells were determined by indirect immunofluorescence and *in situ* hybridization techniques as described in Figs. 5 and 6.

cells do influence the activity of neurons in the head ganglia, it is by way of the hormonal action of the bag-cell peptides released into the circulation and not by synaptic or other local interactions. Another possibility is that the ELH-expressing neurons in the head ganglia directly release peptides onto cell bodies at regions of close contact with particular interneurons and motor neurons in these ganglia. Alternatively, the behavioral components of egg laying mediated by neurons in the head ganglia may be controlled by other unidentified peptides or neurotransmitters.

4. Development of the Neurons Expressing the ELH Gene Family

Expression of the ELH gene family during *Aplysia* development has been examined using DNA probes encoding ELH genes, *in situ* hybridization, and antibodies to ELH (McAllister *et al.*, 1983). Since the bag cells and atrial gland both arise after metamorphosis, they are convenient structures to examine developmentally. The bag cells are particularly suitable for this type of study because they are the last neurons to develop. In 10 to 40-day postmetamorphic animals, a number of cells expressing the ELH gene family can be found along the inner lining of the caudal region of the body wall (Fig. 8). Additional cells appear at locations rostral to this area, and as the region of the abdominal ganglion is approached, the cells are located on fibrous strands connecting the body wall with the abdominal ganglion or the large connectives running between the abdominal and pleural ganglia. Although it is not possible to follow the path of migration of individual neurons using this methodology, these data suggest that the bag cells originate in proliferative zones in the caudal ectoderm and migrate to their positions on the rostral portion of the abdominal ganglion.

The development of cells expressing the ELH gene family in the atrial gland is quite different. Cells do not appear until somewhat later, about 40–60 days after metamorphosis. At this time a highly infolded region of the lumen of the large hermaphroditic duct appears. Interestingly, the ELH probe hybridizes to various cells in this area in a patchy distribution, and no cells appear to be present in the tissues surrounding the duct. Thus, it appears that the atrial gland arises from a thickened epithelium of the large hermaphroditic duct.

The data developed to date indicate that two distinct modes of proliferation give rise to the bag cells and atrial gland. In bag-cell development, the neurites arise from the ectoderm in the caudal region of the animal and migrate into the central nervous system. Since these cells are expressing the ELH genes while they are migrating along the body wall, it appears their fate is determined prior to migration. In other words, the fate of these cells is determined by lineage and not by induction. There is no evidence that the cells migrate to the abdominal ganglion, sense their position, and undergo differentiation into functional bag cells. It appears that their development is determined before they leave the ectoderm. How the proper members of the gene family are expressed in different tissues and in what manner expression is related to the developmental origin of the cells is not known.

5. *Aplysia* Genes Encode a Number of Neuropeptide Precursors

The *Aplysia* neuropeptide precursors have a number of common structural features, many of which are consistent with those of other secretory proteins. Neuropeptide precursors are somewhat unique in that several physiologically active peptides may be released from a single polypeptide precursor. Individual peptides are normally flanked on either side by paired basic lysine and/or arginine residues, or by a single arginine. Processing enzymes recognize these basic amino acids and cleave out the active molecule from the precursor polypeptide. Why this arrangement of multiple peptides on a large precursor has evolved is certainly related to the function of these peptides in the nervous system. By examining the structure of individual precursor peptides, one can speculate about the functional advantage of this arrangement.

5.1. The Bag Cell and Atrial Gland Precursors

Detailed analysis of the nucleotide sequences of the various members of the ELH gene family suggests that they encode three classes of neuropeptide precursors (Scheller *et al.*, 1982, 1983a). Two of these precursors encode the A and B peptides expressed in the atrial gland, and the other encodes the ELH precursor expressed in the bag cells (Fig. 9A). All of the precursors begin with a methionine which is followed by a number of hydrophobic residues composing a "signal sequence." This sequence is thought to function in most membrane-bound or secreted proteins to facilitate entrance of the nascent protein chain into the endoplasmic reticulum (reviewed by Docherty and Steiner, 1982).

The ELH precursor contains four pairs of basic amino acid residues flanking two related peptide sequences called β-bag cell factor (β-BCF) and γ-bag cell factor (γ-BCF) and an unrelated heptapeptide sequence known as δ-bag cell factor (δ-BCF). Further toward the carboxyl end of the precursor is a nine amino acid stretch encoding α-bag cell peptide (α-BCP), a neuropeptide released from the bag cells (Rothman *et al.*, 1983b) whose activity is discussed in Section 6. Interestingly, α-bag-cell peptide (BCP) corresponds to the nine amino acids at the carboxyl end of the A and B peptides. As illustrated in Fig. 9A the β, γ, and δ peptides lie within an insert 80 amino acids in length that is not present in either the A or B precursor genes.

It is significant that the 47 amino acid sequence immediately following the carboxy-terminus cleavage site of α-BCP and the A and B peptides contains a number of amino acid changes that result in different potential enzymatic cleavage sites of the precursor. The clustering of these amino acids changes suggests that differential cleavages in this region are functionally significant.

Still further toward the carboxy-terminus of the precursor is a 36 amino acid sequence encoding ELH, which in turn is followed by a 27 amino acid sequence corresponding to the acidic peptide, another previously identified neuropeptide. This peptide is coreleased with ELH upon depolarization of the bag cells.

The A and B peptide precursors are considerably smaller than the ELH precursor. Just to the carboxyl end of the signal sequence of both precursors lie the A and B peptides. In the B peptide precursor, a single base deletion alters the reading frame placing a stop codon after the sixth amino acid of ELH. However, this deletion is not found in the A peptide gene. Further to the carboxyl end of the A precursor is a single arginine within the region homologous to ELH which generates an additional potential cleavage site. Further toward the carboxyl end are two cysteine residues which may form a disulfide bridge at the terminal of the precursor.

The numerous peptides cleaved from the three precursors can be grouped into three major classes according to their primary sequences: the ELH peptide homology, the A/B peptide homology, and the acidic peptide homology. At least

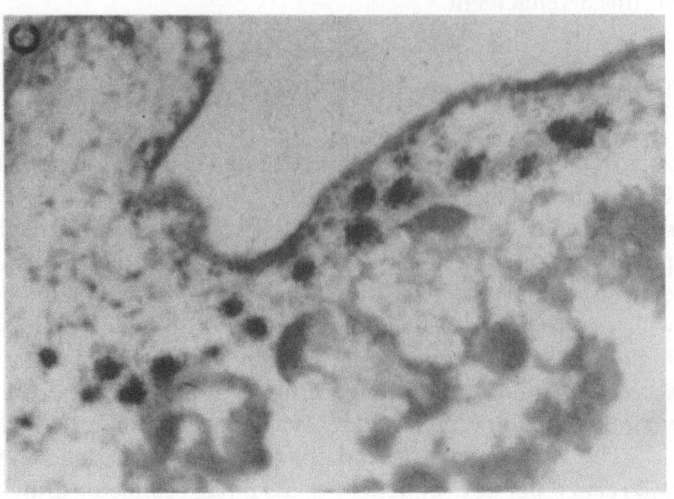
C

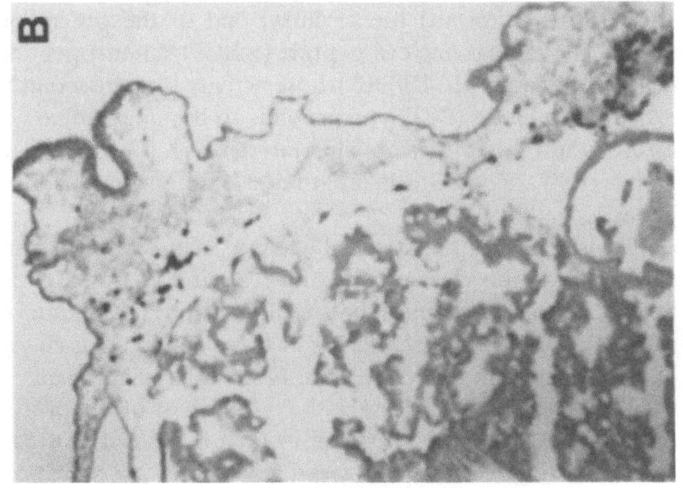
B

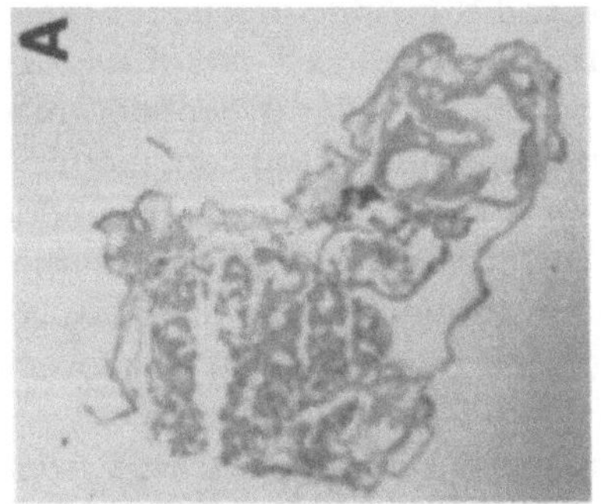
A

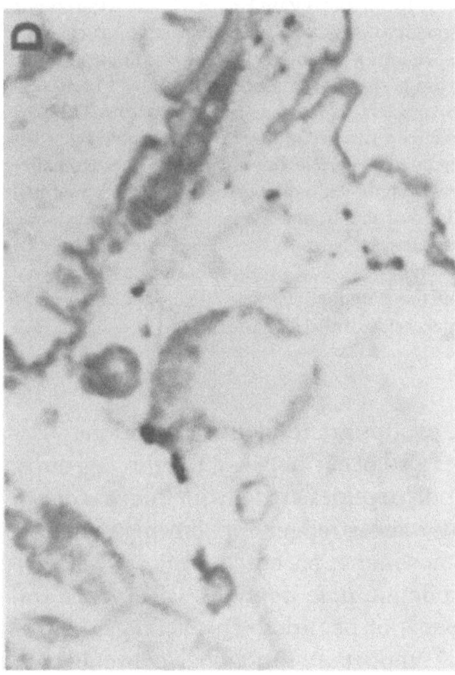

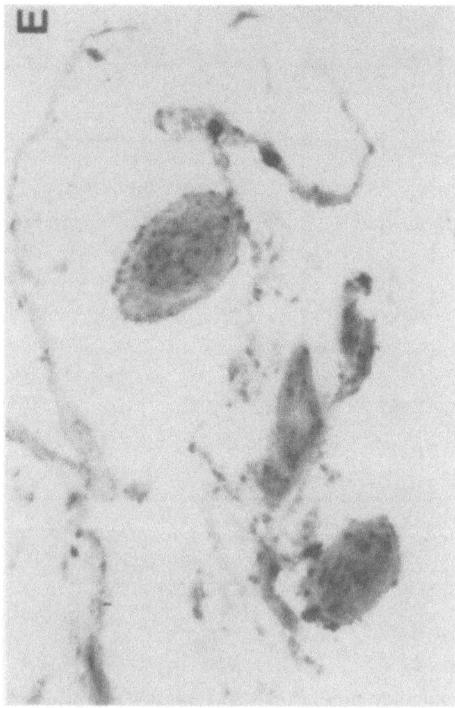

FIGURE 8. Cells expressing the ELH genes originate in the body wall. Neurons expressing the ELH gene family originate in the body wall and migrate to the central ganglia. (A) An *in situ* hybridization to a section through a whole animal 10-days postmetamorphosis. Cells expressing the gene family are located along the length of the body wall; magnification is X 15. (B) A portion of A magnified X 50. (C) A portion of A magnified X 230. (D) An *in situ* hybridization to a developing organism 35 days after metamorphosis. Neurons expressing the ELH genes appear to migrate from the body wall along connective tissue fibers to their ultimate positions in the ganglia. Cells are shown migrating along fibrous threads to a primitive ganglion in the center of the section. Magnification is X 92. (E) An *in situ* hybridization to a transverse section through a developing pleural abdominal connective. Fibers containing cells expressing the ELH genes are evident. Magnification is X 92.

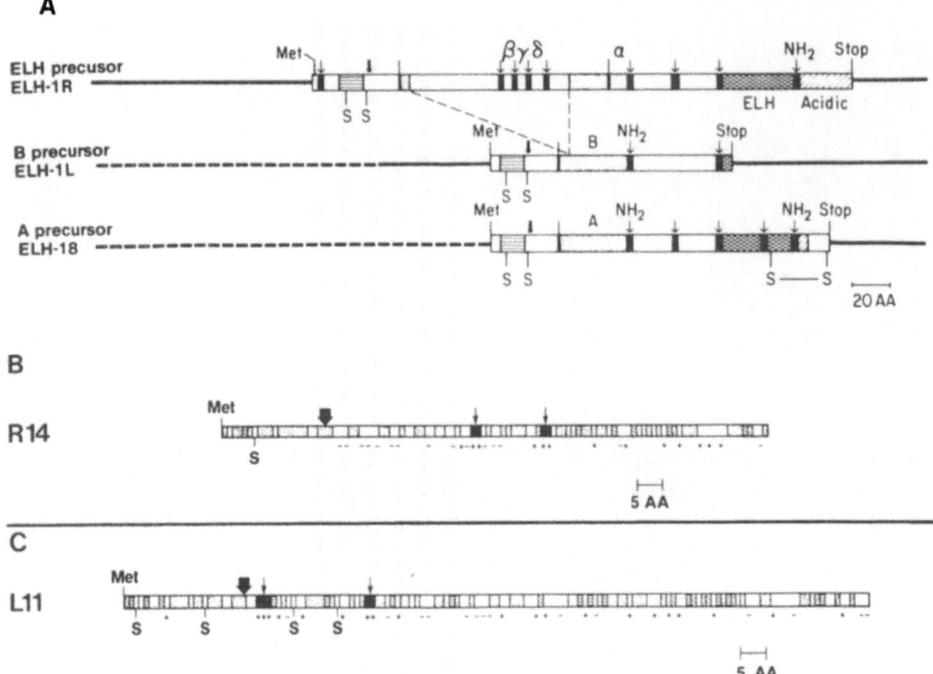

FIGURE 9. Schematic representation of *Aplysia* neuropeptide precursors. (A) Indicates the principal peptides and potential cleavage sites (arrows) of the ELH, B, and A precursor polypeptides. Each of the three peptides is initiated by a methionine followed by a hydrophobic region (≡). An S below the coding region indicates the location of a cysteine residue. Large arrows indicate the putative site of cleavage of the signal sequence (⬇). A line above the sequence represents a potential cleavage at a single arginine residue, while arrows represent potential or known cleavage sites at sequences of 2, 3, or 4 basic residues. If carboxyterminal amidation is believed to occur, an NH_2 is written above the arrow. The A/B peptide homology is represented by stippled boxes ( ). The ELH homology is represented by crosshatched boxes ( ). The acidic peptide homology is designated by parallel lines ( ). Solid lines symbolize sequenced noncoding regions and dotted lines depict regions not yet sequenced. (B) Indicates potential cleavage sites (small arrows), the putative cleavage position of the "signal sequence" (large arrow), and the charged amino acids (+, − signs) of the R3-14 precursor. Hydrophobic regions are indicated by the stippled pattern ( ), histidine residues by crosshatching ( ), and proline residues by horizontal lines (≡). "S" indicates cysteine residues. (C) Indicates the potential cleavage sites, putative cleavage position of the "signal sequence," and the charged amino acids of the L11 precursor.

70% of the amino acids within each of these groups are in identical positions. Less obvious but greater than random homologies also occur between the three groups. These homologies imply that the precursor polypeptides arose from internal duplications within the transcription unit of a single ancestral gene. Following the internal duplications, large regions of the chromosome apparently duplicated to give rise to a family of genes, each of which encode multiple peptides. Divergence following the duplications then generated the sets of peptides. This arrangement of multiple peptides on a large precursor has important functional consequences which are discussed in Section 7.

5.2. The R3-14 Precursor

The gene encoding neuropeptides in neurons R3-14 gives rise to a 13.5-kd precursor (Fig. 9B) (Nambu *et al.*, 1983). As with the ELH gene family precursors, the initiator methionine is immediately followed by the characteristic signal sequence. Beyond this sequence are 88 amino acids containing two paired basic residues 12 amino acids apart, which are potential enzymatic cleavage sites. In addition, there are several other single arginines that may also serve as cleavage sites. Processing of the R3-14 precursor by proteolytic enzymes in intact cells results in the release of several peptides (R.-R. Kaldany *et al.*, 1984, Aswad, 1978). The identity of these peptides and the mechanism by which the precursor is processed is currently being studied in our laboratory.

5.3. The L11 Precursor

The L11 gene encodes a 14.7-kd protein that is thought to be the precursor of one or more neuroactive peptides (Taussig *et al.*, 1983b). As shown in Fig. 9C, the nucleotide sequence of the gene suggests a neuropeptide precursor 125 amino acids in length that is similar in size to a prevalent protein seen *in vivo.* The precursor contains two major potential cleavage sites at a pair of basic residues (28–29) and at a sequence of three basic residues (50–52). Proteolytic cleavage at these positions would liberate a peptide 19 residues in length. This potential cleavage product has two cysteine residues that may function in determining the secondary and tertiary structure of the protein. Since the precursor contains a number of single arginine residues including one within the 19 amino acid peptide, the precise number of neuroactive peptides released from the precursor is not known.

6. Neuropeptides Mediate Discrete Physiological and Behavioral Units

Relatively little is known about the function of the individual *Aplysia* neuropeptides. Several clues about their role in egg laying have come from studies of the effects of bag-cell discharge, bag-cell extracts, and purified peptides on individual cells. Many of the behavioral components of the egg-laying behavioral repertoire can be elicited by injection of bag-cell extract into the animal's circulatory system. Furthermore, discharge of the bag cells always precedes spontaneous egg laying. An important question that we and others have not been able to answer is what triggers discharge of the bag cells. It is known that extracts of the atrial gland initiate egg laying when injected into mature animals, and the A and B peptides depolarize the bag cells when applied to the clusters. However, if the A and B peptides do in fact trigger egg laying, it is not clear how they enter the circulatory system. The atrial gland is an exocrine gland and releases peptides into the lumen of the large hermaphroditic duct. The gland does not secrete hormones directly into the circulation. Since the atrial gland lies in the wall of the gonopore, it is tempting to devise schemes incorporating the assumption that mating triggers release of the A and B peptides, which in turn leads to discharge of the bag cells and egg laying. But in the absence of a mode of transport of the A and B peptides into the circulation, these mechanisms remain speculative.

Bag-cell discharge results in long-term changes in the activity of specific neurons in the abdominal ganglion (Mayeri *et al.*, 1979a,b). These effects include burst augmentation, slow inhibition, transient excitation, and prolonged excitation. Bag-cell discharge also causes long-term excitation of gill and siphon motor neurons in this ganglion which leads to enhanced respiratory pumping (Brownell and Schaefer, 1982; Schaefer and Brownell, 1984), a behavioral change apparently designed to allow greater oxygen uptake by the gills.

Although little is known about the impact of particular peptides on behavior, their influence on identified cells is being documented. For example, when ELH is infused into the vasculature, it causes prolonged excitation of the LC cells, a cluster of neurons on the dorsal surface of the abdominal ganglion (Scheller *et al.*, 1983b). In addition, there is considerable evidence that when α-bag cell peptide α-BCP) is released from the bag cells, it induces slow inhibition of neuron L6 (Rothman *et al.*, 1983b), a neuron in the left upper quadrant of the abdominal ganglion. Long-term changes in the excitability of neuron L6 due to bag cell discharge and a phosphodiesterase inhibitor correlate with changes in protein phosphorylation in this neuron (Schaefer *et al.*, 1984). Further studies are required to determine whether protein phosphorylation and dephosphorylation mediate neuropeptide-induced long-term changes in the electrical activity of neurons.

Neurosecretory cells R3-14 are particularly well suited for studies of the role of neuropeptides in modulating physiological activities. Each neuron has a single large axon that exits the ganglion via the branchial nerve and innervates the efferent vein of the gill at the base of the heart. Neuron R14 also sends an axon to the arteries close to the abdominal ganglion. These cells contain a large number of vesicles in which both peptides and the amino acid glycine are stored (Price *et al.*, 1979; Price and McAdoo, 1981). Which peptides are actually stored and secreted in these cells is not known. We are currently identifying the secreted peptides and determining their role in cardiovascular physiology. We are especially interested in the function of the peptides in relation to glycine, the other transmitter in these neurons.

The cells and processes containing the R3-14 neuropeptides have been identified using immunohistochemical techniques (Kreiner *et al.*, 1984). Antibodies directed against synthetic peptides corresponding to three regions of the precursor were generated and reacted to sectioned tissues as described earlier. These studies indicate that specific cells in the right upper quadrant of the *Aplysia* abdominal ganglion as well as R14 in the lower right quadrant synthesize these peptides. Processes arising from these cells extend to the efferent vein of the heart and other portions of the vascular system via the branchial and genital-pericardial nerves. Processes containing numerous varicosities also extend up the pleuroabdominal connectives. This immunohistochemical study provides evidence that one or more of the R3-14 neuropeptides function to regulate cardiovascular physiology.

Once the function(s) of each of the peptides secreted by the various *Aplysia* neurosecretory cells is determined, a cohesive picture of how neuropeptides and classical neurotransmitters interact to control behavioral and physiological events should emerge. It is evident at this stage that even the relatively simple *Aplysia* nervous system employs a wide array of neurochemical messengers to modulate physiology and behavior.

7. Single Neurons Utilize Multiple Chemical Messengers

Why do neurons such as the R3-14, L11, and the bag cells use multiple chemical messengers? Presumably, this arrangement has evolved because a single transmitter does not meet the needs of these cells for intercellular communication. In the case of the bag cells, several peptides appear to be utilized to achieve a number of coordinated behavioral and physiological events. Depolarization of these electrically coupled cells results in the release of multiple peptides that apparently act both locally and hormonally to influence target cells.

Neurons R3-14 utilize both the simple amino acid glycine and one or more peptides as intercellular messengers. Similarly, acetylcholine and one or more peptides function as messengers in neuron L11. Although we have much to learn about the function of multiple chemical messengers, it appears that peptides and classical neurotransmitters perform duties that are at least partly complementary. Neurotransmitters act over short distances and their effects on the postsynaptic cells are normally of short duration (milliseconds). Peptides, on the hand, may act by way of synaptic or other local interactions, or hormonally, exerting their effects over greater distances. In addition, in contrast to classic neurotransmitters, neuropeptides may affect the excitability of target cells for minutes to hours. The specificity for synaptic communcation lies in the spatial interactions of the presynaptic and postsynaptic cells. Close juxtaposition of the two neurons is required because receptors on postsynaptic cells are normally not unique, but are common to numerous cells. In the case of neuropeptides acting hormonally, unique receptors must be present on target cells and not on other cells. Here the source of specificity is structural and not spatial. Thus, the diversity of chemical messengers provides neurons with the versatility to communicate by way of messages of short duration at specific local sites, and/or by messages of longer duration at multiple, spatially diverse locations.

8. Common Characteristics of the *Aplysia* and Mammalian Neuroendocrine Systems

The neuroendocrine systems of invertebrates and higher organisms including mammals share a number of common structural and functional characteristics, particularly in the fundamental mechanisms they employ in intercellular communication. The strategies used by higher organisms to transfer information both synaptically and hormonally are highly conserved. In particular, the neurotransmitters commonly used in the *Aplysia* nervous system such as serotonin, acetylcholine, and dopamine are identical to those found in higher systems. However, the existing data suggest that this is not the case for neuropeptides. Although the peptide transmitters as a class of molecules are used similarly in both invertebrates and mammals, it appears that their structures have diverged significantly. A number of *Aplysia* neurons are immunoreactive with antibodies to Leu-enkephalin (Hopkins *et al.*, 1982) and to neurotensin (Price *et al.*, 1983). However, recent attempts to locate genes similar to proopiomelanocortin in *Aplysia* have not proved to be fruitful (M. Uhler, personal communication); nor have efforts to isolate mammalian genes homologous to the ELH gene family (J. Nambu, personal communication). The extent of homology between other invertebrate and mammalian neuropeptides is an area of active research in numerous laboratories. Regardless of the extent of homology, it is well established that there are striking similarities in the organization of the genes that encode neuropeptides, the way they are synthesized and secreted, and their mode of action on target tissues. It is these characteristics that make studies of *Aplysia* relevant to higher organisms, and it is the relative simplicity and accessibility of the *Aplysia* neuroendocrine system that allows one to examine these kinds of fundamental questions.

8.1. *Aplysia* and Mammalian Neuropeptides Function Similarly

A prominent similarity between *Aplysia* and mammalian neuropeptides is their mode of action. In both systems they may act locally as neurotransmitters or over longer distances as hormones to influence both neuronal and nonneuronal target cells. In addition, these peptides are apparently always cleaved from a large precursor protein. Egg-laying hormone, for example, is released from the bag cells along with other peptides from a common precursor, just as adrenocorticotropin, melanophore-stimulating hormone, and the endorphins are released from proopiomelanocortin in the mammalian brain. Finally, there are significant homologies in the amino acid sequences of the ELH family peptides secreted from both neural and nonneural tissues of *Aplysia* (atrial gland), just as there are similarities in numerous mammalians peptides originating in different tissues. This conservation of structure and function may represent a common mechanism by which organisms coordinate related behavioral functions (Bloom, 1980; Krieger, 1983).

9. The Link Between Gene Expression and Behavior

The *Aplysia* neuroendocrine system provides the opportunity to investigate the function of neuropeptides from the level of gene expression to neurosecretion, action on target neurons, and ultimately, modulation of behavior. This allows the neural circuitry underlying certain behavioral and physiological events to be determined and analyzed in detail. The *Aplysia* system is composed of numerous motor, sensory, and interneurons that are organized into multiple neural pathways and utilize a large number of chemical messengers. A single cell may employ multiple transmitters, some acting locally and others hormonally, to influence the activity of both neural and nonneural cells. Thus, although the *Aplysia* neuroendocrine system is relatively simple, it is highly sophisticated in that it utilizes the same fundamental mechanisms of communication used in higher systems. It is the balance between simplicity and complexity that makes *Aplysia* a useful model neural system.

How are genes encoding neuropeptides organized and when and where are these genes expressed? Why are multiple peptides encoded by a single gene? How have these genes evolved? What is the mechanism by which neuropeptides are secreted and in what manner do they influence target cells to modulate behavior? These are all questions that can be answered by examining the fundamental genetic and neuroendocrine processes of this marine mollusk.

ACKNOWLEDGMENTS. The authors wish to express their appreciation to John R. Nambu and Rashad-Rudolph Kaldany for critically reading the manuscript. In particular, we wish to acknowledge the work of the numerous individuals cited in the chapter whose efforts made this review possible.

References

Arch, S., 1972, Biosynthesis of the egg laying hormone (ELH) in the bag cell neurons of *Aplysia californica, J. Gen. Physiol.* **60:**102.

Aswad, D. W., 1978, Biosynthesis and processing of presumed neurosecretory proteins in single identified neurons of *Aplysia californica, J. Neurobiol.* **9** (4) :267.

Bloom, F. E., 1980, *Peptides: Integrators of Cell and Tissue Function.* Raven Press, New York.

Brownell, P. H., and Schaefer, M., 1982, Activation of a long lasting motor program by the bag cell neurons in *Aplysia, Soc. Neurosci. Abstr.* **8:**736.

Docherty, K., and Steiner, D. F., 1982, Post-translational proteolysis in polypeptide hormone biosynthesis, *Annu. Rev. Physiol.* **44:**625–638.

Dudek, F. E., and Tobe, S. S., 1979, Bag cell peptides act directly on ovotestis of *Aplysia californica:* Basis for an *in vivo* bioassay, *Gen. Comp. Endocrinol.* **36:**618.

Hopkins, W. E., Stone, L. S., Rothman, B. S., Basbaum, A. I., and Mayeri, E., 1982, Egg-laying hormone, leucine-enkephalin and serotonin immunoreactivity in the abdominal ganglion of *Aplysia:* A light microscopic study, *Soc. Neurosci. Abstr.* **8:**587.

Kaldany, R.-R., Schaefer, M., Evans, C., Mak, G., and Scheller, R. H., 1984, Processing of a neuropeptide precursor in the R3-14 cells of *Aplysia, Soc. Neurosci. Abstr.* **10:**285.

Kandel, E. R., 1979, *Behavioral Biology of Aplysia,* W. H. Freeman and Co., New York.

Kreiner, T., Rothbard, J., Schoolnick, G. K., and Scheller, R. H., 1984, Antibodies to synthetic peptides defined by cDNA cloning reveal a network of peptidergic neurons in *Aplysia, J. Neruosci.,* **4:**2581–2589.

Krieger, D. T., 1983, Brain peptides: What, where, and why? *Science* **222:**975.

Kupfermann, I., 1970, Stimulation of egg-laying by extracts of neuroendocrine cells (bag cells) of abdominal ganglion of *Aplysia californica, J. Neurophysiol* **3:**877.

Mackey, S., and Carew, T. J., 1983, Locomotion in *Aplysia:* Triggering by serotonin and modulation by bag cell extract, *J. Neurosci.* **3**(7):1469.

Mahon, A. C., and Scheller, R. H., 1983, The moleculer basis of a neuroendocrine fixed action pattern: Egg laying in *Aplysia, Cold Spring Harbor Symp. Quant. Biol.* **48:**405–412.

Mayeri, E., Brownell, P., and Branton, W. D., 1979a, Multiple, prolonged actions of neuroendocrine bag cells on neurons in *Aplysia.* I. Effects of bursting pacemaker neurons, *J. Neurophysiol.* **42:**1165.

Mayeri, E., Brownell, P., and Branton, W. D., 1979b, Multiple, prolonged actions of neuroendocrine bag cells on neurons in *Aplysia.* II. Effects on beating pacemaker and silent neurons, *J. Neurophysiol.* **42:**1185.

McAllister, L. B., Scheller, R. H., Kandel, E. R., and Axel, R., 1983, *In situ* hybridization to study the origin and fate of identified neurons, *Science* **222:**800.

Nambu, J. R., and Scheller, R. H., 1983, Molecular cloning and characterization of neuropeptide genes from identified *Aplysia* neurons, in: *Molecular Approaches to the Nervous System,* 1983 Short Course Syllabus, Society for Neuroscience, Bethesda, Maryland, pp. 110–121.

Nambu, J., Taussig, R., Mahon, A. C., and Scheller, R. H., 1983, Gene isolation with cDNA probes from identified *Aplysia* neurons: Neuropeptide modulators of cardiovascular physiology, *Cell* **35:**47.

Price, C. H., and McAdoo, D. J., 1981, Localization of axonally transported ^{3}H-glycine in vesicles of identified neurons, *Brain Res.* **219:**307.

Price, C. H., McAdoo, D. J., Farr, W., and Okuda, R., 1979, Bidirectional axonal transport of free glycine in identified neurons R3-R14 of *Aplysia, J. Neurobiol.* **10:**551.

Price, C. H., Ruane, S. E., and Carraway, R. E., 1983, Neurotensin-like peptides in the brain and gut of *Aplysia:* Studies on biological and chemical properties, in: *Molluscan Neuroendocrinology. Proceedings of the International Minisymposium on Molluscan Endocrinology* (J. Lever and H. H. Boer, eds.), North-Holland Publishing Co., Amsterdam, pp. 14–20.

Rothman, B. S., Weir, G., and Dudek, R. E. 1983a, Egg laying hormone: Direct action on the ovotestis of *Aplysia, Gen. Comp. Endocrinol.* **52:**134.

Rothman, B. S., Mayeri, E., Brown, R. O., Yuan, P.-M., and Shively, J., 1983b, Primary structure and neuronal effects of α-bag cell peptide, a second candidate neurotransmitter encoded by a single gene in bag cell neurons of *Aplysia, Proc. Natl. Acad. Sci. U.S.A.* **80:**5753.

Schaefer, M., and Brownell, P., 1984, Modulation of a respiratory motor program by peptide-secreting neurons in *Aplysia, J. Neurobiol.* (In Press).

Schaefer, M. Shirk, P. D., Roth, D. R., and Brownell, P. H., 1984, Activity-related changes in protein phosphorylation in an identified *Aplysia* neuron (submitted).

Scheller, R. H., Jackson, J. F., McAllister, L. B., Schwartz, J. H., Kandel, E. R., and Axel, R., 1982, A family of genes that codes for ELH, a neuropeptide eliciting a stereotyped pattern of behavior in *Aplysia, Cell* **28:**707.

Scheller, R. H., Jackson, J. F., McAllister, L. B., Rothman, B. S., Mayeri, E., and Axel, R., 1983a, A single gene encodes multiple neuropeptides mediating a stereotyped behavior, *Cell* **32:**7.

Scheller, R. H., Rothman, B. S., and Mayeri, R., 1983b, A single gene encodes multiple peptide transmitter candidates involved in a stereotyped behavior, *Trends Neurosci.* **6** (8) **:**340.

Scheller, R. H., Kaldany, R.-R., Kreiner, T., Mahon, A. C., Nambu, J. R., Schaefer, M., and Taussig, R., 1984, Neuropeptides: Mediators of Behavior in *Aplysia, Sci.* **225:**1300–1308.

Taussig, R., Picciotto, M. R., and Scheller, R. H., 1984a, Two introns define functional domains in an *Aplysia* neuropeptide precursor, in: *Molecular Biology of Development,* Vol. 19 (E. H. Davidson and R. I. Firtel, eds.), Liss, New York, pp. 551–560.

Taussig, R., Kaldany, R.-R., and Scheller, R.H., 1984b, A cDNA clone encoding neuropeptides isolated from Aplysia neuron L11. *Proc. Natl. Acad. Sci. U.S.A.* **81:**4988–4992.

27

The *Drosophila* Thorax as a Model System for Neurogenetics

ROBERT J. WYMAN, JOHN B. THOMAS, LAWRENCE SALKOFF, AND WALTER COSTELLO

1. Introduction

Each scientific question requires careful choice of the experimental system in which an attempt to answer the question will be made. If a familiar system can be used, this has the tremendous advantage of allowing the researcher to capitalize on all the information about the system that previous researchers have gained. Developing a new experimental system may take many scientist-years of effort before the rewards can be reaped. There are outstanding success stories in using both old and new systems, e.g., the continuing excellence of the cat as a neurophysiological system (Hubel, 1982; Wiesel, 1982) and the emergence of *Aplysia* as a superb animal for cellular neurophysiology (Kandel, 1976).

Neurogenetics may be defined as the study of the genetics of the nervous system or the use of genetics to study the nervous system. Here again some classical systems have been exploited (mouse cortex: Caviness and Rakic, 1978), some new systems have been attempted and abandoned (locust: Goddman, 1978; cricket: Bentley, 1975), and some new systems have been developed with marvelous success (nematodes: Brenner, 1973; Horvitz, *et al.*, 1983). The metazoan animal whose genetics is most thoroughly known is, of course, *Drosophila melanogaster* (Sturtevant and Beadle, 1939; Ashburner and Novitski, 1976–1983). If neurobiological techniques could be adapted to *Drosophila,* then one would have available this vast store of genetic information for application to the nervous system. In the early 1970s, a number of laboratories started working on the nervous system of *Drosophila* with a

ROBERT J. WYMAN, JOHN B. THOMAS, LAWRENCE SALKOFF, AND WALTER COSTELLO • Yale University, Department of Biology, New Haven, Connecticut 06511.

variety of questions in mind: the study of visual transduction (Pak *et al.*, 1969), the study of electrophysiology and neural networks (Ikeda and Kaplan, 1970; Levine and Wyman, 1973), or the genetic dissection of behavior (Benzer, 1973).

One of the major stumbling blocks to the success of *Drosophila* as a system has been the perception that it is very difficult to do sophisticated electrophysiology on this very tiny animal. Nevertheless, a number of techniques were rapidly established: single-unit muscle recording (Levine and Wyman, 1973), nerve–muscle preparations in the larva (Jan and Jan, 1976) and adult (Harcombe and Wyman, 1977), and even intracellular recording (Ikeda and Kaplan, 1970; Alawi and Pak, 1971). Over the past several years, our laboratory has been developing the *Drosophila* thorax as a system in which sophisticated electrophysiology could be done and in which important neurobiological questions could be addressed. The hope was that if the thoracic nervous system could be made into a feasible system for electrophysiolgoy, then it could become a perfect blend of a classical system for genetics and a new system for neurobiology.

We found that the thorax contained large cells that could be individually and repeatedly identified in preparation after preparation. These cells included particular muscle cells that had dimensions from 100 to 300 μm and nerve cells with somas of 15–30 μm and axons of 5–8 μm. We developed techniques so that these identified cells could be individually stimulated, recorded from, voltage clamped, and filled with dyes. We succeeded in elucidating neural circuits in which the component cells could be described physiologically and anatomically. The identified cells could be recognized in light micrographs, and electron micrographs could be obtained from the same identified synapses that had been studied physiologically. In short, most of the techniques of invertebrate identified cell neurophysiology could be applied to *Drosophila*. This chapter will summarize some of the progress and promise in these areas.

2. Cast of Characters

The thoracic ganglion is composed of three neuromeres, one each for the prothoracic, mesothoracic, and metathoracic segment (Fig. 1). These neuromeres derive from ganglia that are separate in the embryo, but which are condensed in the larva and adult. The ganglion has the typical invertebrate structure of peripherally placed cell bodies and an interior neuropil. From each soma there is a thin process that enters the neuropil and arborizes there. A single neuron may arborize extensively in the thoracic and brain neuropil (Fig. 2). With the exception noted below, all the synapses between central neurons are in the neuropil.

The neurons that we study are those that are involved in an escape response. We chose this behavior because the cells involved are all large and accessible and it is easy to find mutants that do not escape. Flies will show a startle response to a variety of stimuli. By recording the electrical activity of the muscles during these responses, it was found that the pattern of activation of the various muscles involved varied in response to different stimuli (Wyman *et al.*, 1984). Out of this

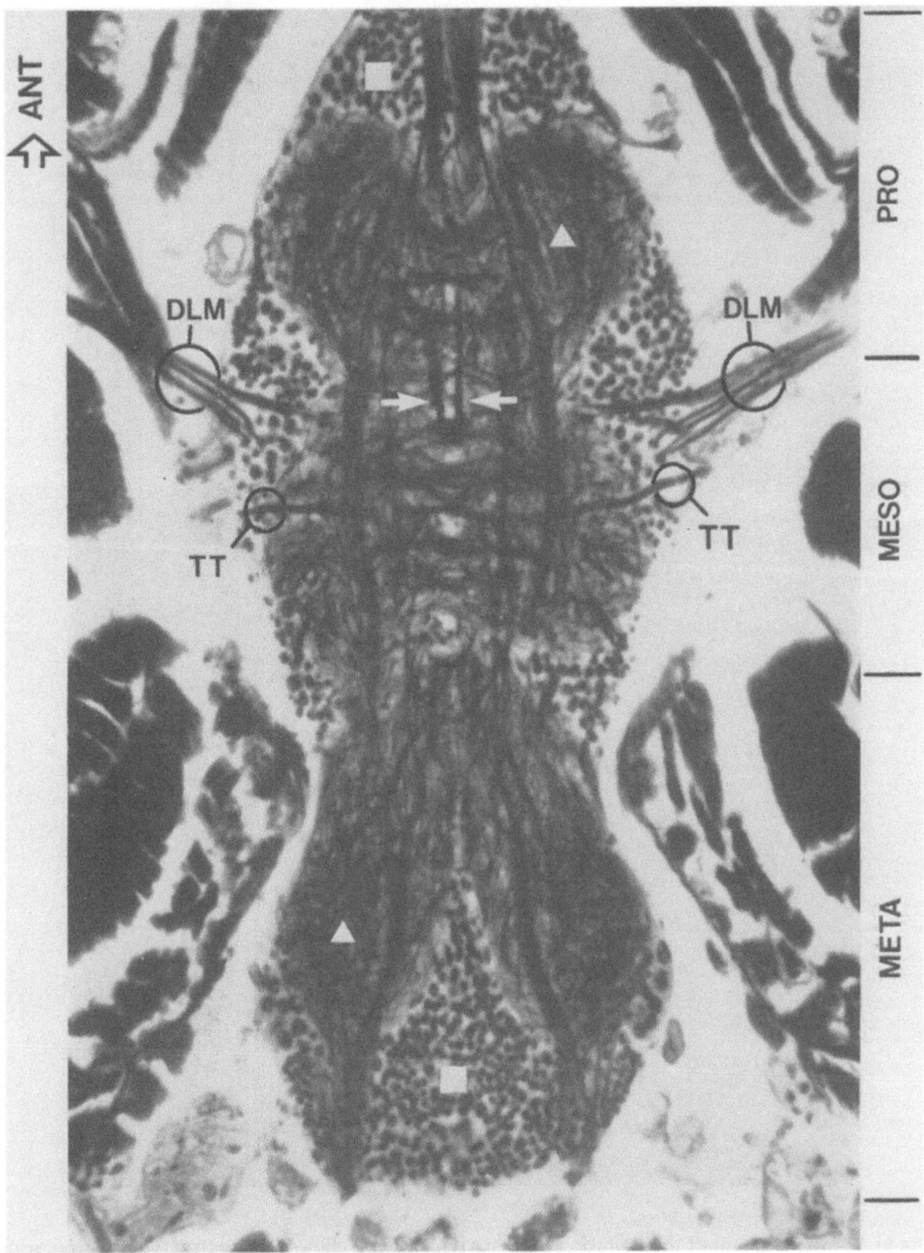

FIGURE 1. Horizontal section through adult thoracic ganglion. The ganglia for each of the three thoracic segments and the abdominal segments are fused, in the adult, into one thoracic ganglion. The portion of the ganglion deriving from each segment is termed a neuromere. The prothoracic, mesothoracic, and metathoracic neuromeres are visible in this section. The centrally located neuropil (white triangles) is surrounded by a cortex of cell bodies (white squares). The two giant fibers (arrows), the jump motoneurons (TTM), and the wing depressor motoneurons (DLM) can be identified. (From Power, 1948.)

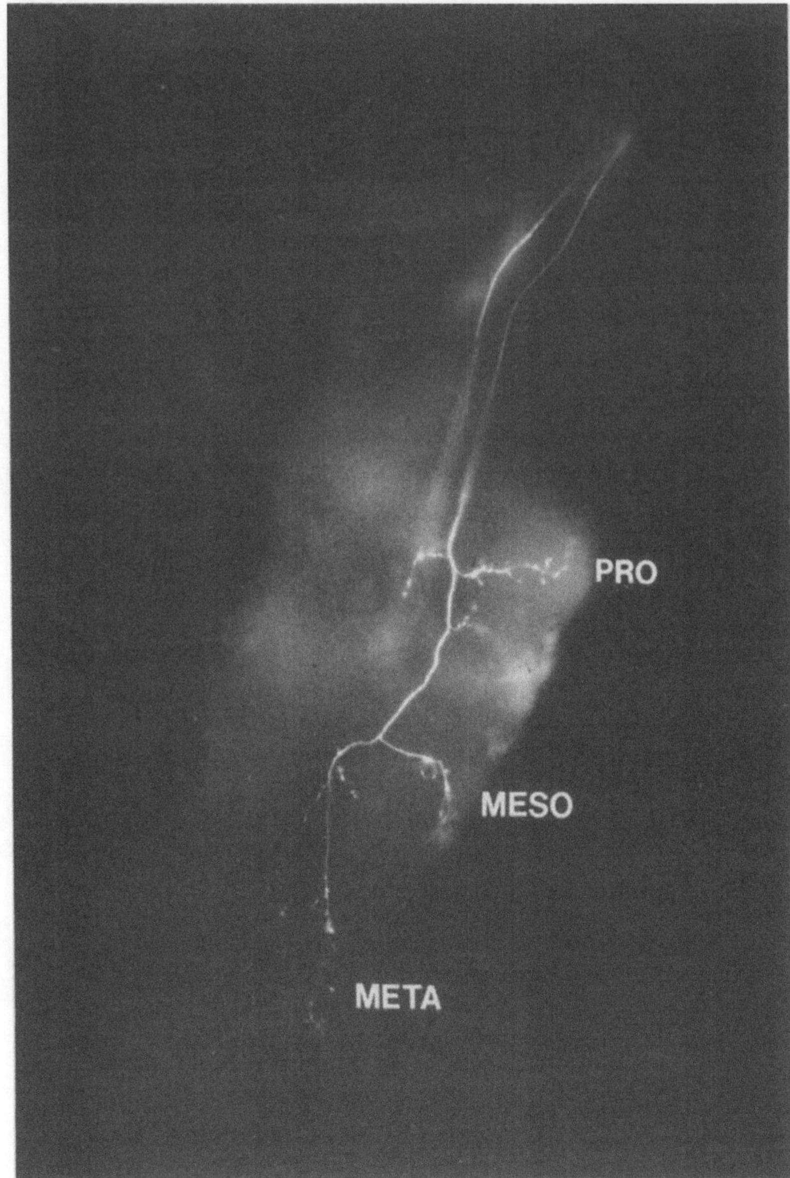

FIGURE 2. Whole mount of the thoracic ganglion with Lucifer-Yellow-filled neuron. This neuron, with its soma in the brain, ramifies in the region of the leg motoneurons in each of the three thoracic neuropils. A portion of the axon of another descending neuron is visible before it dives out of the plane of focus.

diversity of responses, we have focused on one fixed action pattern: the jump response elicited by a rapid light-off stimulus.

A startled fly jumps and initiates flight. The mesothoracic legs provide the majority of thrust during a jump (Mulloney, 1969). The tergotrochanteral muscle (TTM) is the largest twitch muscle in the fly, and acts as the main extensor muscle of the mesothoracic leg. It is innervated by one large axon (Koto and Wyman, 1984). Flight is maintained by the alternate contraction of two opposing sets of indirect flight muscles. The wing elevators (dorsoventral muscles, DVMs) and wing depressors (dorsal longitudinal muscles, DLMs) are composed of 13 bilaterally paired giant muscle fibers. Each fiber is innervated by a single axon.

Escape may be studied in the laboratory by suspending a fly glued to a pin and then inserting recording electrodes into particular muscles. An electrode in a flight muscle fiber records only the spikes elicited by the single axon innervating that fiber. An electrode in the jump muscle records only the spikes of its axon. A light-off visual stimulus initiates a fixed pattern of activity in the flight and jump muscles. An electrical stimulus to the brain elicits the same pattern of activity. A single spike is elicited in the TTM muscle and the resultant twitch causes the jump. Flight then ensues by activity of the wing depressors and elevators. However, after the initial one spike, the TTM is silent. It takes less than 2 msec from the time the brain is shocked until spikes start in the muscles. There must be a very rapidly conducting pathway from the brain to the escape muscles.

In other organisms, these rapid escape responses are coordinated by large axons called giant fibers (Eaton, 1984). *Drosophila* has two axons that (seen as they run through the cervical connective) are much larger than any others (Coggshall *et al.*, 1973). These have been named the cervical giant fibers (GFs) (Power, 1948; King and Wyman, 1980). These GFs, although only 5–8 μm in diameter, can be reliably impaled with a microelectrode (Tanouye and Wyman, 1980). When the GF is stimulated by an intracellular electrode, it is seen that a single spike is sufficient to drive the same stereotyped muscle response that is elicited by either a light-off stimulus or by extracellularly stimulating the GF in the brain. When the response is elicited by light-off or by brain stimulation, the threshold for the muscle response is also the threshold for a spike in the GF (as seen by the intracellular electrode). Thus the GF is a command neuron for this escape response: a single spike in the GF is sufficient and apparently necessary to elicit the stereotypic pattern of muscle activity.

The morphology of the GF can be visualized by filling it with Lucifer Yellow (Koto *et al.*, 1981). The GF cell body lies posteriorly in the brain (Fig. 3) where it has a number of branches. It sends an axon through the cervical connective to the thoracic ganglion. The GF is unbranched in its course through the prothoracic neuromere. Within the mesothoracic neuromere, it sends out a small tuft of branches (Fig. 4A) and then bends laterally and ends abruptly (Fig. 4B). The motoneuron for the jump muscle (TTM, Fig. 1) can be studied (Koto, 1983) by filling the muscle with horseradish peroxidase (HRP) and allowing retrograde transport of the enzyme (Fig. 5). It has two major branches, one posteriorly directed to the region of the leg motoneurons and a centrally directed branch that goes to the same region as the lateral bend of the giant fiber (Fig. 4B). At no point does it

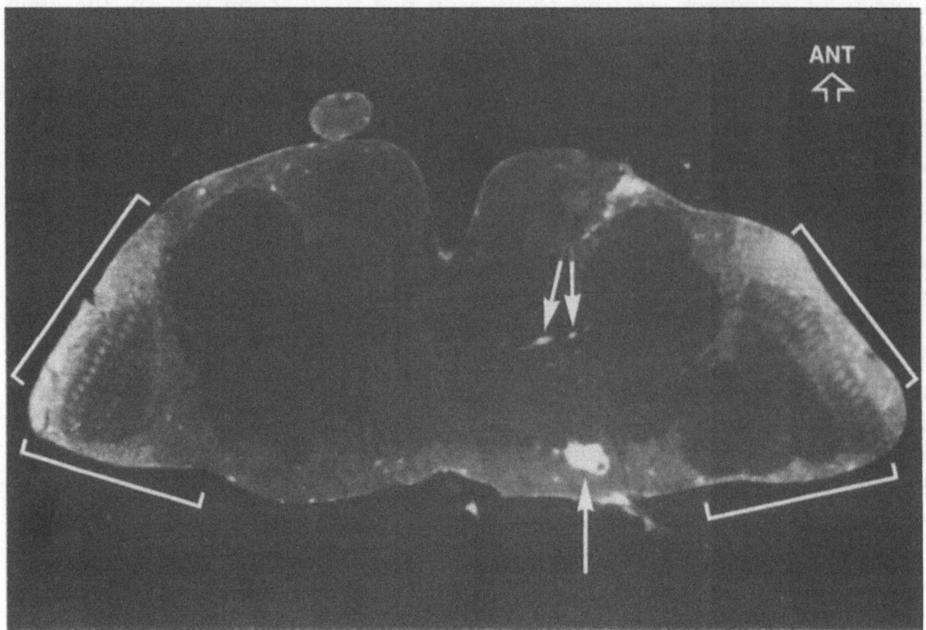

FIGURE 3. Horizontal section through the brain with Lucifer-Yellow-filled soma of the giant fiber (single arrow). This section also passes through one branch of the GF in the brain (double arrow). The optic lobes are also apparent (brackets). (From Koto *et al.*, 1981.)

extend to the contralateral side. When the laterally bending region of the GF was serially sectioned for the electron microscope, it was found that the GF makes a direct contact with the jump motoneuron (King and Wyman, 1980). Lucifer Yellow passes through this junction indicating an electrical synapse (Fig. 6). Thus, starting at the GF, the jump response involves only one neuron-to-neuron synapse and a single spike in each element (Tanouye and Wyman, 1980). The simplicity of this circuit has made it possible to identify mutations that affect this synapse.

Serial sections through the small tuft of branches (just anterior to the bend) reveal contacts between the GF and an interneuron, the peripherally synapsing interneuron (PSI) (King and Wyman, 1980). Lucifer Yellow is also passed at this junction (Fig. 6) indicating another electrical synapse (Wyman and Thomas, 1983). From this synapse, the PSI crosses the ganglion and leaves the ganglion in the contralateral posterior dorsal mesothoracic nerve. In the nerve, the PSI synapses on the five axons of the wing depressor muscle (Fig. 1) and then terminates (King and Wyman, 1980). To our knowledge, this synapse in *Drosophila* and closely related species (King and Valentino, 1983) is the only reported example in the literature where central neurons synapse with each other along the course of a peripheral nerve. The convenient location of this identified synapse makes it possible to readily determine whether any mutation that affects synaptic function also alters synaptic ultrastructure.

The wing depressor muscle is composed of six giant ($65 \times 150 \times 300$ μm)

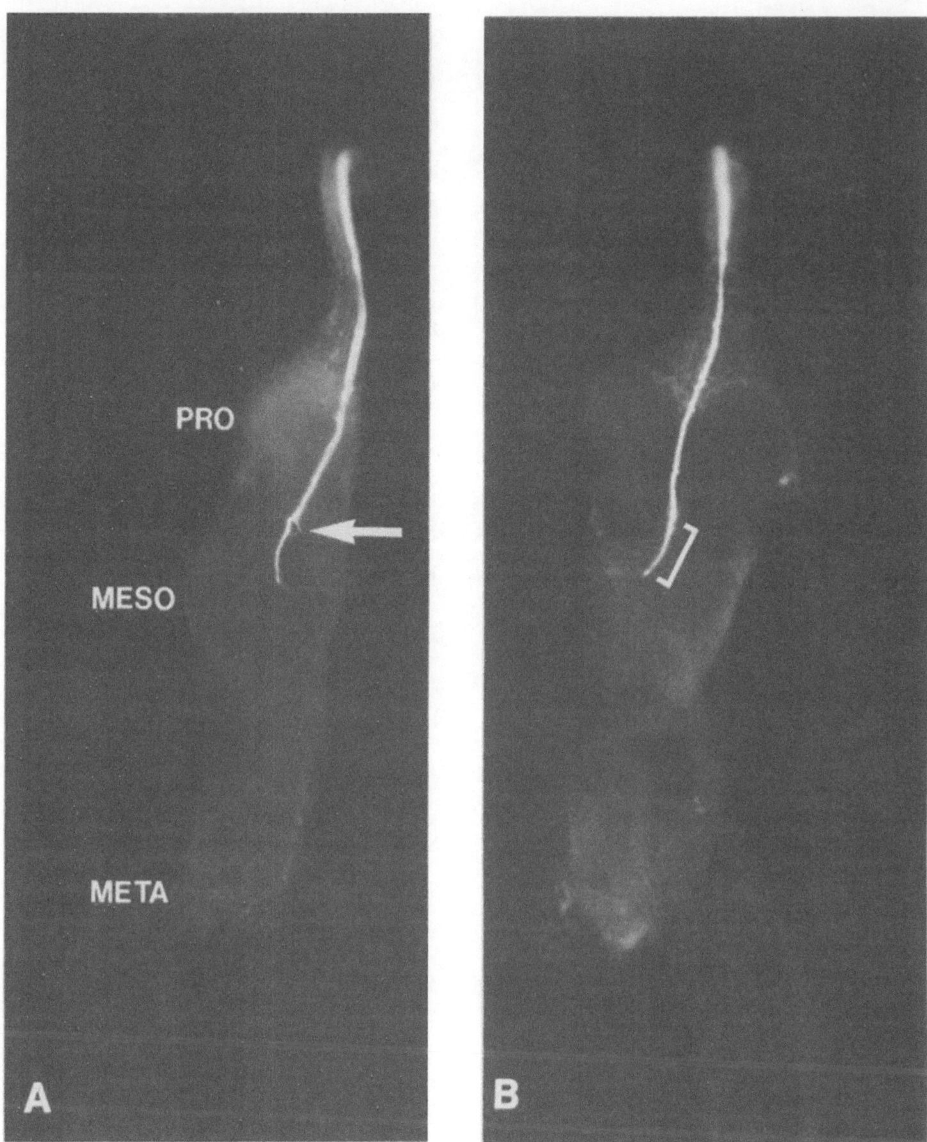

FIGURE 4. Whole mount of thoracic ganglion with Lucifer-Yellow-filled giant fiber. (A) Sagittal view. The ventral protruberances of the prothoracic, mesothoracic, and metathoracic neuromeres are visible from this angle. The tuft region of the GF, where it contacts the PSI, (arrow), can also be seen. (B) Horizontal view. The lateral bend of the GF in the mesothoracic neuropil is marked (bracket).

muscle fibers. Each muscle fiber is innervated by only a single axon (Ikeda *et al.*, 1980). One axon innervates both of the top two fibers (named a and b). The four other axons each innervate one of the four ventral fibers (named c through f). These motoneurons may be studied by injection of Lucifer Yellow into the cell body (Koto, 1983). Alternatively, HRP may be injected into any one of these mus-

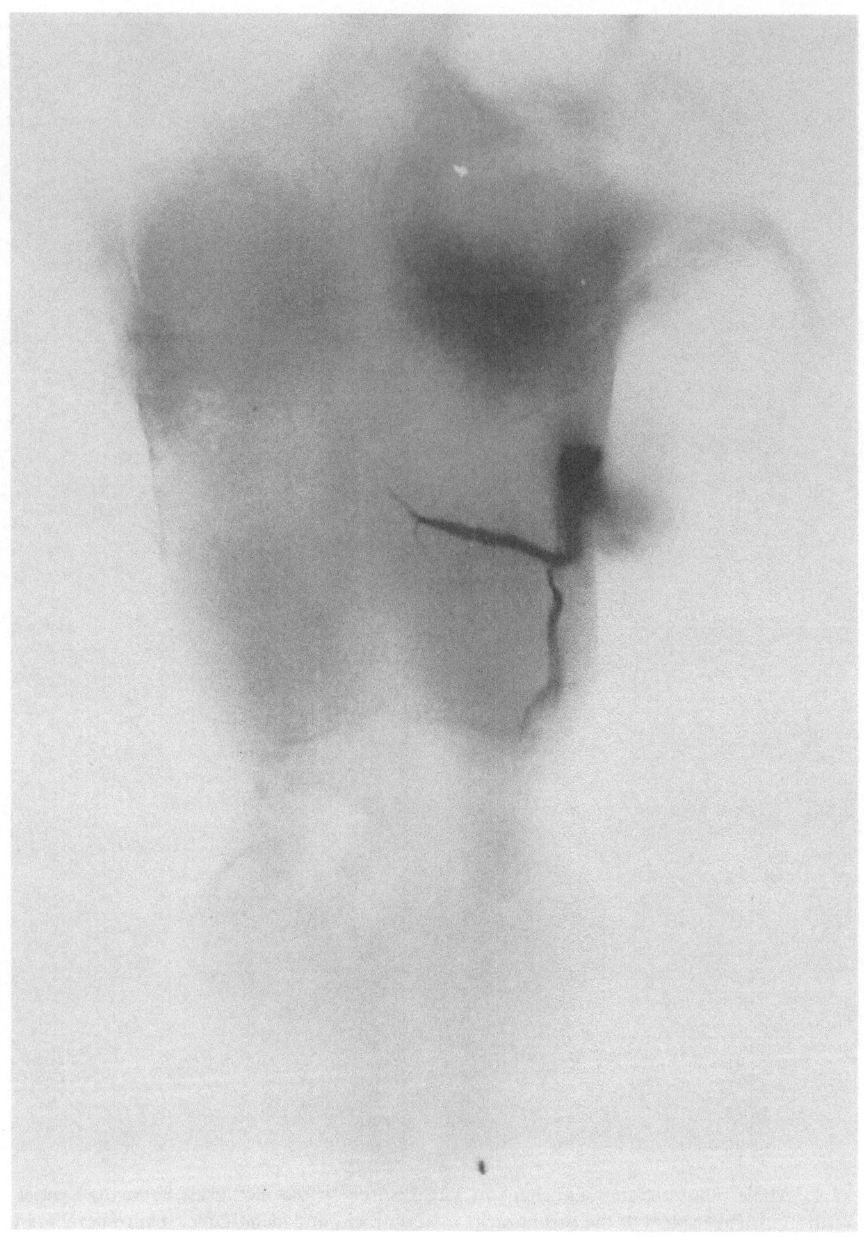

FIGURE 5. Whole mount of thoracic ganglion. The jump motoneuron (TTM) has been filled with HRP.

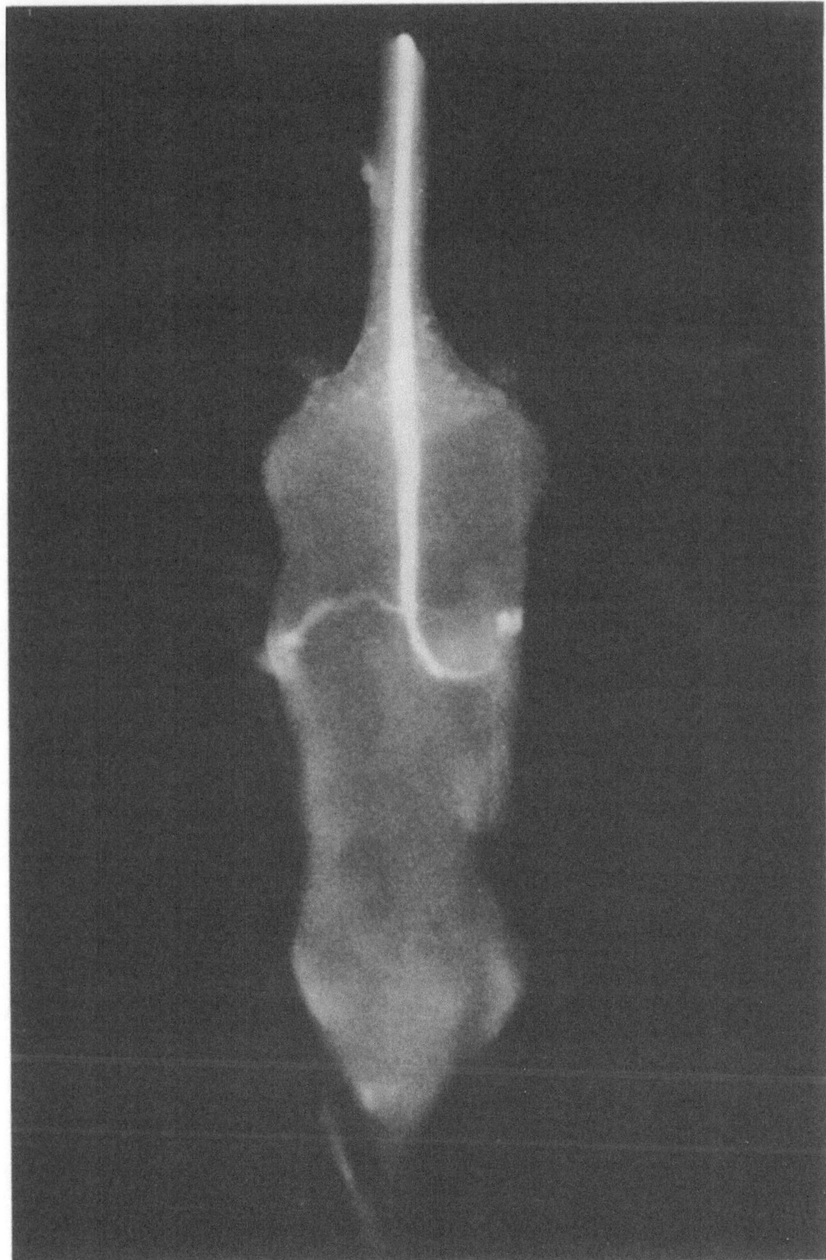

FIGURE 6. Whole mount of thoracic ganglion. The GF was filled with Lucifer Yellow. The dye has passed into the ipsilateral jump motoneuron (right side) and the contralateral PSI (left side).

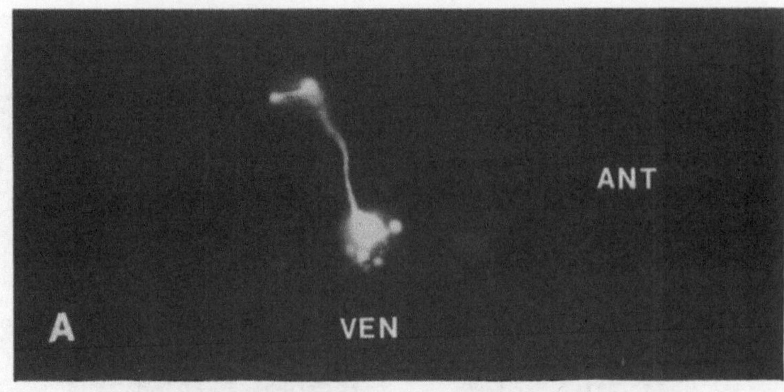

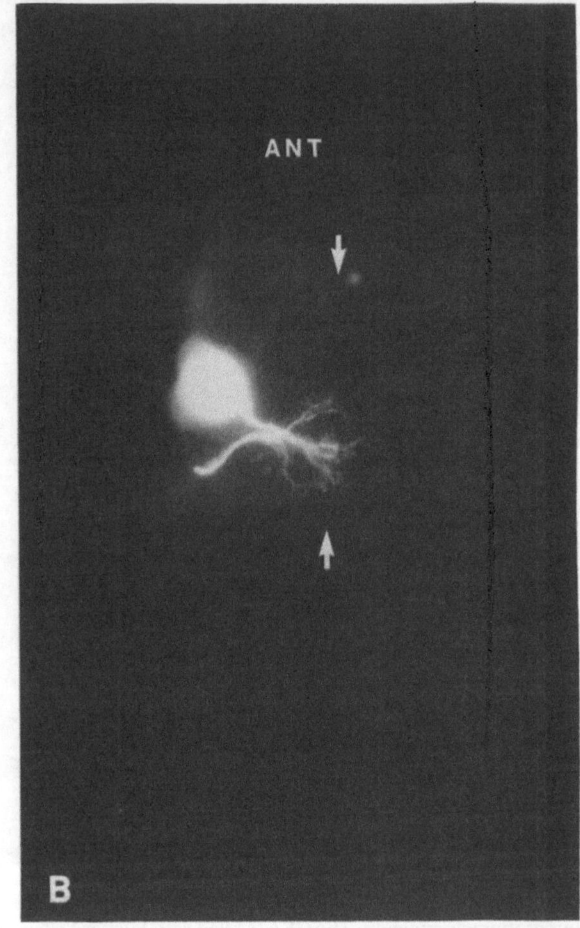

FIGURE 7. Whole mount of thoracic ganglion showing Lucifer-Yellow-filled wing depressor (DLM) motoneuron. (A) Sagittal view. The cell body is ventral; the primary neurite ascends dorsally and ramifies in the dorsal plane. (B) Horizontal view. The cell body is now out of focus; it appears large and blurred. The midline is marked with arrows. The dendrites extend to and just across the midline. The main axon has been cut where it leaves the ganglion to the left.

cle fibers and it will be taken up by that fiber's motoneuron (Coggshall, 1978). Figure 7A,B shows the structure of one of the motoneurons.

3. A Study of the Specificity of Connectivity

During development, neurons actively respond to some kind of guidance cues in their local environment in order to reach and finally synapse with their target cells. The main concern of our laboratory is to discover the molecular basis of these cell guidance and recognition cues. In this era of molecular genetics, possibly the easiest way to find a molecule about which we know nothing (except some idea about its function) is to first find the gene that makes the molecule. Then through molecular biological techniques, it should be possible to use the gene to find the molecule. Our research strategy is to create mutants in which the specificity of neural connectivity among our identified cells is disrupted.

3.1. Mutant Selection

For selecting jumpless mutants, we start with a stock of white-eyed flies since they jump more readily to the light-off stimulus than do red-eyed flies (Thomas and Wyman, 1984a; Wyman *et al.*, 1984). We feed the males a mutagen (ethyl methane sulfonate) to induce mutations in their meiosing sperms. The progeny of these flies are then screened by a variety of methods for their ability to jump to the light-off stimulus. In one protocol, the mutant progeny are introduced into the bottom of an inverted glass flask. Those that are capable of walking crawl up the sides. Severely debilitated mutants are left at the bottom and are discarded. The lights are then flicked out and most flies jump and fall a small distance. After a few stimuli, those that can jump emerge from the bottom of the flask. Nonjumpers remain in the flask. Over 50,000 F1 males carrying mutagenized X chromosomes were screened. From these, 57 mutant strains were derived which are defective in jumping, but appear normal in their other behaviors.

Mutations in the flight system are detected in a similar way. The selecting arena is a large hollow box. Mutagenized flies are introduced through a funnel at the top. As they are falling, most flies fly away and land on the sides of the box. Flightless flies fall straight down into a collecting tube at the bottom of a box. About 50,000 flies were screened in this way and 40 mutant strains were derived which cannot fly, but appear normal in their other behaviors.

The mutants isolated in the behavioral screens can be defective in either jumping or flying. There are many possibilities for the focus of the defect (e.g., visual system, muscular system, skeletal system). Consider the jump response for instance. There are three stages to the jump response: (1) processing of visual information presynaptically to the GF; (2) transmission from the GF to the motoneurons; and (3) transmission from the motoneurons to the muscles and transduction into motion. We are primarily interested in defects in stage 2, because these entail defects in the identified neurons. Stage 1 can be bypassed by direct stimulation of the GF while recording from postsynaptic cells. The GF can be easily stimulated extracellularly by passing current across the brain or the cervical connective. Nonjumpers that have a normal motor output after GF stimulation are apparently defective in visual reception or integration. We do not study these. Nonjumpers with abnormal motor output after brain stimulation are then checked to see if neuromuscular transmission and muscle contraction (stage 3) are normal. This is done by stimulating the motoneurons directly and recording muscle spikes and leg motions. Flies with a normal neuromuscular system must then have defects involving transmission from the GF to the TTM motoneuron, or transmission from the GF to the wing depressor motoneurons via the PSI.

3.2. Mutants

As examples of the types of mutants that we find, the two mutants that we have studied most will be described. Both of these mutations disrupt the connection between the GF and the jump motoneuron. One mutant affects a branch of the GF, the other mutant affects a branch of the jump motoneuron.

3.2.1. Mutation Affecting the GF

One mutant, called *bendless (ben),* isolated in the screen for jumpless flies, was found by electrophysiology to lack direct transmission from the GF to the TTM

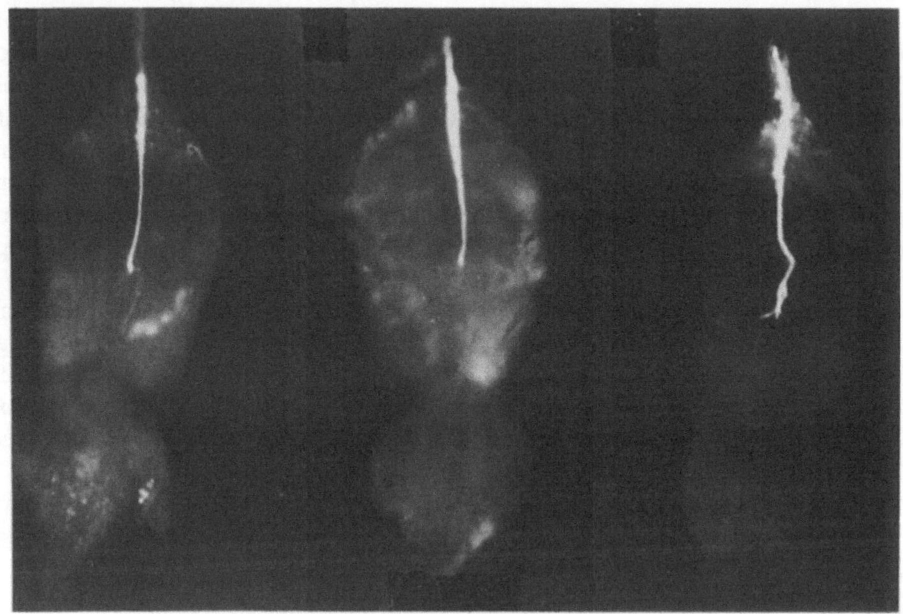

FIGURE 8. Whole mounts of the thoracic ganglion in three *bendless* flies. The GF has been filled with Lucifer Yellow in each. These fibers do not form the lateral bend region. Instead, the GF ends in posteriorly directed tendrils of various extents. The right-hand individual shows some of the largest supernumerary tendrils.

motoneuron (Thomas and Wyman, 1984a). The output from the GF to the PSI functions normally: when the GF is stimulated, the flight muscles are activated normally. Thus, the GF is present, can carry spikes, and contacts and makes an electrical synapse with the PSI. The PSI, in turn, must make its chemical synapses onto the flight motoneurons. However, the jump muscle is not activated normally. These facts suggest that the GF is normal down to its synapse with the PSI, but that there is a defect somewhere distal to this point. The jump motoneuron, when stimulated directly, proves to be present and drives the leg kick normally. Thus, the defect should be proximal to the jump motoneuron axon. When the mutant GF was filled with Lucifer Yellow, it was found that the lateral bend of the fiber (which normally makes the connection to the motoneuron) was missing (Fig. 8); hence the name *bendless* (Thomas and Wyman, 1982). The structure of the GF in the brain (cell body and dendrites) of normal flies is indistinguishable from that in mutant flies. The jump motoneuron is unaffected: its morphology is normal.

In the mutant, the GF terminates in very fine processes or sometimes in larger tendrils (Fig. 8C). One interpretation of this is that these processes are the result of searching attempts by the GF to make contact with the jump motoneuron. It is possible that, due to the lack of some molecular recognition signal, the connection was never made or made but not maintained.

3.2.2. Mutation Affecting the Jump Motoneuron

A second mutant called *passover (pas)* was also found by electrophysiology to lack direct transmission between the GF and the jump motoneuron (Thomas and Wyman, 1984a). The GF is present, conducts spikes, and can make output synapses, but not the normal one to the jump motoneuron. The jump motoneuron is present and carries spikes; neuromuscular transmission to the jump muscle is normal. When the GF in the mutant was filled with Lucifer Yellow, it was seen to be normal. The lateral bend is present and extends to the region where it normally makes contact with the TTM motoneuron. However, the jump motoneuron of the mutant is abnormal. One such mutant motoneuron is shown in Fig. 9. Normally the medial branch of the motoneuron (which makes the synapse with the GF) does not extend over the midline. In Fig. 9 it can be seen that this branch grows to the midline and sends a branch up the midline. It then crosses over the midline and grows down what appears to be the mirror-image path on the opposite side.

When the mutant motoneuron is viewed at high magnification, it can be seen that there are tendrils branching off from the main neurite all along the abnormal part of the growth path. The branches of the motoneuron must come very close to, indeed may touch, the ipsilateral GF and they may also contact the GF on the contralateral side. However, a normal synapse is not made. (The name *passover* refers to the fact that the motoneuron passes right over the region of normal synaptic contact with the GF.) It is again possible that a recognition signal is missing, and that lacking this, the motoneuron extends its processes beyond its normal limits.

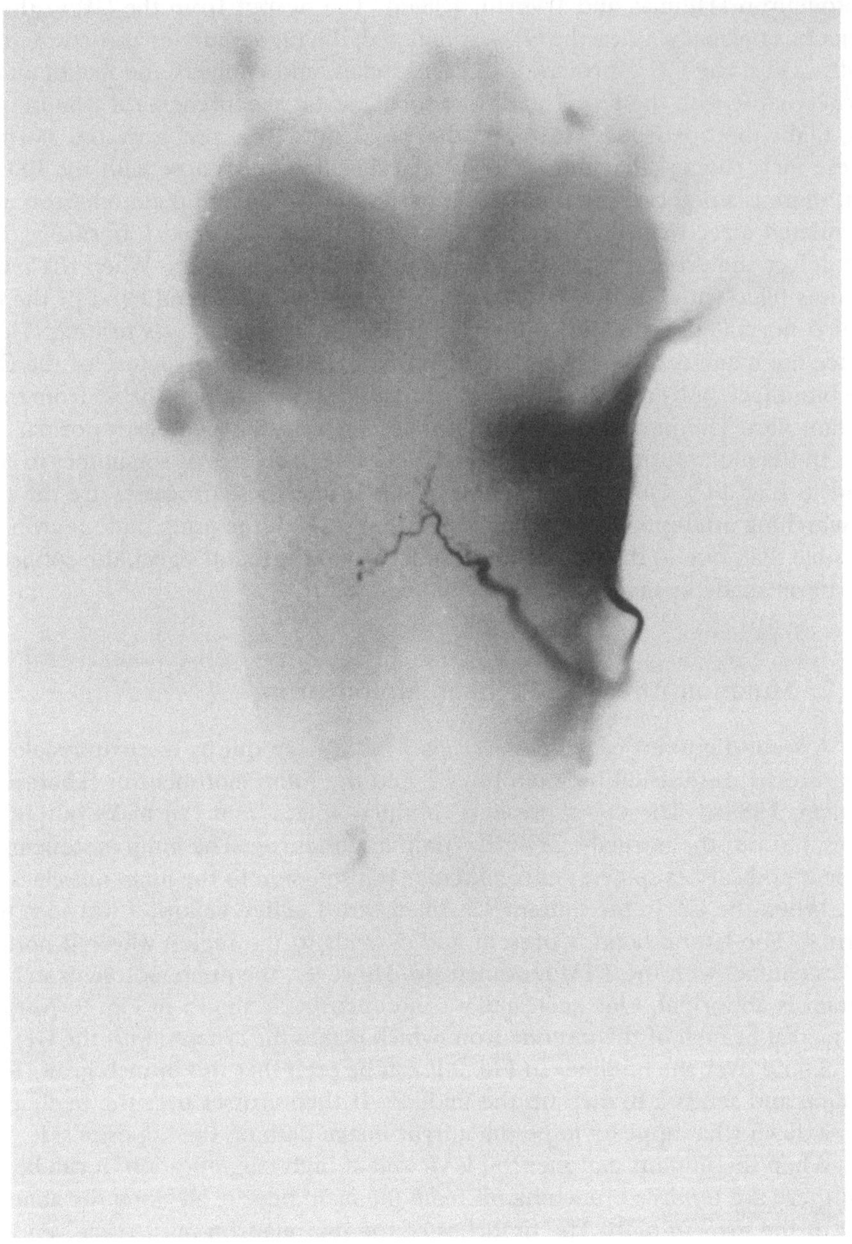

FIGURE 9. Whole mounts of thoracic ganglion of *passover* mutant fly with HRP-filled jump motoneuron. The central branch, instead of being smooth surfaced and stopping short of the midline, has small tendrils emanating from it and supernumerary processes that grow up the midline and down an approximately mirror-image path on the contralateral side. Posterior branch is out of plane of focus.

3.3. Future Directions

We have found mutations in genes that appear to be necessary for specific aspects of neural connectivity. It is quite striking to have found two mutants that both disrupt the same synaptic connection, yet one affects the branching pattern of the presynaptic cell while the other affects the shape of the postsynaptic cell. In both cases the affected cell sends out many supernumerary processes that might be indications of the failure of a specific recognition process.

A number of lines of investigation must be followed before one can do more than guess what the significance of these results are. We would like to know in which cells the gene products act to allow normal connectivity. We should be able to ascertain this by making mosaics in which the GF is mutant and the jump motoneuron is normal and by making the reverse mosaic in which the GF is normal, but it grows through mutant thoracic terrain.

We plan to study the development of these cells in normal and mutant flies. We know the time period, during metamorphosis, when the connections start functioning. By injecting dyes into the GF and motoneuron during metamorphosis, we should be able to visualize the cells before they make contact and during the period of synaptogenesis. It would be interesting to see what, if any, searching activities are taken by these cells as they grow to contact each other. It will be crucial to know how and when the mutant developmental program differs from the normal.

We will also extend the genetic analysis of this system. It is important for us to find more alleles of these genes. It is possible that different alleles have different phenotypes. Null alleles (which make no gene product whatsoever) would allow us to see the full range of defects caused by a total absence of the gene product. Temperature-sensitive alleles would give us information on when in development the normal gene product is necessary for normal connectivity. We would also like to find the full set of genes similar to *ben* and *pas* in order to determine how many genes are necessary to form this synapse that are not also necessary for the general development and functioning of the nervous system. Eventually, when we are convinced that these, or other genes, do indeed code for "guidance" or "recognition" factors, we will clone the genes and study their molecular functions.

4. Plasticity and Ion Channels

In the course of doing experiments with the flight muscle, we noticed a very intriguing case of plasticity. We were performing experiments on a nerve–muscle preparation utilizing the large cells of the dorsal longitudinal flight muscle. A small amount of current injected into one of these muscle cells elicits only a passive depolarization. However, if the nerve is stimulated once causing a single spike in the muscle, then the same amount of current injected into the muscle cell causes a train of spiking events (Fig. 10). This was one of the most striking cases of plasticity we had seen. The single spike had switched the muscle membrane from a passive state into a hyperexcitable state (Salkoff and Wyman, 1980). The switching phenomenon could be shown to last for up to 2 sec.

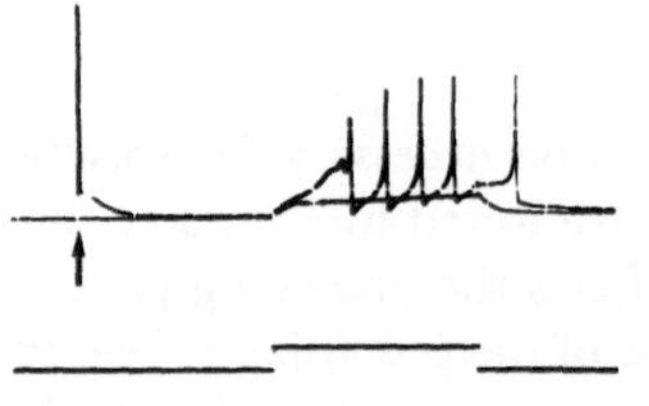

FIGURE 10. Muscle membrane responses to two pulses of current. In the first sweep a small amount of current (lower trace) was passed into the muscle cell. The membrane voltage (top trace) depolarized and repolarized in a passive manner. In the second sweep (superimposed) the motoneuron was shocked once causing one spike in the muscle membrane (arrow). Then the same amount of current was injected into the muscle (lower trace) and the membrane responded (upper trace) with a series of spikelike events. (From Salkoff and Wyman, 1980.)

Further investigation led to the conclusion that the phenomenon was due to the properties of the postsynaptic membrane: a current pulse delivered directly to the muscle by an intracellular electrode could throw the switch as well as a nerve-driven spike. To push the work further, the biophysics of the muscle cell membrane would have to be investigated. The muscle cell could be voltage clamped, but the response of the membrane was a complex mixture of the contribution of many channels (Salkoff and Wyman, 1983a,b). Luckily, this muscle cell develops during pupation, in the middle of the fly's life span, and it was possible to voltage clamp the cell during its development (Salkoff and Wyman, 1981a). At about 55 hr into pupation, there are no electrically excitable channels in the membrane. This gives one the chance to study the properties of the underlying membrane without excitable channels. The membrane shows an interesting form of nonlinearity called instantaneous rectification (Fig. 1a in Salkoff and Wyman, 1981a).

At 72 hr the first excitable channels appear. These channels carry the A current. They are a voltage-sensitive, fast-activating, fast-inactivating potassium channel. Later on, the delayed rectification channel and a calcium channel appear. As soon as the calcium channel is functional, one can see that there is a second type of A-current channel that is *not* primarily activated by voltage, but by calcium entering through the calcium channel (Salkoff, 1983a).

One can use the 72-hr pupa to obtain records of the A current alone, without contamination from current passing through other channels. At this early stage, it could be seen that the A current, aside from being rapidly inactivated, is very slow to recover from inactivation. This turned out to be the basis for the plasticity phenomenon. When the muscle is stimulated the first time, the calcium channel and the A-current channel are activated almost simultaneously and the inward Ca^{2+} current is balanced by the outward K^{+} current, hence the membrane appears passive. The nerve-driven muscle spike activates both the calcium and the A current. The calcium current recovers from inactivation rapidly, whereas it can take the A current up to 2 sec to recover. Thus a second current pulse into the muscle, following the nerve-driven muscle spike, elicits a calcium current with little countervailing A current and the muscle membrane depolarizes into regenerative spikes. Thus this striking example of plasticity had a simple explanation: the slow time course of recovery of the A current. A similar regulation of neural excitability by an A current in the mammalion CNS has recently been discovered (Segal *et al.*, 1984).

These results set the stage for a genetic analysis of membrane excitability in *Drosophila.* Realizing that this preparation was ideal for determining whether mutations affected the A current, we started a mutagenesis. We guessed that the phenotype of an A-current-defective mutant would be hyperexcitability, so we also examined various hyperexcitable (and other neurological) stocks that were available (e.g., *Hyperkinetic, bang-sensitive, ether-a-gogo, Shaker*). It was the last named of these, *Shaker* (Kaplan and Trout, 1969), which in fact affected the A current. One allele was found to double the speed of inactivation of the current and another allele was found to cause a complete absence of the current (Salkoff and Wyman, 1981b). Other channels were not affected by the mutations. By cytogenetics, the gene was localized to band 16F on the X chromosome (Salkoff and Wyman, 1983b; Tanouye *et al.,* 1981).

Meanwhile, and quite independently, Mark Tanouye, who had developed the technique for recording from the GF in our laboratory, was studying the genetics of *Shaker* and its effect on the spikes recorded and in the GF. He found that some alleles caused a prolongation of the repolarization after spikes, and other alleles caused multiple firing. He also suggested that a potassium current was involved (Tanouye *et al.,* 1981).

These mutations had also been studied in the larva by Jan *et al.,* (1977), who found prolonged transmitter release at the larval neuromuscular junction, and since this phenotype could be mimicked by application of 4-AP (a known potassium-channel blocker), they suggested that a potassium channel was involved.

Thus it appears that the *Shaker* gene codes for the A-current channel in both adult and larval nerve and in adult flight muscle. Great excitement has attended the discovery that *Shaker* controls the A-current channel. This is the first *Drosophila* gene that has been unambiguously attached to a particular channel. Several laboratories are racing to clone the gene. We expect that the clone and then the amino acid sequence of the channel protein will be available in the near future (Salkoff, 1983b).

5. Neural Circuits and Pattern Generation

During flight, the DLM motoneurons in *Drosophila* (Harcombe and Wyman, 1978) and other Diptera (Wyman, 1969a,b) are activated in an interesting cyclic pattern. Pattern generation in the system has been studied by recording from the individual motor units (Wyman, 1965) or by antidromic stimulation of the axons (Mulloney, 1970; Harcombe and Wyman, 1977).

Ikeda (1976) found that when flies homozygous for the temperature-sensitive mutation *shi* are subjected to a temperature of 29°C, the flight motor pattern is automatically turned on. Using this mutation, Koenig and Ikeda (1983) have been able to elicit flight motor output in the highly dissected preparations necessary for intracellular recording. This technique has allowed intracellular recording from the motoneuron's axon (Ikeda *et al.,* 1980) or soma (Koenig and Ikeda, 1983) and is yielding very valuable new information.

The pattern is generated by direct interaction of the motoneurons (Harcombe

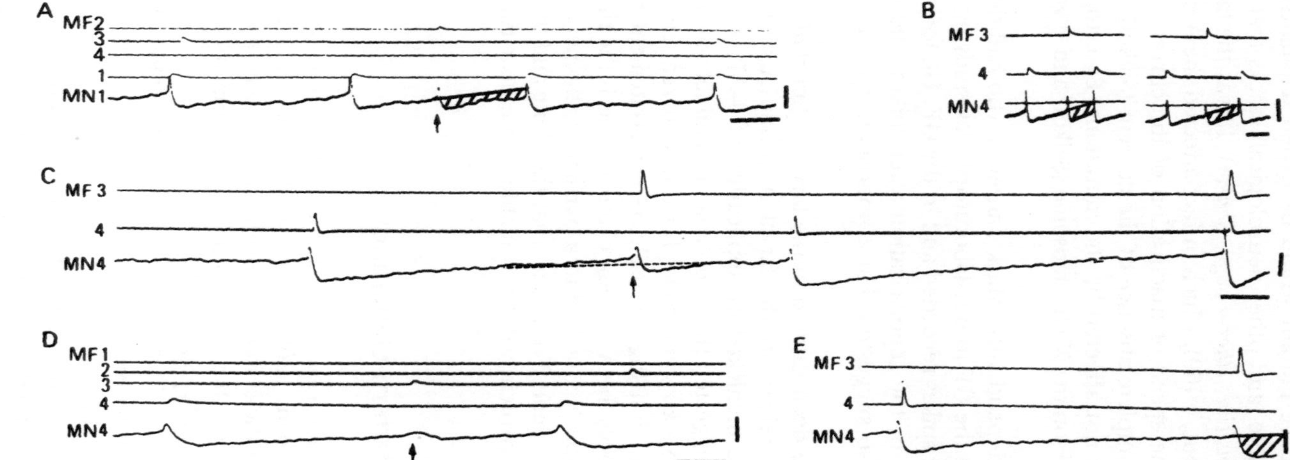

FIGURE 11. Recordings of DLM muscle potentials (top traces) and intracellular recording from a DLM motoneuron (lowest trace). Each record shows the effect of a spike in one DLM motoneuron when another DLM motoneuron fires. (A) Response in MN1 when MN2 fires. (B–E) Response in MN4 when MN3 fires. In A, B, and E, we (Wyman *et al.*) have drawn in a baseline that might have been expected if the heteronymous neuron had not fired. The effect of the firing is biphasic: an initial depolarization (extending above the baseline) followed by a longer-lasting hyperpolarization (extending below the baseline and crosshatched). In C, Koenig and Ikeda have drawn in an expected baseline showing the same biphasic effect. (From Koenig and Ikeda, 1983.)

and Wyman, 1977; Tanouye and Wyman, 1981; Koenig and Ikeda, 1983). Harcombe and Wyman (1977) proposed that when a motoneuron fires, it resets its own firing rhythm, possibly because "the spikes invade the cell's integrating zone and repolarize it." Koenig and Ikeda (1983), in intracellular recordings from the motoneuron soma, have seen that the spike invades the soma and repolarizes it. Assuming that the repolarization occurs also at the integrating zone, this substantiates the self-resetting hypothesis.

Each motoneuron fires once per cycle. The firing times of the different motoneurons are separated so that only rarely does a motoneuron fire synchronously or near synchronously with another motoneuron. Wyman and co-workers have proposed that this spacing is due to mutual inhibition between the motoneurons (Wyman, 1969a,b; Harcombe and Wyman, 1977; Tanouye and Wyman, 1981; Wyman and Tanouye, 1982). Harcombe and Wyman (1977) showed that an antidromic spike, set up in one motoneuron, inhibited the firing of the other motoneurons.

Although the great majority of spikes occur asynchronously, about 1 to 5% of spikes of one motoneuron will occur synchronously with the spike of a second motoneuron (Harcombe and Wyman, 1978, p. 274; Koenig and Ikeda, 1980, Figs. 1, 2c). Because of these synchronies, Koenig and Ikeda (1980) have proposed that mutual excitation is also important in pattern generation. In intracellular recordings they find that when one motoneuron fires, the spike waveform is transmitted (probably via an electrical synapse) to another motoneuron. They believe that excitation transmitted in this way can explain all observations and they feel their results are "incompatible with a system of mutual inhibition" (Koenig and Ikeda, 1983). However, their records (Koenig and Ikeda, 1983) clearly show that the hyperpolarizing (inhibitory) phase, as well as the depolarizing (excitatory) phase, of the spike is transmitted to the impaled cell (Fig. 11). Depending on how one projects the baseline that would be expected if the driving motoneurons had not fired, either the depolarization or the hyperpolarization can look like the larger effect. Thus, the intracellular data currently available can be used to support either hypothesis.

It seems that if one motoneuron is very close to firing when it receives the depolarizing synaptic potential, then it can be excited into a synchronous firing (1–5% of cases). If it does not fire synchronously (95–99% of cases), then its membrane is hyperpolarized by the transmitted spike's after potential (Koenig and Ikeda, 1983, Figs. 1, 4, 6) and will thereby be inhibited.

The foregoing suggests an intriguing and simplifying hypothesis that could explain all the extant data. The spike waveform itself, by invading the integrating zone of the firing motoneuron and by being passed through an electrical synapse to other motoneurons, could be the major pattern-generating signal.

6. Homeotic Mutations

In recent molecular biological work, it has been possible to isolate messages transcribed from the homeotic genes in the bithorax and antennapedia regions.

Recessive mutations of the bithorax complex change anatomical features of more posterior segments toward those of the mesothorax. Several dominant mutations do the reverse. Dominant mutations of antennapedia change parts of the antenna into parts of the mesothoracic leg; recessive mutations do the opposite. Surprisingly, the major tissue of synthesis of the messages so far isolated from these gene complexes is the ventral ganglion. Hafen *et al.* (1984) found in advanced embryos that antennapedia transcripts were predominantly located in the mesothoracic neuromere. They also found, in embryos carrying bithorax complex mutations, that the neuromeres in more posterior, transformed segments also accumulated antennapedia transcripts. Akam (1983) showed that ultrabithorax transcripts, in the late embryo, were most abundant in the metathoracic and first abdominal neuromere.

Other studies of the nervous system in homeotics have largely focused on the effect of mutations of these genes on the ingrowing sensory fibers from the transformed cuticle (e.g., Palka and Ghysen, 1982). However, the clearest information on the effect of homeotic genes on the nervous system should come from the study of their effects on identified neurons. Consequently, we have embarked on a study of the effect of mutations of the bithorax complex on the identified neurons of the escape system.

In a bithorax fly where the cuticle of the metathorax is fully transformed into a mesothoracic cuticle, the GF duplicates its mesothoracic branches in the metathorax (Thomas and Wyman, 1984b). Since the GF cell body is in the brain, and since it is generally thought that the bithorax genes are not active in the head (Lewis, 1978), it is unlikely that the reiteration of GF branching is due to autonomous expression of the mutated genes in the GF itself. Instead, it is more likely that at least some elements of the metathoracic ganglion have been transformed and that the GF is acting as a probe for these changes. The GF reiteration is then a response to the duplication in the metathoracic neuromere of the normal mesothoracic cues involved in GF branching. That the GF repeats its mesothoracic morphology exactly, without addition, deletion, or variation, suggests that the cues are identical in both the mesothorax and homeotic metathorax.

7. Conclusion

In this chapter we have discussed four lines of research that have been pursued successfully by using the *Drosophila* thorax. There are quite a number of other areas of neurobiological research that have also centered on the *Drosophila* thorax. Just a few more of these will be mentioned as further examples of the range of research on this structure: muscle development (Deak, 1978; Costello and Thomas, 1981; Lawrence, 1982); the endocytosis of synaptic vesicles (Costello and Salkoff, 1982; Kosaka and Ikeda, 1983); and the embryology of the nervous system (Thomas, *et al.*, 1984).

We have demonstrated that many of the sophisticated neurobiological techniques are adaptable to *Drosophila*. Combining these methods with the range of techniques available in *Drosophila* genetics and molecular genetics is now leading to rapid progress. It is clear that advances in this area have only just begun to accelerate.

Support: NIH:NS–07314 and NS–14887.

References

Akam, M. E., 1983, The location of *Ultrabithorax* transcripts in *Drosophila* tissue sections, *EMBO J* **2:**2075–2084.

Alawi, A. A., and Pak, W. L., 1971, On-transient of insect electroretinogram: Its cellular origin, *Science* **172:**1055–1057.

Ashburner, M., and Novitski, E. (eds.), 1976–1983, *The Genetics and Biology of Drosophila,* Volumes 1a–3c, Academic Press, New York.

Bentley, D., 1975, Single gene cricket mutations: Effects on behavior, sensilla, sensory neurons, and identified interneurons, *Science* **187:**760–764.

Benzer, S., 1973, Genetic dissection of behavior, *Sci. Am.* **229:**24–37.

Brenner, S., 1973, The genetics of behavior, *Br. Med. Bull.* **29:**269–271.

Caviness, V. S., Jr., and Rakic, P., 1978, Mechanisms of cortical development: A view from mutations in mice, *Annu. Rev. Neurosci.* **1:**297–326.

Coggshall, J. C., 1978, Neurons associated with the dorsal longitudinal flight muscles of *Drosophila melanogaster, J. Comp. Neurol.* **177:**707–720.

Coggshall, J. C., Boschek, C. B., and Buchner, S. M., 1973, Preliminary investigations on a pair of giant fibres in the central nervous system of dipteran flies, *Z. Naturforsch.* **28:**783–4.

Costello, W. J., and Salkoff, L., 1982, Suppression of abnormal synaptic vesicle depletion by divalent cations in the *Drosophila* mutant *shibire, Soc. Neurosci. Abstr.* **8:**494.

Costello, W. J., and Thomas, J. B., 1981, Development of thoracic muscles in muscle-specific mutant and normal *Drosophila melanogaster, Soc. Neurosci. Abstr.* **7:**543.

Deak, I. I., 1978, Thoracic duplications in the mutant *wingless* of *Drosophila* and their effect on muscles and nerves, *Dev. Biol.* **66:**422–441.

Eaton, R. C., 1984, *Neural Mechanism in Startle Behavior,* Plenum Press, New York.

Goodman C. S., 1978, Isogenic grasshoppers: Genetic variability in the morphology of identified neurons, *J. Comp. Neurol.* **182:**681–706.

Hafen, E., Levine, M., and Gehring, W. J., 1984, Regulation of *Antennapedia* transcript distribution by the *bithorax* complex in *Drosophila, Nature (London)* **307:**287–289.

Harcombe, E. S., and Wyman, R. J., 1977, Output pattern generation by *Drosophila* flight motoneurons, *J. Neurophysiol.* **40:**1066–1077.

Harcombe, E. S., and Wyman, R. J., 1978, The cyclically repetitive firing sequences of *Drosophila* flight motorneurons, *J. Comp. Physiol.* **123:**171–179.

Horvitz, H. R., Sternberg, P. W., Greenwald, I. S., Fixsen, W., and Moyed Ellis, H., 1983, Mutations that affect neural cell lineages and cell fates during the development of *Caenorhabditis elegans, Cold Spring Harbor Symp. Quant. Biol.* **48:**453–462.

Hubel, D. H., 1982, Exploration of the primary visual cortex, 1955–78, *Nature (London)* **299:**515–524.

Ikeda, K., 1976, Temperature controlled release of flight pattern by a single-gene mutant in *Drosophila melanogaster, Fed. Proc.* **35:**642.

Ikeda, K., and Kaplan, W. D., 1970, Patterned neural activity of a mutant *Drosophila melanogaster, Proc. Natl. Acad. Sci. U.S.A.* **66:**765–772.

Ikeda, K., Koenig, J. H., and Tsuruhara, T., 1980, Organization of identified axons innervating the dorsal longitudinal flight muscle of *Drosophila melanogaster, J. Neurocytol.* **9:**799–823.

Jan, L. Y., and Jan, Y. N., 1976, Properties of the larval neuromuscular junction in *Drosophila melanogaster, J. Physiol.* **262:**189–213.

Jan, L. Y., Jan, Y. N., and Dennis, M. J., 1977, Two mutations of synaptic transmission in *Drosophila, Proc. R. Soc. Ser. B* **198:**87–108.

Kandel, E. R., 1976, *Cellular Basis of Behavior: An Introduction to Behavioral Neurobiology,* W. H. Freeman, San Francisco, California.

Kaplan, W. D., and Trout, W. E., III, 1969, The behavior of four neurological mutants of *Drosophila, Genetics* **61:**399–409.

King, D. G., and Valentino, K. L., 1983, On neuronal homology: A comparison of similar axons in *Musca, Sarcophaga,* and *Drosophila* (Diptera: Schizophora), *J. Comp. Neurol.* **219:**1–9.

King, D. G., and Wyman, R. J., 1980, Anatomy of the giant fiber pathway in *Drosophila.* I. Three thoracic components of the pathway, *J. Neurocytol.* **9:**753–770.

Koenig, J. H., and Ikeda, K., 1980, Neural interactions controlling timing of flight muscle activity in *Drosophila, J. Exp. Biol.* **87:**121–136.

Koenig, J. H., and Ikeda, K., 1983, Reciprocal excitation between identified flight motor neurons in *Drosophila* and its effect on pattern generation, *J. Comp. Physiol.* **150:**305–317.

Kosaka, T., and Ikeda, K., 1985, Possible temperature-dependent blockage of synaptic vesicle recycling induced by a single gene mutation in *Drosophila, J. Neurobiol.* **14:**207–225.

Koto, M. L., 1983, Morphology of giant fiber system neurons in wild-type and mutant *Drosophila melanogaster,* Ph.D. Thesis, Yale University, New Haven, Connecticut.

Koto, M. L., Tanouye, M. A., Ferrus, A., Thomas, J. B., and Wyman, R. J., 1981, The morphology of the cervical giant fiber neuron of *Drosophila, Brain Res.* **221:**213–217.

Lawrence, P. A., 1982, Cell lineage of the thoracic, muscles of *Drosophila, Cell* **29:**493–503.

Levine, J. D., and Wyman, R. J., 1973, Neurophysiology of flight in wild-type and a mutant *Drosophila, Proc. Natl. Acad. Sci. U.S.A.* **70:**1050–1054.

Lewis, E. B., 1978, A gene complex controlling segmentation in *Drosophila, Nature* **276:**565–570.

Mulloney, B., 1969, Interneurons in the central nervous system of flies and the start of flight, *Z. Vergl. Physiol.* **64:**243–253.

Mulloney, B., 1970, Organization of flight motoneurons of *Diptera, J. Neurophysiol.* **33:**86–95.

Pak, W. L., Grossfield, J., and White, N. V., 1969, Nonphototactic mutants in a study of vision of *Drosophila, Nature, (London)* **222:**351–354.

Palka, J., and Ghysen, A., 1982, Segments, compartments and axon paths in *Drosophila, Trends Neurosci.* **5:**382–386.

Power, M. E., 1948, The thoracico-abdominal nervous system of an adult insect, *Drosophila melanogaster, J. Comp. Neurol.* **88:**347–409.

Salkoff, L., 1983a, *Drosophila* mutants reveal two components of fast outward current, *Nature, (London)* **302:**249–251.

Salkoff, L., 1983b, Genetic and voltage clamp analysis of a *Drosophila* potassium channel, *Cold Spring Harbor Symp. Quant. Biol.* **48:**221–231.

Salkoff, L., and Wyman, R. J., 1980, Facilitation of membrane electrical excitability in *Drosophila, Proc. Natl. Acad. Sci. U.S.A.* **77:**6216–6220.

Salkoff, L., and Wyman, R. J., 1981a, Outward currents in developing *Drosophila* flight muscle, *Science,* **212:**461–463.

Salkoff, L., and Wyman, R. J., 1981b, Genetic modification of potassium channels in *Drosophila Shaker* mutants, *Nature (London)* **293:**228–230.

Salkoff, L., and Wyman, R. J., 1983a, Ion currents in *Drosophila* flight muscles, *J. Physiol.* **337:**687–709.

Salkoff, L., and Wyman, R. J., 1983b, Ion channels in *Drosophila* muscle, *Trends Neurosci.* **6:**128–133.

Segal, M., Rogawski, M. A., and Barker, J. L., 1984, A transient potassium conductance regulates the excitability of cultured hippocampal and spinal neurons, *J. Neurosci.* **4:**604–609.

Sturtevant, A. H., and Beadle, G. W., 1939, *An Introduction to Genetics,* W. B. Saunders, Philadelphia.

Tanouye, M. A., and Wyman, R. J., 1980, Motor outputs of the giant nerve fiber in *Drosophila, J. Neurophysiol.* **44:**405–421.

Tanouye, M. A., and Wyman, R. J., 1981, Inhibition between flight motor neurons in *Drosophila, J. Comp. Physiol* **144:**345–355.

Tanouye, M. A., Ferrus, A., and Fujita, S. C., 1981, Abnormal action potentials associated with the *Shaker* complex locus of *Drosophila, Proc. Natl. Acad. Sci. U.S.A.* **78:**6548–6552.

Thomas, J. B., and Wyman, R. J., 1982, A mutation in *Drosophila* alters normal connectivity between two identified neurons, *Nature (London)* **298:**650–651.

Thomas, J. B., and Wyman, R. J., 1983, Normal and mutant connectivity between identified neurons in *Drosophila, Trends Neurosci.* **6:**214–219.

Thomas, J. B., and Wyman, R. J., 1984a, Mutations altering synaptic connectivity between identified neurons in *Drosophila, J. Neurosci.* **4:**530–538.

Thomas, J. B., and Wyman, R. J., 1984b, Duplicated neural structure in Bithorax flies, *Dev. Biol.* **102:**531–533.

Thomas, J. B., Bastiani, M. J., Bate, M., and Goodman, C. S., 1984, From grasshopper to *Drosophila:* A common plan for neuronal development, *Nature (London)* **310:**203–207.

Weisel, T. N., 1982, Postnatal development of the visual cortex and the influence of environment, *Nature, (London)* **299:**583–591.

Wyman, R., 1965, Probabilistic characterization of simultaneous nerve impulse sequences controlling dipteran flight, *Biophys. J.* **5:**447–471.

Wyman, R. J., 1969a, Lateral inhibition in a motor output system. I. Reciprocal inhibition in dipteran flight motor system, *J. Neurophysiol.* **32:**297–306.

Wyman, R. J., 1969b, Lateral inhibition in a motor output system. II. Diverse forms of patterning, *J. Neurophysiol.* **32:**307–314.

Wyman, R. J., and Tanouye, M. A., 1982, *Drosophila* flight motor pattern: The evidence from interspike intervals, *J. Exp. Biol.* **96:**413–416.

Wyman, R. J., and Thomas, J. B., 1983, What genes are necessary to make an Identified synapse?, *Cold Spring Harbor Symp. Quant. Biol.* **48:**641–652.

Wyman, R. J., Thomas, J. B., Salkoff, L., and King, D. G., 1984, The *Drosophila* giant fiber system, in: *Neural Mechanisms of Startle Behavior* (R. Eaton, ed.), Plenum Press, New York, pp. 133–161.

28

Genetic and Molecular Studies of a Potassium Channel Gene in *Drosophila*

YUH NUNG JAN AND LILY YEH JAN

1. Introduction

Molecular studies of functionally important elements in the nervous system have advanced rapidly in certain areas (e.g., studies of acetylcholine receptors and Na^+ channels), because there are specific high-affinity toxins and antibodies as well as organisms with specialized tissues that contain a large number of these molecules. Unfortunately, this is not true for most molecules that are important for neural function or development, and purification and biochemical studies of these molecules have been very difficult. Various alternative approaches have been proposed. For instance, "transport specificity fractionation" has been used successfully in the purification of an ATP-dependent calcium transport protein (Papazian *et al.*, 1979). One general approach to studying molecules that is important for neuronal function or neural development is to make use of genetics and molecular biology that has been well developed in certain organisms like *Caenorhabditis elegans* and *Drosophila.* If one can identify genes that are important for the function or development of the nervous system, one should be able to isolate them by molecular cloning and then begin to study these genes and their gene products in molecular terms. In this chapter, we will use the current studies of a gene in *Drosophila* for a voltage-sensitive K^+ channel as an example to illustrate this general approach.

Several different types of voltage-sensitive potassium channels have been characterized in recent biophysical studies (for review, Adams *et al.*, 1980). Different

YUH NUNG JAN AND LILY YEH JAN • Department of Physiology, University of California, San Francisco, California 94143.

neurons have been shown to have rather different makeup of these K^+ channels, which to a large extent determines the firing pattern and shape of action potentials in neurons (Byrne, 1980). Since the amount of transmitter released from a neuron can be drastically altered by slight alterations of the shape or duration of the action potential (Klein and Kandel, 1980; Llinas *et al.*, 1982), conceivably K^+ channels may play an important role in the control of synaptic efficacy. In fact, some K^+ channels are known to be influenced not only by membrane potential, but also by transmitter molecules such as acetylcholine or serotonin, and intracellular messengers such as Ca^{2+} or cAMP (e.g., Adams *et al.*, 1982; DePeyer *et al.*, 1982). For instance, the mechanism underlying sensitization in *Aplysia* appears to involve the inactivation of a particular K^+ channel by serotonin (Klein *et al.*, 1982; Siegelbaum *et al.*, 1982). To better understand how the expression of different K^+ channels may be controlled by various chemical messengers, one needs to characterize K^+ channels in molecular terms. Unfortunately, little information is available concerning the structure of K^+ channels or their genes, because there are no antibodies against K^+ channels and, with the exception of Ca^{2+}-activated K^+ channels (Hugues *et al.*, 1982), no toxins with high affinity have been found for K^+ channels.

One possible approach to a molecular study of K^+ channels is to clone the genes for K^+ channels. This involves the traditional genetic approach for identifying genes that code for protein molecules and can be done with animal systems with well-developed genetic tools, such as *Drosophila.* The first step in this approach is the induction of a mutation that gives a clear indication of affecting a K^+ channel. The mutant gene that is shown to alter normal K^+ conductance by physiological tests can be mapped genetically to a particular locus. Once the locus is known, the gene can then be cloned by a variety of techniques. In *Drosophila,* cloning may be done in a "brute force" manner by "walking" along the chromosome. Alternatively, transposable elements may be used to "tag" the gene and facilitate cloning (Bingham *et al.*, 1981; Modolell *et al.*, 1983; Searles *et al.*, 1982). Transposable elements are DNA fragments that apparently can generate and insert copies of themselves elsewhere in the genome, often causing mutations at the sites of insertion. The frequency of transposition of a particular transposable element, the p factor, is drastically increased by mating appropriate strains of fruit flies (Engels, 1981; Rubin *et al.*, 1982) (see Section 2). This "hybrid dysgenesis" phenomenon may then be used systematically for the cloning of genes in *Drosophila.*

In *Drosophila,* mutations at the *Shaker* locus affect a particular K^+ channel, the A channel, thereby causing prolonged action potentials and transmitter release (Jan *et al.*, 1977; Tanouye *et al.*, 1981; Salkoff, 1983). Thus studies of the *Shaker* gene(s) and gene products might lead to molecular studies of K^+ channels. Because *a priori* we do not know just how much DNA is included in the *Shaker* locus, both cloning strategies, namely transposon tagging and chromosome walking, are being used so that potentially cloning of the *Shaker* locus can be initiated at more than one entry point. In this chapter, we will first review evidence for the *Shaker* locus in *Drosophila* being the site for the structural gene of a K^+ channel, and discuss the strategies to be used for the molecular cloning of the *Shaker* locus. Then, we will describe the isolation of hybrid dysgenesis-induced *Shaker* mutants (Jan *et al.*, 1983), which may be useful in the cloning and analysis of DNA from the *Shaker* locus, and the preliminary results on the cloning of the *Shaker* locus by chromosome walking (Jan *et al.*, 1984).

2. Background

2.1. The *Shaker* Locus Probably Contains the Structural Gene for a K^+ Channel—A Review

2.1.1. Identification of *Shaker* as a Locus Important for K^+ Channel Function

Mutations affecting nervous function most likely would cause behavioral abnormality, paralysis, or lethality. With this expectation, over 100 existing behavioral mutants were screened electrophysiologically for possible defects in larval neuromuscular transmission. From that search, mutations of the *Shaker* locus were found to cause abnormally large and prolonged transmitter release, especially when the extracellular Ca^{2+} concentration was low (Jan *et al.*, 1977). Apparently the nerve action potentials in *Shaker* mutants were prolonged and carried recurrent spikes (Jan and Jan, 1980). This caused prolonged elevation of calcium conductance at the nerve terminal and prolonged transmitter release (Jan *et al.*, 1977). The cause of the *Shaker* phenotype was attributed to a defect in K^+ channel function, because the behavioral and physiological phenotype of *Shaker* can be mimicked by treating wild-type larvae with 4-aminopyridine (4-AP), a K^+ channel blocker, whereas blocking Na^+ or Ca^{2+} channels did not correct for the abnormality (Jan *et al.*, 1977). Recent voltage-clamp studies of *Drosophila* pupal muscles showed that 4-AP specifically blocked one type of K^+ channels, the A channel (Salkoff, 1983). Moreover, various *Shaker* mutations were found to affect the A channel function specifically (Salkoff, 1983), as discussed in Section 2.1.2.

2.1.2. Evidence for the *Shaker* Locus to Contain the Structural Gene for a K^+ Channel

In addition to abnormal synaptic transmission, *Shaker* mutants showed prolonged action potential duration in the cervical giant fibers of adult flies (Tanouye *et al.*, 1981), as well as altered A current in pupal muscles (Salkoff and Wyman, 1981) (Fig. 1). These studies have raised the possibility that the *Shaker* locus contains the structural gene for the A channel. A summary of existing evidence is given as follows.

2.1.2.a. The Null Phenotype Caused by Deletion in Shaker Locus. Deleting a small fragment within the *Shaker* locus [between the breakpoints of T(X;Y) B55 and T (X;Y) W32] eliminates the A current in pupal muscles (Salkoff, 1983) and results in prolonged transmitter release from larval nerve terminals, similar to what was found in the Sh^{KS133} mutant. This suggests that genes in the *Shaker* locus are necessary for the expression of functional A channels.

2.1.2.b. Gene Dosage Studies. Although the Sh^{KS133} mutant exhibits the same

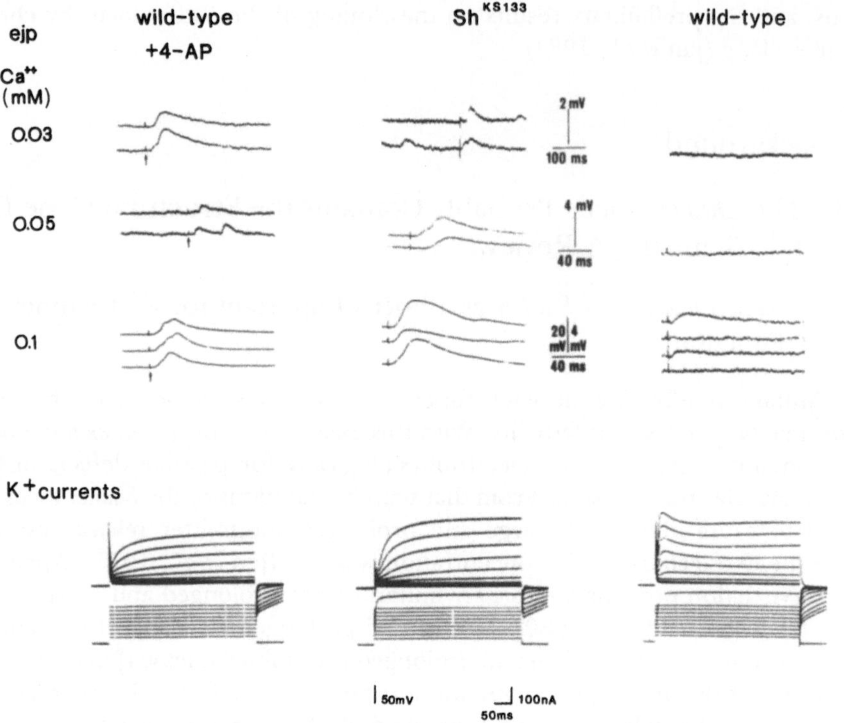

FIGURE 1. (Upper three rows) Intracellular recordings from ventral lateral longitudinal muscle fibers of third-instar larvae. Cut motor nerves were stimulated with a suction electrode; arrows point to stimulus artifacts. In 0.03 mM and 0.05 mM Ca^{2+}, motor-nerve stimulation failed to elicit a synaptic potential (excitatory junctional potential) in the wild-type fly, whereas stimulation succeeded in the Sh^{KS133} mutant and in the wild-type larva that was treated with 6 mM 4-AP. In 0.1 mM Ca^{2+}, motor-nerve stimulation evoked a much larger and prolonged response in Sh^{KS133} and in the 4-AP-treated wild-type larva than in wild-type larvae without 4-AP. (Bottom row) Currents from dorsal flight muscles of pupae, measured with a two-electrode voltage clamp. The voltage-jump protocol, the same for each experiment, is displayed beneath the current traces. The transient outward current, or A current, was present in the wild-type pupa missing in the SH^{KS133} pupa, and blocked by 4-AP in the wild-type pupa. The delayed rectifying K^+ current was not affected by the Sh^{KS133} mutation or by low concentrations of 4-AP. (From Jan *et al.*, 1983.)

phenotype as the deletion ($B55^D/W32^P$) (the null phenotype), the mutation apparently causes the production of an abnormal gene product, because the *Shaker* phenotype is not totally suppressed even in flies carrying enough copies of the Sh^+ gene to make close to 100% of the normal gene product (Tanouye *et al.*, 1981; Salkoff, 1983). This indicates that the gene affected by the Sh^{KS133} mutation is a structural gene.

2.1.2.c. Altered Voltage Dependence of A Channel Inactivation in the Sh^5 mutant. Study of a less severe *Shaker* mutant, Sh^5, has provided additional evidence for this hypothesis (Salkoff, 1983). In pupal muscle of Sh^5 flies, the A current has normal amplitude and rise time, but it inactivates more abruptly than in wild-type. Although normally the time constant for inactivation shows strong voltage dependence, in Sh^5 it is small and independent of membrane potential. Although the phenotype of Sh^5 differs from that of Sh^{KS133}, the two mutations are probably allelic; no recombination was observed between these two mutations among 8000 flies scored (Salkoff, 1983). The fact that one mutation of the *Shaker* locus alters the voltage dependence of A channel inactivation, whereas another mutation eliminates the A current, suggests that the *Shaker* locus contains the structural gene of the A channel.

2.2. Use of Transposable Elements in the Cloning of Genes of *Drosophila*

The cloning of genes whose product is either not defined or not available because of difficulties in isolation can represent a formidable problem. In *Drosophila,* two general approaches have been developed to isolate DNA sequences of genes that have been defined only by genetic criteria.

Most frequently these genes have been cloned by first genetically defining the position of the locus of interest on the polytene chromosome map, then obtaining a nucleic acid sequence (usually a piece of cloned DNA or a mRNA) that hybridizes to the polytene chromosomes at a cytological position that is near the locus of interest. Using a process referred to as "chromosome walking," it is possible to isolate adjacent cloned sequences overlapping the start clone and by a reiterative process "walk down the chromosome" to the region of interest. This process is often laborious and because of inaccuracies of cytological localization, the length of a walk cannot be reliably predicted.

An alternative proposed by Bingham *et al.,* (1981) has been called "transposon tagging." Should transposition lead to a damaging insertion within a gene, the mutant gene sequences will be "tagged" by the presence of the transposable element. If the transposable element DNA has been isolated, clones of "tagged" sequences can be identified by hybridization between the linked tag and labeled transposable element DNA. This approach has been used to clone the genes for *white, scute,* and *RNA polymerase* (Bingham *et al.,* 1981; Modolell *et al.,* 1983; Searles *et al.,* 1982).

The p element of *Drosophila* is a transposon that is particularly useful for transposon tagging because its frequency of transposition can be induced to levels that result in useful frequencies of transposon-induced mutations. Transposition is drastically increased by crosses between males from a strain containing p elements (P strain) and females from a strain lacking p elements (M strain) (Engels, 1981; Rubin *et al.,* 1982). The progeny of these "dysgenic crosses" can be screened for any particular mutant phenotype, and these will usually be due to a damaging insertion of p-element sequences into the gene of interest. If a library of clones is produced from the new mutant, those clones carrying tagged sequences can be identified by hybridization to labeled p-element DNA.

3. Hybrid Dygenesis-induced *Shaker* Mutants

Knowing that the *Shaker* locus is likely to contain the structural gene for the A channel (see Section 2), we undertook the isolation of DNA from the *Shaker* locus for molecular studies (Jan *et al.*, 1983, 1984). The strategy used was to first isolate *Shaker* mutations induced by hybrid dysgenesis, presumably caused by the insertion of p factors into the *Shaker* locus (see Background). Such mutants will allow the use of a cloned p element to isolate DNA that is contiguous with the p factor within the *Shaker* locus. In the meantime, chromosome walking was also initiated. Results of this latter cloning procedure will be described in the next section.

3.1. The Physiological Abnormalities of Sh^{SBn} Are Similar to those of Existing *Shaker* Mutants

Having found several dysgenesis-induced mutants that shake under ether anesthesia (isolated by Sandra Barbel), we first wanted to know whether they showed physiological abnormalities characteristic of *Shaker* mutants: namely, an altered A current and/or prolonged transmitter release caused by prolonged nerve action potential. We found that some Sh^{SBn} mutants showed the extreme, null phenotype of *Shaker*, whereas other dysgenesis-induced Sh^{SBn} mutants showed a weaker *Shaker* phenotype (Jan *et al.*, 1983).

3.2. Sh^{SBn} Mutations Appear to Affect the *Shaker* Locus

Knowing that the Sh^{SBn} mutations caused physiological abnormalities similar to those of *Shaker* mutants, we then asked whether these various Sh^{SBn} mutations mapped to the same region on the X chromosome as the *Shaker* locus. Using both genetic recombination and complementation tests involving chromosomal rearrangements that result in the duplication or deletion of a fragment of X chromosome containing the *Shaker* locus, we mapped all dysgenesis-induced Sh^{SBn} mutations roughly to the *Shaker* locus (Jan *et al.*, 1983).

3.3. A P Factor Is Present near the *Shaker* Locus of Most Sh^{SBn} Mutants

The purpose of generating *Shaker* mutations via hybrid dysgenesis was to insert a p factor into the *Shaker* locus so that the *Shaker* DNA sequences tagged with the p factor could be identified and isolated by hybridization with the labeled p element (see Section 2). Having isolated and mapped dysgenesis-induced mutations that showed the typical *Shaker* phenotype, we then did *in situ* hybridization experiments to test whether the radioactive p element probe would hybridize to the 16F region of the salivary chromosome, which contained the *Shaker* locus (Tanouye *et al.*, 1981).

The radioactive p element probe was made from a recombinant plasmid, p 25.1 (O'Hare and Rubin, 1983), provided by G. M. Rubin. Radioactive label was found around the 16F region in several Sh^{SBn} mutants examined, indicating that a p factor is inserted near the *Shaker* locus in these *Shaker* mutants (Jan *et al.*, 1983).

To summarize, we have isolated *Shaker* mutants through dysgenic crosses that promoted the transposition of p elements. These mutatins showed both behavioral and physiological phenotypes of *Shaker* mutations and mapped to roughly the region of the *Shaker* locus. Finally, a p factor was found near the *Shaker* locus by *in situ* hybridization with the salivary chromosomes. These Sh^{SBn} mutations may prove very useful in the cloning and molecular analysis of the *Shaker* locus.

4. Cloning of the *Shaker* Locus by Chromosome Walking

There are two indications that the *Shaker* locus may be genetically complex. Rearrangement breakpoints covering a substantial region around 16F give some evidence of a shaking phenotype (Tanouye *et al.*, 1981). Additionally, there is more than one phenotypic consequence of mutations at the *Shaker* locus. These mutations may be different alleles of a complicated gene. Alternatively, this locus may contain multiple structural genes with related phenotype. Because a number of *Shaker* mutations are not completely recessive, complementation studies are unable to resolve this issue. We hope to address this question by cloning the entire *Shaker* region and defining its detailed genetic make up by correlating coding regions with the molecularly mapped positions of various *Shaker* mutations of known phenotype.

In order to ensure a speedy isolation of DNA from the entire *Shaker* region, we have begun chromosome walking. For chromosome walking, a cDNA clone of Mariana Wolfner (1980) that maps to the 16F/17 region of the X chromosome was used as the starting point. From the Canton-S (wild-type) embryonic library of Maniatis *et al.* (1978), we have isolated about 70–80 kb of DNA that encompass rearrangements of certain *Shaker* mutations like Sh^5, a mutation that affects the voltage dependence of A channel inactivation (Salkoff, 1983). These rearrangements fall within 7 kb of DNA, which is duplicated once, in tandem, in the fly's genome. Detailed descriptions of the cloning and physical mapping of mutations will be reported elsewhere. Possible implications of these results are discussed in Section 5.

5. Discussion

Molecular studies of K^+ channels are important because these channels are likely to play important roles in the control of neuronal activity and synaptic efficacy. The genetics of *Drosophila* and mutations of the *Shaker* locus offer an alternative approach for cloning K^+ channels, in the absence of high-affinity toxins or antibodies against K^+ channels. Genetic analyses using electrophysiological assays including voltage clamping have provided strong evidence that the *Shaker* locus contains the structural gene for a K^+ channel, the A channel (Tanouye *et al.*, 1981;

Salkoff, 1983). At the same time, cloning strategies using either transposon tagging or chromosome walking can be applied to the cloning of the *Shaker* gene(s).

We have isolated and analyzed over 70 kb of DNA from the *Shaker* locus. Certain *Shaker* mutations show rearrangements to within a 7-kb fragment that is duplicated once in tandem. Whether this region is indeed important for a channel function is being tested in transformation studies (D. M. Papazian and Y. N. Jan, unpublished studies) where pieces of *Shaker* DNA are individually inserted within a p element containing the *alcohol dehydrogenase* gene, injected into embryos, and tested in subsequent generations for alcohol resistance and A-channel function. Previous studies (Spradling and Rubin, 1983; Scholnick *et al.*, 1983; Goldberg *et al.*, 1983) have shown that with this transformation procedure the exogenous DNA is not only integrated singly and stably, it is expressed properly in the correct tissue, at the correct developmental stages, and for approximately the correct amount. Thus, it should be possible to apply the available electrophysiological techniques to analyze muscles and neuromuscular junctions of transformants, and to test whether a piece of wild-type *Shaker* DNA can restore the A-channel function in a mutant with no endogenous A current.

How does one demonstrate that the *Shaker* locus contains the structural genes for A channel? If the A channel were composed of a single subunit, one might expect an expression system such as the *Xenopus* oocytes to be useful, as has been the case for acetylcholine receptors (Barnard *et al.*, 1982). In this case, one would need to use cloned *Shaker* DNA to isolate the appropriate messenger RNA so that the latter may be injected into cells like *Xenopus* oocytes. If the A channel were composed of more than one subunit and not all subunits were encoded by DNA from the *Shaker* locus, one might first identify the DNA sequences in the *Shaker* locus that codes for one of the subunits. This might be done by identifying sequences that could hybridize to and remove a messenger RNA that is necessary for the formation of functional A channels. DNA transformation of *Drosophila* (Rubin and Spradling, 1982) provides another powerful method that could be used to show that a particular DNA sequence includes the information damaged by *Shaker* mutations. The DNA segment that restores A channel in the transformed fly with no endogenous A current is likely to contain the structural gene and be used to generate large amounts of its gene products. By using specialized cloning vectors that will allow expression of the *Shaker* DNA that encodes a subunit of the A channel, one may then generate antibodies to the protein product and use them in the purification and molecular characterization of A channel polypeptides. Analysis of the molecular defects in certain mutants, such as Sh^5, that produce altered channels should help define functional domains in the channels. These analyses might also be extended to determine whether different K^+ channels in *Drosophila* share structural homology and whether their evolutionary conservation permits the identification of other channel genes in other organisms.

ACKNOWLEDGMENT. We with to thank our collaborators, Drs. P. H. O'Farrell, D. M. Papazian, and L. Salkoff, whose efforts made this review possible.

References

Adams, D. J., Smith, S. J., and Thompson, S. H., 1980, Ionic currents in molluscan soma, *Annu. Rev. Neurosci.* **3**:141.

Adams, P. R., Brown, D. A., and Constanti, A., 1982, Pharmacological inhibition of the M-current, *J. Physiol.* **332**:223.

Barnard, E. A., Mileda, R., and Sumikawa, K., 1982, Translation of exogenous messenger RNA encoding for nicotinic acetylcholine receptors produces functional receptors in *Xenopus* oocytes, *Proc. R. Soc. London B.* **215**:241.

Bingham, P. M., Levis, R., and Rubin, G. M., 1981, Cloning of DNA sequences from the *white* locus of *D. melanogaster* by a novel and general method, *Cell* **25**:693.

Byrne, J. H., 1980, Analysis of ionic conductance mechanisms in motor cells mediating inking behavior in *Aplysia californica, J. Neurophysiol.* **43**:630.

DePeyer, J. E., Cachelin, A. B., Levitan, I. B., and Reuther, H., 1982, Ca^{2+}-activated K^+ conductance in internally perfused neurons is enhanced by protein phosphoyrlation, *Proc. Natl. Acad. Sci. U.S.A.* **79**:4207.

Engels, W. R., 1981, Hybrid dysgenesis in *Drosophila* and the stochastic loss hypothesis, *Cold Spring Harbor Symp. Quant. Biol.* **45**:561.

Goldberg, D. A., Posakony, J. W., and Manialis, T., 1983, Correct developmental expression of a cloned alcohol dehydrogenase gene transduced into the Drosophila germ line, *Cell* **34**:59–73.

Hugues, M., Duval, D., Kitabgi, P., Lazkunski, M., and Vincent, J. P., 1982, Preparation of a pure monoiodo derivative of the bee venom neurotoxin apamin and its binding properties to rat brain synaptosomes, *J. Biol. Chem.* **257**:2762.

Jan, L. Y., Barbel, S., Timpe, L., Laffer, C., Salkoff, L., O'Farrell, P., and Jan, Y. N., 1983, Mutating a gene for a K^+ channel by hybrid dysgenesis: An approach to the cloning of the *Shaker* locus in *Drosophila, Cold Spring Harbor Symp. Quant. Biol.* **48**:233–245.

Jan, L. Y., Papazian, D. M., Jan, Y. N., and O'Farrell, P. H., 1984, Cloning of potassium channel gene(s) in the *Shaker* locus of *Drosophila, Neurosci. Abstr.* **10**:1089.

Jan, Y. N., and Jan, L. Y., 1980, Genetic disection of synaptic transmission in *Drosophila melanogaster,* in: *Insect Neurobiology and Pesticide Action* (F. E. Rickett, ed.), The Society of Chemical Industry, London, pp. 161–168.

Jan, Y. N., Jan, L. Y. and Dennis, M. J., 1977, Two mutations of synaptic transmission in *Drosophila, Proc. R. Soc. London B.* **198**:87.

Klein, M., and Kandel, E. R., 1980, Mechanism of calcium current modulation underlying presynaptic facilitation and behavioral sensitization in *Aplysia, Proc. Natl. Acad. Sci. U.S.A.* **77**:6912.

Klein, M., Camardo, J. and Kandel, E. R., 1982, Serotonin modulates a specific potassium current in the sensory neurons that show presynaptic facilitation in *Aplysia, Proc. Natl. Acad. Sci. U.S.A.* **79**:5713.

Llinas, R., Sugimori, M., and Simon, S. M., 1982, Transmission by presynaptic spike-like depolarization in the squid giant synapse, *Proc. Natl. Acad. Sci. U.S.A.* **79**:2415.

Maniati, S. T., Hardison, R. C., Lacy, E., Lauer, J., O'Connell, C., Quon, D., Sim, D. K., and Efstratiadis, A., 1978, The isolation of structural genes from libraries of eucaryotic DNA, *Cell* **15**:687.

Modolell, J., Bender, W., and Meselson, M., 1983, *Drosophila melanogaster* mutations suppressible by the suppressor of *Hairy-wing* are insertions of a 7.3-kilobase mobile element, *Proc. Natl. Acad. Sci. U.S.A.* **80**:1678.

O'Hare, K., and Rubin, G. M., 1983, Structures of P transposable elements and their sites of insertion and excision in the *Drosophila melanogaster* genome, *Cell* **34**:25–35.

Papazian, D. M., Rahamimoff, H., and Goldin, S. M., 1979, Reconstitution and purification by "transport specificity fractionation" of an ATP-dependent calcium transport component from synaptosome-derived vesicles, *Proc. Natl. Acad. Sci. U.S.A.* **76**:3708.

Rubin, G. M., and Spradling, A. C., 1982, Genetic transformation of *Drosophila* with transposable element vectors, *Science* **218**:348.

Rubin, G. M., Kidwell, M. G., and Bingham, P. M., 1982, The molecular basis of P-M hybrid dysgenesis: The nature of induced mutations, *Cell* **29**:987.

Salkoff, L., 1983, Genetic and voltage-clamp analysis of a *Drosophila* K^+ channel, *Cold Spring Harbor Symp. Quant. Biol.* **XLVIII:**221–231.

Salkoff, L., and Wyman, R., 1981, Genetic modification of potassium channels in *Drosophila Shaker* mutants, *Nature (London)* **293:**228.

Scholnick, S. B., Morgan, B. A., and Hirsh, J., 1983, The cloned dopa decarboxylase gene is developmentally regulated when reintegrated into the *Drosophila* genome, *Cell* **34:**37–45.

Searles, L. L., Jokerst, R. S., Bingham, P. M., Voelker, R. A., and Greenleaf, A. L., 1982, Molecular cloning of sequences from a *Drosophila* RNA polymerase II locus by p element transposon tagging, *Cell* **31:**585.

Siegelbaum, S. A., Camardo, J. S., and Kandel, E. R., 1982, Serotonin and cyclic AMP close single K^+ channels in *Aplysia* sensory neurons, *Nature* (*London*) **299:**413.

Spradling, A. C., and Rubin, G. M., 1983, The effect of chromosomal position on the expression of the *Drosophila* xanthine dehydrogenase gene, *Cell* **34:**47–57.

Tanouye, M. A., Ferrus, A., and Fujita, S. C., 1981, Abnormal action potentials associated with the *Shaker* locus of *Drosophila, Proc. Natl. Acad. Sci. U.S.A.* **78:**6548.

Wolfner, M., 1980, Ecdysone responsive genes of the salivary gland of *Drosophila melanogaster,* Ph.D. Thesis, Stanford University.

Index